Receptor Pharmacology and Function

CLINICAL PHARMACOLOGY

Series Editor

MURRAY WEINER

Division of Clinical Pharmacology and Toxicology
University of Cincinnati
College of Medicine
Cincinnati, Ohio

Volume 1 Nicotinic Acid: Nutrient-Cofactor-Drug, *by Murray Weiner and Jan van Eys*

Volume 2 Drugs in Central Nervous System Disorders, *edited by David C. Horwell*

Volume 3 Clinical Application of Interferons and Their Inducers, Second Edition, *edited by Dale A. Stringfellow*

Volume 4 Beta-Lactam Antibiotics for Clinical Use, *edited by Sherry F. Queener, J. Alan Webber, and Stephen W. Queener*

Volume 5 Receptor Binding in Drug Research, *edited by Robert A. O'Brien*

Volume 6 Drug Therapy in Hypertension, *edited by Jan I. M. Drayer, David T. Lowenthal, and Michael A. Weber*

Volume 7 Adverse Reactions to Muscle Relaxants, *edited by Isaac Azar*

Volume 8 Clinical Pharmacology in the Elderly, *edited by Cameron G. Swift*

Volume 9 Anti-Inflammatory Compounds, *edited by W. R. Nigel Williamson*

Volume 10 Nonsteroidal Anti-Inflammatory Drugs: Mechanisms and Clinical Use, *edited by Alan J. Lewis and Daniel E. Furst*

Volume 11 Drug Stereochemistry: Analytical Methods and Pharmacology, *edited by Irving W. Wainer and Dennis E. Drayer*

Volume 12 Drug Interaction Compendium, *by Murray Weiner*

Volume 13 Receptor Pharmacology and Function, *edited by Michael Williams, Richard A. Glennon, and Pieter B. M. W. M. Timmermans*

Additional Volumes in Preparation

Receptor Pharmacology and Function

edited by

Michael Williams
CIBA-GEIGY Corporation
Summit, New Jersey

Richard A. Glennon
Medical College of Virginia
Virginia Commonwealth University
Richmond, Virginia

Pieter B. M. W. M. Timmermans
E. I. du Pont de Nemours & Company, Inc.
Wilmington, Delaware

New York and Basel MARCEL DEKKER, INC.

Library of Congress Cataloging-in-Publication Data

Receptor pharmacology and function / edited by Michael Williams,
Richard A. Glennon, Pieter B. M. W. M. Timmermans.
p. cm. -- Clinical pharmacology ; 13)
Based on a symposium held in Boston in August 1985; sponsored by the American Society for Pharmacology and Experimental Therapeutics and the Medicinal Chemistry Division of the American Chemical Society.
Includes index.
ISBN 0-8247-7841-3
1. Drug receptors--Congresses. 2. Drugs--Structure-activity relationships--Congresses. I. Williams, Michael.
II. Glennon, Richard A. III. Timmermans, P. B. M. W. M.
IV. American Society for Pharmacology and Experimental Therapeutics.
V. American Chemical Society. Medicinal Chemistry Division.
VI. Series: Clinical pharmacology ; v. 13.
[DNLM: 1. Ligands--physiology--congresses. 2. Receptors, Drug--physiology--congresses. 3. Receptors, Endogenous Substances--physiology--congresses. 4. Structure-Activity Relationship--congresses. W1 CL764H v. 13 / QH 603.C43 R2947 1985]
RM301.41. R42 1989
615'.7--dc19
DNLM/DLC
for Library of Congress 88-22937
CIP

MARCEL DEKKER, INC.
270 Madison Avenue, New York, New York 10016

Current printing (last digit):
10 9 8 7 6 5 4 3 2 1

PRINTED IN THE UNITED STATES OF AMERICA

Series Introduction

The receptor-ligand concept is basically as old as the lock and key. As each specialty rediscovers the concept, researchers develop nomenclature and images consistent with their perspectives. Immunologists used the lock-and-key analogy in early concepts of antibody-antigen interactions. They noted the role antigen (key) could play in determining the specificity of the antibody (lock). Biochemists started with soluble enzyme-substrate systems, though they now recognize that the enzyme (lock) can be fixed to a surface membrane and that soluble enzymes can act on fixed substrates. The fact that exposure to substrate can "induce" enzyme production is reminiscent of the immune system. Vital transport functions, such as the delivery of oxygen by hemoglobin from lung to distant tissues and the movement of metals and other critical materials by highly specific apoproteins, may also be looked upon as "lock and key." Pharmacologists found the terms "receptor" and "agonist" particularly suitable to describe their locks and keys. Pharmacologic receptors are usually fixed structures that "receive" specific ligands that are widely dispersed via blood or highly localized. Fixed receptors can also be up- or down-regulated as a function of exposure to the mobile agonist ligand.

Lock-key interactions can have a wide variety of effects. In some instances a lock destroys the key as enzymes do substrates. The key may distort the lock so as to trigger a signal as in many pharmacologic receptor-ligand reactions. The triggering can result in a cascade of widely scattered events, as exemplified by the effects of some antigen-antibody complexes or the very local activation of second or third messengers, so important to pharmacology. Cofactors and allosteric effectors can alter receptor configuration and reactivity. The receptor-ligand interaction can be reversible or irreversible, with some ligands serving as competitive, noncompetitive, or suicidal inhibitors. The same key (ligand) may have both agonist and antagonist effects, depending on circumstances. Each component of the receptor-ligand complex may distort the form of the other.

Thus, the lock-and-key concept is an excellent example of an expanded version of an old proverb: The truth is simple, but the whole truth is very complicated. The receptor-ligand concept is indeed a fundamental, simple truth; however, to understand receptors in pharmacology completely, much more must be mastered.

There have been leaps and bound of progress in receptor technology and understanding in the last two decades, and this volume by Williams, Glennon, and Timmermans is a remarkable testament to that progress. But there is still a long way to go. These authors recognize that complex receptor phenomena are only one of several sets of important factors that determine what a drug will do in a particular therapeutic setting. Their awareness of this fact makes for a balanced and therefore more valuable text in this specialized area.

Murray Weiner, M.D.

Foreword

Over the last two decades our concept of receptors has changed dramatically. No longer is a receptor a hypothetical construct resulting from studies on isolated tissue preparations; in more and more cases it is an actual protein molecule of known sequence. Although recent studies have demonstrated that receptor molecules exhibit an unforseen degree of heterogeneity, some unifying principles have emerged. Indeed, it now appears that there are basically two types of receptors. The first type consists of a single-membrane-spanning polypeptide chain that contains approximately seven α-helical regions. Information resulting from agonist binding to such a receptor is "transduced" to an effector molecule (e.g., adenylate cyclase, phospholipase C, or an ion channel) by means of a G protein. A second type of receptor consists of several membrane-spanning subunits arranged to form an ion channel. Agonist binding to this kind of receptor gates the ion channel directly. Some agonists, such as acetylcholine, GABA, or glutamate can act on receptors from both categories. Thus, muscarinic cholinergic receptors are G-protein linked, whereas nicotinic cholinergic receptors direct the opening and closing of a cation-specific ionophore. Receptors within each category exhibit interesting sequence homologies, a fact that is not surprising considering their overall similarity in structure. For example, the GABA-A and glycine receptors, which both gate ion channels, are obviously related to the nicotinic receptor. For each specific type of receptor within a category, subtle alterations in sequence are also found that give rise to further variations, e.g., multiple types of muscarinic and nicotinic receptors are now known to exist. Thus, receptors represent several themes with variations of symphonic proportions.

All this progress is the result of several technological innovations. Molecular biological approaches have proved to be particularly powerful and revealing when applied to investigations of receptors. However, molecular pharmacological techniques such as ligand binding studies have probably been at least as important and continue to be so. Indeed it was the application of ligand binding techniques to entities such as the insulin, opiate, and nicotinic receptors that opened up the receptor field to the immense quantities of data now available.

Where do we go from here? Obviously the molecular characterization of receptor structure and mechanism will continue apace. On another tack, the realization that considerable receptor heterogeneity exists offers the molecular pharmacologist new and unique opportunities. Clearly, it should be possible to design novel agents specific for the newly emerging receptor subtypes. In addition to their potential therapeutic utility, such molecules will become important probes for the identification of these receptors in vivo and in vitro, thereby providing further insights into their regulation, physiology, and functions. Clearly the diverse techniques for the identification of receptors are assuming a more and more important role. It is essential that we are able to design novel, specific radioligands and understand how to interpret binding data in the most accurate and meaningful fashion. The time is clearly right for a review of these issues. In the current volume, the authors have dealt with these various technical points and put them in perspective by examining how they have been used to clarify the structure and function of several key receptor types.

For those of us concerned with receptors these are exciting times. It is hard to believe that so much has been discovered so quickly. The matters discussed in this volume therefore provide essential information for any modern pharmacologist, physiologist, or biochemist.

R. J. Miller, Ph.D.

Department of Pharmacology
University of Chicago

Preface

The receptor concept, as it relates to the understanding of tissue function and the discovery of new therapeutic entities, is almost a century old, originating in the seminal work of Langley and Ehrlich (1). It is only in the past 20 years that the technology to facilitate testing many of the theories and hypotheses put forward over the years has been developed. Increased sophistication in electronic recording devices, computerization, and molecular biological techniques in the broadest sense of the term have allowed the study of receptor function to advance exponentially.

For the classically trained pharmacologist and medicinal chemist, the concepts of the dose-response curve, the Schild plot, parallel shifts and pA_2 values, of efficacy and affinity, of competitive and noncompetitive inhibition, are second nature; however, in recent years scientists from many other disciplines, including biochemistry, biophysics, and molecular genetics, have become intimately involved in the process of drug discovery. As a result of this influx of diverse talents the receptor is often regarded as a distinct molecular entity, amenable to isolation and subject to analysis from many vantage points. The intrinsic physiological role of this entity, however, is often overlooked. Whether research endeavors related to providing this functional component are basic (related to an improvement in understanding of the physiology, pathology, and pharmacology of receptor function) or pragmatic (related to the discovery of new therapeutic entities), it is of crucial importance that information flow should be maximized between different disciplines and integrated to facilitate the understanding of tissue function and thus provide the basis for the design of new drugs (2). While pharmacology in turn has become a molecular discipline, it still provides the major integrative focus for data derived from these diverse disciplines.

The theme of the present volume has been to present the interrelationships of the various disciplines, melding chemistry with biology at the level of the structure-activity relationship (SAR) in a manner that is neither overly simplistic nor complex, thus providing a synthesis of current knowledge for the novice and the expert to aid in their understanding of receptor function and its relationship to basic pharmacological and physiological processes.

The ligand/drug-receptor interaction comprises two major processes: recognition and transduction. Prior to 1973, studies of the recognition event required the use of classical tissue and animal models to define SARs for the receptor-ligand interaction. However, with the development of radioligand binding by

Solomon Snyder and his colleagues at Johns Hopkins Medical School, it has now become possible to define the SAR in terms of the recognition process, rather than indirectly via effector systems as is the case with many traditional systems. Thus it is now possible to derive structure-affinity (SAFIR) relationships.

While the epithet "grind and bind" has often been used in regard to this technology (especially when in vitro data has been overinterpreted without reference to function), it has nevertheless added considerably to our knowledge of receptor function at the molecular level. In concert with electrophysiological studies and studies of second messenger systems fueled by the Nobel prize-winning work of Sutherland and his colleagues in the 1960s and the work of Berridge and associates in the 1980s, in vitro analysis of receptor-mediated events has provided important conceptual changes in the way that receptors are thought to function. Indeed, studies on receptors for inhibitory (3) and excitatory (4) amino acid neurotransmitters that have provided evidence for allosterically modulated receptor complexes have challenged the basic tenets of receptor theory (5,6).

In the present volume, Hollenberg provides an overview of the receptor concept with reference to receptor-receptor interactions. After Titeler's consideration of the practical aspects of receptor binding, McPherson addresses the theory of data analysis. Kinnier provides a framework for the effector/second messenger systems related to receptor activation that is echoed throughout most of the remainder of the volume, and Filer and colleagues point out the practical aspects of radioligand synthesis and use. In a timely overview, Martin and Danaher, using the dopamine receptor as a primary example, discuss molecular modeling of the receptor-ligand interaction.

Following from this basic background applicable in a generic sense to most receptors, individual receptors are discussed by recognized authorities. Adrenoceptors are reviewed by Timmermans and Brodde; serotonin receptors by Glennon; histamine receptors by Buschauer and co-workers; GABA receptors by Krogsgaard-Larsen; benzodiazepine receptors by Williams and Olsen; excitatory amino acid receptors by Ferkany; cholinergic receptors by Kaiser and Rzeszotarski, and purine receptors by Williams. The rapidly expanding subject of peptide receptors is reviewed in separate chapters by Moody and Perry and by Tam. Eicosanoid receptors are discussed by Crooke and his colleagues, and Martin reviews "drug"-related receptors with particular reference to cannabinoid, phencyclidine, and antidepressant receptors. This theme is extended by Triggle and Janis in their discussion of receptors for calcium entry blockers.

El-Fakahany addresses differences in receptor characteristics in comparing homogenate data with that derived from intact cells, and Niznik and Jarvie review dopamine receptor function.

The emphasis throughout the text is on the history of each receptor, receptor classification, techniques for receptor identification and classification, second messenger/effector systems related to receptor activation, and the pathophysiological states related to receptor malfunction. It is hoped that the present work provides a unifying framework whereby the specialists in the many different disciplines involved in the study of receptor function may integrate their particular efforts with a more generalized understanding of receptor function.

The present volume is the outcome of a successful symposium held in Boston in August 1985, sponsored by the American Society for Pharmacology and Experimental Therapeutics and the Medicinal Chemistry Division of the American Chemical Society. This symposium, the brainchild of Dr. John Neumeyer of Northeastern University, was supported by the Biomedical Products Division of E. I. du Pont de Nemours, Nova Pharmaceuticals, the Pharmaceuticals Division of CIBA-GEIGY, and Research Biochemicals, Inc.

This volume is designed as a valuable sourcebook for chemists and biologists involved in understanding receptor function, particularly as it relates to the discovery of new therapeutic entities. It is also hoped that the volume will be useful to the expert seeking information outside his area of expertise, to those interested in drug discovery and development, and to graduates and undergraduates interested in keeping up-to-date on developments in this most exciting area of biomedical research.

1. P. M. Dean, *Molecular Foundations of Drug-Receptor Interaction*. Cambridge University Press, London, 1987.
2. J. W. Black, Scrip *1309*, 24 (1988).
3. W. Haefley, E. Kyburz, M. Gerecke, and H. Mohler, Adv. Drug Res. *14*, 166 (1985).
4. J. A. Kemp, A. C. Foster, and E. H. K. Wong, Trends Neurosci. *10*, 294 (1987).
5. M. Williams and M. A. Sills, in *Comprehensive Medicinal Chemistry, Vol. 3* (J. D. Emmett, ed.) in press (1989).
6. T. A. Kenakin, Trends Pharmacol. Sci. *8*, 423 (1987).

Michael Williams
Richard A. Glennon
Pieter B. M. W. M. Timmermans

Contents

Contributors

Jean-Michel Arrang, Ph.D., Paul Broca Center of INSERM, Paris, France

Otto-Erich Brodde, University of Essen, Essen, Federal Republic of Germany

Armin Buschauer, Free University of Berlin, Berlin, Federal Republic of Germany

*Mike Clark, Ph.D.,** Smith Kline and French Laboratories, Philadelphia, Pennsylvania

Stanley T. Crooke, M.D., Ph.D., Smith Kline and French Laboratories, Philadelphia, Pennsylvania

Elizabeth Danaher, Ph.D., Abbott Laboratories, Abbott Park, Illinois

Esam E. El-Fakahany, Ph.D., University of Maryland School of Pharmacy, Baltimore, Maryland

John W. Ferkany, Ph.D., Nova Pharmaceutical Corporation, Baltimore, Maryland

Crist Filer, Ph.D., E. I. du Pont de Nemours & Company, Inc., Boston, Massachusetts

Monique Garbarg, Ph.D., Paul Broca Center of INSERM, Paris, France

John Gleason, Smith Kline and French Laboratories, Philadelphia, Pennsylvania

Richard A. Glennon, Ph.D., Medical College of Virginia, Virginia Commonwealth University, Richmond, Virginia

G. Kurt Hogaboom, Ph.D., Smith Kline and French Laboratories, Philadelphia, Pennsylvania

Morley D. Hollenberg, D.Phil., M.D., University of Calgary, Calgary, Alberta, Canada

Stephen Hurt, Ph.D., E. I. du Pont de Nemours & Company, Inc., Boston, Massachusetts

---

**Present affiliation*: Washington University Medical School, St. Louis, Missouri

R. A. Janis, Ph.D., Miles Institute for Preclinical Pharmacology, West Haven, Connecticut

K. R. Jarvie, Ph.D., University of Toronto, Toronto, Ontario, Canada

Carl Kaiser, Ph.D., Nova Pharmaceutical Corporation, Baltimore, Maryland

William J. Kinnier, Ph.D., Nova Pharmaceutical Corporation, Baltimore, Maryland

Povl Krogsgaard-Larsen, D.Sc., Ph.D., The Royal Danish School of Pharmacy, Copenhagen, Denmark

Monica Lewis, Ph.D., Smith Kline and French Laboratories, Philadelphia, Pennsylvania

*Grant A. McPherson, Ph.D.,** University of Vermont, Burlington, Vermont

Billy R. Martin, Ph.D., Medical College of Virginia, Virginia Commonwealth University, Richmond, Virginia

Yvonne C. Martin, Ph.D., Abbott Laboratories, Abbott Park, Illinois

Seymour Mong, Smith Kline and French Laboratories, Philadelphia, Pennsylvania

Terry W. Moody, Ph.D., The George Washington University School of Medicine and Health Sciences, Washington, D.C.

H. B. Niznik, Ph.D., University of Toronto, Toronto, Ontario, Canada

Richard W. Olsen, Ph.D., University of California School of Medicine, Los Angeles, California

David C. Perry, Ph.D., The George Washington University School of Medicine and Health Sciences, Washington, D.C.

W. Janusz Rzeszotarski, Ph.D., Nova Pharmaceutical Corporation, Baltimore, Maryland

Walter Schunack, Free University of Berlin, Berlin, Federal Republic of Germany

Jean-Charles Schwartz, Ph.D., Paul Broca Center of INSERM, Paris, France

S. William Tam, Ph.D., E. I. du Pont de Nemours & Company, Inc., Wilmington, Delaware

Pieter B. M. W. M. Timmermans, Ph.D., E. I. du Pont de Nemours & Company, Inc., Wilmington, Delaware

Milt Titeler, Ph.D., Albany Medical College, Albany, New York

---

**Present affiliation*: Baker Medical Research Institute, Prahran, Victoria, Australia

D. J. Triggle, Ph.D., State University of New York, Buffalo, New York

Yieh-Ping Wan, Ph.D., E. I. du Pont de Nemours & Company, Inc., Boston, Massachusetts

Michael Williams, Ph.D., D.Sc., CIBA-GEIGY Corporation, Summit, New Jersey

J. Michael Young, Ph.D., University of Cambridge, Cambridge, England

Receptor Pharmacology and Function

1

Receptors: Their Reactions and Interactions

Morley D. Hollenberg / University of Calgary, Calgary, Alberta, Canada

The concept that "receptive substances" mediate the action of a variety of drugs, toxins, and other agents was explicitly developed at the turn of the century, largely from the independent studies of Langley (1) and Ehrlich (2). It is of some interest that Langley was focused on the action of nicotine and curare at what is now recognized as the "traditional" nicotinic cholinergic receptor in striated muscle, whereas Ehrlich's deductions were based on studies of the trypanocidal properties of triphenylmethane dyes. What both Langley and Ehrlich had in common was the use of a responsive bioassay system in which to evaluate the action of the compounds of interest. What resulted from their insights was a concept that was of inestimable use for the study of agents that may act via fundamentally different mechanisms. Thus, in general terms, a receptor may represent any cellular constituent (e.g., enzyme, ion channel, or surface oligosaccharide) that exhibits chemical specificity and that plays a key role in regulating the cellular effects of a particular active substance (for example, the Na,K-ATPase in the case of digitalis; the ligand-responsive ion channel in the case of acetylcholine; or the GM_1 ganglioside binding site for cholera toxin). Thus, it is entirely appropriate that this volume, devoted to receptor pharmacology, contains information not only about a large number of well-recognized cell surface "receptors," but also about calcium channels and enzymes; for enzymes and metabolite channels in many ways can fulfill the criteria required of a receptor as envisioned by Ehrlich and Langley. In large part, however, this volume is devoted to cell surface binding sites (or receptors) that play simultaneously both a recognition and an activation role in terms of regulating cell function. It is on these cell surface recognition-activation constituents that this chapter focuses; for information concerning intracellular recognition sites (e.g., steroid hormone receptors, cytoplasmic enzymes, and chromatin) the reader is obliged to look elsewhere (e.g., References 3-6).

The diversity of information presented in this volume attests to how fruitful the receptor concept has been in terms of leading to the detailed characterization of such a variety of chemically specific cell surface signaling elements. As recently as 5 years ago, what might be called the "grind and bind" era of receptor studies was still in full bloom, and a large number of studies were focused on determining receptor affinities and numbers under a variety of physiological conditions. Such studies indicated that receptors are not static elements fixed in the membrane at a single site. Rather, it has become fully appreciated that receptors for neurotransmitters, hormones, and other active agents are dynamic elements both in the plane of the plasma membrane and in the cytoplasmic compartment. Also, as described in the chapters that follow this one, much information is now known about the detailed physicochemical (e.g., molecular weights and hydrodynamic parameters) and chemical (e.g., subunit composition and amino acid sequence) properties of a number of receptors, and reliable data are now available concerning the cell surface mobility and aggregation properties of many receptors. Structural and functional relationships between receptors of different types are now recognized (for instance, receptors for growth factors, like insulin, insulinlike growth factor I, colony stimulating factor I, platelet-derived growth factor, and epidermal growth factor, are all known to contain a tyrosine kinase domain), and a number of distinct receptor-mediated transmembrane signaling reactions have now been described. This information has now set the stage for considering the following questions. How is receptor function regulated; by what mechanism(s) might the activation of one receptor regulate the behavior of a second receptor? This chapter tries to provide a perspective for a partial answer to these questions. In the chapters that follow, there is specific information about individual receptors that can be subjected to regulation by the mechanisms discussed below.

RECEPTORS, ACCEPTORS, CHANNELS, AND MECHANISMS OF TRANSMEMBRANE SIGNALING

Receptors Versus Acceptors and Channels as Signaling Elements

Although the overall function of the plasma membrane is to maintain an effective barrier between the intracellular and extracellular milieu, the membrane contains highly specialized structures (e.g., ion channels, nutrient transporters, and histocompatibility determinants) that play pivotal roles in the selective transfer of information from the external to the internal cell environment (and in some cases, vice versa). In some cases, information is transferred simply by the selective uptake of ions or nutrients via specific membrane channels. In such cases the chemical specificity of the channel and the energy provided by a concentration gradient or by cellular energy (e.g., ATP) are the principal factors governing information transfer, in which the transported ligand (for example, sodium ion) is both the messenger and the message to which the cell must respond. Because of their specificity, these channels can serve as "receptors" for both drugs and toxins. For instance, the voltage-sensitive sodium channel is a receptor for the frog-derived poison, batrachotoxin, and the voltage-sensitive calcium channel as discussed in Chapter 20 can be seen as a receptor for the antiarrhythmic agents, like verapamil. It is now evident that the membrane constituents that act as selective transport molecules are highly complex transmembrane proteins that may be subject to the same kind of regulatory processes (e.g., phosphorylation-dephosphorylation and allosteric inhibition) that affect either enzymes or pharmacological receptors, as discussed below. Nonetheless, the channels themselves do not exhibit the kind of signaling properties expected of pharmacological receptors, as outlined below.

In a number of cases, substances that convey information for the regulation of intracellular processes (for example, vitamin B_{12} or cholesterol) are associated with carrier proteins in the bloodstream (e.g., transcobalamin II for B_{12} and low-density lipoprotein, LDL, for cholesterol). The blood-borne carrier proteins can be recognized in a very specific high-affinity manner by specialized transporters responsible for the selective adsorptive pinocytosis of the regulatory ligands (the examples used here are B_{12} and cholesterol). Here again, as in the case of the sodium ion, the ligand transported (i.e., cholesterol or B_{12}) is the message being taken into the cell. The binding constituents (either the carrier proteins or the cell surface uptake proteins) perform a message-carrying function and do not, of themselves, generate a transmembrane signal. Thus the binding constituents for substances like the LDL-cholesterol complex or the transcobalamin II-B_{12} complex, although commonly referred to as receptors, do not truly function as receptors in the pharmacological sense. It has proved convenient to refer to such highly specific transporters as either "acceptors" (7) or receptors of the class II type (8).

In contrast to acceptors or metabolite channels, pharmacological receptors, located in the plasma membrane, exhibit not only the ability to recognize a ligand with high affinity and selectivity but also the capacity, once combined with the specific ligand, to participate in the process of cell activation. It is this dual recognition-activation property, by which the receptor per se acts in part as a message-generating system, that distinguishes receptors from other cell surface recognition and transport constituents. It is to be noted that many pharmacological receptors may be closely associated with ion channels in a supramolecular complex, like the nicotinic receptor-sodium channel or the GABA-benzodiazepine receptor-chloride channel. As is seen in the following section, ligand-regulated ion channel activity comprises one of several transmembrane signaling processes mediated via pharmacological receptors.

General Mechanisms of Transmembrane Signaling

Basic Mechanisms

In order to fulfill their recognition-activation function, membrane receptors must be able to generate an intracellular signal that can be greatly amplified. Structural data that have been obtained to date indicate that a physical contiguity between an extracellular and intracellular receptor domain may be required for receptor-mediated signaling. The mechanisms whereby the transmembrane receptor domain generates the intracellular signal may turn out to be few in number. Based on data accumulated to date, one can single out three basic processes, as illustrated in Figure 1: (1) ligand-modulated ion channel activity, (2) ligand-regulated enzymatic activity, with the catalytic domain situated in the intracellular portion of the receptor, and (3) ligand-regulated liberation of cryptic mediators via interactions of intracellular receptor domains with other submembraneous constituents, such as the so-called G- or N-regulatory components of the adenylate cyclase system (see below). The process of receptor internalization, in which the receptor-bound ligand finds its way into intracellular organelles (9,10), provides an intriguing mechanism whereby one of the three basic reactions performed by receptors can be translocated to a targeted intracellular environment. A key element of receptor function, irrespective of which of the above-cited mechanisms is used, is the basic requirement of the ligand (or its surrogate) to act as an allosteric regulator of the receptor mechanism.

Signal Amplification

A fundamental question to answer is, How does the cell amplify the signal generated by each of the three basic processes just outlined? Fortunately, information acquired to date provides a number of answers to this question for each basic mechanism. For instance, a change in membrane potential resulting from a ligand-induced change in ion flux (Figure 1a) would, as elaborated upon by Zierler and colleagues (11,12), have a profound effect on the orientation of many membrane proteins, thereby rapidly generating an overall intracellular signal. The voltage-sensitive sodium channel (13,14) provides an interesting example of the effect of membrane potential on membrane protein function, with consequent signal amplication. Alternatively, a ligand-regulated channel for an enzyme regulatory cation like calcium could rapidly control intracellular calcium-modulated enzymes that participate in cascade reactions vital to cell regulation. Calcium channels are discussed at greater length in Chapter 20. In principle, ligand-modulated channels for either anions or cations could lead to such amplified intracellular signals. The distinction between the receptor channel activity and the activities of other ion or metab-

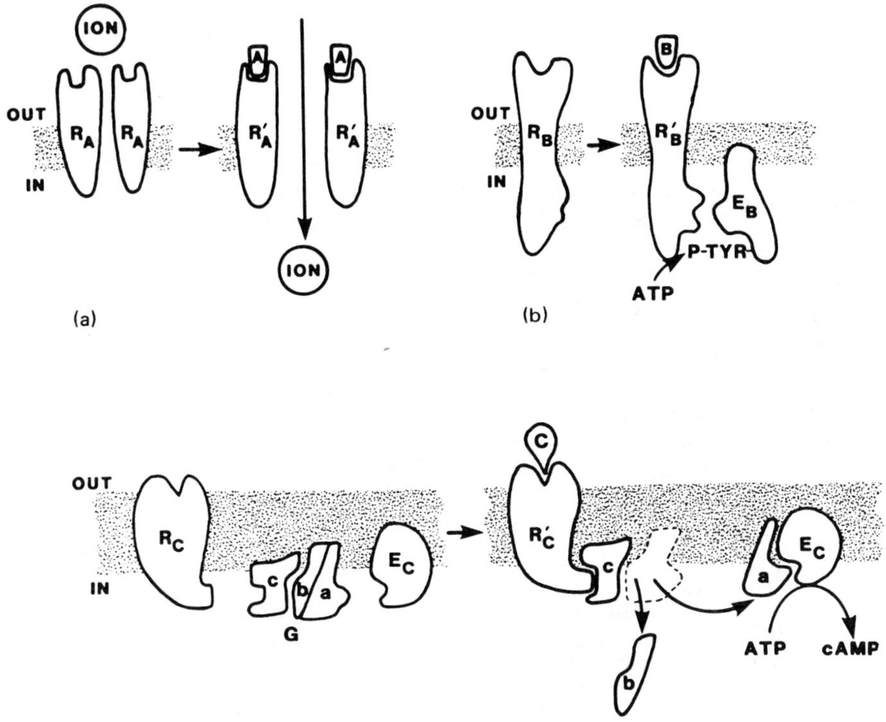

(a) (b)

(c) (d)

Figure 1 Hypothetical models of ligand-activated transmembrane signaling using three basic mechanisms. (a) Ligand-regulated ion channel, like the nicotinic cholinergic receptor. (b) Ligand-regulated enzyme activity. Receptor tyrosine kinase is used as the example in which phosphorylation of a membrane effector E_B plays a role in signaling. (c) Ligand-modulated release of cryptic mediators. Upon interaction with a ligand-occupied receptor, a hypothetical guanine nucleotide regulator complex (G) is shown to release an effector-regulatory component (a) and at least one additional polypeptide (b) from the G-protein oligomer (a, b, c). The b subunit is thought to be common to a variety of G proteins and to be a key modulator of the activity of the a subunits. (From Reference 31.)

olite channels rests entirely on the ligand-regulated properties of the receptor channels. The nicotinic cholinergic receptor represents the best understood ligand-regulated channel of this kind (15-19). The F_c receptor for immunoglobulins can also be seen to function as a ligand-regulated channel for monovalent cations (20,21).

It is comparatively easy to visualize how the second basic receptor process, ligand-regulated enzymatic activity, could yield an amplified receptor signal. The discovery that receptors for epidermal growth factor-urogastrone (EGF-URO) (22,23), insulin (24-27), insulinlike growth factor I (IGF-I, or somatomedin), platelet-derived growth factor (PDGF), and colony-stimulating factor 1 (CSF-1) are ligand-regulated tyrosine kinases (for a brief review, see Reference 28), indicates that phosphorylation-dephosphorylation cascade reactions, known to play a vital role in many metabolic pathways, probably also play a central role in receptor function. Such cascade reactions, initiated by ligand-triggered tyrosine kinase activity (Figure 1b), could readily amplify the initial receptor stimulus. Apart from kinase activities, any other enzymatic activity (e.g., proteolytic activity and phospholipase activity) could in principle be used by a receptor system to initiate a cascade signal amplification process. To date, there is no evidence for or against the existence of receptors with such intrinsic enzymatic activities. Such ligand-driven enzymatic processes could readily generate compounds (e.g., peptides, diacylglycerol, and arachidonic acid) that might serve as intracellular "second messengers." It will be of great interest in the future to look for receptors with intrinsic enzymatic activities other than that of the kinase class.

Those receptors that regulate the production of the second messenger, cyclic AMP, represent well-know examples of receptors that utilize the third basic process outlined previously. Only recently has it been appreciated that the catalytic activity of the adenylate cyclase system is not controlled directly by the receptor itself, but rather by an indirect process, whereby the receptor liberates a separate cyclase-regulatory polypeptide (so-called α_s) (29). The liberation of such cryptic regulatory polypeptides from the so-called guanine nucleotide binding regulatory proteins (G proteins or N proteins) (30,31) can provide the concerted regulation of a variety of cell activities apart from adenylate cyclase. A generalized case for this mechanism is illustrated in Figure 1c. The oligomeric structure of the G proteins (known to contain α, β, and γ subunits) and the discovery of several distinct α substituents (29-31) provides enormous flexibility in the use of such transmembrane signaling process. In terms of the cyclic AMP system, the amplification reactions involving cyclic AMP-dependent protein kinases have been widely studied. This represents perhaps the best understood (biochemically) receptor-mediated signaling system. However, because the common β substituent can combine with (and presumably regulate) a variety of the family of α substituents and because the metabolic functions of the α subunits other than the cyclase-regulatory α_s substituent have yet to be determined, it appears likely that the cyclase system may represent only the tip of the iceberg in terms of the regulation of cell function by this third distinct basic receptor-signaling process. One looks forward eagerly to interesting developments in this area of study.

Messengers

As pointed out in an earlier section, in the situation in which information transfer is mediated via ion or metabolite channels, the ion itself represents the message. In contrast, receptor-mediated information transfer involves the generation of at least one and perhaps multiple diffusible messengers. In one sense, for the action of hormones like insulin, the receptor per se can be called the second messenger (the hormone itself can be thought of as the first messenger in the communication chain leading from hormone binding to cellular activation). For some time, cyclic AMP was designated as a second messenger for the action of such hormones as glucagon or epinephrine. Now, however, it can be appreciated that the so-called α_s-stimulatory subunit of the cyclase-regulatory guanine nucleotide binding complex (29-31) also participates as a messenger in the course of the action of either glucagon or epinephrine. Further, it is not unlikely that a single hormone-receptor combination may initiate not one, but perhaps many membrane-localized reactions. Thus, it may be inappropriate to single out any particular diffusable low-molecular-weight compound as a primary messenger for any hormone; rather, it may be more fruitful to think in terms of a characteristic matrix of diffusable messengers generated by the combination of a single hormone with its receptor. Compounds that have been singled out for particular attention in terms of transmembrane signaling processes are (1) sodium ion (nicotinic cholinergic receptor); (2) cyclic AMP (cyclase-related hormones); (3) cyclic GMP (several neurotransmitter receptors, including the muscarinic cholinergic receptor); (4) calcium (for neurotransmitter receptors, including muscarinic cholinergic as well as for other peptide receptors, such as those for angiotensin and pancreozymin); (5) diacylglycerol (32,33); (6) inositol trisphosphate (32,33); (7) α substituents of the G-protein complex (29-31); (8) as yet unidentified peptide regulators related to the action of insulin (34-38); and (9) an unidentified glucosamine or inositol-containing phospholipid related to the action of insulin (39). It is important to note that in the case of many neurotransmitters and hormones, not one, but three of the identified messengers, namely, calcium, diacylglycerol, and inositol trisphosphate, may be involved in a complex bifurcating signal pathway involving the hydrolysis of membrane phosphoinositides (32,33). In selected cases, two messengers, which alone may not trigger a response, may have to work together to generate an appropriate intracellular signal (e.g., the ability of inositol-2,4,5-trisphosphate and inositol-1,3,4,5-tetrakisphosphate to act in concert to activate sea urchin eggs) (40,41). A critical factor in understanding the chain of information transfer via the receptor to the cell relates to the identification of the key targets of the messengers, namely, the protein kinases regulated by cyclic AMP; kinase C regulated by diacylglycerol (42); intracellular calcium binding proteins from which calcium is released in the presence of inositiol triphosphate; and calcium-modulated regulatory enzymes or proteins like calmodulin that may have a widespread effect on overall cell regulation. Possibly, only a handful of such messengers, generated via membrane-localized reactions, are involved in the action of most neurotransmitters and hormones.

Transmembrane Signaling Mechanisms and Receptor Regulation

A main hypothesis put forward in this chapter is that the same set of reactions triggered by receptors to generate a transmembrane signal is also involved in receptor regulation. As outlined in the following this regulation may in some cases affect only the receptor that initiates the signal; alternatively, the receptor-triggered reactions may coordinately affect a variety of other receptors present in the membrane. Thus, the transmembrane signaling process has built into it a feedback loop that can affect specifically those constituents involved in initiating the signal. The regulatory feedback reactions may affect not only the pharmacological receptors but also the other elements already discussed (i.e., channels and acceptors) that are generally involved in transmitting information across the cell membrane.

RECEPTOR REGULATION AND RECEPTOR-RECEPTOR INTERACTIONS

Levels of Control of Receptor Function

There are many levels at which receptors may be controlled, both with respect to receptor number and with respect to overall receptor activity. In principle, as outlined in Figure 2, control could be exerted at the level of the receptor, at the level of the mediators generated by receptor activity, or at some point farther along in the reaction pathway. Quite removed from the initial signal, for instance, could be the triggering of the biosynthesis of new receptor or of other membrane constituents. For this chapter, attention is focused on the receptor per se, not on control points that might occur distal to the receptor-generated signal.

Receptor control may occur at the level of the gene, where the stage of the cell cycle, the state of cell differentiation, or the influence of such agents as glucocorticoids may change receptor expression, or receptor regulation may occur at an epigenetic level in response both to nonspecific (e.g., change in cell substratum, cell-cell contact, a chemical insult) and receptor-triggered stimuli. When thinking about the possible mechanisms that may be involved in regulating receptors, one must consider not only the ligand binding domain

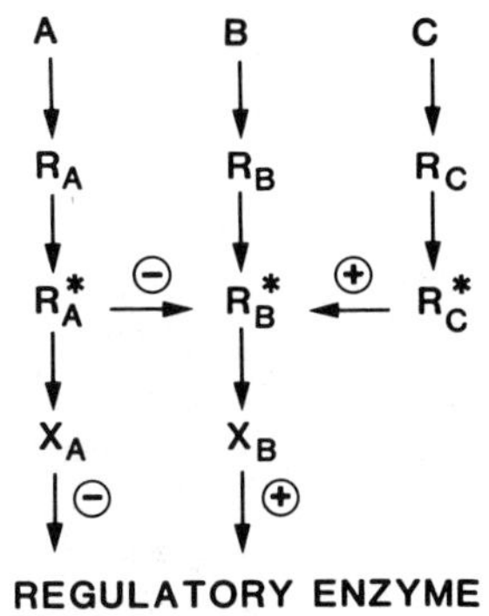

Figure 2 Levels of control of receptor function. The positive cellular signal mediator X_B resulting from the activation of receptor R_B by ligand B is shown to be modulated either at the receptor level via receptors R_A and R_C or at the cellular level by the negative cellular mediator X_A generated by the triggering of receptor R_A by ligand A. The asterisks denote the ligand-activated states of the receptor.

of the receptor but also other receptor domains that may be involved in overall receptor function. Receptor domains apart from the ligand binding region that may be targets for receptor-regulatory processes could include: (1) a receptor catalytic site (e.g., tyrosine kinase); (2) a specific ion binding site (e.g., for sodium, calcium, or chloride); (3) phosphate acceptor sites (serine, threonine, or tyrosine) that control overall receptor conformation; (4) oligosaccharide attachment sites; (5) sites involved in clustering and internalization; and (6) a cryptic peptide-messenger domain that could be liberated proteolytically from the receptor. In summary, irrespective of the stimulus that triggers receptor regulation (i.e., nonspecific or ligand-triggered events), mechanisms that involve the receptor per se can be thought of at the level of specific receptor domains that can be targets for the reactions to be discussed in the following sections.

Mechanisms of Receptor Regulation

As discussed in the second section, a number of receptor-triggered reactions involved in transmembrane signaling may affect receptor function in the course of cell activation. Each of the three basic processes outlined in Figure 1 could ultimately lead to a change in receptor activity via reactions that may or may not involve the formation or breaking of a covalent bond.

Noncovalent Interactions

Noncovalent mechanisms that might regulate receptor function could include (1) changes in membrane potential, (2) changes in receptor distribution (patching or capping, for example), (3) allosteric interactions of the receptor with other membrane protein (e.g., G proteins) or with smaller ligands (cations, anions, and nucleotides), and (4) changes in the membrane lipid environment. For instance, the change in membrane potential generated by receptor-mediated events causes profound changes in the behavior of the voltage-sensitive sodium channel (13,14). As repeatedly pointed out by Zierler and colleagues (11,12), changes in membrane potential could also have a profound effect on the function of many membrane proteins. Thus a change in membrane potential caused by one membrane receptor could readily change the function of other receptors on the same cell surface. Among the other mechanisms already outlined, the interaction of receptors with G-protein complexes can be singled out as potentially playing a critical role in regulating receptor activity. As envisioned by the mobile or floating receptor model developed by a number of investigators (e.g., Reference 43) and summarized in some detail elsewhere (44,45), the noncovalent interaction of the ligand-occupied receptor with a membrane effector, such as a G protein, would be expected to change the receptor's affinity for the ligand. In summary, all four of the noncovalent mechanisms outlined previously could have broad implications for the control of receptor function.

Covalent Modification of Receptors

Several covalent reactions can be singled out for their potential to play a role in the rapid membrane-localized regulation of receptor function. In principle, the potential reactions very likely reflect the same reactions used for the regulation of cytoplasmic enzymes. Thus, it comes as no surprise that phosphorylation-dephosphorylation reactions, involving the receptor as either enzyme or substrate, are prime candidates for the control of receptor activity. Two other covalent reactions that merit attention are (1) disulfide-sulfhydryl interchange and (2) limited proteolysis.

Examples of all three of these mechanisms can now be cited in the context of receptor studies. For instance, the phosphorylation of the EGF-URO receptor caused by the activation of the phorbol ester receptor (i.e., kinase c) is thought to regulate both the binding activity and kinase activity of the EGF-URO receptor (46). Alternatively, desensitization of the β-adrenergic receptor is thought to involve phosphorylation mediated either via a cyclic-AMP-dependent kinase or via a novel kinase that can phosphorylate the β receptor primarily when the receptor is occupied by an agonist (47). Disulfide-sulfhydryl exchange reactions may be important not only for the regulation of the structure and oligomeric forms of the receptors, such as that for insulin (48), but also for the control of the sulfhydryl-reagent-sensitive tyrosine protein kinase activity of growth factor receptors (EGF-URO, insulin, IGF-I, and PDGF). Finally, proteolytic mechanisms may not only control receptor binding (49) but may also liberate limited receptor domains that could regulate intracellular processes (38). Potentially, many of the rapid membrane-localized receptor-receptor interactions that have now been documented (reviewed in References 45 and 50) could be rationalized in terms of the noncovalent or covalent reactions just outlined.

Receptor Homology and Coordinate Receptor Regulation

It has been appreciated for many years that there can be a crossover between some receptor types in terms of the recognition of a single compound, such as epinephrine, which can activate both β_1- and β_2-adrenergic receptors as well as the α receptor. It would come as no surprise, therefore, if structural homologies (e.g., antigenic epitopes or common amino acid sequences) were to be found between, say, the β_1 and β_2 receptors. A priori, however, homologies between receptor domains other than the ligand recognition domain may not have been expected. Now, however, accumulating evidence indicates that many receptor systems may exhibit domains that are homologous with respect to function if not with respect to regions of amino acid sequence. For instance, it is now clear that all receptor species that regulate adenylate cyclase (e.g., β-adrenergic and glucagon) must possess a domain capable of interacting with the regulatory G_s subunits of the cyclase system. Similarly, as mentioned, such receptors as those for insulin, insulinlike growth factor I, epidermal growth factor-urogastrone, and platelet-derived growth factor appear to have a common catalytic domain conferring tyrosine kinase activity on the receptor species. Other functional properties of a variety of receptor and acceptors, such as the ability to internalize via "coated pit" regions of the cell or to migrate to lysosomal structures, imply that there may be a number of common functional domains shared by receptors or acceptors with quite distinct recognition properties.

Perhaps the most striking example of receptor homology that has come to light to date can be seen between the receptors for insulin and IGF-I. These two receptors, which possess selective (but cross-reactive) recognition sites for insulin and IGF-I, have both been found to possess two principal subunits: the α (135 kD) and the β (90 kD) that form the receptor heterotetramer $\alpha_2\beta_2$. The tyrosine kinase activity of both receptors resides in the β subunit, which also possesses the transmembrane domain of the receptor; the α subunits represent extracellular receptor domains linked to the β subunit via disulfide bonds (51-56). The immunologic and enzymatic relationship initially found between the two receptors (57-60) heralded the remarkable amino acid sequence homology that has been observed between the two receptors (53-56). It is highly likely that close structural and functional homologies will also be found between receptors of other classes (e.g., β_1 and β_2 and α_1 and α_2).

This discussion indicates clearly that there may be common domains found in receptors of different classes as well as highly homologous regions in receptors of the same class. Thus, reactions that are found to be involved in the regulation of an individual receptor may also simultaneously regulate many receptors in an individual cell. For instance, the elevation of cyclic AMP caused by activating the β-adrenergic receptor could lead to a concerted control of all receptors that possess a regulatory cyclic AMP kinase substrate domain. Similarly, receptor kinases, such as that discovered for the β-adrenergic receptor (47), may be found to regulate a variety of receptors that act via G proteins by phosphorylating a domain that has been found in common between rhodopsin and the β receptor (61). The "code" for receptor regulation would thus be built into the regulatory reaction and its domain targets, irrespective of the agonist that may have triggered the cell initially. In theory, most of the specific receptor-regulatory reactions outlined here could be used for coordinated receptor control. These reactions, comprising a unique "receptor code," may in principle be either unidirectional or reversible. In practice, the challenge is to identify those receptor-targeted reactions that can be seen to play a role in the physiologic setting of an intact cell.

RECEPTOR CASCADES

Although most of this chapter is focused on receptor regulatory mechanisms situated at the inner aspect of the plasma membrane, it is instructive to consider examples in which the activation of one receptor can, via the secretion of an extracellular messenger, activate a second cell surface receptor. This process can be thought of as a cascade mechanism leading to a positive amplification via a matrix of receptor-mediated signals triggered by the activation of a single receptor. In a sense this kind of cascade can be thought of as an autocrine system that can function within a selected microenvironment for somatic cells. In the nervous system, corelease of a variety of transmitters may play a similar role. Drawing from work by my colleagues and myself, one can point to the generation of IGF-I-like material by the action of EGF-URO on fibroblasts (62,63), to the likely participation of prostaglandins in the contractile response of rat arterial strips to EGF-URO (64), and to the possible participation of surface-localized interferon-related constituents in the course of a mitogenic response to EGF-URO (64). Other and perhaps better examples of this kind of receptor cascade situation are found in the literature; it is not as yet certain as to how widespread this mechanism of receptor-receptor communication may be.

SUMMARY AND PERSPECTIVES FOR FUTURE DEVELOPMENTS

Using the same set of reactions that are designed for communicating with intracellular reaction pathways, receptors can also communicate with each other. This communication can take place via a variety of intramembranous interactions that may or may not result in a covalent modification of the receptors themselves. In some cases, receptors may also communicate with each other via a cascade of secreted messenger molecules that act at the cell surface in an autocrine fashion. The existence of common receptor polypeptide domains and a limited number of cell activation pathways raises the possibility that certain classes of receptors (e.g., those activating adenylate cyclase) may modulate overall cell function by using a common code of receptor-regulatory reactions.

In the chapters that follow, a great deal of information is presented, with a focus on a number of individual receptors. This kind of information sets the stage for exploring receptor-receptor interactions, as outlined in this chapter. On the one hand, coordinate receptor regulation and similar cell responses triggered by different receptors can be rationalized in terms of common receptor domains. On the other hand, receptor-specific reaction pathways and receptor-specific cell responses may be understood in terms of distinct receptor domains and differential substrate specificity of the various receptor kinases. In the future, it should be possible with the use of site-directed mutagenesis, site-specific antibodies, and hybrid synthetic receptor genes to explore in molecular detail the function of specific receptor polypeptide domains. The keys to such studies will be found not only in the new recombinant technologies becoming generally available, but also in the choice of an appropriate propagable cell system in which one may study both receptor biochemistry and a receptor-triggered cellular response. It is hoped that this chapter and this volume provide some stimulus for such studies in the future.

ACKNOWLEDGMENTS

Work in the author's laboratory, summarized in part in this chapter, is supported largely by funds from the Medical Research Council of Canada, the Alberta Heritage Foundation for Medical Research, the Alberta Heritage Savings Trust Fund—Cancer, and the Alberta Heart and Stroke Foundation.

REFERENCES

1. P. Ehrlich, in *The Collected Papers of P. Ehrlich* (F. Himmelweit, M. Marquardt, and H. Dale, eds.), Vol. 3, Pergamon, Oxford, 1956, p. 183.
2. J. N. Langley, *Proc. R. Soc. Lond. [Biol.]* *78*:170 (1906).

3. K. K. Kontula, O. A. Janne, and C. W. Bardin, in *The Receptors* (P. M. Conn, ed.), Vol. IV, Academic Press, New York, 1986, p. 37.
4. R. M. Riehl and D. O. Toft, in *The Receptors* (P. M. Conn, ed.), Vol. IV, Academic Press, New York, 1986, p. 149.
5. M. A. Venis, in *The Receptors* (P. M. Conn, ed.), Vol. IV, Academic Press, New York, 1986, p. 275.
6. W. V. Welshons and J. Gorski, in *The Receptors* (P. M. Conn, ed.), Vol IV, Academic Press, New York, 1986, p. 97.
7. M. D. Hollenberg, *Pharmacol. Rev. 30*:393 (1979).
8. J. Kaplan, *Science 212*:14 (1981).
9. I. H. Pastan and M. C. Willingham, *Science 214*:504 (1981).
10. B. I. Posner, M. N. Khan, and J. J. M. Bergeron, in *Polypeptide Hormone Receptors* (B. I. Posner, ed.), Marcel Dekker, New York, 1985, p. 61.
11. K. Zierler, in *Insulin, Its Receptor and Diabetes* (M. D. Hollenberg, ed.), Marcel Dekker, New York, 1985, p. 141.
12. K. Zierler and E. M. Rogus, *Fed. Proc. 40*:121 (1981).
13. W. A. Catterall, *Trends Neurosci. 5*:303 (1981).
14. M. Noda, S. Shimizu, T. Tanabe, T. Takai, T. Kayano, T. Ikeda, H. Takahashi, H. Nakayama, Y. Kanaoka, N. Minimino, K. Kanagawa, H. Matsuo, M. A. Raftery, T. Hirose, S. Inayama, H. Hayashida, R. Miyata, and S. Numa, *Nature 312*:121 (1984).
15. B. M. Conti-Tronconi and M. A. Raftery, *Ann. Rev. Biochem. 51*:491 (1982).
16. J. L. Marx, *Science 219*:1055 (1983).
17. M. Mishina, T. Tobimatsu, K. Imoto, K.-I. Tanaka, Y. Fujita, K. Fukuda, M. Kurasaki, H. Takahashi, Y. Morimoto, T. Hirose, S. Inayama, T. Takahashi, M. Kuno, and S. Numa. *Nature 313*:364 (1985).
18. M. Noda, H. Takahashi, T. Tanabe, M. Toyosato, S. Kikyotani, Y. Furutani, T. Hirose, H. Takashima, S. Inayama, T. Miyata, and S. Numa, *Nature 302*: 528 (1983).
19. P. Taylor, R. D. Brown, and D. A. Johnson, *Curr. Top. Membr. Transport 18*:407 (1983).
20. I. S. Mellman and J. C. Unkeless, *J. Exp. Med. 152*:1048 (1980).
21. J. D.-E. Young, J. C. Unkeless, H. R. Kaback, and Z. A. Cohn, *Proc. Natl. Acad. Sci. USA 80*:1636 (1983).
22. G. Carpenter, in *Tissue Growth Factors, Handbook of Experimental Pharmacology* (R. Baserga, ed.), Springer Verlag, New York, 1981, p. 89.
23. G. Carpenter and S. Cohen, *Ann. Rev. Biochem. 48*:193 (1979).
24. M. Kasuga, Y. Zick, D. L. Blithe, M. Grettaz, and C. R. Kahn, *Nature 298*: 667 (1982).
25. O. M. Rosen, R. Herrera, Y. Olowe, L. M. Petruzzelli, and M. H. Cobb, *Proc. Natl. Acad. Sci. USA 80*:3237 (1983).
26. R. A. Roth and D. J. Cassell, *Science 219*:299 (1983).
27. E. Van Obberghen and A. Kowalski, *FEBS Lett. 143*:179 (1982).
28. M. D. Hollenberg, *Trends Pharmacol. Sci. 3*:271 (1982).
29. A. G. Gilman, *J. Clin. Invest. 73*:1 (1984).

30. M. Rodbell, *Nature 284*:17 (1980).
31. J. K. Northup, in *Molecular Mechanisms of Transmembrane Signaling* (P. Cohen and M. Houslay, eds.), Elsevier (Biomedical Division), Amsterdam, 1985, p. 91.
32. M. J. Berridge and R. F. Irvine, *Nature 312*:315 (1984).
33. M. J. Berridge, *Biochem. J. 220*:345 (1984).
34. L. Jarett and J. R. Seals, *Science 206*:1407 (1979).
35. J. Larner, G. Galasko, K. Cheng, A. A. DePaoli-Roach, L. Huang, P. Daggy, and J. Kellogg, *Science 206*:1408 (1979).
36. J. R. Seals and M. P. Czech, *J. Biol. Chem. 255*:6529 (1980).
37. A. R. Saltiel, M. I. Siegel, S. Jacobs, and P. Cuatrecasas, *Proc. Natl. Acad. Sci. USA 79*:3513 (1982).
38. F. L. Kiechle and L. Jarett, in *Insulin, Its Receptor and Diabetes* (M. D. Hollenberg, ed.), Marcel Dekker, New York, 1985, p. 181.
39. A. R. Saltiel, J. A. Fox, P. Sherline, and P. Cuatrecasas, *Science 233*:967 (1986).
40. R. F. Irvine and R. M. Moore, *Biochem. J. 240*:917 (1986).
41. B. Michell, *Nature 324*:613 (1986).
42. Y. Nishizuka, *Nature 308*:693 (1984).
43. S. Jacobs and P. Cuatrecasas, in *Polypeptide Hormone Receptors* (B. Posner, ed.), Marcel Dekker, New York, 1985, p. 39.
44. M. D. Hollenberg and H. J. Goren, in *Mechanisms of Receptor Regulation* (G. Poste and S. T. Crooke, eds.), Plenum Press, New York, 1985, p. 323.
45. M. D. Hollenberg, *Experientia 42*:718 (1986).
46. J. Downward, M. D. Waterfield, and P. J. Parker, *J. Biol. Chem. 260*:14538 (1985).
47. J. L. Benovic, R. H. Strasser, M. G. Caron, and R. J. Lefkowitz, *Proc. Natl. Acad. Sci. USA 83*:2797 (1986).
48. J. M. Maturo III, M. D. Hollenberg, and L. S. Aglio, *Biochemistry 22*:2579 (1983).
49. P. Cuatrecasas, *J. Biol. Chem. 246*:6522 (1971).
50. M. D. Hollenberg, in *Mechanisms of Receptor Regulation* (G. Poste and S. T. Crooke, eds.), Plenum Press, New York, 1985, p. 295.
51. J. Massague and M. P. Czech, *J. Biol. Chem. 257*:5038 (1982).
52. M. D. Lane, in *Insulin, Its Receptor and Diabetes* (M. D. Hollenberg, ed.), Marcel Dekker, New York, 1985, p. 237.
53. A. Ullrich, J. R. Bell, E. Y. Chen, R. Herrera, L. M. Petruzzelli, T. J. Dull, A. Gray, L. Coussens, Y.-C. Liao, M. Tsubokawa, A. Mason, P. H. Seeburg, C. Grunfeld, O. M. Rosen, and J. Ramachandran, *Nature 313*:756 (1985).
54. Y. Ebina, L. Ellis, K. Jarnagin, M. Edery, L. Graf, E. Clauser, J.-H. Ou, F. Maslarz, Y. W. Kan, I. D. Goldfine, R. A. Roth, and W. J. Rutter, *Cell 40*:747 (1985).
55. Y. Fujita-Yamaguchi, T. R. LeBon, M. Tsubokawa, W. Henzel, S. Kathuria, D. Koyal, and J. Ramachandran, *J. Biol. Chem. 261*:16727 (1986).
56. A. Ullrich, A. Gray, A. W. Tam, T. Yang-Feng, M. Tsubokawa, C. Collins,

W. Henzel, T. LeBon, S. Kathuria, E. Chen, S. Jacobs, U. Francke, J. Ramachandran, and Y. Fujita-Yamaguchi, *EMBO J.* *5*:2503 (1986).
57. S. D. Chernausek, S. Jacobs, and J. J. Van Wyk, *Biochemistry 20*:7345 (1981).
58. B. Bhaumick, G. D. Armstrong, M. D. Hollenberg, and R. M. Bala, *Can. J. Biochem. 60*:923 (1982).
59. G. D. Armstrong, M. D. Hollenberg, B. Bhaumick, and R. M. Bala, *J. Cell. Biochem. 20*:283 (1983).
60. G. D. Armstrong, M. D. Hollenberg, B. Bhaumick, R. M. Bala, and J. M. Maturo III, *Can. J. Biochem. Cell Biol. 61*:650 (1983).
61. B. K. Kobilka, R. A. F. Dixon, T. Frielle, H. G. Dohlman, M. A. Bolanowski, I. S. Sigal, T. L. Yang-Feng, U. Francke, M. G. Caron, and R. J. Lefkowitz, *Proc. Natl. Acad. Sci. USA 84*:46 (1987).
62. P. R. Atkison, R. M. Bala, and M. D. Hollenberg, *Can. J. Biochem. Cell Biol. 62*:1335 (1984).
63. P. R. Atkison, L. J. Hayden, R. M. Bala, and M. D. Hollenberg, *Can. J. Biochem. Cell Biol. 62*:1343 (1984).
64. I. Muramatsu, M. D. Hollenberg, and K. Lederis, *Can. J. Physiol. Pharmacol. 63*:994 (1984).
65. S. Lin, P. O. P. Ts'o, and M. D. Hollenberg, *Life Sci. 32*:1479 (1983).

2

Receptor Binding Theory and Methodology

Milt Titeler / Albany Medical College, Albany, New York

This chapter is intended to be a brief guide to understanding the theoretical basis of research concerning the direct interaction of neurotransmitters, hormones, and/or drugs ("ligands") with their biologic sites of action. A powerful approach to studying ligand-receptor interactions is to label the ligand radioactively and study the radioactive signal produced when the radioactive ligand is allowed to interact with biologic tissues. The first major aim of this chapter is to bridge the gap between pure ligand binding site theory and radioligand binding site theory. The second aim of the chapter is to discuss the major methodologies employed; this section explains in general terms, the techniques that are commonly used. The theoretical discussion should help the reader understand the rationale behind the methodologies used in radiolabeling studies of receptors; that is, it should become clear that under the appropriate conditions it is inevitable that reliable receptor radiolabeling studies have been and will continue to be developed. In the final sections of this chapter the flexibility of the general strategies involved in radiolabeling receptors is illustrated by briefly discussing two subspecialities of radiolabeling studies, autoradiography and receptor-GTP binding protein interactions. The goal of these later sections is to provide the reader with a perspective on these facets of radiolabeling studies, that is, why these techniques and strategies may be of interest to a researcher concerned with a specific question.

In summary this chapter is intended to answer the following questions, among others:

What is an IC_{50}?

How does one differentiate artifactual binding of a radioligand from receptor-associated binding?

Should a planned radiolabeling study be done in homogenates or in tissue-mounted slices (autoradiography)?

Can intracellular events coupled to receptors be monitored with radiolabeling studies, or is the technique useful only for counting the number of sites?

SIMPLE BINDING THEORY

The theoretical basis of the radioreceptor assay is derived from the law of mass action as applied to a ligand and a receptor. The following derivations involve the general thermodynamic and kinetic laws governing the interactions of ligands and binding sites adapted to the specific case of a radioligand.

$$D^*R \underset{k_1}{\overset{k_2}{\rightleftharpoons}} D^* + R \tag{1}$$

where D^* = radioactive drug and R = receptor;

$$\frac{d(D^*R)}{dt} = k_1(D^*)(R) \qquad -\frac{d(D^*R)}{dt} = k_2(D^*R)$$

where D^*R = drug-receptor complex; at equilibrium,

$$\frac{d(D^*R)}{dt} = -\frac{d(D^*R)}{dt}$$

$$\frac{k_2}{k_1} = K_{D^*} = \frac{(D^*)(R)}{(D^*R)} \tag{2}$$

where K_{D^*} = equilibrium dissociation constant of D^* for R. Experimentally, (D^*) and (D^*R) can be measured (see below) but (R) generally cannot:

$$(R_T) = (D^*R) + (R) \qquad (R) = (R_T) - (D^*R)$$

where R_T = total receptor population = $(D^*R)_{max}$; substituting

$$K_{D^*} = \frac{(D^*)[(D^*R)_{max} - (D^*R)]}{(D^*R)} \tag{3}$$

and rearranging

$$\frac{(D^*R)}{(D^*R)_{max}} = \frac{1}{(1 + K_{D^*}/D^*)} = \frac{(D^*)}{[(D^*) + K_{D^*}]} \tag{4}$$

$(D^*R)/(D^*R)_{max}$ is simply the fraction of total receptors occupied by radioactive ligands, that is, the fractional occupancy.

SATURATION CURVE AND SCATCHARD ANALYSIS

In order to understand fully the meaning of Equation (4), let D* range from extremely high concentrations to extremely low concentrations. At concentrations of D* that are so high as to make K_{D*}/D^* approach 0, fractional occupancy approaches 1. This means the receptors have become fully occupied or saturated by radioactive ligand. At concentrations of D* that are so low that $K_{D*}/D^* \gg 1$, fractional occupancy approaches 0. At a concentration of $D^* = K_{D*}$, fractional occupancy = 1/2 = 50%. In fact, a table of fractional occupancies calculated from various values of K_{D*}/D^* can be produced and a theoretical saturation curve plotted (Table 1 and Figure 1).

The amount of radioactive drug that produces 50% occupancy is by definition the experimentally determined K_{D*} [see Equation (4)]. As it is experimentally difficult to create conditions in which exactly 50% of the receptors are occupied, transformation of the binding equation to a form that allows an estimate of the K_{D*} and the $(D^*R)_{max}$ is commonly performed [Equation (5)].

$$\frac{(D^*R)}{(D^*)} = -\frac{(D^*R)}{(K_{D*})} + \frac{(D^*R)_{max}}{K_{D*}} \tag{5}$$

Table 1 Effect on the Fractional Occupancy of a Binding Site by Increasing Radioactive Ligand Concentrations Using Equation (5)[a]

KD*/D*	Fractional occupancy (%)	$\log \frac{\%\ \text{bound}}{100 - \%\ \text{bound}}$
0.01	99	1.99
0.05	95	1.28
0.10	91	1.00
0.50	67	0.31
1.00	50	0.00
5.00	17	−0.69
10.00	9	−0.95
50.00	2	−1.89
100.00	1	−1.99

[a]In the calculation the radioactive drug D* ranges from 1/100 the equilibrium dissociation constant K_{D*} to 100 times the K_{D*}. The far right column represents calculations of the left-hand side of Equation (6) for the determination of the Hill coefficient.

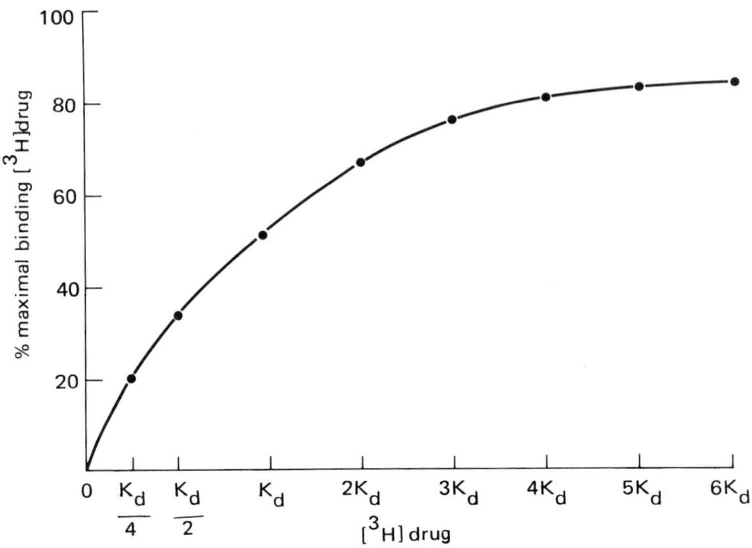

Figure 1 Theoretical saturation curve. Radioactive drug concentration $[^3H]$ drug is expressed in terms of K_D (equilibrium dissociation constant). The points on the curve are from the calculations presented in Table 1.

where (D^*R) = bound (B in Figure 2) and (D^*) = free drug concentration (F in Figure 2). The graph created by this equation is known as a Scatchard plot (1). Figure 2 is a Scatchard plot of the theoretical saturation curve (Figure 1).

From this discussion it is apparent that by varying the amount of radioactive ligand in the tissue incubation solution, the K_{D^*} and total receptor population $(R_T) = (D^*R)_{max}$ can be determined. This is generally the first series of experiments performed in radioreceptor assays.

The amount of information about a receptor that can be determined by this type of experiment is limited. A saturation curve indicates that a radioactive drug apparently binds with a characteristic affinity to a binding site or sites. This does not prove that the binding site or sites are to physiologically relevant receptor(s) or that the interaction(s) being measured is of biologic significance.

One of the major technical limitations of the typical radioligand receptor assay is the limited levels of radioactive ligand that can be used effectively (see Technical Considerations). Therefore, if the ligand is binding to a site with a relatively low affinity, significant fractional occupancy of the binding site is impossible to monitor accurately. Recently there have been strong criticisms of the use (or misuse) of the Scatchard transformation of saturation curve data (2).

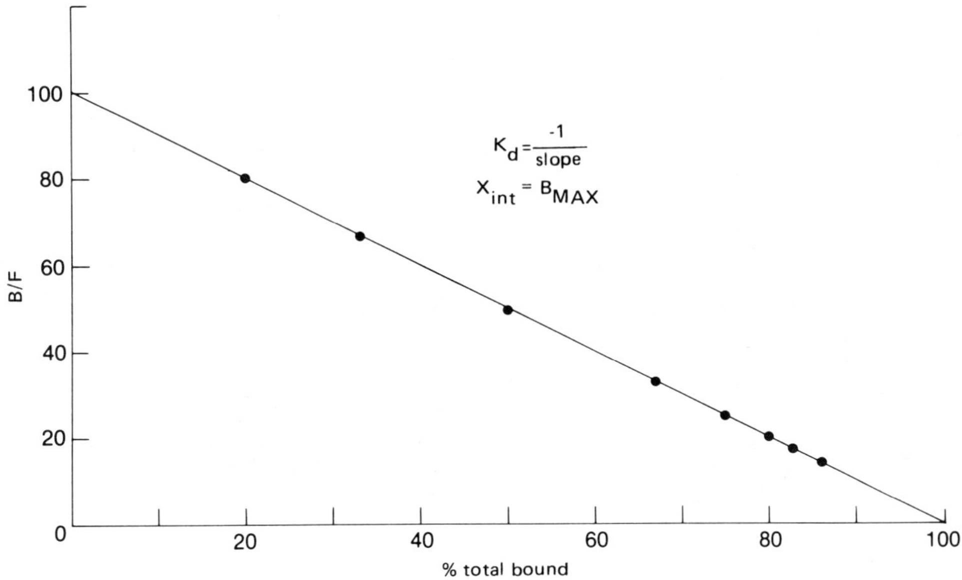

Figure 2 Scatchard analysis of saturation curve data (Figure 1). Bound drug B is represented as a percentage of maximal competible binding. The points on the line were derived from the calculations in Table 1.

Basically the point of the criticism is that more than 50% occupancy of a binding site by a radioactive ligand is necessary to determine reliable parameters of the interaction of that site by the ligand (with or without computer-assisted analysis); see Chapter 3 (McPherson). The scientific literature is replete with Scatchard transformations of insufficient saturation curve data. Thus the problem may not only be with the Scatchard transformation but also with the investigators performing the insufficient saturation curve experiments It is hoped that chapters such as this will lead to an improvement in the design, execution, and interpretation of all aspects of radioligand binding experiments.

HILL PLOTS

The binding of a radioactive drug to a binding site should follow the law of mass action and produce a hyperbolic function, as displayed in Figure 1. The fractional occupancy of a binding site by a drug should range from 2 to 95% over approximately four orders of magnitude of drug concentration (Table 1). In general, deviations from the expected hyperbolic function for the binding to one ligand to one site are not readily observed from the hyperbolic saturation curve, and the Scatchard transformation also has severe limitations in detecting deviations from simple one-site interactions (see Multiple Binding Sites; Technical Considerations). In order to quantitate the mode of saturation (i.e., is the saturation curve consistent with a simple one-site interaction?), the Hill plot (3) is generally employed. Equation (6) is the Hill equation adapted for binding results:

$$\log \frac{\text{fractional occupancy}}{1 - \text{fractional occupancy}} = \text{Hill coefficient} \times \log[\text{D}^*] \qquad (6)$$

where $[\text{D}^*]$ = free concentration of radioactive drug.

By plotting log $[\text{D}^*]$ on the x axis and the left side of the equation (Table 1), for each value of log $[\text{D}^*]$ on the y axis a plot is obtained in which the slope of the line is the Hill coefficient (Figure 3). A Hill coefficient approximately equal to unity indicates independent binding of a radioactive drug to a site or sites with one apparent affinity. A Hill coefficient significantly greater

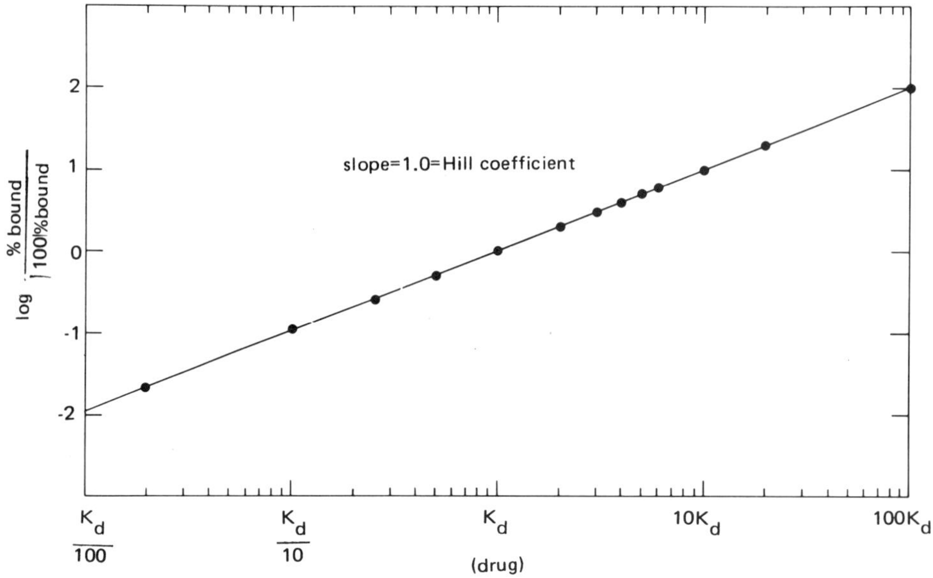

Figure 3 Theoretical Hill plot of the saturation data presented in Figure 1. Drug concentrations are presented in terms of K_D. The slope of the line represents the Hill coefficient. The points on the line were derived from the calculations in Table 1.

than unity indicates positive cooperativity among the binding sites, that is, the binding of one drug molecule to a site in some way increases the affinity of another drug molecule for the next unoccupied site (rarely observed in receptor pharmacology). A Hill coefficient significantly less than unity indicates negative cooperativity or multiple binding sites for the radioactive compound. The presence of drug interactions with receptors producing Hill coefficients less than unity is not uncommon. Generally these types of interactions have been found to be caused by either multiple receptor interactions or multiple receptor-state interactions. These issues are pursued further in later sections.

COMPETITION CURVES

A great deal of information can be obtained by adding various concentrations of nonradioactive drugs to the incubation medium and observing the effects they have on the binding properties of the radioactive drug. This type of study is known as a competition experiment. The results of competition studies are routinely used to identify the pharmacological nature of the radioactively labeled receptor. These types of studies also provide the strongest evidence for the selectivity or nonselectivity of a radioligand (see Multiple Binding Sites). Computer-assisted analyses of competition curves have become important tools for studying multiple-state receptors (see the last section). If the nonradioactive drug binds to the same site(s) as the radioactive drug, it will interfere with the binding of the radioactive drug. This is the molecular basis of direct competitive pharmacological antagonism. It is unfortunate that the word "displacement" has become so widely used in discussing competition studies. Typically there is no displacement of one drug by another at a receptor binding site; when the binding site is occupied by one drug there is no attraction for the receptor from another drug. When the first drug dissociates from the binding site then the other drug may bind to the now unoccupied site. Displacement implies some sort of influence by one drug on the drug-receptor interaction of another drug; this kind of molecular mechanism is not part of classical competitive antagonism. However, if an allosteric mechanism is suspected in the inhibition of the specific radioactive drug signal by a nonradioactive drug, then displacement may be a more appropriate term.

In the typical competition experiment the concentration of radioactive drug and the amount of tissue are kept constant while the nonradioactive, competing drug concentrations are varied. The equation describing this type of experiment is as follows:

$$\frac{(D^*R)}{(D^*R)_{max}} = \frac{1}{1 + [(K_{D^*}/D^*)(1 + D/K_D)]} \qquad (7)$$

where D = nonradioactive drug and K_D = equilibrium dissociation constant of the nonradioactive drug for the receptor.

At low concentrations of nonradioactive drug, D/K_D approaches zero and little or no competition is observed. At high concentrations of nonradioactive drug, $D/K_D \gg 1$ and radioactive drug fractional occupancy is reduced.

A table of the fractional occupancy of the radioactive drug to the receptor population can be calculated from various values of D/K_D (Table 2). Figure 4 is a theoretical competition curve based on the theoretical data presented in Table 2.

A great deal of information about a binding site can be determined from a series of competition curves. The concentration of a nonradioactive drug that inhibits 50% of the radioactive drug's "specific" binding is called the IC_{50}. "Specific" binding is more precisely defined later (see Technical Considerations); for this discussion specific binding refers to the binding of the radioactive ligand to sites for which nonradioactive drugs with selectivity for the given recognition site can compete. At this concentration of nonradioactive drug (the IC_{50}), one-half of the binding sites originally occupied by the radioactive drug are occupied by the nonradioactive drug. Although by definition the concentration of a drug that occupies 50% of a population of binding sites is the K_D, this definition applies only when the drug is free to interact with the receptor in the absence

Table 2 Effect of a Competing Drug on the Binding of a Radioactive Drug to One Site Using Equation (7)[a]

D/K_D	Fractional occupancy (%)
0.01	99
0.05	95
0.10	91
0.50	67
1.00	50
5.00	17
10.00	9
50.00	2
100.00	1

[a]The competing drug concentration ranges from 1/100 the K_D to 100 times the K_D. The concentration of D^* is assumed to be equal to the K_{D^*}, and thus the fractional occupancy in the absence of competing drug is 50%. This initial fractional occupancy is normalized to 100% for the competition calculation.

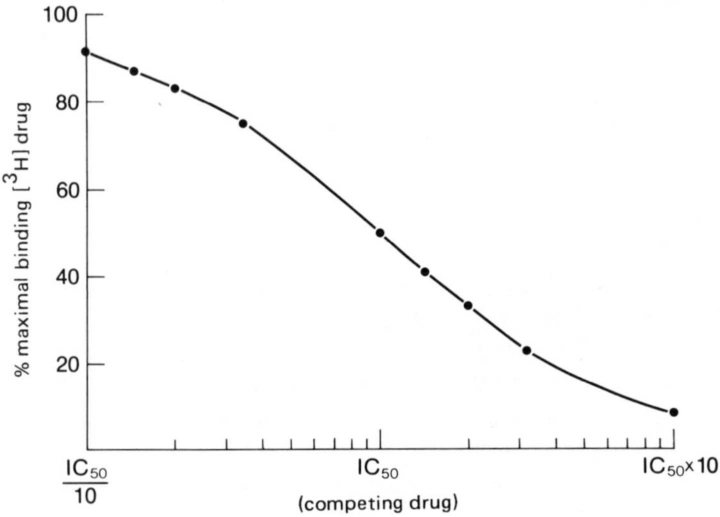

Figure 4 Theoretical competition curve. The data are from the calculation in Table 2. The IC_{50} represents the concentration of drug that inhibits specific radioactive drug [^{3}H] drug by 50%.

of any other interfering drug. In the case of the radioligand binding assay, there is always radioactive drug present in concentrations sufficient to bind to the receptor. The radioactive drug interferes with nonradioactive drug binding just as nonradioactive drug binding interferes with radioactive drug binding. In order to ascertain the K_i from the IC_{50}, the effect produced by the radioactive drug must be included. Equation (8), described by Cheng and Prusoff (4), converts the IC_{50} values of a competition curve to an apparent K_i:

$$K_i = \frac{IC_{50}}{1 + (D^*/K_{i^*})} \tag{8}$$

where IC_{50} = concentration of nonradioactive drug that inhibits 50% of the radioactive drug binding, K_i = apparent equilibrium dissociation constant of a nonradioactive drug for the radioactive drug binding site, D^* = free concentration of the radioactive drug in the incubation solution, and K_{i^*} = equilibrium dissociation constant of the radioactive drug for the binding site.

In theory this equation only holds for competition curves producing Hill coefficients of unity; however it is commonly employed in the receptor literature in situations involving competition curves with Hill coefficients less than unity. Presumably the reason for this is that most authors and referees prefer to see tables with "K_i" values as headings, rather than "IC_{50}" values as headings.

A reasonable compromise in such a situation is to use the Cheng-Prusoff equation but note which drugs did not produce Hill coefficients close to unity (i.e., the K_i value for that drug should be interpreted with caution).

Hill coefficients can be calculated from competition curves just as they can be calculated for saturation curves (see above). Equation (9) is the Hill equation adapted for the competition type of experiment:

$$\log \frac{\%\ \text{inhibition}}{100\%\ \text{inhibition}} = \text{Hill coefficient} \times \log\,[\text{D}] \tag{9}$$

where D = nonradioactive competing drug. Just as with saturation experiments, a Hill coefficient greater than 1 indicates positive cooperativity in the inhibition of the radioligand binding. A Hill coefficient significantly less than 1 indicates negative cooperativity or that the radioligand is labeling multiple sites and the competing radioligand is competing for these sites (see below). A Hill coefficient of 1 indicates the competing drug is inhibiting the radioligand binding to one site or that both the radioligand and the competing drug are interacting with multiple sites with similar affinities for the radioligand and similar affinities for the competing drug. These issues are dealt with in more detail in the next section. It is not uncommon in the receptor binding literature to see phases of competition that appear to begin and end over one logarithmic unit of drug concentration. A Hill coefficient calculation of that phase of inhibition would reveal an extremely high number. Generally such reports have been the result of improper technique and interpretation, and when the experiments have been repeated with improved technique (better radioligands and more competing drug concentrations; see Technical Considerations), a "shallower" competition curve is reported with competition actually occurring over at least four logarithmic units of competing drug concentration.

MULTIPLE BINDING SITES

It is well known that drugs can have high affinity for more than one receptor. It is not surprising that, when a drug is made radioactive (see Chapter 5) and applied to a biologic tissue preparation, more than one binding site is labeled. This situation presents theoretical and technical problems to the investigator interested in identifying the sites labeled and/or examining the properties of these sites individually. Of course, one way to avoid such a situation is to use a radioligand that labels only the receptor of interest. However, in order for the responsible investigator to establish (from his or her own work or from the literature) that the "specific" radioligand chosen possesses the properties desired, the concepts and strategies discussed in this section need be understood. In other words, in order to prove that a radioligand labels only one site it is important to establish that it does not label multiple sites.

Table 3 Effect on the Fractional Occupancy of Two Binding Sites by Increasing Radioactive Drug Concentration

[D*] (nM)	Site A[a] (fmol/mg)	Site B[b] (fmol/mg)	Percentage[c]
0.2	17	4	10.5
0.5	33	9	21
1.0	50	17	33
2.0	67	29	48
3.0	75	37	56
4.0	80	44	62
5.0	83	50	67
10.0	91	67	79
20.0	95	80	88

[a]100 fmol/mg; K_{D*} = 1 nM.
[b]100 fmol/mg; K_{D*} = 5 nM.
[c]Percentage of total available sites occupied by D* at the given concentrations.

The first experimental step in determining whether the tritiated ligand with which one is dealing is binding to more than one site is to produce a saturation curve and a Scatchard analysis of these data. As an example, let us imagine a preparation containing two binding sites of equal densities and fivefold differences in affinities for a tritiated drug. Table 3 lists the calculated binding of the tritiated drug to each site and to the total population of sites.

The theoretical saturation curve produced in such a situation is depicted in Figure 5. The Scatchard analysis of a single binding site produces a straight line whose slope equals the negative reciprocal of the K_{i*} (Figure 2). Figure 6 is a Scatchard analysis of Figure 5. There are actually two lines of fivefold differences in slope produced by the binding of the radioactive drug to the two binding sites. Yet the Scatchard analysis reveals a curvilinear plot. It is more often than not impossible for an experimenter to decide whether one line or two lines form a curved line. The points depicted in Figures 5 and 6 are ideal in that there are no error bars included. Actual experimental data produce a degree of variability that further complicates interpretations. If the two sites have closer affinities and/or different densities than the example illustrated here, the interpretation becomes even more difficult. It should also be pointed out that because of technical considerations (see below), the concentration of radioactive drug than can be effectively applied in the typical receptor binding assay is limited. Thus relatively low affinity sites (K_{i*} values of 10^{-8} M or greater) cannot be significantly labeled and thus not reliably assayed. Taking these

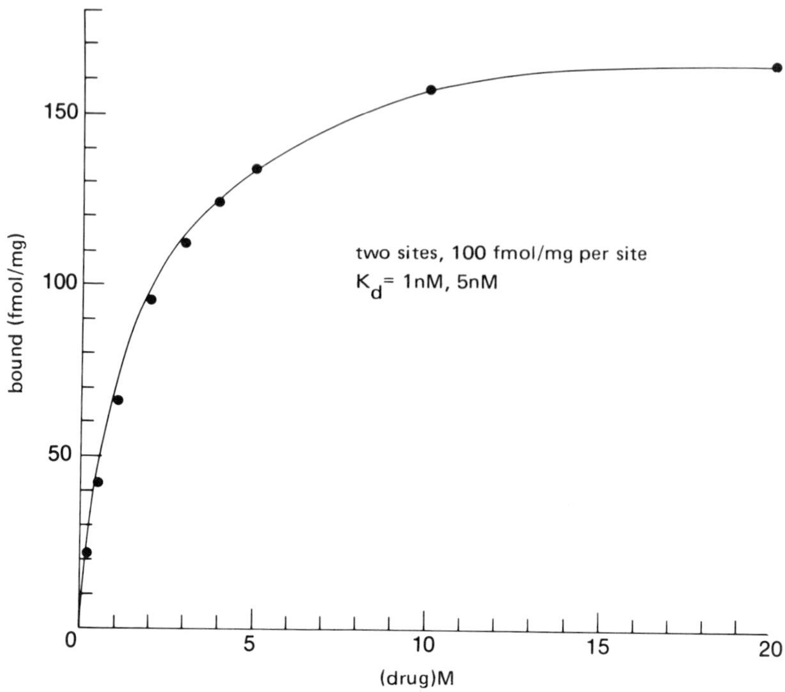

Figure 5 Theoretical saturation curve obtained assuming two binding sites of equal densities and fivefold differences in affinities (see text).

factors into consideration it becomes clear that, for the task of establishing multiple site binding of a radioactive ligand and determining the parameters (the B_{max} and K_{i*}) of the ligand for the multiple sites, the Scatchard analysis is a relatively weak tool.

The Hill plot is somewhat helpful (Figure 7). The saturation curve data from Figure 5, replotted as a Hill plot in Figure 7, produce a Hill coefficient of 0.83. A Hill coefficient less than unity may result from either multiple sites or negative cooperativity. Generally, even with the variability introduced in actual experimental situations, it is possible to statistically establish a Hill coefficient below unity in cases of multiple radioligand site binding. Unfortunately, the Hill plot does not allow a quantitation of the sites or an examination of their properties, such as their respective affinities for the radioligand. Thus Hill plots are also of limited value.

Competition curve experiments represent an additional tool for the examination of multiple sites. If there are two high-affinity binding sites for the

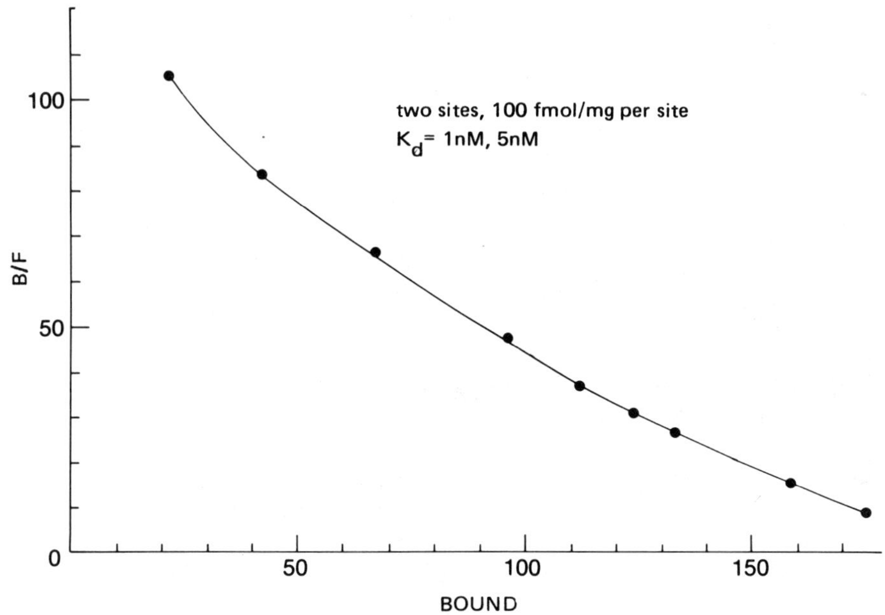

Figure 6 Theoretical Scatchard analysis of the saturation data presented in Figure 5.

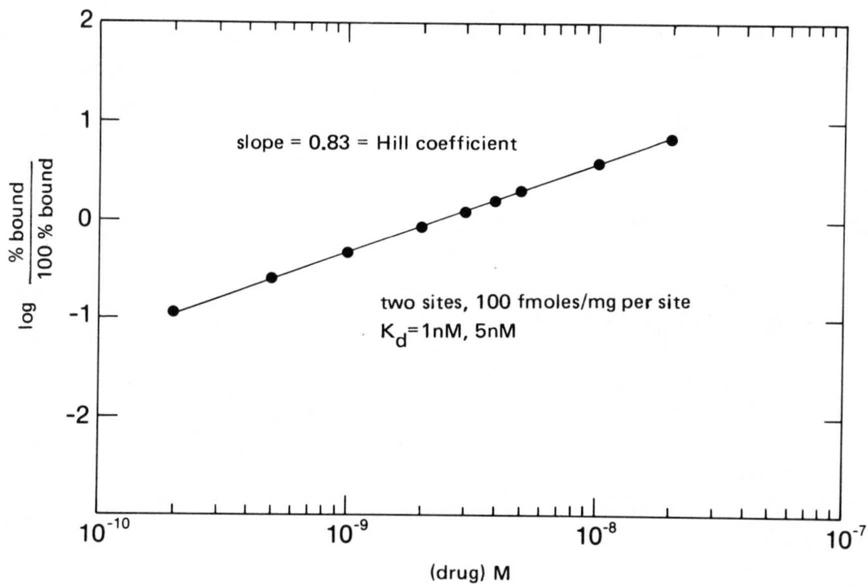

Figure 7 Theoretical Hill plot of the saturation data presented in Figure 5.

radioactive drug, it is very likely that nonradioactive drugs are available that display very different affinities for the two sites. By allowing these drugs to compete with the radioactive drug for the binding sites, "flat" or "shallow" competition curves may be produced (Figure 8). If the competing drug is an antagonist, the presence of a "shallow" competition curve is strong evidence for multiple sites; if the competing drug is an agonist, the situation is more complex (see the last section). If the affinity differences are substantial enough (>1000-fold), there will be a concentration of the nonradioactive drug that essentially saturates the high-affinity site (20 nM in Figure 8) and essentially avoids the lower affinity site. By including this concentration of drug in all subsequent experiments, the properties of the second site can be examined selectively. To determine the properties of the first binding site, one can either subtract the second-site data from the data produced by total binding, or one can find a drug that selectively blocks the second site. Alternatively, computer-assisted analyses of the competition curves can be used to determine the properties of the nonradioactive drug interactions with the labeled sites. Note in Figure 8 that the nonradioactive form of the radioactive drug produced a "steep" competition curve. This is generally the case because if multiple sites are labeled by the radioligand, the sites must have similar affinities for the radioligand (see Technical Considerations). Thus the results using the radioactive ligand and its nonradioactive form usually appear to indicate the labeling of one site.

Generally, if there are multiple binding sites for the radioactive drug, most drugs produce competition curves with Hill coefficients of less than 1. Drugs that possess the same basic structure as the radioactive drug are the most likely candidates to produce Hill coefficients close to 1, indicating an equal affinity for both sites. As just mentioned, in multiple-site situations the radioligand itself generally has similar affinities for the multiple sites; it is not surprising that molecules strongly resembling the radioligand would also have similar high affinities for the sites labeled. Therefore it is always advisable to test drugs from other structural families when multiple-site binding is being investigated; these structurally dissimilar drugs are the most likely to display different affinities for the binding sites and thus produce "shallow" competition curves clearly revealing the presence of multiple sites.

KINETICS OF BINDING

The equilibrium between the drug and the receptor is based on two rates, the dissociation rate $k_2(D^*R)$ and the association rate $k_1(D^*)(R)$. Experimentally it is extremely difficult to measure association rates directly as high-affinity drugs have very rapid rates of association and dissociation of the drug from the receptor will occur simultaneously. In fact, accurate measurement of association rates involves a *Catch-22* situation. The very early time points (i.e., low fractional

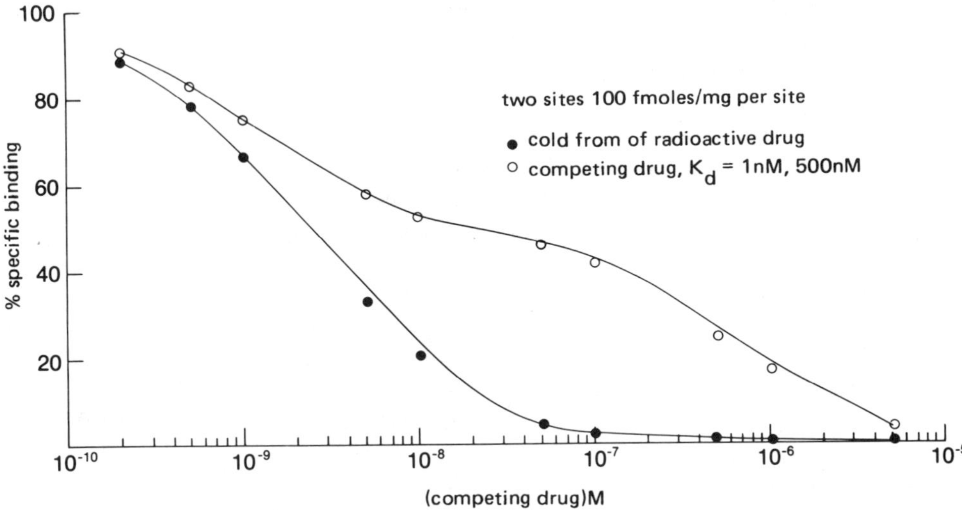

Figure 8 Competition curves produced by using the nonradioactive form of the radioactive drug (filled circles) that has equal affinity for two binding sites, or a drug (open circles) with a 500-fold difference in affinity for the two binding sites in competition experiments. The biphasic nature of the curves is strong evidence for two binding sites.

occupancy phases of the association curve) reflect the actual rate of association of the radioligand to the receptor. However, the measurement of such low fractional occupancies is difficult and these data tend to have large variability. As significant fractional occupancy occurs and more reliable data are achievable, dissociation of the radioligand from the receptor becomes a complicating factor. If a very accurate measurement of the association rate (k_1) is needed, it is best to perform the association and dissociation rate experiments using varying concentrations of radioligand.

$$k_{obs} = k_1(D^*) - k_2 \tag{10}$$

Equation (10) is derived from Equation (1), where k_{obs} equals the observed association rate constant. By plotting k_{obs} versus (D^*) for varying (D^*), a straight line with a slope equal to k_1 and a y intercept equal to $-k_2$ can be created. In practice most researchers simply report $k_{obs/D*}$ for the association rate constant.

Dissociation rates for high-affinity drugs can be measured directly with less theoretical and technical difficulties to overcome relative to association rates. One reason for this is that by definition the radioactive ligand will have high affinity for the receptor ($K_{D*} \cong 10^{-9}$ M); this means the dissociation rate will be in the order of 10^9 times slower than the association rate. Another reason for the ease of measuring dissociation rates is that there is a simple strategy for observing the dissociation of the radioligand from the receptor without any association to complicate the measurement. This procedure involves allowing the radioactive drug to equilibrate with the receptor and then instituting a procedure that prevents reassociation of the radioactive drug onto the receptor after it has dissociated. The easiest and theoretically most satisfactory method is the addition of a large excess of nonradioactive competing drug to the equilibrated radioactive drug-membrane solution. As labeled drug dissociates from the receptor, nonradioactive drug occupies the receptor. The rate of reduction of specific binding over time after addition of nonradioactive drug is directly related to the dissociation rate of the radioactive drug from the receptor. Figure 9 illustrates a theoretical association and dissociation rate experiment. The dissociation rate is determined from the plot illustrated in the inset in Figure 9 (for a typical pseudo-first-order reaction). An indirect but theoretically sound method of determining the association rate constant is to divide the dissociation rate constant by the K_{D*} (determined independently by Scatchard analysis).

$$K_{D*} = \frac{k_2}{k_1} \qquad k_1 = \frac{k_2}{k_{D*}} \tag{11}$$

Many investigators have been using the dissociation rate type of experiment to determine if a nonradioactive ligand that inhibits radioactive ligand binding is doing so through a competitive or allosteric mechanism. Competitive inhibition of radioligand binding should not affect the dissociation rate of an equilibrated radioligand-receptor interaction, whereas an allosteric mechanism of inhibition usually involves an increase in the dissociation rate.

TECHNICAL CONSIDERATIONS

There are three major technical problems in developing a successful assay procedure. The first problem is determining the free (unbound) drug D* and the bound drug D*R. The second problem is that almost all radioactive drugs adsorb nonspecifically to lipids, proteins, etc. and this type of drug binding interaction must be differentiated from the binding of radioactive drug to specific sites. The third problem is the low density of neurotransmitter receptors in tissues. A step-by-step description of the strategies and procedures that may overcome these problems follows.

Tissue Preparation

A range of commonly employed membrane preparations is used in receptor radiolabeling studies. The easiest preparation involves simply homogenizing the tissue, washing it several times in an appropriate buffer, and resuspending it

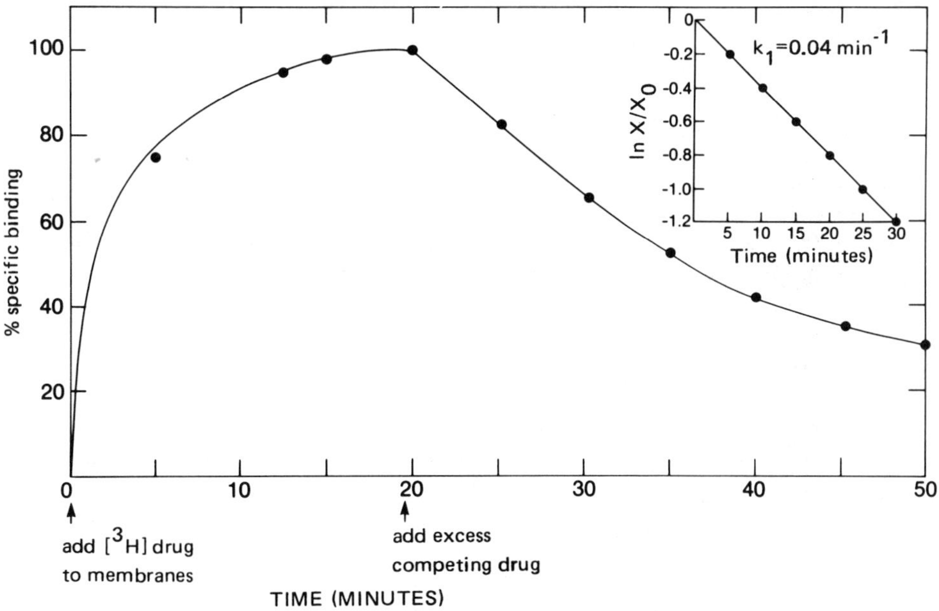

Figure 9 Association and dissociation rate experiment. At time 0 the radioactive drug is applied to the membranes and equilibration is allowed to occur. Excess nonradioactive drug is then applied, and the binding is assayed at various time intervals. The inset is the determination of the first-order dissociation rate constant.

for incubation with the appropriate radioactive and nonradioactive substances. Such preparation is satisfactory if a sufficient number of receptors are available to provide a workable radioactive signal in spite of the possibility that the cell type contain the receptor of interest may represent a small fraction of the total tissue in the homogenate. It is gratifying, albeit somewhat intuitively surprising, how often this simple homogenate preparation proves adequate. A slightly more sophisticated preparation involves initially suspending the tissue in an isotonic sucrose solution, gently homogenizing, and gently centrifuging. The pellet formed from this preparation is composed of nuclei and "cellular debris," that is, heavy material that sediments in the sucrose under relatively low centrifugal force. The supernatant, which should contain the tissue of interest, is then resuspended in the appropriate buffer and washed several times. This preparation is theoretically "cleaner" and may provide a slightly more reliable radioactive signal. For specialized investigations, such as subcellular localization of receptors, there are more complex centrifugation procedures involving varying densities of sucrose solutions and varying centrifugation steps (5,6). The vast majority of in vitro binding experiments utilize the simplest tissue preparation; it is beyond the scope of this chapter to detail the various tissue preparations.

Bound Drug Quantitation

Once the tissue preparation and the radioactive drug have been equilibrated, it becomes necessary to separate the unbound radioactive drug in the incubation medium from the bound radioactive drug. Several methods are currently used, the most popular being vacuum filtration. In this technique the incubation solution is applied to a filter apparatus connected to a vacuum pump. The fluid containing the unbound radioactive drug passes through the filter, while the membrane containing the bound radioactive drug is trapped on the filter. Glass fiber filters are especially well suited for this purpose.

Another popular technique for separating membranes from their aqueous environment is centrifugation. In this procedure the membrane suspension is spun down, forming a pellet containing the membranes with bound radioactive drug. The supernatant containing the unbound radioactive drug is discarded. The pellet is solubilized, and liquid scintillation fluid is applied for counting. This approach is considerably more labor and time intensive.

Specific Binding Determination

By trapping the membranes on a filter or by centrifuging the membranes into a pellet the bound radioactive drug has been separated from the free (unbound) radioactive drug. The critical next step in the assay requires the differentiation of the "specifically" bound radioactive drug from the "nonspecifically" bound drug on the filter (or in the pellet). In order to understand this process it is important to examine the differences in the properties of the binding of drugs to specific sites (a saturable phenomena) and the adsorption of drugs onto lipids, proteins, and the filter (nonsaturable phenomena). Specific binding sites, such as receptors, exist in relatively small finite numbers in tissues and therefore can be saturated with drug molecules (see Figure 10). The number of nonspecific sites is usually very large compared with the number of specific binding sites and often appears to increase linearly over the concentration range of the radioactive drug used. This difference in specific and nonspecific sites allows the experimenter to differentiate the two types of interactions of radioactive drugs with membranes.

Figure 10 diagrammatically describes the technical "trick" used to quantitate specific binding. In one set of test tubes the membranes are incubated with radioactive drug. The amount of counts per minute (cpm) detected on filters that have been used to trap the membranes is considered the total binding and includes the binding of the radioactive drugs to specific and

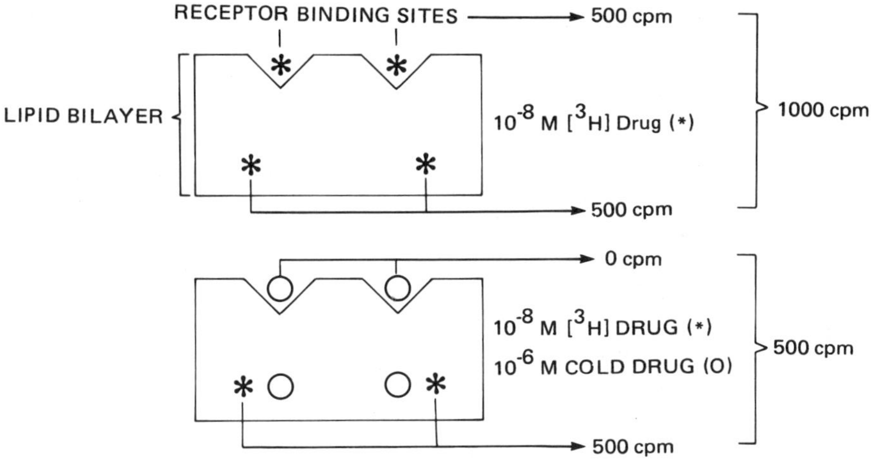

Figure 10 Specific and nonspecific binding determination; the asterisks represent radioactive drug molecules. By substracting the amount of binding in the presence of excess nonradioactive drug from the amount of binding of radioactive drug in the absence of excess competing drug, specific binding can be monitored.

nonspecific sites. In another set of tubes, radioactive drugs and membranes are incubated along with a large excess of nonradioactive drug (in this case a 100-fold excess). The excess nonradioactive drug reduces the specific radioactivity of the radioactive drug to 1% of the original drug and increases the total drug concentration by 100-fold. The specific sites can bind a limited amount of drug as they become saturated at concentrations of nonradioactive drug 10-fold higher than the K_D, according to the law of mass action. The major effect of the 100-fold excess drug on the specifically bound drug is to reduce the specific radioactivity of the bound drug to approximately 1% of the original. The amount of nonspecifically bound drug is increased 100-fold because of the increased total drug concentration and because the nonspecific adsorption is an essentially nonsaturable phenomenon in this concentration range. However, this increase in total amount of drug adsorbed nonspecifically to the membranes is due to nonradioactive drug; therefore the amount of radioactivity in the nonspecific compartment of membrane bound radioactivity is unchanged by the addition of excess nonradioactive drug. The difference between the nonspecific and total binding is assumed to represent the saturable, specific, or "competible" binding to the membranes. All radioligand binding assays are based on this general strategy.

It is important to be aware that the definition of "specific binding" at this point in the discussion is operational; it is the difference in radioactive ligand binding between membranes incubated with and without excess nonradioactive drug. A common mistake is to assume that "specific binding" and "receptor binding" are synonymous. It would be wonderful were this true; however, the situation is often more complex than this. At this point in a radioligand study (having established a "specific signal") the data are consistent, but not sufficient, to claim that the radioligand is specifically labeling a given receptor. In Receptor Criteria (see below), the steps necessary to define the specific binding as receptor binding are discussed.

Although this technique allows differentiation between specific ("competible") and nonspecific binding, other procedures are involved in order to improve the proportion of specific binding. This is very important as the relibility of the signal is strongly related to the proportion of specific relative to nonspecific binding. One of the most important properties of a radioactive drug is its affinity for a receptor. If the affinity is very high ($K_D < 10^{-9}$ M), then very low concentrations of the drug can be used in the test tube to label the receptors. The low concentrations of the drug tend to minimize the nonspecific binding. If the radioactive drug has a lower affinity for the receptor, then higher concentrations of radioactive drug are necessary for detectable labeling, and this usually results in a higher proportion of nonspecific binding. Another important attribute of a radioactive drug is its membrane-buffer partition coefficient. The higher the coefficient, the more the radioactive drug will be adsorbed nonspecifically into the lipid bilayer membrane. The ideal radioactive drug possesses high affinity for the receptor and a low partition coefficient.

The nonspecific binding can be reduced by washing the drug-receptor complex after collection on a filter, preferably with cold buffer. This wash should remove a significant portion of the nonspecific binding without interfering significantly with the specific binding. The reason for this selective effect of a rapid wash is that the high-affinity binding of a radioligand involves a slow dissociation rate while the nonspecifically adsorbed radioligand rapidly washes off the membranes. The buffer can be passed through the filter in seconds, leaving the high-affinity radioactive drug binding intact while removing at least some of the nonspecific binding.

RECEPTOR CRITERIA

In order to conclude that a drug is binding to a neurotransmitter receptor, several criteria must be met (Table 4). The biochemical criteria, which derive from our understanding of the physical nature of hormone and neurotransmitter receptors, include stereospecificity and heat lability. The latter criterion applies because receptors are well accepted as proteins with three-dimensional structure; boiling should destroy the three-dimensional structure, eliminating

Table 4 Basic Criteria for Identifying a Specific Radiolabeled Binding Site as a Receptor

1.	Saturability, competibility, stereospecificity
2.	Heat lability, irreversible thermal denaturation
3.	Appropriate pharmacology, stereospecificity
4.	Appropriate distribution

the specific radioligand signal. The stereospecific nature of radioligand binding is also derived from the three-dimensional nature of the binding site of the prototypical receptor; stereoisomers of drugs should interact with different affinities with a receptor binding site. The degree of stereospecificity may vary from receptor and from drug to drug; if the asymmetric carbon is in a position on the ligand that is crucial to the ligand-receptor interaction, then a high degree of stereospecificity may be observed for that drug. If the asymmetric carbon is somewhat distant from the critical areas of the ligand for receptor binding, then little or no stereospecificity may be observed. Because one pair of stereoisomers does not display differing affinities for a given radiolabeled site, one should not rule out the possibility that this site is indeed a receptor. However, if numerous stereospecific pairs are tested and none shows any appreciable affinity differences for the radiolabeled site, then the receptor nature of the specifically labeled site becomes questionable.

The issue of stereospecificity may become pharmacological if the question is whether the radioligand is labeling a receptor that has an already well-delineated pharmacology. If such a receptor has been shown to display a stereospecific interaction with a pair of stereoisomeric drugs in, for instance, behavioral studies, then the burden of proof is on the radioligand receptor labeling laboratory to demonstrate a similar stereospecific interaction with the radiolabeled site.

The criterion of "appropriate distribution" is necessarily vague and should be applied with a great deal of flexibility. In studies of neurotransmitter receptors for which detailed distribution of the neurotransmitters is known, it is expected that some sort of correlation should exist between the levels of neurotransmitter and the levels of the radiolabeled receptor in a given tissue (or brain area). Often, however, tissue (or areas of the brain) may be supplied by a neurotransmitter and yet have very low (i.e., undetectable) levels of the receptor although these few receptors are of critical functional importance. The most sensitive technique for detecting low levels of receptors is radioautography of receptors using radioiodinated ligands (see Receptor Autoradiography). Distribution can become an important criterion if a radiolabeled site is proposed to represent a "new" receptor (for which there are no pharmacological data available). If this site is truly a receptor, it should be found in different levels in different tissues or in various brain regions.

The criterion of "appropriate pharmacology" can only be applied to radiolabeled sites proposed to represent receptors already studied using techniques less direct than radiolabeling studies. In practice, researchers often attempt to correlate the pharmacology (i.e., order of potency of drugs) determined in clinical, behavioral, and biochemical studies with the order of potency of drugs in interacting with a radiolabeled site proposed to represent the relevant receptor. Experience has demonstrated that when a large series (a dozen or more) of drugs, agonists or antagonists, displaying varying potencies in biologic systems, correlate strongly with affinities for a radiolabeled site, then this site is probably the relevant receptor. The debatable point is what constitutes a strong correlation. A correlation coefficient of 0.75 or above is not unusual even when clinical events are correlated with radioligand binding results. In studies comparing in vitro results, such as the potencies of agonists in producing receptor-mediated stimulation of adenylate cyclase activity, with the affinities of the same agonists as determined by competition studies using radiolabeling techniques, it is not unusual to produce correlation coefficients of 0.90 or higher. However, as the potency of an agonist may depend on its efficacy as well as its affinity, stronger correlations are generally found between the potencies of antagonists in blocking receptor-mediated responses and their receptor affinities as determined through radioligand competition studies.

It is not unusual to find reports in the literature of high-affinity, heat-labile, saturable binding for radioligands (drugs, neurotransmitters, ion channel effectors, toxins, growth factors, hormones, and others) for which no pharmacological correlation can be found. This may be regarded as a "good news, bad news" situation. The bad news is that, lacking a strong pharmacological correlation, the investigator may have a difficult time establishing the biologic relevance of this site, that is, establishing that this binding phenomenon is something more than just an artifact. The good news is that the investigator may have discovered a "new" receptor. Generally, in situations in which the binding site has turned out to represent a functional newly discovered receptor, the pharmacological data from the radiolabeling studies have been extremely valuable in guiding subsequent research into the function of the receptor. Radiolabeling studies have revealed the existence and pharmacology of heretofore undetected opioid, serotonin, and adrenergic receptors, to name a few.

RECEPTOR AUTORADIOGRAPHY

The purpose of this section is not to detail the methodology involved in receptor autoradiography, but rather to compare this technique with the standard receptor radiolabeling techniques involving tissue homogenates. It is hoped that this section, in conjunction with the previous discussions, will allow the reader to compare the strengths and weaknesses of receptor labeling of tissue homogenates and slide-mounted tissues. For excellent detailed reviews and step-by-step discussions of the receptor autoradiography procedures, the reader is referred to References 7 and 8.

Radiolabeling receptors in tissue homogenates is obviously a very powerful technique. However, there are some severe inherent technical limitations. The precise anatomic localization of receptors is impossible to determine, owing to the disruption of the tissue. If the receptors are present in relatively low amounts, the dilution of these receptors in the total tissue homogenate may result in an unworkably low specific radioactive signal. If the tissue in which the receptors are localized happens to be a small tissue (or brain region) in a small animal (rat or mouse), it may be impractical or impossible to collect enough tissue to study the receptors reliably. Sometimes there are solutions to these problems. For instance, the specific radioactive signal of a preparation containing a low density of receptors may be greatly improved by substituting a radioiodinated ligand for a tritiated ligand; radioiodinated ligands tend to have specific radioactivities in the order of 2000 Ci/mmol, but useful tritiated ligands tend to have specific radioactivities in the order of 10-100 Ci/mmol. To get around the problem of a very small tissue or brain region, many researchers simply procure tissue from larger species (i.e., pig, cow, sheep, and human). Sometimes none of these strategies reveals a reliable radioactive signal in spite of overwhelming evidence that a given functional receptor is present in the tissue in question. Certainly there is no strategy that gets around the problem of the lack of precise anatomic localization owing to the disruption of the tissue.

Autoradiography of radiolabeled receptors avoids the disruption of the tissue so that the precise localization is possible and at the same time avoids the dilution of the radioactive signal inherent in the tissue homogenate procedure. The key discovery in the history of the development of this technique was the realization that perfusing a rat with saline or sucrose solutions, followed by fixing, freezing, slicing, and mounting of tissue (brain) slices, did not significantly alter the binding properties of receptors (9). This can be demonstrated in two ways. The slide-mounted tissues can be dipped in solutions containing radioactive ligands with or without appropriate nonradioactive ligands. This procedure is analogous to incubating membrane homogenates with radioactive ligands with or without appropriate nonradioactive ligands. The radioactivity labeled tissue slices can then be blotted onto filter paper, the filters placed into scintillation vials, and the radioactive signal analyzed for its pharamacological properties. Invariably, the receptors are found to have retained their binding properties in spite of the tissue mounting procedures. The radioautographic procedure involves exposing radioactive-sensitive film or coverslips layered with radioactive-sensitive emulsions to the radioactively labeled slide-mounted tissue. After a suitable length of exposure time (weeks to months for tritiated ligands; hours to days for radioiodinated ligands), the film or coverslip can be examined with a microscope to reveal the areas in which the radioactivity has been of sufficient quantity to produce "grains." By using the same technical tricks to separate nonspecific binding (nonspecific "grain production") from total binding (total "grain production"), specific labeling can be identified. The power of the technique resides in the nondisruptive treatment of the tissue, allowing a much more precise anatomic localization of receptors as well as increasing the likelihood of reliably detecting receptors in very small regions of a tissue. Above and beyond the allure of the receptor autoradiographic procedure because of its technical advantages, for certain types of studies, the technique can be mastered without becoming involved in the sometimes perplexing world of receptor biochemistry. For scientists from such disciplines as histology, neuroanatomy, and neurophysiology, the basic receptor autoradiographic procedures are at one time familiar and intuitively comprehensible. The same cannot be said for radiolabeling studies of receptors in homogenates. On the other hand, biochemists, biochemical pharmacologists, and physiologists may be more comfortable with homogenate binding and may find the larger degree of control and manipulability of the incubation conditions in the homogenate procedures more critical for answering their research questions.

Quantitating receptor autoradiograms involves determining the density of grains in the areas of the slide-mounted tissue exposed to radioligand that appear to have higher densities of grains than appear on slide-mounted tissues that have been exposed to radioligand and excess nonradioactive ligand. This is a rather labor-intensive chore. Recently, computer-assisted densitometry has been introduced into the field of receptor autoradiography. Although the computer approach does save time, a thorough quantitative receptor autoradiographic study is still a very time-consuming, tedious job. However, the ability to map the density of a given receptor throughout a complex tissue, such as the brain, is a remarkable technical achievement; the successful conclusion of such a study is always a remarkable contribution. Computerized color coding of the different densities of the grains has resulted in the production of eye-pleasing multicolored receptor distribution displays. Although the scientific value of these color-coded images relative to the display of the original grain production is questionable, the color-enhanced displays have sparked more interest and excitement in the overall scientific and nonscientific community than might have occurred without the color enhancement.

RADIOLABELING STUDIES OF RECEPTOR-GTP BINDING PROTEIN COMPLEXES

Receptors influence intracellular events; therefore, receptors must be chemically and/or physically associated with intracellular components. An important question is whether radiolabeling studies of receptors can be used to probe the interactions of the receptor with the receptor-coupled intracellular components. An aspect of this question is whether radiolabeling studies of receptors can be used to distinguish agonist-receptor interactions from antagonist-receptor interactions.

Cyclic AMP (cAMP) is the most studied "second messenger" system (for reviews, see References 10 and 11) and Chapter 4 (Kinnier). The levels of cAMP (which activates cAMP-dependent protein kinase) are modulated by agonist stimulation of cell surface receptors in essentially all types of cells. Receptors that, when stimulated by an agonist interaction, produce elevated intracellular levels of cAMP include glucagon, β_1- and β_2-adrenergic, H_2 histamine, adenosine A_2, and D_1 dopamine receptors, to name just a few. Receptors that, when stimulated by an agonist interaction, inhibit the production of cAMP include α_2-adrenergic, opiate (μ subtype), and D_2 dopamine receptors. The receptor-mediated effects upon cAMP levels are now well accepted as GTP dependent. The GTP is obligatory because it is an endogenous activating intracellular ligand for a protein (GTP binding protein, heretofore referred to as G) that interacts with both the plasma membrane receptor and the intracellular adenylate cyclase enzyme. Whether a receptor-mediated influence on adenylate cyclase activity will be stimulatory or inhibitory appears to depend upon with which of two types of G the receptor interacts (12,13). If the agonist-receptor complex is coupled to G_s, then that receptor will stimulate adenylate cyclase activity when activated by an agonist. If the receptor is coupled to G_i, then adenylate cyclase activity will be inhibited when the receptor is activated.

There has been an explosion of interest during the last several years (14,15) in the relationship between receptor activation and stimulation of the breakdown of the phospholipid phosphatidylinositol (PI). It has been definitely established that many hormone and neurotransmitter receptors, when activated by agonists, are capable of stimulating the activity of the enzyme phospholipase C, which is the rate-limiting enzyme in the production of inositol phosphates from the parent PI molecule (16). It is also currently believed that the resulting inositol phosphates are involved in releasing intracellular stores of calcium and the resulting diacylglycerol is involved in stimulating a non-cAMP-dependent protein kinase. Thus the PI metabolic system provides two second messengers. In the last two years several lines of evidence have indicated the likelihood that receptors coupled to the PI response interact with G subunit, which is also a GTP binding protein. It has been shown that in neutrophils the chemotactic receptor-mediated PI response is inhibited by pertussis toxin (a specific G_i inhibitor) and that the toxin can catalyze the ADP ribosylation of G_i coupled to the receptor (17). In membrane preparations, Litosch et al. (18) have shown a GTP dependence for serotonin receptor-mediated stimulation of the PI response in blowfly salivary glands (a classical PI response system). Titeler and coworkers (19,20) have used radiolabeling techniques (see below) to demonstrate that brain 5-HT_2 receptors are coupled to G protein; the 5-HT_2 receptor has been demonstrated to be coupled to the PI response in the rat brain (21,22) and in the human platelet (23).

The a priori observations that have been made involving radiolabeling studies of G-protein-coupled receptors are listed below. The molecular mechanisms probably underlying these observations are discussed following this list.

1. Radioactive antagonist ligands specific to the given receptor type appear to label a homogeneous population of sites; competition studies using nonradioactive antagonist ligands to compete for the specific binding of the radioactive antagonist ligand support the apparent homogeneity of the sites labeled; that is, competition curves produce Hill coefficients close to unity (Figure 11). In Figure 11, the drug cinanserin is a prototypical 5-HT_2 receptor antagonist; it competes for [^{3}H]ketanserin (an antagonist radioligand)-labeled 5-HT_2 receptors with a Hill coefficient not significantly less than 1 (19).
2. Competition studies using nonradioactive agonist ligands to compete for the specific binding of the radioactive antagonist ligand produce competition curves that have Hill coefficients significantly less than unity; that is, the agonists reveal what appears to be heterogeneity in the radioactivity labeled receptors (Figure 11). In Figure 11, serotonin is the agonist used that reveals an apparent heterogeneity in the radiolabeled 5-HT_2 receptor population (19).
3. Guanyl nucleotides, but not adenyl nucleotides, change the nonradioactive agonist competition curves (for radioactive antagonist-labeled receptors) from "shallow" to "steep"; that is, the Hill coefficients in the presence of guanyl nucleotides are close to unity (Figure 11). In Figure 11 the effects of the nonhydrolyzable analog of GTP, GppNHp, are illustrated. The apparent affinity of the agonist for the radioactive antagonist-labeled receptor will appear to be lower as the IC_{50} is "shifted" to the right on the log concentration-inhibition curve. The guanyl nucleotides have no effect on the competition curves produced by nonradioactive competing antagonists (Figure 11).
4. The competition curves produced by the agonists in competing for the antagonist-labeled receptor can be analyzed by computer and will best fit a two-site model as compared with a one-site model. Computer programs can generate the proportion of the sites that seem to have high affinity for agonists (R_H) and the proportion of the sites that seem to have low affinity

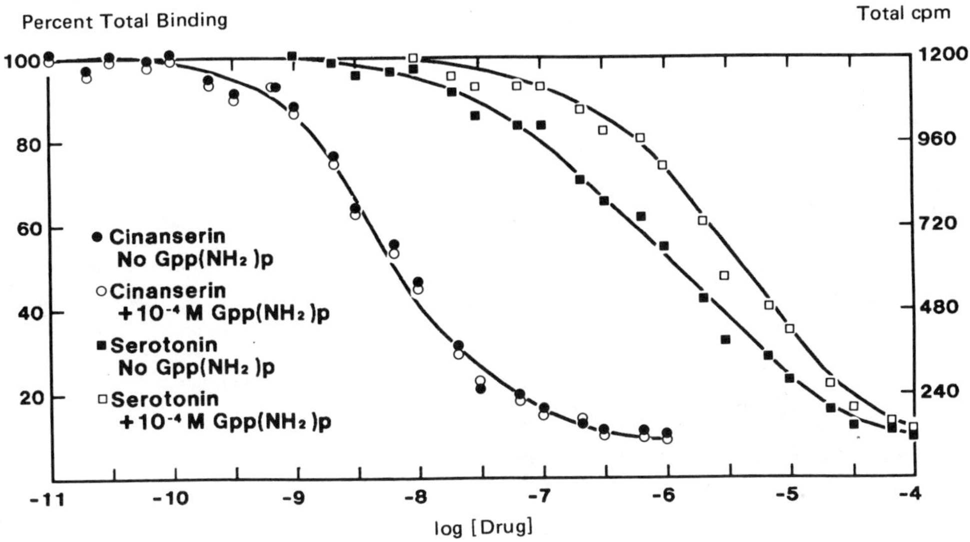

Figure 11 Competition curves produced by coincubating increasing concentration of nonradioactive cinanserin or serotonin with $4 \times 10^{-10} \times$ [^{3}H]ketanserin in homogenates of rat prefrontal cortex tissue. (From Reference 19).

for the sites (R_L). Also, the apparent affinities the agonists have for the two sites (or "states" of the receptor) are generated by the two-site computer-assisted analyses (Chapter 3, McPherson).

5. Divalent cations appear to increase the proportion of the sites in the agonist high-affinity form; divalent cations have little or no effect on antagonist binding. Monovalent cations tend to lower agonist affinity for both states of the receptor and tend to increase the affinity of antagonists for the receptor (although this is not definitively established as a general rule).
6. Radioactive agonist-labeled receptors appear to be a homogeneous population of sites with B_{max} values significantly less than the B_{max} values determined with a radioactive antagonist ligand. Nonradioactive agonist and antagonist competition curves tend to be "steep," that is, produce Hill coefficients not significantly different from unity. The binding is inhibited by guanyl nucleotides and monovalent cations. Divalent cations increase the B_{max} of the radioactive agonist-labeled receptor.

All these observations are easily explained with a simple model, usually referred to as the "ternary complex model" (Figure 12). In this model the receptor may exist in two "states." The free receptor (usually the majority

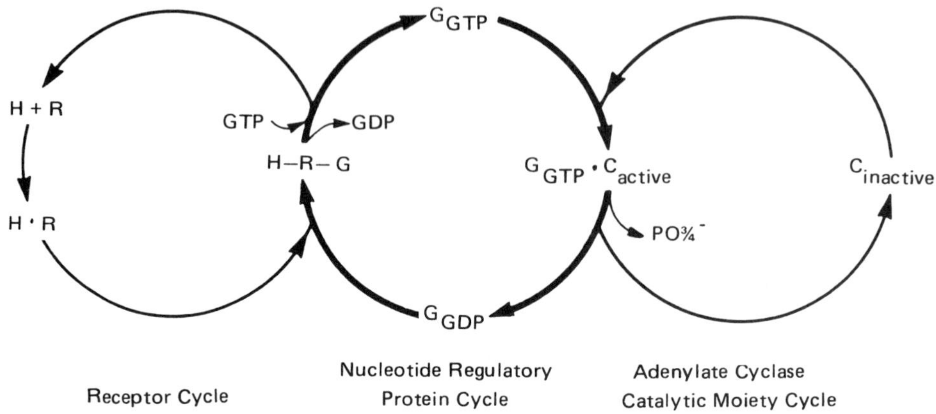

Figure 12 Schematic depiction of the "ternary complex model." (Adapted from Reference 10.)

of the sites) exhibits relatively low affinity for the agonist. The receptor-G protein complex (apparently present in tissue homogenates in detectable levels) exhibits relatively high affinity for the agonist. In Figure 12 the relevant components are R (free receptor), which has high affinity for antagonists and relatively low affinity for agonists, and R–G (receptor-G protein complex), which has a high affinity for antagonists and agonists. In the scheme depicted in Figure 12, R–N does not appear; it is a subject of some debate whether R–G (as opposed to H–R–G) does exist in vivo. Guanyl nucleotides dissociate the receptor-G protein complex; divalent cations stabilize the complex. This simple model adequately explains most of the results of radiolabeling studies of G protein-associated receptors and does not disagree with any of the significant observations on receptor-G protein-adenylate cyclase or phospholipase C interactions. However, models based on some sort of cooperative phenomenon to explain the "shallow" agonist competition curves have not been well explored and may actually better explain some of the observations.

REFERENCES

1. G. Scatchard, *Ann. N.Y. Acad. Sci. 51*:660 (1949).
2. I. M. Klotz, *Science 217*:1247 (1982).
3. A. V. Hill, *J. Physiol.* (*Lond.*) *40*:4 (1910).
4. Y. C. Cheng and W. H, Prusoff, *Biochem. Pharmacol. 22*:3099 (1973).
5. V. P. Whittaker, I. A. Michaelson and R. J. A. Kirkland, *Biochem. J. 90*: 293 (1964).

6. P. Laduron, M. Verwimp, P. F. M. Janssen, and J. Leysen, *Life Sci. 18*: 433 (1976).
7. M. J. Kuhar, in *Neurotransmitter Receptor Binding* 2nd ed. (H. Yamamura, S. Enna, and M. Kuhar, eds.), 1985, p. 153.
8. M. J. Kuhar and J. R. Unnerstall, *Trends Neurosci. 8*:49 (1985).
9. W. S. Young and M. J. Kuhar, *Brain Res. 179*:255 (1979).
10. R. J. Lefkowitz, J. M. Stadel and M. G. Caron, *Annu. Rev. Biochem. 52*: 159 (1983).
11. A. G. Gilman, *J. Clin. Invest. 73*:1 (1984).
12. L. E. Limbird, *Biochem. J. 195*:1 (1981).
13. D. M. Cooper, *FEBS Lett. 138*:157 (1982).
14. R. H. Michell, *Trends Biochem. Sci. 4*:128 (1979).
15. M. J. Berridge, *Mol. Cell. Endocrinol. 24*:115 (1981).
16. Y. Nishizuka, *Nature 312*:315 (1984).
17. C. D. Smith, B. C. Lane, I. Kusaka, M. W. Verghese, and R. Snyderman, *J. Biol. Chem. 260*:5875 (1985).
18. I. Litosch, C. Wallis, and J. N. Fain, *J. Biol. Chem. 260*:5464 (1985).
19. M. Titeler, G. Battaglia, and M. Shannon, *J. Rec. Res. 4*:705 (1984).
20. G. Battaglia, M. Shannon, and M. Titeler, *J. Neurochem. 43*:1213 (1984).
21. J. Conn and E. Sanders-Bush, *J. Pharmacol. Exp. Ther. 234*:195 (1985).
22. D. A. Kendall and S. R. Nahorski, *J. Pharmacol. Exp. Ther. 233*:473 (1985).
23. D. de Chaffoy de Courcelles, J. E. Leysen, F. De Clerck, H. Van Belle, and P. A. J. Janssen, *J. Biol. Chem. 260*:7603 (1985).

3

A Mathematical Approach to Receptor Characterization

Grant A. McPherson* / University of Vermont, Burlington, Vermont

A quantitative approach to receptor theory was developed when Clark (1-3) and Gaddum (4) found that concentration-response data could be described by the adsorption isotherm of Langmuir (5) in some experimental situations. More specifically, there appeared to be a hyperbolic relationship between agonist concentration and the appropriate biologic response. After this time, and before the development of specific radioligands, receptor characteristics were inferred from concentration-response data. Clark's treatment assumed that the response was directly proportional to the number of receptors occupied and that the maximum tissue response was only obtained with complete receptor activation. Limitations of this "occupancy" model were apparent, and many alternatives have been proposed [for extensive reviews, see Ruffolo (6) and Kenakin (7)]. With respect to receptor pharmacology, an inherent problem with all such models is the unknown relationship between receptor occupancy and pharmacological response. Although methods have been developed that supposedly eliminate this complication [so-called null methods (8)], at a practical level it is more difficult to achieve this (9). In any case, the advent of radiolabeled drugs has allowed direct assessment of receptor pharmacology and function without the need to involve the stimulus-response relationship.

Independent of the final purpose of the radioligand binding study, an essential prerequisite is that the correct model be selected to describe the radioligand-binding site interaction. Obviously, the informativeness of such studies, in terms of calculation of equation constants, is limited by the model's applicability. In addition, the model selected regulates, to some extent, various ex-

---

**Present affiliation*: Baker Medical Research Institute, Prahran, Victoria, Australia

perimental conditions that must be met if the experimental results are to be considered valid. The purpose of this chapter is to define the mathematical models commonly adopted to describe radioligand-binding site interactions. It should be stressed that the models that have been developed are complex and, by necessity, require some form of computing facility for two reasons. First, to adequately characterize the system, nonlinear curve-fitting techniques are required to obtain accurate estimates of the constants defined in the models. Second, the choice of one model over another can have profound effects on the final interpretation of the data. Computerized statistical analyses are an important component of any analytic technique (10,11). Despite these requirements, commonly used graphic procedures are still useful for displaying the data. Consequently, these methods are also discussed.

THE BIMOLECULAR INTERACTION BETWEEN DRUG AND RECEPTOR

Much of the modeling involved in radioligand binding studies has stemmed from experiments performed in the chemical and biochemical sciences. Langmuir (5) considered the relationship between the partial pressure of a gas and its adsorption onto a solid surface. It was thought that adsorption could only occur in a monolayer and that there were a limited number of active sites at which the adsorption could take place. The binding of drug to receptor has been considered in an analogous manner, in which a discrete number of macromolecular receptors are distributed over the surface of the plasma membrane, or perhaps present throughout the cytosol, which can bind drugs present in the biophase (Figure 1). In accordance with Langmuir's (5) adsorption isotherm and considering the law of mass action, the rate of formation of the drug-receptor complex would be proportional to both the fraction of sites not already occupied and the concentration of unbound drug. The process is reversible, and dissociation would be proportional to the amount of drug bound. A mathematical model can be constructed that describes this bimolecular interaction:

$$[L] + [R] \underset{K_{-1}}{\overset{K_1}{\rightleftharpoons}} [RL] \tag{1a}$$

where [R] is the unbound (free) concentration of receptor, [L] the unbound (free) concentration of drug (henceforth designated [F]), [RL] is the concentration of drug-receptor complex (i.e., the amount of drug bound, henceforth designated [B]), and K_1 and K_{-1} are the proportionality constants of association (association rate constant) and dissociation (dissociation rate constant), respectively.

The model given by Equation (1a) represents the simplest relationship describing drug (radioligand)-receptor (binding site) formation. Many receptors

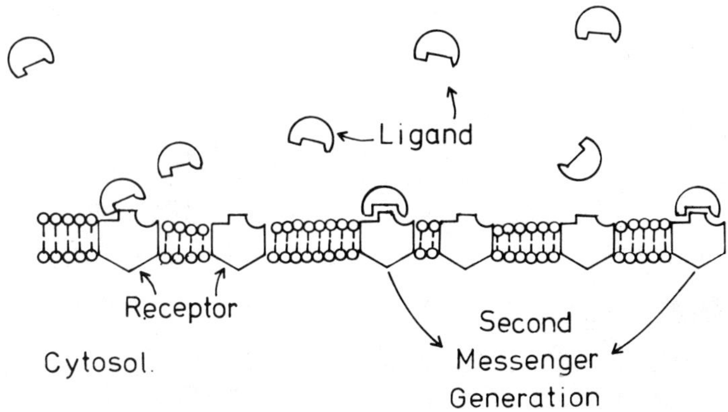

Figure 1 Diagramatic representation of the bimolecular model described by Equation (1a). Receptor macromolecules embedded in the cell's plasma membrane interact independently with ligand molecules in the surrounding biophase. Kinetic and equilibrium equations can be developed from Langmuir's adsorption isotherm and the law of mass action to quantify both ligand affinity and density of sites.

have been subdivided (e.g., α_1 and α_2; H_1 and H_2) and although specific agonists and antagonists interact in a qualitatively similar manner, there are quantitative differences (i.e., differences in affinity and efficacy). Furthermore, there are many instances in which receptor subtypes coexist in the same tissue. Given the limited selectivity of most drugs, it is necessary to consider the binding of a ligand to two or more receptors simultaneously. Thus:

$$[L] + [R_1] \underset{K_{-1}}{\overset{K_1}{\rightleftharpoons}} [R_1 L] \qquad \text{site 1}$$

$$[L] + [R_2] \underset{K_{-1}^*}{\overset{K_1^*}{\rightleftharpoons}} [R_2 L] \qquad \text{site 2} \tag{1b}$$

where $[R_1]$ and $[R_2]$ are the free concentrations of the two receptor subtypes, which also have differing rate constants.

Two aspects of the bimolecular model are described separately: first, the equations that describe the kinetic properties of a ligand; second, those that describe binding characteristics when the process has reached equilibrium. In both cases, it is necessary to examine both the binding of the radioligand and also that observed when the radioligand and an unlabeled drug compete for the same site simultaneously. Although it is not feasible to radiolabel all drugs, binding constants can be calculated by examining the way they compete with the radioligand for the binding site.

BIMOLECULAR KINETIC STUDIES

The concentration of a radioligand-binding site complex at any one time is related to its rate of formation (association) and reduction (dissociation). Each aspect is considered separately.

Association of a Single Radioligand

The association of a radioligand to a binding site is a second-order process since it depends on the concentrations of both the radioligand and the binding site [Equation (1a)]. In practice it is not possible to isolate the association process for separate study since concurrent dissociation always occurs. This is not the case when studying dissociation, since association can be experimentally prevented by infinitely diluting the incubation medium prior to studying the dissociation time course. In association studies, therefore, the equations relating the amount of radioligand bound at a given time must in some way account for both association and dissociation.

The rate of radioligand-binding site formation, assuming the simple bimolecular interaction described in Equation (1a) and the law of mass action, can be described by Equation (2).

$$\frac{d[B_t]}{dt} = [F]\,[R]\,K_1 - [B_t]\,K_{-1} \tag{2}$$

where $[B_t]$ is the amount of ligand bound at time t. The free radioligand concentration can be expressed as

$$[F] = [T] - [B_t]$$

where [T] is the total radioligand concentration. Also, the concentration of receptors:

$$[R] = B_{max} - [B_t]$$

where B_{max} is the maximum number of binding sites. Substituting into Equation (2) gives

$$\frac{d[B_t]}{dt} = ([T] - [B_t])(B_{max} - [B_t])K_1 - [B_t]\,K_{-1} \tag{3}$$

At equilibrium, association and dissociation proceed at the same rate; that is, $d[B_t]/dt = 0$. At this time, $[B_t]$ is equal to $[B_E]$ (the amount of radioligand bound at equilibrium). Thus,

$$([T] - [B_E])(B_{max} - [B_E])K_1 = [B_E]K_{-1} \tag{4}$$

or

$$K_{-1} = \frac{K_1([T] - [B_E])(B_{max} - [B_E])}{[B_E]} \tag{5}$$

or rearranging in terms of $[B_E]$ gives

$$[B_E] = \frac{B_{max}([T] - [B_E])}{K_{-1}/K_1 + ([T] - [B_E])} \tag{6}$$

Substitution of Equation (5) into Equation (3) gives

$$\frac{d[B_t]}{dt} = ([T] - [B_t])(B_{max} - [B_t])K_1 - \frac{K_1([T] - [B_E])(B_{max} - [B_E])[B_t]}{[B_E]} \tag{7}$$

After integration (12), the following equation is produced:

$$\ln \frac{[B_E]([T] - [B_t][B_E]/B_{max})}{[T]([B_E] - [B_t])} = K_1 t \frac{([T]B_{max} - [B_E])}{[B_E]} \tag{8}$$

Rearranging in terms of $[B_t]$ gives

$$[B_t] = \frac{K[T][B_E] - [B_E]}{K[T] - [B_E]^2/B_{max}^2} \tag{9}$$

where $K = e^{K_1 Zt}$ and $Z = [T]B_{max}/[B_E] - [B_E]$.

Equation (9) represents the complete model describing the time course of a simple bimolecular interaction defined by Equation (1a).

Alternatively, Equation (8) can be simplified by performing the experiment under pseudo-first-order conditions. In this situation, only a small amount (<10%) of the radioligand added is assumed to be bound. Under such circumstances, the system is said to exist in “zone A” (13) conditions. Consequently the following approximations can be made:

$$[T] - [B_E] \cong [T]$$

that is,

$$[T] >> [B_E]$$

Consequently,

$$[T] >> [B_t]\frac{[B_E]}{B_{max}}$$

and

$$[T]\ \frac{B_{max}}{[B_E]} >> [B_E]$$

Thus, under pseudo-first-order conditions, Equation (8) reduces to

$$\ln \frac{[B_E]}{[B_E] - [B_t]} = K_1 t\ \ [T]\ \frac{B_{max}}{[B_E]} \tag{10}$$

Expressing in terms of $[B_t]$,

$$[B_t] = [B_E](1 - e^{-K_{obs}t}) \tag{11a}$$

where K_{obs} is the pseudo-first-order rate constant. Equation (11a) can be rearranged into various forms to allow graphic representation. Thus,

$$[B_t] = [B_E] - [B_E]e^{-K_{obs}t} \tag{11b}$$

Therefore,

$$[B_E] - [B_t] = [B_E]e^{-K_{obs}t} \tag{11c}$$

Therefore,

$$\ln([B_E] - [B_t]) = \ln[B_E] - K_{obs}t \tag{11d}$$

or

$$\ln \frac{[B_E]}{[B_E] - [B_t]} = K_{obs}t \tag{11e}$$

The pseudo-first-order association rate constant is

$$K_{obs} = K_1 \left([T]\ \frac{B_{max}}{[B_E]} \right) \tag{12}$$

Given the pseudo-first-order conditions in which $[T] \cong [F]$ and using a rearrangement of Equation (6), where

$$\frac{B_{max}}{[B_E]} \cong \frac{K_{-1}/K_1 + [T]}{[T]}$$

Then

$$K_{obs} = K_1[T] + K_{-1} \tag{13}$$

The half-life $t_{1/2}$ for this pseudo-first-order reaction is equal to $0.693/K_{obs}$.

It is possible for the radioligand to associate with more than one site simultaneously [Equation (1b)]. Under these circumstances, the total amount of radioligand bound at any time (assuming pseudo-first-order conditions) is equal to the sum of the binding to each individual site [Equation (11a)]. Thus,

$$[B_t] = \sum_{i=1}^{i=n} [B_{Ei}](1 - e^{-K_{obsi}t}) \tag{14}$$

where n = the maximum number of sites.

Dissociation of a Single Radioligand

The process of calculating the dissociation rate constant of a radioligand is simplified since the process is truly first order. In addition, concurrent association can be prevented. Thus Equation (2) can be simplified to

$$\frac{d[B_t]}{dt} = -[B_t]K_{-1} \tag{15}$$

Integrating gives

$$[B_t] = [B_0]e^{-K_{-1}t} \tag{16a}$$

where $[B_0]$ is the amount bound at time t = 0. Taking the natural logarithm gives

$$\ln[B_t] = \ln[B_0] - K_{-1}t \tag{16b}$$

If the radioligand dissociates from more than one site simultaneously, then the total amount of radioligand bound at any time would be given by

$$[B_t] = \sum_{i=1}^{i=n} [B_{0i}]e^{-K_{-1i}t} \tag{17}$$

where n = the maximum number of sites.

Graphic Analysis of Kinetic Studies

Graphic methods of analysis generally rely on a linear relationship between ordinate and abscissa values. This permits simple linear regression techniques to be employed to gain estimates of the model constants. If there is no linear relationship in the original data, then the data need to be transformed in some way.

Association

Two equations have been developed that would allow calculation of the association rate constant. The bound-time data can be transposed according to Equation (8). A plot of the left-hand versus the right-hand side should yield a straight line with a slope equal to K_1. Alternatively, a general nonlinear curve-fitting program can be employed to fit Equation (9) directly. In either case, however, prior knowledge of the maximum number of binding sites B_{max} is required.

More commonly, the experiment is performed under pseudo-first-order conditions in which case the association rate constant can be calculated after rearranging Equation (13) to give

$$K_1 = \frac{K_{obs} - K_{-1}}{[T]} \tag{18}$$

This method requires prior knowledge of the dissociation rate constant (see B_{max} in the first method). However, it is essential that the experiments be carried out under pseudo-first-order conditions, and it would be prudent to calculate K_1 at a number of different ligand concentrations. A plot of K_{obs} versus the total radioligand concentration [T] should yield a straight line with a slope equal to K_1 and y intercept of K_{-1}. Points deviating from linearity at high [T] values may indicate that pseudo-first-order conditions have not been met. In addition, this method does not require the independent calculation of K_{-1} (in fact, it is a calculated value). Consequently, this method also allows verification of the K_{-1} constant calculated separately in dissociation studies. In either case, K_{obs} can be calculated from a plot of [see Equation (11d)]

$$\ln([B_E] - [B_t]) \text{ versus time} \tag{19}$$

The plot should yield a straight line in the case of a monoexponential model with the slope equal to $-K_{obs}$ and a y intercept of $\ln [B_E]$. In the case of a multisite association, graphic procedures become complex. Figure 2a represents theoretical association time course data (using a biexponential model) carried out under pseudo-first-order conditions. Also included are the graphically transformed data given by Equation (19) (Figure 2b). In the situation in which the radioligand associates to more than one site simultaneously, the curve must be "feathered" (see Appendix) to obtain estimates of K_{obs} and $[B_E]$ for each site. However, given the complexities of the pursuant calculations, the subjectiveness of the process, and the absence of statistical tests to aid in selection of the appropriate model (i.e., mono-, biexponential, and so on), it would be necessary to employ some form of nonlinear curve fitting that uses the exact equation [Equation (14)] (10,14).

Dissociation

Monoexponential dissociation time data can be linearized by plotting [see Equation (16a)]

$$\ln [B_t] \text{ versus time} \tag{20}$$

The slope of the resultant straight line is equal to $-K_{-1}$ and the y intercept equal to $\ln [B_0]$. If the radioligand is dissociating from more than one site simultaneously, then the curve again must be feathered to calculate K_{-1} values for each site. Figure 3a shows predicted dissociation data using a biexponential

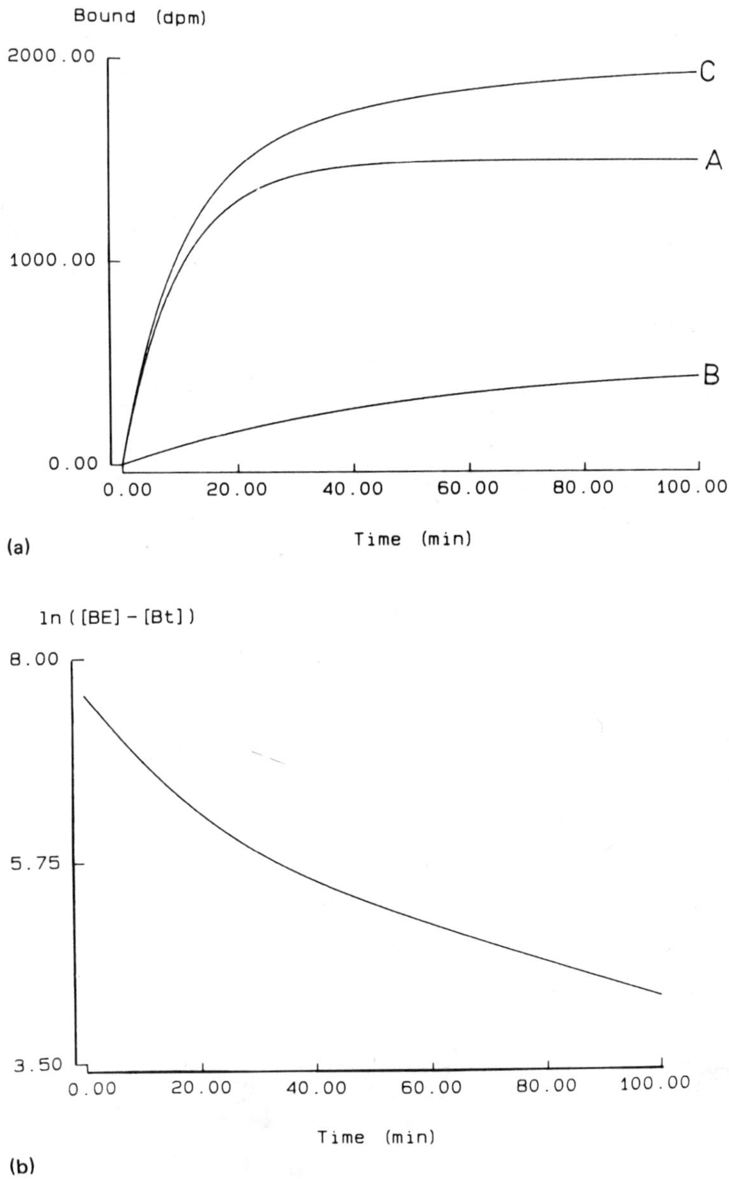

Figure 2 (a) Computer-generated association bound-time data using Equation (14). The data are plotted as bound versus time for the radioligand binding to two sites individually (A, site 1, $K_{obs_1} = 0.1\ min^{-1}$; $[B_E] = 1500$ dpm: B, site 2, $K_{obs_2} = 0.02\ min^{-1}$, $[B_E] = 500$ dpm) and in a combined (biexponential) model (C). (b) Transposition of C (C = A + B) using Equation (19) yields a line that is curvilinear. Actual values for the respective model parameters can be calculated graphically using a feathering technique described in the Appendix.

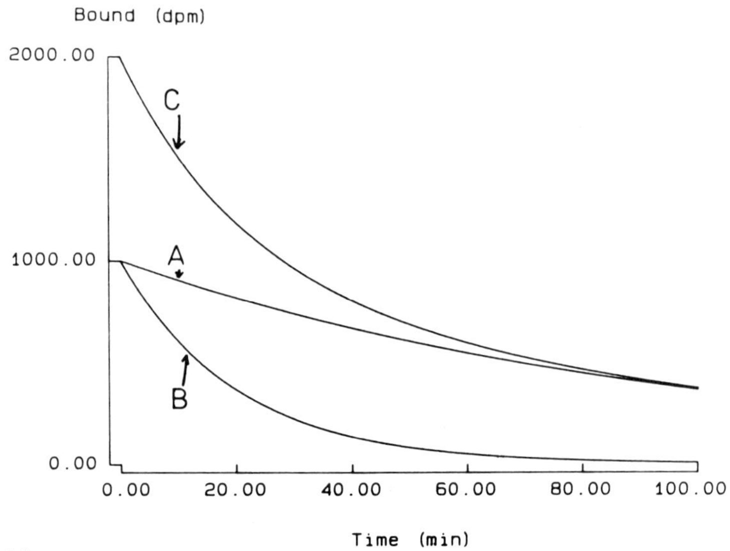

(a)

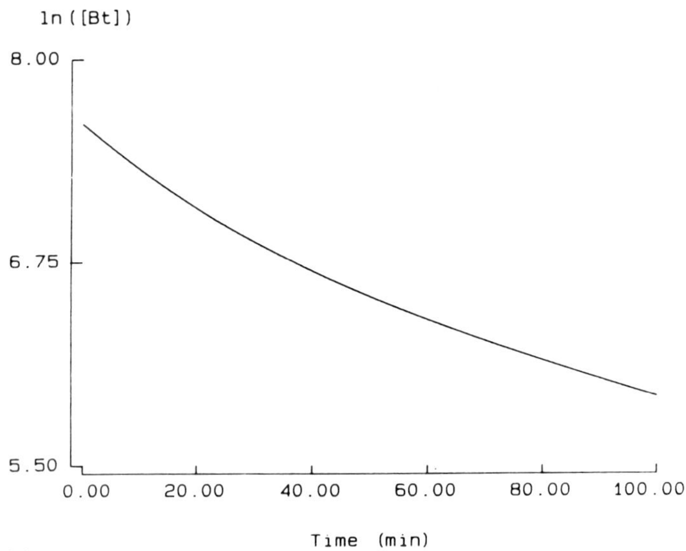

(b)

Figure 3 (a) Computer-generated dissociation bound-time data using Equation (17). The data are represented as bound versus time for the radioligand binding to two sites individually (A, site 1, $K_{-1} = 0.01\ min^{-1}$; $[B_0] = 1000$ dpm; B, site 2, $K_{-1} = 0.05\ min^{-1}$; $[B_0] = 1000$ dpm) and in the actual biexponential model (C). (b) Transposition of C using Equation (20) yields a line that is curvilinear. Actual values for the respective model parameters can be calculated graphically using a feathering technique described in the Appendix.

model. Also included are the same data transformed according to Equation (20) (Figure 3b). Alternatively, and preferably, a nonlinear curve-fitting program could be employed to directly fit the untransformed data (10,14).

Kinetic Studies Involving Two Ligands

To fully characterize a receptor system, it is essential to examine the binding characteristics of a full range of compounds (both agonists and antagonists) that have biologic activity. It is not feasible to radiolabel all drugs. Consequently, binding constants for these drugs must be calculated by examining their interaction with the radioligand for a particular binding site. The following sections deal with equations describing the kinetic properties of two drugs (one of which is radiolabeled) binding simultaneously to a given receptor population. A generalized model, similar to the situation for a single ligand [Equation (1a)], can be developed:

$$[R] + [L] \underset{K_{-1}}{\overset{K_1}{\rightleftharpoons}} [RL] \tag{21a}$$

$$[R] + [U] \underset{K_{U-1}}{\overset{K_{U1}}{\rightleftharpoons}} [RU] \tag{21b}$$

All symbols for the labeled ligand [Equation (21a)] have been previously described. [U] represents the free concentration, [RU] is the concentration of unlabeled drug bound, and K_{U1} and K_{U-1} are the association and dissociation rate constants, respectively. For simplicity, the model again assumes that less than 10% of either drug is bound to the binding site by the time equilibrium is reached. Given the law of mass action, the following equations can be written.

$$\frac{d[RL]}{dt} = [L][R]K_1 - [RL]K_{-1} \tag{22}$$

$$\frac{d[RU]}{dt} = [U][R]K_{U1} - [RU]K_{U-1} \tag{23}$$

where the free concentration of receptor is

$$[R] = B_{max} - [RL] - [RU] \tag{24}$$

Solving Equations (22) through (24) simultaneously (15,16) and expressing in terms of $[B_t]$ = [RL] (i.e., the amount of radioligand bound) gives

$$[B_t] = \frac{B_{max}K_1[L]}{A-B}\left(\frac{K_{U-1}(A-B)}{AB} + \frac{K_{U-1}-A}{A}e^{-At} - \frac{K_{U-1}-B}{B}e^{-Bt}\right) \tag{25}$$

where

$A = 0.5(C + D + SQR[(C - D)^2 + 4K_1K_{U1}[L][U]])$
$B = 0.5(C + D - SQR[(C - D)^2 + 4K_1K_{U1}[L][U]])$
$C = K_1[L] + K_{-1}$
$D = K_{U1}[U] + K_{U-1}$

Given the complexities of this equation, it is not possible to use graphic analytic techniques. However, the common availability of nonlinear curve-fitting programs offers the means of calculating the kinetic constants of unlabeled drugs (see Weighted Nonlinear Curve Fitting). The only requirements are that the experiments be performed under pseudo-first-order conditions and that the kinetic constants for the labeled ligand be determined in separate experiments.

Although the model described by Equation (25) allows the calculation of kinetic constants for unlabeled drugs, it also has important uses in studies performed at equilibrium. More specifically, it provides the means of determining when equilibrium is reached. This aspect is discussed at length below.

BIMOLECULAR EQUILIBRIUM STUDIES

As alluded to earlier, the binding of a ligand at "equilibrium" simply means that the rates of association and dissociation are equal. In theory, however, equilibrium is never reached; it is approached asymptotically. In practice, equilibrium conditions can be considered met when the amount bound deviates by a small amount (less than 3%) from the true equilibrium value (16). The model for studies performed at equilibrium should be considered a unique case of the kinetic models just discussed; that is, when time approaches infinity, or in practice, at a time when the amount bound deviates less than 3% from the true equilibrium value.

Saturation Studies

It has already been shown that at equilibrium [from Equation (6)]

$$[B_E] = \frac{B_{max}[F]}{K_{-1}/K_1 + [F]}$$

A ligand's dissociation constant K_D is defined as

$$K_D = \frac{K_{-1}}{K_1}$$

Thus

$$[B_E] = \frac{B_{max}[F]}{K_D + [F]} \tag{26}$$

This equation is identical in form to the Michaelis-Menten equation developed to describe enzyme substrate kinetics. The assumptions inherent in this model are as follows:

1. Each molecular interaction is independent.
2. Measurement of binding is taken at equilibrium.
3. [F] represents the free concentration of ligand.

In practical terms, the use of this model involves the calculation of the amount of radioligand bound at equilibrium for a given free concentration over a wide range of concentrations of the ligand. Equation (26) describes the hyperbolic relationship between bound and free ligand shown in Figure 4a. Two points should be apparent. First, there are a finite number of binding sites (i.e., the system is saturable; hence saturation studies). Also, the K_D is numerically equal to the concentration of free drug resulting in 50% occupancy of the total receptor population (B_{max}). Equation (26), in its various forms, is the basic mathematical model adopted to describe the majority of radioligand-binding site interactions at equilibrium.

Under conditions in which the ligand is binding to two or more sites simultaneously, the equation relating the total amount of radioligand bound is given by

$$[B_E] = \sum_{i=1}^{i=n} \frac{B_{maxi}[F]}{K_{Di} + [F]} \tag{27}$$

where K_{Di} and B_{maxi} are the constants for the ligand binding to site i of n possible sites. In practice, it is pössible for a radioligand to bind, not only to specific receptor sites, but also to others distinct from the receptor (nonspecific binding). Nonspecific binding generally results from the low-affinity association of the radioligand to other components of the assay system. Nonspecific binding can generally be determined experimentally and subtracted from total binding to reveal the specific component. In the kinetic models just described, nonspecific binding was not incorporated into the related models. Consequently, all equations relate to the specific binding of the radioligand. In studies performed at equilibrium, there can be a predictable (linear) relationship between nonspecific binding and the radioligand concentration (see Figure 4b). In this case, therefore, a nonspecific binding component can be incorporated into the generalized model. The appropriate equation is

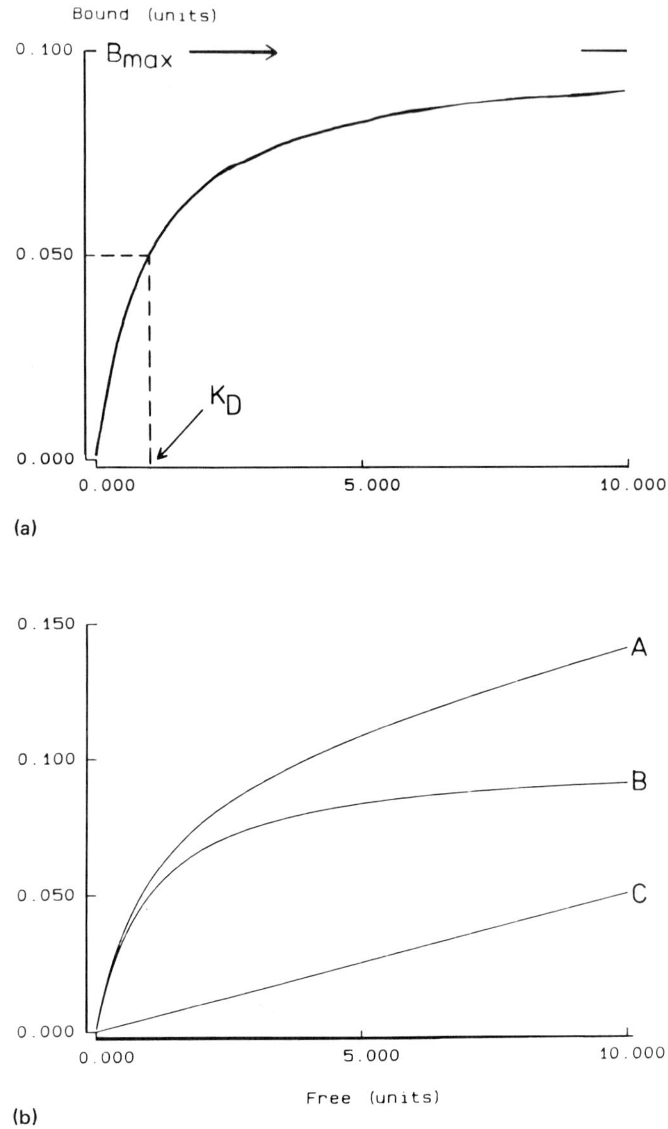

Figure 4 (a) Computer-generated saturation data using Equation (26), which describes a hyperbolic relationship between free ligand concentration and the amount bound. The dissociation constant K_D was assigned a value of 1 unit and the maximum number of sites B_{max}, 0.1 units. The K_D (K_{-1}/K_1) value represents the free radioligand concentration required to produce 50% occupancy of the total number of sites. At a free concentration 10-fold higher than the K_D value, the system is only approximately 90% saturated. (b) In this case, the model has been extended to include a nonspecific component (N = 0.01), as described by Equation (28). Total binding (A) results from both the hyperbolic specific component (B) and a linear nonspecific component (C). Experimentally, specific binding can be calculated by subtracting nonspecific binding from total binding.

$$[B_T] = \sum_{i=1}^{i=n} \frac{B_{max}[F]}{K_{Di} + [F]} + N[F] \tag{28}$$

where $[B_T]$ is the total observed binding (i.e., specific plus nonspecific) and N is the ratio of bound to free at infinite free concentration and effectively represents the gradient of the line relating nonspecific binding with free concentration of radioligand as expressed in Figure 4b. Thus, multiplying the N value by the free concentration $N[F]$ gives the amount of nonspecific binding at that particular point.

Graphic Analysis of Saturation Data

As mentioned, the basic equations describing ligand binding at equilibrium are similar to those for Michaelis-Menten enzyme kinetics. Consequently, the analytic techniques used initially to process radioligand binding data are derived from those developed to process enzyme-substrate data. If the specific amount of radioligand bound is plotted against the amount free, then accurate estimates of the required constants (K_D and B_{max}) cannot be calculated, since it is difficult to estimate the asymptotic value of B_{max}. There are, however, several rearrangements of the data that can be made to linearize the curve. One such manipulation was described by Lineweaver and Burke (17) (the double reciprocal plot) but one more widely used in radioligand binding studies is that described by Scatchard (18):

$$[B_E] = B_{max} - K_D \frac{[B_E]}{[F]} \tag{29}$$

Thus a plot (a "Scatchard plot") of the bound/free ratio versus bound should yield a straight line with a slope = $-1/K_D$ and an x intercept of B_{max}. Linear Scatchard plots suggest that the ligand is binding to a homogeneous population of receptors. However, many tissues are not homogeneous with respect to receptor subtype [Equation (1b)]. Consequently, a radioligand may bind simultaneously to two or more sites, with differing affinities. The graphic methods just described are inadequate for analyzing complex binding isotherms (10,11, 19-21). This point cannot be stressed too heavily. To demonstrate this, saturation binding data have been generated using a two-site model [Equation (27)]. In such a case the total amount of radioligand specifically bound is

$$[B_E] = \underbrace{\frac{B_{max_1}[F]}{K_{D_1} + [F]}}_{\text{Site 1}} + \underbrace{\frac{B_{max_2}[F]}{K_{D_2} + [F]}}_{\text{Site 2}} \tag{30}$$

where K_{D1}, B_{max_1} and K_{D2}, B_{max_2} are the dissociation constants and maximum number of binding sites for sites 1 and 2, respectively. From the model

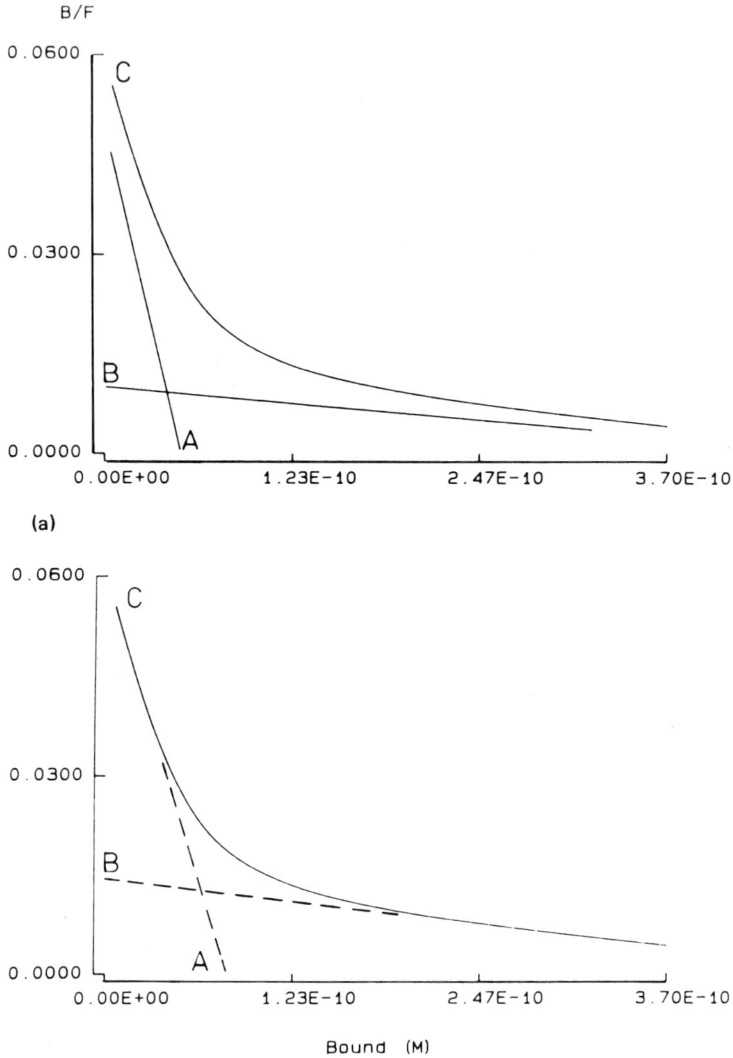

(b)

Figure 5 (a) Scatchard plot representation of a two-site model produced by Equation (27) (model parameters were site 1, K_D - 0.1 nM, B_{max} = 50 pM; site 2, K_D = 50 nM, B_{max} = 500 pM). Line C represents the actual isotherm, which is biphasic. Lines A and B are the resultant lines for sites 1 and 2, respectively. As expected these are linear. (b) Past use of the Scatchard plot as an analytic tool relied on simple linear regression through the linear portions of the biphasic curve (C) to gain estimates for potential high- (A) and low-affinity (B) sites. Obviously, estimates of the K_D and B_{max} for both sites are different from the actual values if this latter technique is utilized.

constants defined in Figure 5, Scatchard plots can be constructed (Figure 5). Also presented on the plot are the individual lines representing binding to each site. When a ligand binds to two or more sites simultaneously, the resultant Scatchard plot is curvilinear. In this case there appear to be two linear portions at each end of the line. However, these do not correspond exactly to those that would be obtained if binding were restricted to either site alone. The true line describing each binding isotherm falls inside that obtained if simple linear regression is used on the two aspects of the actual curve (see Figure 5).

Deviations from the Law of Mass Action

Linear Scatchard plots imply that the ligand is binding to a single homogeneous population of binding sites. In a number of cases, Scatchard plots are curvilinear. There are many reasons for this effect, but commonly these can be narrowed to two alternatives. First, the ligand may be binding to two or more sites simultaneously with differing affinities as described earlier. Second, there are binding site interactions resulting in either positive or negative cooperativity [for extensive reviews, see De Meyts and Rousseau (22) and Levitzki (23)]. In this latter case, the affinity of the unoccupied receptor for the ligand is progressively altered by the binding of the ligand to the receptor. A progressive reduction in affinity, producing a concave Scatchard plot, is indicative of negative cooperativity. A progressive increase in affinity, producing a convex Scatchard plot, is indicative of positive cooperativity. In the form of Hill (24,25), Equation (1a) can be rewritten as

$$[R]_n + n[L] \rightleftharpoons [RL]_n \tag{31}$$

where the overall dissociation constant is

$$K_{D'} = \frac{[R][L]^n}{[RL]_n}$$

and n (the Hill coefficient nH) is the number of binding sites per receptor molecule (in the classic, noninteractive case: n = 1).

At equilibrium,

$$[B_E] = \frac{B_{max}[L]^n}{K_{D'} + [L]^n} \tag{32}$$

Equation (32) can be rearranged to give

$$\text{Log} \frac{[B_E]}{B_{max} - [B_E]} = n \log [L] - \log K_{D'} \tag{33}$$

A plot of $\log [[B_E]/(B_{max} - [B_E])]$ versus $\log [L]$ should yield a straight line with a slope equal to the Hill coefficient nH. Values of n greater than unity indicate positive cooperativity; values less than unity indicate either negative cooperativity or the binding of the ligand to two or more sites.

At very low and high concentrations of the radioligand, the Hill coefficient always approaches unity. At these concentrations the Hill plot corresponds to the radioligand binding to the first and last receptors, respectively, which effectively do not interact cooperatively. Consequently, Hill plots should only be constructed of data resulting from free concentrations of $K_D/10$ to $K_D \times 10$.

Competition Studies

Competition studies involve simultaneously assessing the binding of a single radioligand and an unlabeled drug. Equation (25) gave the kinetic equation relating the amount of radioligand bound in the presence of unlabeled drug. At equilibrium (i.e., as t approaches infinity), this equation reduces to (16)

$$[B_E] = \frac{B_{max} K_1 K_{-1}}{A + B}$$

where A and B have been defined previously. Expanding gives

$$[B_E] = \frac{B_{max} [L]}{K_D(1 + [U]/K_{DU} + [L]/K_D)}$$

which can be rearranged to

$$[B_E] = \frac{B_{max} [L]}{K_D(1 + [U]/K_{DU}) + [L]} \qquad (34)$$

From the derivation of Equation (25), Equation (34) is only valid if the experiment is performed under zone A conditions (<10% bound at equilibrium): that is when the total amount of radioligand added [T] can be used to approximate the free concentration [L] at all concentrations of unlabeled drug utilized.

If the radioligand and unlabeled drug bind to two or more sites simultaneously, then the amount of radioligand bound at a given concentration of unlabeled drug is

$$[B_E] = \sum_{i=1}^{i=n} \frac{B_{max} [L]}{K_{Di}(1 + [U]/K_{DUi}) + [L]} \qquad (35a)$$

where i is the number of sites of a maximum of n sites.

Two experimental approaches can be utilized to calculate the K_{DU} for the unlabeled drug. First, saturation experiments using the radioligand can be performed in the presence of an increasing concentration of the unlabeled drug. Equation (34) is still in the form of a rectangular hyperbola and can therefore be graphically represented by a Scatchard plot. The effect of adding increasing concentrations of an unlabeled competitor would be to lower the apparent affinity of the radioligand without affecting the maximal number of binding sites [Equation (35b)]. Thus,

$$[B_E] = \frac{B_{max}[L]}{K_{Dapp} + [L]} \tag{35b}$$

where the apparent K_D (K_{Dapp}) is related to the true K_D of the radioligand by the relationship

$$\begin{aligned} K_{Dapp} &= K_D\left(1 + \frac{[U]}{K_{DU}}\right) \\ &= K_D + [U]\,\frac{K_D}{K_{DU}} \end{aligned} \tag{36}$$

Thus a plot of K_{Dapp} versus [U] should yield a straight line with a y intercept of K_D and a slope of K_D/K_{DU} from which K_{DU} can be calculated. This procedure is diagramatically represented in Figure 6.

More commonly, a single concentration of radioligand is utilized and the concentration of competitor required to inhibit 50% of radioligand binding ($[U] = [IC_{50}]$) is calculated. Under these circumstances, radioligand bound is half of that bound with no competitor.

$$\frac{B_{max}[L]}{K_D(1 + [IC_{50}]/K_{DU}) + [L]} = 0.5\,\frac{B_{max}[L]}{K_D + [L]}$$

The $B_{max}[L]$ cancels out and after rearrangement gives

$$2\left[K_D\left(1 + \frac{[IC_{50}]}{K_{DU}}\right) + [L]\right] = K_D + [L]$$

Rearranging gives

$$K_{DU} = \frac{[IC_{50}]}{1 + [L]/K_D} \tag{37}$$

This equation is identical to that developed by Cheng and Prusoff (26) to calculate the inhibition constant K_i for a competitive inhibitor of an enzyme-substrate reaction. In the more general case, Linden (27) has developed equations that allow the K_{DU} to be calculated from the IC_{50} but with corrections made for the amount of radioligand bound (i.e., in non-zone A conditions). In this case, the free concentration of unlabeled drug displacing 50% of total binding IC_{50} (free) is given by

$$IC_{50}(\text{free}) = IC_{50} - B_{max} + \frac{B_{max}}{2}\left(\frac{[L]}{K_D + [L]} + \frac{K_D}{K_D + [L] + B_{max}/2}\right) \tag{38a}$$

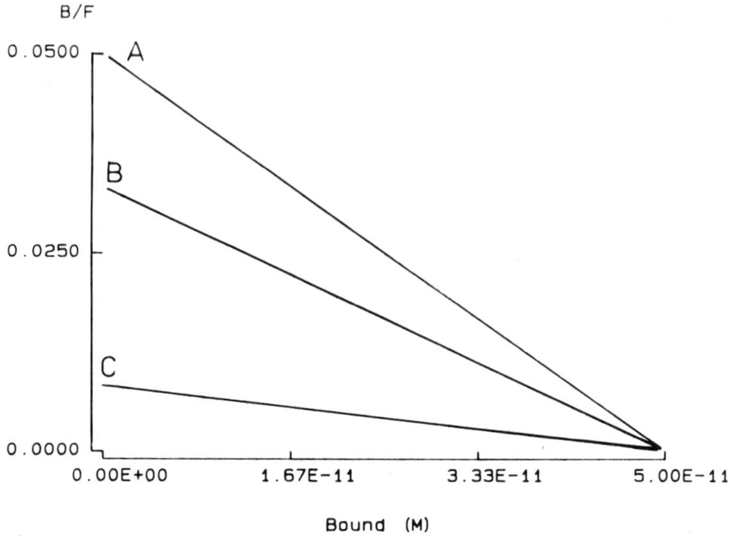

(a)

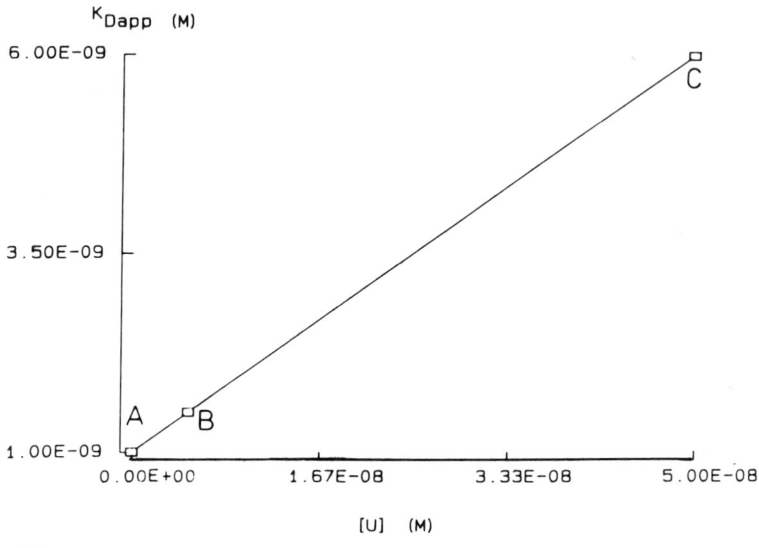

(b)

Figure 6 (a) Scatchard plot of saturation data obtained for a radioligand (K_D = 1 nM, B_{max} = 50 pM) in the absence (A) and presence of a competitive inhibitor (K_{DU} = 10 nM) at a concentration of 5 nM (B) and 50 nM (C). As the concentration of the competitive inhibitor is increased, the apparent affinity of the radioligand K_{Dapp} decreases. (b) The plot of K_{Dapp} versus concentration of competitor [U]. The dissociation constant of the competitor can be calculated from the slope of the resultant line.

in which case the corrected K_{DU} is given by Equation (38b):

$$K_{DU} = \frac{IC_{50}(\text{free})}{[1 + [L]/K_D + (B_{max}/K_D)(K_D + [L]/2)/(K_D + [L])]} \tag{38b}$$

where K_D is the dissociation constant for the radioligand, [L] is the free concentration of the radioligand in the absence of any inhibitor, and B_{max} is the maximum density of binding sites. Under zone A conditions, B_{max}/K_D is small and Equation (38b) condenses to the Cheng-Prusoff equation [Equation (37)]. In either case, the Cheng-Prusoff approach for calculating an unlabeled compound's dissociation constant is not recommended for a number of reasons. First, the Cheng-Prusoff method is virtually impossible to apply in systems involving multiple binding sites for the radioligand and unlabeled drug. Second, the availability of several computer programs that use the exact model (see Computer-Assisted Nonlinear Curve Fitting) makes this method of data processing redundant. Despite these reservations, the Cheng-Prusoff method has gained popular use in the analysis of radioligand binding experiments and for data presentation.

In the general case (in zone A conditions), the fraction of radioligand bound in the presence of increasing concentrations of a competitive inhibitor is given by

$$f = \frac{[B_{EU}]}{[B_E]}$$

where $[B_{EU}]$ is the amount of radioligand bound in the presence of the competitor and $[B_E]$ the amount bound in the competitor's absence. Thus,

$$f = \frac{B_{max}[L]/[K_D(1 + [U]/K_{DU}) + [L]]}{B_{max}[L]/(K_D + [L])}$$

$$= \frac{K_D + [L]}{K_D(1 + [U]/K_{DU}) + [L]} \tag{39}$$

But from Equation (37), $[IC_{50}] = K_{DU}(1 + [L]/K_D)$. Substituting into Equation (39) and rearranging gives

$$f = \frac{[IC_{50}]}{[U] + [IC_{50}]} \tag{40}$$

or alternatively,

$$f = \frac{[U]^{-1}}{[IC_{50}]^{-1} + [U]^{-1}} \tag{41}$$

If the fraction of radiologand displaced (1 – f) is defined as d, then

$$d = \frac{[U]}{[IC_{50}] + [U]} \tag{42}$$

Thus there is a hyperbolic relationship between the fraction of radioligand displaced (d) and the competing ligand's concentration [U]. Consequently, many of the graphic principles just applied to saturation data can also be applied to competition studies. In this respect, Equation (42) can be rearranged in a similar fashion to that described for the Scatchard plot [Equation (29)]. Thus,

$$d = 1 - [IC_{50}]\,\frac{d}{[U]} \tag{43}$$

Graphic Analysis of Competition Studies

Competition curves are generally plotted as the percentage of radioligand bound (amount of radioligand bound in the absence of any drug = 100%) versus the logarithm of the concentration of displacing drug (log [U]). Figure 7a displays this form. Figure 7a can be linearized by transformation to a Hofstee plot (28). In this transformation, the percentage of radioligand displaced is plotted against the percentage of radioligand displaced divided by [U] according to Equation (43). In this case, the $[IC_{50}]$ of the unlabeled drug is given by the slope (Figure 7b) (29). The Hofstee plot is identical in form to the Scatchard plot used to analyze saturation data, with the exception of the transposition of the axes.

Curvilinear Hofstee plots can occur experimentally in a fashion similar to that described for saturation studies. Hill coefficients can be calculated for the competitor using Equation (44a):

$$\log \frac{\%B}{100 - \%B} = -n \log [U] + n \log [IC_{50}] \tag{44a}$$

where %B is the percentage of radioligand bound ($[B_{E0}] = 100$).

A plot of log [%B/(100 – %B)] versus log [U] should yield a straight line with a slope equal to the Hill coefficient nH. A coefficient less than unity may indicate either a form of negative cooperativity in binding or binding of the competitor to two or more sites. If a heterogeneous binding site population does underlie the multiphasic Hofstee plot, then graphic procedures are, again, an inadequate means of calculating the affinity constants for the unlabeled drug. Theoretical displacement curves and a corresponding Hofstee plot for a two-site model using Equation (35a) are given in Figure 8.

Many computer programs have been written (30,31) that use a logistic function to calculate the $[IC_{50}]$ and the slope factor (equivalent to the Hill coefficient). The function takes the form

$$[\%B_E] = \frac{100[U]^n}{[IC_{50}]^n + [U]^n} \tag{44b}$$

where $[\%B_E]$ is the percentage of radioligand bound in the presence of increasing concentrations of [U], and n is the slope factor.

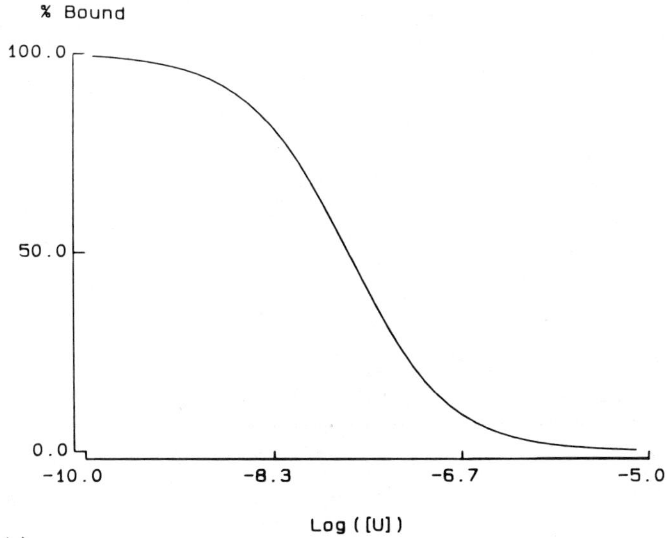

(a)

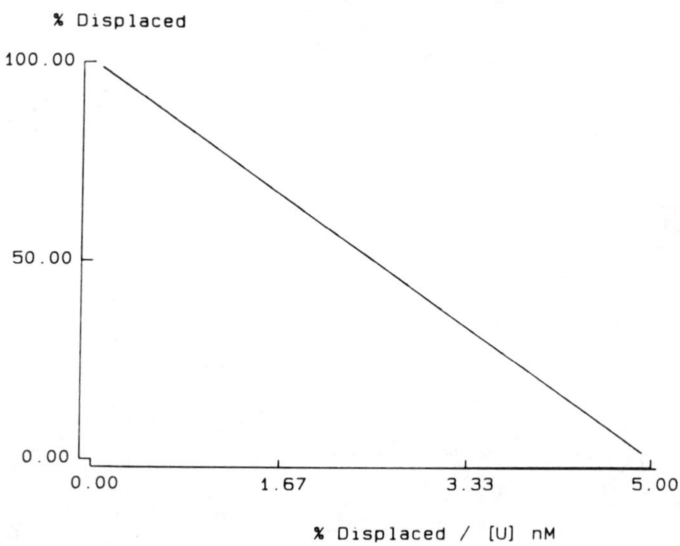

(b)

Figure 7 (a) Computer-generated displacement curve created using Equation (41) with an $[IC_{50}]$ value of 20 nM. (b) Hofstee transformation of the data presented in (a) using Equation (43), which linearizes the displacement curve. The $[IC_{50}]$ value can then be calculated from the slope of the resultant line.

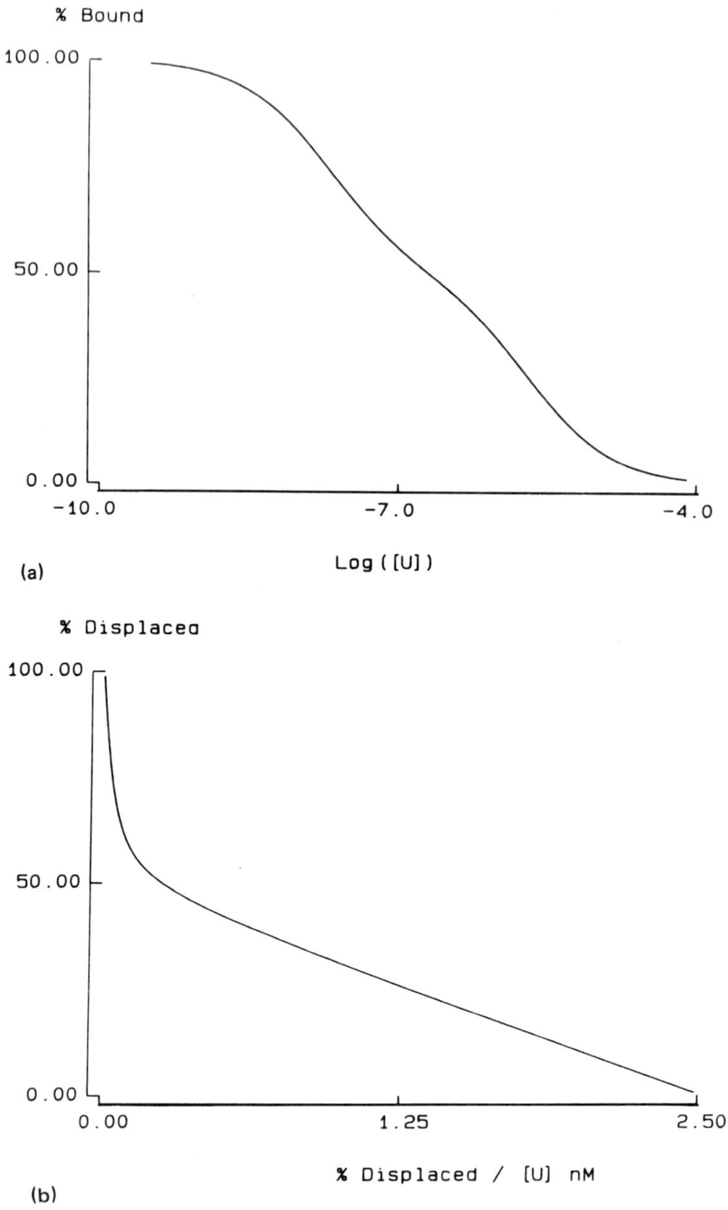

Figure 8 Computer-generated two-site displacement curve created using Equation (35a). It was assumed that the radioligand binds with a similar affinity to both sites (K_D = 1 nM, B_{max} = 50 pM) and the unlabeled drug distinguished between the two (K_{DU} values of 10 and 1000 nM for sites 1 and 2, respectively). The resultant curve has a low Hill coefficient (slope factor) and can be described as biphasic. (b) Transformation of the data presented in (a) to a Hofstee plot. As in saturation studies, the resultant line is biphasic and cannot be readily used to calculate values of the binding constants.

Some Practical Implications of the Bimolecular Model

When is Equilibrium Reached?

An important assumption made during the performance of saturation and competition studies is that "equilibrium" (see Saturation Studies) has been reached. Failure to ensure this, or to take into account nonequilibrium conditions, can have profound effects on subsequent parameter calculations (12,15,16,32). By the use of the equations developed here, it is possible to calculate times of incubation that would ensure that the system is at, or acceptably close to, the true equilibrium value.

Saturation studies As previously mentioned, the amount of radioligand bound at true equilibrium, although never reached, can be approximated by a value that is trivially different (<3%) (16). As is shown below, this would be achieved if the incubation proceeded for five half-lives. Thus, for a given concentration of radioligand, the amount of radioligand bound at any time is [from Equation (11a)]

$$[B_t] = [B_E](1 - e^{-K_{obs}t}) \tag{45}$$

where the pseudo-first-order rate constant [Equation (13)] is given by

$$K_{obs} = K_1[L] + K_{-1} \tag{46}$$

After five half-lives (i.e., 5 × 0.698/K_{obs}), the amount bound would be

$$\begin{aligned}[B_t] &= [B_E](1 - e^{-K_{obs} 5 \times 0.693/K_{obs}}) \\ &= [B_E](1 - 0.031) \\ &\cong 0.97[B_E]\end{aligned}$$

In the following discussion, the time required to reach this value is defined as the time to equilibrium T_e. From Equation (46) it is apparent that (1) the K_{obs} value (and consequently the T_e value) obtained from such a study is dependent on the radioligand concentration utilized; and (2) as [L] approaches very large values, that is, [L] >> K_D, the radioligand effectively binds instantaneously.

For example, consider a radioligand with an association rate constant of 2 × 10^8 M^{-1} min^{-1} and a dissociation rate constant of 0.02 min^{-1}. The radioligand therefore has a K_D value of 0.1 nM. If an association experiment were performed under zone A (<10% bound) conditions using a concentration equivalent to its K_D value (0.1 nM), the observed association rate constant would be

$$\begin{aligned}K_{obs} &= (2 \times 10^8)(1 \times 10^{-10}) + 0.02 \\ &= 0.04\ min^{-1}\end{aligned}$$

Therefore, to ensure that at least 97% of the radioligand had bound at this concentration, the incubation time would be

$$T_e = 5\,\frac{0.693}{0.04}$$

$$= 87 \text{ min}$$

At lower concentrations, the amount bound would not have yet reached this value, but at higher concentrations the amount bound would be in excess of the 97% value. For example, if the concentration used was 0.01 nM ($K_D/10$), then the amount bound at 87 min would be

$$K_{obs} = (2 \times 10^8)(0.01 \times 10^{-9}) + 0.02$$

$$= 0.022 \text{ min}^{-1}$$

Therefore,

$$[B_t] = [B_E](1 - e^{-0.022 \times 87})$$

$$= 0.85[B_E]$$

That is, only 85% is bound at this time.

What model parameter actually determines the time to equilibrium in a saturation experiment? The time to equilibrium is given by

$$T_e = \frac{3.5}{K_{obs}}$$

$$= \frac{3.5}{K_1[L] + K_{-1}} \tag{47}$$

At very low concentrations (those that take the longest to reach equilibrium), [L] approaches zero. Consequently, Equation (47) reduces to

$$T_e = \frac{3.5}{K_{-1}} \tag{48}$$

The time to reach equilibrium is therefore dependent on the dissociation rate constant of the radioligand. This may be considered somewhat surprising since it is common practice to determine T_e through an association experiment. More appropriately, a dissociation experiment should be performed and T_e calculated using Equation (48). In the preceding theoretical example, a time of incubation that would ensure all concentrations are at equilibrium would be

$$T_e = \frac{3.5}{0.02}$$

$$= 175 \text{ min}$$

If a saturation experiment is performed well short of the T_e value, then complex (curvilinear) Scatchard plots can result. This is diagramatically represented in Figure 9, which uses the parameter values previously defined and Equation (45) to generate apparent bound values for an 87-min incubation period. The resultant Scatchard plot is convex (13).

Competition studies When a radioligand and an unlabeled drug compete simultaneously for the same site, the time to reach equilibrium is greater than, or equal to, the time required for the radioligand alone (16). Consequently, it is possible that the incubation time utilized in the saturation studies may be less than that required for a competition study to reach effective equilibrium. To demonstrate this, simple modeling can be performed using Equation (25), which describes concurrent binding of two drugs (one of which is radiolabeled) to a homogeneous population of sites. The modeling can be simplified by assuming that the concentration of the unlabeled drug is at its $[IC_{50}]$ value. In addition, the association constants K_1^* for both drugs are equivalent and equal to $2 \times 10^8\ M^{-1}\ min^{-1}$, a value that approximates the value experi-

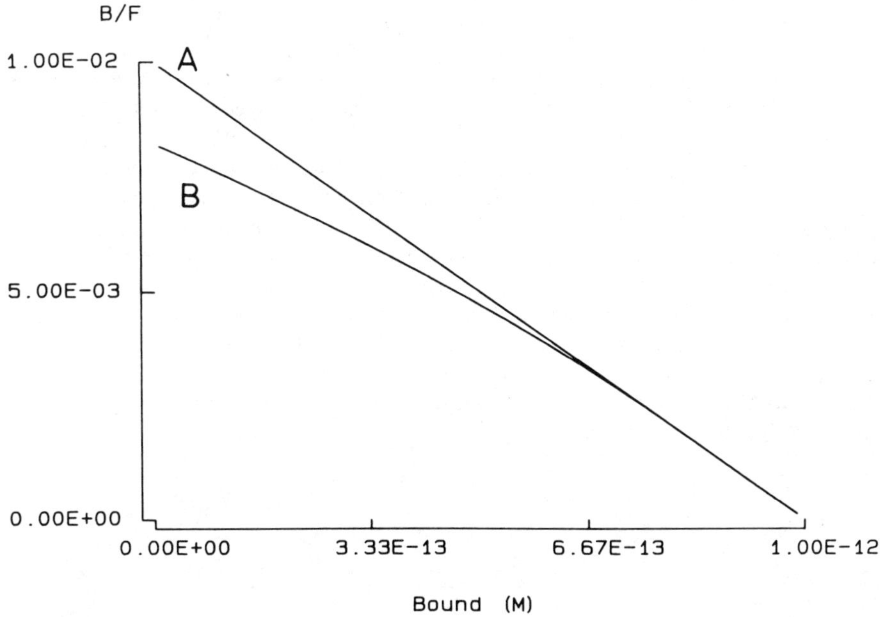

Figure 9 Scatchard plot of saturation data (B_{max} = 1 pM, K_D = 0.1 nM, K_{-1} = 0.02 min^{-1}, $K_1 = 2 \times 10^8\ M^{-1}\ min^{-1}$) using an incubation time of 87 min. (B) Based on the dissociation rate constant, the actual time should have been at least 175 min (A). As a result of the short incubation time, the plot (B) is curvilinear.

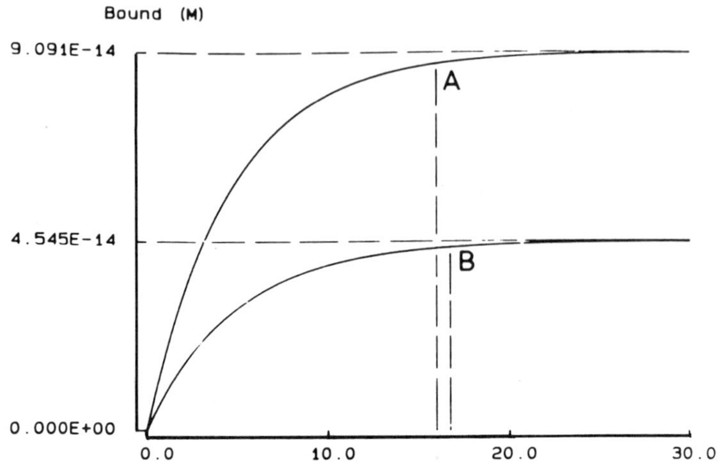

(a)

Figure 10 (a) Case 1. Association time course data generated using Equation (25) in the absence (A) and in the presence (B) of a competitive inhibitor at its [IC_{50}] value. Horizontal dashed lines represent the final amount of radioligand bound at equilibrium. Vertical dashed lines represent the time required to reach 97% of the final equilibrium value. In this case, where the concentration of radioligand is low ($K_D/10$; see Table 1) and the competitor dissociates much more rapidly, the time to reach equilibrium is the same as for the radioligand alone. (b) Case 2. The same case model as described in (a) except that the radioligand concentration is greater ($K_D \times 10$). Although increasing [L] decreases the time to equilibrium for the radioligand alone (A), in the presence of the competitor (B) the reaction cannot proceed faster than $1.75/K_{-1}$. (c) Case 3. Association data for a radioligand alone (A) and in the presence of a competitive inhibitor (B) that dissociates more slowly than the radioligand (see Table 1). The amount of radioligand bound actually overshoots its final equilibrium value. The time to equilibrium is equal to $1.75/K_{U-1}$ and is independent of the radioligand concentration.

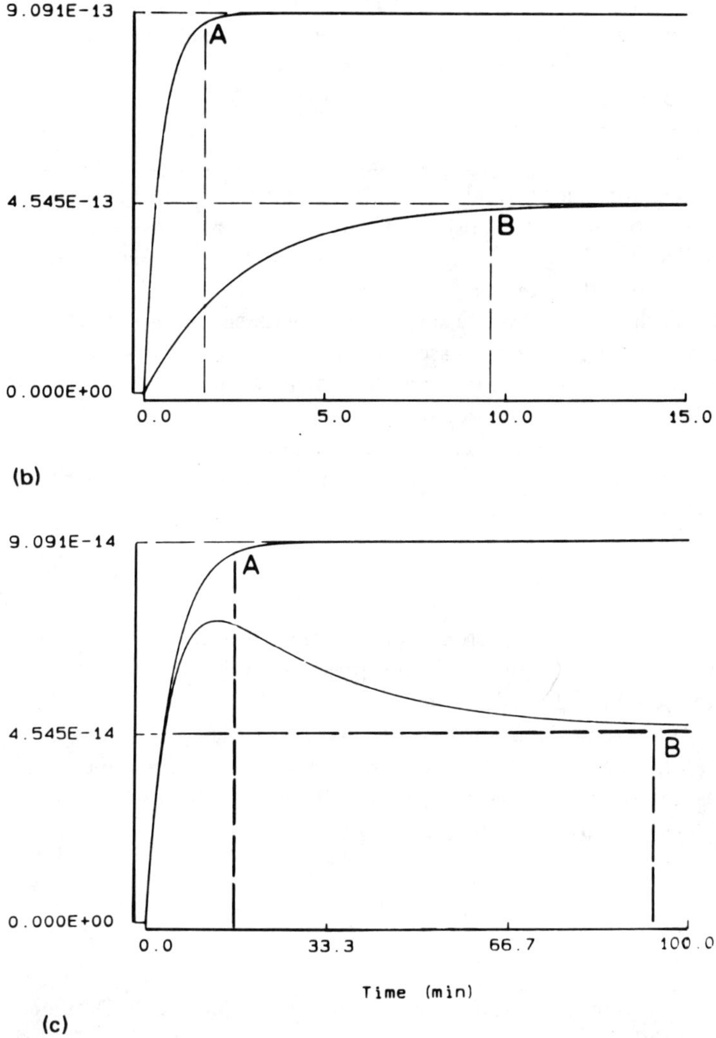

Figure 10 (Continued)

mentally obtained using a number of radioligands and receptor systems. The time required to reach equilibrium is largely dependent on the dissociation constants of each drug. In this respect, two extreme cases can be examined to determine an approximate time to reach T_e: when the radioligand has a much higher affinity than the unlabeled drug (i.e., $K_{-1} << K_{U-1}$), and second, when the unlabeled drug has a higher affinity than the radioligand (i.e., $K_{-1} >> K_{U-1}$). The mathematical derivations have been extensively described elsewhere (16). It is important to realize the profound effect that the affinity of the unlabeled drug can have on the time to reach equilibrium, which in all cases will be greater than that for the radioligand alone.

The time to reach equilibrium in a saturation experiment is dependent on the radioligand concentration. In the case of competition studies in which the radioligand has a higher affinity for K_D than the unlabeled drug, the radioligand concentration also needs to be considered. There are two extremes.

When $[L] << K_D$ (case 1), the time to reach equilibrium is the same as that required by small concentrations of the radioligand alone (i.e., it is dependent on K_{-1}). Thus,

$$T_e = \frac{3.5}{K_{-1}} \tag{49}$$

Figure 10a diagramatically represents this situation, substituting the values presented in Table 1 into Equation (25). Also presented in Figure 10a is the kinetic profile expected for the radioligand alone.

When $[L] >> K_D$ (case 2), in the case of radioligand alone, increasing the concentration results in decreasing T_e values to the point at which equilibrium is reached almost instantaneously (i.e., T_e approaches zero). However, in the presence of an unlabeled drug the shortest incubation period is

$$T_e = \frac{1.75}{K_{-1}} \tag{50}$$

Table 1 Theoretical Kinetic Constants for Radioligand Binding in the Presence of a Competitive Inhibitor[a]

Case	[L] (nM)	K_{-1} (min^{-1})	K_D (nM)	K_{U-1} (min^{-1})	K_{DU} (nM)
1	0.1	0.2	1	200	1000
2	10	0.2	1	200	1000
3	0.1	0.2	1	0.02	0.1

[a]Values are given for the three examples discussed in the text. In all cases the association rate constant (K_{-1} and K_{U-1}) was $2 \times 10^8\ M^{-1}\ min^{-1}$ with a B_{max} value of 1 pM.

This is equal to half the maximum time required to reach equilibrium for the radioligand alone. However, if one compares the T_e values obtained in the presence and absence of the drug, it is apparent that it takes much longer to reach equilibrium when the competing drug is present. This is diagramatically represented in Figure 10b.

When $K_{-1} >> K_{U-1}$ (case 3)—when the competing drug has a higher affinity than the radioligand—the time to reach equilibrium is dependent on the dissociation constant of the unlabeled drug. Thus the time to equilibrium is

$$T_e = \frac{1.75}{K_{U-1}} \tag{51}$$

In this situation, the time to equilibrium is independent of the radioligand concentration. It should also be noted that the amount of radioligand bound actually exceeds its final equilibrium value at some points in time. An example of this phenomenon is displayed in Figure 10c.

In all three cases, the calculated $[IC_{50}]$ value varies with time and only reaches its true value once equilibrium has been reached (16). Subsequent K_{DU} and Hill coefficient values can be erroneous if these conditions are not met. In the situation in which the competing drug dissociates more slowly than the radioligand, the time to reach equilibrium is determined by the dissociation constant of the competing drug. The converse holds true if the competing drug dissociates more rapidly. To avoid the possibility of nonequilibrium conditions, one can use the kinetic equation [Equation (25)] rather than one specifically applicable at equilibrium [Equation (26)] to calculate the model constants.

Calculation of the Unbound Radioligand Concentration

Another assumption in the bimolecular model is that the free concentration of the ligand is known or can be accurately determined. In the case of a saturation experiment, free radioligand concentration can be calculated by subtracting the amount bound from the total radioligand added. This procedure suffices under some circumstances, but at very low or high concentrations of the radioligand, the relatively uncertain calculations of the bound value can give erroneous free concentrations. The error is then transferred when the data are transformed to the bound/free values required for Scatchard plot analysis.

An alternative approach, already eluded to, is to ensure that less than 10% of the radioligand is bound at equilibrium (zone A). In this situation the total radioligand concentration can be used to approximate the free concentration at all concentrations utilized. If this assumption is made but zone A conditions are not met, then the resultant Scatchard plot has a lower slope (K_D) than expected owing to the overestimation of the free value (see below) (33). Similar arguments apply to both competition and kinetic studies.

A practical problem involved in the calculation of free radioligand results from the usual procedure of determining receptor-bound radioligand under nonequilibrium conditions (14). That is, the mixture is rapidly filtered, the bound radioligand quantified, and the free concentration calculated by difference. Binding of the radioligand just prior to filtration is likely to involve not only high-affinity (receptor) components, but also a low-affinity-high-capacity portion resulting from binding to other cellular structures distinct from the receptor. Although this form of binding could be classified as nonspecific, it should be included in the model as an additional binding site. However, it is unlikely that this low-affinity component will be represented in experimentally determined nonspecific binding since it dissociates so readily during the filtration process. This form of undetected nonspecific binding results in a miscalculation (overestimation) of the free concentration. Consequently, bound/free values will be lower and the radioligand will appear to bind with an affinity less than the actual value. This phenomenon is diagrammatically represented in Figure 11.

This form of undetected nonspecific binding may have important ramifications in all aspects of radioligand binding studies. Many studies designed to elucidate receptor pharmacology and function rely on assessing changes in binding characteristics produced by inclusion in the assay, for example, of divalent cations and guanine nucleotides, in addition to a multitude of unlabeled drugs. Clearly, changes in the apparent affinity of binding of a radioligand and/or unlabeled drugs can be achieved by alteration of the degree of binding to the undetected nonspecific component. To ensure that the filtration technique does not artifactually distort the true binding constant values, it may be prudent to use an independent procedure, such as equilibrium dialysis, to determine the true free radioligand concentration (14).

COMPUTER-ASSISTED NONLINEAR CURVE FITTING

The preceding sections briefly describe the types of equations that can be utilized to allow practical assessment of a particular model, in this case the simple bimolecular interaction. Largely for historical reasons, methods for graphically displaying, and in some instances analyzing, such data have also been presented. It should be apparent, however, that these methods are highly subjective and in some cases are inappropriate for analyzing the more complex situations.

Computer-assisted, nonlinear curve-fitting (NLCF) techniques provide a means of accurately analyzing all forms of radioligand binding experiments. Once the appropriate model has been defined, NLCF techniques rely on iterative procedures to modify initial estimates of model parameters (e.g., K_D, B_{max}, K_1, and K_{-1}) until the predicted line varies by a minimal amount from the experimental data. Several advantages are associated with NLCF. (1) No transformation of the data is necessary since the exact model can be used. (2) Sub-

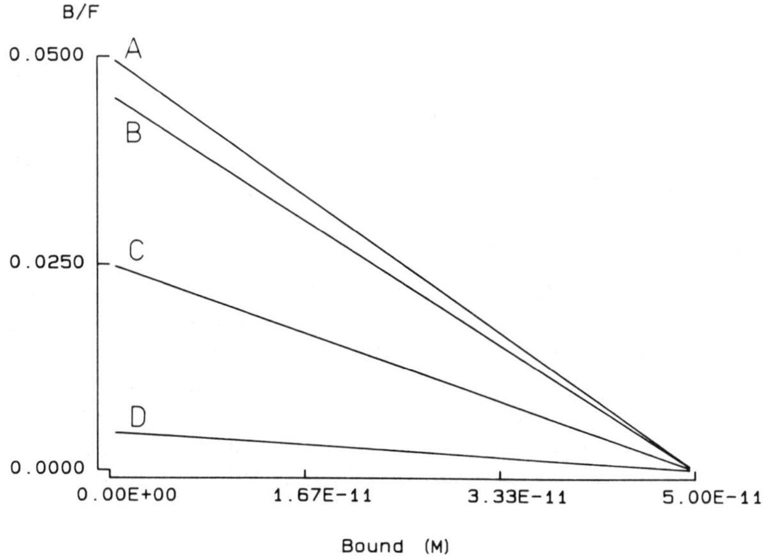

Figure 11 Scatchard plot of saturation data generated using Equation (28) (K_D) = 1 nM, B_{max} = 50 pM). Presented are the data obtained in the absence (A) and presence of an undetected low-affinity site (K_D = 1 μM) with B_{max} values of 0.1 (B), 1 (C), and 10 (D) μM. The presence of the second site may not be detected when filtration is used to separate bound and free. Consequently, the affinity of the radioligand for the high-affinity site is underestimated. The B_{max} calculation is not affected.

jective analysis, which is present in graphic procedures, is eliminated. (3) A number of statistical tests can be performed on the data to determine the suitability of the chosen model compared with other proposed models (34).

Many generalized nonlinear curve-fitting programs are readily available (e.g., MLAB, CONSAAM, NONLIN, and FUNFIT) that can be used to analyze complex models. Other nonlinear curve-fitting programs have been written to analyze specific aspects of radioligand binding experiments. For example, Feldman (35) has developed a technique for analyzing complex ligand-binding interactions at equilibrium. Feldman's general model involves the simultaneous binding of several ligands (some of which may be radiolabeled) to several independent receptor populations, assuming the law of mass action as described in Equation (1b). This technique has been implemented in such programs as SCATFIT (36) and LIGAND (21). An additional advantage of both these programs is that the free radioligand concentration is calculated by a more accurate iterative procedure (36) and therefore does not limit experiments to zone A conditions. Such programs are easily adapted to operate on microcomputers (10,34, 37).

CONCLUSION

The use of radiolabeled drugs as molecular probes for putative biologic receptors can be a powerful investigative procedure, although the results are meaningful only if due care is used in performing the experiments and analyzing the data. First, the correct model must be selected prior to any serious attempt to characterize the ligand-binding system, since the model may dictate some experimental condition that must be met. Second, the values for the model parameters obtained in subsequent analysis are only as meaningful as the model itself. A way to ensure valid results is to perform the experiments using a number of different techniques that should, if the model is correct, arrive at the same final result. Although this approach is somewhat more tedious, it can eliminate possible subsequent confusion regarding the characteristics of a particular receptor system.

APPENDIX: EXPONENTIAL CURVE FEATHERING

The technique of curve feathering (also referred to as curve peeling, curve stripping, or the residuals technique) is commonly used in the pharmacokinetic sciences in which drug absorption and elimination generally follow first-order kinetics (38). The technique relies on graphically subtracting the influence of each exponential (each binding site) from total binding. The procedure is best described by an example.

Dissociation

Consider a biexponential dissociation model having the following characteristics.

$$[B_t] = \underset{\text{Site 1}}{1000e^{-0.1t}} + \underset{\text{Site 2}}{1000e^{-0.01t}} \qquad (52)$$

Thus there are two sites. For each site, the amount bound at time (0) = 1000 (i.e., $[B_{01}]$ and $[B_{02}]$ = 1000). The rate constant for site 1 equals 0.1 min^{-1} ($t_{1/2}$ = 0.693/0.1 = 6.9 min). For site 2, the dissociation rate constant is 0.01($t_{1/2}$ = 69 min). A total of 10 data points were generated using Equation (52) for times between 0 and 100 min. The resultant data are summarized in Table 2 and displayed in Figure 12. Rather than convert the bound data to actual logarithm values, it is customary to plot the data using semilogarithmic graph paper (Figure 12). The terminal phase of the resultant plot is linear and

Table 2 Theoretical Dissociation Data Generated Using Equation (52)

Time	$[B_t]$	$E[B_t]^a$	Residual[b]
0	2000	1000	1000
0.5	1946	998	948
1	1784	995	899
2	1798	990	808
4	1631	960	671
10	1272	900	372
20	954	820	134
40	688	—	
80	449	—	
100	367	—	

[a]Extrapolated $[B_t]$ values obtained graphically.
[b]Residual is equal to $[B_t] - E[B_t]$.

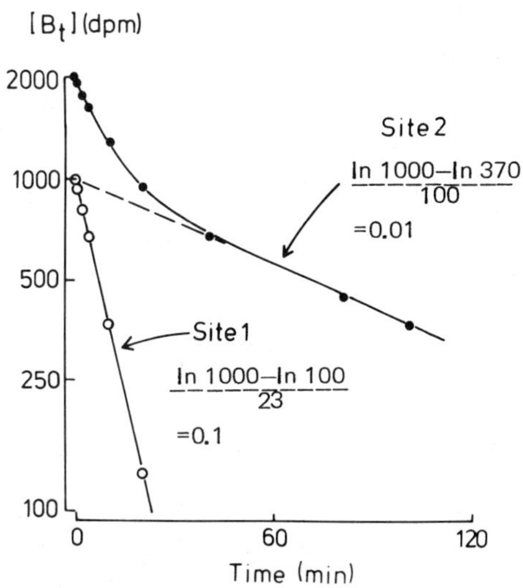

Figure 12 $[B_t]$ versus time plot for a biexponential dissociation model (●). The most slowly dissociating site, dominant in the terminal portion of the curve, is subtracted from the total curve. Subtraction of extrapolated $[B_t]$ values from the actual data yields a new set corresponding to site 1 (○). As with the association example, K_{-1} values for each site can be calculated from the slope of the logarithmically transformed data.

Table 3 Theoretical Biexponential Association Data Generated Using Equation (53)

Time	$[B_t]$	$[B_E] - [B_t]$[a]	E[NB][b]	Residual[c]
0	0	2000	1000	1000
1	114	1886	900	896
2	220	1780	980	800
4	406	1594	950	644
10	813	1187	820	367
20	1194	806	680	126
40	1532	486	–	
80	1797	203	–	
100	1864	136	–	

[a] $[B_E]$ was determined at time points greater than 100. Points close to $[B_E]$ are generally excluded since, experimentally, they can distort the resultant linear transformation. The value $[B_E] - [B_t]$ represents the amount of ligand not bound [NB] at time t.
[b] E[NB] represents the extrapolated [NB] value using the terminal portion of the curve presented in Figure 13.
[c] Residual refers to the value [NB] – E[NB].

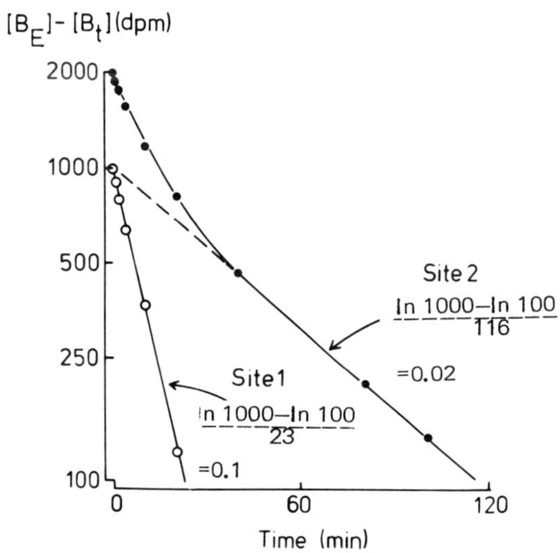

Figure 13 $[B_E] - [B_t]$ versus time plot for a biexponential association model (●). The most slowly associating site (site 2) is represented by the terminal portion of the plot. The last three points are used to extrapolate back to the y axis. The corresponding $[B_E] - [B_t]$ points are subtracted from the remaining points, which are then replotted (○). These data correspond to the radioligand binding to site 1. Slope values (K_{obs}) are calculated by taking the natural logarithms of the y-axis values before dividing by the corresponding time range.

has a slope value proportional to K_{-1} for the most slowly dissociating site (site 2). Extrapolation of this line produces a y intercept of $[B_{02}]$. To calculate the K_{-1} value (the slope), the natural logarithm values of the bound values must first be taken (see Figure 12). Subtracting the calculated bound values on the extrapolated line from the actual data (see Table 2) produces a new set of $[B_t]$ values, which are then replotted (Figure 12). Since a biexponential model was considered, the residual values fall on the actual line for site 1. However, if another site were present, there would again be a terminal phase, corresponding to the next most slowly dissociating site, for which the procedure could be repeated.

Association

The curve peeling technique for analyzing association studies is identical for that just described for dissociation studies. Consider a biexponential model:

$$[B_t] = \underbrace{1000(1 - e^{-0.1t})}_{\text{Site 1}} + \underbrace{1000(1 - e^{-0.02t})}_{\text{Site 2}} \tag{53}$$

where 0.1 and 0.02 are the K_{obs} values for sites 1 and 2, respectively. The corresponding $t_{1/2}$ values for each site would be 6.9 and 34 min^{-1}, respectively. Site 2 therefore represents the most slowly associating site. A total of 10 points were generated using Equation (53) and are presented in Table 3 and graphically represented in Figure 13. The data have also been transposed according to Equation (11d). The linear terminal phase represents association to site 2 (the slowest). As with dissociation study analysis, actual data ($[B_E] - [B_t]$) could be plotted on semilogarithmic paper. Extrapolation of the data would produce a y intercept of $[B_E] - [B_{E2}]$. K_{obs} can be calculated (see Figure 13) by first taking the natural logarithms of selected y-intercept values before calculating the slope value.

From the extrapolated line, $[B_E] - [B_t]$ values can be calculated and subtracted from the original data to yield a new set of values (see Table 3 and Figure 13). From these, values for K_{obs_1} and $[B_{E1}]$ can be calculated.

REFERENCES

1. A. J. Clark, *J. Physiol. (Lond.) 61*:530 (1926).
2. A. J. Clark, *J. Physiol. (Lond.) 61*:547 (1926).
3. A. J. Clark, *Handb. Exp. Pharmakol. 4*:61 (1937).
4. J. H. Gaddum, *J. Physiol. (Lond.) 61*:141 (1926).
5. I. Langmuir, *J. Am. Chem. Soc. 40*:1361 (1918).
6. R. R. Ruffolo, *J. Auton. Pharmacol. 2*:277 (1982).
7. T. P. Kenakin, *Pharmacol. Rev. 36*:165 (1984).

8. D. Colquhoun, in *Drug Receptors* (H. P. Rang, ed.), Macmillan, New York, 1973, p. 149.
9. M. Stone, *J. Theor. Biol. 111*:261 (1984).
10. G. A. McPherson, *J. Pharmacol. Methods 14*:213 (1985).
11. P. J. Munson, in *Brain Receptor Methodologies* (P. J. Marangos, I. C. Campbell, and R. M. Cohen, eds.), Academic Press, New York, 1984, p. 33.
12. G. A. Weiland and P. B. Molinoff, *Life Sci. 29*:313 (1981).
13. J. M. Boeynaems and J. E. Dumont, in *Outlines of Receptor Theory*, Elsevier/North Holland, Amsterdam, 1980.
14. P. K. Siiteri, *Science 223*:191 (1984).
15. P. Aranyi, *Biochim. Biophys. Acta 628*:220 (1980).
16. H. J. Motulsky and L. C. Mahan, *Mol. Pharmacol. 25*:1 (1984).
17. H. Lineweaver and D. Burke, *J. Am. Chem. Soc. 56*:658 (1934).
18. G. Scatchard, *Ann. N.Y. Acad. Sci. 51*:660 (1949).
19. J. G. Nørby, P. Ottolenghi, and J. Jensen, *Anal. Biochem. 102*:318 (1980).
20. I. M. Klotz, *Science 217*:1247 (1982).
21. P. J. Munson and D. Rodbard, *Anal. Biochem. 107*:220 (1980).
22. D. D. De Meyts and G. G. Rousseau, *Circ. Res. 46*:13 (1980).
23. A. Levitzki, in *Quantitative Aspects of Allosteric Mechanisms*, Springer-Verlag, New York, 1978.
24. A. V. Hill, *Biochem. J. 7*:471 (1913).
25. A. V. Hill, *J. Physiol. (Lond.) 40*:4 (1910).
26. Y. Cheng and W. H. Prusoff, *Biochem. Pharmacol. 22*:3099 (1973).
27. J. Linden, *J. Nucleotide Res. 8*:163 (1982).
28. B. H. J. Hofstee, *Science 116*:329 (1952).
29. E. L. Rugg, D. B. Barnett, and S. R. Nahorski, *Mol. Pharmacol. 14*:996 (1978).
30. G. A. McPherson, *Comput. Programs Biomed. 17*:107 (1983).
31. R. B. Parker and D. R. Waud, *J. Pharmacol. Exp. Ther. 177*:1 (1971).
32. F. J. Ehlert, W. R. Roeske, and H. I. Yamamura, *Mol. Pharmacol. 19*:367 (1981).
33. K.-J. Chang, S. Jacobs, and P. Cuatrecasas, *Biochim. Biophys. Acta 406*: 294 (1975).
34. G. A. McPherson, *Analysis of Radioligand Binding Experiments; A Collection of Computer Programs for the IBM PC.* Biosoft, Cambridge, England, 1986.
35. H. A. Feldman, *Anal. Biochem. 48*:317 (1972).
36. A. De Lean, A. A. Hancock, and R. J. Lefkowitz, *Mol. Pharmacol. 21*:5 (1982).
37. G. A. McPherson, *Trends Pharmacol. Sci. 4*:369 (1983).
38. M. Gibaldi and D. Perrier, in *Pharmacokinetics*, Marcel Dekker, New York, 1982.

4

Receptor-Linked Second Messenger Systems

William J. Kinnier / Nova Pharmaceutical Corporation, Baltimore, Maryland

Second messenger systems are traditionally thought of as intracellular chemicals, cations, cyclic nucleotides, and phospholipids that are effectors within the cell for events initiated by receptor activation. Calcium, which was first discovered in 1808 by Humphrey Davy (1), became important in animal cell function and survival in 1882, when it was found to be required for normal contraction of frog heart (2). Using the calcium-selective dye, alizarium sulfonate, Heilbrunn (3) was able to study the action of and the role of intracellular calcium on heart function. The importance of intracellular calcium as a regulator or second messenger was received with little enthusiasm at this time. Work by Rall and Sutherland (4) in 1958 on the heat-stable "adenine ribonucleotide," later called cyclic AMP (3,5-cyclic adenosine monophosphate) (Figure 1), provided a fresh perspective to the second messenger concept. Piecing together the interrelationships of extracellular events of the hormone (epinephrine) and such intracellular events as the production of cyclic AMP, Sutherland and colleagues elegantly integrated the initial, intermediate, and final events to one of the most comprehensive interrelationships between biochemistry and physiological responses. This was termed the "two-messenger hypothesis". (5) to describe hormone action and, it was postulated that the hormone epinephrine is the first messenger circulating in the blood. It then binds to the plasma membrane of the target cell in the muscle and activates the enzyme adenylate cyclase. Cyclic AMP, the "second messenger," is then generated on the inner surface of the cell membrane from the substrate; ATP diffuses through the cell and brings about the appropriate physiological response, in this instance glycogen metabolism. This was the first time that the concept of a second messenger was used and the insight into the interrelationship of hormone/neurotransmitter $\longrightarrow$ second messenger

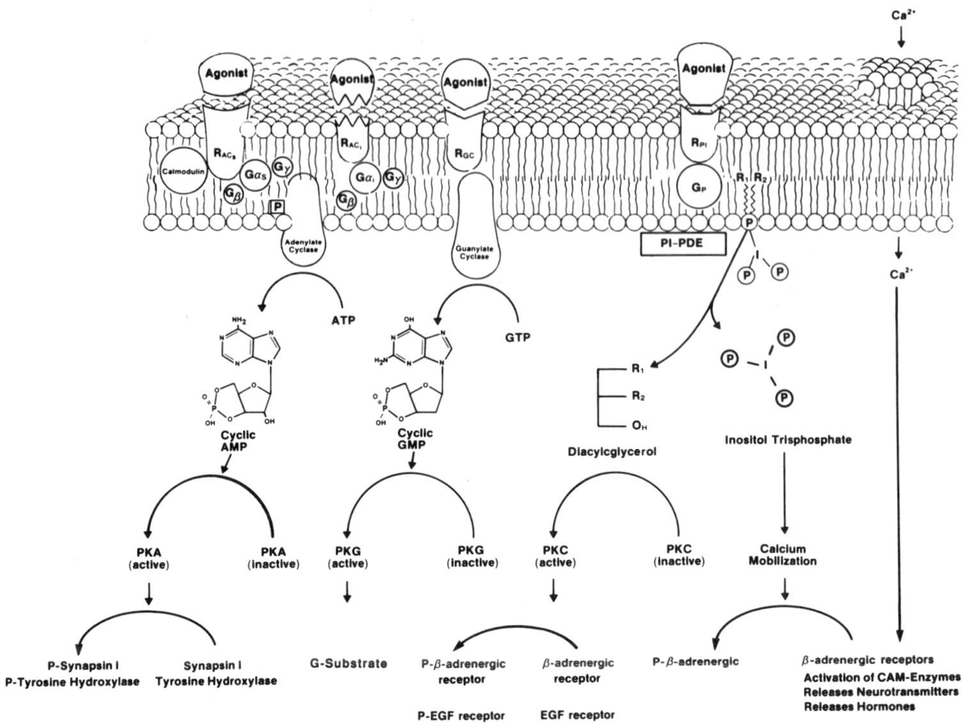

Figure 1 Schematic lipid bilayer of a cell membrane with the adenylate cyclase, guanylate cyclase, PI, and Ca^{2+} system components.

⟶ effector ⟶ physiological response that would influence, guide, and advance the understanding of other second messenger systems. Cyclic AMP pandemonium followed for the next two decades with an intensity of research activity that few other subjects have ever experienced (19,000 papers since 1966).

Shortly after the discovery of cyclic AMP came the identification 3′,5′-cyclic guanine monophosphate (cyclic GMP; Figure 1) (6). This cyclic nucleotide did not experience quite the research popularity (3500 papers since 1966) as cyclic AMP, and a comprehensive functional role of cyclic GMP has remained elusive. Cyclic GMP research paralleled most aspects of the cyclic AMP work, revealing more differences than similarities. Some notable differences of the guanylate cyclase system are the comparative insensitivity to hormonal agents while being stimulated by lipids, peroxides, free radicals, and oxidants (7).

The initial discovery that hormones had effects on phosphoinositide metabolism was first described by Hokin and Hokin (9) in 1953, several years before the pioneering work of Rall and Sutherland (4); however, the enzyme systems that mediate the phosphoinositide turnover took many years to characterize. One of these enzymes, myoinositol-1-phosphatase, which rapidly liberates free myoinositol from inositol-1-phosphate (Figures 1 and 2), made research on the phosphoinositides very difficult until it was discovered (10) that myoinositol formation was lower in lithium-treated rats. It was subsequently noted (11) that lithium inhibited the enzyme myoinositol-1-phosphatase. Lithium, in the incubation medium, facilitated the accumulation of inositol-1-phosphate. This observation represented a crucial event in the understanding of the physological role of phosphatidylinositol in tissue function and a clear understanding of the interrelationships of intracellular Ca^{2+} fluxes and inositol-1,4,5-trisphosphate. Although as early as 1975 workers (12) proposed that inositol phosphate was responsible for mobilizing calcium, Berridge and coworkers (13) in 1984 clearly

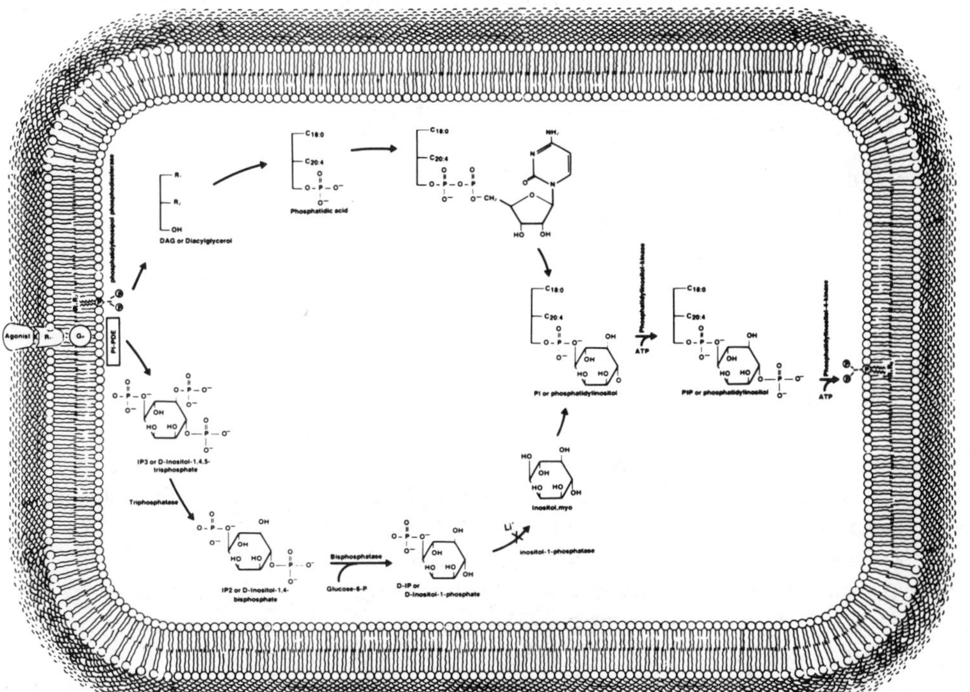

Figure 2 Schematic cross section of a cell with the substrates and enzymes involved in the PI and DAG cycle.

showed that interrelationship from hormone (epinephrine) ⟶ second messenger (inositol-1,4,5-trisphosphate) ⟶ intracellular calcium mobilization.

Diacyclglycerol's (DAG) role as a second messenger also ran into scientific resistance until recent years, but thanks to the persistent work of Takai et al. (14) it was demonstrated that DAG activated the enzyme protein kinase C, which could then phosphorylate various substrate target proteins.

RECEPTOR-COUPLED SYSTEM

The most widely used analogy of receptor-neurotransmitter interactions that generate second messengers is that of a lock (the receptor) and key (the neurotransmitter). This mechanistic view, although crude, nevertheless has been very helpful in conveying the biochemical events—of course assuming that scientists understand the workings of a lock. Many receptors transfer their signal inside the cell through a second messenger system.

The rest of this chapter briefly reviews some of the major components (i.e., external signal, receptor, transducer, amplifier, and internal effector) of the second messenger systems of cyclic AMP, cyclic GMP, calcium, IP_3, and DAG, which will be defined and delineated further by specific receptors in the other chapters.

CYCLIC AMP GENERATION

The cyclic AMP generating system is made up of receptor site (R), nucleotide regulatory protein (G), and adenylate cyclase (C) (Figure 1). It may or may not have calmodulin (CAM) and an adenosine-sensitive site (P) (16,17).

Receptor Site

The receptor site imparts the selectivity, sensitivity, and specificity needed for the cyclic AMP response. Some receptors increase cyclic AMP production; others inhibit it. As new chemical entities with increased selectivity, sensitivity, and specificity became available, the subcategorization of the receptor system has occurred. The production of cyclic AMP can either be stimulated by adenosine A_2 (18), β-adrenergic (19), dopamine D_1 (20), serotonin 2 (21), histamine H_2 (22), and vasopressin 2 (23) receptors (R_{AC_a}) or inhibited by adenosine A_1 (26), α_2 adrenergic (25), dopamine D_2 (26), muscarinic 2 (27), and δ opiate receptors (26) (R_{AC_i}) (Figure 1, Table 1). The binding of ligand (neurotransmitter or drug; L) is the first in the sequence of molecular events required in the transmembrane signal to effect cyclic AMP production.

The L–R complex is thought, then, to bind to the nucleotide regulatory protein (L–R–G). The L–R–G complex favors the binding of GTP to G.

Table 1 Receptor Classification with Second Messenger Systems

Receptor	Subtype	Second messenger
ACTH		PI, Ca^{2+} (103)
Adenosine	A_1	cAMP (inhibits) (24)
	A_2	cAMP (stimulated) (18)
Adrenergic	α_1	PI (76), Ca^{2+}, cGMP (69)
	α_2	cAMP (inhibits) (25)
	β_1	cAMP (stimulates) (19)
	β_2	cAMP (stimulates) (19)
Angiotensin II		cAMP (inhibits) (104) PI, Ca^{2+} (105)
Atrial naturietic factor (ANF)		cAMP (inhibits) (106) cGMP (104)
Bradykinin		PI (77), Ca^{2+}
Cholecystokinin		PI, Ca^{2+} (108)
Corticotropin releasing hormone		PI, Ca^{2+} (109) AC (stimulation) (109)
Dopamine	D_{-1}	cAMP (stimulation) (20)
	D_{-2}	cAMP (inhibits) (26) PI (77)
Excitatory amino acid	NMDA	PI (78), Ca^{2+}, cGMP (74)
	Quisqualate	cGMP (74)
	Kainate	cGMP (74)
Follicle stimulating hormone		cAMP (stimulates) (110)
GABA	$GABA_B$	cAMP (enhances) (111)
Glucagon		cAMP (stimulates) (112)
Histamine	H_{-1}	PI (79), Ca^{2+}
	H_{-2}	cAMP (stimulates) (22)
Human chorionic gonadotropin		cAMP (stimulates)
Insulin		cAMP (inhibits) (113), PI (114)
Luteinizing hormone		cAMP (stimulates) (115)
Luteinizing hormone releasing hormone		cAMP (stimulates)
Melanocyte stimulating hormone		cAMP (stimulates) (116)
Muscarinic cholinergic	M_{-1}	PI (80), Ca^{2+}, cGMP (70)
	M_{-2}	cAMP (inhibits) (27)
Nerve growth factor (NGF)		PI, CA^{2+} (117)
Neuropeptide Y		cAMP (inhibits) (118)

Table 1 (Continued)

Receptor	Subtype	Second messenger
Neurotensin		PI, Ca^{2+} (119) cAMP (inhibits) (120)
Opiates	δ	cAMP (inhibits) (16)
Parathyroid hormone		cAMP (stimulates) (121)
Platelet activating factor (PAF)		PI, Ca^{2+} (122) cAMP (inhibits) (123)
Prostaglandin E_1		cAMP (stimulates) (124)
Prostaglandin E_2		cAMP (inhibits)
Serotonin	$5-HT_1$ $5-HT_2$	PI (81), Ca^{2+} cAMP (stimulating) (21)
Somatostatin		cAMP (inhibits) (125)
Substance P		PI, Ca^{2+} (126)
Thrombin		PI, Ca^{2+} (127)
Thyrotropin releasing hormone		PI, Ca^{2+} (128)
Thyroid stimulating hormone		PI, Ca^{2+} (129)
Vasointestinal peptide		cAMP (stimulates) (130)
Vasopressin	V_1 V_2	PI (82), Ca^{2+} cAMP (stimulates) (23)

The binding of GTP–G decreases the affinity of R for L (agonist) and G, which causes dissociation. GTP–G is then free to interact with the adenylate cyclase. This "ternary complex" model has been very useful in examining the sequence of molecular events but is still the subject of considerable study.

The total number of active or sensitized receptor sites can also be increased (supersensitized or up-regulated) or decreased (desensitized or down-regulated). Although the desensitization of the receptor is induced by the agonist, there is clearly no single mechanism but various mechanisms, such as internalization (29), phosphorylation (30), methylation (31), uncoupling (32), and protein synthesis (33), that can cause desensitization (32).

Nucleotide Regulatory Protein

The nucleotide regulatory protein (also known as G, G/F, N, and GTP binding protein) binds GTP. This G–GTP complex acts as an intramembrane transducing mechanism to transfer the signal from the L–R complex to the adenylate cyclase. The concept of nucleotide regulatory protein and its importance to the hormone sensitivity of adenylate cyclase was first suggested by Pfeuffer (35) in 1975. Without GTP or a nonhydrolyzable form of GTP, such as GppNHp or GTP–γ–S, the receptor becomes uncoupled to the adenylate cyclase. Two distinct G proteins have been demonstrated, the guanine nucleotide regulatory protein that mediate stimulation of adenylate cyclase (G_s) and the protein that mediate the inhibition of adenylate cyclase (G_i) (36,37).

The G_s protein is composed of three homologous α, β, and γ subunits. The subunit has a M_r of 52,000 or 45,000 daltons and has a single high-affinity site for guanine nucleotides. It is also the site for cholera toxin ADP-ribosylation. The β subunit has a M_r of 35,000 daltons; the γ subunit has an M_r of approximately 10,000 daltons. Fluoride and GTP appear to cause the dissociation of the α_s unit from the β unit. The dissociation of the β subunit increases the activity of the α subunit. Other characteristics of G_s are a low-K_m GTPase, the enzyme that metabolizes GTP to GMP, which also requires one order magnitude lower GTP concentrations than G_i. Moreover, GppNHp, a nonhydrolyzable form of GTP, can substitute.

The G_i protein is also composed of three homologs. The β and γ subunits are identical to the G_s subunits, but the α_i subunit is unique. The α_i subunit has a M_r of 41,000 daltons, which is ADP-ribosylated by pertussis toxin and islet activating protein (38). Like the G_s system, ribosylation also changes the activation of the α_i subunit by dissociation of the β subunit. Other characteristics of G_i are its requirement for a higher concentration of GTP for activity and a low K_m GTPase, that GppNHp often cannot substitute, and that NaCl enhances activity (38).

Adenylate Cyclase

Cyclic AMP is generated inside cells by adenylate cyclase [ATP pyrophosphate-lyase (cycling) enzyme E.C. 4.6.1.1], which catalyzes the conversion of ATP to 3′,5′-cyclic AMP and pyrophosphate. This reaction requires the presence of magnesium ions that complex with the ATP.

The use of cell mutants such as S49 mouse lymphoma variant which are phenotypically deficient in the catalytic subunit but which still retain the β adrenoceptor, enables the definition and delineation of the separate nature of these proteins (39).

The adenylate cyclase (C) has been more difficult to study in a purified state. It is very labile when dissociated from its other components, and only recently has the partially purified catalytic subunit been solubilized from bovine cerebral cortex (40). The resolved catalytic activity does not respond to GppNp and sodium fluoride (NaF); however, when the G_s protein is added back, GppNp and fluoride sensitivity are restored.

The addition of sulfhydryl agents can destroy the ability of the catalytic subunit to interact with the G_s protein. It has been interpreted that a cysteine moiety mediates the coupling of the G_s and C subunits (41,42). Also associated with the catalytic subunit is a regulatory divalent cation binding site where Mn^{2+} is thought to activate and stabilize the enzyme (43). Moreover, the diterpene forskolin, isolated from the plant *Coleus forskohlii*, can cause a greater activation of the enzyme than NaF. It will synergize the stimulation by neurotransmitter agents, and its mode of action is believed to be directly on the catalytic subunit (44-48). Other notable properties of the enzyme are its large hydrophobic surface, which facilitates its localization in the lipid bilayer, and its sensitivity to proteases, which demonstrates its protein nature. It is a large protein, target size analysis from liver having led to an estimate of its mass at 1×10^6 daltons (49).

Calmodulin

A number of different names for calmodulin appeared in the literature since its discovery in 1970 (50,51), including calcium dependent activator, calcium dependent regulator (CDR), modulatory protein, and troponin C-like protein, before the term calmodulin was accepted.

It is heat stable, monomeric, globular protein of 17,000 daltons. The amino acid composition is acidic in nature with an isoelectric point of 4.0. Calmodulin is so abundant in eukaryotic cells that it is a problem to remove it when trying to purify or study enzymes that are activated by CAM.

Calmodulin contains four cation binding sites with affinities for calcium in the micromolar range. When calcium is bound to CAM, it can activate adenylate cyclase preparations, such as cerebral cortex (52), adrenal medulla (53), pancreatic islets (54), and enterocytes (55), while having no effect on other adenylate cyclase systems. The presence of GTP is not required for calmodulin stimulation of adenylate cyclase, suggesting a direct interaction with the catalytic subunit (56,57).

Inhibitors of calmodulin, such as calmidazolium, have been synthesized but to date hold little therapeutic application owing to a lack of specificity (58,59).

Purinergic Site

A purine-specific site on the catalytic subunit can also lead to inhibition of cyclic AMP. It is distinct from the adenosine receptor, per se, in that it is intracellular. It is independent of guanine nucleotide action and is selectively acted upon by 2′,5′-dideoxyadenosine (61-62).

Cyclic AMP-Dependent Protein Kinase

Cyclic AMP is believed to exert the majority of its biologic effects via cyclic AMP-dependent protein kinase (also known as protein kinase A) that can exist as two isozymes (types I and II) that are made up of regulatory and catalytic subunits. Binding of cyclic AMP to the regulatory subunit causes dissociation of the catalytic subunit, which increases the activity of the latter. The phosphorylation of the specific proteins by the catalytic subunit causes a change in the level of biologic activity of those proteins. Some well-studied examples of cyclic AMP-dependent phosphorylated substrates are synapsin I (63), the nicotinic acetylcholine receptor (64), β-adrenoceptor (59), and the neurotransmitter synthesizing enzyme tyrosine hydroxylase (66). The phosphorylation of the β adrenoceptor leads to desensitization of the receptor. Other examples as well as mechanisms are described in detail elsewhere (67,68).

Cyclic Nucleotide Phosphodiesterase

The previous sections deal with the formation of cyclic AMP from the adenylate cyclase. The steady-state level of cyclic AMP is also controlled by its degradation by the cyclic nucleotide phosphodiesterases. There are multiple forms of the cyclic nucleotide phosphodiesterase (PDE), which differ in their substrate specificity and tissue distribution. On this basis, three major forms of the enzyme have been identified. PDE I is responsible primarily for the hydrolysis of cyclic GMP, although it will hydrolyze cyclic AMP at higher concentrations. PDE II preferentially hydrolyzes cyclic GMP and is calcium dependent, and. PDE III is relatively specific for cyclic AMP.

CYCLIC GMP SYSTEM

Cyclic GMP levels are affected by receptor-mediated stimulation of guanylate cyclase (69), as well as nonhormonal stimulation of guanylate cyclase (Figure 1) (70). Unlike cyclic AMP, a definitive role of cyclic GMP is biologic regulation has not been fully substantiated. The cyclic GMP effector system consists primarily of a receptor site and guanylate cyclase (71-73).

Receptor Site

In contrast to adenylate cyclase, receptors linked to guanylate cyclase maintain only a certain amount of control over the enzyme. α_1-Adrenergic (69), muscarinic (70), and excitatory amino acids (74) receptors (RG_c) are some of the more notable whose activation is linked to the stimulation of guanylate cyclase (Figure 1); however, these stimulatory effects require Ca^{2+} and are lost in cell-free systems. Nonhormonal mediation by the free radical of guanylate cyclase has been the subject of much more research.

Guanylate Cyclase

Guanylate cyclase exists in soluble and particular forms (69,70). The M_r of the soluble form ranges from 150,000 to 200,000 daltons and the particulate enzyme ranges from 300,000 to 400,000 daltons. Guanylate cyclase can be activated by nitrogen oxides, but the protein heme is required for this activity. The nitric oxides that are thought to act primarily through guanylate cyclase may have therapeutic potential (75), being related to endothelially derived relaxation factor (EDRF).

Link to Inositol Triphosphate and Calcium

It has long been known that receptor sites that increased intracellular calcium showed a concomitant increase in cyclic GMP. Although the link is still a mystery, certain researchers speculate that arachidonic acid, a coproduct of phosphatidylinositol turnover, is converted to thromboxane, prostaglandins, and leukotrienes, which can activate the guanylate cyclase.

Cyclic GMP-Dependent Protein Kinase

Cyclic GMP is believed to exert its primary biologic effects via a cyclic GMP-dependent protein kinase (protein kinase G). It exists as a dimer composed of a regulatory and a catalytic subunit. The association of cyclic GMP with the regulatory subunit causes dissociation of the catalytic subunit. This dissociation increases the activity of the catalytic subunit. The substrates for the protein kinase G have not been well characterized (64). The best studied is the G substrate, also known as the ≅ 23,000 M substrate. This protein, which is isolated from rabbit cerebellum, is not phosphorylated by the protein kinase A. The phosphorylated form of the G substrate can then inhibit a protein phosphatase that dephosphorylates proteins also present in rabbit cerebellum.

INOSITOL TRISPHOSPHATE AND DIACYCLGLYCEROL SYSTEMS

The inositol triphosphate (IP_3) system is actually only part of a multiple second messenger system. The stimulation of α_1-adrenergic (76), dopamine (77), excitatory amino acid (78), histamine 1 (79), muscarinic 1 (80), serotonin 1 (81), vasopressin 1 (82), and bradykinin (77) receptors and others (Table 1) (77) initiates the production of inositol trisphosphate that mobilizes calcium as well as liberates diacylglycerol (DAG), which stimulates protein kinase C. The major components of the IP_3 and DAG systems are the receptor site, guanine nucleotide regulatory protein, and phosphoinositide phosphodiesterase (PI–PDE) (Figures 1 and 2) (83-85).

Recently, the generation of inositol tetrakisphosphate (IP_4) by thyrotropin releasing hormone acting on GH_4 pituitary cell line has been reported (86). Although the role of IP_4 has yet to be established, it has been shown that it can act to facilitate the conversion of IP(1,3,4) to IP.

Receptor Sites

As in the cyclic AMP system, the binding of agonist to the receptor site initiates the cascade. The receptor (R_{Pi}) also provides the selectivity, sensitivity, and specificity (Figure 1). The number of receptor sites that have been discovered to initiate phosphatidylinositol turnover is growing rapidly. The same ternary complex model used to explain receptor site-ligand interaction is used for the inositol trisphosphate system. It has only been recently that the existence of a G_p protein and its coupling to the receptor sites were known. In addition, the question of desensitization and supersensitization of receptors coupled to phosphatidylinositol turnover has just begun to be addressed (87). Fortunately, advances in our understanding of these receptors are exceptionally rapid as well as exciting, which has helped to maintain a high profile for this second messenger system.

Nucleotide Regulatory Protein

The recent description of a nucleotide regulatory protein, termed N_p/G_p (88,89) or G_o (90), showed characteristics similar to those of other G proteins. It has GTPase activity, binds GTP, and appears to be pertussis toxin sensitive; the existence of subunits has not been reported.

Phosphoinositide Phosphodiesterase

Phosphoinositide phosphodiesterase catalyzes the release of inositol triphosphate from the diacylglycerol backbone and is calcium dependent (90). This enzyme is termed phospholipase C phosphodiesterase (91). Although it may exist as several isozymes and is localized in the cytosol, little is known about how the receptor site and G_p interact with this enzyme.

Actions of Diacylglycerol

The primary role of DAG appears to be activation of protein kinase C, which in turn phosphorylates specific proteins, such as myosin light chain (92). Also of interest is that the tumor-promoting phorbol esters appear to substitute for DAG to turn on the protein kinase C (92).

Protein Kinase C

Protein kinaase C (also known as calcium/phospholipid-dependent protein kinase) is stimulated by Ca^{2+}, DAG, and phospholipid. Unlike the cyclic nucleotide counterpart, it exists as a monomer. It has a broad spectrum of substrates. β-Adrenergic and epidermal growth factor (93) receptors are two of its substrates. The phosphorylation of the β adrenoceptor (94) by protein kinase C, like protein kinase A, causes desensitization (67,95). Other protein kinase C effects center around the ability of phorbol esters to substitute for DAG and thereby activate the protein kinase C. This activation has been hypothesized to be the tumor-promoting mode of action for the phorbol esters (96).

Since the activity of the enzyme myoinositol-1-phosphatase, which liberates free myoinositol-1-phosphate, is inhibited by lithium, it has been suggested (96) that this mechanism of action may account for the ability of the cation to control manic conditions. Other inhibitors of this enzyme may also prove to be therapeutic.

Antagonists of the DAG site on protein kinase C may have antitumor activity. The receptor sites coupled to phosphatidylinositol turnover, like the previously discussed systems, may also serve as a good test system for screening compounds for receptor site subtype selectivity (83,84,96).

CALCIUM SYSTEM

Calcium has the most far reaching effect of all the second messengers, yet, as has been noted, none exist independently to the others. Many of the same receptors that mediate cyclic GMP, IP_3, and DAG also regulate calcium, and these receptor sites appear to be distinct from those that regulate cyclic AMP. Calcium is made available from two sources: movement from outside the cell through calcium channels and intracellular mobilization from endoplasmic reticulum and mitochondria. Free calcium then binds to calcium binding proteins that act on various enzyme systems (Figure 3) (97-99).

Physiology

Because of the major roles that calcium plays in cellular function, the mobilization of calcium elicits a diverse range of activity dependent upon the tissue examined: in smooth muscle and skeletal muscle, it causes contraction; in nerve cells, it causes depolarization and neurotransmitter release; in liver, it activates phosphorylase kinase activity for glycogen metabolism; and in parotid glands and the pancreas, it stimulates fluid secretion (98). Calcium is not, therefore, a selective messenger, this role being served by the particular receptor coupled to calcium fluxes.

Action Potentials

In excitable cells, such as neurons and muscle cells, a membrane potential E_r (difference in electrical charge on the inside versus outside of the cell) is main-

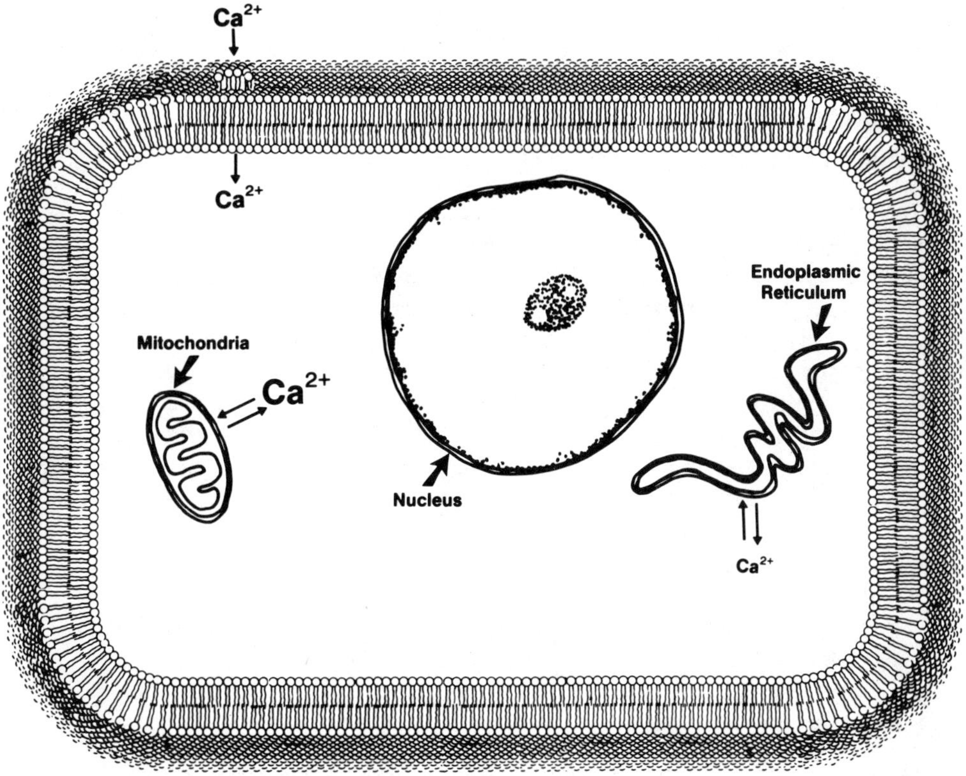

Figure 3 Schematic cross section of a cell with the major sites for Ca^{2+} mobilization.

tained at about 60-90 V positive outside (98). The inside of the cell is polarized in comparison with the outside of the cell. When the cell is stimulated by certain events, the cell depolarizes where there is a reversal of the membrane potential and an influx of calcium. This process lasts only several milliseconds and is defined as an action potential (98).

During those milliseconds, about 3000 calcium ions are admitted through the Ca^{2+} channels. The depolarization can be initiated by electrical or chemical transmission. An example of electrical transmission can be observed in the crayfish, with the rapid transmission from the spinal cord to the motor axons that causes the tail to flex. Here the electric signal transfers across the membranes from one cell to the other (98). Chemical transmission is best exemplified by the classic work with the vagus nerve innervation of the heart. Acetylcholine is released from the vagus nerve endings, crosses the synapse, and act on the cardiac muscle receptors, slowing the beating of the heart muscle (99).

Calcium Channels

The calcium channel is responsible for large changes in intracellular calcium levels, as in nerve and muscle cells, which repeatedly experience influxes of calcium from action potentials (Figures 1 and 3). When chemical transmission is involved, the neurotransmitters bind to the receptor sites, which causes a depolarization or change in the voltage gradient while electrical transmission transfer the depolarization across the membranes directly. This activates voltage-sensitive Na^+ channels with an influx of Na^+, followed by the opening of voltage-sensitive Ca^{2+} channels (VSCC) with an influx of Ca^{2+}. As the action potential peaks at about +40 mV, compensatory K^+ is pumped out and Cl^- is pumped in. Resting polarization is restored by Na^+,K^+ pumps and the process is repeated. Little is known about how the channel responds to voltage changes, but it is speculated that a protein component is probably voltage sensitive.

In nonexcitable cells there are also calcium channels, but they are not voltage sensitive. Protein phosphorylation, by one of the previously discussed kinases, may control calcium channel opening.

Intracellular Mobilization

The location of intracellular calcium stores is still somewhat controversial, but the consensus is that the endoplasmic reticulum is the primary source (Figure 3). In fact, it has been speculated that the endoplasmic reticulum also has calcium channels for the release of calcium into the cytosol. Whatever the mechanism of release, it has been elegantly demonstrated that IP_3 stimulates this release (13).

Mitochondria may also play an important role in calcium regulation. They serve to help regulate the larger fluxes in calcium levels above and below, with the smaller normal fluxes controlled by the endoplasmic reticulum.

Calcium/Calmodulin-Dependent Protein Kinase

This is not included in the schematic diagram (Figure 1) because it is not necessarily the primary site of action of calcium, which should not detract from its importance. Calcium exerts its effects on many enzymes via the ubiquitous calmodulin. Synapsin I and tyrosine hyroxylase are two of the better known substrates for this enzyme (67).

CONCLUSION

It has been a little over two decades since the proposal of the second messenger hypothesis for the cyclic nucleotides. It has not only survived the test of time but has served to develop the concept of other second messengers. Although a great deal is known about the mechanism of action of receptors and adenylate cyclase, there are still many unanswered questions. How are the receptor G proteins and adenylate cyclase coupled, and how do they affect transition and conformational state changes? How are the complex interactions among second messenger systems accomplished?

Although a basic understanding of the role of the guanylate cyclase system is still elusive, the calcium and phosphatidylinositol fields are exploding with newfound knowledge. Much of this chapter has been concerned with familiarizing the reader with the detailed components involved in second messenger systems. There are often multiple receptor systems on cells that when stimulated affect the levels of second messengers. Furthermore, this is probably a dualism in which each second messenger sometimes opposes the actions of the other and at other times synergizes it. The net effect is an integration of signals to produce the desired physiological response. A monodirectional system can exist when multiple second messengers have their effects in one direction, as do cyclic AMP and cyclic GMP in the release of cortisol, whereas a bidirectional system is the result of opposing actions of multiple second messengers, such as cyclic AMP inhibition and Ca^{2+} stimulation of platelet aggregation. This concept was suggested over a decade ago by Goldberg and colleagues (102) as the "yin-yang" hypothesis, based on the ancient Oriental philosophy of forces that control life (130).

In addition to the systems discussed, arachidonic acid, prostaglandins, thromboxanes, and leukotrienes may also function as second messengers (131). One issue is whether arachidonic acid and the others act only intracellularly, making them true second messengers, or whether they act extracellularly, which also makes them secretagogues. Furthermore, some substances, such as glucose and certain amino acids influence cellular events that do not involve any of the second messengers. The second messenger hypothesis will always be subject to the criticism of oversimplification, but when kept in perspective as merely a small piece of nature's integrated system it provides a great deal of insight into other homeostatic systems.

REFERENCES

1. H. Davy, Philos. Trans. R. Soc. Lond. *98*, 1 (1808).
2. S. Ringer, J. Physiol. (Lond.) *3*, 380 (1882).
3. L. V. Heilbrunn, Physiol. Zool. *13*, 88 (1940).

4. T. W. Rall and E. W. Sutherland, J. Biol. Chem. *232*, 1065 (1958).
5. E. W. Sutherland, I. Oye, and R. W. Butcher, Recent Prog. Horm. Res. *21*, 623 (1965).
6. D. F. Ashman, R. Lipton, M. M. Melicow, and T. D. Price, Biochem. Biophys. Res. Commun. *11*, 330 (1963).
7. E. Bohme, G. Grossman, J. Herz, A. Mulsch, C. Spies, and G. Schultz, Adv. Cyclic Nucleotide Prot. Phosp. Res. *17*, 259 (1984).
8. L. J. Ignarro, K. D. Hood, and M. S. Holin, Adv. Cyclic Nucleotide Prot. Phosp. Res. *17*, 267 (1984).
9. M. R. Hokin and L. E. Hokin, J. Biol. Chem. *203*, 967 (1953).
10. J. H. Allison and M. A. Stewart, Nature (New Biol.) *233*, 267 (1971).
11. L. M. Hallcher and W. R. Sherman, J. Biol. Chem. *255*, 10896 (1980).
12. R. H. Michell, Biochim. Biophys. Acta *415*, 81 (1975).
13. G. M. Burgess, P. P. Godfrey, J. S. McKinney, M. J. Berridge, R. F. Irvine, and J. W. Putney, Nature *309*, 63 (1984).
14. U. Takai, A. Kishimoto, U. Kikkama, T. Mori, and Y. Nishizuka, Biochem. Biophys. Res. Commun. *91*, 1218 (1979).
15. J. N. Langley, Proc. R. Soc. Lond. [Biol.] *78*, 170 (1906).
16. M. Berridge, Sci. Am. *253*, 142 (1985).
17. M. D. Smigel, K. M. Ferguson, and A. G. Gilman, Adv. Cyclic Nucleotide Prot. Phosp. Res. *19*, 103 (1985).
18. G. Guroff, G. Dickens, D. End, and C. Londos, J. Neurochem. *37*, 1431 (1981).
19. E. M. Ross and A. G. Gilman, Annu. Rev. Biochem. *49*, 533 (1980).
20. J. W. Kebabian and P. Greengard, Science *174*, 1346 (1971).
21. D. L. Nelson, A. Herbert, A. Enjalbert, J. Bockaert, and M. Hamon, Biochem. Pharmacol. *29*, 2445 (1980).
22. C. V. Perrier and M. Griessen, Eur. J. Clin. Invest. *6*, 113 (1976).
23. C. Londos, D. Cooper, W. Schlezel, and M. Rodbell, Proc. Natl. Acad. Sci. USA *75*, 5362 (1978).
24. J. Mooney, W. Horne, R. Handin, J. Schildkraut, and R. Alexander, Mol. Pharmacol. *21*, 600 (1982).
25. M. Munemura, T. E. Cote, K. Tsuruta, R. L. Eskay, and J. W. Kebabian, Endocrinology *107*, 1676 (1980).
26. M. Olianas, P. Onali, N. Neff, and E. Costa, Mol. Pharmacol. *23*, 393 (1982).
27. A. J. Blume, D. Lichtshtein, and G. Boone, Proc. Natl. Acad. Sci. USA *76*, 5626 (1979).
28. D. M. Chuang, L. Farber, W. J. Kinnier, and E. Costa, Adv. Biochem. Psychopharmacol. *21*, 143 (1980).
29. A. Goldstein, S. Tachibana, L. I. Lowney, M. Hunkapillar, and L. Hood, Proc. Natl. Acad. Sci. USA *76*, 6666 (1979).
30. W. J. Strittmatter, F. Hirata, and J. Axelrod, Adv. Cyclic Nucleotide Res. *14*, 83 (1980).
31. T. K. Harden, Y. F. Su, and J. P. Perkins, J. Cyclic Nucleotide Res. *5*, 99 (1979).

32. G. Nickols and G. Brooker, J. Cyclic Nucleotide Res. *5*, 435 (1979).
33. T. Pfeuffer and E. J. Helmreich, J. Biol. Chem. *250*, 867 (1975).
34. S. H. Snyder, Sci. Am. *253*, 132 (1985).
35. C. Londos, D. Cooper, and M. Rodbell, Adv. Cyclic Nucleotide Res. *14*, 163 (1981).
36. J. D. Hildebrandt, R. D. Sekura, J. Codina, R. Iyengar, C. Manclark, and L. Birnbaumer, Nature *302*, 706 (1983).
37. M. Rodbell, P. M. Lad, T. B. Nielsen, D. M. F. Cooper, W. Schlegel, M. S. Preston, C. Londos, and E. S. Kempner, Adv. Cyclic Nucleotide Res. *14*, 3 (1981).
38. A. G. Gilman, Cell *36*, 577 (1984).
39. P. A. Insel, M. E. Maguire, A. G. Gilman, H. R. Bourne, P. Coffino, and K. L. Melmon, Mol. Pharmacol. *12*, 1062 (1976).
40. S. Strittmatter and E. J. Neer, Proc. Natl. Acad. Sci. USA *77*, 6344 (1980).
41. E. M. Ross, A. C. Howlett, K. M. Ferguson, and A. G. Gilman, J. Biol. Chem. *253*, 6401 (1978).
42. K. Barovsky, C. Pedone, and G. Brooker, Mol. Pharmacol. *25*, 256 (1983).
43. D. L. Garbeis and R. A. Johnson, J. Biol. Chem. *250*, 8449 (1975).
44. C. O. Brostrom, Y. C. Huang, B. M. Breckenridge, and F. J. Wolff, Proc. Natl. Acad. Sci. USA *72*, 64 (1975).
45. J. W. Daly, W. Padgett, and K. B. Seamon, J. Neurochem. *38*, 532 (1982).
46. K. B. Seamon and J. W. Daly, J. Cyclic Nucleotide Res. *7*, 201 (1981).
47. K. B. Seamon and J. W. Daly, J. Biol. Chem. *256*, 9799 (1981).
48. K. B. Seamon, W. Padgett, and J. W. Daly, Proc. Natl. Acad. Sci. USA *78*, 3363 (1981).
49. L. E. Limbrid and R. J. Lefkowitz, J. Biol. Chem. *252*, 5776 (1977).
50. W. Y. Cheung, Biochem. Biophys. Res. Commun. *90*, 533 (1970).
51. S. Kakiuchi and R. Yamazaki, Biochem. Biophys. Res. Commun. *41*, 1104 (1970).
52. N. C. LeDonne, Fed. Proc. *40*, 1791 (1981).
53. I. Valverde, A. Vandermurs, R. Anganetulur, and W. J. Malaisse, Science *206*, 225 (1979).
54. L. M. Pinkus, S. Sulimovici, F. I. Susser, and M. S. Roginsky, Biochim. Biophys. Acta (1983).
55. N. D. Goldberg and M. K. Haddox, Annu. Rev. Biochem. *46*, 823 (1977).
56. A. S. Manahan and C. B. Klee, Adv. Cyclic Nucleotide Prot. Phosp. Res. *18*, 227 (1984).
57. W. Y. Cheung, Science *207*, 19 (1980).
58. G. C. Palmer, Life Sci. *36*, 1995 (1985).
59. M. Kaliner, J. Allergy Clin. Immunol. *76*, 321 (1985).
60. J. W. Daly, J. Med. Chem. *25*, 197 (1982).
61. J. W. Daly, R. F. Bruns, and S. H. Snyder, Life Sci. *28*, 2083 (1981).
62. J. W. Phillis and P. H. Wu, Prog. Neurobiol. *16*, 187 (1981).
63. W. Sieghart, J. Forn, R. Schwarcz, J. T. Coyle, and P. Greengard, Brain Res. *156*, 345 (1978).
64. R. L. Huganir and P. Greengard, Proc. Natl. Acad. Sci. USA *80*, 1130 (1983).

65. J. M. Stadel, P. Nambi, R. G. Schorr, D. F. Sawyer, M. G. Caron, and R. J. Lefkowitz, Proc. Natl. Acad. Sci. USA *80*, 3173 (1983).
66. J. W. Haycock, J. A. Meligini, W. F. Bennett, and W. F. Waymire, J. Biol. Chem. *257*, 12641 (1982).
67. A. C. Nairn, H. C. Hemmings, and P. Greengard, Annu. Rev. Biochem. *54*, 931 (1985).
68. P. Cohen, Eur. J. Biochem. *151*, 123 (1985).
69. L. J. Ignarro, K. S. Wood, and M. S. Wolin, Adv. Cyclic Nucleotide Res. *17*, 267 (1984).
70. L. E. White, L. J. Ignarro, and W. J. George, Pharmacologist *15*, 157 (1973).
71. U. Walter, Adv. Cyclic Nucleotide Prot. Phos. Res. *17*, 249 (1984).
72. E. A. Schwartz, Annu. Rev. Neurosci. *8*, 339 (1985).
73. W. Hoss and J. Ellis, Int. Rev. Neurobiol. *26*, 151 (1985).
74. J. W. Ferkany, R. Zaczek, and J. T. Coyle, Nature *298*, 757 (1982).
75. G. C. Blank and N. B. Giao, Adv. Biosci. *24*, 15 (1979).
76. V. Prpec, P. F. Blackmore, and J. H. Exton, J. Biol. Chem. *257*, 11323 (1982).
77. R. H. Michell, Cell Calcium *3*, 285 (1982).
78. F. Nicoletti, M. I. Iadarola, J. T. Wroblewski, and E. Costa, Neuroscience Abstracts *241*, 4 (1985).
79. L. M. Jones, S. Cockcroft, and R. H. Michell, Biochem. J. *182*, 669 (1979).
80. S. J. Weiss and J. W. Putney, Biochem J. *194*, 463 (1981).
81. J. N. Fain and M. J. Berridge, Biochem. J. *180*, 655 (1979).
82. A. P. Takhar and C. J. Kirk, Biochem. J. *194*, 167 (198).
83. L. Hokin, Annu. Rev. Biochem. *54*, 205 (1985).
84. M. J. Berridge, Biochem. J. *220*, 345 (1984).
85. R. V. Farese, Endocrinol. Res. *10*, 515 (1985).
86. J. P. Heslop, R. F. Irvine, A. Tashjian, and M. J. Berridge, J. Exp. Biol. *119*, 395 (1985).
87. C. P. Downes and M. M. Wusterman, Biochem. J. *216*, 633 (1983).
88. C. D. Smith, C. C. Cox, and R. Snyderman, Science *232*, 97 (1986).
89. M. W. Verghese, C. D. Smith, and R. Snyderman, Biochem. Biophys. Res. Commun. *127*, 450 (1985).
90. R. F. Irvine, Cell Calcium *3*, 295 (1982).
91. S. K. Fisher, L. A. Van Rooijen, and B. W. Agranoff, Trends Neurosci. *8* (1984).
92. M. Naka, M. Nishikawa, R. S. Adelstein, and H. Hedaka, Nature *206*, 490 (1983).
93. C. Cochet, G. N. Gill, J. Meisenhilder, J. A. Cooper, and T. Hunter, J. Biol. Chem. *259*, 2553 (1984).
94. D. J. Kelleher, J. E. Pessin, A. E. Ruoho, and G. L. Johnson, J. Biol. Chem. *81*, 4316 (1984).
95. H. Rasmussen, I. Kojina, W. Zawalich, and W. Apfeldorf, Adv. Cyclic Nucleotide Prot. Phosp. Res. *18*, 159 (1984).
96. Y. Nishizuka, Nature *308*, 693 (1984).

97. D. Gill, Adv. Cyclic Nucleotide Prot. Phosp. *19*, 307 (1985).
98. E. Carofoli and J. T. Penniston, Sci. Am. *253*, 70 (1985).
99. E. J. Furshpan and D. D. Potter, J. Physiol. (Lond.) *145*, 289 (1959).
100. O. Loewi, Pflugers Arch. *189*, 239 (1921).
101. S. Ebaski, Experientia *41*, 978 (1985).
102. N. D. Goldberg, M. K. Haddox, E. Dunham, C. Lopez, and J. W. Haddox, Adv. Cyclic Nucleotide Res. *5*, 307 (1975).
103. L. H. Schrama, P. N. DeGraan, J. Eichberg, and W. H. Gispen, Eur. J. Pharmacol. *121*, 401 (1986).
104. S. Gunther, J. Biol. Chem. *259*, 7622 (1984).
105. H. Okamoto, N. Sone, H. Hirata, M. Yoshida, and Y. Kagawa, J. Biol. Chem. *252*, 6125 (1979).
106. M. A. Dillingham and R. J. Anderson, Science *231*, 1572 (1986).
107. S. Hersher, J. Simard, E. Assayag, Y. Mehri, and F. Labrie, Mol. Cell. Endocrinol. *44*, 125 (1986).
108. C. W. Len, B. R. Bianchi, D. Grant, T. Miller, E. A. Danaker, M. D. Tufano, H. Kopecha, and A. M. Nadzen, J. Pharmacol. Exp. Ther. *236*, 729 (1986).
109. S. Husler, J. Simard, E. Assayag, Y. Mehri, and F. Labrie, Mol. Cell. Endocrinol. *44*, 125 (1986).
110. E. A. Soto, N. J. Kliman, J. F. Strauss, and L. G. Parvola, Biol. Reprod. *34*, 559 (1986).
111. D. R. Hill, Br. J. Pharmacol. *84*, 249 (1985).
112. E. W. Sutherland and T. W. Rall, Pharmacol. Rev. *12*, 265 (1960).
113. M. Barbaccia, D. M. Chuang, E. Costa, *Regulatory Peptides from Molecular Biology Function*, Raven Press, New York, 1982, p. 511.
114. R. V. Farese, J. S. Davis, D. E. Barnes, M. L. Standaert, J. S. Babischkin, R. Hock, N. J. Rosic, and R. J. Pollet, Biochem. J. *231*, 269 (1985).
115. J. M. Marsh, R. W. Butcher, K. Savard, and E. W. Sutherland, J. Biol. Chem. *241*, 5436 (1966).
116. K. Abe, G. A. Robinson, G. W. Liddle, R. W. Butcher, W. E. Nicholson, and C. E. Baird, Endocrinology *85*, 674 (1969).
117. A. E. Traynor, D. Schubert, and W. R. Allen, J. Neurochem. *39*, 1677 (1982).
118. B. B. Fredholm, I. Jansen, and L. Edvensson, Acta Physiol. Scand. *124*, 467 (1985).
119. P. C. Canonico, M. A. Sortino, C. Speciale, and V. Scapagnini, Mol. Cell. Endocrinol. *42*, 215 (1985).
120. J. C. Bozou, S. Amar, J. P. Vincent, and P. Kitabgi, Mol. Pharmacol. *29*, 489 (1986).
121. L. R. Chase and G. D. Aurbach, Proc. Natl. Acad. Sci. USA *58*, 518 (1967).
122. S. D. Shukla, Arch. Biochem. Biophys. *240*, 674 (1985).
123. M. A. Proll, R. B. Clark, and R. W. Butcher, Mol. Pharmacol. *29*, 307 (1986).

124. A. C. Howlett, J. M. Qualy, and L. L. Khachatrion, Mol. Pharmacol. *29*, 307 (1986).
125. S. J. Weiss, J. S. McKinney, and J. W. Putney, Biochem. J. *206*, 555 (1982).
126. S. Nakashima, T. Tohmatsu, H. Hattori, Y. Okano, and Y. Nozawa, Biochem. Biophys. Res. Commun. *135*, 1099 (1986).
127. T. F. Martin, J. Biol. Chem. *258*, 14816 (1983).
128. I. Pastan and S. H. Wollman, J. Cell Biol. *35*, 262 (1967).
129. P. Onali, J. P. Schwartz, and E. Costa, Proc. Natl. Acad. Sci. USA *78*, 6531 (1981).
130. R. Wilhelm (ed.), *The I Ching or Book of Changes*, Routledge and Kegan Paul, London (1951).
131. P. F. Worley, J. M. Baraban, C. Van Dop, E. J. Neer, and S. H. Snyder, Proc. Natl. Acad. Sci. USA *83*, 4561 (1986).

5

Radioligands: Synthesis and Handling

Crist Filer, Stephen Hurt, and Yieh-Ping Wan / E. I. du Pont de Nemours & Co., Boston, Massachusetts

Radiolabeled ligands have been instrumental in the study of cellular receptors since the first demonstration of ligand binding to the nicotinic cholinergic receptor in the late 1960s. Ligand binding experiments were uniquely exploited for studies of the opiate receptor using [^{3}H]naloxone (1). With the exception of naloxone, which had been commercially available in tritiated form for many years, ligands for early work had to be prepared individually by the researcher, purified by chromatographic procedures in the researcher's laboratory, and used directly in binding assays. Although this is still an important process and is a relatively inexpensive way to investigate the value of a new ligand, it requires a range of capabilities that would limit the ligand binding technique to the relatively few laboratories with both biological and chemical expertise.

As receptor binding techniques became more widely used, radiochemical companies began to supply labeled ligands, the first being [^{3}H]dihydroalprenolol from NEN in 1975. At that time, NEN scientists assisted Lefkowitz in proving the structure of this tritiated ligand (2). As with many firsts, naloxone and dihydroalprenolol proved to be enduring standards. In addition, they demonstrated the value of these techniques for labeling ligands using exchange or reduction with tritium gas or other simple chemical reactions to yield biologically active analogs for binding studies.

Following dihydroalprenolol, diazepam, spiperone, dihydroergocryptine, the enkephalins, and other "classic" ligands became commercially available in tritiated form as quickly as their value to neuroscience research was identified. New product introduction continues to be an essential researcher-industry collaboration that is highly valued. In providing new ligands, the radiochemist has been an essential element in the progress of neurochemistry.

This chapter presents information intended to assist the researcher in purchasing, stabilizing, and handling radiochemicals. The section on synthesis of radiochemicals may help the scientist appreciate the intricacies of preparing these products or to plan a synthetic project particularly suited for their laboratory. The authors hope that by using the information in this chapter, research can be accomplished more efficiently and productively.

PURCHASE OF RADIOCHEMICALS

As an investigator plans a project and determines the required radiochemicals, there are several options available. Specific radiochemicals can be obtained in several forms:

Crude, unpurified products via a radiolabeling service arrangement
Standard, purified, radiolabeled chemicals from radiochemical catalogs
Unique research chemicals prepared and purified by means of a custom synthesis contract

The purpose of this section is to review these options, making it easier for a neuroscientist to select one that best fits his or her technical and financial situation.

The Catalog

A scientist interested in acquiring a radiolabeled neurochemical should first consult the radiochemical catalogs. Current catalogs contain thousands of radiolabeled items and literally hundreds of these are ^{3}H, ^{125}I, and ^{35}S radioligands specifically marketed as tools to study various receptor families. Table 1 highlights a representative listing of currently available radioligands for each receptor class. At present, 20-30 new radiolabeled products are introduced each year to this area by radiochemical firms. Because of this commitment to neuroscience research, investigators can satisfy their research needs from readily available catalog items.

Commercial suppliers put fresh lots of catalog compounds into stock on a timely basis, and their purity is checked at regular, predetermined intervals by a quality control group. With the purchase of each catalog item is included a specification sheet detailing storage and analytic evaluation data for that particular production lot.

On occasion, a supplier finds it necessary to delete or change the specifications of an item in its catalog for various reasons. Possibly, a compound's decreased use makes it impossible to justify retaining it as a catalog item, or perhaps a superior product has been introduced. So that investigators can smoothly adjust to such changes, most companies alert them several months in advance

Table 1 Currently Available Ligands for Each Receptor Class

Receptor	Compound	Pharmacological Action of Ligand
Adenosine	N^6 [adenine-2,8-^{3}H] cyclohexyladenosine	A_1 Agonist
	5′-*N*-[adenine-2,8-^{3}H] ethylcarboxamidoadenosine (NECA)	A_2 Agonist
Adrenergic	[Benzene ring-10,11-^{3}H] desmethylimipramine,	Uptake
	L-[*N*-methyl-^{3}H] Epinephrine	Endogenous
α-Adrenergic	p-[3,5-^{3}H] Azidoclonidine	α_2 Photoaffinity
	[Benzene ring-^{3}H] clonidine	α_2 Agonist
	[^{125}I] HEAT	α_1 Antagonist
	[^{125}I] Iodoazidoarylprazosin	α_1 Photoaffinity
	[7-Methoxy-^{3}H] Prazosin	α_1 Antagonist
	[Methyl-^{3}H] Rauwolscine	α_2 Antagonist
	[Imidazolyl-4,5-^{3}H] UK-14, 304	α_2 Agonist
	[Methyl-^{3}H] Yohimbine	α_2 Antagonst
β-Adrenergic	[^{125}] 15-(4′azido-3′-iodobenzyl) carazolol	Photoaffinity
	L-[propyl-1,2,3-^{3}H] Dihydroalprenolol	Antagonist
	(–)-[^{125}I] iodocyanopindolol	Antagonist
Amino acids	DL-2-[3,4-^{3}H] amino-4-phosphonobutyric acid	Glutamate antagonist
	[5-Mehthyl-^{3}H] AMPA	Glutamate agonist
	D-[2,3-^{3}H] aspartic acid	Aspartate agonist
	[Vinylidene-^{3}H] kainic acid	Glutamate agonist
	[Glutamyl-3,5-^{3}H(N)] NAAG	Endogenous, glutamate
	CPP	NMDA antagonist
	MK 801	NMDA antagonist
Benzodiazepine	[Methyl-^{3}H] flunitrazepam	Agonist
	[*N*-methyl-^{3}H] PK11195	Peripheral antagonist
	[7,9-^{3}H] RO15-4513	Photoaffinity antagonist
	[*N*-methyl-^{3}H] RO15-1788	Central antagonist
	[*N*-methyl-^{3}H] RO5-4864	Peripheral agonist
Calcium channel	[*N*-methyl-^{3}H] D600	Phenylalkylamine antagonist

Table 1 (Continued)

Receptor	Compound	Pharmacological Action of Ligand
	cis-(+)-[*N*-methyl-^{3}H] diltiazem	Benzothiazepine antagonist
	Isopropyl-1,3-^{3}H] nimodipine	Dihydropyridine antagonist
	[5-Methyl-^{3}H] nitrendipine	Dihydropyridine antagonist
	(+)-[5-Methyl-^{3}H] PN 200-110	Dihydropyridine antagonist
	[*N*-methyl-^{3}H] verapamil	Phenylalkylamine antagonist
Cholinergic		
Muscarinic	[Methyl-^{3}H] hemicholinium-3	Uptake
	[Methyl-^{3}H] oxotremorine-M	Agonist
	[*N*-methyl-^{3}H] pirenzepine	M-1 antagonist
	[Propyl-2,3-^{3}H] propylbenzilylcholine mustard HCl	Irreversible
Nicotinic	L-[benzilic-4,4′-^{3}H(N)] QNB	Antagonist
	L-[N-methyl-^{3}H] nicotine	Agonist
	[^{125}I] α-bungarotoxin	Antagonist
Dopamine	[Ring-2,5,6-^{3}H] dopamine	Endogenous
	[Ethanol-1,2-^{3}H] fluphenazine	D-1 antagonist
	[^{3}H]GBR 12935	Uptake
	[^{3}H(G)] haloperidol	D-2 antagonist
	[^{125}I] 4,-iodospiperone	D-2 antagonist
	[4′-^{3}H] mazindol	Uptake
	[Methyl-^{3}H] MPP(+)	Neurotoxin
	[Methyl-^{3}H] MPTP	Neurotoxin
	L(−)-[propyl-^{3}H(N)] Propylnorapomorphine	Agonist
	[Methyl-^{3}H] SCH 23390	D-1 antagonist
	[Benzene ring-^{3}H] spiperone	D-2 antagonist
	(−)-[Methoxy-^{3}H] sulpiride	D-2 antagonist

Table 1 (Continued)

Receptor	Compound	Pharmacological Action of Ligand
GABA	[2,3-^{3}H]γ-aminobutyric acid	Endogenous
	(–)-[Butyl-4-^{3}H(N)] Baclofen	$GABA_B$ agonist
	(–)-[Methyl-^{3}H] Bicuculline methyl chloride	$GABA_A$ antagonist
	[^{35}S] Butylbicyclophosphorothionate	Cl channel antagonist
	[Methylene-^{3}H(N)] Muscimol	Agonist
Histamine	[Methylene ring-^{3}H] histamine	Endogenous
	[Methyl-^{3}H] tiotidine	H_2 antagonist
Opiate	(–)-[9-^{3}H(N)] bremazocine	κ Agonist
	[*N*-Methyl-^{3}H] dihydromorphine	Agonist
	[Tyrosyl-3,5-^{3}H(N)] DAGO	μ Agonist
	[^{125}I]-endorphin	Analog
	[D-Ala^2-D-Leu^5]-[tyr-3,5-^{3}H(N)] enkephalin	δ Agonist
	[D-pen^2,D-pen^5]-^{3}H-[tyr-2,6-^{3}H(N)] enkephalin	δ Agonist
	(–)-[9-^{3}H] ethylketocyclazocine	κ Agonist
	[*N*-methyl-^{3}H] morphine	Agonist
	[*N*-allyl-2,3-^{3}H] naloxone	Antagonist
	[Piperidyl-3,4-^{3}H(N)] phencyclidine	σ-PCP agonist
	(+)-[*N*-allyl-2,3-^{3}H] SKF-10,047	σ Agonist
	[^{3}H] TCP	σ-PCP agonist
Serotonin	[^{3}H] 8-hydroxy-DPAT	$5HT_{1A}$ agonist
	[^{3}H] hydroxytryptamine binoxalate	Endogenous
	[*N*-methyl-^{3}H] Imipramine	Uptake
	2-[^{125}I] Iodo-LSD	Agonist
	[Ethylene-^{3}H] ketanserin	$5HT_2$ antagonist
	[*N*-methyl-^{3}H] LSD	Agonist
	[^{3}H] Paroxetine	Uptake
Sodium channel	[^{3}H] batrachotoxinin 20-α-benzoate	Antagonist
Peptide	(5-Ile) [tyrosyl-3,5-^{3}H(N)] angiotensin II	Endogenous
	[2,3-Prolyl-3,4-^{3}H(N)] bradykinin	Endogenous

Table 1 (Continued)

Receptor	Compound	Pharmacological Action of Ligand
	N-[Phe-2,6-^{3}H(N)] Formyl Met-Leu Phe OH	Agonist
	[^{3}H] leuteinizing hormone releasing hormone	Endogenous
	[^{3}H] neurotensin	Endogenous
	[^{3}H] oxytocin	Endogenous
	[^{3}H] substance P	Endogenous
	[^{3}H] thyrotropin releasing hormone	Endogenous
	(3-Methyl-his)[^{3}H] thyrotropin releasing hormone	Analog
	[^{3}H] vasopressin-arg	Endogenous

and also may retain the compound in a nonroutinely available (NRA) listing. This latter status may allow former catalog compounds to be purchased on an "as is" basis. Because no minimal purity level is guaranteed, a purity determination can be requested from the supplier's quality control group.

Radiolabeling Services

Most industrial suppliers have tritium and ^{125}I radiolabeling services. These services are an economical way to obtain radiolabeled compounds for those who have the license required to receive large quantities of radioactivity and facilities to purify the product.

Because of the industries' extensive experience in tritiating a wide variety of compounds, investigators are encouraged to contact suppliers for consultation regarding the tritiation service. Once needs have been determined, that is, total radioactivity and specific activity for a compound, one of several tritiation strategies may be recommended. It is not uncommon for this technical consultation, price quotation, and order placement to be accomplished during the same telephone call. The investigator would then send the agreed-upon precursor to the tritium laboratory, and the crude tritiated product with labile tritium removed would be shipped typically within 4-5 weeks. Virtually all structural types of neurochemicals have been successfully tritiated by such tritiation services. In consideration of license restrictions, arrangements can usually be made to make partial shipments to the investigator over an extended period of time.

It should be emphasized that for most companies it is the tritiation service that is purchased, not the compound. The investigator should be aware that tritiation requires exposure of a precursor to a destructive radioactive environment and, therefore, should never expect the resulting product to be pure. The exact amount of decomposition depends upon the stability of the precursor and/or final product and the method of tritiation. Upon request, the service laboratory will run a thin-layer chromatographic (TLC) analysis of the crude tritiated product for the customer. Given the potential of this option and positive comments by satisfied scientists, the success rate of such tritiation services is relatively high.

A completely analogous arrangement usually exists in most companies for an ^{125}I radiolabeling service, and all the comments and caveats already expressed apply to it. The methods that are currently employed to radiolabel with ^{125}I are (1) chloramine T-$Na^{125}I$; (2) ^{125}ICl; (3) exchange labeling with $Na^{125}I$; (4) iodination with lactoperoxidase, and (5) labeling with Bolton-Hunter reagent (3). As one may imagine, amines, peptides containing tyrosine, and phenolic compounds are especially frequent candidates for such ^{125}I radiolabeling, but nearly every type of chemical has been labeled by these methods.

Custom Synthesis

Although more expensive, the custom synthesis route is one in which the customer can exert the most control and obtain a radiochemical best suited to meet the technical requirements. Not only the position of radiolabeling, but also the purity criteria, special packaging arrangements, and shipping instructions can be specified. Since the vast majority of custom-synthesized radiolabeled neurochemicals are tritiated, we restrict our discussion exclusively to this radioisotope.

After the tritiated compound has been successfully prepared at the appropriate specific activity and, radiochemical purity, it is packaged according to the customer's specifications. Most often for mass and stability reasons, the compound is packaged in solution. Ethanol is often employed, and most laboratories prefer to package the product solution in a sealed glass ampule under an inert atmosphere, such as argon. However, if the customer desires a crimp-sealed vial can be used in lieu of a glass ampule. The compound is then shipped in the most appropriate manner. Almost routinely the product is shipped in dry ice (−78°C). Under separate cover, a specification sheet, and results of high-pressure liquid chromatography (HPLC), TLC, and (if requested) a ^{3}H nuclear magnetic resonance (NMR) are sent.

Although hypothetical, the following example is representative of the issues attending a custom tritiation. A custom synthesis laboratory may be approached to quote on the tritiation of pumiliotoxin A [1] (4). The neuroscientist is willing to have the substance radiolabeled in any position but suggests that it be prepared at $\geqslant$ 15 Ci/mmol. Having reviewed the possible strategies to prepare the tritiated toxin, it might be concluded that the most reasonable approach would be a NaB^3H_4 reduction of ketone intermediate [2] to afford [15-^{3}H] pumiliotoxin A [3]. Although this method would theoretically achieve a specific activity of 29 Ci/mmol, the most that a laboratory could confidently assure the customer would be ≥10 Ci/mmol. Assume, however, that after talking to the investigator, the custom synthesis group learns that this lower specific activity range would be acceptable.

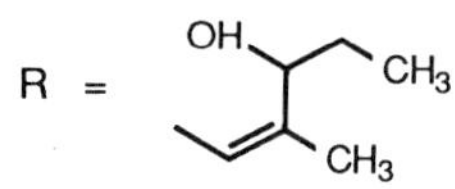

1

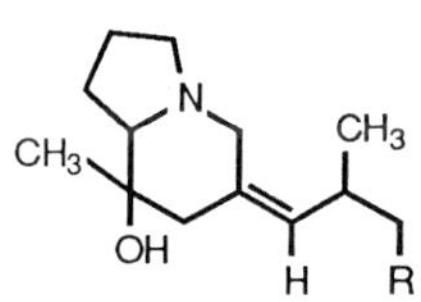

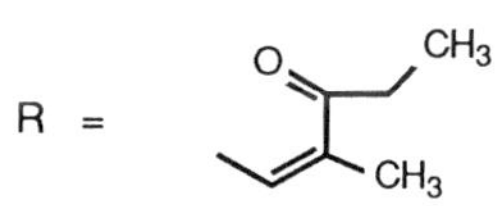

2

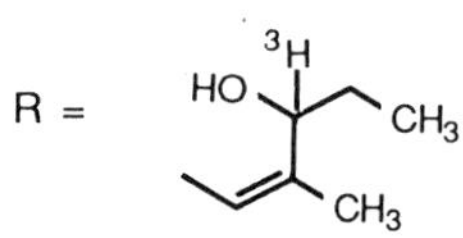

3

In all likelihood, the investigator would be given a price quotation based on his providing either cold precursor [2] or only toxin [1]. The latter quotation would be more expensive since the custom synthesis group would have to prepare precursor [2]. Also assume, in this instance, that the investigator has no synthetic expertise but a generous colleague has provided 200 mg of a 1:1 mixture of [1] and [2]. In this case, the custom group would quote an additional charge for the separation of [1] from [2] by HPLC, thereby providing both sufficient tritiation precursor and cold standard. To our knowledge, this example is totally hypothetical, but it does illustrate the flexibility that exists for collaborating with scientists to custom tritiate compounds necessary for their research efforts.

SYNTHESIS OF RADIOCHEMICALS

General Exchange

The ultimate source of tritium in all tritiated radioligands is tritium gas. Consequently, the most direct method of tritiation is exposure of a compound to allow hydrogens in chemically labile positions to exchange for tritium atoms. Although this technique has been used extensively in the past, it suffers from limitations that rarely make it the method of choice. Its principal drawback is that the resultant specific activity is typically only a few curies per millimole. By the nature of the process, the product may also contain tritium in positions sufficiently labile that back-exchange with the storage solvent is a problem. From an practical standpoint, the technique often results in complex product mixtures, requiring time-consuming purifications that nevertheless result in products with poor stability.

The applicability of general exchange depends on the presence of structural features in the target compound that make certain of the hydrogens susceptible to exchange. For example, compounds containing aromatic rings can usually be labeled by general exchange with either tritium gas or ^3H_2O. Specific activity is generally increased by using a transition metal catalyst, such as platinum (Pt), palladium (Pd), or ruthenium (Rh). Examples of currently available ligands so produced include [^{3}H]amphetamine [4] and [^{3}H]*p*-hydroxyacetanilide [5].

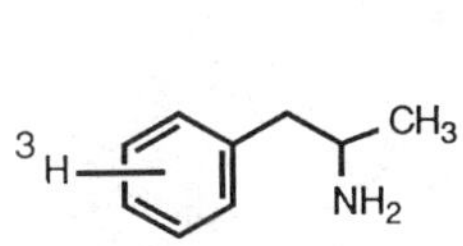

4

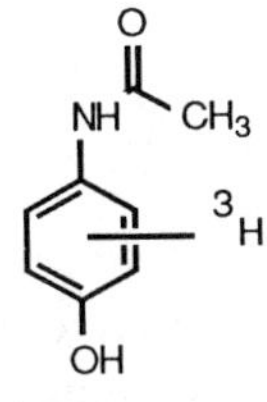

5

The other most commonly employed approach is the exchange of enolizable hydrogens adjacent to a carbonyl group, such as a ketone or a carboxylic acid derivative. In this case, ^3H_2O is almost always used, and the reaction is catalyzed by either acid or base. The chief disadvantage of this method is that tritium resides in labile positions that can back-exchange into the storage medium. The product should be stored at as low a temperature as is practical in order to minimize this problem. Notable illustrations of ^3H_2O exchange labeling are [^{3}H]haloperidol [6] and [^{3}H]saxitoxin [7].

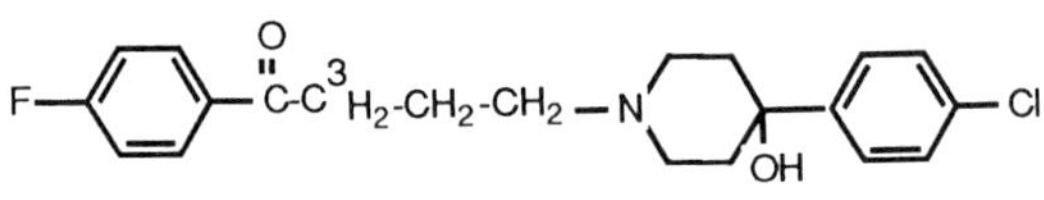

6

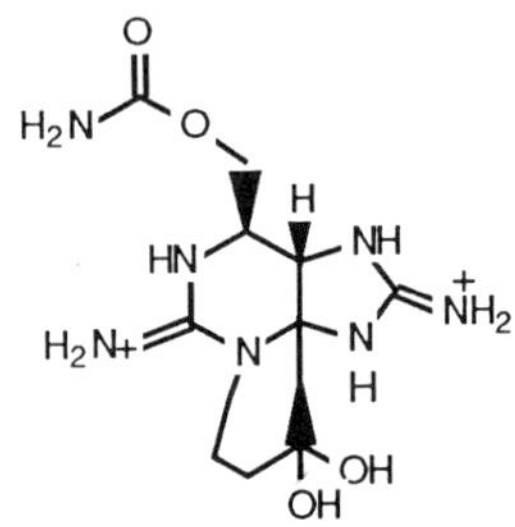

7

Other examples of exchange labeling generally involve more unique strategies and are therefore applicable in limited instances. A particularly interesting approach was taken in labeling the GABA analog THIP [8] (5). In this case, tritium was incorporated into a nitroso precursor [9] by exchange with ^3H_2O, since the nitroso group renders the adjacent methylene hydrogens considerably more exchangeable than in THIP itself. Furthermore, once the nitroso group is removed to yield [^{3}H] THIP, the tritium is in nonlabile positions.

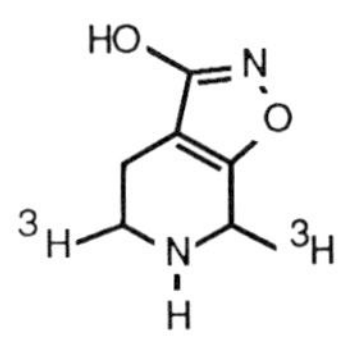

8

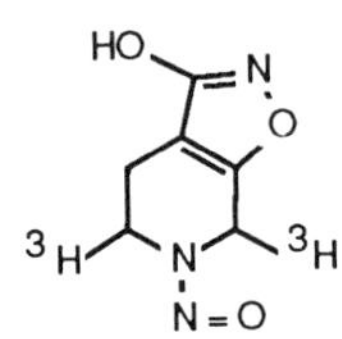

9

A final illustration of exchange labeling is cited, because the product, [^{3}H]-muscimol [10], is a commonly used radioligand. In this instance, the precursor, ibotenic acid [11], is decarboxylated in the presence of ^3H_2O. Tritium is thereby specifically incorporated into the methylene position.

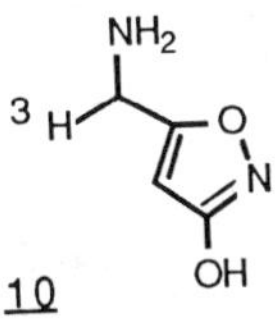

10

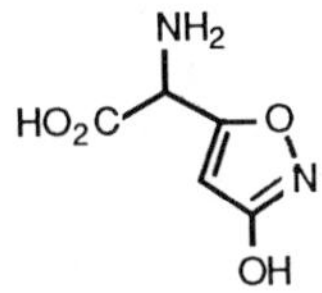

11

In recent years, well over 100 new radioligands have become commercially available. Few have been prepared by general exchange, but rather by more contemporary methods that have been developed to provide higher specific activity and more stable products.

Synthetic Methods

A number of standard strategies for introducing tritium into a molecule have been developed. Owing to their versatility, they have been exploited as a means of labeling nearly all the radioligands available at present. These approaches have been widely adopted because they have a number of characteristics that make them particularly attractive for radioligand synthesis:

1. The specific activity achieved is usually sufficient for ligand binding studies.
2. Reaction conditions are relatively mild, and yields are typically high.
3. The necessary precursors are usually available or can be readily synthesized.

In this section, examples of these synthetic methods are discussed.

The radiolabeling of peptides is based on many of the same reactions but often involves unique concerns owing to the functional complexity of peptides and is therefore discussed separately.

Catalytic Reduction of Multiple Bonds

Olefins The catalytic reduction of multiple bonds with tritium gas is a very effective means of tritiation since the resultant specific activity and radiochemical yield are generally high. The most frequently used precursors for catalytic reductions are olefins owing to the large number of synthetic methods available for forming carbon-carbon double bonds. In recent years, many of the new radioligands have been compounds developed as potential new drugs by the pharmaceutical industry. It is not uncommon to find that unsaturated analogs have also been prepared in the course of the drug development program and that this analog can serve as a precursor for the desired radioligand. However, if such a precursor is not already available, in many cases it can be readily synthesized.

Either heterogeneous or homogeneous catalysts can be used, with Pd or Pt heterogeneous catalysts the most frequent choice because of convenience. Heterogeneous catalysts can be removed from the reaction mixture by simple filtration, whereas homogeneous catalysts must be separated from the product chromatographically and this may complicate purification protocols. The maximum specific activity theoretically attainable by reduction of an olefin with carrier-free tritium gas is approximately 58 Ci/mmol (29 Ci per milliatom of tritium). In practice, using a heterogeneous catalyst, the specific activity usually falls short of this mark into the range of 30-50 Ci/mmol. There are two principal sources of hydrogen responsible for this specific activity dilution: hydrogen preabsorbed on the catalyst surface and hydrogen abstracted by the catalyst from the reaction solvent. Examples of radioligands prepared by catalytic reduction, illustrating the structural diversity that can be accomodated by this approach, include [^{3}H] dihydroalprenolol [12], [^{3}H] GBR12935 [13], and [^{3}H] TCP [14].

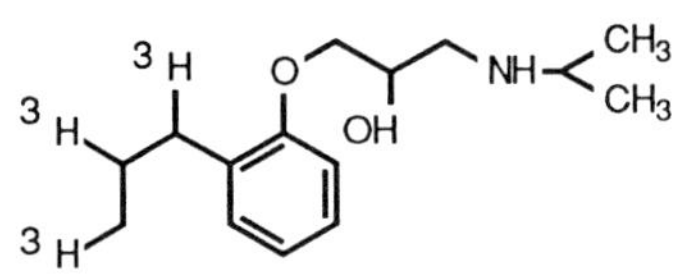

12

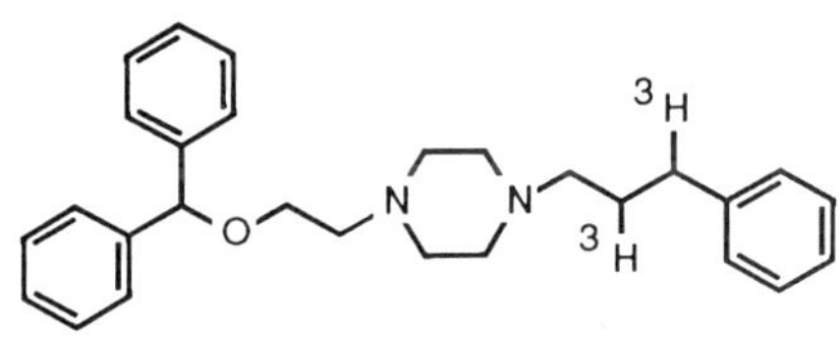

13

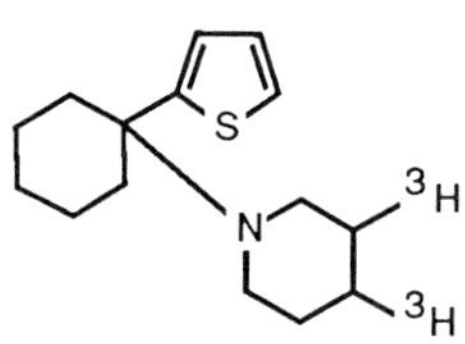

14

Although less frequently employed, homogeneous catalysts, such as Wilkinson's catalyst, offer certain advantages that recommend their use. Experience has shown that homogeneous catalysts generally give specific activities closer to the theoretical limit. They also tend to be more selective for reducing olefins in the presence of other easily reducible functionalities, such as nitro groups and aryl halides. Finally, homogeneous catalysts will reduce an isolated double bond, whereas hindered double bonds will not react under such conditions.

Acetylenes A carbon-carbon triple bond can also be catalytically reduced with tritium gas. This approach is utilized less often than the reduction of olefins since there are fewer synthetic methods available for preparing acetylenes. The reduction of an acetylene proceeds through the corresponding olefin to the saturated product and can be stopped at the intermediate olefin stage by using Lindlar catalyst. This strategy has been used to prepare tritiated olefins; the outstanding examples are [^{3}H]naloxone [15] and [^{3}H]SKF 10,047 [16], obtained from the corresponding N-propargyl precursors (6).

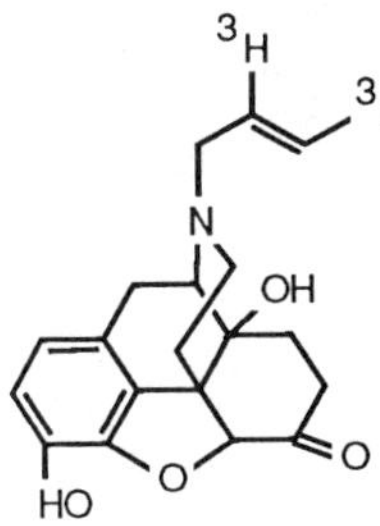

15

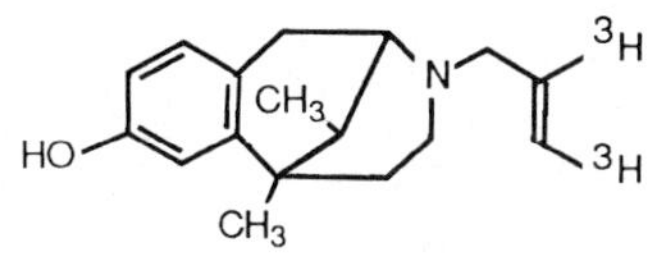

16

Since this process theoretically introduces four tritium atoms, the resultant specific activity exceeds that obtained from olefin reductions. Some dilution usually occurs, so the typical specific activity range obtained is 60-90 Ci/mM. Complete reduction of acetylene precursors with Pt or Pd catalysts affords the radioligands [^{3}H]buspirone [17] and [^{3}H]3-PPP [18].

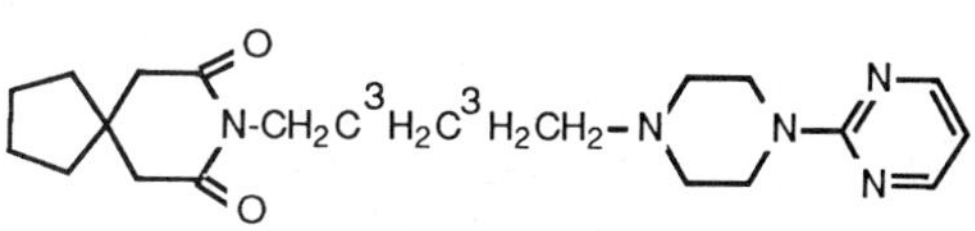

17

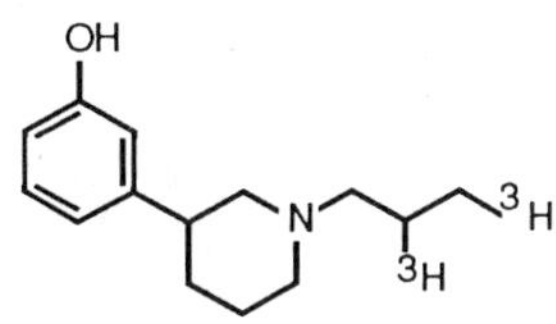

18

Nitriles The catalytic reduction of nitriles to amines has been exploited as a means of preparing tritiated ligands, generally with a specific activity of 20-40 Ci/mmol. The limiting factor in this approach may be in obtaining the required nitrile precursor, which can involve a multistep synthesis. Examples of radioligands produced in this manner include [^{3}H]baclofen [19] and [^{3}H]melatonin [20].

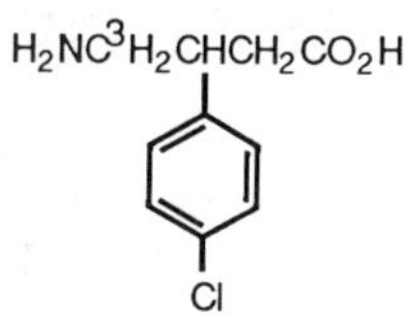

19

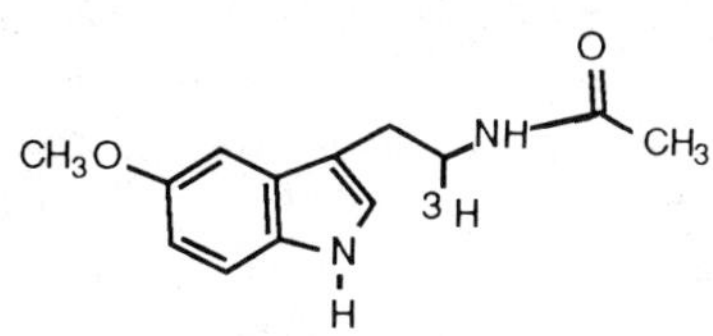

20

Enamines and nitrostyrenes Tritiated amines can also be synthesized by reducing the corresponding enamine precursor. This strategy is most appropriate for cyclic tertiary amines since the enamine is usually accessible by dehydrogenation of the parent amine with $Hg(OAc)_2$ (7). The preparation of [^{3}H]naltrexone [21] from its precursor [22] illustrates this approach (8).

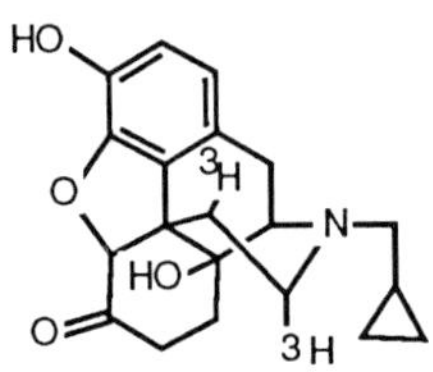

21

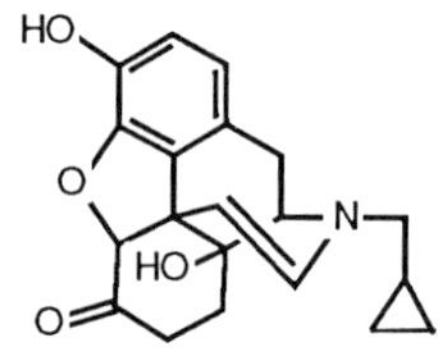

22

In an analogous fashion, nitrostyrenes can serve as precursors to tritiated amines since the nitro group is simultaneously converted to an amine, [23] → [24]. Both [^{3}H]dopamine [25] and [^{3}H]serotonin [26] have been prepared from nitrostyrene precursors.

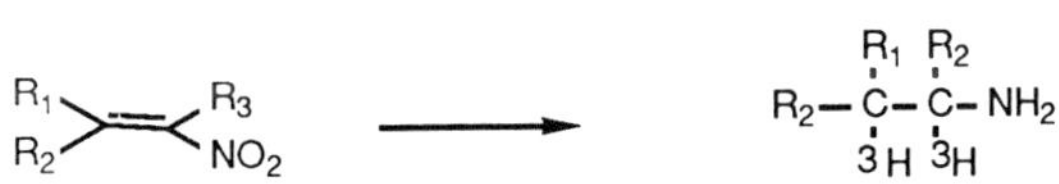

23 24

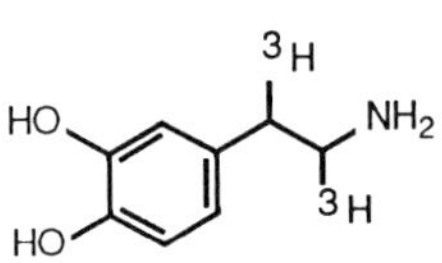

25

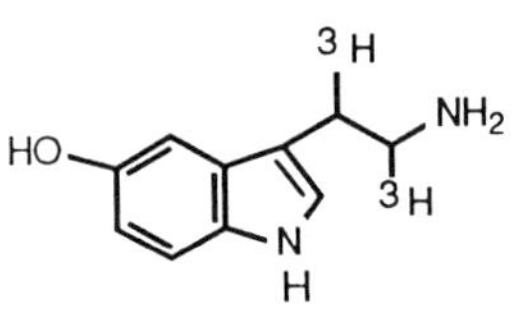

26

Reductive dehalogenation with tritium Reductive dehalogenation of an aryl halide with tritium gas is a generally useful labeling method that places tritium in the position formerly occupied by the halogen. This reaction is catalyzed by Pd, Pt, or Rh heterogeneous catalysts, with Rh/Al_2O_3 the usual choice. As long as there are no other easily reducible groups present to interfere, this approach usually gives excellent yields.

The relative reactivity of halides toward reductive dehalogenation follows the order I > Br > Cl >> F. Although iodides are the most reactive, bromides are generally more stable and easier to synthesize and are thus the most frequent choice. Bromination is usually accomplished by treating a compound with Br_2 in such solvents as glacial acetic acid, chloroform, or carbon tetrachloride. Substituents that activate aromatic rings toward electrophilic substitution (e.g., $-NH_2$, $-OH$, and $-OCH_3$) facilitate the reaction so that it is possible to obtain polybrominated products. This is advantageous because the resultant specific activity is generally 20-30 Ci/mmol for each bromine present in the precursor. The dehalogenation of bromo precursors has been used to label such ligands as [^{3}H] spiperone [27], [^{3}H] imipramine [28], and [^{3}H] -bremazocine [29].

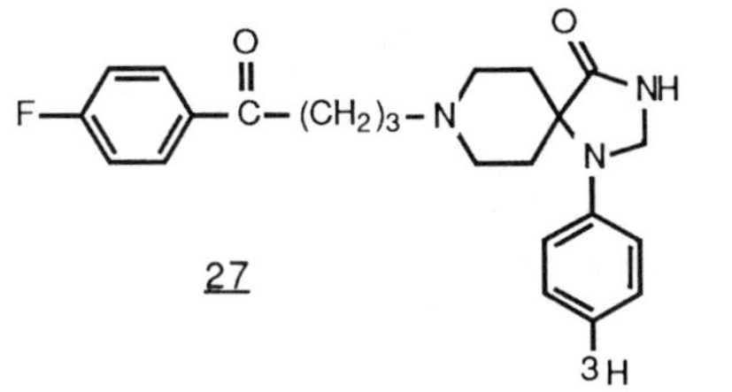

27

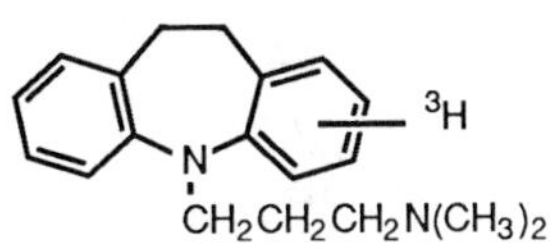

28

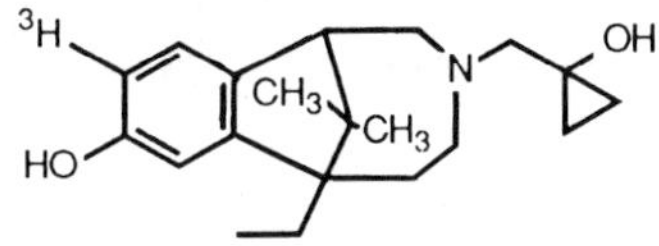

29

The most frequently used reagent for preparing iodides is iodine monochloride (ICl) since molecular iodine (I_2) is usually unreactive in electrophilic substitution reactions. Iodination is most successful for aromatic rings containing activating substituents, but polyiodinated products are rarely obtained, thus placing a limitation on specific activity. Dehalogenation of the appropriate iodide precursor has produced such radioligands as [^{3}H] strychnine [30] and [^{3}H] dextrorphan [31].

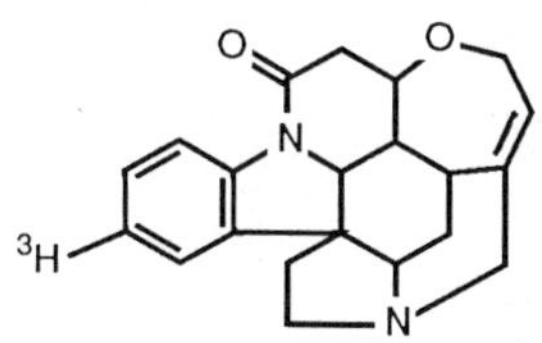

30

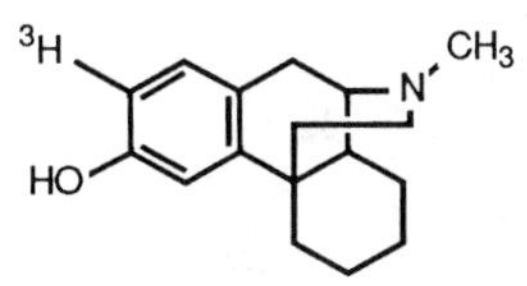

31

Direct chlorination of a drug to give a chlorinated precursor is rarely, if ever, feasible. In those instances in which chloro derivatives have been utilized, they are usually synthesized from commercially available starting materials.

Dichloro-QNB was prepared in this fashion, yielding [^{3}H]QNB [32] upon reductive dechlorination.

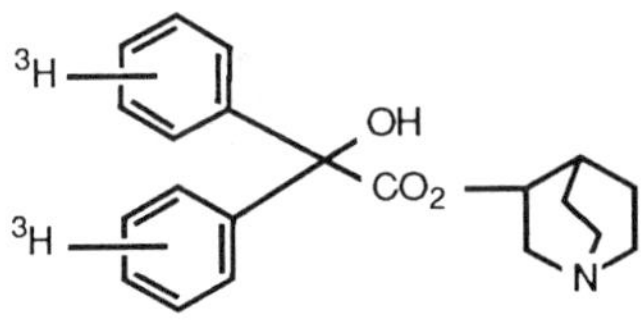

32

The order of reactivity of the various halogens, as previously described, has been used to advantage. Since iodine and bromine can be cleaved much more readily than chlorine or fluorine, it is possible to selectively replace a reactive halide in the presence of one less so. For example, dehalogenation of bromo-chloro (or fluoro) precursors gives the chloro- or fluoro-containing radioligands [^{3}H]chlorpromazine [33], [^{3}H]aminoclonidine [34], and [^{3}H]paroxetine [35].

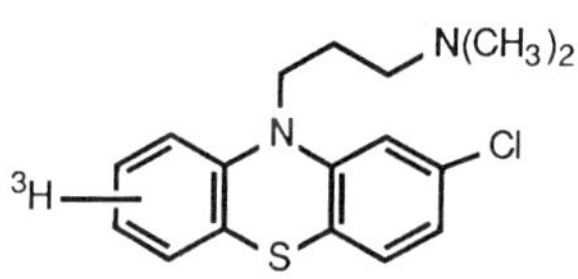

33

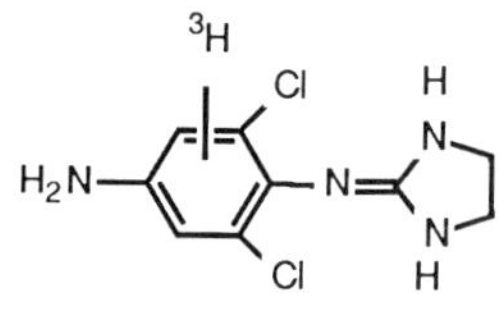

34

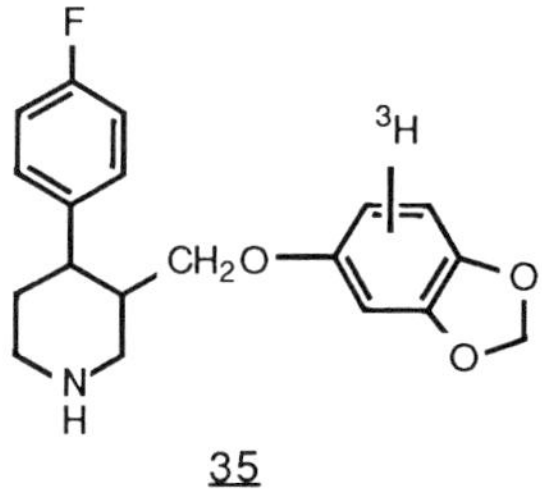

35

Carbonyl Reductions with Sodium Borotritide

The reduction of carbonyl-containing compounds with sodium borotritide (NaB^3H_4) is a valuable method for preparing tritiated ligands. This reagent is available at >90 Ci/mmol, in contrast to other hydride reducing agents, such as $LiAlH_4$, which, so far, have only been prepared at low specific activity. The most frequently utilized precursors are aldehydes and ketones, which yield tritiated alcohols. Two features of this method tend to limit its applicability:

1. Only a single tritium atom is introduced, limiting the specific activity.
2. Racemic products may result when one would like to obtain a single optical isomer; such is the case when noradrenalone is reduced to give [^{3}H] D,L-norepinephrine [36].

Nevertheless, in many cases this approach is a viable alternative, as in the preparation of [^{3}H]phorbol dibutyrate [37] from its aldehyde precursor.

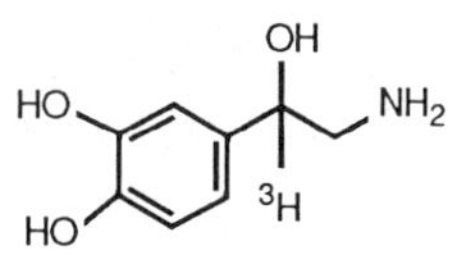

36

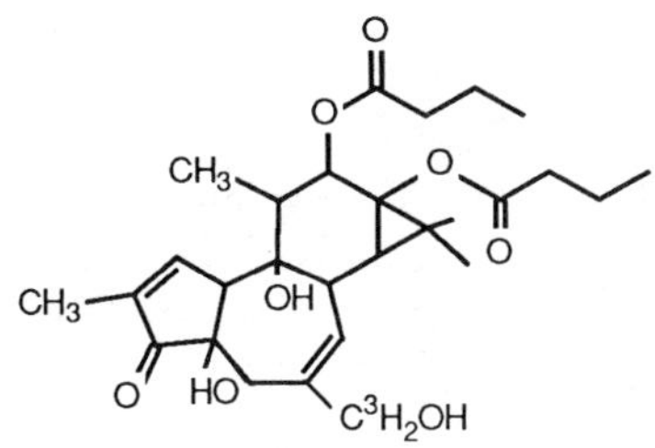

37

Incorporation of High-Specific-Activity Precursors

Each of the aforementioned labeling techniques, with the exception of general exchange, share a common strategic theme: a precursor compound is subjected to reductive conditions to introduce tritium and convert it to the desired radiochemical. Although these methods have proven to be very powerful and have been used to synthesize numerous radioligands, they cannot be applied in every case. For instance, the requisite precursor may not be available or is unstable under the reaction conditions. Radiochemists have therefore expended considerable effort developing additional reagents to widen the useful scope of synthetic methods.

In recent years, the principal advance in radiochemical synthesis has been the preparation of high-specific-activity tritiated precursors, which are then incorporated into a synthetic pathway. These precursors are typically small alkyl halides, the chief example being [^{3}H] CH_3I, which can be prepared at >80 Ci/mM. The utility of this reagent stems from the fact that many neurochemicals are amines with N-methyl substituents. Alkylation of the corresponding N-desmethyl derivative with [^{3}H] CH_3I yields the chemical of interest. The desmethyl precursors are usually readily available, since in cases in which the desired product is a compound developed by a pharmaceutical company, the desmethyl counterpart may also have been prepared in the course of structure-activity studies. If not, there are a number of efficient synthetic methods to effect N-demethylations. In this manner such radioligands as [^{3}H] pirenzepine [38] and [^{3}H] MPTP [39] have been prepared.

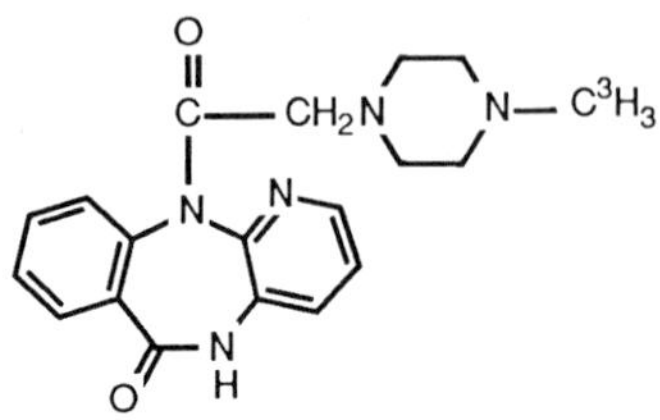

38

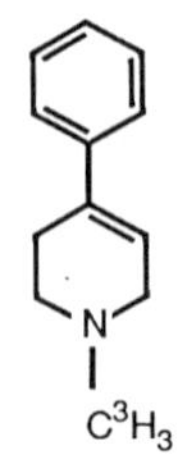

39

[^{3}H] Methyl iodide can also serve as the precursor to tritiated methyl esters. The corresponding carboxylic acid is alkylated with [^{3}H] CH_3I to yield such radioligands as [^{3}H]yohimbine [40] and [^{3}H]nitrendipine [41].

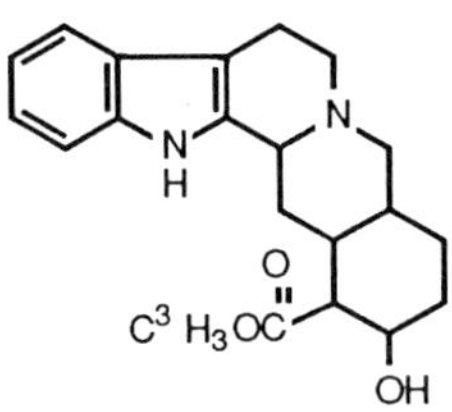

40

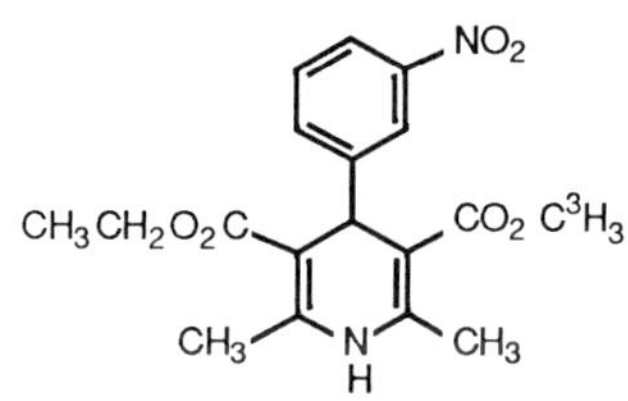

41

Although they are used less frequently than [^{3}H] CH_3I, it is possible to generate [^{3}H]ethyl iodide and [^{3}H]isopropyl iodide at high specific activity. Analogous alkylation reactions with these two precursors give labeled amines and esters, such as [^{3}H] chloroquine [42] and [^{3}H]nimodipine [43].

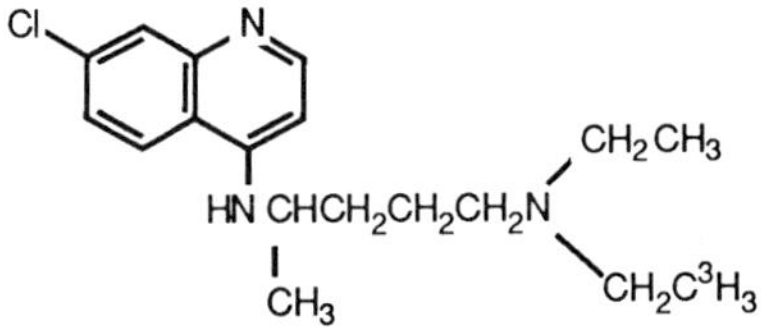

42

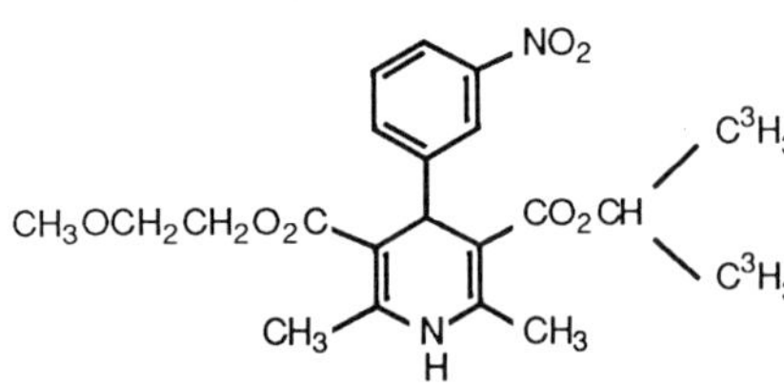

43

Finally, a proprietary synthesis of [^{3}H]ethylene dibromide at 60-90 Ci/mM has recently been developed. An attractive feature of this reagent is that it can be incorporated into internal portions of a molecule, since both its carbons are functionalized. As an illustration, [^{3}H] UK 14,304 [44] and [^{3}H] ketanserin [45] are prepared using [^{3}H] ethylene dibromide as the labeled starting material.

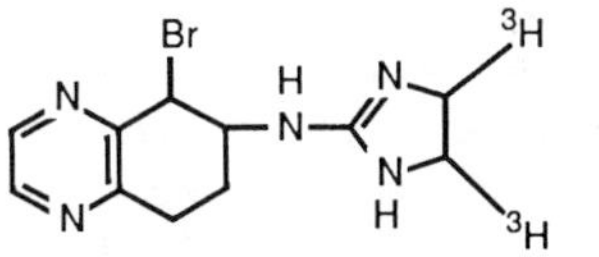

44

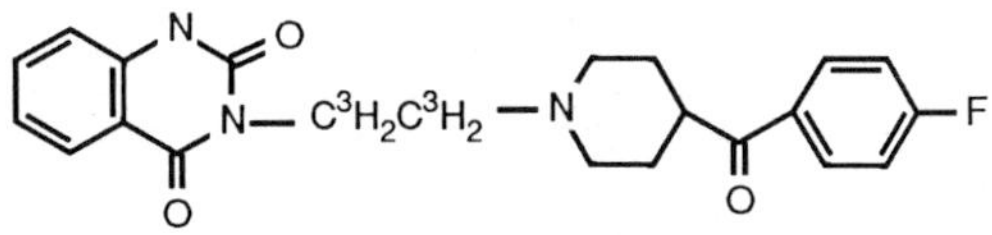

45

RADIOLABELED PEPTIDES

Synthesis

Recent interest in peptides as possible neurotransmitters or neuromodulators has resulted in an increased demand for labeled peptides. Many different methods for peptide labeling have been reported in the literature; however, selection of the appropriate method for labeling a particular peptide is complicated by the structural diversity of peptides. The purpose of this section is to review the synthetic options and the advantages of each so that one can select the procedure that best fits the peptide.

Exchange Labeling

Exposure of a peptide to either tritium gas or tritiated water, usually in the presence of catalyst, achieves a general exchange of the chemically labile hydrogen atoms of histidine, tyrosine, phenylalanine, tryptophan, and methionine in the peptide sequence. The advantage of this method is simplicity; however, it usually leads to significant degradation of the peptide and produces randomly labeled products of low specific activity.

Catalytic exchange reactions have also been applied to protected peptides. Brundish et al. (9) reported that moderate specific activities could be achieved for histidine-containing peptides. For example, a specific activity of 8 Ci/mM was obtained when the tripeptide thyrotropin releasing hormone (TRH, pyro-Glu-His-Pro-NH_2) was labeled by catalytic exchange. They also reported that the degree of labeling decreased sharply with increasing molecular weight of the peptide.

Catalytic Reduction

Reductive dehalogenation Catalytic dehalogenation with tritium is probably the most frequently used method in preparing [^{3}H] peptides. Peptides containing halogenated aromatic amino acids, such as diiodotyrosine, dibromotyrosine, chlorophenylalanine, or iodohistidine, are commonly used as precursors. Halogen-containing peptides have been obtained by conventional peptide synthesis or solid-phase peptide synthesis or by converting a tyrosine- or histidine-containing peptide to its iodo derivative.

In an effort to develop more selective ligands for the different opiate receptors, Gacel et al. (10), using classic methods, prepared the 3,5-dibromotyrosyl precursors of three peptides: Tyr-D-Ala-Gly-NH-$CH(CH_3)$-CH_2-$CH(CH_3)_2$, Tyr-D-Ser-Gly-Phe-Leu-Thr, and Tyr-D-Thr-Gly-Phe-Leu-Thr. These precursors were then tritiated in the presence of palladium oxide as catalyst to produce the desired [^{3}H] peptides with specific activities between 45 and 60 Ci/mmol. Precursor peptides can also be synthesized by the solid-phase peptide procedure.

Houghten and Li (11) incorporated a 3,5-diiodotyrosine residue into the sequence of human β endorphin to form [3,5-diiodo-tyr^{27}]-β-endorphin using solid-phase peptide synthesis. The iodinated peptide, after catalytic tritiation in a solvent mixture of hexamethylphosphoramide and dimethylacetamide, yielded [^{3}H] human β endorphin with a specific activity of 50 Ci/mmol.

Catalytic dehalogenation of protected peptide precursors has been used to prepare [^{3}H]-labeled peptides. Allen et al. (12) synthesized [4-^{3}H-Phe^8] substance P by catalytic deiodination of Boc-Arg-Pro-Lys(Boc)-Pro-Gln-Gln-Phe-Phe-(I)-Gly-Leu-Met-NH_2 followed by acidolytic deprotection. They claimed that catalytic dehalogenation of a protected peptide precursor, rather than free peptide, affords the advantage of protection of ionizable amino functions and allows the use of an aprotic solvent. Aprotic solvents minimize dilution through exchange of tritium with hydrogen and produce products with higher specific activity. In addition, the use of a protected peptide reduces the loss of product that, as a free peptide, would be absorbed irreversibly by the catalyst. Thus, [^{3}H] substance P was prepared with high specific activity (23 Ci/mmol) and in good yield (17-38%).

Probably the simplest method of preparing an authentic tritiated peptide of biologic origin without chemical modification, with an acceptable specific activity, and without resorting to the great labor of synthesizing a sophisticated precursor is to prepare a halogenated precursor from the parent peptide. This is classically done by iodination and incorporation of tritium by catalytic dehalogenation [Equation (1)].

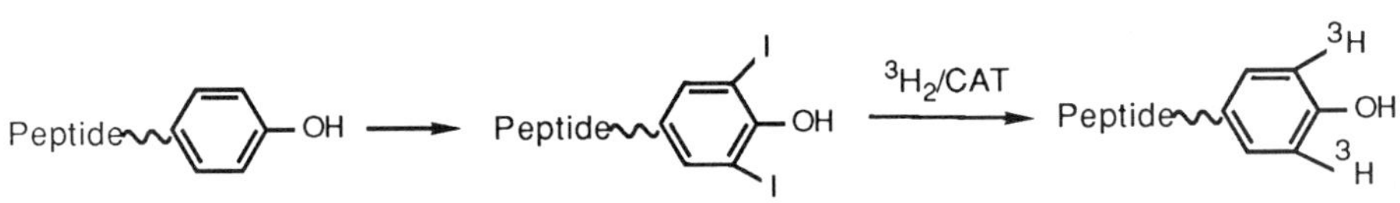

(1)

Morgat et al. (13) reported the iodination of angiotensin II, Asp-Arg-Val-Tyr-Val-His-Pro-Phe, using iodine monochloride. The diiodotyrosyl derivative of angiotensin II was isolated, characterized, and then tritiated to produce [^{3}H]-angiotensin II with a specific activity of 56 Ci/mM. [^{3}H] histidine-containing peptides have also been synthesized by this approach. Pradelles et al. (14) prepared [^{3}H]TRH(pyro-Glu-His-Pro-NH_2) with a specific activity of 60 Ci/mmol. Lin et al. (15) prepared [^{3}H]secretin (7 Ci/mmol), a 37 amino acid peptide. Molecular iodine has also been used as the iodination reagent. Flouret and his coworkers (16) prepared the 3,5-diiodotyrosyl derivatives of arginine vasopressin and oxytocin using iodine. After tritiation, [^{3}H]arginine vasopressin and [^{3}H]oxytocin with specific activities between 17 and 31 Ci/mM were obtained.

Although this two-step tritium-labeling procedure for peptides appears to be simple and convenient, it is important to be aware that side reactions may occur during the procedures commonly employed for iodination. Oxidative reactions can convert methionine residues into methionine sulfoxide and possibly attack the disulfide bridge within the peptide chain. An excess of the iodination reagent can cause degradation of the imidazole ring of histidine and modification of tryptophan or serine.

Reduction of peptides containing unsaturated amino acids The catalytic reduction with tritium gas of a peptide that contains an unsaturated amino acid has been widely used as a means of introducing tritium into peptides. Thus, peptide analogs containing unsaturated amino acids, such as dehydroproline, dehydroleucine, and 2-amino-4-hexynoic acid, have been prepared and reduced to produce tritiated peptides. The reduction has been carried out in acetic acid, methanol, or aprotic solvents. Pd and PtO_2 are common catalysts in this type of reduction. Depending upon the length and amino acid composition of the peptide precursor, as well as the conditions of the reduction, products with various specific activities have been obtained. Felix et al. (17) reported the synthesis and reductive tritiation of several 3,4-dehydroproline-containing peptides. Small peptides, such as [dehydro-Pro^3]-TRH, and [dehydro-Pro^1]-MRIH, were successfully reduced to yield [^{3}H]TRH and [^{3}H]MRIH with very high specificities—58 and 55 Ci/mM, respectively. Under the same mild conditions, [dehydro-Pro^3]-bradykinin, a nonapeptide, was not fully reduced. No specific activity of the product was given, and the authors stated that complete reduction of a dehydroproline residue in longer peptides may require more forcing conditions.

Perrin et al. (18) prepared [^{3}H]gonadotropin releasing hormone by tritiating [dehydro-D,L-Pro^9]-GnRH, which was synthesized by solid phase-peptide synthesis. The tritiated material, containing a mixture of [^{3}H] [D-Pro^9]-GnRh and [^{3}H] [Pro^9]-GnRH, was purified by high-pressure liquid chromatography to obtain [^{3}H] [Pro]-GnRH with a specific activity of 16 and 25 Ci/mM in two preparations. Hardy et al. (19) prepared [4,5-^{3}H-Leu]- and [3,4-^{3}H-Pro] locust adipokinetic hormones (LAKH) by the catalytic tritiation of [4,5-dehydro-Leu] and [3,4-dehydro-Pro]-LAKH and obtained specific activities of 115 and 12 Ci/mM, respectively. The authors suggested that the low specific activity obtained for [3,4-^{3}H-Pro] LAKH was probably a reflection of the slow rate of reduction of the precursor with concomitant dilution of specific activity from the proton-tritium exchange. The extremely high specific activity of [4,5-^{3}H-Leu] LAKH was the result of substantial isotopic exchange during saturation of the alkene bond of 4,5-dehydroleucine. They concluded that the 4,5-dehydroleucine residue may be generally useful in the preparation of specifically tritiated peptides of high specific activity.

Tritiated peptides containing methionine are sometimes more difficult to prepare by catalytic tritiation procedures because of possible catalyst poisoning by the sulfur function of methionine. Also, purified methionine-containing [^{3}H] peptides tend to be much less stable and are easily converted to methionine sulfoxide analogs under normal storage conditions. For these reasons, many [^{3}H] peptide analogs with a substitution of methionine by its isostere, norleucine, have been synthesized and used as radioactive ligands for receptor binding work. The substitution of methionine by norleucine in the peptide not only provides a compound that is more suitable for tritiation and more stable as a final product, but also provides an additional opportunity for labeling.

L-2-amino-4-hexynoic acid (Aha), an acetylenic norleucine was incorporated into enkephalin (20), pentagastrin (20), and cholecystokinin (21) sequences. The catalytic tritiation of these [Aha] -containing precursor peptides gave [^{3}H] [Nle^5] -enkephalin, [^{3}H] [Nle^3] -pentagastrin, and [^{3}H] *p*-Boc-[Nle^{28,31}] -CCK (27-33) with the specific activities of 120, 80-90, and 150 Ci/mM, respectively.

Labeling of Tryptophan-Containing Peptides

A method of tritiating tryptophan-containing peptides was developed by Marche and his coworkers (22). The procedure consists of formation of an *o*-nitrophenylsulfenyl tryptophan derivative by reacting the peptide with *o*-nitrophenylsulfenyl chloride. The sulfenylated peptide is then catalytically reduced in the presence of tritium to give the [^{3}H] -labeled product [Equation 2]. A [^{3}H]$^-$ pentagastrin analog, Boc-Gly-[^{3}H] -Try-Met-Asp-Phe-NH_2, specific activity 5-7 Ci/mM, was prepared by this procedure.

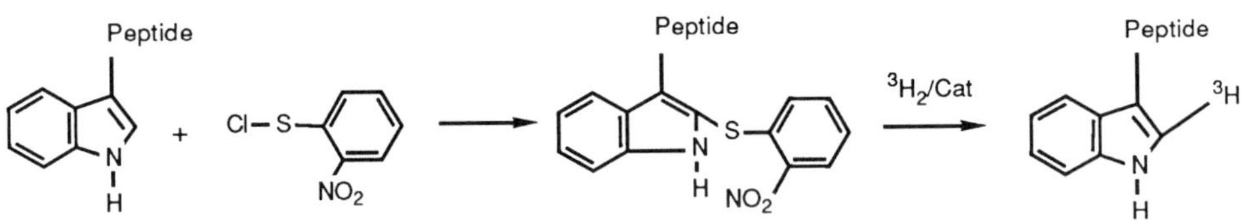

(2)

Labeling by Methylation

It is known that peptides and protein that contain methionine residues can be methylated with methyl iodide to the corresponding sulfonium derivatives.

The resulting methyl sulfonium derivatives can subsequently be demethylated at alkaline pH in the presence of a thiol to regenerate the methionine function by the following scheme [Equation (3)].

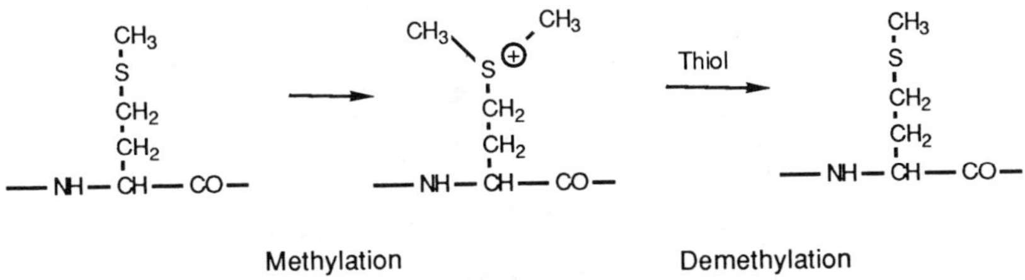

(3)

Recently this technique has been applied to the synthesis of [^{3}H] methionine-containing peptides by reaction with [^{3}H] methyl iodide (23). The specific activity of the products obtained by this methyl exchange procedure is usually one-half of the specific activity of the starting [^{3}H] methyl iodide owing to lack of discrimination between methyl groups in the demethylation step.

Several high-specific-activity [^{3}H] peptides containing a C-terminal methionine amide residue, such as substance P and neurokinins A and B, have been prepared by methylation of the homocysteine thiolactone derivatives of the peptide in liquid ammonia with [^{3}H] methyltosyl ester [Equation (4)] (24).

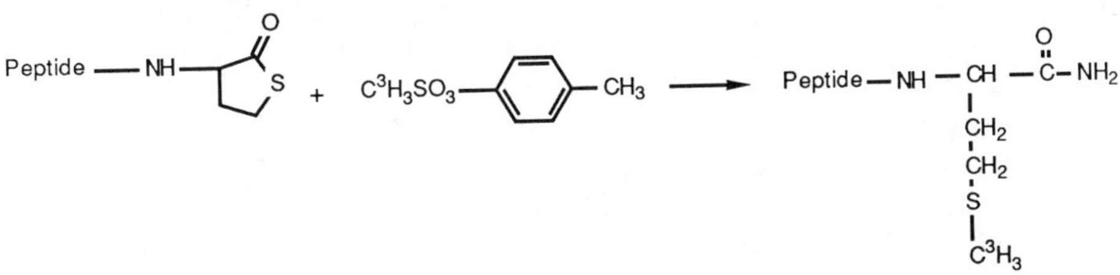

(4)

Semisynthesis

It is often impractical and sometimes almost impossible to prepare an appropriate tritiation precursor that contains either unsaturated or halogenated amino acids for a large peptide with a complex sequence. Semisynthesis, in which fragments of naturally occurring peptides and proteins are used as reaction intermediates for the preparation, is probably the best method for labeling such substances. [^{3}H] insulin has been prepared by semisynthesis (25). Porcine insulin was first converted to di-protected des-Phe(B1) insulin [Equation (5)]. The di-protected des-Phe(B1) insulin was then coupled with Boc-[^{3}H] phenylalanine active ester to give protected [^{3}H-Phe(B1)] insulin, which was deprotected to give [^{3}H-Phe(B1)] insulin with a specific activity as high as 20 Ci/mmol.

Insulin ⟶ [Nα -Boc-Gly^{A-1},N^{t}-Boc-Lys^{B-29}]-des-phe^{B-1}-insulin

↓ Boc-phe-[^{3}H]-active ester

[Nα -Boc-Gly^{A-1}, N^{a}-Boc-phe^{B-1}[^{3}H],^{t}-Boc-Lys^{B-29}]-insulin (5)

↓

phe^{B-1}[^{3}H]-insulin

By a similar approach, [^{3}H-Gly(A1)] insulin with specific activities ranging from 1 to 44 Ci/mM has also been synthesized (26).

Total Synthesis

[^{3}H] Peptides have also been prepared by the incorporation of [^{3}H] amino acids into the peptide sequences using standard peptide synthesis procedures. This synthetic approach provides the opportunity to maximize the specific activity of the product, in contrast to catalytic reduction in which specific activity is almost impossible to predict with any degree of certainty. This is true especially when the reduction is carried out with a precursor containing a methionine or cysteine residue. Since the tritium is introduced into the peptide as a tritiated amino acid, the regiospecificity of the label in the product is well defined. This is particularly important for peptides used for metabolic studies. The major disadvantage of this approach is the safety and potential health physics hazards involved in the work. A high level of radioactivity is always employed in order to yield sufficient product since such syntheses usually involve multiple chemical reaction steps.

Parnes and Shelton (27) reported the synthesis of [D-3-(2-naphthyl)-2,3-^{3}H-Alanine^{6}] LHRH at 43 Ci/mM by conventional peptide synthesis. Boc-D-[^{3}H] Na was first coupled to the C-terminal tetrapeptide of LHRH to give the desired pentapeptide sequence, which was subsequently coupled to the N-terminal pentapeptide to produce [D-3-(2-naphthyl)-2,3-^{3}H-Ala^{6}] LHRH. The syntheses of tritiated oxytocin (28) and [Nva^{13}] α-MSH (29) have also been reported. Tritiated peptides have also been prepared by solid-phase peptide synthesis. Holton et al. (30)prepared tritiated arginine vasopressin, and Wilkes et al. (31) reported the synthesis of [^{3}H] -Ac-[Nle^{4},D-Phe^{7}] α-MSH(4-11)-NH$_{2}$.

Labeling Reagents

Labeled peptides have been prepared by reacting peptides with an appropriate labeling reagent. These reagents, containing tritium or radioiodine, react with specific functional groups in the peptide to produce labeled derivatives of peptides that often retain high biological activity.

Several [^{3}H]-labeling reagents for peptides and proteins are now commercially available. Among them, succinimidyl [2,3-^{3}H]propionate [46], an active ester of [^{3}H]propionic acid with high specific activity, is probably the most frequently used reagent. This reagent acylates an amino group of the peptide to form a [^{3}H]propionyl peptide derivative. Dolly et al. (32) prepared tritiated α-bungarotoxin using this reagent. The advantages of this reagent are the relative small size of the propionyl group, which gives less structural alteration to the labeled product and also the relative ease of use.

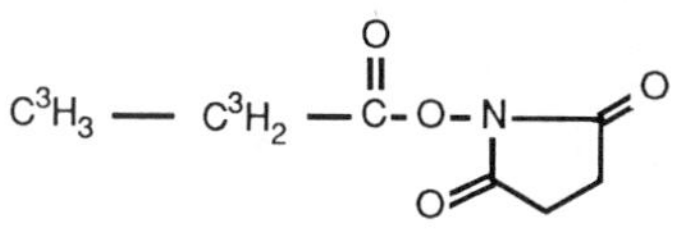

46

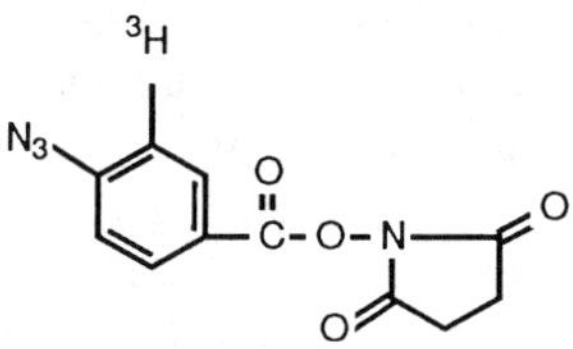

47

A similar labeling reagent, [^{3}H] succinimidyl 4-azidobenzoate [47], which has the added capability for photoaffinity labeling, is also available commercially.

It is known that peptides with free amino groups can react with formaldehyde to form Schiff bases. The Schiff base can subsequently be reduced with sodium borohydride to a stable amine, as shown in Equation (6):

$$\text{Peptide-NH}_2 \xrightarrow[\text{NaBH}_4]{\text{HCHO}} \text{Peptide-NH-CH}_3 \xrightarrow[\text{NaBH}_4]{\text{HCHO}} \text{Peptide-N(CH}_3)_2 \qquad (6)$$

Taking advantage of the commercially available high-specific-activity sodium borotritide, tritiated peptides and proteins with high specific activities have been prepared. Hammonds et al. (33) prepared [^{3}H-Me]β-endorphin with a specific activity of 10 Ci/mM using this approach.

Radioiodination

Radioiodine-labeled peptides are increasingly used in receptor binding studies. The commonly used isotope is iodine 125, which has a half-life of 60 days. An advantage of radioiodine labeling is that it provides a much higher specific activity than tritium, 2200 versus 30 Ci/mM. However, as with other derivatizing reagents, one must be cautious in utilizing radioiodine-labeled peptides, since the introduction of the relatively large atom(s) or group(s) into the peptide can sometimes significantly affect the structural orientation and hence the biological properties of the compound. The use of elemental radioiodine or radioiodine monochloride as the iodinating agent has been known for many years; however, because of their volatile nature these agents are now rarely used. The most widely used method today is iodination with chloramine-T (34) and sodium radioiodide. Chloramine-T, acting as an oxidizing agent, oxidizes iodide to iodine in situ. The procedure is relatively straightforward and usually gives a high efficiency of utilization of the radioactive iodine.

There are structural requirements for the peptide to be iodinated; the peptide needs to possess a tyrosine or histidine residue. Leucine enkephalin, Tyr-Gly-Gly-Phe-Leu, was iodinated using this method to give [monoiodinated-Tyr]-leucine enkephalin with a specific activity of 2200 Ci/mmol, and monoiodinated thyrotropin releasing hormone, pGlu-[^{125}I] His-Pro-NH_2, was obtained with a specific activity of 2000 Ci/mM.

For peptides that do not contain tyrosine or histidine residues or peptides that become biologically inactive after being iodinated by the chloramine-T method, labeling with Bolton-Hunter reagent is a potential choice. Bolton-Hunter reagent is the N-hydroxysuccinimide ester of iodinated *p*-hydroxyphenylpropionic acid [48]. As an active ester, the reagent acylates the free amino groups of the peptide to form the iodinated p-hydroxyphenylpropionyl peptide. Miller et al. (35) prepared N-^{125}I-labeled CCK-8 using this method. Besides Bolton-Hunter reagent, Wood's reagent, methyl-3-iodo-4-hydroxybenzimidate [49], is also used for labeling peptide ligands in a similar fashion. A radioiodine labeling reagent, MAHIB, methyl-3-azido-4-hydroxy-5-iodo-benzimidate [50], which has the added capability for photoaffinity labeling, is also available commercially.

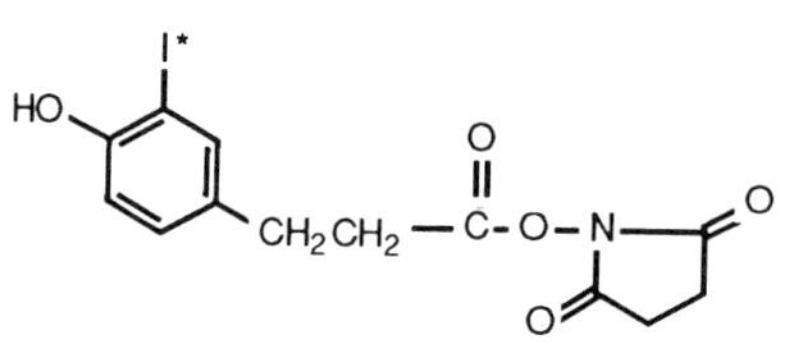

48

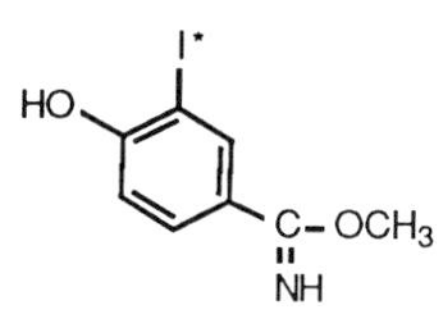

49

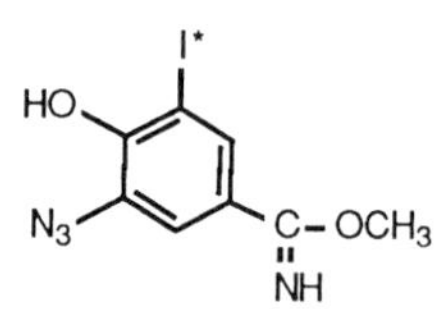

50

RADIOCHEMICALS: TECHNICAL CONSIDERATIONS

The correct radiochemical is an experimental tool of superb analytic sensitivity and consequent value to the researcher. This value is compromised if the radiochemical is not handled properly to retain the quantitative and chemical integrity necessary for the experiment. In the following section the stability and practical aspects of handling radiochemicals are discussed with the expectation that the intrinsic value of these materials can be retained with proper care.

Stability

The stability of all chemicals depends on the environment in which they are stored and their inherent stability. Radiochemicals are inherently less stable than their unlabeled counterparts for the following reasons:

1. They undergo nuclear decay to an extent inversely proportional to the half-life of the isotope.
2. Nuclear decay leads to generation of free radicals and/or other high-energy particles that degrade the chemical purity of the radiochemical and its surrounding environment.
3. The nature of the free radical process is such that a single radical event can cause multiple subsequent events.

In addition to being inherently less stable, the high-specific-activity ligands used in neuroscience are necessarily handled and stored in dilute solutions, which, although increasing their stability through dilution, presents the opportunity for decomposition through reactions with dissolved gases, solvent, solvent impurities, and even the container walls and caps.

Commercial radiolabeled ligands have a wide range of stabilities and useful shelf-lives depending on their chemical nature and the experiment in which they are used. Because the shelf-life is application dependent and there may be hundreds of uses for a single product, most radiochemicals are initially prepared and evaluated on the basis of radiochemical purity. In the majority of cases, the rate of decomposition begins as a linear decrease in radiochemical purity but increases rapidly after 10-15% of impurities accumulate. For this reason, although many radiochemicals are stable for many months, close attention to purity and its effect on experimental results must be carefully monitored during the course of experimental work.

The researcher can determine the radiochemical purity by careful chromatographic determinations as suggested on the technical data sheets provided by the manufacturer. HPLC, if available, is the method of choice, but each method depends on excellent quantitative technique and attention to the following details;

Manipulations should be done as quickly as is consistent with good technique under an inert gas, at room temperature or lower, and away from light.

The choice of solvent and glass or plasticware should be consistent with the properties of the radiochemical. Polar materials, such as peptides or strongly basic ligands, should be contained in silanized glassware or plastic. Very nonpolar and organic soluble materials should be handled in glass as they may adhere to or dissolve plastic. Pipettes and all other equipment contacting the sample should be consistent with these considerations.

Concentration of the solution may best be done under a gentle stream of nitrogen, taking care not to entrain air into the atmosphere over the chemical or to physically blow away the solvent and hence the compound. A slight dimple in the surface of the solvent is usually sufficient to effect rapid evaporation.

Avoid reducing the solution to complete dryness. A few microliters of solvent remaining often does not affect analytic or biological results. If necessary, addition of 100 μl of alternate solvent and a second evaporation is preferable to total solvent removal.

Injection into the HPLC is best done in the mobile phase of the chromatography solvent. In order to avoid irreversible surface effects due to the small mass, unlabeled compound should be added to the sample. When using TLC, it is advisable to apply unlabeled compound to the plate under the labeled aliquot to avoid irreversible adsorption or decomposition at the point of application. The authentic material also assists in detecting and validating the identity of the material.

Radiochemical purity is often evaluated from the results of biological experiments. Increasing or high background or blank determinations, poor binding, or metabolic conversions in control experiments may be the most pragmatic determination of suitability for continued use of the radiochemical. A scientifically valid experimental result is a fundamental purity criterion since radiochemical determinations of purity do not always detect experimentally important impurities present in minute amounts.

Handling

In addition to the suggestions just outlined there are preferred techniques for maintaining purity of the radiochemical during the experimental manipulations.

Minimize the number of times the primary container is opened, and maintain a reduced temperature consistent with the storage solvent.

Avoid repetitive freeze-thaw cycles. Except at extremely low temperatures, radiochemicals should not be stored frozen. The crystallization and continuing recrystallization of the solvent concentrates the compound into the solvent crystal interstices and thus accelerates decomposition.

Although many buffers containing proteins, bipolar salts, and so on stabilize radiochemicals, such buffers or solvents often accelerate decomposition. Stability tests should be used to determine the advisability of preparing multiple aliquots for future use.

Many ligands require special techniques to overcome poor solubility characteristics in biologically compatible buffers. During each new application or change in procedure the researcher should validate the quantitative transfer of the radiochemical into the solvent medium.

Radiochemical decomposition to an unusable extent is an inevitable end to each purchase. In order to minimize the economic consequences, experiments and purchases should be carefully coordinated so that the radiochemical is as fresh as possible and experiments are consummated within its usable shelf-life. Information on stability, usable shelf-life, and the rate of radiochemical decomposition is usually available from the manufacturer at the time of purchase. Newer products, however, may not have been available long enough to allow the accumulation of this information.

Repurification

Repurification of small, often microgram, amounts of a radiochemical is a challenging undertaking that is best avoided by prior planning. If necessary, HPLC is often the best tool. All the precautions of analysis and handling must be followed. Unless the eluting solvent can be used directly in the experimental procedure, volatile buffers should be used and removed quickly and carefully under vacuum. These buffers should be prepared from carefully redistilled reagents. Common volatile aqueous buffers contain

Triethylamine-acetic acid
Diethylamine
Ammonium acetate
Trifluoroacetic acid and/or acetic acid

Pyridinium acetate is less desirable as it oxidizes readily and is not as volatile as the others.

Some radiochemicals can be successfully purified and recovered from paper or thin-layer chromatographic procedures. In these instances careful preliminary experiments should validate the procedure. Some possible techniques to preserve fragile radiochemicals are as follows.

Apply the compound to an area of the plate (paper) that has been previously wet with solvent and partially or completely air dried.

The radioactive compound may be identified and located by chromatographing it between spots of authentic material, which are more easily visualized at their higher concentration. Avoid chromogenic sprays if possible, and expose the chromatogram to ultraviolet light as briefly as possible. If a spray is essential, plan to break a TLC plate along a scratch or cut strips from a paper to avoid spraying or heating the product.

Use cellulose plates if possible since silica gel plates are the most difficult to use successfully owing to the surface effects of the highly reactive adsorbent.

It is vital to elute the compound from the plate before allowing the plate to dry completely.

Wrap a partially dry plate or paper chromatogram in plastic wrap to prevent overdrying and exposure to oxygen in the air while visualizing the unlabeled spots or while other manipulations are being carried out.

This discussion of radiochemical handling and stability should assist the researcher in obtaining the best possible results from carefully planned and executed experimental protocols.

REFERENCES

1. C. B. Pert and S. H. Snyder, Proc. Natl. Acad. Sci. USA *70*, 2243 (1973).
2. M. H. Randall, L. J. Altman, and R. J. Lefkowitz, J. Med. Chem. *20*, 1090 (1977).
3. A. E. Bolton and W. M. Hunter, Biochem. J. *133*, 529 (1973).
4. L. E. Overman and K. L. Bell, J. Am. Chem. Soc. *103*, 1851 (1981).
5. P. Krogsgaard-Larsen, J. S. Johansen, and E. Falch, J. Labelled Comp. Radiopharm. *19*, 689 (1982).
6. C. Filer, D. G. Ahern, R. Fazio, and R. J. Sequin, J. Org. Chem. *46*, 4968 (1981).
7. N. J. Leonard, A. S. Hay, R. W. Fulmer, and V. W. Gash, J. Am. Chem. Soc. *77*, 439 (1955).
8. G. A. Brine and J. A. Kepler, J. Labelled Comp. Radiopharm. *12*, 401 (1976).
9. D. E. Brundish, M. G. Combe, and R. Wade, J. Labelled Comp. Radiopharm. *20*, 869 (1983).
10. G. Gacel, P. Dodey, B. P. Roques, J. L. Morgat, J. Roy, and P. Fromageot, J. Labelled Comp. Radiopharm. *20*, 719 (1983).
11. R. A. Houghten and C. H. Li, J. Pept. Protein Res. *12*, 325 (1978).
12. M. C. Allen, D. E. Brundish, R. Wade, B. E. B. Sandberg, M. R. Hanley, and L. L. Iversen, J. Med. Chem. *25*, 1209 (1982).
13. J. L. Morgat, L. T. Hung, and P. Fromageot, Biochim. Biophys. Acta *207*, 374 (1970).
14. P. Pradelles, J. L. Morgat, P. Fromageot, C. Oliver, P. Jacquet, D. Gourdji, and A. Tixier-Vidal, FEBS Lett. *22*, 19 (1972).
15. J.-T. Lin, S. Milutinovic, and F. Fahrenholz, Anal. Biochem. *88*, 587 (1978).
16. G. Flouret, S. Terada, F. Yang, S. H. Nakagawa, T. Nakahara, and O. Hechter, Biochemistry *16*, 2119 (1977).
17. A. M. Felix, C.-T. Wang, A. A. Liebman, C. M. Delaney, T. Mowles, B. A. Burghardt, A. M. Charnecki, and J. Meienhofer, Int. J. Pept. Protein Res. *10*, 299 (1977).
18. M. H. Perrin, J. E. Rivier, and W. W. Vale, Endocrinology *106*, 1289 (1980).
19. P. M. Hardy, P. W. Sheppard, D. E. Brundish, and R. Wade, J. Chem. Soc. Perkin Trans. I 731 (1983).
20. N. A. Sasaki, S. Funakoshi, R. Michelot, P. Potier, J.-L. Morgat, J. M. Zajac,

and B. P. Roques, *Peptides 1984: Proceedings of the 18th European Peptide Symposium* (V. Ragnarsson, ed.), Almquist and Wiksell International, Stockholm, 1985, p. 255.

21. N. A. Sasaki, S. Funakoshi, P. Potier, J.-L. Morgat, R. Genet, G. Gacel, B. Charpentier, and B. P. Roques, J. Labelled Comp. Radiopharm. *22*, 1123 (1985).
22. P. Marche, J.-P. Girma, J.-L. Morgat, P. Fromageot, C. Ghelis, M. Dubrasquet, and S. Bonfils, Eur. J. Biochem. *50*, 375 (1975).
23. P. M. Barling, J. R. Preston, N. J. Bibby, and T. Wilson, Anal. Biochem. *144*, 542 (1985).
24. J.-L. Morgat, R. Genet, P. Fromageot, G. Chassaing, S. Lavielle, and A. Marquet, Abstract 9th American Peptide Symposium, June 23-28, 1985, p. 37.
25. P. A. Halban and R. E. Offord, Biochem. J. *151*, 219 (1975).
26. J. G. Davies and R. E. Offord, Biochem. J. *231*, 389 (1985).
27. H. Parnes and E. J. Shelton, J. Labelled Comp. Radiopharm. *21*, 263 (1984).
28. V. du Vigneaud, C. H. Schneider, J. E. Stouffer, V. V. S. Murti, J. P. Aroskar, and G. Winestock, J. Am. Chem. Soc. *84*, 409 (1962).
29. A. N. Eberle and A. Zeller, Helv. Chim. Acta *68*, 1880 (1985).
30. H. H. Holton, L. A. Branda, and B. M. Ferrier, Can. J. Chem. *51*, 1910 (1973).
31. B. C. Wilkes, V. J. Hruby, H. I. Yamamura, K. Akiyama, A. M. L. Castrucci, M. E. Hadley, J. R. Andrews, and Y.-P. Wan, Life Sci. *34*, 977 (1984).
32. J. O. Dolly, E. A. V. Nockles, M. M. S. Lo, and E. A. Barnard, Biochem. J. *193*, 919 (1981).
33. R. G. Hammonds, Jr., N. Ling, and D. Puett, Biochem. Biophys. Res. Commun. *98*, 211 (1981).
34. W. M. Hunter, and F. C. Greenwood, Nature (Lond.) *194*, 495 (1962).
35. L. J. Miller, S. A. Rosenzweig, and J. D. Jamieson, J. Biol. Chem. *256*, 12417 (1981).

6

Molecular Modeling of Receptor-Ligand Interactions

Yvonne C. Martin and Elizabeth Danaher / Abbott Laboratories, Abbott Park, Illinois

It is a fundamental tenet of chemistry that the covalent structure of a molecule and its associated solvent and reagents contain all the information necessary to understand the molecule's reactivity and physical properties. The study of chemistry is the attempt to decode the properties encoded in the chemical structure. A generation ago chemists came to realize that for such understanding one must examine the three-dimensional structure, the conformation, of the molecules of interest. Most molecules have many possible conformations of different stability. The tool of single-crystal x-ray diffraction has provided a low-energy conformation of thousands of small molecules (1) and hundreds of proteins (2). In addition, a variety of spectroscopic and chemical probes have provided insight into molecular flexibility or conformational change as well as minimum energy conformation (3,4). Such experimental results plus the increasing power of quantum chemistry have produced a steady increase in the precision and scope of our understanding of the relationship between three-dimensional structure and molecular properties. The 1980s has seen the fruits of these previous labors transformed with the aid of computers into tools for use by ordinary scientists (5-12).

COMPUTER TOOLS FOR MOLECULAR MODELING

Molecular Graphics Display of Molecular Structure and Motion (6,8,10,12,13)

Molecular graphics is the displaying of three-dimensional structure of a molecule on a computer screen. Displays are typically in colors that are used to distinguish different atom types or chemical properties of one molecule or different molecules in a multimolecule display. The graphics device can range from still pictures of molecular backbones on a black-and-white graphics terminal or personal computer all the way to color images of molecular backbones and/or surfaces that can be rotated in real time.

The structure may be the result of experimental observations or theoretical optimization, or it may be built up from pieces as one would build a ball-and-stick model. Usually one can perform bond rotations on a dial or a function key and receive feedback of the energy. Text and other nonstructural information may also be displayed.

Rotation of the whole molecule about an axis in real time gives an impression of the three-dimensional shape. Various devices are also used to deliver different images to each eye and so produce a more pronounced effect of three dimensions. (Most figures in this chapter are presented as stereo pairs. To view them one must decouple closely focusing the eyes from convergence of the two eyes on the same image. In other words, look straight ahead but focus. After a minute or so a third image will appear between the two printed ones: this center image will have three-dimensional properties. The viewing may be facilitated by placing one's nose on a sheet of paper between the images or using special glasses.*)

Most molecular graphics programs have the facility to show molecular surfaces. These pictures may be space filling or show the surface as dots or a mesh over the stock background. On an individual molecule color is used to show the atom type or to show a chemical property. An example of a single molecule display is shown in Figure 1.

Software designed for structure-activity analysis has provisions for the display of several molecules simultaneously (8,10). Molecules are superimposed by minimizing the root-mean-square difference of distances between specified atoms. In addition, one usually has the provision to display a single surface that encloses several molecules, a surface that encloses only the space common to all of a set of molecules, and regions in space uniquely occupied by one or a subset of molecules (10). For example, Figure 2 shows the superposition of apomorphine over dopamine and the extra volume occupied by apomorphine.

As the examples discussed here illustrate, these molecular graphics tools have advantages over mechanical molecular models in accuracy, versatility, superposability, retrievability, and expanded capabilities. They are, however, no more accurate than the science on which they are built. Indeed, it is to be expected that the availability of modern computational tools will catalyze the critical evaluation of many of the concepts of current chemical theory.

---

*These glasses may be purchased from Reed 3-D Enterprises, P. O. Box 2368, Culver City, California 90231.

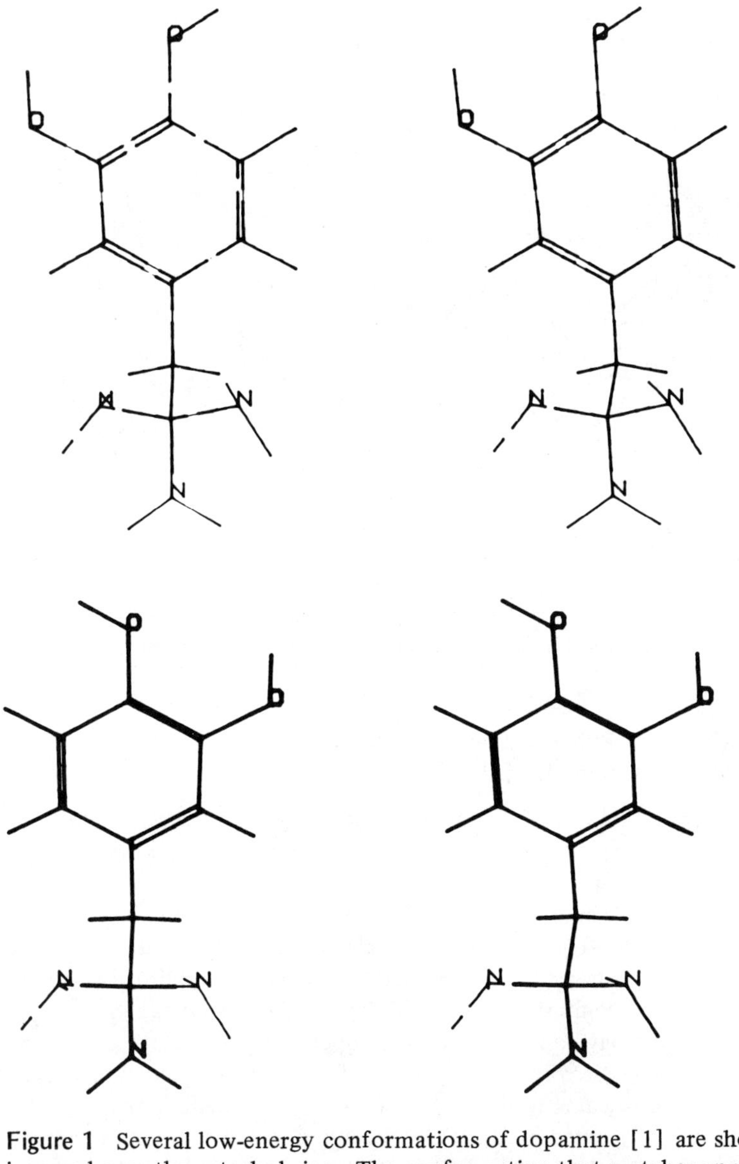

Figure 1 Several low-energy conformations of dopamine [1] are shown superimposed over the catechol ring. The conformation that matches apomorphine is shown as a heavy solid line; the others were derived by rigid rotation of its side-chain bonds.

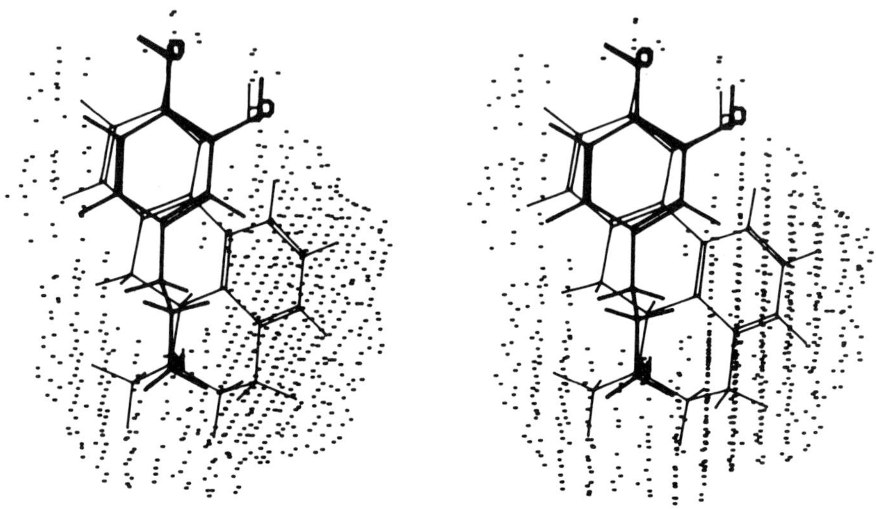

Figure 2 The low-energy conformation of apomorphine [2] is shown as a fine solid line. The structure shown in molecular mechanics (MMP2) minimized from the single-crystal x-ray diffraction results (14). The conformation of dopamine that best matches apomorphine is shown as a heavy line. Finally, the dots indicate regions in space that are occupied by apomorphine, not dopamine.

Molecular Mechanics or Potential Energy Calculations of the Energy of a Molecular Structure or Ensemble (5)

This type of computation is used to minimize the energy of a structure and so find the closest preferred conformation, to search conformational space for all low-energy conformations (9,15), to examine molecular flexibility and atomic motion by a molecular dynamics calculation (16-18), and to calculate the change in a particular free energy when one member of a molecular ensemble is altered (19). Minimization of a small molecule may be accomplished in minutes even on a personal computer: the calculation of the free energy of a protein-ligand interaction may consume days on a super-computer.

In molecular mechanics the relative energy of a particular three-dimensional arrangement of atoms is equal to the sum of several independent contributions. Each contribution is evaluated by the use of an equation and parameters that have been derived empirically, principally by fit to experimental data but also to high-quality quantum mechanical results. The exact parameters used and the form of the relationships are termed the force field.

Thus, the bond stretch-compression and bond angle bend contributions to energy are a consequence of the deviation of the bond lengths and angles from an ideal value. Torsional contributions to energy are a consequence of deviations from the ideal arrangement of four bonded atoms in space. This ideal arrangement is determined by quantum mechanical factors, such as the tendency of double bonds to be planar. The van der Waals contribution to energy reflects that atoms associate but cannot interpenetrate. The electrostatic contribution to energy arises from the attraction of opposite charges and the repulsion of like ones. Some force fields also include an explicit solvation contribution to energy; for the others, solvation is ignored or treated by including solvent molecules in the calculation. Some force fields include cross-terms, for example between bond stretch and angle bend. Others keep bond lengths and angles fixed at identical values and rotate about bonds.

The principal advantage of molecular mechanical compared with quantum mechanical calculation of energy is that the calculation is much faster with no loss in accuracy.

Very good molecular mechanics parameters are available for hydrocarbons and simple substituents. For example, Figure 3 shows a comparison of the crystal and MM2 (20) structures of apomorphine. Fairly reliable parameters are also available for water, proteins, and nucleic acids (15-17,22,23).

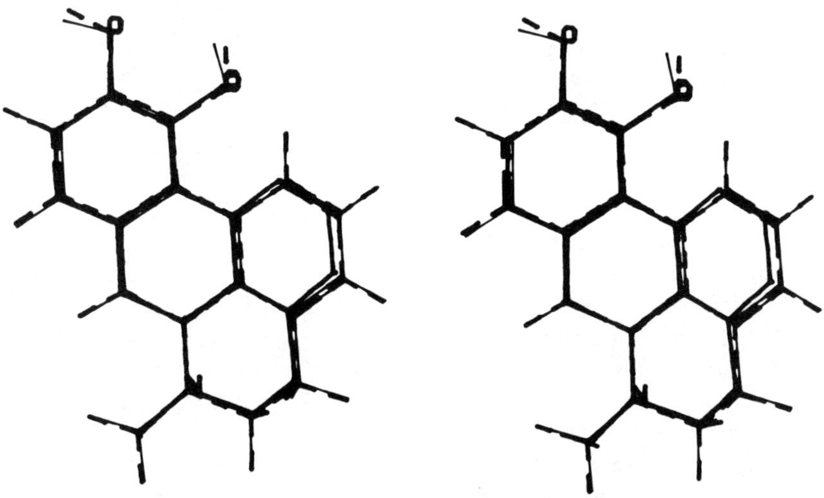

Figure 3 The molecular mechanics (MMP2) (20) minimized (fine line) structure of apomorphine is superimposed over that determined by single-crystal x-ray diffraction (14) (heavy solid line) and that minimized with the quantum mechanics program MNDO (22) (heavy dashed line).

The greatest deficiency in molecular mechanics is that one must have values of the empirical parameters: they simply do not exist for many drug molecules. Sometimes one can make a reasonable guess for a parameter value, but not always. Molecular mechanics methods will need special extentions in order for them to be useful in calculating reaction pathways.

Quantum Mechanical Calculation of the Energy and Chemical Properties of Molecules and Molecular Ensembles (24,25)

Quantum chemical calculations have become practical for medicinal chemists because of the increase in available computer power and because of the power of molecular graphics that makes input easier and output more easily interpreted. They may be used to calculate the relative energy and chemical properties of a molecular system, to optimize the three-dimensional structure, and to investigate the structures of meta-stable intermediates in reaction pathways (11,21,26-39). The advantage of quantum mechanics over molecular mechanics is that those few parameters needed are already available. The disadvantage is the large or enormous amounts of computer time and memory required for accurate calculations.

Semiempirical methods (such as CNDO/2, INDO, MINDO/3, MNDO, and AM1) use a variety of assumptions and parameterizations to reduce the size of the calculation. They generally suffer in accuracy and speed compared with molecular mechanics, but are fast enough (a drug molecule might take 10 hr to minimize a structure) and accurate enough to provide useful information. They are currently parameterized for all except certain transition metal and high atomic number atoms. Figure 3 shows a comparison of the x-ray, molecular mechanics, and quantum mechanics structure of apomorphine.

Ab initio methods in principle are very accurate: the accuracy is attained with the massive use of computer resources. A compromise is chosen with the use of minimal basis set (STO-3G) for the description of the atomic orbitals (39). A single energy calculation, not structure minimization, can take a day or longer. Such calculations produce fairly reliable semiquantitative chemical property information. For such purposes calculations are often done on structure optimized by molecular mechanics or semiempirical quantum mechanics.

The electrostatic interactions between the atoms appears to be the single most important factor in molecular recognition: steric fit can spoil or augment an interaction, but the electrostatics must be correct for an interaction to occur at all (40). Thus, scientists are eager to calculate the electrostatic potential at various points around a molecule (29,34,36-38). The regions of minima in the electrostatic potential are regions very attractive to an H^+; these are the sites of H-bond accepting ability. The regions of maxima are repulsive to an H^+. The repulsion may indicate a positively charged region in space and so a site of H-

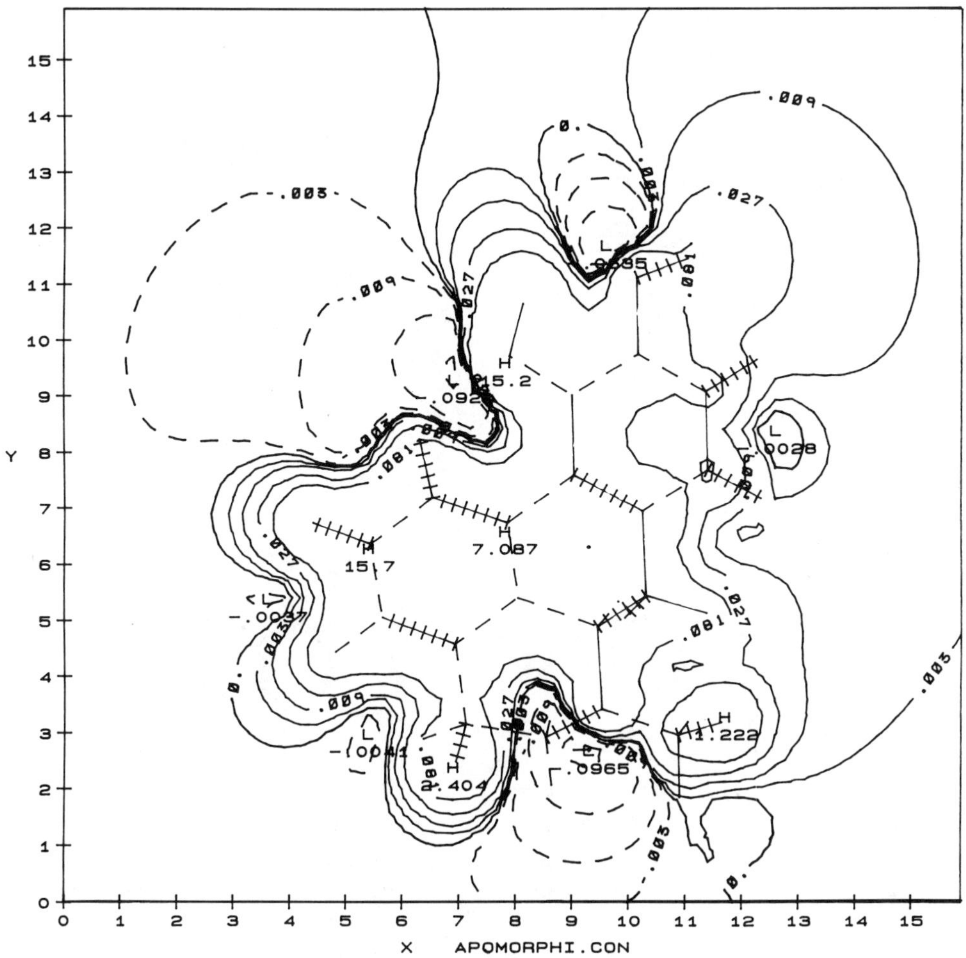

Figure 4 The molecular electrostatic potential of apomorphine [2] in the plane of the catechol ring. The electrostatic potential was calculated from the ab initio STO-3G wavefunction for the MMP2-optimized crystal structure. Dashed lines represent negative potential and solid lines, positive potential.

bond donation. However, positive values also arise from steric repulsion. Steric repulsion is not a factor when one is beyond the van der Waals radius.

The molecular graphics display of electrostatic potential uses either contour lines or colors to distinguish chemically different regions in space. Figures 4 and

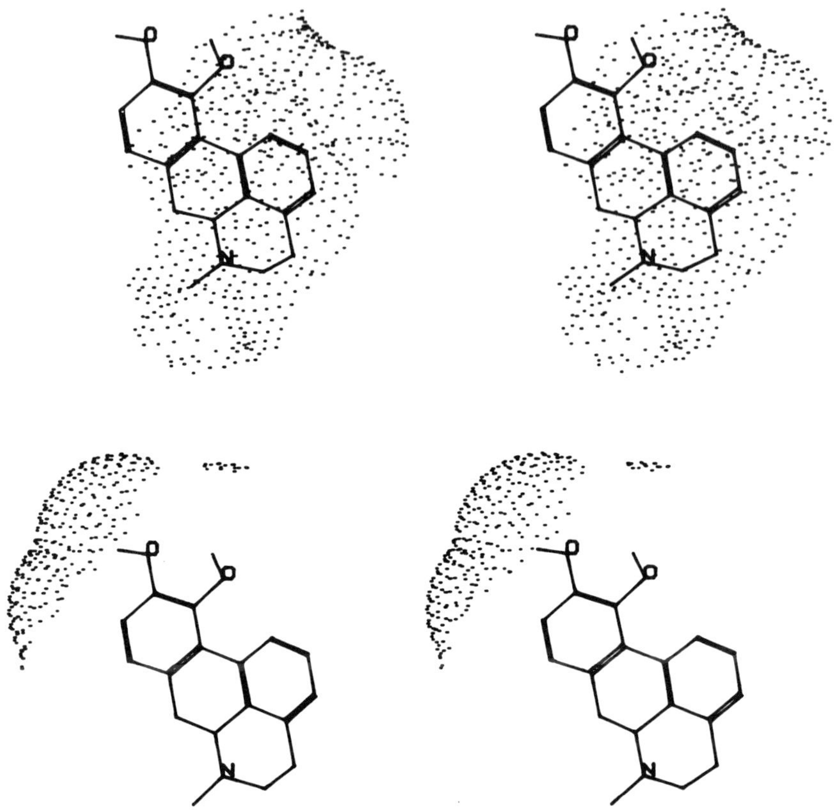

Figure 5 The top figure shows regions on the molecular surface of apomorphine [2] (calculated at the van der Waals radius plus 1.4 Å) at which an H^+ is attracted with an energy of −3.0 kcal/mol or greater. The bottom figure shows regions at which an H^+ is repelled with an energy of +3.0 kcal/mol or greater. The electrostatic potential was calculated as in Figure 4.

5 show such displays for apomorphine [2]. (Notice that the electrostatic potential in the region of the catechol would change sign if the –OH groups were rotated 180°.) This type of calculation is typically done on the neutral molecule even though the ion may be considered the biologically active species since such an ion does not exist in solution or the receptor but would be neutralized by electrostatic interactions.

Quantum mechanical calculations cannot today be done on a whole protein. However, various approximations can be made to allow one to investigate such questions as enzyme mechanisms or proton transfer upon binding of the ligand to an enzyme (26,33,35).

DETERMINATION OR MOLECULAR MODELING OF COMPLETE THREE-DIMENSIONAL STRUCTURE OF A LIGAND-RECEPTOR COMPLEX

Crystallography of Proteins and Protein-Ligand Complexes

The ligand binding site of a protein is typically made up of residues distant in the amino acid sequence. Thus experimental and theoretical methods to establish the three-dimensional structure of receptors is of great interest.

Experience has shown that because of the large amount of cocrystallized solvent, the three-dimensional structure of a protein in a crystal is often very similar to that in solution. At this time, no one has experimentally determined the three-dimensional, atomic-level resolution structure of a receptor. However, the recent isolation and cloning of several receptors suggests that a sufficient supply of material for crystallography will soon be available (41-43). In addition, the recent breakthrough in crystallization techniques for membrane-bound proteins (4) offer the promise that soon a three-dimensional receptor structure will be known. The solution of a protein structure is a great task. It may take several years' work to arrive at the structure. In addition, one is not guaranteed a successful outcome because the protein may not form suitable crystals or it may prove impossible to convert the observed x-ray diffraction pattern into the electron density distribution. However, if one does obtain a protein structure and if the binding of a ligand does not change the conformation of the protein dramatically or change the packing of the protein molecules in the crystal, then it is quite straightforward to solve for the differences between the free and liganded protein (3,4). For example, see Figure 8. One would thus directly observe the receptor-bound conformation of the ligand and the (small) changes in receptor conformation or solvent binding that accompany ligand binding.

Of course, ligand binding may produce a large change in the receptor conformation. If this happens, then the difference method cannot be used, but the new structure must be solved de novo.

In any event, the prospect of having the three-dimensional atomic-level structure of a liganded receptor is very exciting. Such a structure would be useful directly for the design of potential agonists and antagonists of novel structure (7,44-46) but also for the design of biochemical and biophysical experiments to probe the structural basis for receptor function.

Nuclear Magnetic Resonance Spectroscopy of Proteins

Nuclear magnetic resonance (NMR) techniques tell us which atoms are close in space. From this information and the amino acid sequence of the protein, one may use computer graphics, distance geometry, and molecular mechanics to propose a three-dimensional structure of the protein (47).

The advantage of NMR over x-ray diffraction is that one does not have to search empirically for suitable crystals. A disadvantage of NMR is that the protein may have several conformations observed simultaneously. If this is true, the computer modeling of NMR data may provide no solution. A more serious current limitation of NMR is that it is useful only for proteins of less than 100 amino acids. Intact receptors sites are much larger than this. For all these reasons NMR spectroscopy does not appear to be the method that will provide the first three-dimensional atomic-level resolution structure of a receptor.

Modeling a Protein Structure from that of Homologous Proteins

Of course, we will not soon have an experimentally determined three-dimensional structure of every receptor discussed in this book. Will knowing one such structure help us know the others? Experience in protein crystallography suggests that there are a limited number of structural motifs from which three-dimensional protein structures are constructed. Even more importantly, one frequently sees a strong three-dimensional homology between proteins with similar function (3,48,49). For example, Figure 6 shows the superposition of the α carbons of the x-ray structures of three hydrolytic enzymes of the serine protease class. The close similarity of the structures in the region of the active site is especially noteworthy. This similarity was used in modeling the structure of the blood clotting protein factor Xa when bound to its substrate (Figure 7).

Figure 6 The crystal structures of chymotrypsin, trypsin, rat mast cell protease, and elastase have been superimposed to maximize correspondence of the active site. Only the α carbons are shown. The view of the molecules is looking into the serine protease active site with the specificity pocket positioned to the right. [After J. Greer (50).]

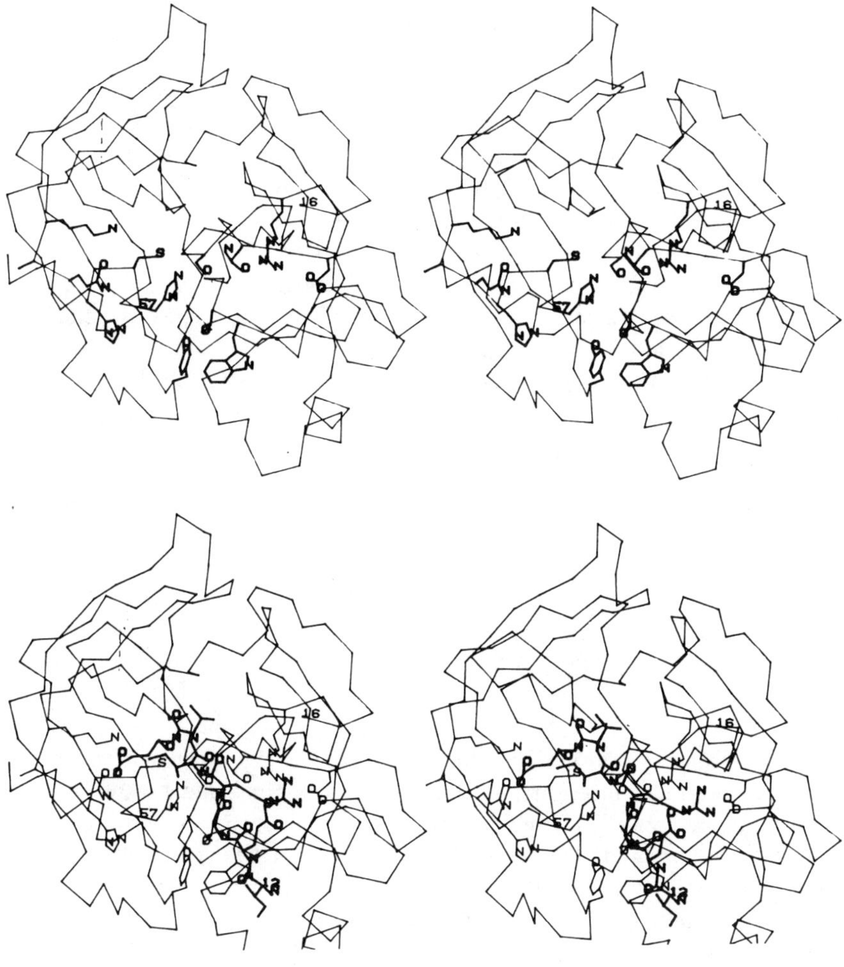

Figure 7 (a) The structure of the serine protease blood-clotting factor Xa modeled from the structures of the three serine proteases shown in Figure 6 (49). The side chains that interact with the prothrombin substrate are shown in heavy lines. (b) The modeled structure of the complex between factor Xa (fine lines) and prothrombin (heavy lines). The serine and imidazole of the catalytic site are in the center of the figure. Notice the two guanidium-carboxylate salt bridges on the right and the ammonium-carboxylate on the left.

Such three-dimensional homologies have led to procedures whereby one proposes a three-dimensional structure of a protein from its sequence and the three-dimensional structure of one or several proteins with which it is homologous (48-50). One first aligns the amino acid sequences of the unknown with one of the known proteins and then changes the side chains of the closest known into those of the unknown structure. Molecular graphics may help one see how to position the side chains of the new structure. Energy minimization may then be used to refine the proposed structure of the unknown protein.

There is pharmacological structure-activity evidence that the ligand binding sites of receptors will be found to be structurally similar. For example, dopamine [1] binds to at least six different receptors; D_{-1}, D_{-2}, α_1, α_2, β_1, and β_2; apomorphine [2] binds to D_{-1}, D_{-2}, and α_2 receptors; 7,8-dihydroxy-3-benzazepine [3] binds to D_{-1}, D_{-2}, α_1, and α_2 receptors; and B-HT 920 [4] binds to

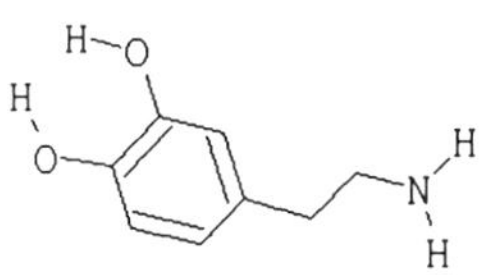

1

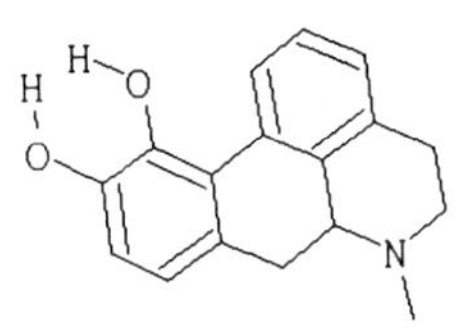

2

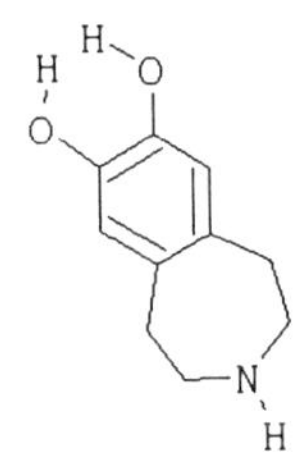

3

D_{-2} and α_2 receptors. On the other hand, SKF-38393 [5] shows 100 times enhanced binding to D_{-1} receptors, no change in D_{-2} binding, and a loss of α_2 binding compared with [3]. Such results suggest that the receptors for catecholamines will be found (1) to contain identical atoms involved in binding the basic nitrogen atom and the catechol group or surrogate nitrogen atom and (2) to vary in the structure of nearby accessory binding sites.

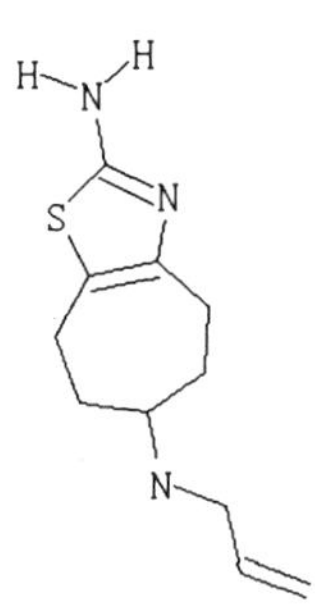

4

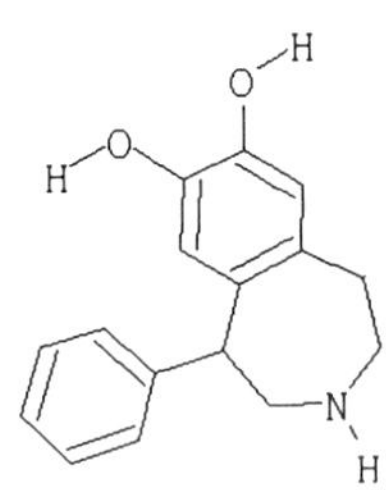

5

There is also biochemical evidence of similarities in the structure of the domains on the receptor that are responsible for the signal transduction. For example, one would expect that the part of any receptor structure that binds to the N_i protein should be similar in three-dimensional structure to any other receptor that binds to N_i, and so on. Just as the binding site for NADPH is similar in all NADPH-binding proteins studied (3), so we expect to find similarities within the classes of binding sites for the protein that couple receptor occupancy to some biophysical event.

The final evidence that three-dimensional homology will be found between receptor structures is to be seen in the resemblance between rhodopsin, the receptor for light, and the β_2-adrenergic receptor (43). There is observable homology in the amino acid sequence of these two proteins. Each sequence suggests that there are (seven) helical domains of a length that spans a membrane bilayer. Since rhodopsin and the β_2-adrenergic receptor may be so similar in three-dimensional properties, we expect that there will be even more resemblance between the three-dimensional structures of receptors that recognize more similar ligands.

The advantage of the modeling by homology methods is that one is able to predict a provisional three-dimensional structure from amino acid sequence without understanding the nuances of how and why proteins fold the way that they do. Such proposed structures may then be used to examine the reason for the observed structure-activity relationships of ligands and to propose the structure of novel ligands. By consideration of the structure of several receptors, one may design ligands that possess the required selectivity or broad spectrum of activity.

Examination of Protein-Ligand Complexes with Computer Graphics and Molecular Mechanics

Computer techniques are often used to refine the initial x-ray or NMR structure to increase the degree of fit between the observable and the calculated atomic positions. However, the real power of computer modeling is in providing a tool to probe the basis of structure-function relationships and to examine molecular dynamics. As noted earlier, potential energy calculations can be used to further optimize the protein structure; to stimulate motions of the protein, associated solvent, and ligand; and to estimate changes in the strength of binding or the structure of the complex that might arise from changes in the covalent structure of the ligand, the protein, or the solvent.

Although in principle one could examine the details of receptor function with these techniques, with current computers the simulation of protein dynamics typically covers a time period of no more than 10^{-9} sec range. On the other hand, the half-time for the conversion of the resting to the open ion channel state of the nicotinic acetylcholine receptor (the fastest event measured with this receptor) is in the range of 10^{-4} sec, and it is 10^{-2} for conversion of the fast-onset desensitized to the slow-onset desensitized form of this receptor (51). Even if these events require no change in the covalent structure of the receptor, molecular dynamics simulation of the pathway of the change in conformation between these states is not something that can be done today because of the enormous amount of time that would be required.

A special application of potential energy calculations and molecular graphics is in the design of novel ligands that affect protein function (7,44,45,46,52). An example is seen in Figure 8, which shows the basic residues in *Escherichia coli* dihydrofolate reductase that interact with the carbonyl groups of the inhibitor methotrexate. From a molecular graphics analysis it was decided to add a carboxylate to the inhibitor trimethoprim to form the additional salt bridge with the enzyme. Inhibitor potency increased 50-fold (44). To aid in such a task a special computer program has been written to identify the three-dimensional position of favorable binding sites on a protein for various types of ligand atoms (53).

Quantitative structure-activity relationship (QSAR) analysis is a useful tool for probing the molecular properties in a series of related ligands that result in enhanced binding to a protein. Molecular graphics analysis of systems for which there is both a protein structure and a QSAR reveal that the QSAR generally detected the important features of the binding site (54). Since many ligands are

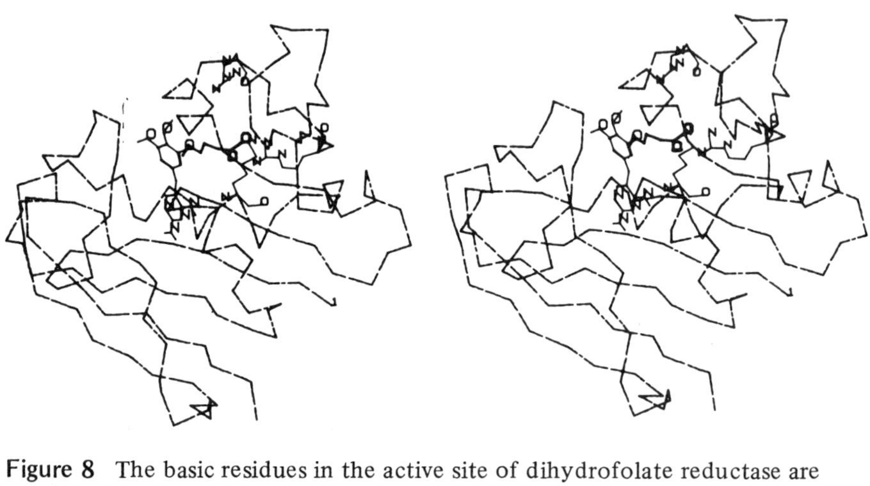

Figure 8 The basic residues in the active site of dihydrofolate reductase are shown as bound to an analog of trimethoprim designed from computer modeling (44). The added side chain is shown in heavy lines.

included in a QSAR, it may give an indication of the ability of the protein to accommodate ligands of different size and shape. The complementary nature of protein crystallography and QSAR suggests that these methods will be used in concert in drug design problems in the future.

COMPUTER MOLECULAR MODELING OF THE LIGAND BINDING SITE OF THE RECEPTOR FROM THE STRUCTURE-ACTIVITY RELATIONSHIPS OF LIGANDS

Objectives of Receptor Mapping

Receptor-binding or radioligand binding assays (Chaper 2) can be used to provide a wealth of structure-activity relationship (SAR) information. Medicinal chemists use this SAR information to design more potent or selective compounds, structural biochemists use it to probe the nature of the binding site for the small molecule on the receptor, and molecular pharmacologists and biophysists use it to investigate the details of receptor function and differences. The aims of all these groups are served at least partly in a family of techniques loosely grouped under the term "receptor mapping" (8,9,10,30,55). The goal of receptor mapping is to derive a three-dimensional model of the ligand binding site on the receptor from the SAR information on binding and response. Whatever computer tools are necessary are used. Such a model contains locations of primary recognition sites, an outline of the explored boundaries of the binding site, and regions in space that cannot be occupied by a ligand. Although such models will never contain the information of an x-ray crystallographic structure of such a receptor, they are powerful tools in organizing SAR data and suggesting new compounds for synthesis. The positions of the atoms in a protein determined by x-ray crystallography are located with an accuracy of about 10 times less than with small-molecule crystallography. Accordingly, theoretical molecular modeling of the ligand SAR is a complement to the experimental determination of receptor structure.

In order for one to expect such a receptor map to have some validity, it is important that one use the correct structure for each molecule used to build the model. Radioligand binding assays are ideal in this respect because the possibilities of metabolism are remote. However, since the receptor may have a different conformation when an agonist is bound than when an antagonist is bound, one must also have functional information to know whether the compound is an agonist or antagonist and be catious about mixing the two classes of compounds. Of course, partial agonists present a special problem for which no universal solution is yet known.

Since the whole methodology of radioligand binding assays is the displacement of one ligand by another, such experience prepares us to believe that two mutually displaceable ligands occupy the same site on a protein. The two ligands thus are expected to have some chemical and shape features in common. Thus the first step of receptor mapping is to use computer graphics and energy calculations to tease out of the structures of a series of active molecules those features essential for activity (8,10,30). These features form the pharmacophore. All active molecules are then superimposed in such a way and in such a conformation as to optimize the match between corresponding pharmacophore atoms in the different molecules. Finally, the surface that encloses all active molecules may be calculated as a proposed space available to be occupied by any proposed analog (10). Frequently one observes that inactive analogs protrude outside this volume (10).

Selection of the Pharmacophore Atoms

The selection of the pharmacophore atoms ranges from the obvious to the hopeless depending on the information available. First, one needs data from a well-designed structure-activity or molecular dissection analysis. For example, if deletion of a certain functional group destroys activity, one may hypothesize that this group is part of the pharmacophore. However, such a group may be required in a particular structure only because it influences the conformational or electronic properties of the groups that bind strongly to the receptor.

For example, in the case of D_{-2} agonists, the various hydroxylated 2-amino tetralins have provided definitive structure-activity information that led to the identification of the pharmacophore (56,57). The order of binding affinity is [6=8 > 7=9 > 10 > 11] ; [6-9] are full agonists (58). The corresponding methoxy analogs are not D_{-2} agonists, nor is the 8-hydroxy analog [11] or the desoxy analog [12] (57).

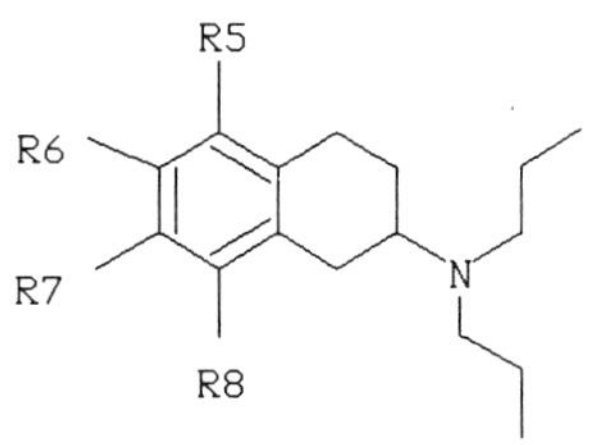

6 R5= R6= OH

7 R6= R7= OH

8 R5= OH

9 R7= OH

10 R6= OH

11 R8= OH

12 R5= R6= R7= R8= H

The D_{-2}-agonist pharmacophore is further refined with the observation that B-HT 920 [4] and many ergolines (for example, [13-17]) are also D_{-2} agonists (57).

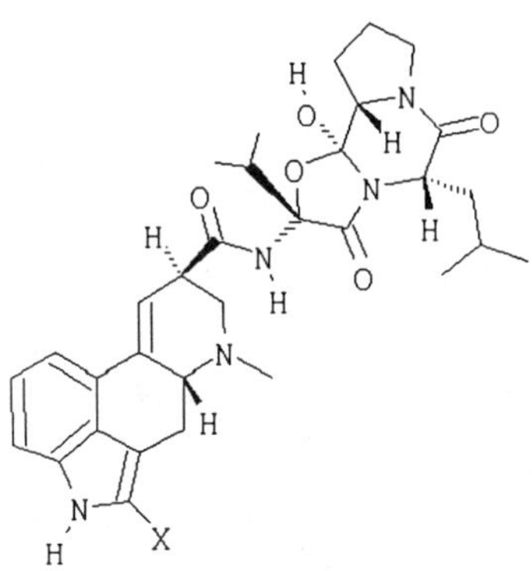

13 X= H

14 X= Br

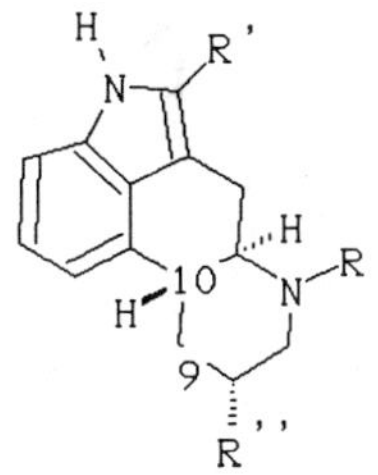

15 R= Me, R'= Cl, R''= CH_2CN

16 $\Delta^{9,10}$, R= Me, R'= R''= H

C8= —— NHC(=O)N(ET)$_2$

17 R= nPr, R'= R''= H, 9= O

Finally, a number of compounds have been synthesized to show that the distance of the basic nitrogen from the meta substituent must be between 6.5 and 7.3 Å (59,60). For example, [11] is also a meta analog, but the N to 0 distance is shorter than the active agonists. Thus the pharmacophore for the D_{-2}-agonist receptor is a basic nitrogen 6.5-7.3 Å away from an electronegative atom that is attached to a hydrogen atom.

The selection of the pharmacophore may be aided by a comparison of the electrostatic potential maps of different molecules. For example, Figure 9 shows the electrostatic potential of two noncatechol dopamine agonists. In comparison with Figure 4, all three have strong hydrogen bond donating potential toward the top of the figure. In most other respects, the maps of the three molecules are very different. It should be noted that our interpretation differs from that of others who emphasized the negative potential above the plane of the aromatic ring (37). The SAR analysis does not support the latter since the methoxy analogs have a similar potential above the plane of the ring. This latter interaction may contribute to binding, but it does not define the pharmacophore.

One may also propose a pharmacophore from the location of atoms that bind strongly to the analogs. For example, Figure 10 shows the regions in space at which an 0^- would bind to each of the three dopaminergic agonists. Superposition may be better done on these putative receptor atoms than on the atoms of the ligand (27). Ghose and Crippen carry this one step further and actually assign an energy value to each interaction (55).

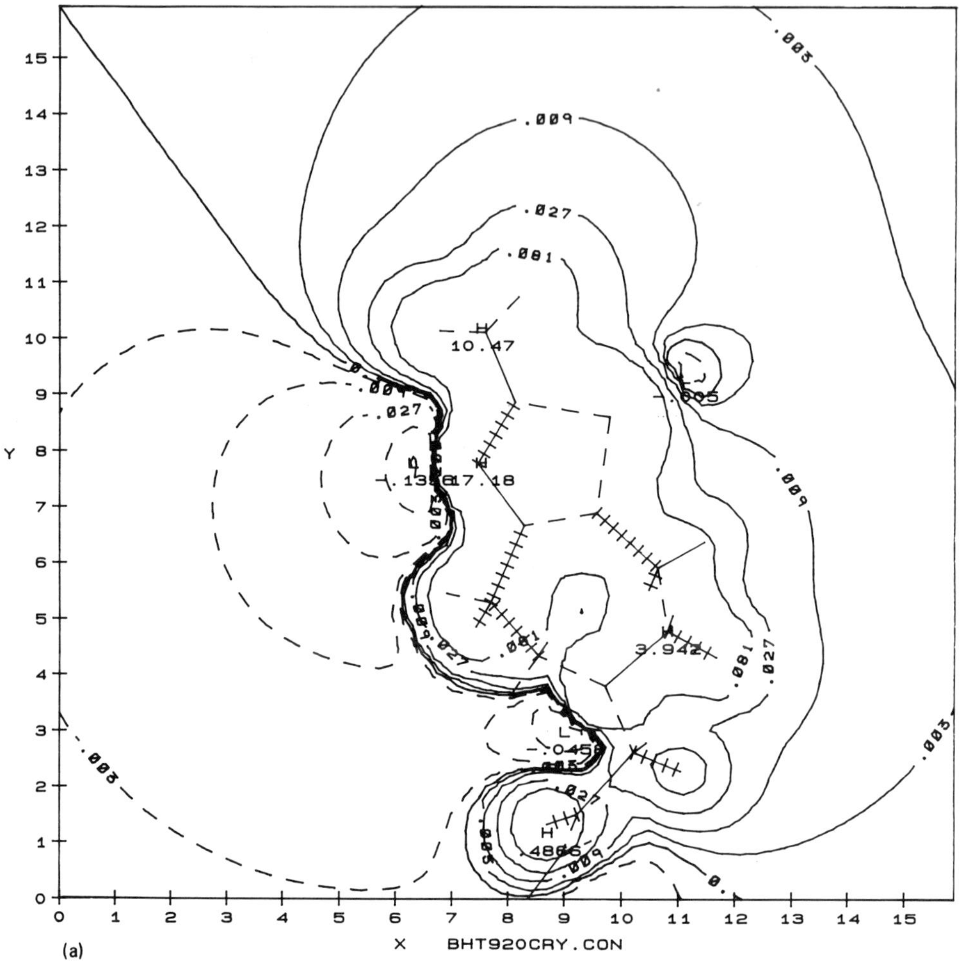

Figure 9 The electrostatic potential in the plane of the aromatic ring calculated from the STO-3G wavefunction for the x-ray structure of (a) B-HT 920 (61) [4] and (b) the MMP2-minimized structure of lergotrile [15]. (Y. C. Martin, unpublished observations.)

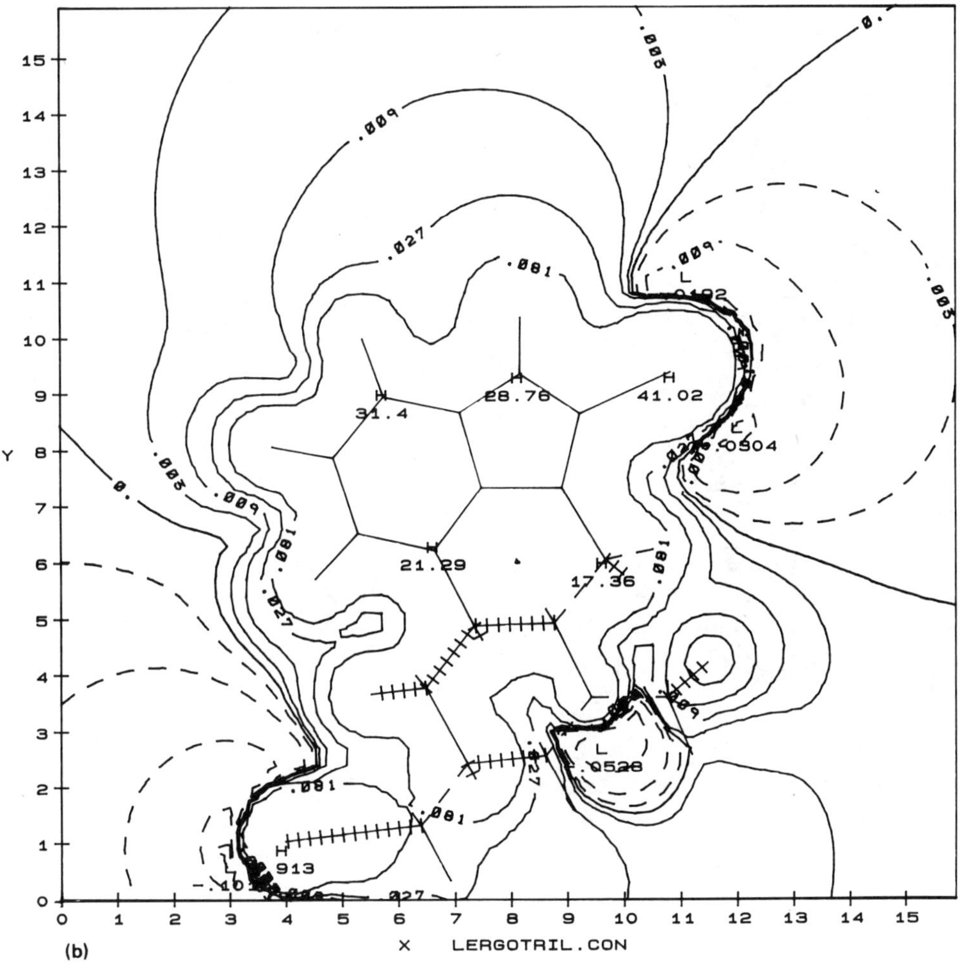

(b)

Figure 9 (Continued)

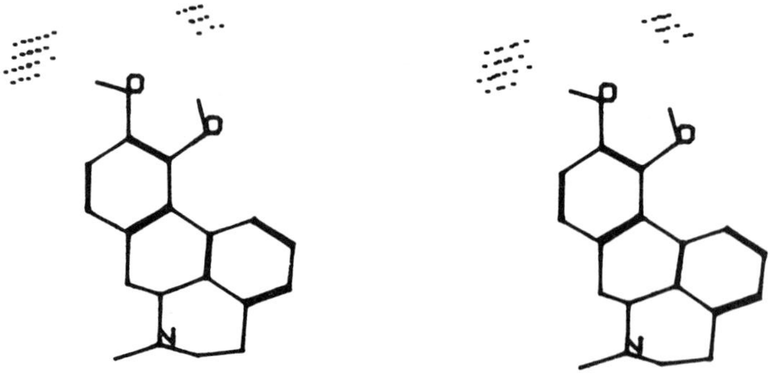
O
O
N
O
O
N

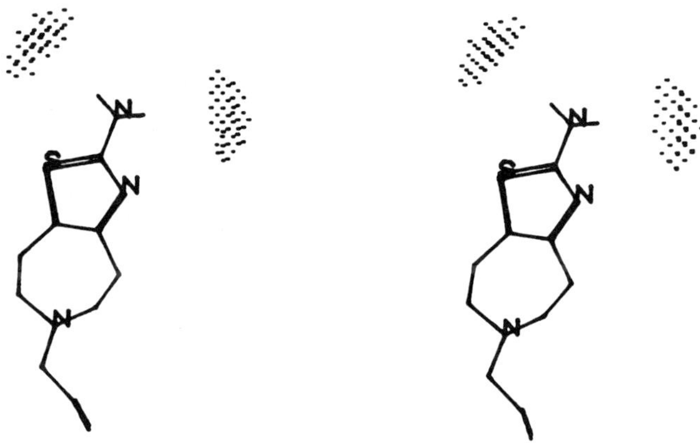
N
S
N
N
N
S
N
N

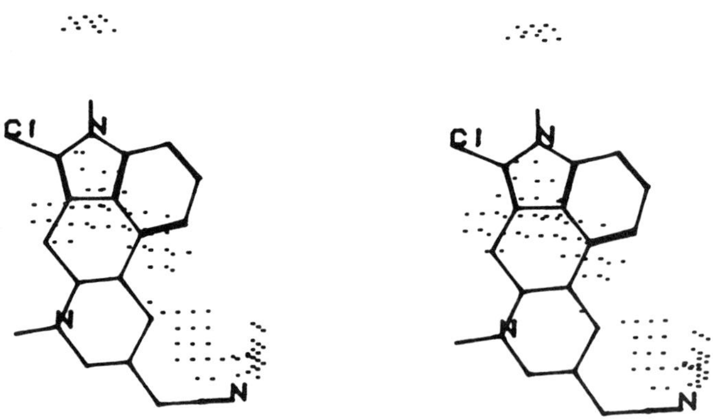
Cl
N
N
N
Cl
N
N
N

In its most useful role, receptor mapping is not a post facto analysis of SAR but an interactive process with synthesis of new probes. If the receptor mapping analysis suggests that there are ambiguities in the data, then new molecules may be designed and tested to further explore the definition of the pharmacophore. Such molecules themselves may show increased selectivity or potency.

The separation of ligand atoms into pharmacophore and nonpharmacophore atoms is artificial. For example, Figures 7 and 8 show that many atoms of the ligand participate in the ligand-receptor interaction. In the case of D_{-1} agonists, the pendant phenyl of SKF-38393 [5] increases binding affinity 100 times (62). Since [3] has measurable activity, the phenyl does not fit the pharmacophore definition. However, it clearly adds to potency and so one cannot forget about this feature in molecular comparisons.

Because the pharmacophore atoms are required for activity, they are used for the superposition of molecules (Figure 11). The active enantiomer of apomorphine and the ergoline are superimposed to take this into account (63). Disagreements as to how to superimpose molecules (64) are really disagreements about the features of the pharmacophore. Close attention to the known SAR may help settle such problems.

Selection of the Biologically Active Conformation of a Molecule

One must also decide what conformation and stereoisomer to use in the receptor map. A rigid compound of known stereochemistry provides the most definitive answer: the ultimate test of a receptor map based on flexible molecules is the design, synthesis, resolution, and testing of a more rigid compound.

Every degree of conformational freedom must be considered. For example, in apomorphine the –OH groups can rotate to reverse the direction of potential hydrogen bonds. However, only in the rotation shown would the hydrogen of the meta hydroxyl superimpose with the indole (N)H of the ergolines and one exocyclic (N)H of B-HT 920.

Since binding to the receptor is energetically favorable, a molecule need not bind in its low-energy conformation. The active conformation must be energetically possible, however. If the pharmacophore is well established and if one or more molecules of the set has a rather fixed distance between pharmacophore atoms, then this distance may be used as the target to which more flexible mole-

---

Figure 10 Stereoplots of the regions in space around apomorphine [2], B-HT 920 [4], and lergotrile [15] that are attractive to an O^- with an energy of –2.0 kcal/mol. These points were calculated with the potential energy program GRID (53). Point charges on the ligand atoms were calculated by a fit to the molecular electrostatic potential. (Y. C. Martin, unpublished observations.)

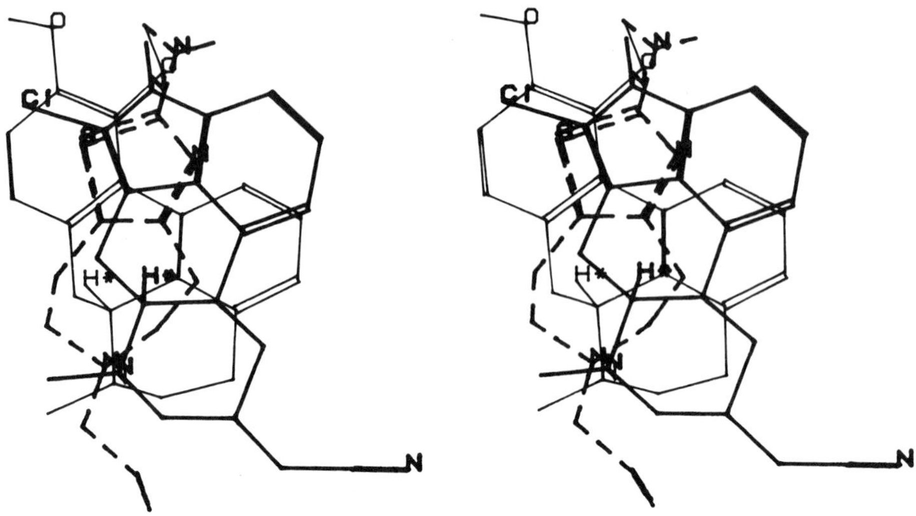

Figure 11 The structure of lergotrile [15] (heavy solid line) and of B-HT 920 [4] (heavy dashed line) is superimposed over apomorphine [2] (fine line). The lergotrile-apomorphine superposition is the second of the Camermans (64).

cules are matched to find the active conformation. Otherwise some sort of search must be used (10,55).

Candidate conformations are then superimposed over some template molecule. If inactive compounds that have the correct pharmacophore atoms are known, for example, if the "wrong" enantiomer of a ligand does not bind, then these compounds are also considered in the selection of the active conformation and enantiomer.

For example, let us consider again the D_{-2} agonists. The unnatural apomorphin enantiomer binds 800 times less strongly and it is not an agonist (57). The absolute stereochemistry of the active enantiomer of lergotrile is also known; it binds 10 times more strongly than its enantiomer (57). Accordingly, these features must be preserved in the comparisons (63) (Figure 11), not violated to gain more overlap of the total structure as in Figure 12 (64).

Now consider compounds [7] and [8]. Compound [8] is a substructure of apomorphine [2] and so it is straightforward to model and to select the active enantiomer. However, in order to superimpose the meta-oxygen and the nitrogen of [7] with the corresponding atoms of apomorphine, one must select the enantiomer with the opposite stereochemistry. Figure 13 illustrates this point. McDermed et al. recognized this, resolved and established the absolute stereochemistry of [7] and [8], and found the expected result (65). Any receptor

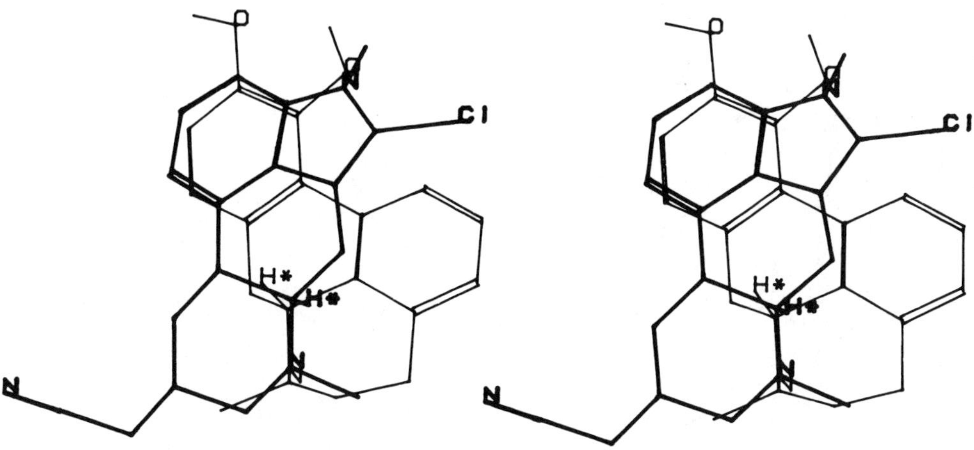

Figure 12 The superposition of lergotrile [15] (heavy solid line) and apomorphine [2] (fine solid line) that maximizes molecular overlap but violates stereochemical constraints (64).

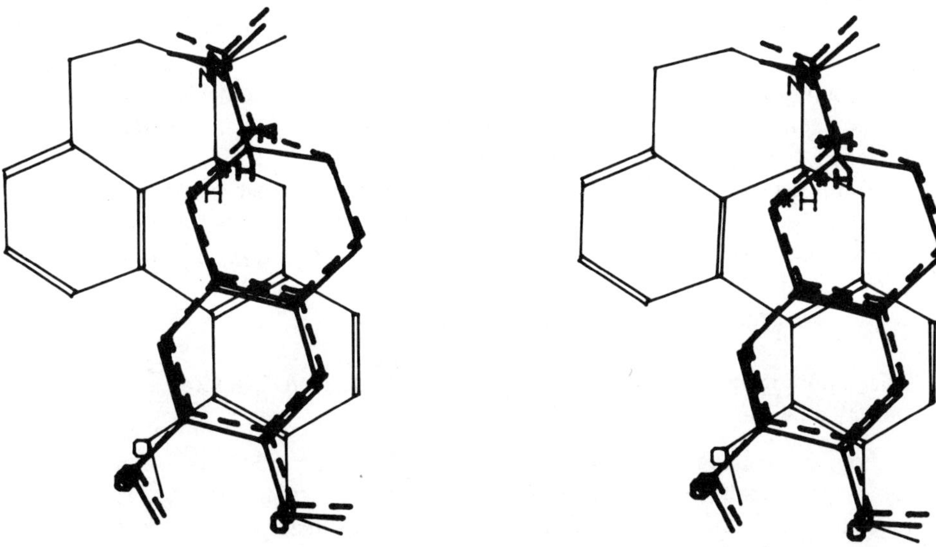

Figure 13 The active enantiomer of [7] (solid heavy line) superimposed over apomorphine [2] in a fine solid line. The inactive enantiomer of [7] is shown as a dashed heavy line.

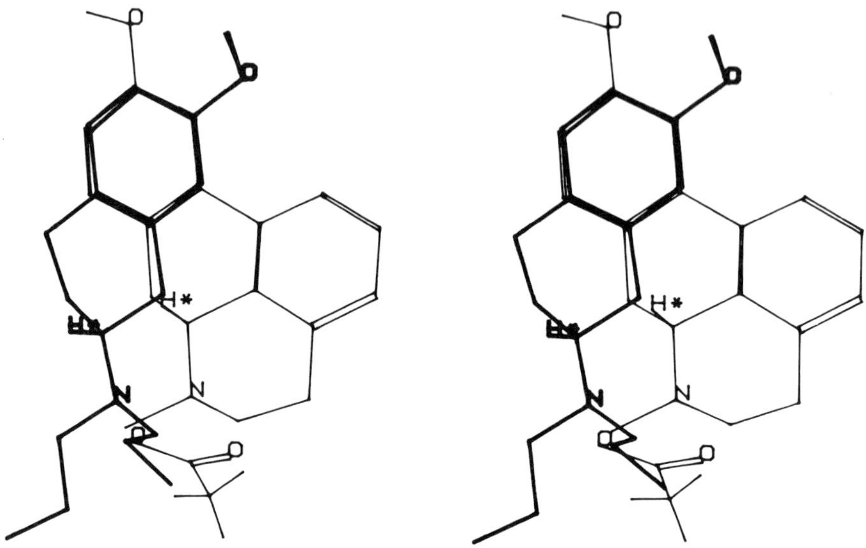

Figure 14 The superposition of [9] (heavy solid line) over apomorphine (light solid line) according to the recent proposal (66). The possible receptor carboxylate is shown at the bottom of the figure.

map is a hypothesis. Frequently alternative hypotheses can be made. For example, on the basis of the different effects on potency of N-propylation of [6] and [7] (58), recently it has been proposed that [8] and [9] should be superimposed over the phenol only (Figure 14) (66). Although now the nitrogen atoms do not superimpose, in their protonated forms their N-H are positioned to hydrogen bond to opposite oxygen atoms of a carboxylate group. With this superposition there is no requirement for the same chirality at the asymmetric centers of the two analogs.

Calculation of the Binding Site Boundaries and the Regions in Space Occupied by the Receptor

In receptor mapping, molecules that displace one another in radioligand binding assays are assumed to occupy the same site on the receptor protein. It is further assumed that any region in space occupied by an active molecule defines an open space in the ligand binding site (10). Thus the volume occupied by the union of all active molecules correctly superimposed defines the volume of the ligand binding site on the receptor. For example, the volume enclosed by the dots in Figure 15 shows the boundary of the D_{-2} receptor as it is currently

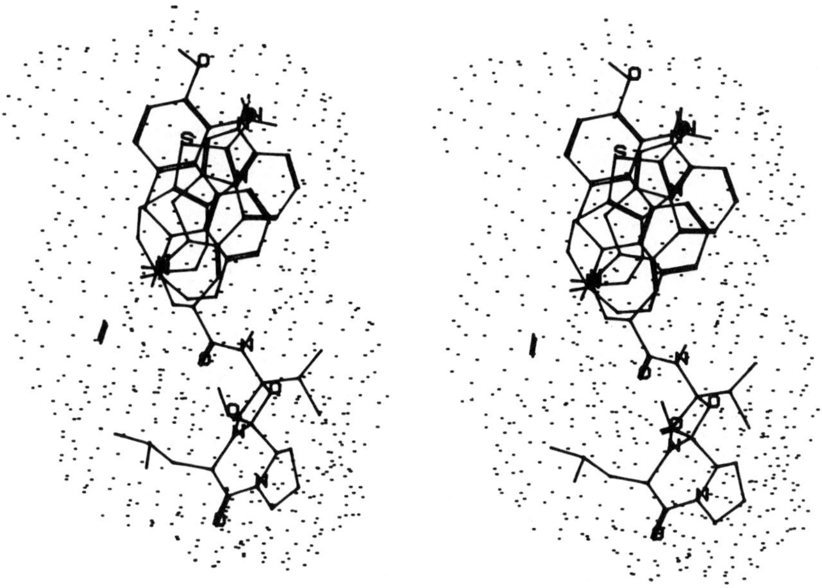

Figure 15 Representative compounds in the volume occupied by the 27 D_{-2} agonists (catechols and ergolines) that bind with a K_D^{high} of <100 nM (57). The compounds are B-HT 902 [4], ergocriptine [13], and *n*-propyl apomorphine [27].

viewed. The molecules included are those that bind with a K_D^{high} of <100 nM. They are superimposed as by others; see next section (57,60,65,67-69).

The three semirigid catecholamines [18-20] bind with a K_D of >15000 nM. The drastically reduced affinity of [20] compared with that of [7] is explained by proposing that the rightmost ring of [20] interferes with its proper orientation on the receptor (65,67). Figure 16 shows the new volume occupied by these three inactive catecholamines.

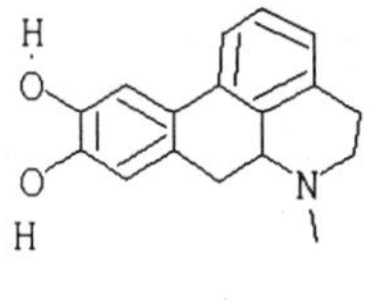

18

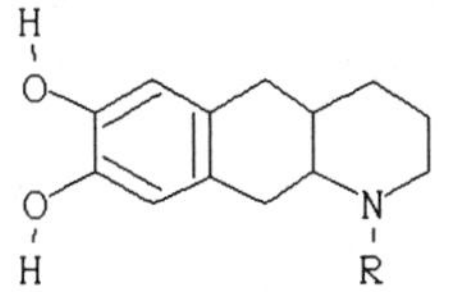

TRANS

19 R= Me

20 R= Pr

Such receptor boundaries and steric obstacles are hypotheses. They ignore the likelihood that the receptor changes conformation to maximize favorable

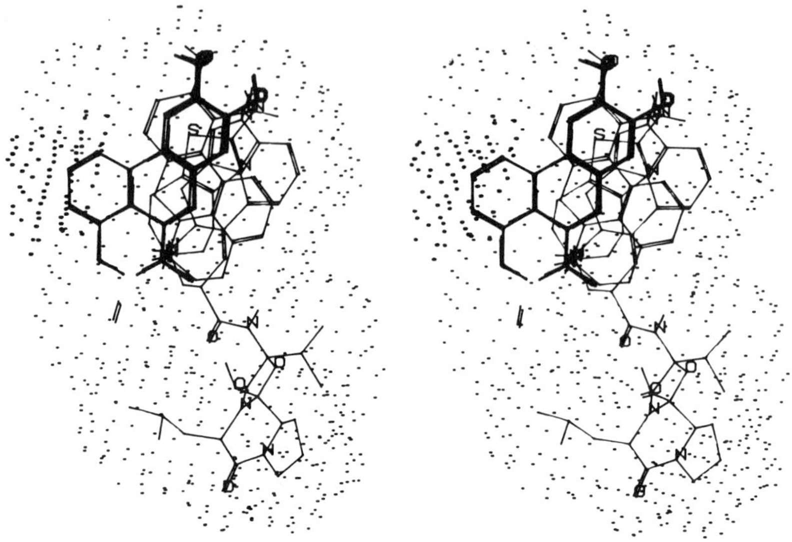

Figure 16 As Figure 15 (all fine lines) with the addition of the inactive compounds [19], [20], and [28] in solid heavy lines and the new volume occupied in heavy dots.

interactions with the ligand. Thus the concepts described here assume that a given change in ligand structure will always result in quantitatively the same change in binding affinity. If exceptions to such regularities occur, then more complex models are needed.

Detailed Example: The Agonist Binding Site of the Dopamine Receptor(s)

Many workers have contributed key observations to our knowledge of the structure of this binding site and proposed receptor models: only the most dramatic are noted here. As early as 1974, the importance of a certain N to 0 distance was realized (59,60,70), although this work was based on stimulation of adenylate cyclase. In 1977 the specific receptor binding of 6,7-ADTN, the analog of [7] in which $N(Pr)_2$ is replaced by NH_2, was reported (71). In the sâme year conformational calculations on dopamine and molecular models led to the postulate of the steric obstacle on the dopamine receptor (67). McDermed et al.

followed this up by the resolution of [7] and [8] and a more elaborate model (65).

In 1978, the N-0 distance requirement was further validated in that trans and not cis [21] displaces [^{3}H] dopamine from synaptosomes (72). Resolved

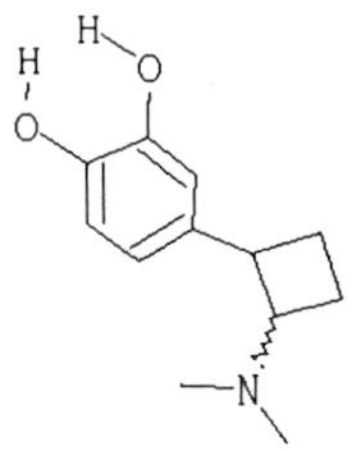

21

[8] and racemic [9] and [10] were tested in a binding assay in 1979 (56). Work on catechol derivatives of octahydrobenzo-[f]-quinolines in 1979 further refined the N-0 distance requirements and binding site shape albeit with a cat rather than receptor assay (59). This work was extended to the resolved phenols by others (68) and in 1986 elaborated into a molecular mechanics-based computer model (69) indistinguishable from our own on catechols and phenols. Between 1982 and 1984 three groups reported additional analogs that are inactive because of a steric obstacle; compounds [22-26] (73-75). Mechanical models were superimposed to derive a receptor model for D_{-1} agonists based on [28] (76).

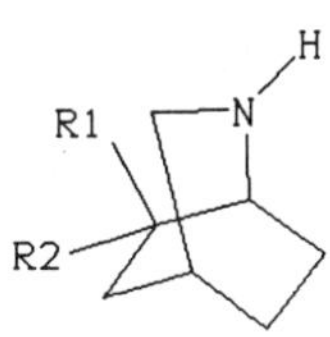

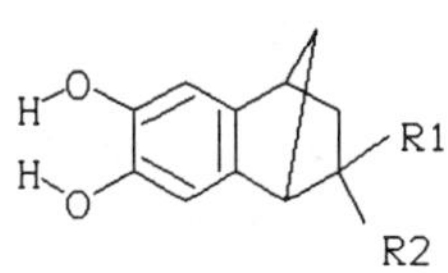

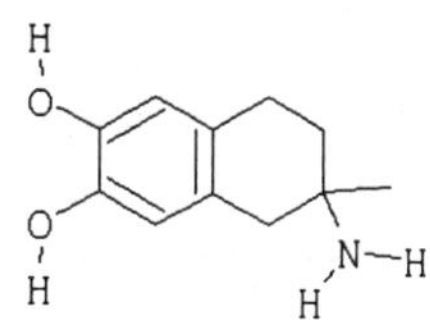

22 R1= Catechol, R2= H

23 R1= H, R2= Catechol

24 R1= NH2, R2= H

25 R1= H, R2= NH2

26

Do such models suggest novel compounds for synthesis? Yes, receptor models of the resemblance between apromorphine and ergolines led to the synthesis of several noncatecholic dopamine agonists [29-34] (77-79). Partial ergoline structures that lack the pyrrole were shown to be inactive (80).

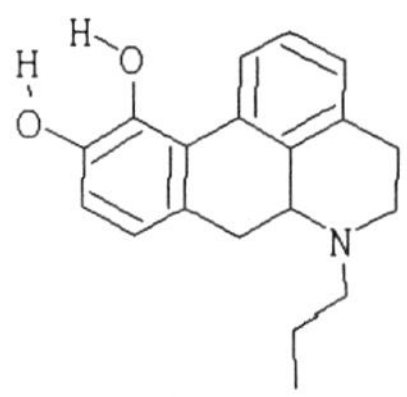

27

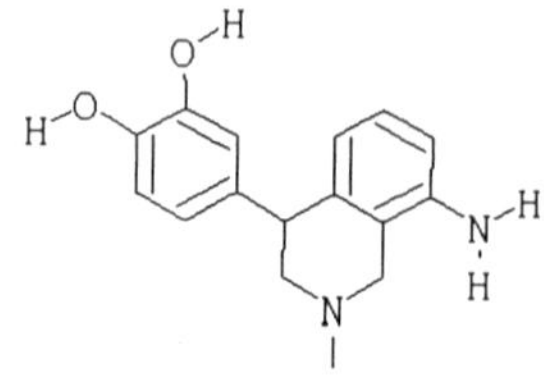

28

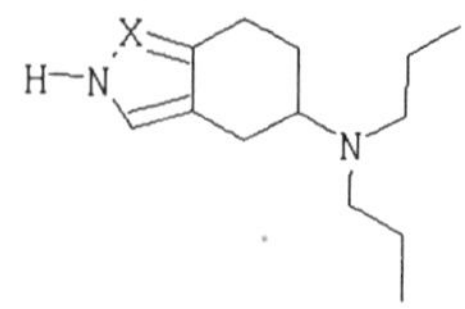

29 X= CH

30 X= N

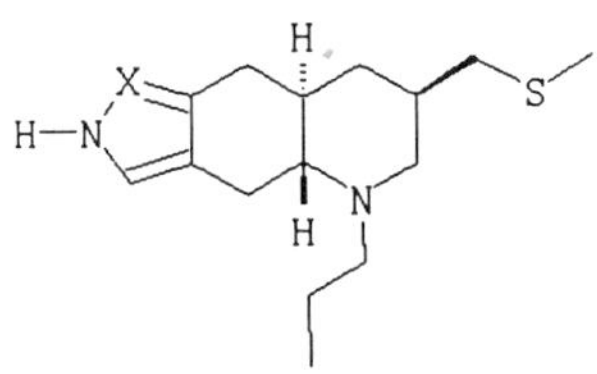

31 X= CH

32 X= N

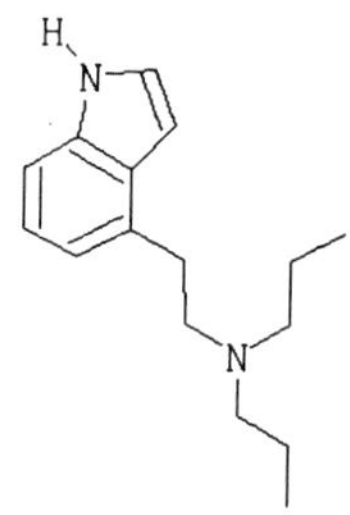

33

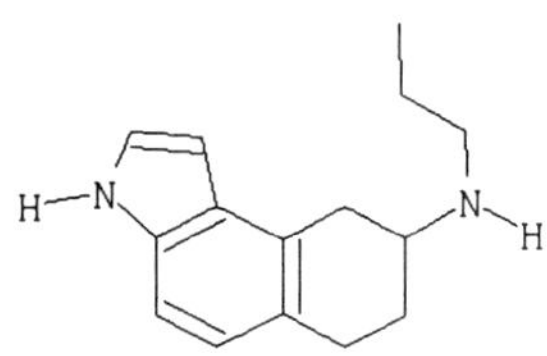

34

The current receptor maps for dopamine agonists are not the final word. In particular, the superposition suggested by Grol has not been evaluated for consistency with all SAR. There are no publications on the possible differences between the agonist binding sites on D_{-1} and D_{-2} receptors, although two groups derived models from renal blood flow measurements on semirigid molecules (81,82). Finally, several groups have addressed the enantioselectivity of apomorphine, but no definitive answer is known (79,83,84).

Receptor Maps of Dopamine Antagonist Sites

Very early in dopamine receptor work it was realized that neuroleptics are dopamine antagonists. As early as 1975, receptor models for phenothiazine drugs were derived from potential energy calculations (85) and x-ray crystal structures (86). In the same year the first of a series of investigations of the relationship between the structure of apomorphine and the novel antipsychotic butaclamol [35] was reported based on crystallographic measurement and potential energy

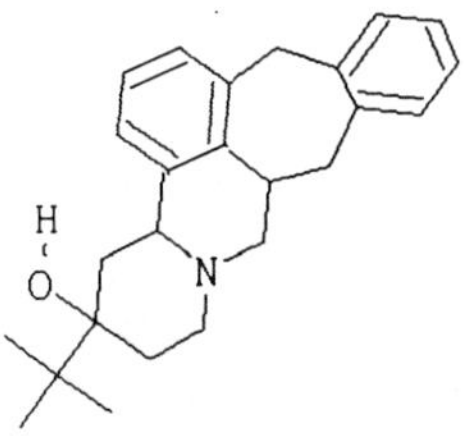

35

calculations (87,88). This was extended to butyrophenones (89) in a study that discussed the advantages of rigid analogs. It was implied that conformational considerations led to the synthesis of a novel series of γ-carboline neuroleptics (90). From an analysis of intermolecular interactions of known neuroleptics in crystals, plus more elaborate model building, Olson et al. proposed a structure of the neuroleptic ligand binding site (91). This led to the synthesis of a novel class of antipsychotic agent.

Extension of Receptor Maps to Multiple Receptors: Comparison of D_{-2} and 5-HT_{1A} Binding Sites

Many molecules interact with more than one receptor. Thus the ergoline LSD [36] is not only a dopaminergic but also a serotonergic [37] agonist (63,92).

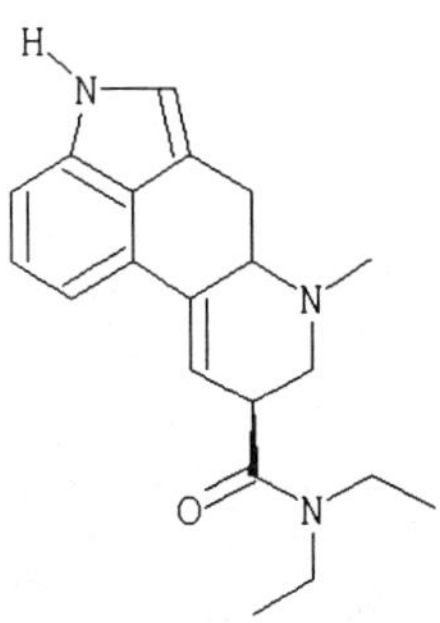

36

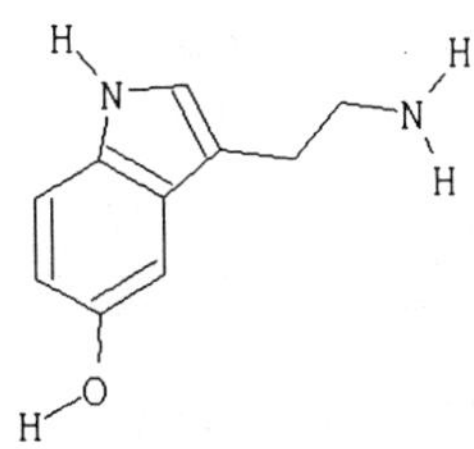

37

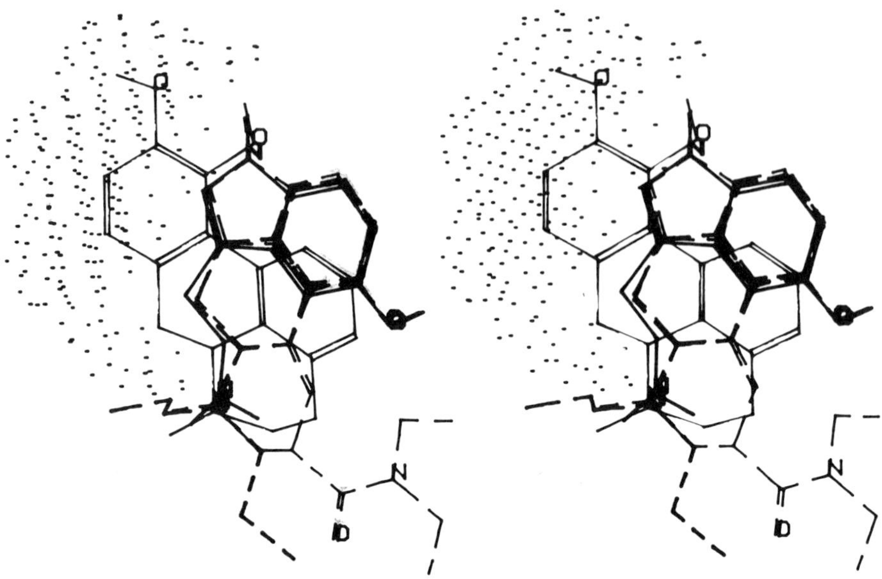

Figure 17 The $5HT_{1A}$ agonists [8], [36], and [37] in heavy lines; the prototype D_{-2} agonist apomorphine [2] in fine lines; and the extra volume it requires in fine dots.

However, *N,N*-dipropyl-8-hydroxy-2-aminotetralin [11] is only serotonergic, not dopaminergic (93), whereas the opposite is true of apomorphine. Molecular superpositions have been proposed (93,94). Figure 17 shows a comparison of the volumes available to D_{-2} agonists compared with that available to $5\text{-}HT_{1A}$ agonists. The implication of such a comparison is that from it one should be able to design the desired specificity into another molecule.

ADVANTAGES OF MODERN COMPUTER TOOLS IN DRUG RESEARCH

The objective of structural biochemistry (and of medicinal chemistry) is an understanding of the three-dimensional structure (of a drug) and a macro-molecular target. Computer graphics offers the advantage of data integrity, ease of visualization and comparison, availability of color-coding chemical properties, and an easy tie-in to energetics and dynamics calculations as well as crystallographic or spectroscopic measurements. The more complex the comparison of molecules to be viewed, the more advantages are seen in the computer graphics display over mechanical models. The tie to computational programs is essential for molecules that are conformationally flexible. For the design of any but the most obvious analogs one needs a graphics display of all known active compounds, the volume tolerated, and the regions forbidden to be occupied.

These tools have associated costs: money for equipment and software and time to construct the models. In addition, the available tools have not been perfected and some of those needed do not exist. For example, the various molecular mechanics force fields lack parameters for many functional groups; it is not clear how to treat solvent; and calculating free energy changes is extremely time consuming. We still need tools for the design of a ligand from a complete or model structure of a binding site. Because of the lack of understanding of the relationship between the functional and structural consequences of binding a ligand to a protein, the design of a drug from the structure of its target biomolecule is not necessarily straightforward. However, the concepts and approaches described in this chapter promise to offer new insight into ligand-receptor interaction and its consequences and to be a substantial aid to new drug design.

ACKNOWLEDGMENTS

All figures shown in this chapter were prepared with the Abbott Laboratories molecular graphics software system INTERACT/CMD,PDS/GRAMPS (95). Principal contributors to this system are J. Greer, S. Harrod, C. Haseman, T. Koschmann, K. Mitchell, T. J. O'Donnell, F. Thomas, J. Wakat, and H. J. R. Weinbraub. The figures were drawn on a HP7550A Graphics Plotter from images seen on Evans and Sutherland MPS devices or on office graphics terminals through the intermediacy of TEMPLATE software. The installation and integration of the ab initio quantum chemistry software into our modeling system was the work of C. Chabalowski and J. Schubert. L. Sanathanan devised the algorithms for surface comparisons and for locating the local maxima and minima in molecular electrostatic potential surfaces.

REFERENCES

1. F. H. Allen, O. Kennard, and R. Taylor, Acc. Chem. Res. *16*, 146 (1983).
2. Brookhaven Laboratories, Data Base of Protein Structures, New York, updated twice a year.
3. C. R. Cantor and P. R. Schimmel, Biophysical Chemistry, D. H. Freeman, San Francisco, 1980.
4. F. Richards, R. Baldwin, G. R. Galluppi, R. Griffin, E. T. Kaiser, B. Matthews, J. A. McCammon, A. Redfield, B. Reid, R. Sauer, A. Schecter, P. Sigler, P. von Hippel, and D. Wiley, Research Briefings 1986, Vol. 38, National Academy Press, Washington, D.C., 1986.
5. U. Burkert and N. L. Allinger, Molecular Mechanics, American Chemical Society, Washington, D.C., 1982.

6. R. Feldmann, in Computer Applications in Chemistry (S. R. Heller and R. Potenzone, Jr., eds.), Vol. 9. Elsevier, Amsterdam, 1983.
7. P. J. Goodford, J. Med. Chem. *27*, 557 (1984).
8. P. Gund, J. D. Andose, J. B. Rhodes, and G. M. Smith, Science *208*, 1425 (1980).
9. A. J. Hopfinger, J. Med. Chem. *28*, 1133 (1985).
10. C. Humblet and G. R. Marshall, Drug. Dev. Res. *1*, 409 (1981).
11. P. Kollman, Acc. Chem. Res. *18*, 105 (1985).
12. R. Langridge, T. E. Ferrin, I. D. Kuntz, and M. L. Connolly, Science *211*, 661 (1981).
13. T. J. O'Donnell and A. J. Olson, Computer Graphics *15*,133 (1981).
14. J. Giesecke, Acta Crystallogr. *33*, 302 (1977).
15. H. A. Scheraga, Ann. N.Y. Acad. Sci. *439*, 170 (1985).
16. P. Dauber, D. J. Osguthorpe, and A. T. Hagler, J. Am. Chem. Soc. *10*(5), 312 (1982).
17. B. R. Brooks, R. E. Bruccoleri, B. D. Olafson, D. J. States, S. Swaminathan, and M. Karplus, J. Comput. Chem. *4*, 187 (1983).
18. M. Levitt, Annu. Rev. Biophys. Bioeng. *11*, 251 (1982).
19. B. L. Tembe and A. McCammon, Computers Chem. *8*, 281 (1984).
20. N. L. Allinger, MM2 Reference Guide, Molecular Design, San Leandro, Calif., 1987.
21. A. Warshel and R. M. Weiss, Ann. N.Y. Acad. Sci. *367*, 370 (1981).
22. W. K. Olson, J. Am. Chem. Soc. *104*, 278 (1982).
23. P. K. Weiner and P. A. Kollman, J. Comp. Chem. *2*, 287 (1981).
24. W. G. Richards, Quantum Pharmacology, Butterworth, Woburn, Mass., 1977.
25. H. Weinstein and J. P. Green (eds.), Ann. N.Y. Acad. Sci. *367* (1981).
26. M. S. Dewar and W. Thiel, J. Am. Chem. Soc. *99*, 4899 (1977).
27. P. R. Andrews and D. A. Winkler, in Drug Design: Fact or Fantasy? (G. Jolles and K. R. H. Wooldridge, eds.), Academic Press, London, 1984.
28. D. B. Boyd, Ann. N.Y. Acad. Sci. *367*(6), 364 (1981).
29. J. J. Kaufman, P. C. Hariharan, H. E. Popkie, and C. Petrongolo, Ann. N.Y. Acad. Sci. *367*, 452 (1981).
30. L. B. Kier, Molecular Orbital Theory in Drug Research, Academic Press, New York, 1971.
31. G. H. Loew, Z. S. Herman, M. M. Rohmer, A. Goldblum, and A. Pudzianowski, Ann. N.Y.Acad. Sci. *367*, 192 (1981).
32. M. Martin, F. Sanz, M. Campillo, L. Pardo, J. Perez, J. Turmo, and J. M. Aullo, Int. J. Quant. Chem. *23*, 1643 (1983).
33. R. Osman, H. Weinstein, and S. Topiol, Ann. N.Y. Acad. Sci. *367*, 356 (1981).
34. P. Politzer and D. G. Truhlar (eds.), in Chemical Applications of Atomic Molecular Electrostatic Potential, Plenum Press, New York, 1981.
35. A. Pullman, Ann. N.Y. Acad. Sci. *367*, 340 (1981).
36. H. van de Waterbeemd, P. Carrupt, and B. Testa, J. Med. Chem. *29*, 600 (1986).

37. D. Kocjan, M. Hodoscek, and D. Hadzi, J. Med. Chem. *29*, 1418 (1986).
38. H. Weinstein, R. Osman, S. Topiol, and J. Green, Ann. N.Y. Acad. Sci. *367*, 434 (1981).
39. W. J. Hehre and J. A. Pople, J. Am. Chem. Soc. *92*, 2191 (1970).
40. M. F. Perutz, Science *201*, 1187 (1978).
41. M. Mishina, T. Kurosaki, T. Tobimatsu, Y. Morimoro, M. Noda, T. Yamamoto, M. Terao, J. Lindstrom, T. Takahashi, M. Kuno, and S. Numa, Nature *307*, 604 (1984).
42. S. Green, P. Walter, V. Kumar, A. Krust, J. M. Bornert, P. Argos, and P. Chambon, Nature *320*, 134 (1986).
43. R. A. Dixon, B. K. Kobilka, D. J. Strader, J. L. Benovic, H. G. Dohlman, T. Frielle, M. A. Bolanowski, C. D. Bennett, E. Rands, R. E. Diehl, R. A. Mumford, E. E. Slater, I. S. Sigal, M. G. Caron, R. J. Lefkowitz, and C. D Strader, Nature *321*, 75 (1986).
44. L. F. Kuyper, B. Roth, D. P. Baccanari, R. Ferone, C. Beddell, J. Champness, D. Stammers, J. Dann, F. Norrington, D. Baker, and P. Goodford, J. Med. Chem. *28*, 303 (1985).
45. C. R. Beddell, Chem. Soc. Rev. *13*, 279 (1984).
46. W. G. Hol, Angew. Chem. Int. Ed. Eng. *25*, 767 (1986).
47. K. Wuthrich, M. Billeter, and W. Braun, J. Mol. Biol. *180*, 715 (1984).
48. R. J. Feldmann, D. H. Bind, M. Potter, C. Mainhart, B. Furie, B. C. Furie, and L. H. Caporale, Ann. N.Y. Acad. Sci. *439*, 12 (1985).
49. J. Greer, Ann. N.Y. Acad. Sci. *439*, 44 (1985).
50. J. Greer, J. Mol. Biol. *152*, 1027 (1981).
51. A. Karlin, P. N. Kao, and M. DiPaola, TIPS 304 (1986).
52. J. M. Blaney, P. K. Weiner, A. Dearing, P. A. Kollman, E. C. Jorgensen, S. J. Oatley, J. M. Burridge, and C. C. F. Blake, J. Am. Chem. Soc. *104*, 6424 (1982).
53. P. J. Goodford, J. Med. Chem. *28*, 849 (1985).
54. C. Hansch and J. M. Blaney, in Drug Design: Fact or Fantasy (G. Jolles and K. R. H. Woolridge, eds.), Academic Press, London, 1984, p. 185.
55. A. K. Ghose and G. M. Crippen, J. Comput. Chem. *6*, 350 (1985).
56. J. L. Tedesco, P. Seeman, and J. D. McDermed, Mol. Pharmacol. *16*, 369 (1979).
57. P. Seeman, M. Watanabe, D. Grigoriadis, J. L. Tedesco, S. R. George, U. Svensson, J. L. G. Nilsson, and J. L. Neumeyer, Mol. Pharmacol. *28*, 391 (1985).
58. M. P. Seiler and R. Markstein, Mol. Pharmacol. *26*, 452 (1984).
59. J. G. Cannon, Prog. Drug Res. *29*, 303 (1985).
60. C. Kaiser and T. Jain, in Highlights in Receptor Chemistry (C. Melchiorre and M. Giannella, eds.), Elsevier, Amsterdam, 1984, p. 29.
61. P. Luger, G. Griss, R. Hurnaus, and G. Trummlitz, Acta Crystallogr. *42*, 359 (1986).
62. S. T. Ross, R. G. Franz, J. W. Wilson, M. Brenner, R. M. DeMarinis, J. P. Hieble, and H. M. Sarau, J. Med. Chem. *29*, 733 (1986).

63. D. E. Nichols, J. Theor. Biol. *59*, 167 (1976).
64. N. Camerman and A. Camerman, Mol. Pharmacol. *19*, 517 (1981).
65. J. D. McDermed, H. S. Freeman, and R. M. Ferris, in Catecholamines: Basic and Clinical Frontiers (E. Usdin, I. J. Kopin, J. D. Barchas, eds.), Pergamon Press, New York, 1978, p. 568.
66. C. J. Grol, L. J. Jansen, and H. Rollema, J. Med. Chem. *28*, 679 (1985).
67. C. J. Grol and H. Rollema, J. Pharm. Pharmacol. *29*, 153 (1977).
68. H. Wikstrom, B. Andersson, D. Sanchez, P. Lindgerg, L. Arvidsson, A. M. Johansson, J. L. G. Nilsson, K. Svensson, S. Hjorth, and A. Carlsson, J. Med. Chem. *28*, 215 (1985).
69. T. Liljefors and H. Wikstrom, J. Med. Chem. *29*, 1896 (1986).
70. H. Sheppard and C. R. Burghardt, Mol. Pharmacol. *10*, 721 (1974).
71. P. J. Roberts, G. N. Woodruff, and J. A. Poat, Mol. Pharmacol. *13*, 541 (1977).
72. H. L. Komiskey, J. F. Bossart, D. D. Miller, and P. N. Patil, Proc. Natl. Acad. Sci. USA *75*, 2641 (1978).
73. S. Law, J. M. Morgan, L. W. Masten, R. F. Borne, G. W. Arana, N. S. Kula, and R. J. Baldessarini, J. Med. Chem. *25*, 213 (1982).
74. D. E. Nichols, J. N. Jacob, A. J. Hoffman, J. K. Kohli, and D. Glock, J. Med. Chem. *27*, 1701 (1984).
75. D. I. Schuster, H. E. Katerinopoulos, W. L. Holden, A. P. S. Narula, R. B. Libes, and R. B. Murphy, J. Med. Chem. *25*, 850 (1982).
76. C. Kaiser, in Dopamine Receptors (C. Kaiser and J. W. Kebabian, eds.), Vol. 224, American Chemical Society, Washington, D.C., 1983, p. 223.
77. N. J. Bach, E. C. Kornfeld, N. D. Jones, M. O. Chaney, D. E. Dorman, J. W. Paschal, J. A. Clemens, and E. B. Smalstig, J. Med. Chem. *23*, 492 (1980).
78. J. G. Cannon, B. J. Demopoulos, J. P. Long, J. R. Flynn, and F. M. Sharabi, J. Med. Chem. *24*, 238 (1981).
79. A. A. Asselin, L. G. Humber, K. Voith, and G. Metcalf, J. Med. Chem. *29*, 648 (1986).
80. N. J. Bach, E. C. Kornfeld, J. A. Clemens, and E. B. Smalstig, J. Med. Chem. *23*, 812 (1980).
81. P. W. Erhardt, J. Pharm. Sci. *69*, 1059 (1980).
82. P. H. Volkman, J. D. Kohli, L. I. Goldberg, J. G. Cannon, and T. Lee, Proc. Natl. Acad. Sci. USA *74*, 3602 (1977).
83. M. Froimowitz, J. L. Neumeyer, and R. J. Baldessarini, J. Med. Chem. *29*, 1570 (1986).
84. J. L. Neumeyer, D. Reischig, G. W. Arana, A. Campbell, R. J. Baldessarini, N. S. Kula, and K. J. Watling, J. Med. Chem. *26*, 516 (1983).
85. P. Feinerg and S. H. Snyder, Proc. Natl. Acad. Sci. USA *72*, 1899 (1975).

86. A. S. Horn, M. L. Post, and O. Kennard, J. Pharm. Pharmacol. *27*, 553 (1975).
87. L. G. Humber, F. T. Bruderlein, and K. Voith, Mol. Pharmacol. *11*, 833 (1975).
88. L. G. Humber, A. H. Philipp, F. T. Bruderlein, M. Gotz, and K. Voith, in Computer-Assisted Drug Design (E. C. Olson and R. E. Christoffersen, eds.), American Chemical Society, Washington, D.C., 1979, p. 227.
89. H. Moereels and J. P. Tollenaere, Life Sci. *23*, 459 (1978).
90. C. A. Harbert, J. J. Plattner, W. M. Welch, A. Weissman, and B. K. Koe, Mol. Pharmacol. *17*, 38 (1980).
91. G. L. Olson, H. Cheung, K. D. Morgan, J. F. Blount, L. Todaro, L. Berger, A. B. Davidson, and E. Boff, J. Med. Chem. *24*, 1026 (1981).
92. R. A. Glennon, J. Med. Chem. *30*, 1 (1987).
93. L. Arvidsson, U. Hacksell, A. M. Johansson, J. L. G. Nilsson, P. Lindberg, D. Sanchez, H. Wikstrom, K. Svensson, S. Hjorth, and A. Carlsson, J. Med. Chem. *29*, 45 (1984).
94. D. E. Nichols, W. R. Pfister, and G. K. Kim, Life Sci. *22*, 2165 (1978).
95. Y. C. Martin, K. H. Kim, T. Koschmann, and T. J. O'Donnell, in Computer Applications in Chemistry (S. R. Heller and R. Potenzone, eds.), Elsevier, Amsterdam, 1983, p. 285.

7

α Adrenoceptors

Pieter B. M. W. M. Timmermans / E. I. du Pont de Nemours & Company, Inc., Wilmington, Delaware

HISTORICAL SURVEY

Although it may seem that the high interest in α adrenoceptors and in the various chemical structures interacting with this class of adrenergic receptors is fairly recent, some of these drugs were already identified at the beginning of this century. In 1906, ergotamine and its congeners were reported by Dale (1) to block stimulator but not inhibitory effects of epinephrine. In 1925, the alkaloid yohimbine was found to cause "epinephrine reversal" (2). Phentolamine and phenoxybenzamine, two well-known α-adrenoceptor blockers, were synthesized in the 1940s.

Renewed interest in the field of α adrenoceptors has occurred especially during the last decade. Numerous explanations can be given for this renaissance, for example, the observation that centrally located α adrenoceptors are the main targets for the clinically important antihypertensive drugs clonidine and α-methyl-DOPA (see below), the distinction between pre- and postsynaptic (pre- and postjunctional) α adrenoceptors, as well as the general acceptance of the most workable subdivision of α adrenoceptors into two prototypes, α_1 and α_2 adrenoceptors. The rediscovery of α adrenoceptors as attractive areas of research has been stimulated further by the introduction of new, more selective compounds, the detection and characterization of novel locations and functions of α adrenoceptors, the availability of "easy" techniques, like radioligand binding, and the increased understanding of the processes linking α-adrenoceptor activation to the pharmacological effect. In pharmacological history it has invariably been observed that a better knowledge of receptors, their subtypes, locations, and functions and the development of selective agonists and antagonists have resulted in the introduction of clinically relevant and interesting novel compounds. For this reason it can be expected that new drugs based on the interaction with various subtypes of the α adrenoceptors will become available in the forthcoming years.

CLASSIFICATION

The alpha (α) beta (β) classification of adrenergic receptors was proposed by Ahlquist (3) to explain the actions of six sympathomimetic amines. In 1958, the dual-receptor hypothesis was put on firm ground by Powell and Slater (4) and by Moran and Perkins (5) when they showed that β-adrenoceptor stimulatory actions of catecholamines could be antagonized by dichloroisoproterenol. Previously, only α-adrenoceptor responses had been susceptible to blockade. The subdivision of β-adrenoceptors into β_1 and β_2 types (see Chapter 5), originally demonstrated by Furchgott (6) and Lands et al. (7), is generally accepted and is also therapeutically relevant. The need for an analogous subdivision of α adrenoceptors was first felt after the discovery and establishment of α adrenoceptors present at the varicosity membrane of noradrenergic nerve endings. Until then, the classic view considered only the receptors located at the membrane of the target organ, the so-called postsynaptic or postjunctional receptors, which upon stimulation induce the pharmacological effect. Several lines of experimental evidence have demonstrated that presynaptic α adrenoceptors take part in a negative feedback mechanism of transmitter (e.g., norepinephrine) release (Figure 1). Activation by appropriate exogenous or endogenous (autoinhibition) agonists causes an inhibition, whereas blockade induces a facilitation of the release. Transmitter norepinephrine released by nerve stimulation not only controls further release but also diffuses through the synaptic cleft and can stimulate the postsynaptic α adrenoceptors (e.g., vasoconstriction) (Figure 1). The concept of presynaptic control of neurotransmitter release through α adrenoceptors (and other receptors) has been the subject of a number of reviews (8-13) and need not be treated in too much more detail here.

Ample experimental findings support the concept that the prejunctional α adrenoceptors involved in the regulation of norepinephrine release are distinguished from those situated postjunctionally, mediating smooth muscle constriction with respect to the relative activities of agonists and antagonists (see reviews 12-15 and references quoted there). The most elaborate differentiation has been summarized in Figure 2 (15). The higher the column, the greater is the selectivity of the particular antagonist (left part) or agonist (right part) to block or stimulate the postsynaptic α adrenoceptors. The presynaptic-postsynaptic ratios vary about 500-fold for agonists, and the affinity ratios differ more than 1000-fold for antagonists Prazosin (Pra) has been shown to be a very selective antagonist of postjunctional α adrenoceptors, whereas yohimbine (Yoh) and

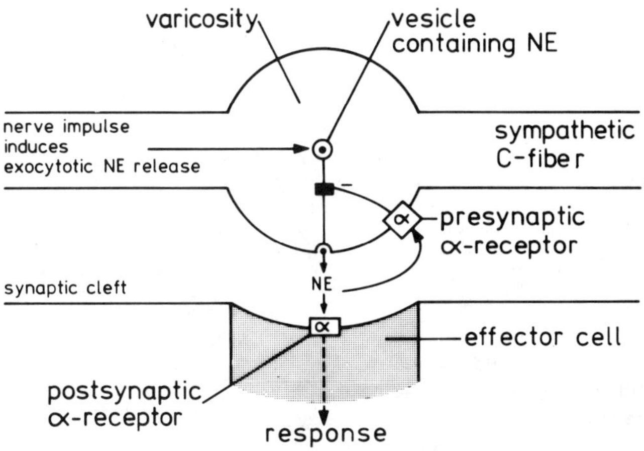

Figure 1 Localization of pre- and postsynaptic α adrenoceptors in a noradrenergic synapse. Transmitter norepinephrine (NE) stimulates postsynaptic α adrenoceptors after its exocytotic release from the noradrenergic varicosity. This gives rise to a physiological effect, such as vasoconstriction. However, norepinephrine can also activate a negative feedback mechanism via stimulation of presynaptic α adrenoceptors on the varicosity membrane. This inhibits the release of transmitter norepinephrine per nerve impulse.

rauwolscine (Rauw) have been characterized as preferential blocking agents of presynaptic α adrenoceptors. In addition, methoxamine (Meth) and phenylephrine (Pheny) behave as preferential agonists of postjunctional α adrenoceptors, and α-methylnorepinephrine and tramazoline preferentially activate the presynaptic population. These observations, together with the obvious differences in location and function among pre- and postsynaptic α adrenoceptors led Langer (16) to the proposal that postsynaptic α adrenoceptors, which mediate excitatory responses, be nominated α_1 and presynaptic α adrenoceptors at noradrenergic nerve terminals, which mediate inhibitory effects, α_2. However, this classification could not be adhered to. Although at first the terms α_1 and α_2 receptors were proposed synonyms for post- and presynaptic α adrenoceptors, respectively, the detection of α_2 adrenoceptors outside noradrenergic terminal axons, on some organelles lacking synapses, and even at postsynaptic sites possessing the general characteristics of those identified presynaptically, made it necessary to refine the classification (15,17-19). At present, the terms pre- and postsynaptic refer solely to the location of the particular α adrenoceptor with respect to the synapse. The designation α_1 or α_2 is exclusively based upon the drug selectivities of the α adrenoceptor in question, irrespective of its function or anatomic location (20-22).

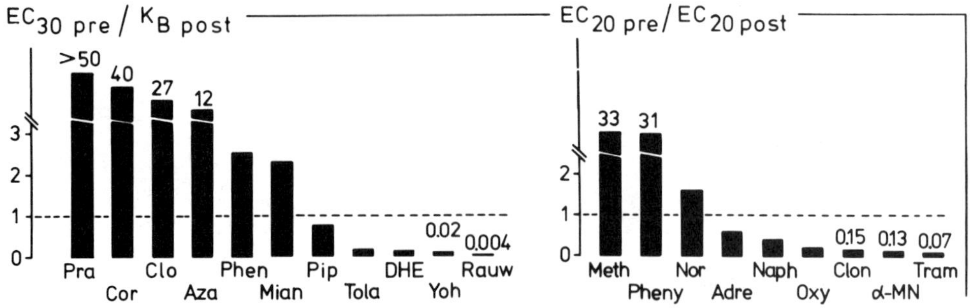

Figure 2 Pre/postsynaptic selectivity ratios for various agonists (right) and antagonists (left) established in the rabbit isolated pulmonary artery. EC_{30} (pre) = concentration facilitating [^{3}H] norepinephrine release by 30% (presynaptic effect); K_B (post) = antagonism against phenylephrine- and norepinephrine-induced contraction (postsynaptic effect). From left to right: prazosin, corynanthine, clozapine, azapetine, phentolamine, mianserin, piperoxan, tolazoline, dihydroergotamine, yohimbine, and rauwolscine. EC_{20} (pre) = concentration inhibiting [^{3}H] norepinephrine overflow by 20% (presynaptic effect). EC_{20} (post) = concentration inducing 20% of the maximal contraction (postsynaptic effect). From left to right: methoxamine, phenylephrine, norepinephrine, epinephrine, naphazoline, oxymetazoline, clonidine, α-methylnorepinephrine, and tramazoline. The chemical structures of some of the above-mentioned compounds are given in the Figures 3, 5, and 8. (Data taken from Reference 15.)

AGONISTS AND ANTAGONISTS

At present, various agonists and antagonists are available that show a high degree of selectivity for α_1 or α_2 adrenoceptors. The compounds that are currently recognized as preferential agonists of α_1 adrenoceptors are depicted in Figure 3. (±)-Erythro-methoxamine (23-25), (–)-amidephrine (27,28), (+)- and (–)-SKF 89748 (29,30), St 587 (31), and Sgd 101/75 (indanidine) (32), and cirazoline (24) fulfill the criteria for selective α_1-adrenoceptor stimulants. In addition to its selective α_1-adrenoceptor agonist activity, cirazoline also behaves as a potent antagonist of α_2 adrenoceptors (33). A hydroxyl group has been found to be necessary for α_1-adrenoceptor agonist activity for open-chain phenethylamines related to methoxamine in which this amine-containing side chain can freely adopt various conformations (34). More sterically defined structures, such as SKF 89748, retain activity without this hydroxyl substituent. Within this series of 2-aminotetralins, substitutent alterations on the ring, as well as on the nitrogen, change the α_1-adrenoceptor agonist activity by over three orders of magnitude, producing optimum activity by placing substituents at both the 5

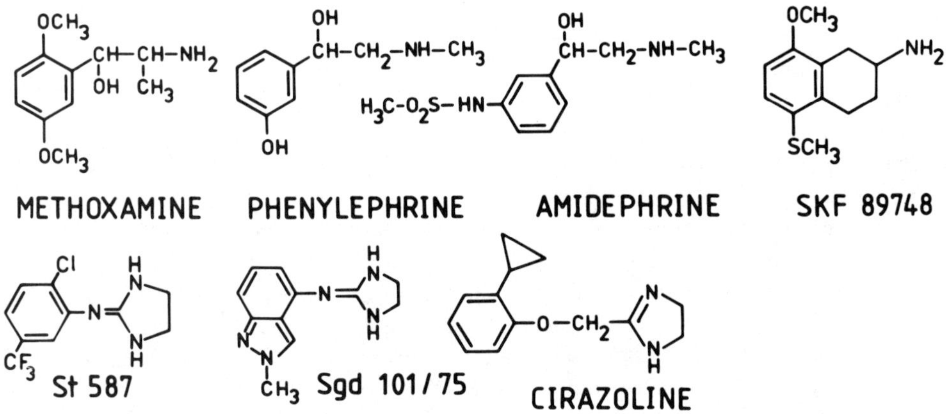

Figure 3 Selective agonists of α_1 adrenoceptors.

and 8 position (35). The rationale behind St 587 has been the observation that the bulk of the 5-substitutent in 2-chloro-substituted 2-(phenylimino)imidazolidines determines the selectivity to stimulate α_1 adrenoceptors (36).

The chemical structures of the currently classified selective α_2-adrenoceptor agonists are shown in Figure 4. High α_2-adrenoceptor agonist activity has been observed for B-HT 920 (alefexole) (24,25). B-HT 933 (azepexole) (24,25,37), B-HT 958 (38), guanabenz (39), UK-14,304 (24,40), xylazine (39), TL-99 (41), M-7 (42,43), DP-6,7-ADTN (44), and some exoisomers of 2-amino-6,7-dihy-

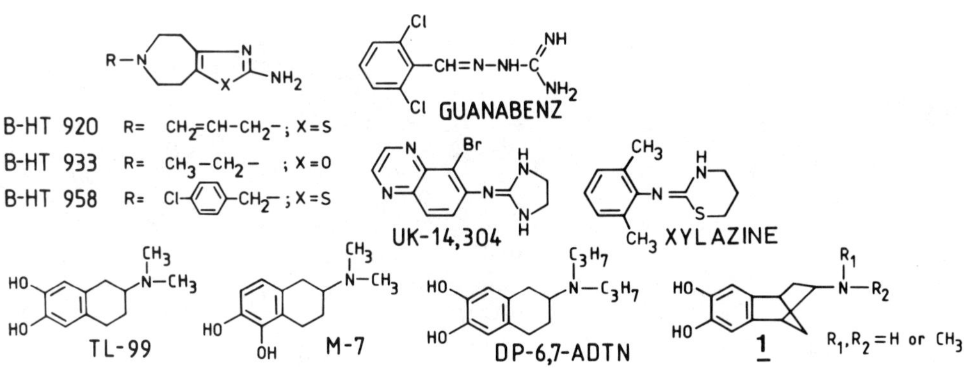

Figure 4 Selective agonists of α_2 adrenoceptors.

droxybenzonorbornene [1] (45). It should be stressed that there is as yet no rationale for the design of selective α_2-adrenoceptor agonists.

In comparison with the preferential α_1- and α_2-adrenoceptor stimulants just listed, the more commonly known α-adrenergic stimulants, clonidine, tramazoline, naphazoline, and oxymetazoline, as well as the endogenous agonists norepinephrine and epinephrine, are considered mixed agonists of α_1 and α_2 adrenoceptors (14). In addition, the relatively newer agents, guanfacine, lofexidine, and tiamenidine, also behave as nonselective stimulants (14).

Substances characterized as preferable α_1-adrenoceptor antagonists are shown in Figure 5. Prazosin (46) is, to date, the α_1-adrenoceptor blocker possessing the highest activity and selectivity, although corynanthine (47), labetalol, which also exhibits β-adrenoceptor antagonist activity (48,49), WB 4101 (50), and AR-C 239 (51) all selectively block α_1 adrenoceptors. The basic structure of prazosin has been a lead for other selective α_1-adrenoceptor antagonists, such as terazosin (the tetrahydrofuran analog of prazosin), doxazosin, and trimazosin (Figure 6) (52). Like prazosin, UK-18,596 (53), OR H3088 (54), E-643 (55), and alfuzosin (56) preferentially interact with α_1 adrenoreceptors.

The benzodioxanes represent one of the oldest groups of competitive antagonists of α adrenoceptors (57). In WB 4101-like molecules, a di-ortho substitution with methoxyl groups affords high affinity for α_1 adrenoceptors (58,59); replacement of the methoxyl groups by methyl decreases affinity to the level of the nonmethoxylated molecule. Repositioning of a methoxy group to the meta or para position reduces, but introduction of an hydroxy functionality increases, potency. Although the distance between the benzodioxane and the 2,6-

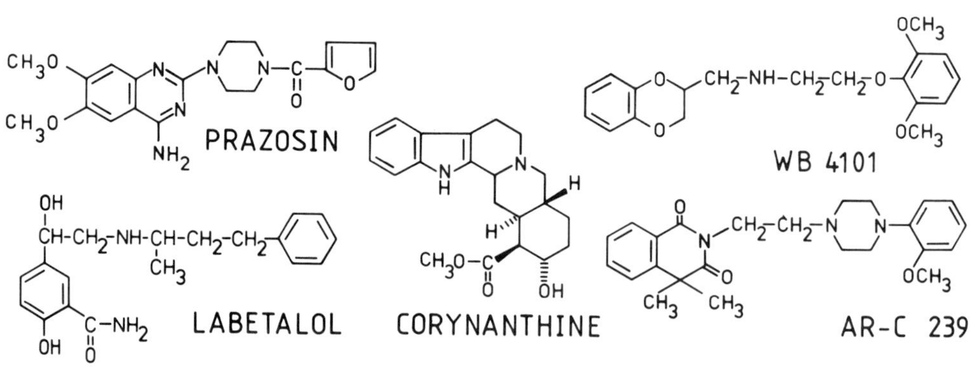

Figure 5 Selective antagonists of α_1 adrenoceptors.

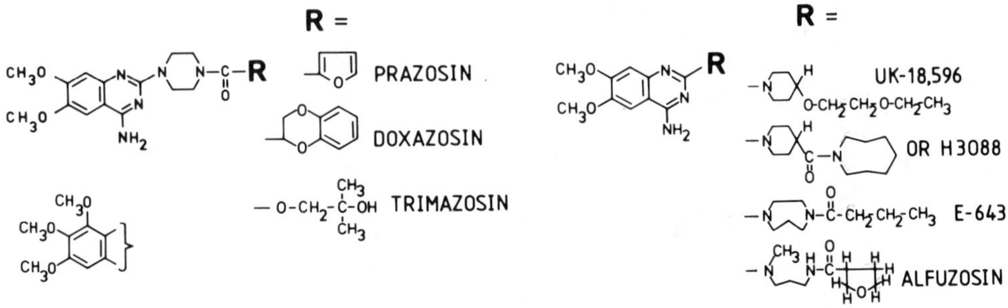

Figure 6 Newer selective antagonists of α_1 adrenoceptors structurally related to prazosin.

dimethoxyphenyl nuclei is important, the oxygen atom in the side chain appears not essential for affinity. The benzodioxane moiety itself may not be critical for high affinity, either (60), and substitution of the oxygen at position 4 in this nucleus with a sulfur has recently yielded a more potent and selective α_1-adrenoceptor antagonist (61). The importance of the secondary amine structure in the side chain is emphasized by the dramatic loss of affinity for α_1 adrenoceptors upon alkylation. The S isomer of WB 4104 exhibits the highest activity at and selectivity for α_1 adrenoceptors (62,63). The introduction of an additional chiral center in the side chain also demonstrated the pronounced dependency of the affinity on the configuration (59).

Figure 7 shows some additional compounds that, by virtually lacking affinity for α_2 adrenoceptors, selectively block α_1 adrenoceptors (although some of them possess other properties). Tibalosine has been suggested to lower arterial pressure owing to blockade of vascular α_1 adrenoceptors (64). In addition, a central mechanism has also been suggested (65). Although ketanserin (see Chap. 7) is a selective serotonin (5-HT_2) antagonist (66), it also exerts significant selective affinity for α_1 adrenoceptors (67,68). Urapidil is known to block postsynaptic α_1 adrenoceptors in a similar manner as has been established for prazosin, but also a central action unrelated to α_1-adrenoceptor blockade has been advocated (69-71). BE 2254 (HEAT) and some of its congeners act as rather selective antagonists of α_1 adrenoceptors (72,73). This series of compounds has provided [^{125}I] HEAT, which finds application in direct labeling of α_1 adrenoceptors (see below) (74). Indoramin is a clinically effective antihypertensive agent; it selectively interferes with vascular α_1 adrenoceptors and possesses a myocardial membrane stabilizing effect (75).

Numerous, structurally different compounds have been characterized as selective antagonists of α_2 adrenoceptors (Figure 8). Yohimbine and rauwol-

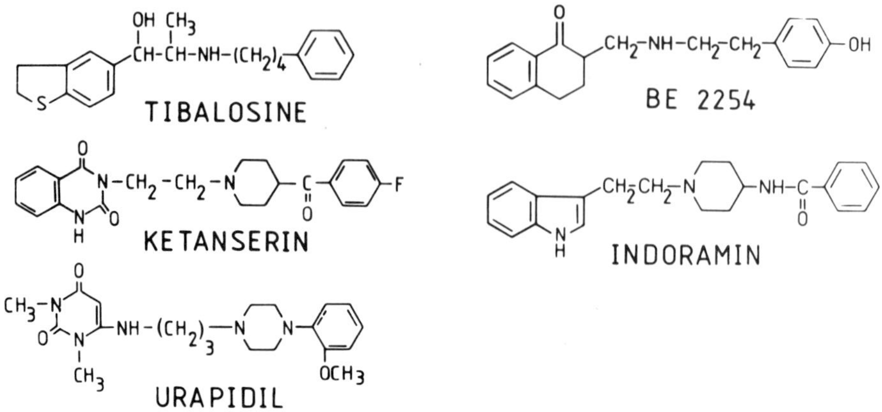

Figure 7 Additional compounds acting as selective antagonists of α_1 adrenoceptors.

scine (Figure 8), as well as corynanthine (Figure 5), are diastereoisomers. In contrast to corynanthine, which selectively blocks α_1 adrenoceptors (see above), yohimbine and especially rauwolscine show high selectivity toward α_2 adrenoceptors (47,76-78). The rationale behind the design of idazoxan, currently considered one of the most potent and selective α_2-adrenoceptor antagonists (79,80), has been outlined (81). More recently, the observation has been made that the selectivity and potency of this class of compounds is markedly enhanced by certain substituents located at the 2-position of the benzodioxane ring (82,83). Although imiloxan is 30- to 100-fold less potent than idazoxan, it

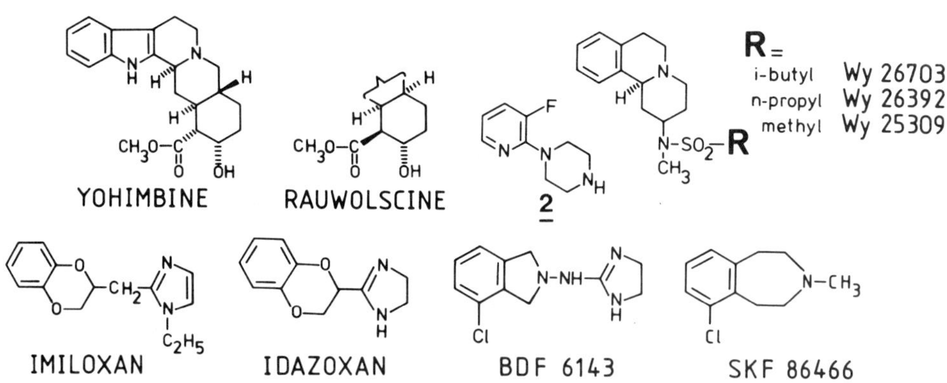

Figure 8 Selective antagonists of α_2 adrenoceptors.

shows high preference in blocking α_2 adrenoceptors (80,84). Imiloxan turned out to be the most selective compound within a large series of 1-alkylimidazoles (85). Several members of a series of 1-(2-pyridinyl)piperazines were found to be potent and selective antagonists of α_2 adrenoceptors. In various assays, the 3-fluoro analog [2] showed enhanced or comparable potency and selectivity as rauwolscine (86). BDF 6143 (87) and SKF 86466 (88) represent rather novel chemical entities possessing favorable preference for antagonizing α_2-adrenoceptor-mediated effects. Extreme α_2-adrenoreceptor blocking selectivity exceeding that of yohimbine (90), has been identified for some benzoquinolizines, such as Wy 25309, Wy 26392, and Wy 26703 (89). It is not known which structural features determine whether a compound exhibits α_1- or α_2-adrenoceptor blocking selectivity, since the structure-activity and selectivity relationships for α_1- and α_2-adrenoceptor antagonists are largely unexplored. The classic antagonists of α-adrenoceptors, phentolamine, tolazoline, and piperoxan, have been recognized as nonselective for either subtype. Phenoxybenzamine is considered α_1 selective, although its selectivity is less pronounced than that of, for example, prazosin (14).

Radioligand binding studies have supported the concept of α_1 and α_2 subtypes and significantly extended the knowledge about the occurrence and distribution of α_1 and α_2 adrenoceptors (91-95). The partial agonist $[^3H]$dihydroergocryptine was the first ligand available. Although at first the two binding sites detected with this radioligand were interpreted as agonist and antagonist α-adrenoceptor binding sites (96,97), subsequent studies made it very plausible that $[^3H]$dihydroergocryptine labels α_1- as well as α_2-adrenoceptor sites with equal affinity (92,98,99). At present, tritium-labeled clonidine is widely used for a direct identification of α_2 adrenoceptors. Numerous reports have been published on the specific high-affinity binding to α_2 adrenoceptors in brain (100,101) and various peripheral tissues (91,92,95,99). *p*-Amino$[^3H]$clonidine (102) and $[^3H]$UK-14,304 (103) have also proven very useful probes for labeling α_2 adrenoceptors. These two ligands bind to α_2 sites with appreciably higher affinity than $[^3H]$clonidine. The pharmacologically and structurally related agonists $[^3H]$guanfacine and $[^3H]$lofexidine occupy the same α_2-adrenoceptor population as $[^3H]$clonidine (104,105). In view of the higher specificity of yohimbine and rauwolscine over dihydroergocryptine, $[^3H]$yohimbine and $[^3H]$rauwolscine are excellent additional tools in binding studies involving α_2 adrenoceptors (106,107). More recently, $[^3H]$idazoxan has become available for direct binding to α_2 adrenoceptors (108). α_1 Adrenoceptors can be most conveniently labeled by $[^3H]$WB 4101 (100,109), $[^3H]$prazosin (110,111), and particularly by $[^{125}I]$HEAT (112).

Table 1 Distribution, Localization, and Function of α_1 Adrenoceptors

Organ/tissue (location)	Activation causes
Blood vessels (postsynaptic)	Contraction
Smooth muscle (postsynaptic)	Contraction
Heart (postsynaptic)	Positive inotropy/chronotropy (negative chronotropy)
Eye (postsynaptic)	Mydriasis, ocular hypertension
Liver (postsynaptic)	Glycogen phosphorylase activation
CNS (postsynaptic)	Stimulation; inhibition of baroreceptor afferent inputs
Sympathetic neurons (presynaptic?)	Inhibition of norepinephrine release (?)

DISTRIBUTION, LOCALIZATION, AND FUNCTION

Table 1 lists the various tissues and organs in which α_1 adrenoceptors have been identified. Indicated also are the effects caused by stimulation of these receptors by endogenous neurotransmitters or exogenous drugs. α_1 Adrenoceptors are mainly found at postsynaptic sites mediating a variety of (stimulatory) physiological and biochemical actions. All organs subject to neuronal control by the sympathetic nervous system contain postsynaptic α_1 adrenoceptors (8-15). Functionally most important are probably the α_1 adrenoceptors present in the vascular wall; there they play a pivotal role in the regulation of vascular tone. The existence of a minor population of α_1 adrenoceptors on the varicosity membrane of noradrenergic nerve terminals and their inhibitory role in neurotransmission has been suggested (113,114).

The distribution, localization, and function of α_2 adrenoceptors in the periphery and central nervous system (CNS) are reported in Table 2. Presynaptic α_2 adrenoceptors located at the noradrenergic nerve endings mediating the (auto)inhibition of norepinephrine release are considered the prototypes of α_2 adrenoceptors. Inhibitory prejunctionally located α_2 adrenoceptors have been identified at almost all noradrenergic axons in the periphery and in the CNS (8-15). In all tissues examined so far, the presynaptic α_2 adrenoceptors appear to be similar. In addition to noradrenergic nerve terminals, presynaptic α_2 adrenoceptors are found at central and peripheral cholinergic and serotonergic neurons where they are involved in the modulation of acetylcholine and serotonin release, respectively (9,11,15). Stimulation of α_2 adrenoceptors can hyperpolarize noradrenergic cell bodies in sympathetic ganglia, and similar recep-

Table 2 Distribution, Localization, and Function of α_2 Adrenoceptors

Organ/tissue (location)	Activation causes
Sympathetic, cholinergic, and serotonergic neurons (presynaptic)	Inhibition of norepinephrine, acetylcholine, and serotonin release
CNS (postsynaptic)	Hypotension, bradycardia
CNS (presynaptic?)	Sedation/behavioral depression, analgesia/psychotropic activity, endocrinologic effects
Sympathetic ganglia	Hyperpolarization
Somadendrites in CNS	Inhibition of firing
Blood vessels (postsynaptic)	Contraction
Platelets	Aggregation
Fat cells	Inhibition of lipolysis
Pancreatic islets	Inhibition of insulin secretion
Eye	Ocular hypotension
Intestinal epithelial cells	Inhibition of intestinal secretions

tors inhibit the firing of some somadendrites in the CNS (11-15). Postsynaptic α_2 adrenoceptors have been identified on platelets where they participate in aggregation upon stimulation (115). Inhibition of lipolysis by α-adrenoceptor agonists in fat cells is mediated by α_2 adrenoreceptors (116). The α adrenoceptors responsible for the agonist-induced inhibition of the release of immunoreactive insulin from pancreatic islets resemble α_2 adrenoceptors (117,118). In the eye, α_2 adrenoceptors, the synaptic location of which is not known, have been identified; they participate in a reduction of intraocular pressure exerted by α_2-adrenoceptor agonists (119). A postsynaptic location of α_2 adrenoceptors has been demonstrated on intestinal epithelial cells. Stimulation is though to inhibit intestinal secretions (120).

It is generally accepted that the centrally mediated hypotension and bradycardia elicited by clonidine, guanfacine, and similar compounds, as well as by α-methyl-DOPA (via α-methylnorepinephrine), is brought about by stimulation of α adrenoceptors located at pontomedullary sites. The basis for the formulation of the mode of action of clonidine and α-methyl-DOPA need not be described in full detail here, since these principles have been reviewed extensively (121-125). In addition, there is ample experimental evidence to support the view that the central hypotension (bradycardia) induced by clonidine and related drugs is brought about by stimulation of α_2 adrenoceptors (20,125-127). Arguments on which this classification is based are (1) selective antagonists of α_2 adrenoceptors, such as yohimbine, rauwolscine, and idazoxan, are very effective blockers compared with preferential α_1-adrenoceptor antagonists, such as prazosin and corynanthine (see Figure 9); (2) selective agonists of α_2 adrenoceptors, such as alefexole, azepexole, UK-14,304, guanabenz, and xylazine, all show pronounced centrally induced hypotensive effects; (3) agonists of α_1 adrenoceptors, such as cirazoline and St 587, which are clearly lipophilic enough to enter the brain, lack hypotensive activity (see Reference 125 for review). Since multiple sites seem to be involved, it is presumably not correct to define a single neuron or brain area as the sole and major target of centrally acting hypotensive α_2-adrenoceptor agonists. However, the nucleus of the solitary tract (NTS)

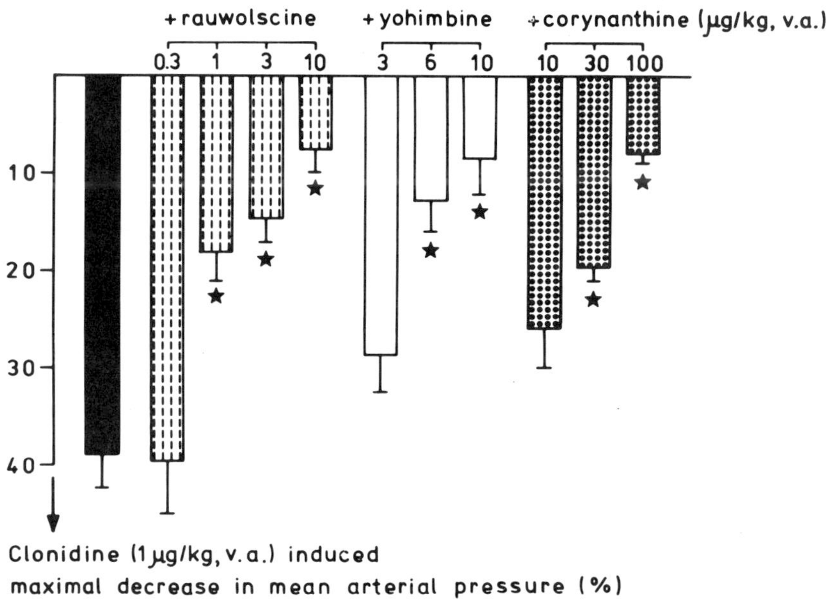

Figure 9 The antagonism by rauwolscine and yohimbine (selective α_2-adrenoceptor antagonists) and corynanthine (selective α_1-adrenoceptor antagonist) of the central hypotensive effect of clonidine (1 µg/kg) in chloralose-anesthetized cats. Clonidine was infused into the left thoracic vertebral artery. The antagonists were injected 15 min prior to clonidine via the same route. The maximal hypotensive effect of clonidine was expressed as a percentage decrease in mean arterial pressure. Means ± SEM (n = 5-7). Star indicates $p < 0.05$ (Student's t-test) with respect to nonpretreated control group (black column). (From Reference 126, with permission.)

looks attractive as an important site of action (125,128). A high density of α_2 adrenoceptors is found in this nucleus (129). Local injections of α-adrenoceptor agonists into the NTS elicit a decrease in arterial pressure and heart rate (130), although an unequivocal classification of these hypotensive α adrenoceptors in the NTS in terms of α_1 or α_2 subtypes has not been possible. The question of whether the central cardiovascular inhibitory α_2 adrenoceptors are present on the terminals of noradrenergic neurons (presynaptic) or on the cell bodies or dendrites of noradrenergically innervated neurons (postsynaptic) cannot be answered with certainty mainly because of methodological limitations. Support for a presynaptic location is based on the observation that one of the most prominent CNS effects of clonidine and related α_2-adrenoceptor agonists is a decrease in the turnover and rate of synthesis of norepinephrine (131). This action is most likely due to a negative feedback in noradrenergic neurons involving presynaptic α_2 adrenoceptors. On the other hand, depletion of norepinephrine by reserpine or 6-hydroxydopamine and subsequent inhibition of synthesis did not reduce clonidine's depressant effect on sympathetic nervous activity and did not affect its vagally mediated bradycardia (124,125,128). Consequently, central endogenous transmitter norepinephrine is apparently not required for the central cardiovascular response to clonidine, indicating a postsynaptic location of the α_2 adrenoceptors. Furthermore, it is likely that these α_2 adrenoceptors virtually lack a noradrenergic input, since α-adrenoceptor antagonists are generally inactive at the central level to change cardiovascular parameters and elimination of central endogenous transmitters only moderately and transiently affects the peripheral cardiovascular system.

Stimulation of central α adrenoceptors lead to behavioral depression in laboratory animals (123,124,132,133). In fact, sedation is a prominent side effect in the antihypertensive therapy using centrally acting α-adrenoceptor agonists. With respect to the clonidine-induced sedation, the involvement of α_2 adrenoceptors has been made plausible (125,126,134). Some findings favor the view that the central α_2 adrenoceptors initiating sedation of α-adrenoceptor stimulants are inhibitor presynaptic α_2 adrenoceptors at noradrenergic neurons. Depletion of catecholamine stores and/or destruction of noradrenergic neurons prevented the sedation produced by clonidine and that of the selective α_2-adrenoceptor agonist alefexole (135,136).

Apart from the central α_2 adrenoceptors already discussed, similar receptors in the CNS presumably located presynaptically are involved in the analgesia produced by α_2-adrenoceptor stimulants. In this respect, the antinociceptive actions of clonidine, which are distinct from those of morphine, have been explored extensively (137). A variety of psychotropic activities is observed upon activation of central α_2 adrenoceptors. These studies, the majority of which have used clonidine, indicate that central α_2 adrenoceptors could play a role in anxiety, learning and memory, heroin and alcohol withdrawal, depression, thermoregulation, schizophrenia, hunger drives, dementia, and psychosis (14,133). Certain endocrinologic effects are elicited by stimulation of central α_2 adrenoceptors. Again, most of these effects have been demonstrated with clonidine, which, for example, increases plasma levels of growth hormone (138) and controls ACTH release (139).

Undoubtedly, the discovery of a simultaneous occurrence of postsynaptic α_1 and α_2 adrenoceptors in vascular smooth muscle, both involved in drug-induced vasoconstriction, has definitely established the classification of α adrenoceptors into two distinct functional classes. The first indications for a heterogeneity of postsynaptic vascular α adrenoceptors came from observations with isolated strips of human palmar digital arteries showing a prazosin-resistant component of the contraction to norepinephrine (140). Following these first observations, subsequent studies in vivo presented very clear evidence of the existence of functional α_1 and α_2 adrenoceptors as postjunctional sites in the intact circulatory system of various animal species (see Figure 10) (for review see Reference 141).

The remarkable convenience of separate stimulation of postsynaptic α_1 and α_2 adrenoceptors to elicit a vasopressor response in vivo has been applied to differentiate α-adrenoceptor antagonists with respect to selectivity for either class. In pithed normotensive rats, pressor responses to α_1-adrenoceptor stimulation was activated by phenylephrine and azepexole was used to produce α_2-adrenoceptor-mediated vasoconstriction. As a measure of the selectivity for either postsynaptic α_1 or α_2 adrenoceptors in vascular smooth muscle, the ratio of K_B (post α_2) to K_B (post α_1) was determined. As summarized in Figure 11, the α-adrenoceptor antagonists were found to cover an approximately 30,000-fold range of activity ratios. The antagonists most selective for either type were prazosin (α_1) and rauwolscine (α_2). The selectivity thus established for α-adrenoceptor antagonists at postsynaptic vascular α_1 and α_2 adrenoceptors agree very well with those involving presynaptic α_2 adrenoceptors (compare with Figure 2), indicating that pre- and postsynaptic α_2 adrenoceptors have similar drug specificities (142).

The evidence for functional constrictor α_2 adrenoceptors is largely based on experiments performed in vivo. It has been rather difficult to convincingly detect and satisfactorily characterize postsynaptic α_2 adrenoceptors in vitro by functional tests, although direct binding studies could generally demonstrate the presence of these receptors. At present, it is recognized that in vascular smooth muscle the postsynaptic α_1 adrenoceptor is the predominant vasoconstrictor type, but stimulation of postsynaptic α_2 adrenoceptors can also induce vasoconstriction. The relative contribution of α_2 adrenoceptors to drug-induced vasoconstriction depends heavily on the particular vessel or vascular bed and animal species and also on the diameter of the blood vessel (143). In the resistance vessels more α_2 adrenoceptors seem to contribute. Evidence is accumu-

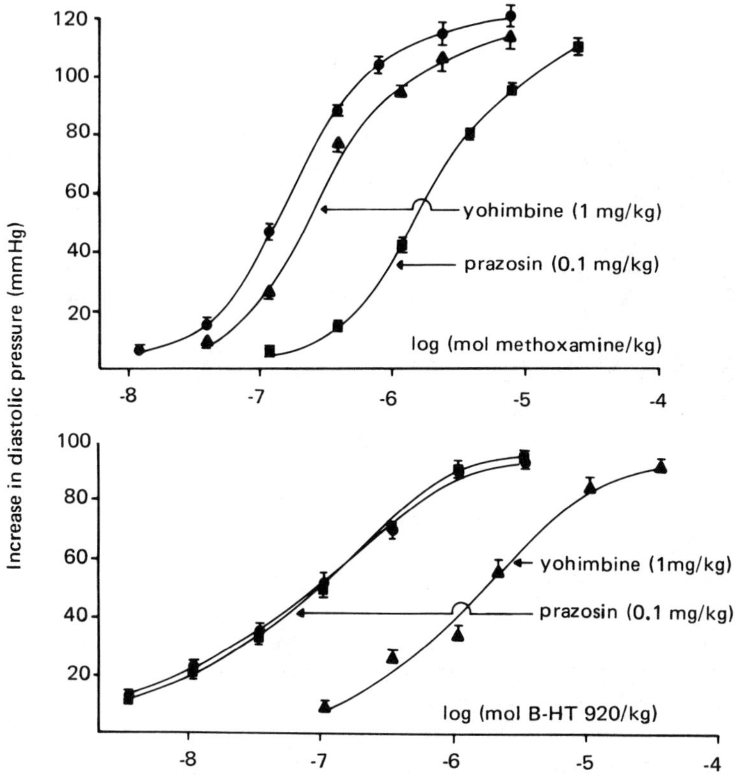

Figure 10 Log dose-vasopressor response curves of the α_1-adrenoceptor agonist methoxamine (above) and the α_2-adrenoceptor agonist B-HT 920 (alefexole, below) with respect to the increase in diastolic pressure in pithed normotensive rats after pretreatment (IV, 15 min) with saline (●–●), yohimbine (1 mg/kg, ▲–▲), or prazosin (0.1 mg/kg, ■–■). Symbols represent mean values ± SEM (n = 5-7). The data show that distinctly different populations of postsynaptic α_1 and α_2 adrenoceptors can initiate vasoconstriction. (Modified after Reference 24.)

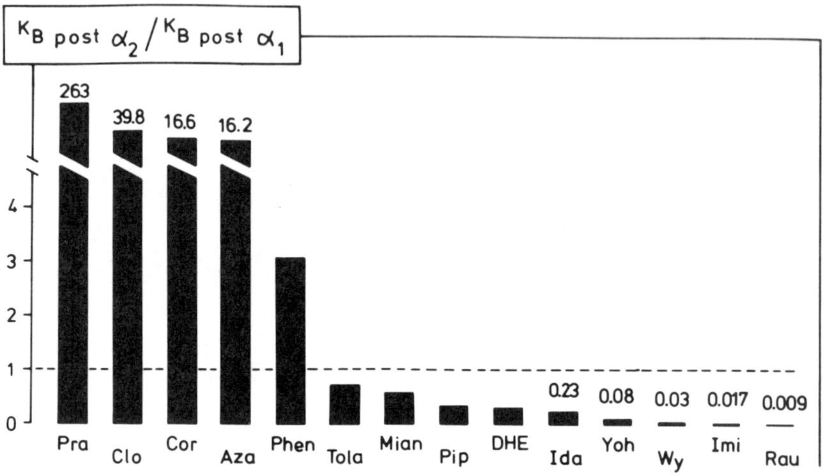

Figure 11 Selectivity ratios of α-adrenoceptor antagonists for postsynaptic α_1 and α_2 adrenoceptors. The ratio of K_B (post α_2) to K_B (post α_1) resulted from antagonism against α_2- and α_1-adrenoceptor-induced vasoconstriction in pithed rats, respectively. From left to right: prazosin, clozapine, corynanthine, azapetine, phentolamine, tolazoline, mianserin, piperoxan, dihydroergotamine, idazoxan, yohimbine, Wy 26703, imiloxan, and rauwolscine. For chemical structures, see Figures 4 and 5. (Data from References 80, 142, and 215.)

lating suggesting that postsynaptic α_2 adrenoceptors are also present in human vasculature (see Reference 144 and References quoted there). Obvious species differences also appear to exist with respect to the functionality of postsynaptic α_2 adrenoceptors in the venous vascular bed. In spinalized ganglion-blocked dogs and pithed cats, α_2-adrenoceptor stimulation did not result in an increase in cardiac output (145,146), in contrast to pithed rats (147).

Since stimulation of both α_1 and α_2 adrenoceptors induces a vasoconstrictor response, the question arises: What is the physiological or pharmacological basis for this differentiation in vasconstrictor α_1 and α_2 adrenoceptors? Ariens and Simonis (148,149) have suggested that the α_2 adrenoceptors are predominantly epinephrinergic in nature and located extrajunctionally, whereas the α_1 adrenoceptors are predominantly norepinephrinergic in nature and located postsynaptically in the sympathetic terminal junction. A differential effectiveness of α-adrenoceptor blocking agents at inhibiting the pressor responses to spinal cord stimulation and intravenously administered norepinephrine in pithed rats has been noticed (150). It has also been found that pressor responses elicited by sympathetic nerve stimulation in the dog were markedly antagonized by prazosin and that in the same species rauwolscine rendered the pressor effects of

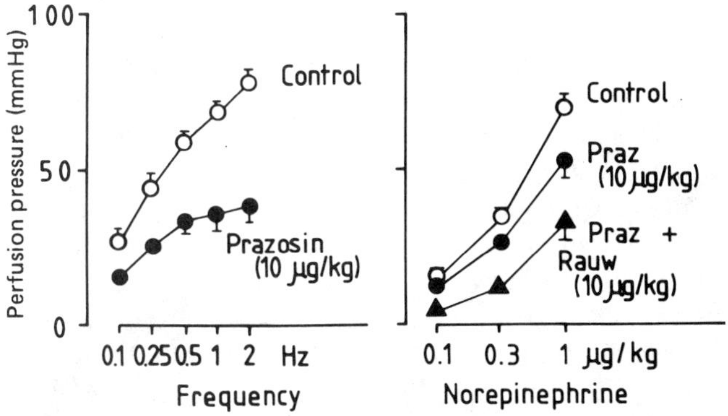

Figure 12 Effect of prazosin on the hindlimb perfusion pressure to lumbar sympathetic stimulation and of prazosin and rauwolscine on responses to exogenous norepinephrine in anesthetized dogs. The effects of norepinephrine were obtained in ganglion- and β-adrenoceptor-blocked animals. (Modified after Reference 151.)

intravenous norepinephrine sensitive to prazosin blockade (see Figure 12) (151). Further studies carried out under in vivo as well as in vitro experimental conditions showed that prazosin was more effective in blocking the responses to norepinephrine released by nerve stimulation than in antagonizing the same end-organ effects induced by exogenous norepinephrine (see Reference 143 and references quoted there). In addition, the pressor effects of norepinephrine released through ganglionic stimulation with DMPP in pithed rats were affected by rauwolscine only at the higher doses of DMPP. After bilateral adrenalectomy, rauwolscine was no longer effective, whereas prazosin invariably exerted a pronounced inhibitor effect (152). These data support the concept that postsynaptic vasoconstrictor α_1 adrenoceptors are part of the synapse or at least close to the nerve terminal and that the corresponding α_2 adrenoceptors have a predominant extrasynaptic location or are at least located farther away from the nerve terminal. However, the hypothesis that vascular postsynaptic α_1 adrenoceptors are innervated in contrast to postsynaptic α_2 adrenoceptors, which are considered hormone receptors, is clearly an oversimplification. Evidence is increasing that postsynaptic α_2 adrenoceptors can be involved to a significant or even predominant degree in the vasoconstrictor response to sympathetic nerve stimulation. Some examples are the tail arteries of spontaneously hypertensive and Sprague-Dawley rats (153,154), the dog (155) and human (156) saphenous veins, the dog (poststenotic) coronary (157) and femoral (158,159) beds, and the rabbit renal vascular bed (160). Indications for postsynaptic vasopressor

α_2 adrenoceptors activated by neurons were also obtained in the pithed rat by using the indirectly acting agent tyramine (161). It has been made plausible that these α_2 adrenoceptors are activated via ganglionic (muscarinic) M-2 receptors (162). As an alternative hypothesis for the extrasynaptic location of vascular α_2 adrenoceptors, it has been suggested that activation of ganglionic nicotinic and muscarinic M-1 receptors leads to stimulation of the nearest varicosities endowed with α_1 adrenoceptors. On the other hand, activation of ganglionic muscarinic M-2 receptors may lead to additional release of neurotransmitter in the more distant varicosities containing α_2 adrenoceptors (see Figure 13) (162). Additional experiments in which autoradiographic studies will play an important role are needed to further clarify and possibly generalize this concept.

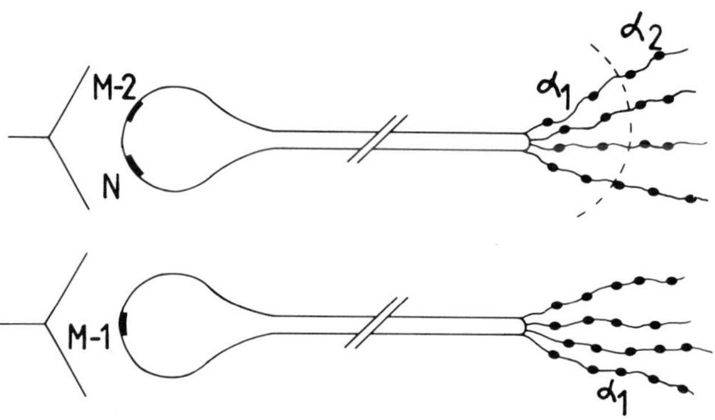

Figure 13 Schematic representation of sympathetic ganglia with their postganglionic neuron terminating in the vascular wall. The nicotine (N), muscarine 1 (M-1), and muscarine 2 (M-2) receptors are located on the cell body of the ganglia. It is suggested that the varicosities of the ganglion endowed with both nicotine and M-2 receptors can roughly be divided over two regions. In the vascular wall the more proximal varicosities interfere predominantly with postsynaptic α_1 adrenoceptors and the more distant with postjunctional α_2 adrenoceptors. A stimulus initiated by activation of the ganglionic nicotine receptors may reach the varicosities in the proximal area, whereas stimulation of the ganglionic M-2 receptors may give rise to a stimulus that reaches the varicosities in the more distant region. Excitation of M-1 receptors elicits activation of postsynaptic α_1 adrenoceptors only. Note that M-1 and nicotine receptors mediate the activation of different populations of α_1 adrenoceptors. This concept is based on data obtained in pithed rats (161,162).

α-ADRENOCEPTOR-EFFECTOR COUPLING

Several mechanisms have been proposed by which presynaptic α_2 adrenoceptors modulate transmitter release. The mechanism by which prejunctional α_2 adrenoceptors regulate transmitter release is only operating on those release pathways that depend on the availability of extracellular Ca^{2+} ions. Thus, K^+, veratrine, and electrical, but not tyramine-induced, norepinephrine release is modulated by α_2-adrenoceptor agonists and antagonists (9,10,13). Consequently, presynaptic α_2 adrenoceptors are presumed to modify transmitter release by regulating the influx of extracellular Ca^{2+} ions. Activation inhibits the transmembrane inward current of Ca^{2+} ions via potential-sensitive permeability channels.

α_2 Adrenoceptors in general may be coupled to adenylate cyclase in an inhibitory manner in that stimulation results in inhibition of cyclic AMP production. Activation of postsynaptic α_2 adrenoceptors located on the cell membranes of platelets, pancreatic islets, adipocytes, and neuroblastoma X glioma hybrid cells particularly responds with an inhibition of cyclic AMP accumulation (163-167). It remains to be determined whether the coupling to adenylate cyclase applies to and is functional for all postsynaptic α_2 adrenoceptors, such as those in the central nervous system and in the vascular wall, and whether stimulation is always effected in suppression of cyclic AMP formulation. It should be added that pertussis toxin, which blocks all known receptor-mediated inhibition of adenylate cyclase activity, has been shown to attenuate several actions of α_2-adrenoceptor agonists, including vasoconstriction (168).

Radioligand binding studies indicate that in all α_2-adrenoceptor-containing tissues, whether or not the inhibition of adenylate cyclase has been demonstrated, the α_2 adrenoceptors complex with a GTP-binding protein similar to, if not the same as, the regulatory subunit of adenylate cyclase. Guanyl nucleotides and Na^+ ions have been found to reduce high-affinity binding of agonists to α_2 adrenoceptors, whereas numerous divalent cations reversed the effects on binding of guanyl nucleotides and Na^+ and potentiated [^3H] clonidine binding to α_2 adrenoceptors. Na^+ ions and guanyl nucleotides convert a large portion of the α_2 adrenoceptors into a low-affinity state. The divalent cations are necessary to form a high-affinity state of an agonist-α_2-adrenoceptor complex. An explanation of the effects of guanyl nucleotides is that agonists promote the association of the α_2 adrenoceptor with a regulatory membrane component. This complex is destabilized by guanyl nucleotides (169,170).

α_1 Adrenoceptors in general are believed not to be coupled to adenylate cyclase, but the current concept is that activation of these receptors leads to an increase in the intracellular free Ca^{2+} concentration (163). Increasing evidence suggests that a widespread, and maybe universal response to stimulation of α_1 adrenoceptors is hydrolysis of inositol lipid. The α_1-adrenoceptor-coupled event is presumably hydrolysis of phosphatidylinositol 4,5-biphosphate generating two second messenger molecules, that is, inositol-1,4,5-triphosphate, which may be responsible for the intracellular mobilization of Ca^{2+}, and 1,2-diacyglycerol, which probably contributes to control of cell function through activation of protein kinase C (171,712). Although the role of inositol lipids in (α_1-adrenergic) receptor-mediated Ca^{2+} mobilization has mainly been tested in nonexcitable mammalian tissues, such as liver, exocrine pancreas, and lacrimal gland, sufficient information is available to indicate that this also applies to excitable (vascular) smooth muscles (173,172).

It has long been recognized that the contractile mechanism in (vascular) smooth muscle is triggered by an increase in intracellular Ca^{2+} concentration (174). The increase in cytoplasmic Ca^{2+} concentration, which activates the contractile process in vascular smooth muscle cells, can be due either to an increased permeability of the cell membrane for extracellular Ca^{2+} (Ca^{2+} influx or entry) or to mobilization of Ca^{2+} from cellular stores (see above). Entry of Ca^{2+} is thought to occur through specific, so-called slow channels activated by either changes in membrane potential (potential-operated channels, POC) or by drug-receptor interactions (receptor-operated channels, ROC) (175,176). Accordingly, α-adrenoceptor-induced vasoconstriction may give rise to an increased cytoplasmic Ca^{2+} concentration by (1) opening of POC, (2) opening of ROC, or (3) release of intracellularly stored Ca^{2+}. Each of these mechanisms contributes to a different extent to the increase in intracellular free Ca^{2+}, depending on the vascular blood vessel and species involved (176-178).

Ca^{2+} has been proposed to play a differential role in the vasoconstrictor processes to stimulation of postsynaptic α_1 and α_2 adrenoceptors (for reviews see References 178 and 179). The general observation has been made that the increase in diastolic pressure in pithed rats as well as in other animal species to all currently available selective α_2-adrenoceptor agonists is markedly sensitive to inhibition by low doses of the so-called Ca^{2+} entry blockers nifedipine, verapamil, and diltiazem (for reviews, see References 178 and 179). Two examples are given in Figure 14. Differences in efficacy and in potency of the α_2-adrenoceptor agonists bear no relationship with the effectiveness of Ca^{2+} slow channel blockers to attenuate these pressor effects. In addition to these experiments in vivo, all in vitro studies involving vascular smooth muscle α_2 adrenoceptors have verified that there are no exceptions to the rule that apparently α_2-adrenoceptor-mediated vasoconstriction heavily, if not exclusively, depends on an influx of extracellular Ca^{2+} (for reviews, see References 178 and 179).

In contrast, the sensitivity of α_1-adrenoceptor-mediated vasoconstriction to blockade by Ca^{2+} slow channel inhibitors has been found to vary depending on the α_1-adrenoceptor agonist and type of vessel used. For example, in pithed rats, the log dose-pressor effect curves to cirazoline (see Figure 15), methoxamine, phenylephrine, amidephrine, and SKF 89748 were but moderately shifted

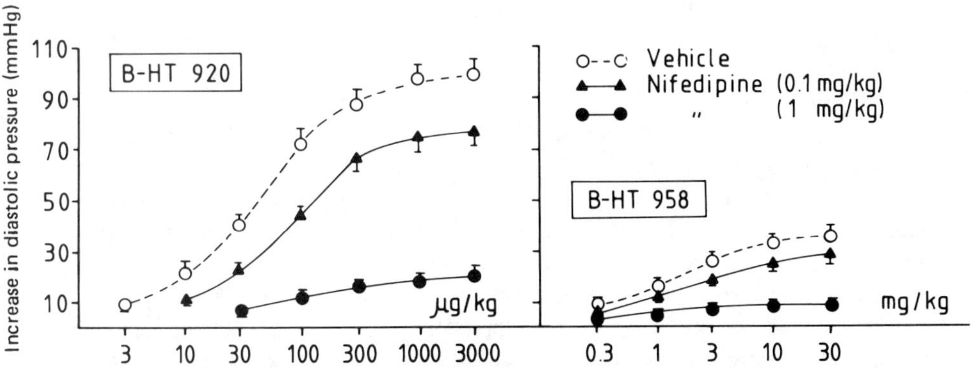

Figure 14 Log dose-vasopressor response curves with respect to the increase in diastolic pressure of pithed normotensive rats brought about by the selective α_2-adrenoceptor agonists B-HT 920 (left) and B-HT 958 (right) following IV administration 15 min after intra-arterial pretreatment with various doses of the Ca^{2+} entry blocker nifedipine. Symbols represent mean values ± SEM (n = 6-7). (Modified after References 216 and 217.)

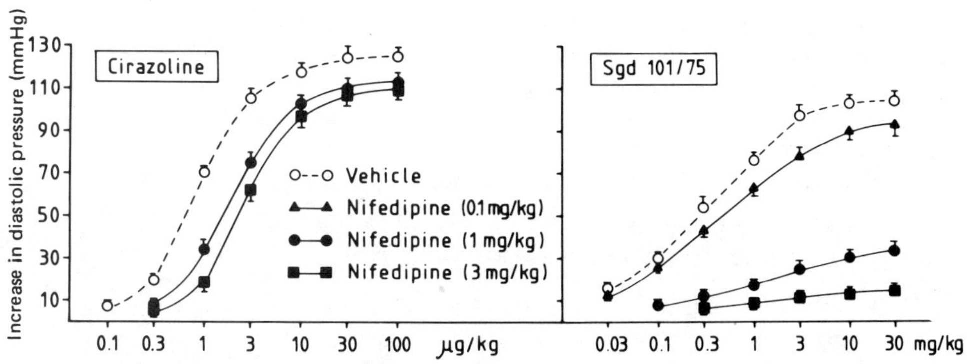

Figure 15 Log dose-vasopressor effect curves with respect to the increase in diastolic pressure of pithed normotensive rats brought about by the selective α_1-adrenoceptor agonists cirazoline (left) and indanidine (Sgd 101/75, right) following IV administration 15 min after intra-arterial pretreatment with various doses of the Ca^{2+} entry blocker nifedipine. Symbols represent mean values ± SEM (n = 6-7). (Modified after References 180 and 182.)

to the right, accompanied by a slight reduction of the maximum following treatment with high doses of nifedipine (180,181). On the other hand, the increase in diastolic pressure of pithed rats produced by Sgd 101/75 (indanidine) (see Figure 15) or St 587 proved sensitive to blockade by nifedipine and verapamil (180,182). Numerous in vitro studies have further demonstrated the variable sensitivity of α_1-adrenoceptor-induced constriction of vascular smooth muscle to inhibition of Ca^{2+} entry. In the rabbit main pulmonary artery, the contractions to α_1-adrenoceptor activation by clonidine were much more sensitive to Ca^{2+} entry blockade than those produced by methoxamine (183). In the rat isolated thoracic aorta, the α_1-adrenoceptor-mediated log dose-constrictor response curves to cirazoline and norepinephrine were shifted in a parallel manner to the right after blockade of Ca^{2+} influx, with but a minor depression of the maximum, whereas both slope and maximum of the concentration-response curves to St 587 and indanidine were markedly reduced (182). Conversely, in the guinea pig aorta the α_1-adrenoceptor-provoked contractions to these stimulants were highly resistant to impairment by Ca^{2+} entry blockade (184). α_1-Adrenoceptor-mediated constriction of various other vessels, such as canine portal and mesenteric veins and coronary and splenic arteries, as well as rat and guinea pig portal veins, have been found considerably susceptible to Ca^{2+} entry blockade or omission of Ca^{2+} from the bath medium, whereas similar contrictions in rabbit aorta, rat tail artery, and anococcygeus muscle are more or less unaffected (for reviews see References 178 and 179). In a comparative study, Muller-Schweinitzer (185) clearly showed that the α_1-adrenoceptor-mediated contractions of the dog isolated circumflex coronary artery, but not the α_1-adrenoceptor-mediated contractions of the dog saphenous vein, were inhibited by nifedipine. It seems safe to conclude that vasoconstriction to α_1-adrenoceptor activation can be governed by Ca^{2+} influx and/or intracellular release of Ca^{2+}. The interaction of the particular α_1-adrenoceptor agonist with the α_1 adrenoceptor per se determines to what extent both processes of Ca^{2+} utilization are activated. In addition, the Ca^{2+} mobilization triggered by the various α_1-adrenoceptor agonists is not uniform among the different tissues.

PHYSIOLOGICAL ROLE AND TARGETS FOR DRUGS

Most experiments aimed at the demonstration of operating presynaptic mechanisms through α_2 adrenoceptors have been performed in vitro or in vivo under electrical stimulation. Consequently, limited information is available about the actual physiological role of presynaptic α_2 adrenoceptors. However, it can be speculated that such a role, if any, may be threefold (13). (1) Presynaptic α_2 adrenoceptors may serve as targets for circulating catecholamines. (2) Presynaptic α_2 adrenoceptors may function as targets for transmitters secreted from neighboring axon terminals. Thus, presynaptic interactions between sympathetic and parasympathetic nervous systems may supplement the classic postsynaptic antagonism between these two parts of the autonomic nervous system. (3) Presynaptic α_2 adrenoceptors may manifest themselves as autoreceptors for endogenous neurotransmitter norepinephrine, although the physiological meaning of autoinhibition of norepinephrine has been criticized (11,186). A functional role of presynaptic α_2 adrenoceptors in peripheral noradrenergic transmission has been suggested in conscious rats (187), but this has not been confirmed in conscious rabbits (188) or humans (189).

Functional and biochemical studies suggest that endogenous depression may be related to impaired central noradrenergic transmission (190). Drug-induced desensitization of central α_2-presynaptic (auto)receptors, stemming from chronic inhibition of neuronal norepinephrine uptake, and hence elevated synaptic levels, has been postulated to play a key role in the mechanism of action of many antidepressants (190). This would lead to yet further enhanced neurotransmission with subsequent down-regulation of central β adrenoceptors, which is possibly a pivotal event in the induction of the desired therapeutic effect (191). However, the situation is far from clear with, on the one hand, conflicting reports about whether chronic desipramine and imipramine do (192) or do not (193,194) down-regulate presynaptic α_2 adrenoceptors (195). These data make it very unlikely that the induction of subsensitive central presynaptic α_2 adrenoceptors represents the fundamental mode of action of antidepressants.

The physiological role of the postsynaptic α_2 adrenoceptors can only be the subject of mere speculation, especially when they are present together with postsynaptic α_1 adrenoceptors. It is an oversimplification to distinguish the α_2 from the corresponding α_1 adrenoceptors in terms of control by the nonadrenergic nervous system or by blood-borne hormones (see above). In the intact circulatory system, postsynaptic α_2 adrenoceptors may contribute to peripheral resistance. Undoubtedly, the α_1 adrenoceptor is the predominant type present in all noradrenergic synapses regulating the responses of the target organs.

Whatever the significance of pre- and postsynaptic α adrenoceptors may be, a number of drug-induced effects that are recognized as of potential therapeutic benefit are presumed to be initiated by pre- or postsynaptic α adrenoceptors. Apart from some use in the symptomatic treatment of rhinitis or conjunctivitis (decongestants), the therapeutic application of drugs interfering with α adrenoceptors is largely limited to their use as antihypertensives. As has already been alluded to (see above), stimulation of central (postsynaptic) α_2 adrenoceptors leads to hypotension and bradycardia. Clonidine and α-methyl-DOPA are the prototypes of these centrally acting antihypertensive drugs (125). Clonidine and related α-adrenoceptor stimulants also activate presynaptic cardiac α_2 adrenoceptors. This possibly contributes to the overall bradycardic action of this class of compounds. The participation of (presynaptic) α_2 adrenoceptors in the production of sedation has already been mentioned. Activation of presynaptic α_2 adrenoceptors inhibiting cholinergic transmission is probably part of the explanation for the occurrence of dry mouth. So far, a separation between the sedative and hypotensive properties of centrally acting α_2-adrenoceptor agonists has not been successfully achieved in therapy.

Presynaptic α_2 adrenoceptors at noradrenergic nerve terminals in vascular smooth muscle are theoretical targets for hypotensive drugs. Stimulation of these α_2 adrenoceptors will impair norepinephrine release, which may give rise to vasodilatation. These drugs must be selective for the presynaptic α_2 type, since stimulation of the postsynaptic α_2-adrenoceptor population in vascular smooth muscle causes vasoconstriction. However, there is as yet no evidence suggesting that such a peripheral presynaptic mechanism is effective enough to lower arterial pressure. Furthermore, it may be very difficult, if not impossible, to develop selective presynaptic α_2-adrenoceptor agonists.

Vascular postsynaptic α_1 adrenoceptors are also important targets for antihypertensive drugs. Prazosin has proven to be very useful in lowering arterial pressure by functional blockade of this subclass of α adrenoceptors (196). Numerous compounds (see above) share the selective α_1-adrenoceptor blocking potency with prazosin. Their place in antihypertensive therapy is not yet firmly established (196). Postsynaptic α_2 adrenoceptors in vascular smooth muscle may offer an additional target to produce vasodilatation. If the activation of these α_2 adrenoceptors contributes to vascular arteriorlar tone, blockade may result in hypotension. It seems possible to lower arterial pressure via this mechanism, since the selective α_2-adrenoceptor antagonists imiloxan (P. M. W. M. Timmermans, unpublished) and SKF 86466 (197) show hypotensive effects in spontaneously hypertensive rats, although the actual involvement of postjunctional α_2 adrenoceptors has not been demonstrated.

Clonidine and some related drugs suppress narcotic withdrawal signs in the rat, monkey, and human (198,199), possibly via stimulation of central presynaptic α_2 adrenoceptors, thereby inhibiting the firing of the locus ceruleus through reduction of norepinephrine release (200). The α_2 adrenoceptors also appear to play a role in the withdrawal syndrome following cessation of long-term antihypertensive treatment with clonidine and some related compounds (201,202). Clonidine has also been found effective in the treatment of alcohol withdrawal states (203). Recently, clonidine has been reported to accelerate growth in children with impaired growth hormone secretion (204).

Elevated central norepinephrine activity (e.g., in the locus ceruleus) may be an important correlate of human anxiety, although it is not the sole one (205, 206). The beneficial effects of clonidine as an antianxiety agent in humans could result from its inhibitory action on norepinephrine activity by stimulation of presynaptic α_2 adrenoceptors (207,208). The observation that yohimbine is anxiogenic in humans is in accordance with this hypothesis (209).

Irrespective of the mechanism of action of tricyclic antidepressants (see above), antagonists of α_2 adrenoceptors per se may have a beneficial effect in depressive illness, since they facilitate noradrenergic neurotransmission. Phenoxybenzamine, dihydroergotamine, and yohimbine, but not the selective α_1-adrenoceptor antagonist prazosin, have been shown to be synergistic with several antidepressants in decreasing β-adrenoceptor density in rat cerebral cortex (210, 211). Imiloxan has been reported to produce a more rapid onset of β-adrenoceptor desensitization by itself than desipramine (212). Preliminary data indicate the effectiveness of idazoxan in hospitalized patients suffering from major primary depression (213).

A potentially attractive therapeutic application of α_2-adrenoceptor agonists may be offered by their profound effect on intraocular pressure (119). Macroscopic ocular side effects or an effect on the pupil size are possibly related to α_1-adrenoceptor stimulation. Thus, selective α_2-adrenoceptor stimulants may represent a group of novel antiglaucomatous agents.

The inhibitory effects of α_2-adrenoceptor agonists on intestinal motility suggest a therapeutic approach for the treatment of excessive gastrointenstinal motility and diarrhea (214).

Under physiological conditions, the net effect of an α-adrenoceptor agonist will be the sum of pre- (inhibition of transmitter release) and postsynaptic (activation of the target organ) actions, since most organs receive a constant noradrenergic input. Obviously the same applies to α adrenoceptor antagonists. The relative proportion of functional postsynaptic α_2 adrenoceptors will then determine to what extent pre (α_2) and postsynaptic (α_1/α_2) effects can be distinguished. Although the general characteristics of pre- and postsynaptic α_2 adrenoceptors are similar (see above), a discrimination between both populations of α_2 adrenoceptors is theoretically still possible. It seems that the development of drugs (agonists and/or antagonists) with a strong preference for either pre- or postsynaptic α_2 adrenoceptors, once met with success, may open new possibilities for novel therapeutic applications of α-adrenoceptor agonists and antagonists.

ACKNOWLEDGMENT

The preparation of the manuscript by Ms. Anna M. Harkins is gratefully acknowledged.

REFERENCES

1. H. H. Dale, J. Physiol. (Lond.) *34*, 163 (1906).
2. C. R. Raymond-Hamet, C. R. Acad. Sci. [III] *180*, 2074 (1925).
3. R. P. Ahlquist, Am. J. Physiol. *153*, 586 (1948).

4. C. E. Powell and I. H. Slater, J. Pharmacol. Exp. Ther. *122*, 480 (1958).
5. N. C. Moran and M. E. Perkins, J. Pharmacol. Exp. Ther. *124*, 223 (1958).
6. R. F. Furchgott, Ann. N.Y. Acad. Sci. *139*, 553 (1967).
7. A. M. Lands, A. Arnold, J. P. McAuliff, F. P. Luduena, and T. G. Brown, Nature (Lond.) *214*, 597 (1967).
8. S. Z. Langer, Br. J. Pharmacol. *60*, 481 (1977).
9. K. Starke, Rev. Physiol. Biochem. Pharmacol. *77*, 1 (1977).
10. T. C. Westfall, Physiol. Rev. *57*, 659 (1977).
11. E. S. Vizi, Prog. Neurobiol. *12*, 181 (1979).
12. S. Z. Langer, Pharmacol. Rev. *32*, 337 (1980).
13. K. Starke, Annu. Rev. Pharmacol. Toxicol. *21*, 7 (1981).
14. P. B. M. W. M. Timmermans and P. A. van Zwieten, J. Med. Chem. *25*, 1389 (1982).
15. K. Starke, Rev. Physiol. Biochem. Pharmacol. *88*, 199 (1981).
16. S. Z. Langer, Biochem. Pharmacol. *23*, 1793 (1974).
17. P. B. M. W. M. Timmermans and P. A. van Zwieten, J. Auton. Pharmacol. *1*, 171 (1981).
18. S. Z. Langer and N. B. Shepperson, Trends Pharmacol. Sci. *3*, 440 (1982).
19. J. C. McGrath, Biochem. Pharmacol. *31*, 467 (1982).
20. S. Berthelsen and W. A. Pettinger, Life Sci. *21*, 595 (1977).
21. J. E. S. Wikberg, Acta Physiol. Scand. (Suppl.) *468*, 1 (1979).
22. K. Starke and S. Z. Langer, in Presynaptic Receptors (S. Z. Langer, K. Starke, and M. L. Dubocovich, eds.), Pergamon Press, Oxford, 1979, pp. 1-3.
23. K. Starke, T. Endo, and H. D. Taube, Naunyn-Schmiedebergs Arch. Pharmacol. *291*, 55 (1975).
24. J. C. A. van Meel, A. de Jonge, P. B. M. W. M. Timmermans, and P. A. van Zwieten, J. Pharmacol. Exp. Ther. *219*, 760 (1981).
25. W. Kobinger and L. Pichler, Eur. J. Pharmacol. *73*, 313 (1981).
26. P. B. M. W. M. Timmermans, H. Y. Kwa, and P. A. van Zwieten, Naunyn-Schmiedebergs Arch. Pharmacol. *310*, 189 (1979).
27. N. A. Flavahan and J. C. McGrath, Br. J. Pharmacol. *72*, 585P (1981).
28. M. J. Mathy, H. N. Doods, M. J. M. C. Thoolen, B. Wilffert, A. de Jonge, P. B. M. W. M. Timmermans, and P. A. van Zwieten, J. Auton. Pharmacol. *3*, 249 (1983).
29. R. M. DeMarinis and J. P. Hieble, J. Med. Chem. *26*, 1215 (1983).
30. P. B. M. W. M. Timmermans, M. J. Mathy, H. N. Doods, M. J. M. C. Thoolen, A. de Jonge, B. Wilffert, P. A. van Zwieten, R. M. DeMarinis, J. Lafferty, R. A. Macia, and W. D. Matthews, Eur. J. Pharmacol. *101*, 45 (1984).
31. A. de Jonge, J. C. A. van Meel, P. B. M. W. M. Timmermans, and P. A. van Zwieten, Life Sci. *28*, 2000 (1981).
32. M. J. Mathy, M. J. M. C. Thoolen, P. B. M. W. M. Timmermans, and P. A. van Zwieten, Br. J. Pharmacol. *81*, 255 (1984).
33. R. R. Ruffolo, Jr., and J. E. Waddell, J. Pharmacol. Exp. Ther. *222*, 29 (1982).

34. R. M. DeMarinis, W. M. Bryan, D. H. Shah, J. P. Hieble, and R. G. Pendleton, J. Med. Chem. *24*, 1432 (1982).
35. R. M. DeMarinis, D. H. Shah, R. F. Hall, J. P. Hieble, and R. G. Pendleton, J. Med. Chem. *25*, 136 (1982).
36. A. de Jonge, F. P. Slothorst-Grisdijk, P. B. M. W. M. Timmermans, and P. A. van Zwieten, Eur. J. Pharmacol. *71*, 411 (1981).
37. P. B. M. W. M. Timmermans and P. A. van Zwieten, Naunyn-Schmiedebergs Arch. Pharmacol. *313*, 17 (1980).
38. L. Pichler, H. Hortnagl, and W. Kobinger, Naunyn-Schmiedebergs Arch. Pharmacol. *320*, 110 (1982).
39. J. R. Docherty and J. C. McGrath, Naunyn-Schmiedebergs Arch. Pharmacol. *312*, 107 (1980).
40. D. Cambridge, Eur. J. Pharmacol. *72*, 413 (1981).
41. P. E. Hicks, and C. Waldron, Br. J. Pharmacol. *74*, 254P (1980).
42. G. M. Drew, Eur. J. Pharmacol. *65*, 85 (1980).
43. P. B. M. W. M. Timmermans, B. Wilffert, D. Davidesko, M. J. Mathy, D. Dijkstra, A. S. Horn, and P. A. van Zwieten, Eur. J. Pharmacol. *225*, 407 (1984).
44. P. B. M. W. M. Timmermans, M. J. Mathy, B. Wilffert, H. O.Kalkman, G. Smit, D. Dijkstra, A. S. Horn, and P. A. van Zwieten, Eur. J. Pharmacol. *97*, 55 (1984).
45. P. E. Hicks, C. Waldron, P. Burn, and P. A. Crooke, J. Pharm. Pharmacol. *35*, 94 (1983).
46. D. Cambridge, M. J. Davey, and R. Massingham, Br. J. Pharmacol. *59*, 514P (1977).
47. R. Weitzell, T. Tanaka, and K. Starke, Naunyn-Schmiedebergs Arch. Pharmacol. *308*, 127 (1979).
48. G. M. Drew, Br. J. Pharmacol. *64*, 292 (1978).
49. R. T. Brittain, G. M. Drew, and G. P. Levy, Br. J. Pharmacol. *73*, 282P (1981).
50. H. Kapur and D. R. Mottram, Biochem. Pharmacol. *27*, 1879 (1978).
51. P. Mouille, A. M. Huchet, J. Chelly, B. Lucet, M. F. Doursout, and H. Schmitt, J. Cardiovasc. Pharmacol. *2*, 175 (1980).
52. P. B. M. W. M. Timmermans and P. A. van Zwieten, in Handbook of Hypertension, Vol. 3, Pharmacology of Antihypertensive Drugs (P. A. van Zwieten, ed.), Elsevier, Amsterdam, 1984, pp. 239-248.
53. P. B. M. W. M. Timmermans, H. Y. Kwa, F. Karamat Ali, and P. A. van Zwieten, Arch. Int. Pharmacodyn. *245*, 218 (1980).
54. L. Eriksson, H. Karappanen, E. Honkanen, and U. M. Kokkonen, Arch. Pharmacol. Toxicol. *54*, 158 (1984).
55. T. Kawasaki, K. Uezono, I. Abe, S. Nakamura, M. Ueno, N. Kawazoe, and T. Omae, Eur. J. Clin. Pharmacol. *20*, 399 (1981).
56. I. Cavero, F. Lefevre-Borg, and P. Manoury, Br. J. Pharmacol. *81*, 13P (1984).
57. C. Melchiorre and B. Belleau, in Adrenoceptors and Catecholamine Action, Vol. I (G. Kunos, ed.), Wiley, New York, 1981, pp. 181-211.

58. H. Kapur, B. Rouot, and S. H. Snyder, Eur. J. Pharmacol. *57*, 317 (1979).
59. P. B. M. W. M. Timmermans, J. E. van Kemenade, H. D. Batink, and P. A. van Zwieten, Pharmacology *26*, 258 (1983).
60. C. Melchiorre, D. Giardina, P. Gallucci, and L. Brasili, J. Pharm. Pharmacol. *34*, 683 (1982).
61. C. Melchiorre, L. Brasili, D. Giardina, M. Pigini, and G. Strappaghetti, J. Med. Chem. *27*, 1536 (1984).
62. D. R. Mottram, J. Pharm. Pharmacol. *31*, 767 (1981).
63. R. R. Ruffolo, Jr., W. L. Nelson, and E. L. Yaden, Naunyn-Schmiedebergs Arch. Pharmacol. *322*, 93 (1983).
64. J. Q. Qian, M. J. Mathy, M. J. M. C. Thoolen, P. B. M. W. M. Timmermans, and P. A. van Zwieten, Arch. Int. Pharmacodyn. *266*, 264 (1983).
65. P. Chatelain, M. Claeys, W. van Dorsser, and J. Roba, Arch. Int. Pharmacodyn. *268*, 271 (1984).
66. J. M. van Neuten, P. A. J. Janssen, J. van Beek, R. Xhonneux, T. J. Verbeuren, and P. M. Vanhoutte, J. Pharmacol. Exp. Ther. *218*, 217 (1981).
67. J. R. Fozard, J. Cardiovasc. Pharmacol. *4*, 829 (1982).
68. H. O. Kalkman, P. B. M. W. M. Timmermans, and P. A. van Zwieten, J. Pharmacol. Exp. Ther. *222*, 227 (1982).
69. W. Schoetensack, P. Bischler, E. C. Dittmann, and V. Steinijans, Arzneimittelforsch. *27*, 1908 (1977).
70. M. Eltze, Eur. J. Pharmacol. *59*, 1 (1979).
71. P. A. van Zwieten, A. de Jonge, B. Wilffert, P. B. M. W. M. Timmermans, J. J. Beckeringh, and M. J. M. C. Thoolen, Arch. Int. Pharmacodyn. *276*, 180 (1985).
72. N. Heinz and E. Hofferber, Arzneimittelforsch. *30*, 2135 (1980).
73. M. Gothert, C. Dieckhofer, and J. Nolte, J. Cardiovasc. Pharmacol. *5*, 13 (1983).
74. G. Engel and D. Hoyer, Eur. J. Pharmacol. *73*, 221 (1981).
75. J. L. Archibald, Br. J. Clin. Pharmacol. *12*, 45S (1981).
76. J. C. Doxey, C. F. C. Smith, and J. M. Walker, Br. J. Pharmacol. *60*, 91 (1977).
77. P. B. M. W. M. Timmermans, J. C. A. van Meel, and P. A. van Zwieten, J. Auton. Pharmacol. *1*, 53 (1980).
78. N. B. Shepperson, N. Duval, R. Massingham, and S. Z. Langer, J. Pharmacol. Exp. Ther. *219*, 540 (1981).
79. J. C. Doxey, A. C. Lane, A. G. Roach, and N. K. Virdee, Naunyn-Schmiedebergs Arch. Pharmacol. *325*, 136 (1984).
80. P. B. M. W. M. Timmermans, J. Q. Qian, R. R. Ruffolo, Jr., and P. A. van Zwieten, J. Pharmacol. Exp. Ther. *228*, 739 (1984).
81. C. B. Chapleo, P. L. Myers, R. C. M. Butler, J. C. Doxey, A. G. Roach, and C. F. C. Smith, J. Med. Chem. *26*, 823 (1983).
82. J. C. Doxey, A. G. Roach, M. R. Stillings, D. A. Strachan, and N. Virdee, Br. J. Pharmacol. *79*, 311P (1983).
83. J. C. Doxey, A. G. Roach, M. R. Stillings, D. A. Strachan, N. Virdee, and A. P. Welbourn, Br. J. Pharmacol. *81*, 181P (1983).

84. A. D. Michel, S. R. Nahorski, and R. L. Whiting, Br. J. Pharmacol. *74*, 845p (1981).
85. J. M. Caroon, R. D. Clark, A. F. Kluge, R. Olah, D. B. Repke, S. H. Unger, A. D. Michel, and R. L. Whiting, J. Med. Chem. *25*, 666 (1982).
86. W. S. Saari, W. Halczenko, S. W. King, J. R. Huff, J. P. Guare, C. A. Hunt, W. C. Randall, P. S. Anderson, V. J. Lotti, D. A. Taylor, and B. V. Clineschmidt, J. Med. Chem. *26*, 1696 (1983).
87. I. B. Armah, D. R. Ferry, and H. Glossmann, Br. J. Pharmacol. *78*, 151P (1983).
88. R. M. DeMarinis, J. P. Hieble, and W. D. Matthews, J. Med. Chem. *26*, 1213 (1983).
89. V. Pierce and J. F. Waterfall, Br. J. Pharmacol. *76*, 263p (1982).
90. P. M. Paciorek, V. Pierce, N. B. Shepperson, and J. F. Waterfall, Br. J. Pharmacol. *82*, 127 (1984).
91. C. L. Wood, C. D. Arnett, W. R. Clarke, B. S. Tsai, and R. J. Lefkowitz, Biochem. Pharmacol. *28*, 1277 (1979).
92. D. C. U'Prichard and S. H. Snyder, Life Sci. *24*, 79 (1979).
93. B. B. Hoffmann and R. J. Lefkowitz, N. Engl. J. Med. *302*, 1390 (1980).
94. H. Glossmann, R. Hornung, and P. Presek, J. Cardiovasc. Pharmacol. (Suppl.) *2*, S303 (1980).
95. D. B. Bylund and D. C. U'Prichard, Int. Rev. Neurobiol. *24*, 343 (1983).
96. D. A. Greenberg and S. H. Snyder, Mol. Pharmacol. *14*, 38 (1978).
97. S. J. Peroutka, D. A. Greenberg, D. C. U'Prichard, and S. H. Snyder, Mol. Pharmacol. *14*, 403 (1978).
98. P. J. Miach, J. P. Dausse, and P. Meyer, Nature *274*, 492 (1978).
99. B. B. Hoffmann, A. DeLean, C. L. Wood, D. D. Schocken, and R. J. Lefkowitz, Life Sci. *24*, 1739 (1979).
100. D. C. U'Prichard, D. A. Greenberg, and S. H. Snyder, Mol. Pharmacol. *13*, 454 (1977).
101. P. J. Miach, J. P. Dausse, A. Cardot, and P. Meyer, Naunyn-Schmiedebergs Arch. Pharmacol. *312*, 23 (1980).
102. B. M. Rouot and S. H. Snyder, Life Sci. *25*, 769 (1979).
103. D. J. Loftus, J. M. Stolk, and D. C. U'Prichard, Life Sci. *35*, 61 (1984).
104. P. B. M. W. M. Timmermans, A. M. C. Schoop, and P. A. van Zwieten, Biochem. Pharmacol. *31*, 899 (1982).
105. P. B. M. W. M. Timmermans, J. E. van Kemenade, Y. M. Harms, G. Prop, E. Graf, and P. A. van Zwieten, Arch. Int. Pharmacodyn. *261*, 23 (1983).
106. M. Daiguji, H. Y. Meltzer, and D. C. U'Prichard, Life Sci. *28*, 2705 (1981).
107. B. D. Perry and D. C. U'Prichard, Eur. J. Pharmacol. *76*, 294p (1982).
108. D. R. Howlett, P. Taylor, and D. S. Walter, Br. J. Pharmacol. *76*, 294P (1982).
109. T. F. Lyon and W. C. Randall, Life Sci. *26*, 1121 (1980).
110. P. Greengrass and R. Bremner, Eur. J. Pharmacol. *55*, 323 (1979).
111. R. Hornung, P. Presek, and H. Glossmann, Naunyn-Schmiedebergs Arch. Pharmacol. *308*, 223 (1979).
112. G. Engel and D. Hoyer, Eur. J. Pharmacol. *73*, 221 (1981).

113. W. Kobinger and L. Pichler, Eur. J. Pharmacol. *82*, 203 (1982).
114. J. R. Docherty, Br. J. Pharmacol. *78*, 655 (1983).
115. P. Lasch and K. H. Jakobs, Naunyn-Schmiedebergs Arch. Pharmacol. *306*, 119 (1979).
116. L. M. Lafontan and M. Berlan, Trends Pharmacol. Sci. *2*, 126 (1981).
117. T. Kakaki, T. Nakadate, and R. Kato, Naunyn-Schmiedebergs Arch. Pharmacol. *313*, 151 (1980).
118. S. Yamazaki, T. Katada, and M. Ui, Mol. Pharmacol. *21*, 648 (1982).
119. H. C. Innemee, A. de Jonge, J. C. A. van Meel., P. B. M. W. M. Timmermans, and P. A. van Zwieten, Naunyn-Schmiedebergs Arch. Pharmacol. *316*, 294 (1981).
120. J. F. DiJoseph, J. A. Taylor, and G. N. Mir, Life Sci. *35*, 1031 (1984).
121. H. Schmitt, Actual. Pharmacol. (Paris) *24*, 93 (1971).
122. P. A. van Zwieten, Prog. Pharmacol. *1*, 1 (1975).
123. H. Schmitt, in Antihypertensive Agents (F. Gross, ed.), Springer-Verlag, Berlin, 1977, pp. 299-396.
124. W. Kobinger, Rev. Physiol. Biochem. Pharmacol. *81*, 39 (1978).
125. P. B. M. W. M. Timmermans, in Handbook of Hypertension, Vol. 3, Pharmacology of Antihypertensive Drugs (P. A. van Zwieten, ed.), Elsevier, Amsterdam, 1984, pp. 102-153.
126. P. B. M. W. M. Timmermans, A. M. C. Schoop, H. Y. Kwa, and P. A. van Zwieten, Eur. J. Pharmacol. *70*, 7 (1981).
127. W. Kobinger and L. Pichler, J. Cardiovasc. Pharmacol. *4*, S81 (1982).
128. G. L. Hauesler, J. Cardiovasc. Pharmacol. *4*, S72 (1982).
129. W. S. Young and M. J. Kuhar, Proc. Natl. Acad. Sci. USA *77*, 1696 (1980).
130. W. de Jong and M. J. Petty, J. Cardiovasc. Pharmacol. *4*, S77 (1982).
131. N. E. Anden, M. Grabowska, and U. Strombom, Naunyn-Schmiedebergs Arch. Pharmacol. *292*, 43 (1976).
132. P. B. M. W. M. Timmermans, W. Hoefke, H. Stahle, and P. A. van Zweiten, Prog. Pharmacol. *3* (1), 1 (1980).
133. H. Lal and S. Fielding (eds.), Psychopharmacology of Clonidine, Alan R. Liss, New York, 1981.
134. G. M. Drew, A. J. Gower, and A. S. Mariott, Br. J. Pharmacol. *67*, 133 (1979).
135. I. Zebrowska-Lupina, E. Przegalinski, M. Sloniec, and Z. Kleinrock, Naunyn-Schmiedebergs Arch. Pharmacol. *297*, 227 (1977).
136. L. Pichler and W. Kobinger, Naunyn-Schmiedebergs Arch. Pharmacol. *317*, 180 (1981).
137. S. Fielding, T. C. Spaulding, and H. Lal, in Psychopharmacology of Clonidine (H. Lal and S. Fielding, eds.), Alan R. Liss, New York, 1981, pp. 226-242.
138. I. Gil-Ad, E. Topper, and Z. Zaron, Lancet *2*, 278 (1979).
139. G. R. van Loon, U. Scapagnini, G. P. Moberg, and W. F. Ganong, Endocrinology *89*, 1464 (1971).
140. R. A. Jauernig, R. F. W. Moulds, and J. Shaw, Arch. Int. Pharmacodyn. *231*, 81 (1978).

141. P. B. M. W. M. Timmermans and P. A. van Zwieten, J. Auton Pharmacol. *1*, 171 (1981).
142. P. B. M. W. M. Timmermans, J. C. A. van Meel, and P. A. van Zweiten, J. Auton. Pharmacol. *1*, 53 (1980).
143. S. Z. Langer and P. E. Hicks, J. Cardiovasc. Pharmacol. *6*, S547 (1984).
144. P. van Brummelen, K. Jie, P. Vermey, P. B. M. W. M. Timmermans, and P. A. van Zwieten, Clin. Sci. *68* (Suppl. 10), 151s (1985).
145. H. O. Kalkman, M. J. M. C. Thoolen, P. B. M. W. M. Timmermans, and P. A. van Zwieten, J. Pharm. Pharmacol. *36*, 265 (1984).
146. P. Zandberg, P. B. M. W. M. Timmermans, and P. A. van Zwieten, J. Cardiovasc. Pharmacol. *6*, 256 (1984).
147. M. Gerold and G. Haeusler, Naunyn-Schmiedebergs Arch. Pharmacol. *322*, 29 (1983).
148. E. J. Ariens and A. M. Simonis, in Beta-Adrenoceptor Blocking Agents (P. R. Saxena and R. P. Forsyth, eds.), Elsevier, Amsterdam, 1976, pp. 3-27.
149. E. J. Ariens and A. M. Simonis, Biochem. Pharmacol. *32*, 1539 (1983).
150. I. Yamaguchi and I. J. Kopin, J. Pharmacol. Exp. Ther. *214*, 275 (1980).
151. S. Z. Langer, R. Massingham, and N. B. Shepperson, Clin. Sci. *59*, 225s (1980).
152. B. Wilffert, P. B. M. W. M. Timmermans, and P. A. van Zwieten, J. Pharmacol. Exp. Ther. *221*, 762 (1982).
153. I. C. Medgett, P. E. Hicks, and S. Z. Langer, J. Pharmacol. Exp. Ther. *231*, 159 (1984).
154. I. C. Medgett, Eur. J. Pharmacol. *108*, 281 (1985).
155. N. A. Flavahan, T. J. Rimele, J. P. Cooke, and P. M. Vanhoutte, J. Pharmacol. Exp. Ther. *230*, 699 (1984).
156. J. R. Docherty and L. Hyland, Br. J. Pharmacol. *84*, 573 (1985).
157. M. Saeed, J. Holtz, D. Elsner, and E. Bassenge, J. Cardiovasc. Pharmacol. *7*, 167 (1985).
158. J. C. Gardiner and C. J. Peters, Eur. J. Pharmacol. *84*, 189 (1982).
159. D. Elsner, M. Saeed, O. Sommer, J. Holtz, and E. Bassenge, Hypertension *6*, 915 (1984).
160. I. F. A. Hesse and E. J. Johns, J. Auton. Pharmacol. *4*, 145 (1984).
161. B. Wilffert, M. A. M. Gouw, A. de Jonge, P. B. M. W. M. Timmermans, and P. A. van Zwieten, J. Pharmacol. Exp. Ther. *223*, 219 (1982).
162. B. Wilffert, D. Davidesko, P. B. M. W. M. Timmermans, and P. A. van Zwieten, J. Pharmacol. Exp. Ther. *226*, 855 (1983).
163. J. N. Fain and J. A. Garcia-Sainz, Life Sci. *26*, 1183 (1980).
164. K. H. Jakobs and G. Schultz, J. Cardiovasc. Pharmacol. *4*, S63 (1982).
165. R. Kato and T. Nakaki, Trends Pharmacol. Sci. *4*, 34 (1983).
166. L. E. Limbird, Trends Pharmacol. Sci. *4*, 135 (1983).
167. S. R. Nahorski, D. B. Barnett, and Y. D. Cheung, Clin. Sci. *68* (Suppl. 10), 39s (1985).
168. J. L. Boyer, C. Cardenas, C. Pasadas, and J. A. Garcia-Sainz, Life Sci. *33*, 2627 (1983).

169. B. B. Hoffman, D. Mullikin-Kilpatrick, and R. J. Lefkowitz, J. Biol. Chem. *255*, 4645 (1980).
170. R. S. Kent, A. DeLean, and R. J. Lefkowitz, Mol. Pharmacol. *17*, 14 (1980).
171. M. J. Berridge, Biochem. J. *220*, 345 (1984).
172. R. H. Michell, Clin. Sci. *68* (Suppl. 10), 43s (1985).
173. T. Takenawa, Cell Calcium *3*, 359 (1982).
174. D. F. Bohr, Circ. Res. *32*, 665 (1973).
175. A. Fleckenstein, Annu. Rev. Pharmacol. Toxicol. *17*, 149 (1977).
176. P. H. Ratz and S. F. Flaim, in Calcium Blockers (S. F. Flaim and R. Zelis, eds.), Urban and Schwartzenberg, Baltimore, 1982, pp. 77-98.
177. C. Cauvin, R. Loutzenhiser, and C. van Breemen, Annu. Rev. Pharmacol. Toxicol. *23*, 373 (1983).
178. P. B. M. W. M. Timmermans and M. J. M. C. Thoolen, in The Alpha-1 Adrenergic Receptors (R. R. Ruffolo, Jr., ed.), Humana Press, Clifton, New Jersey, 1987, pp. 113-187.
179. P. B. M. W. M. Timmermans, M. J. M. C. Thoolen, A. de Jonge, B. Wilffert, and P. A. van Zwieten, in α-Adrenoceptor Blockers in Cardiovascular Disease (H. Refsum and O. D. Mjos, eds.), Churchill Livingstone, London, 1985, pp. 30-50.
180. P. B. M. W. M. Timmermans, M. J. Mathy, B. Wilffert, H. O. Kalkman, M. J. M. C. Thoolen, A. de Jonge, J. C. A. van Meel, and P. A. van Zwieten, Naunyn-Schmiedebergs Arch. Pharmacol. *324*, 239 (1983).
181. P. B. M. W. M. Timmermans, W. D. Matthews, R. M. DeMarinis, J. P. Hieble, M. J. Mathy, H. N. Doods, M. J. M. C. Thoolen, A. de Jonge, B. Wilffert, and P. A. van Zwieten, Eur. J. Pharmacol. *101*, 45 (1984).
182. P. B. M. W. M. Timmermans, M. J. M. C. Thoolen, M. J. Mathy, B. Wilffert, A. de Jonge, and P. A. van Zwieten, Eur. J. Pharmacol. *96*, 187 (1983).
183. M. I. Holck, C. H. M. Jones, and G. Haeusler, J. Cardiovasc. Pharmacol. *5*, 240 (1983).
184. J. J. Beckeringh, M. J. M. C. Thoolen, A. de Jonge, B. Wilffert, P. B. M. W. M. Timmermans, and P. A. van Zwieten, J. Pharmacol. Exp. Ther. *229*, 515 (1984).
185. E. Muller-Schweinitzer, Naunyn-Schmiedebergs Arch. Pharmacol. *324*, 64 (1983).
186. S. Kalsner, M. Suleiman, and R. E. Dobson, J. Pharm. Pharmacol. *32*, 290 (1980).
187. R. M. Graham, W. H. Stephenson, and W. A. Pettinger, Naunyn-Schmiedebergs Arch. Pharmacol. *311*, 129 (1980).
188. C. A. Hamilton, J. L. Reid, and C. Zamboulis, Br. J. Pharmacol. *75*, 417 (1982).
189. C. T. Dollery, G. A. Fitzgerald, and J. Watkins, Br. J. Clin. Pharmacol. *8*, 396P (1979).
190. M. F. Sugrue, Biochem. Pharmacol. *32*, 1811 (1983).
191. F. Sulser, Trends Pharmacol. Sci. *1*, 92 (1979).

192. M. F. Sugrue, Life Sci. *28*, 377 (1981).
193. A. N. M. Schoffelmeer and A. H. Mulder, Naunyn-Schmiedebergs Arch. Pharmacol. *318*, 173 (1982).
194. E. Schlicker, M. Gothert, and R. Clausing, Naunyn-Schmiedebergs Arch. Pharmacol. *320*, 38 (1982).
195. F. Cerrito and M. Raiteri, Eur. J. Pharmacol. *70*, 425 (1981).
196. P. A. van Zwieten (ed.), Handbook of Hypertension, Vol. 3, Pharmacology of Antihypertensive Drugs, Elsevier, Amsterdam, 1984.
197. W. D. Matthews, J. P. Hieble, and R. M. DeMarinis, IUPHAR 9th International Congress of Pharmacology, London, 1984, Abstract 1901P.
198. M. S. Gold, D. E.Redmond, and H. D. Kleber, Lancet *2*, 599 (1978).
199. H. Lal and G. T. Shearman, Prog. Clin. Biol. Res. *71*, 99 (1981).
200. H. Lal and S. Fielding, Trends Pharmacol. Sci. *4*, 70 (1983).
201. M. A. Weber, J. Cardiovasc. Pharmacol. *2* (Suppl. 1), S73 (1980).
202. M. J. M. C. Thoolen, P. B. M. W. M. Timmermans, and P. A. van Zwieten, Gen. Pharmacol. *12*, 303 (1981).
203. J. A. Steiner, A. J. Wilkins, and W. Jenkins, Br. J. Clin. Pharmacol. *15*, 583P (1983).
204. C. Pintor, R.Corda, R. Puggioni, S. Cella, V. Locatelli, S. Loche, and E. Muller, Lancet *1*, 1482 (1985).
205. D. E. Redmond and Y. H. Huang, Life Sci. *25*, 2149 (1979).
206. R. Hoehnsaric, A. F. Merchant, M. L. Keyser, and V. K. Smith, Arch. Gen. Psychiatry *38*, 1278 (1981).
207. D. E. Redmond, Trends Pharmacol. Sci. *3*, 477 (1982).
208. D. S. Charney, G. R. Heninger, D. E. Sternberg, K. M. Hafstad, S. Giddings, and D. H. Landis, Arch. Gen. Psychiatry *39*, 290 (1982).
209. D. S. Charney, G. R. Heninger, and D. E. Redmond, Life Sci. *33*, 19 (1983).
210. T. Reisine, R. Johnson, N. Wiech, R. Ursillo, and H. I. Yamamura, Adv. Biochem. Psychopharmacol. *31*, 63 (1982).
211. J. A. Scott and F. T. Crews, J. Pharmacol. Exp. Ther. *224*, 640 (1983).
212. A. Dye, C. M. Page, and R. L. Whiting, Br. J. Pharmacol. *80*, 665p (1983).
213. D. I. Crossley, IUPHAR 9th International Congress of Pharmacology, London, 1984, Abstract 1724P.
214. N. S. Doherty and A. A. Hancock, J. Pharmacol. Exp. Ther. *225*, 269 (1982).
215. C. Korstanje, H. N. Doods, A. de Jonge, M. J. M. C. Thoolen, B. Wilffert, P. B. M. W. M. Timmermans, and P. A. van Zwieten, J. Auton. Pharmacol. *4*, 287 (1984).
216. P. B. M. W. M. Timmermans, M. J. Mathy, M. J. M. C. Thoolen, A. de Jonge, B. Wilffert, and P. A. van Zwieten, J. Pharm. Pharmacol. *36*, 772 (1984).
217. B. Wilffert, P. N. M. van Heiningen, M. J. Mathy, H. D. Batink, A. de Jonge, M. J. M. C. Thoolen, P. B. M. W. M. Timmermans, and P. A. van Zwieten, Naunyn-Schmiedebergs Arch. Pharmacol. *327*, 90 (1984).

8

β Adrenoceptors

Otto-Erich Brodde / University of Essen, Essen, Federal Republic of Germany

HISTORICAL SURVEY

Adrenoceptors were originally introduced into the pharmacology of the sympathetic nervous system to explain the differences in the action of the endogenous catecholamines norepinephrine and epinephrine. The first evidence that epinephrine may exert its effects at different receptors was presented in the classic work of Sir Henry Dale (1,2) in the years 1906-1913, who demonstrated in the despinalized cat that the blood pressure-increasing effect of epinephrine could be reversed into a blood pressure-decreasing effect after pretreatment of the cats by ergot alkaloids. Dale interpreted his data to mean that the action of epinephrine is composed of a mixture of two components: an excitatory (blood pressure increasing) and an inhibitory (blood pressure decreasing) effect. The conclusion, however, that epinephrine produces its different effects through stimulation of different receptors was drawn by Ahlquist (3) in 1948. He presented evidence for the existence of two different receptors in the sympathetic nervous system by those classic pharmacological methods that are still valid for the identification of different receptors: Ahlquist determined on isolated organ systems the responses to epinephrine, norepinephrine, α methylnorepinephrine, α methylepinephrine, and isoproterenol and found that these agonists produced contraction of smooth muscles of the peripheral blood vessels (vasoconstriction), nictitating membrane, uterus, and musculus dilatator pupillae with an order of potency: epinephrine > norepinephrine > α methylnorepinephrine > α methylepinephrine > isoproterenol. On the other hand, for relaxation of smooth muscle preparations and excitation (increase in force of contraction and rates of beating) of the heart the order of potency was reversed to isoproterenol > epinephrine

> α methylepinephrine > α methylnorepinephrine > norepinephrine. From these different orders of potency for the same agonists, Ahlquist concluded that the effects of sympathomimetic amines are mediated by different receptors, which he named α (for vasoconstriction in smooth muscle cells) (see Chapter 7) and β (for vasodilation in smooth muscle cells or for positive inotropic effects in the heart). This concept could be supported by the synthesis of various specific antagonists, such as phentolamine, phenoxybenzamine (α) or propranolol, pindolol, and alprenolol (β) (Figure 1).

Besides different orders of potency for agonists and specific antagonists, α and β adrenoceptors exert their effects by different mechanisms of action. Although β-adrenoceptor stimulation always leads to an increase in the intracellular level of cyclic AMP (5-8), α-adrenoceptor-mediated effects do not increase (8-10) but (under certain circumstances), rather, decrease the intracellular level of cyclic AMP (11,12). Later it became clear that at least two major subtypes of both β and α adrenoceptors can be distinguished—the β_1 and β_2 and the α_1 and the α_2 adrenoceptors—by a variety of pharmacological criteria. α_1 Adrenoceptors are generally localized postsynaptically; they mediate, for example, the classic smooth muscle contraction. α_2 Adrenoceptors are either localized presynaptically, where they mediate feedback inhibition of norepinephrine release from the sympathetic nerve terminals, or postsynaptically, where they have various effects, for example, constriction of vascular smooth muscle, platelet aggregation, or inhibition of renin release from the kidney. Figure 2 summarizes localization and the functional role of α_1 and α_2 adrenoceptors in the sympathetic nerve endings. This Chapter is concerned with the molecular pharmacology of β adrenoceptors; thus for further detailed information on α-adrenoceptor subtypes the reader is referred to Chapter 7.

α Adrenoceptors

Agonists: epinephrine ⩾ norepinephrine > phenylephrine > isoproterenol
Antagonists: phentolamine, phenoxybenzamine

β Adrenoceptors

Agonists: isoproterenal > epinephrine > norepinephrine > phenylephrine
or isoproterenol > norepinephrine > epinephrine > phenylephrine
Antagonists: propranolol, pindolol, alprenolol

Figure 1 Definition of α and β adrenoceptors according to Furchgott (4).

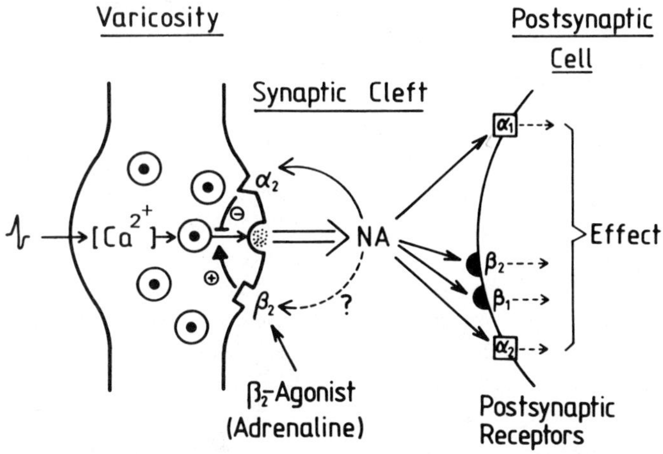

Figure 2 Adrenergic synapse. Stimulation of sympathetic nervous activity evokes release of endogenous norepinephrine (NA). NA reaches postsynaptic α_1 and α_2 as well as β_1 (and in very high concentrations also β_2) adrenoceptors via diffusion; stimulation of these receptors mediates physiological or pharmacological effects. Stimulation of presynaptic α_2 adrenoceptors by endogenous NA or exogenously applied α_2 agonists inhibits NA release (⊖). On the contrary, stimulation of presynaptic β_2 adrenoceptors by endogenous epinephrine (presumably released from the adrenal medulla) or by exogenously applied β_2 agonists facilitates NA release (⊕). Whether NA can stimulate presynaptic β_2 adrenoceptors is still a matter of controversy.

CLASSIFICATION OF β ADRENOCEPTORS

The first suggestion that two distinguished types of β adrenoceptors might exist came from studies on the effects of the α-methyl congeners of methoxamine on β-adrenoceptor-mediated effects in a variety of tissues. Thus, Salvador et al. (13) and Burns et al. (14) observed that isopropyl-methoxamine and butoxamine antagonized catecholamine-evoked increases in plasma glucose, lactic acid, and free fatty acid in anesthetized dogs in concentrations that did not affect cardiac stimulatory effects. Levy (15) and Moran (16) found in dogs and cats, respectively, that butoxamine inhibited isoproterenol-induced vasodilation without affecting cardiac stimulation or intestinal inhibition. In addition, Levy (17) demonstrated that butoxamine in appropriate concentrations inhibited isoproterenol-induced relaxation of the isolated rat uterus, but not of the isolated rabbit jejunum. Lands, Arnold, and coworkers (18-24) assessed the potency of a series of catecholaminergic agonists for producing β-adrenergic effects in both isolated tissues and in intact animals. They found that the rank order of potency for these agonists for various β-adrenoceptor-mediated responses could be subdivided in two major groups; they postulated that these groups reflect the existence of two types of β adrenoceptors, which they named β_1 and β_2. The β adrenoceptors controlling increases in heart rate and force of contraction, lipolysis in rat and guinea pig adipose tissue, inhibition of rabbit small intestine, and calorigenesis in the intact rat were stimulated by catecholamines with a rank order of potency isoproterenol > epinephrine = norepinephrine. They were named β_1 adrenoceptors. The β adrenoceptors controlling vasodepression in the intact dog, bronchodilation in the perfused guinea pig lung, inhibition of rat uterus contraction, facilitation of contraction of the rat diaphragm, and elevation of blood glucose in the dog were stimulated by catecholamines with a rank order of potency isoproterenol > epinephrine > norepinephrine. They were named β_2 adrenoceptors. Concomitantly, Furchgott (25) determined the relative potencies of norepinephrine, epinephrine, isoproterenol and phenylephrine for responses mediated by β adrenoceptors in selected isolated tissues from rat and guinea pig. In some tissues, in addition, the dissociation constant of the β-adrenoceptor antagonist pronethalol was also determined. In all these experiments α adrenoceptors and the adrenergic neuronal uptake were blocked. In accordance with Lands et al. (18,19), Furchgott (25) found two different rank orders of potency for β-adrenoceptor stimulation. On β adrenoceptors in atrial and gastrointestinal smooth muscles (classified by Lands as β_1), it was isoproterenol > norepinephrine = epinephrine > phenylephrine; on β adrenoceptors in aortic smooth muscles classified by Lands as β_2, the rank order of potency was isoproterenol > epinephrine > norepinephrine > phenylephrine. Further experiments by various investigators confirmed the usefulness of this subclassification and extended it to β adrenoceptors in a variety of other tissues and species (see Table 1) (4,26). The most important physiological difference between β_1 and β_2 adrenoceptors is their different sensitivity to norepinephrine. β_1 Adrenoceptors have approximately the same affinity for epinephrine and norepinephrine, and β_2 adrenoceptors have an approximately 10-30 times higher affinity for epinephrine than for norepinephrine. Accordingly, β_1 adrenoceptors can be stimulated by circulating epinephrine or by endogenously released norepinephrine. On the other hand, at β_2 adrenoceptors epinephrine might well be the "natural" agonist. According to these different sensitivities of the endogenous catecholamines, it has been suggested that β_2 adrenoceptors are "hormonal" receptors and β_1 adrenoceptors are "neuronal" receptors (27-29). However, β_1 adrenoceptors can be stimulated both by norepinephrine (i.e., neuronally) and by epinephrine (i.e., hormonally), and β_2 adrenoceptors can be activated by norepinephrine if the concentration is high enough.

In addition to postsynaptic β_1 and β_2 adrenoceptors, the existence of prejunctional β adrenoceptors mediating facilitation of stimulation-induced norepinephrine release has been demonstrated (Figure 2). The first evidence for the

Table 1 β_1- and β_2-Adrenoceptor-Mediated Effects

β_1 Adrenoceptors		β_2 Adrenoceptors	
Effector organ	Response	Effector organ	Response
Heart	Increase in heart rate, increase in contractility glycogenolysis	Heart (human, cat)	Increase in heart rate (increase in contractility?)
Vascular smooth muscle		Vascular smooth muscle	Vasodilation (vasodepression)
(Coronaries)	(Relaxation?)	(Coronaries)	(Relaxation?)
Intestinal smooth muscle	Relaxation	Tracheal and bronchial smooth muscle	Relaxation, antianaphylaxis
Kidney (human)	Renin release	Skeletal muscle	Contraction, glycogenolysis
Adipose tissue			
White	Lipolysis, lipogenesis	Potassium uptake	Hypokalemia
		Urinary bladder	Relaxation
Brown	Calorigenesis (nonshivering thermogenesis)	Uterus	Relaxation
		Liver	Glycogenolysis
		Pancreas	Insulin release
Salivary gland	Secretion	Intestinal sphincter	Relaxation
		Gallbladder	Relaxation
		Vas deferens	Relaxation
		Rat diaphragm	Contraction

presence of prejunctional β adrenoceptors was presented in 1975 by Adler-Graschinsky and Langer (30), who showed that isoproterenol enhanced the stimulation-induced release of norepinephrine from guinea pig isolated atria. In the following years similar observations were made in a variety of isolated cardiac and vascular tissues and other sympathetically innervated tissues from numerous species (for recent reviews, see References 31-33).

These prejunctional β adrenoceptors seem to be of the β_2 subtype: Stjärne and Brundin (34,35) demonstrated on isolated human omental arteries and veins that isoproterenol and the β_2 agonists terbutaline and salbutamol, but not the β_1-selective agonist H110/38, increased stimulation-induced norepinephrine release. On the isolated rat portal vein, terbutaline but not prenalterol enhanced stimulation-induced norepinephrine release; the effect of terbutaline could be inhibited by the β_2-selective antagonist IPS 339 but not by the β_1-selective antagonist practolol (36,37). However, not all studies have shown such clear results; metoprolol especially was found to be an effective antagonist at the presynaptic β adrenoceptors (for review see Reference 32).

Evidence for the existence of prejunctional β adrenoceptors in humans has been obtained from studies on a number of isolated tissues (Table 2). Very recently, Göthert and Hentrich (43) demonstrated on the isolated human pulmo-

Table 2 Human Tissue Containing Facilitatory Prejunctional β Adrenoceptors

Tissue	β-Adrenoceptor subtype determined as to	Reference
Omental arteries and veins	β_2	34,35
Oviduct	–	38
Digital arteries	$(\beta_2{}^a)$?	39
Metatarsal veins	$(\beta_2{}^a)$?	40
Saphenous vein	–[b]	41
Saphenous vein	β_2	(Göthert, personal communication)
Right atrium	β_2	42
Pulmonary artery	β_2	43

[a]Very weak evidence for β_2-adrenoceptor characterization.
[b]β Adrenoceptor not subclassified.

nary artery that isoproterenol and the highly selective β_2-adrenoceptor agonist procaterol, but not prenalterol, enhanced stimulation-induced norepinephrine release. The procaterol-induced effects could be antagonized by the selective β_2 antagonist ICI 118,551, but were not affected by the β_1-selective antagonist atenolol, which strongly supports the view that human prejunctional β adrenoceptors are (predominantly) of the β_2 subtype. The same authors could also demonstrate (44) that cyclic AMP plays a role in the modulation of norepinephrine release in the human pulmonary artery and that presynaptic β_2 adrenoceptors appear to be coupled to the adenylate cyclase system in sympathetic nerve terminals.

AGONISTS AND ANTAGONISTS

The demonstration of the existence of two different types of β adrenoceptors has led to the suggestion that drugs may be developed that show selectivity for each of the two subtypes. In most of the species the β-adrenoceptor mediating increases in heart rate and contractility seem to be predominantly of the β_1 subtype (but see Disease-Induced Changes: Heart Failure), but the β adrenoceptor mediating bronchodilation seems to be predominantly of the β_2 subtype. Thus, β_1-selective drugs may be useful in the treatment of cardiac arrythmias or angina, and β_2-selective drugs may be of beneficial effect in bronchial asthma. During the last 20 years numerous agonists and antagonists have been developed that show physiological selectivity for β-adrenoceptor subtypes in different organs. Several of these agents are now used clinically or as research tools. The structures of commonly used drugs are shown in Figures 3 through 6.

β_1-Selective Agonists

The only β_1-selective full agonist so far available is the endogenous catecholamine norepinephrine; it has an about 10-30 times greater affinity for β_1 than for β_2 adrenoceptors (18,45). In addition, it has been shown in a variety of tissues that the intrinsic activity of norepinephrine both at β_1 and β_2 adrenoceptors is comparable to that of isoproterenol. However, it should be remembered that norepinephrine is a potent α-adrenoceptor stimulant, and, in addition, it is rapidly inactivated by degradation and is a good substrate for neuronal and extraneuronal uptake systems. Thus, in order to get a selective β_1-agonistic effect, norepinephrine must be used in the presence of saturating concentrations of an α antagonist, a carefully calculated concentration of a β_2 antagonist, and blockade of uptake 1 and uptake 2.

A few synthetic drugs have been classified as selective β_1-adrenoceptor agonists by their ability to produce cardiac stimulant effects without concomitant changes in blood pressure: dobutamine, prenalterol, xamoterol (ICI 118,587),

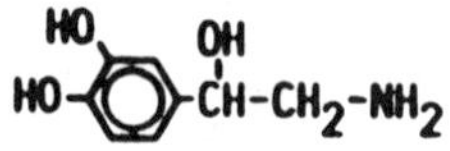

Noradrenaline

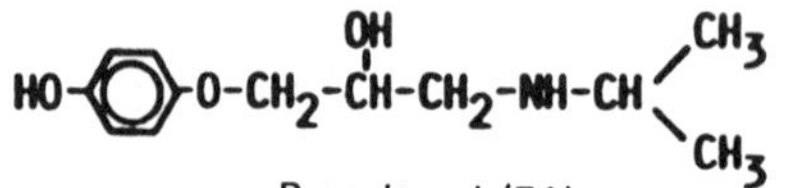

Prenalterol (51)

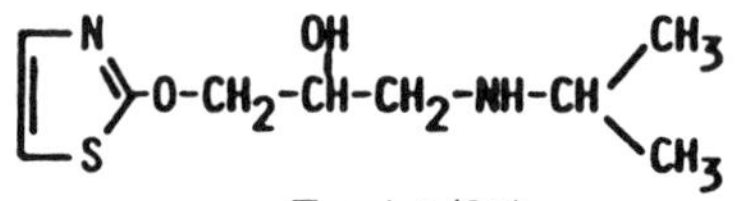

Tazolol (67)

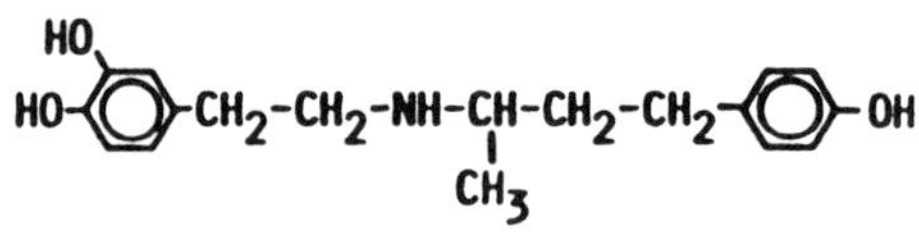

Dobutamine (46,47)

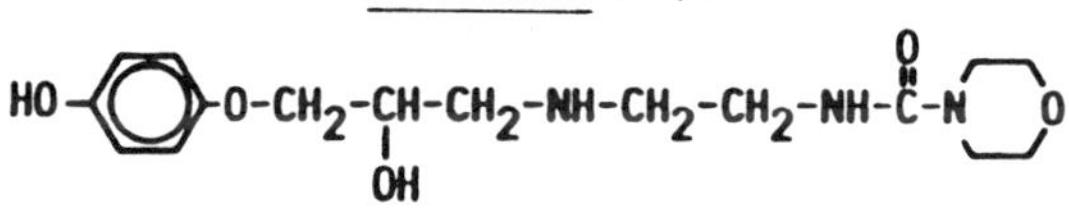

Xamoterol (60)

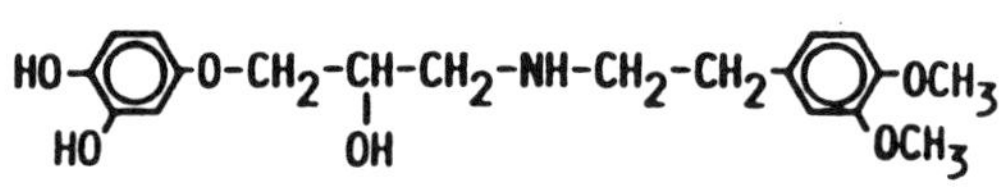

RO-363 (62)

Figure 3 Structural formulas of β_1-adrenoceptor agonists.

and RO-363 (Figure 3). Dobutamine was originally postulated to be a β_1-selective agonist because it produced inotropic effects in concentrations that have little or no effect on heart rate (46,47). However, recent studies have demonstrated that (±)-dobutamine has a preferential agonistic activity at α_1 rather than at α_2 adrenoceptors (48) and that it is an unselective β_1- *and* β_2-adrenoceptor agonist at β adrenoceptors (49,50). (±)-Dobutamine induces α-adrenoceptor actions in blood vessels in a concentration range that is very similar to that stimulating β adrenoceptors; similarly, (±)-dobutamine is also able to stimulate myocardial α and β adrenoceptors in a similar concentration range (50). Both the (–)- and the (+)-isomer of dobutamine are agonists at β_1 and β_2 adrenoceptors, with the (+)-isomer more potent (49,50). Like dobutamine, prenalterol was originally classified as a selective β_1-adrenoceptor agonist based on the findings that it stimulates cardiac β adrenoceptors in experimental animals and in humans at concentrations that do not affect mean arterial blood pressure (51,52). However, studies using isolated tissues have yielded conflicting results on the β_1 selectivity of this drug (53-58). In addition, the maximum β-adrenoceptor-stimulating activity of prenalterol seems to be highly variable. This has led to the conclusion that the apparent β_1 selectivity (cardioselectivity) of prenalterol is caused by tissue rather than receptor selectivity (53,56,57). These properties of prenalterol are due to its ability to cause functional responses by stimulation of either β_1 or β_2 adrenoceptors only if the tissue has a large receptor reserve or a highly efficient stimulus-response coupling mechanism, as indicated by a high sensitivity to the isoproterenol response (for references, see Reference 59). Xamoterol (ICI 118,587) has been shown in vitro and in vivo to be a partial agonist with selective β_1-adrenoceptor activity (60,61). Its intrinsic activity is approximately 50% of that of isoproterenol; however, it seems to be slightly less effective than prenalterol (61). In vitro experiments have provided evidence that xamoterol possesses a 40- to 100-fold higher affinity at β_1 than at β_2 adrenoceptors (58,61). It has been suggested that the selective cardiostimulatory effect of xamoterol is (as in the case of prenalterol) due to tissue rather than receptor selectivity (61).

Very recently, a new β_1-selective stimulant drug has been introduced, RO 363 (62). This agent is a partial agonist; however, at β_1 adrenoceptors its intrinsic activity is very similar to that of isoproterenol (62,63). The selective β_1-adrenoceptor stimulant activity of RO 363 has been shown to be caused by its greater affinity and efficacy for β_1 than for β_2 adrenoceptors (63).

β_1-Selective Antagonists

β_1-Selective antagonists contain an aminopropanolic side chain linked to a phenolic oxygen with a para substituent (Figure 4). Practolol was the first of these drugs to be introduced (64). It was shown to antagonize catecholamine-induced cardiac effects much more potently than the effects on tracheal or bronchial smooth muscles (64). Thereafter, several additional β_1-selective antagonists were developed: metoprolol, atenolol, betaxolol, celiprolol, and, very recently, bisoprolol (for formulas, see Figure 4). All these agents preferentially block cardiac effects mediated by β adrenoceptors and are therefore of beneficial effect in the treatment of cardiac arrhythmias and angina. In addition, they have proven to be very effective as antihypertensive agents (65,66).

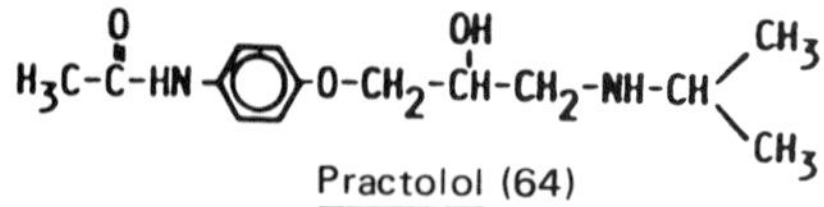

Practolol (64)

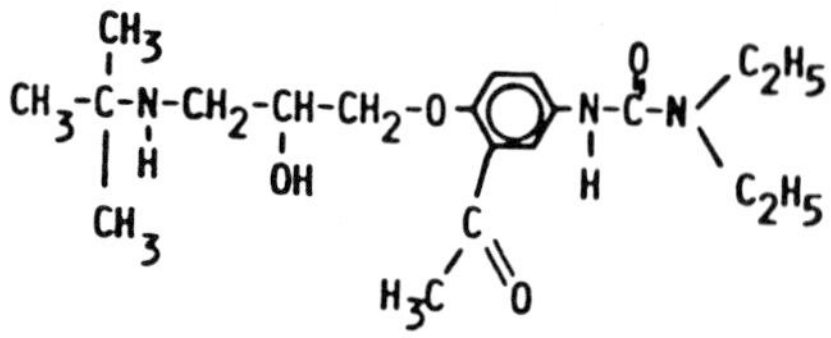

Celiprolol (72)

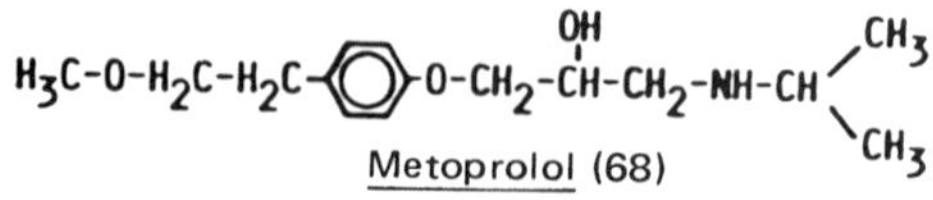

Metoprolol (68)

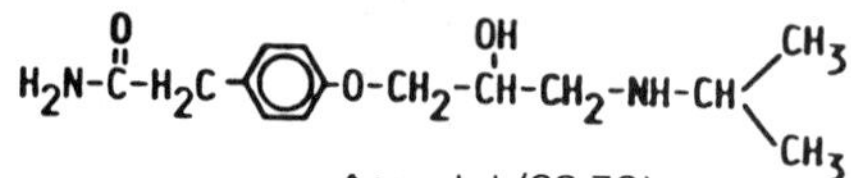

Atenolol (69,70)

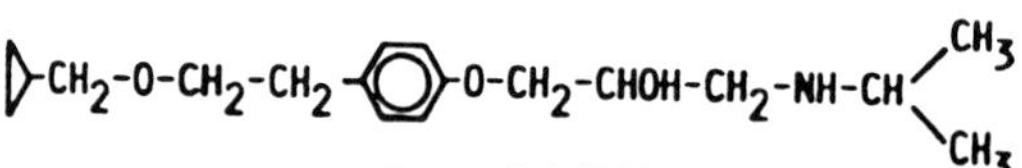

Betaxolol (71)

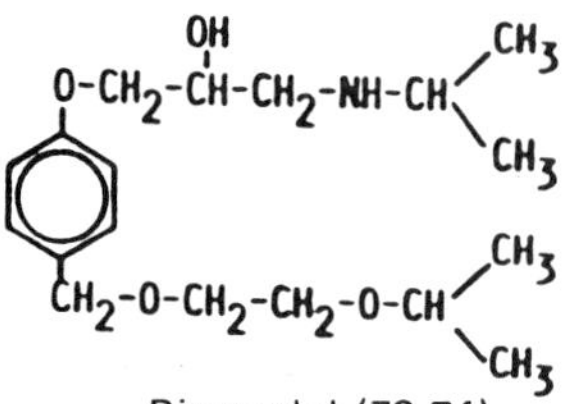

Bisoprolol (73,74)

Figure 4 Structural formulas of β_1-adrenoceptor antagonists.

β_2-Selective Agonists

The β_1-β_2 subclassification by Lands et al. (18,19) and the discovery that non-catechol drugs, such as soterenol or metaproterenol, were useful bronchodilators that caused minimal cardiac stimulation led to the development of numerous β_2-selective agonists (Figure 5). All these agents contain a β-hydroxyphenyl-ethylamine moiety with oxygen or nitrogen substituents at the meta and/or para position of the benzene ring (Figure 5). Detailed structure-activity studies for β_2-adrenoceptor agonists have led to the conclusion that an increase in size of the terminal amine substituent seems to favor β_2-adrenoceptor selectivity (75-77). In addition, introduction of an aromatic ring on the alkyl chain increases the potency. Furthermore, it is worthwhile to note that these agents (except hexoprenaline) do not contain a catechol moiety and are thus resistant to degradation by catechol-*O*-methyltransferase.

β_2-Selective Antagonists

The first rather selective β_2-adrenoceptor antagonists were several congeners of methoxamine, including butoxamine and H-35/25 (Figure 6) (15). These drugs

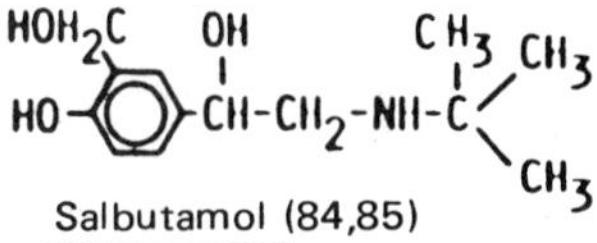

Salbutamol (84,85)

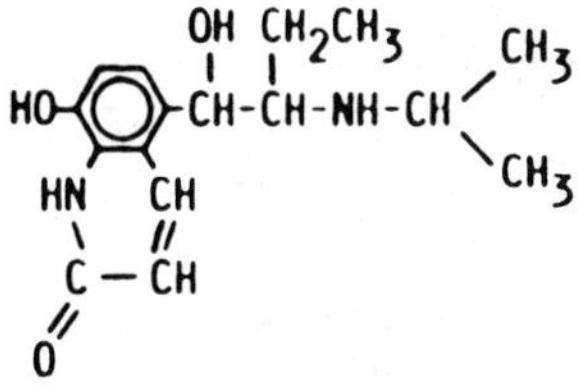

Procaterol (92,93)

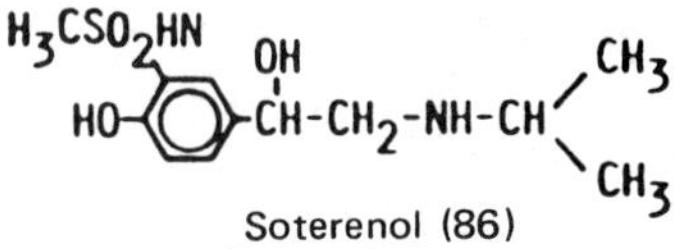

Soterenol (86)

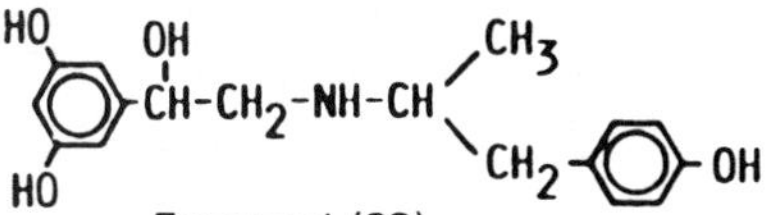

Fenoterol (88)

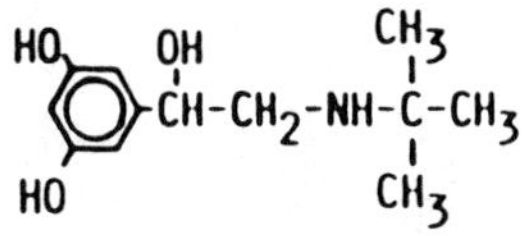

Terbutaline (87)

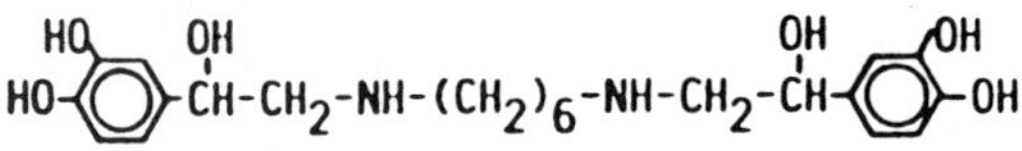

Hexoprenaline (89)

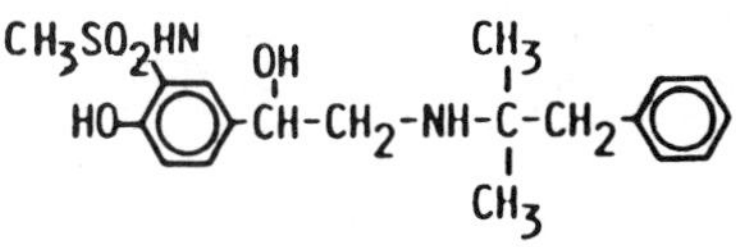

Zinterol (90)

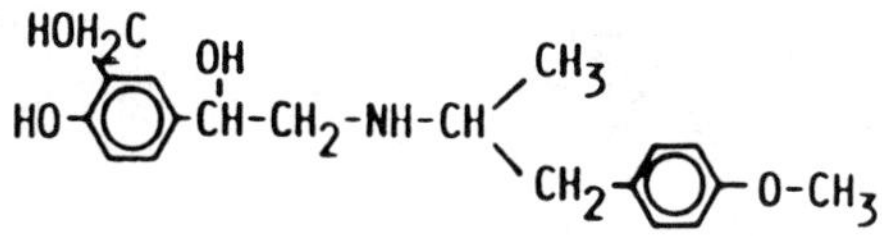

Salmefamol (91)

Figure 5 Structural formulas of β_2-adrenoceptor agonists.

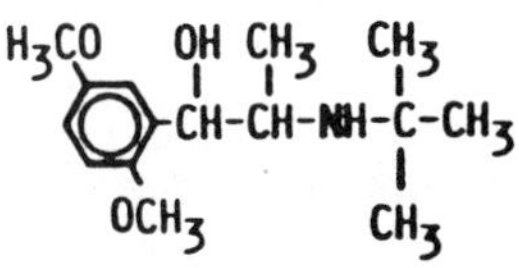

Butoxamine (15)

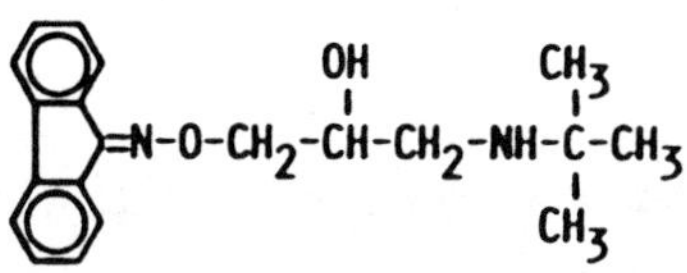

IPS 339 (78)

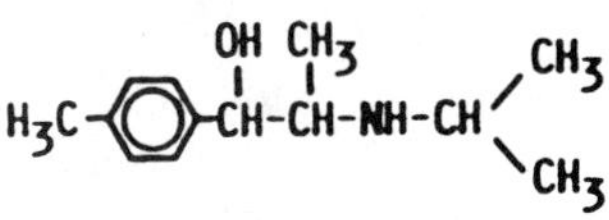

H 35/25 (94)

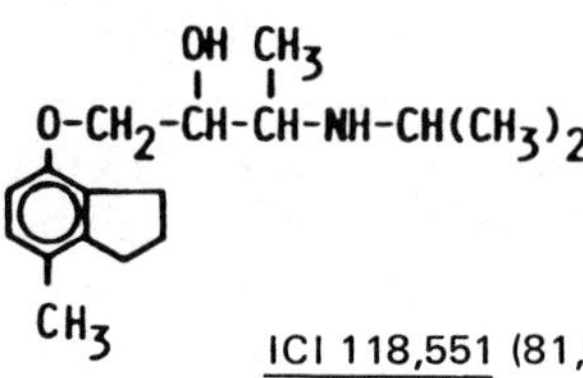

ICI 118,551 (81,82)

Figure 6 Structural formulas of β_2-adrenoceptor antagonists.

did not affect β_1-adrenoceptor-mediated effects, such as inhibition of intestinal contraction, positive inotropic and positive chronotropic effects, or catecholamine-induced increases in plasma free fatty acids. On the other hand, they were very potent in antagonizing vasodilation, uterine relaxation, and increases in blood glucose and lactic acid. Recently, two new β_2-adrenoceptor antagonists have been described—IPS 339 and ICI 118,551 (Figure 6)—that have been demonstrated in numerous in vitro systems to be much more potent and β_2 selective than butoxamine or H 35/25 (45,78-82). In addition, ICI 118,551 has also been shown in humans to be a potent and selective β_2-adrenoceptor antagonist (83).

RADIOLIGAND BINDING STUDIES

Methodological Considerations

During the last decade radioligand binding studies have greatly advanced our knowledge of the molecular pharmacology of β adrenoceptors. This technique has enabled, for the first time, the direct determination of the tissue concentration of β adrenoceptors and, hence, the responsiveness of the tissues to β-adrenergic stimulation.

The first attempt of a direct binding study on β adrenoceptors was performed using labeled catecholamines, such as [^{3}H]norepinephrine or [^{3}H]epinephrine, but such an approach was not suitable to specifically label β adrenoceptors because of high levels of binding to nonreceptor sites (95-97). After the introduction of radioactively labeled high-affinity β-adrenoceptor antagonists, such as (–)-[^{3}H]dihydroalprenolol (98), (±)-[^{125}I]iodohydroxybenzylpindolol (99), (±)-[^{125}I]iodocyanopindolol (100), or (±)-[^{125}I]iodopindolol (101), it was possible to identify β adrenoceptors in a variety of tissues with high specificity. Thereafter, a number of other β-adrenoceptor antagonist radioligands have been developed (Table 3). In addition, a few radiolabeled agonists have also been used for the identification of β adrenoceptors (Table 3).

Coupling β-Adrenoceptor Occupancy to the Adenylate Cyclase System

Radioligand binding studies have provided new detailed information on the mechanism of coupling β-adrenoceptor occupancy to the adenylate cyclase system. It had been consistently found that competition curves for several β-adrenoceptor agonists, but not antagonists, competing with β-adrenoceptor antagonist radioligands, such as (–)-[^{3}H]DHA, IHYP, ICYP, or IPIN, for occupancy of receptors were biphasic and shallow (102-105), indicating that the β adrenoceptor may exist in two affinity states (a high and a low affinity state) for agonists. Typical competition curves for the β-adrenoceptor antagonist propranolol and the β-adrenoceptor agonist isoproterenol for occupancy of the recep-

Table 3 β-Adrenoceptor Radioligands

Radioligand	Maximum specific radioactivity (Ci/mmole)	Specificity
Antagonists		
(–)-[^{3}H] dihydroalprenolol ((–)-[^{3}H] DHA)	110	$\beta_1 = \beta_2$
(–)–[^{3}H] propranolol ((–)-[^{3}H] PRO)	30	$\beta_1 = \beta_2$
(–)-[^{3}H] carazolol ((–)-[^{3}H] CAR)	25	$\beta_1 = \beta_2$
(–)–[^{3}H] bupranolol ((–)-[^{3}H] BU)	18	$\beta_1 = \beta_2$
(–)-[^{3}H] CGP 12177 ((–)-[^{3}H] CGP)	50	$\beta_1 = \beta_2$
(±)-[^{3}H] betaxolol ([^{3}H] BET)	60	β_1
(–)-[^{3}H] bisoprolol ((–)-[^{3}H] BIS)	40	β_1
(±)-[^{3}H] ICI 118,551 ([^{3}H] ICI)	21	β_2
(±)-[^{125}I] iodohydroxybenzylpindolol ((±)-IHYP)	2,200	$\beta_1 = \beta_2$
(–)-[^{125}I]] iodocyanopindolol ((–)-ICYP)	2,200	$\beta_1 = \beta_2$
(–)-[^{125}I] iodopindolol ((–)-IPIN)	2,200	$\beta_1 = \beta_2$
Agonists		
(±)-[^{3}H] hydroxybenzylisoproterenol ((±)-[^{3}H] HBI)	20	$\beta_2 \gg \beta_1$
(±)-[^{3}H] isoproterenol ((±)-[^{3}H] ISO)	15	$\beta_1 = \beta_2$

tors in human right atrium labeled with the β-adrenoceptor antagonist radioligand ICYP are shown in Figure 7. In the presence of guanyl nucleotides, such as GTP or its nonhydrolysable analog Gpp(NH)p, those agonist competition curves were shifted to the right to lower affinities and became monophasic. Thus, guanyl nucleotides converted all high-affinity-state adrenoceptors into low-affinity-state receptors. More detailed studies could show that the ability of an agonist to induce or stabilize this high-affinity coupled receptor is directly related to its intrinsic activity or maximal stimulatory capacity. On the other hand, antagonists always showed a monophasic displacement curve, which was not affected by the addition of guanyl nucleotides to the incubation mixture. Based on these findings, a ternary complex model (106,107) for the coupling of β-adrenoceptor occupancy to the adenylate cyclase has been proposed (Figure 8). In this model, an agonist interacting with the β adrenoceptor induces a (presumably conformational) change in the receptor that promotes the stabilization of a ternary complex consisting of agonist, receptor, and the G protein (AGN). The formation of this complex is associated with the loss of tightly bound GDP from the G protein, which is replaced by GTP. The association of GTP with the ternary complex destabilizes it, and the receptor is reversed into

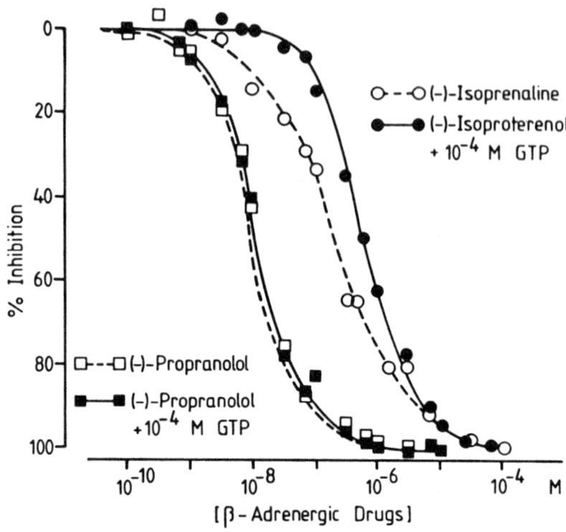

Figure 7 Inhibition of ICYP binding (approximately 50 pM) to β adrenoceptors in human right atrial membranes by (–)-propranolol (□,■) and (–)-isoproterenol (○,●) in the absence (- - - - -) or presence (——) of 10^{-4} M GTP. [Data recalculated from Brodde et al. (132).]

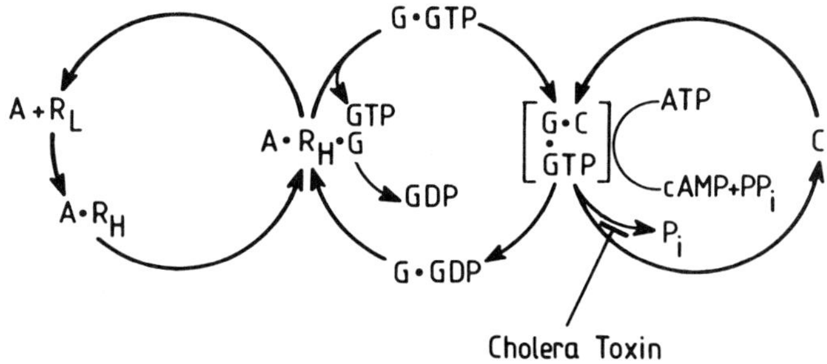

Figure 8 Model for coupling β-adrenoceptor occupany by β agonists to the adenylate cyclase-cAMP system. A = β-adrenoceptor agonist; R = β adrenoceptor; G = guanine nucleotide regulatory protein (G protein); C = catalytic unit of the adenylate cyclase. (|——) cholera toxin inhibits GTPase, thus preventing cleavage of GTP to GDP. [From Lefkowitz et al. (163) with slight modifications.]

its low-affinity form. The G-GTP complex can then interact with the catalytic unit of the adenylate cyclase, thereby activating it to generate cyclic AMP. This activation is terminated by a GTPase (presumably associated with the G protein) that cleaves the terminal phosphate of GTP, converting it to GDP, which does not promote activation. The G protein involved in coupling the stimulatory β adrenoceptors has been shown to contain a GTPase activity that, in fact, terminates each cycle of adenylate cyclase activation by hydrolyzing GTP to GDP (108). Cholera toxin catalyzes an ADP ribosylation of the G protein (presumably at the active side of the GTPase); this ADP ribosylation inactivates the GTPase, thus inhibiting the physiological turnoff mechanism for adenylate cyclase. Under these conditions GTP is an irreversible activator of the adenylate cyclase (109,110).

Coexistence of β_1 and β_2 Adrenoceptors in a Single Organ

The second major insight into the molecular pharmacology of β adrenoceptors to come from radioligand binding studies was that β_1 and β_2 adrenoceptors are not, as was originally postulated by Lands et al. (18,19), organ specifically distributed but coexist in a variety of organs in which, however, one subtype predominates. This was first suggested in 1972 by Carlsson et al. (111), who showed that equal submaximal chronotropic responses of the cat heart to β-adrenergic agonists were antagonized to different extents by the β_1-selective antagonist practolol and by the β_2-selective antagonist H 35/25. Similar effects have been obtained in numerous in vivo as well as in vitro experiments in cat, dog, and guinea pig hearts, as well as in guinea pig and rabbit trachea, thus supporting the concept of the coexistence of β_1 and β_2 adrenoceptors in a single organ (for recent reviews, see References 26 and 112-114). By the use of radioligand binding studies it is now possible to determine the relative proportion of β_1 and β_2 adrenoceptors present in a tissue. The technique most widely applied to quantify the relative amount of β_1 and β_2 adrenoceptors is to assess the inhibition of binding of unselective β-adrenoceptor radioligands, such as (–)-[^{3}H] DHA, IHYP, ICYP, or IPIN, which label β_1 and β_2 adrenoceptors with the same affinity (see Table 3) by unlabeled drugs that have shown in vitro selectivity to β_1 or β_2 adrenoceptors; the resulting competition curves are shallow and biphasic if both β-adrenoceptor subtypes are present in the tissue, but steep and monophasic competition curves are obtained if only one β-adrenoceptor subtype is present. On the other hand, nonselective β-adrenoceptor antagonists always inhibit binding with monophasic competition curves independent of whether one or two subtypes are present. To analyze such competition curves, mainly three different analytic methods have been used: (1) Nahorski and coworkers (115,116) introduced the graphic analysis of such curves by transformation of the binding data into a modified Scatchard or Hofstee plot, (2) Minneman

Table 4 Distribution of β_1 and β_2 Adrenoceptors in Various Tissues

Species and tissue	β_1:β_2 Ratio	Reference
Rat		
Heart		
whole heart	83:17	117
ventricles	100:0	321
left ventricle	74:26	322
right atrium	65:35	124
left atrium	65:35	124
	—[a]	323
Lung	25:75	116
	15:85	117
	20:80	138, 324
Spleen	30-35:65-70	325
Kidney	70:30	275
	63:37	326
	100:0	327
	—[b]	326, 328
Guinea pig		
Heart		
right atrium	77:23	329
left ventricle	100:0	101, 329
	82:18	45
right ventricle	82:18	45
Lung	22:78	101
	26:74	45
trachea	15:85	330
bronchi	12:88	330
parenchyma	15:85	330
Kidney	—[c]	328
Rabbit		
Heart		
right atrium	72:28	331
left atrium	82:18	331
right ventricle	92:8	331
left ventricle	93:7	331
Lung	60:40	116
	80:20	75,138,332
Cat		
Heart		
right atrium	78:22	329
left atrium	79:21	119
left ventricle	98:2	329
	86:14	119

Table 4 (Continued)

Species and tissue	$\beta_1:\beta_2$ Ratio	Reference
right ventricle	82:18	119
Dog		
Heart		
left ventricle	85:15	333
Lung	5:95	333
Human		
Heart		
right atrium	82:18	131, 132
	74:26	133
	75:25	135
	50:50	136
	43:57	137
left atrium	65:35	134
right ventricle	87:13	145
	75:25	135
left ventricle	86:14	133
	80:20	277
	75:25	135
	69:31	322
	65:35	134
	41:59	137
Lung	28:72	147
	29:71	148
Myometrium	13:87	264
Placenta	57:43	334

[a] β_1 Adrenoceptors on myoblasts; β_2 Adrenoceptors on fibroblasts.
[b] Autoradiographic studies using ICYP revealed that β_1 adrenoceptors were localized predominantly on glomeruli and the straight part of the distal tubulus and on the cortical portion of the collecting duct; a few β_2 adrenoceptors were diffusely localized around the corticomedullary junction.
[c] β_1 Adrenoceptors were localized on glomeruli and distal tubulus; high concentrations of β_2 adrenoceptors were localized on the straight part of the proximal tubulus and the cortical portion of the collecting duct.

et al. (117) used an iterative computer analysis of the Hofstee plots, and (3) DeLean et al. (118) introduced the analysis of untransformed binding data with a nonlinear least-squares computer modeling method.

An alternative approach for the *direct* characterization of *one* β-adrenoceptor subtype in tissues containing both β_1 and β_2 adrenoceptors is the use of a radioligand highly selective for a single β-adrenoceptor subtype. The advantage of using such a subtype-selective β-adrenergic radioligand is the *direct* determination of the total amount of one β-adrenoceptor subtype as well as the *direct* determination of the true affinity of β-adrenergic drugs for one β-adrenoceptor subtype. For β_1-adrenoceptor determination, (–)-[^{3}H] bisoprolol (74,119) has been found to be a ligand that selectively labels β_1 adrenoceptors in rabbit lung membranes as well as in kitten and guinea pig heart tissues known to contain β_1 and β_2 adrenoceptors (Table 4).

In addition, for direct determination of β_2 adrenoceptors, Lemoine et al. (45) recently introduced [^{3}H] ICI 118,551. With these radioligand binding techniques the coexistence of β_1 and β_2 adrenoceptors has been demonstrated in the hearts of rats, guinea pigs, cats, and rabbits, as well as in the lung of rats, rabbits, and guinea pigs (Table 4). Although there is a general agreement that in right atria of cats and dogs (51,92,111) both β_1 and β_2 adrenoceptors are involved in the physiological regulation of heart rate and/or contractility, the physiological role of β_2 adrenoceptors present in rat, guinea pig, and rabbit right atria is still not completely understood. In right atria of rats, several authors have found that only β_1 adrenoceptors mediate inotropic and chronotropic responses to β-adrenergic stimulation (113,120-123); however, Kaumann recently found that epinephrine causes positive chronotropic effects via both β_1- and β_2-adrenoceptor stimulation (124). Similar conflicting results were obtained on guinea pig right atria. Dreyer and Offermeier (125) claimed that the β-adrenoceptor subtype mediating positive inotropic effects in guinea pig right atrium is different from that mediating increases in heart rate; on the other hand, several groups found that only β_1 adrenoceptors mediate positive inotropic and chronotropic effects in guinea pig right atrium (123,126,127). Similarly, no evidence for a physiological role of β_2 adrenoceptors present in rabbit right atrium is available at present (123,128).

The first evidence that a heterogeneous population of β_1 and β_2 adrenoceptors may exist in human myocardium was presented in 1974 by Ablad et al. (129), who found, on the isolated electrically driven human right atrium, that the β_1-selective antagonist H 93/26 was more potent in antagonizing the positive inotropic effect of norepinephrine than that of epinephrine, whereas the unselective β-adrenoceptor antagonist propranolol inhibited responses to both catecholamines to about the same degree. Similar results were reported by Bonelli (130), who observed that in humans the β-adrenoceptor antagonist mepindolol inhibited the positive chronotropic effect of isoproterenol much more potently than its positive inotropic effect. Very recently, β_1 and β_2 adrenoceptors have been directly identified in human right and left atria by radioligand binding studies (Table 4). In the majority of these studies it was found that approximately one-third of the total β-adrenoceptor population is of the β_2 subtype (131-135), but in two studies an equal amount of β_1 and β_2 adrenoceptors were found in right atrium (136,137). The reason for these discrepancies is not known at present but may be due to the use of different radioligands and/or different techniques used in preparing the membrane fractions. It should also be mentioned that Dickinson and Nahorski (138) have recently demonstrated that in rat and rabbit lung membranes (after solubilization) β_1 adrenoceptors are much more labile than β_2 adrenoceptors. A similar instability of β_1 adrenoceptors has been described for human right atrial tissue (133). Thus, storage of this tissue results in a loss of β_1 adrenoceptors, thereby apparently increasing the relative amount of β_2 adrenoceptors.

Both β_1 and β_2 adrenoceptors in human right atrium are functionally coupled to the adenylate cyclase (139,140), although this has been recently challenged (141). The functional role of β_1 and β_2 adrenoceptors present in human right atrium is not completely understood at present. In the isolated electrically driven muscle strips of human right atrial appendage, several groups have recently presented convincing evidence that both β_1 and β_2 adrenoceptors mediate the positive inotropic action of β agonists (123,140,142,143). Thus, it may be possible that right atrial β_2 adrenoceptors are involved in the physiological regulation of contractile force. However, they may play an even more important role in regulating heart rate than force of contraction, since it has been shown in the human that isoproterenol-induced tachycardia could be antagonized to a greater extent by propranolol than by the β_1-selective antagonist atenolol (144). Furthermore, Arnold et al. (83) recently demonstrated in humans that ICI 118,551 antagonized the isoproterenol-induced tachycardia much more potently than the β_1-selective antagonist atenolol, whereas propranolol was equipotent to ICI 118,551. In human right (131,135,145) and left ventricular wall (133-135), the coexistence of β_1 and β_2 adrenoceptors could be demonstrated by radioligand binding techniques; the β_1-β_2 ratio was 80:20 (but see Disease-Induced Changes: Heart Failure). The functional importance of the β_2 adrenoceptors present in ventricular myocardium, is not known at present. However, it has been recently demonstrated on isolated electrically driven strips of human right and left ventricular myocardium that stimulation of β_2 adrenoceptors can mediate positive inotropic effects (for references see Ref. 146). Thus it might be possible that also in human ventricular myocardium β_2 adrenoceptors contribute to the positive inotropic effects of β agonists (see also Disease-Induced Changes: Heart Failure). The coexistence of β_1 and β_2 adrenoceptors has been demonstrated in human lung (ratio of β_1 to β_2 = 30:70) (147,148); in human airway smooth muscle, however, there is no evidence for a β_1-adrenoceptor relaxation (149).

Regulation

A third major insight into the molecular pharmacology of β adrenoceptors to come from radioligand binding studies was that the tissue concentration of β adrenoceptors is not a fixed number but rather is dynamically regulated by a variety of drugs, hormones, and pathological and physiological conditions. An important example of a clinically relevant consequence of receptor regulation is the phenomenon of desensitization. In a study with rabbit cerebellar slices, Kakiuchi and Rall (150) first demonstrated that exposure of the tissue to catecholamines resulted in a rapid rise followed by a decline, of the intracellular level of cyclic AMP. Subsequent stimulation of the same tissue with β agonists resulted in a markedly diminished cyclic AMP response. Such a decrease in responsiveness to pharmacological stimuli with time is a general mechanism of cellular adaptation that is referred to as desensitization, tachyphylaxis, or refractoriness. Two major mechanisms of desensitization of the adenylate cyclase-coupled β adrenoceptor have been postulated: "homologous" and "heterologous" desensitization. In homologous regulation, tissue receptor concentration is regulated only by drugs that specifically interact with the receptor. For example, homologous regulation of β adrenoceptors is elicited by β-adrenoceptor agonists such as isoproterenol, epinephrine, or norepinephrine or by β-adrenoceptor antagonists such as propranolol. Heterologous regulation of β adrenoceptors, on the other hand, is brought about by drugs that do not interact with the β adrenoceptor, but with the components of the adenylate cyclase/cyclic AMP system. Cyclic AMP has been postulated to be a mediator of heterologous regulation. Accordingly, any hormone that increases cyclic AMP may induce heterologous desensitization. Catecholamines acting at β adrenoceptors can induce both receptor-specific (homologous) and heterologous types of desensitization in the same cell. Thus, the impaired β-adrenoceptor function following subsequent stimulation is the sum of both processes of desensitization.

Homologous Desensitization

Based on studies of catecholamine-induced desensitization in frog erythrocytes (151-154), various cultured mammalian cell lines (155-159), and very recently in vivo in rat lung (160), the following working model for the sequence of events of the β-adrenoceptor-specific desensitization process has been proposed (Figure 9) (161-163). Very rapidly after occupancy of the β adrenoceptor by a β agonist, there is a functional uncoupling of the receptor from the adenylate cyclase activation; that is, the agonist induces a conversion of the β adrenoceptor into a form that is unable to effectively couple with the G protein of the adenylate cyclase system (presumably by phosphorylation of the receptor by β-adrenoceptor kinase). At this time the density of receptors at the cell

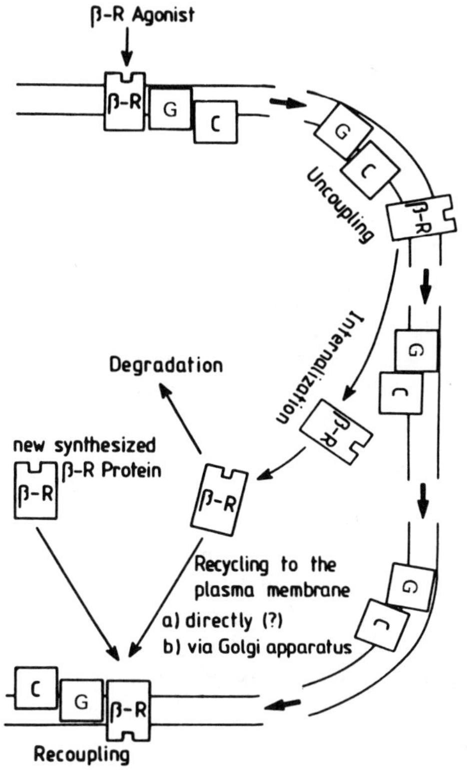

Figure 9 Model for homologous desensitization of β adrenoceptors. β-R = β adrenoceptor; G = guanine nucleotide regulatory protein (G protein); C = catalytic unit of the adenylate cyclase. [From Lefkowitz et al. (163) with slight modifications.]

surface is not altered but the ability of agonist to stimulate adenylate cyclase is diminished. Thereafter, the β adrenoceptors are "internalized" or at least sequestered away from the cell surface. These internalized receptors can be recovered in a light vesicle fraction that sediments at 100,000 × g. It has been shown that such sequestered receptors are functionally normal but are desensitized since they are functionally sequestered away from both cell surface and the physiological effector, that is, the G-protein-adenylate cyclase system. The potential fate of these receptors in the light vesicular fraction is either to recycle via the Golgi apparatus to the surface or to be converted into a form that cannot be detected by radioligand binding studies. Finally, the receptor is degraded; the recovery of such degraded receptors requires the synthesis of new receptor protein.

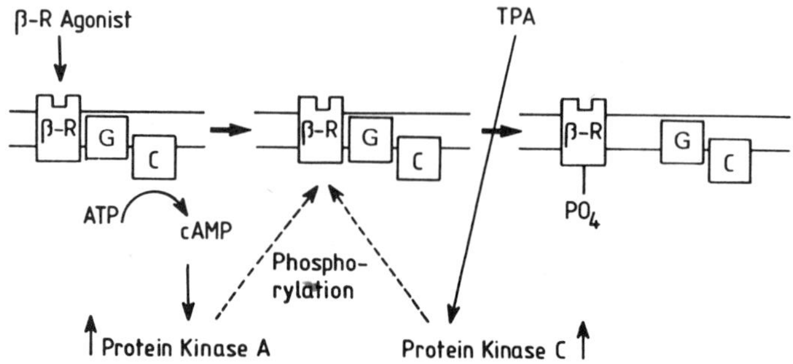

Figure 10 Model of heterologous desensitization of β adrenoceptors. β-R = β adrenoceptor; G = guanine nucleotide regulatory protein (G protein); C = catalytic unit of the adenylate cyclase; TPA = 12-*O*-tetradecanoyl-phorbol-13-acetate. [From Lefkowitz et al. (163) with slight modifications.]

Heterologous Desensitization

In contrast to homologous desensitization, heterologous desensitization seems to be a cyclic AMP-dependent process (164,165). In addition to β-adrenoceptor agonists, phorbol esters can induce desensitization of the β-adrenoceptor-coupled adenylate cyclase (166,167). Since the common effect of β agonists and phorbol esters is to activate protein kinases (protein kinase A via β agonists and protein kinase C via phorbol esters) and since both classes of drugs evoke desensitization, it has been hypothesized that the common link in heterologous desensitization may be phosphorylation of the β adrenoceptor. In fact, Lefkowitz and coworkers (168,169) have recently demonstrated that incubation of turkey and duck erythrocytes with isoproterenol or with the phorbol ester 12-*O*-tetradecanoyl-phorbol-13-acetate (TPA) leads to a marked increase in the amount of phosphates present in the β-adrenoceptor peptides. This is compatible with the view that heterologous desensitization is caused by phosphorylation of the β adrenoceptor. Figure 10 presents a working model of possible mechanisms involved in heterologous desensitization of the β adrenoceptor (163). According to this scheme, protein kinases, such as protein kinase A or C, phosphorylate the β adrenoceptor and directly uncouple it from the G-protein-adenylate cyclase system.

PATHOPHYSIOLOGICAL AND DRUG-INDUCED ALTERATIONS

Desensitization

As mentioned earlier, β adrenoceptors are dynamically regulated by a wide variety of drugs, hormones, and pathological or physiological conditions. To study such alterations in human β adrenoceptor function, circulating lymphocytes containing a homogeneous population of β_2 adrenoceptors (170-172) excitatorily coupled to the adenylate cyclase (173,174) are frequently used. Brodde et al. (175) have recently shown that, in the human, the β-adrenoceptor density in circulating lymphocytes is significantly correlated with the corresponding β-adrenoceptor density and responsiveness (evaluated as positive inotropic responses to isoproterenol) in the right atrial appendages obtained from the same subjects. A detailed analysis of this relationship, however, revealed that lymphocyte β_2 adrenoceptors are significantly better correlated with right atrial β_2 than β_1 adrenoceptors (176). These results demonstrate that it is valid to extrapolate from changes of β adrenoceptors determined in circulating lymphocytes to changes in (at least) β_2 adrenoceptors that may occur in solid human tissues (e.g., the heart).

Agonists acting at β adrenoceptors are commonly used to treat patients suffering from asthma or other pulmonary dysfunctions. An agonist-induced decrease in β-adrenoceptor density and responsiveness to β-adrenergic stimulation has been observed in lymphocytes of healthy volunteers as well as of asthmatic patients following long-term treatment with β agonists (177-187). The resulting tachyphylaxis may markedly limit the therapeutic efficacy of β-adrenergic bronchodilator therapy in asthma.

Glucocorticoid therapy, however, can rapidly reverse β-agonist-induced down-regulation of lymphocyte β_2 adrenoceptors in healthy volunteers and in asthmatic patients (187-190). Down-regulation of β_2 adrenoceptors on lymphocytes has also been demonstrated in vitro after incubation with isoproterenol and epinephrine, but not with norepinephrine (191-193). Although it is now generally accepted that long-term exposure of cells to endogenous catecholamines, or exogenously applied β agonists leads to a reduction in the number of β adrenoceptors on the cell surface, the sequence of events following acute increases in endogenous catecholamines due to acute in vivo stimulation of the sympathetic activity seems to be quite different. Several groups have recently demonstrated that dynamic exercise on a bicycle for 15 min causes about 100% increase in lymphocyte β_2-adrenoceptor density (194-196); this rise in β_2-adrenoceptor density was accompanied by a significant increase in β_2-adrenoceptor responsiveness since, after exercise, isoproterenol produced significantly greater increases in the intracellular level of cyclic AMP. The mechanism of this agonist-induced "acute up-regulation" of β_2 adrenoceptors in human lymphocytes is not known at present. However, it seems to be a β_2-adrenoceptor-dependent process since the exercise-induced increase in lymphocyte β_2-adrenoceptor density was markedly attenuated by propranolol (197) and the β_2 antagonist ICI 118, 551 (Brodde, et al., Am. J. Physiol., in press), but not by the β_1-selective antagonist bisoprolol (197). Similar acute agonist-induced increases in lymphocyte β_2 adrenoceptors have been described by Tohmeh and Cryer (192), who showed that the infusion of isoproterenol or epinephrine in healthy men led to an increase in lymphocyte β_2-adrenoceptor density after 30 min as well as to an increase in β-adrenoceptor responsiveness as measured by heart rate responses to isoproterenol.

Age

It has been reported that the number of β adrenoceptors on lymphocytes derived from neonates at term was only 50% of that in lymphocytes from young adults. In addition, cyclic AMP accumulation following isoproterenol stimulation was diminished (198). Studies on the β-adrenoceptor density on lymphocytes from old subjects have led to contradictory results; one group reported that the β-adrenoceptor density decreased with age (199), but the majority of studies showed that there were no significant age-dependent changes in lymphocyte β_2-adrenoceptor density (193,196,200-202). In a recent study, O'Hara et al. (203) found that the mean number of lymphocyte β_2 adrenoceptors in young adults (19-30 years old) was approximately twice as high as that in lymphocytes derived from neonates. In old subjects (60-86 years), β_2-adrenoceptor density showed a tendency to increase further. Isoproterenol-induced increases in lymphocyte cyclic AMP content were significantly higher in young adults than in neonates or old subjects. The affinity of isoproterenol to the β adrenoceptor was significantly lower in old subjects than in young adults, a finding that is in accordance with data published by Feldman et al. (204). These data convincingly demonstrate that in neonates and old subjects β-adrenoceptor responsiveness is reduced. The mechanism underlying this reduced β_2-adrenoceptor responsiveness in neonates and old subjects, however, is different: in neonates it seems to be due to the reduced β-adrenoceptor density; in old subjects, on the other hand, it is obviously caused by a postreceptor defect. In fact, it has been shown that in the elderly the activity of the catalytic unit of the adenylate cyclase is diminished (205).

Supersensitivity or "Up-Regulation"

In many in vivo studies in animals, depletion of catecholamines (for example by treatment with 6-hydroxydopamine or reserpine or long-term treatment with adrenergic neuron blocking agents like guanethidine) was found to be associated with a supersensitivity of various tissues to β-agonist stimulation. In the overwhelming majority of these studies the supersensitivity was caused by an increased number of β adrenoceptors (206-210).

In humans, supersensitivity and up-regulation have been studied in patients with orthostatic hypotension. Some of these patients had low concentrations of circulating catecholamines and showed enhanced responses to exogenously applied β agonists. In these patients the number of β adrenoceptors on lymphocytes was significantly higher than in normal subjects (211-214).

β Adrenoceptor Changes Evoked by β Adrenoceptor Antagonists

β-Adrenoceptor antagonists (Table 5) are commonly used in the therapy of angina pectoris and hypertension (65,66,215). There are several reports in the literature showing that abrupt cessation of therapy with propranolol (a nonselective β-adrenoceptor antagonist *without* ISA*) can lead to clinical symptoms suggesting adrenergic hypersensitivity ("propranolol-withdrawal phenomenon") (for references, see Reference 216). Such a β-adrenoceptor supersensitivity could be caused by an increase in β-adrenoceptor number as a reflection of the β-adrenoceptor blockade by propranolol (217). In fact, propranolol can lead to

*Intrinsic sympathomimetic activity.

Table 5 β-Adrenoceptor Antagonists

	β_1 Selectivity	Membrane stabilizing effect	Intrinsic sympathomimetic activity
Acebutolol	+	+	+
Alprenolol	–	+	+
Atenolol	+	0	0
Betaxolol	+	±	0
Bisoprolol	+	0	0
Bunolol	–	0	0
Bupranolol	–	+	0
Celipropol	+	0	++
Mepindolol	–	±	++
Metoprolol	+	±	0
Nadolol	–	0	0
Oxprenolol	–	+	+
Penbutolol	–	+	0
Pindolol	–	±	++
Practolol	+	0	++
Propranolol	–	++	0
Sotalol	–	0	0
Timolol	–	0	0

increases in β-adrenoceptor density. This was first demonstrated in the rat, in which chronically administered propranolol produced increases of β-adrenoceptor density in heart (218,219), lung, and lymphocytes (219). Subsequently, several groups have shown that chronic treatment of normotensive volunteers with propranolol leads to a significant increase in β_2-adrenoceptor density in circulating lymphocytes (220-223). After withdrawal, β_2-adrenoceptor density declined slowly, still being significantly increased after 3-5 days (although no propranolol is detectable in plasma after 24 hr). Thus, for 3-5 days there is an increased β-adrenoceptor density not protected by propranolol, which may well be the cause of the β-adrenoceptor supersensitivity following propranolol withdrawal.

On the other hand, when healthy volunteers were treated with β-adrenoceptor antagonists *with* ISA, such as pindolol, mepindolol, celiprolol, or acebutolol, there was a significant reduction in lymphocyte β_2 adrenoceptor density that persisted after withdrawal for at least 4 days (223-227). This β-adrenoceptor reducing effect could be inhibited by propranolol, which strongly suggests that in fact the partial agonistic activity of these β-adrenoceptor antagonists is the cause for the reduction in lymphocyte β_2-adrenoceptor density (226). The magnitude of β-adrenoceptor antagonist-induced reduction in β-adrenoceptor density is obviously dependent on the strength of the ISA. Pindolol produced greater decreases on lyphocyte β_2-adrenoceptor density than mepindolol or celiprolol, as the ISA of pindolol is stronger than that of mepindolol or celiprolol (223,226). On the other hand, alprenolol—a nonselective β-adrenoceptor antagonist with relatively weak ISA—did not significantly alter lymphocyte β_2-adrenoceptor density; that is, it did not reduce, but it also did not, in contrast to propranolol, increase lymphocyte β_2-adrenoceptor density (223). That all β-adrenoceptor antagonists with ISA studied so far decreased, or at least did not affect, β-adrenoceptor density has important clinical implications: since no β-adrenoceptor increases occur, rapid cessation of drug treatment should not induce rebound effects. In fact, pindolol does not show symptoms of adrenergic hypersensitivity after rapid withdrawal (228,229).

HORMONAL REGULATION

Glucocorticoids

It has been well known for a long time that glucocorticoids can modulate responsiveness of tissues to β-adrenergic stimulation. Glucocorticoids can increase β-adrenergic responsiveness as indicated by enhanced inotropic responses in heart muscle, enhanced β_2-adrenoceptor-mediated vasodilation, and enhanced hepatic glucose production after catecholamine stimulation (230-233). These glucocorticoid changes in tissue responsiveness to β-adrenergic stimulation are caused by an increase in β-adrenoceptor density (234). Thus, glucocorticoids can increase β-adrenoceptor density in rat lung (235), rabbit fetal lung (236), cultured human lung cells (237), and cultured human astrocytoma cells (238). Glucocorticoids also appear to increase β-adrenoceptor density and cyclic AMP responses to β-adrenergic stimulation in human lymphocytes and granulocytes (239). Besides this increase in β-adrenoceptor density, glucocorticoids can regulate β-adrenoceptor function by facilitating the formation of the "high-affinity" state of the β adrenoceptor. Thus, it has been shown in healthy volunteers after oral administration of cortisone (240) or prednisone (187) that the ratio of the high- to the low-affinity state of the lymphocyte β_2 adrenoceptor is significantly shifted toward the high-affinity state. Such an increased formation of the agonist-induced high-affinity state enhances the efficiency of coupling receptor occupancy to the adenylate cyclase system, thus leading to exaggerated β-adrenergic responses.

In addition to the "sensitizing" effects of glucocorticoids on β adrenoceptors in normal tissues, it has been shown in vitro as well as in vivo that agonist-induced desensitization of the β-adrenoceptor/adenylate cyclase system can be attenuated (187–190) or rapidly reversed by glucocorticoids. The rapid restoration of down-regulated β-adrenoceptor responsiveness has important clinical implications: asthmatic patients are frequently treated with β_2 agonists. Such a long-term treatment may lead to a desensitization of the bronchial β_2 adrenoceptors, leading to decreased β-agonist-stimulated bronchodilation. Although it is still a matter of controversy whether an impaired β-adrenoceptor function due to a reduced β-adrenoceptor number may be the pathological cause of atopic diseases like bronchial asthma, as originally postulated by Szentivanyi (241), several groups have presented convincing evidence that the number of β_2 adrenoceptors in circulating lymphocytes is significantly reduced in asthmatic patients receiving β_2-agonist therapy. Glucocorticoids can rapidly reverse this reduced lymphocyte β_2-adrenoceptor density. This is in accordance with the observation that the steroids can restore responsiveness to β-adrenergic bronchodilators in tolerant patients (242,243) and animals (244); in addition, hydrocortisone has been shown to accelerate recovery from the desensitized state in isolated human airway smooth muscle (245).

Thyroid Hormones

The clinical symptoms of hyperthyroidism, including tachycardia, enhanced cardiac output, skeletal muscle tremor, hyperkinetic behavior, and increased sweating, resemble those of a hyperadrenergic state. These symptoms can be partly suppressed by application of β-adrenoceptor antagonists. Since the concentration of endogenous catecholamines is not significantly increased or even decreased in hyperthyroid patients, it has been suggested that the cardiovascular changes in hyperthyroidism are due to an exaggerated responsiveness of β-adrenoceptors to stimulation by endogenous catecholamines. On the other hand, in the hypothyroid state cardiovascular alterations occur that are in a direction opposite to those in hyperthyroidism, suggesting a reduced responsiveness of β adrenoceptors to catecholamine stimulation.

There are now numerous in vivo and in vitro studies on the influence of thyroid hormones on the density and responsiveness of β adrenoceptors in a wide variety of animal models, as well as in human lymphocytes. Although the data published so far are rather conflicting (114,246), principally it has been found that β-adrenoceptor density and functional responsiveness (and perhaps coupling of receptor occupancy to the adenylate cyclase) (247,248) is often increased in hyperthyroidism and decreased in hypothyroidism (249-253). A noteworthy exception to this general rule seems to be the liver; it has been observed that, in hyperthyroidism, β-adrenoceptor function in rat liver membranes is decreased but hypothyroidism is associated with an increased β-adrenoceptor function (254-256). Conflicting results have also been obtained in human studies. Although no significant differences in the density of lymphocyte β_2 adrenoceptors were observed between hyperthyroid and euthyroid patients in two studies (257,258), lymphocyte β_2-adrenoceptor density was found to be increased in hyperthyroidism in two other studies (259,260). It is worthwhile to note that in those studies in which differences were found, each subject served as his or her own control. Ginsberg et al. (259) demonstrated in a double-blind study in 24 healthy volunteers that in each individual treated with triiodothyronine, lymphocyte β_2-adrenoceptor density increases, whereas in the placebo group no significant changes occurred. In addition, in 10 patients with hyperthyroidism antithyroid therapy was accompanied by a decrease in leukocyte β_2-adrenoceptor function. Thus, in the euthyroid state these patients had significantly lower β-adrenoceptor density and responsiveness than in the hyperthyroid state (260). The mechanism of this thyroid hormone-dependent regulation of β-adrenoceptor function is not known at present. However, the increase of β adrenoceptors in hyperthyroidism may be due to an enhanced synthesis of new receptor protein, since Gross and Lues (261) have recently shown in rat heart that the triiodothyronine-induced increase in β-adrenoceptor density could be completely prevented by simultaneous application of cycloheximide, which is known to be a potent protein synthesis inhibitor.

Sex Steroids

Little is known about the effects of sex steroids on β adrenoceptors. In animal studies it has been shown that gonadal steroids can alter the α-adrenoceptor responsiveness in myometrial tissues. In rabbits, estrogen treatment or elevated levels of estrogen enhance uterine contractile responses to α-adrenoceptor stimulation by norepinephrine (262). On the other hand, if estrogen treatment is followed by progesterone treatment β-adrenoceptor-mediated relaxation predominates (262). Roberts et al. (263) have demonstrated that estrogen increases the number of α adrenoceptors in rabbit myometrium; in rabbits treated with estrogen followed by progesterone, α-adrenoceptor density was decreased but β-adrenoceptor number was not changed.

In human myometrium, β adrenoceptors coupled to the adenylate cyclase have been identified by (–)-[^{3}H] DHA binding (264,265); in this tissue β_2 adrenoceptors predominate, but β_1 adrenoceptors have been also identified (approximately 15%) (264). In women treated with terbutaline for premature uterine contractions, there was a significant fall in β-adrenoceptor density in myometrial membranes accompanied by a reduced cyclic AMP response to β-adrenoceptor stimulation when compared with an untreated control group (265). Similar effects were also seen in lymphocytes of women treated with hexoprenaline for delaying delivery (Brodde, unpublished observations).

DISEASE-INDUCED CHANGES

Heart Failure

The failing human heart is exposed to increased catecholamine stimulation owing to the deficit in myocardial contractility. Elevated circulating catecholamines in heart failure have been observed in several studies (266-269). As a result of this chronic exposure to endogenous norepinephrine, cardiac β adrenoceptors may be desensitized (270,271). Colucci et al. (272) were the first to demonstrate a (slight) decrease in β-adrenoceptor density in lymphocytes of patients suffering from heart failure when compared with healthy controls. Subsequently, Bristow et al. (273) reported that in end-stage congestive heart failure the β-adrenoceptor density in left ventricles was decreased by about 50%, accompanied by a 45% decrease in maximal isoproterenol-stimulated adenylate cyclase activity when compared with normally functioning hearts. A reduced contractile response of the failing heart to isoproterenol has been observed (274). These results favor the idea that, in heart failure, diminished β-adrenoceptor responsiveness is due to homologous desensitization by endogenous norepinephrine. Norepinephrine, however, has only a very low affinity to β_2 adrenoceptors (18,45). It has been shown by in vitro studies that isoproterenol and epinephrine, but *not* norepinephrine, can "down-regulate" β_2-adrenoceptor number in human lymphocytes (193). Similar results have been obtained in rats transplanted with a norepinephrine-producing pheochromocytoma, in which only β_1 but not β_2-adrenoceptor density was reduced in various tissues (275,276). It may be possible, therefore, that in end-stage congestive heart failure increased norepinephrine levels selectively down-regulate β_1 adrenoceptors. In fact, Bristow et al. (277) have recently found that the β-adrenoceptor density in myocardial membranes obtained from severely failing human left ventricular myocardium was decreased by about 60% compared with that in normal hearts; the ratio of β_1 to β_2 adrenoceptors, however, was changed from 80/20 in normal hearts to

62/38 in the failing heart, indicating a predominant β_1-adrenoceptor down-regulation. Brodde et al. (278) have recently investigated in more detail the regional distribution of β_1 and β_2 adrenoceptors in the human heart. For this purpose competition curves were analyzed of the highly selective β_1 antagonist bisoprolol (73,74) and the highly selective β_2 antagonist ICI 118,551 (81,82) with the non-subtype selective radioligand ICYP in right and left atrium, as well as right and left ventricular wall, obtained from heart transplant recipients who suffered from end-stage heart failure (NYHA class IV). On all four myocardial tissues investigated, β_1 and β_2 adrenoceptors could be identified. The relative amount of β_2 adrenoceptors, however, differed between atria and ventricles: although in both atria approximately 35-40% of the β adrenoceptors were of the β_2 subtype, the amount of β_2 adrenoceptors in ventricular wall was considerably less (approximately 25-30%). The regional distribution of β_1 and β_2 adrenoceptors in human myocardium is obviously different from that in the heart of various other animal species, in which β_1 and β_2 adrenoceptors coexist only in the atria but not in the ventricles (see Table 4). On all four myocardial tissues both β_1 and β_2 adrenoceptors are functionally coupled to the adenylate cyclase system. Norepinephrine, which predominantly stimulates β_1 adrenoceptors (18, 45), was a full agonist in stimulating adenylate cyclase activity on all four tissues. Also, the highly selective β_2 agonist procaterol (92,93) stimulated adenylate cyclase activity on all four tissues; its intrinsic activity, however, was only 55-75% of that of isoproterenol.

The number of β_1 and β_2 adrenoceptors in right atrium and in right and left ventricular wall of these patients with end-stage congestive cardiomyopathy was compared with that in normal hearts. In right atrium the total number of β adrenoceptors was approximately 50-60% less when compared with recently reported data obtained in normal hearts either from patients undergoing elective coronary artery bypass grafting (132,135) or from heart transplant donors (133). This reduction in total β-adrenoceptor density was due to a selective decrease in β_1-adrenoceptor density, but β_2-adrenoceptor density was only marginally reduced (Figure 11). Similar results were obtained on right and left ventricular wall. The total number of β adrenoceptors in membranes obtained from severely failing hearts was approximately 50% less than that determined on ventricular walls obtained from normal hearts (133,135,145,277). As described for right atrium, in the ventricles the decrease in β-adrenoceptor density was also mainly due to a reduction in the density of β_1 adrenoceptors, and β_2-noceptor density was not significantly changed (Figure 11). These data support the hypothesis that one of the pathological mechanisms underlying the decreased catecholamine-induced contractility in end-stage heart failure may be a selective down-regulation of β_1 adrenoceptors by the elevated concentrations of endogenous norepinephrine. That β_1-adrenoceptor responsiveness is markedly reduced and β_2-adrenoceptor function is not altered is further

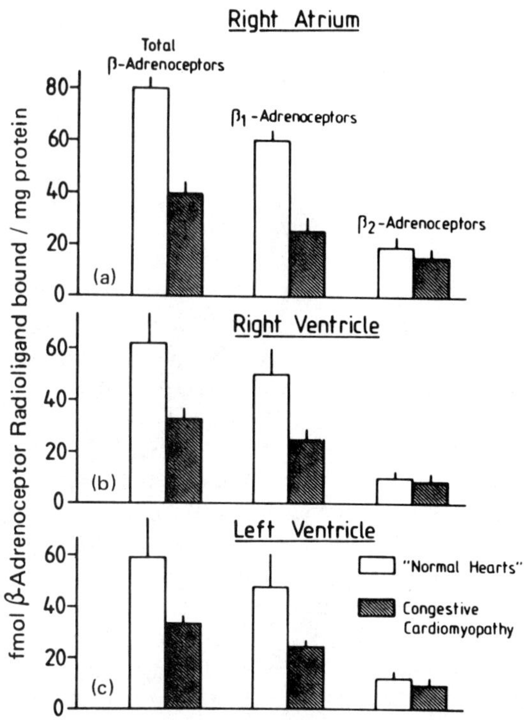

Figure 11 Human myocardial β adrenoceptors: comparison of total β-, β_1-, and β_2-adrenoceptor density in membranes of right atrium (a), right ventricular wall (b), and left ventricular wall (c) from normally functioning hearts (□) with those derived from end-stage congestive cardiomyopathy (■). Ordinates: Total β-, β_1-, and β_2-adrenoceptor density in femtomoles ICYP and (–)-[^{3}H] DHA, respectively, bound per milligram protein. Data from Brodde et al. (278) determined in three explanted hearts of heart transplant recipients who suffered from end-stage congestive cardiomyopathy were compared with data from normally functioning hearts recalculated from (a) Brodde et al. (132), Stiles et al. (133), and Golf et al. (135) for right atrium; (b) Golf et al. (135), and Brodde et al. (145) for right ventricle; (c) Stiles et al. (133), Golf et al. (135), and Bristow et al. (277) for left ventricle.

supported by the recent finding that on isolated, electrically driven right ventricular strips from severely failing hearts the β_2 agonist zinterol produced positive inotropic effects that were not different from those in nonfailing hearts, whereas the positive inotropic effect of isoproterenol was markedly reduced and that of β_1 (partial) agonist denepamine was nearly abolished (277). The unaltered β_2-adrenoceptor function may also explain the beneficial effects of the β_2 agonists fenoterol (279) and salbutamol (280,281) in severe congestive heart failure. Thus, it might be speculated that in the severely failing heart β_2 adrenoceptors may compensate for the loss of β_1 adrenoceptors to maintain (at least a rest of) contractility.

Hypertension

Alterations in sympathetic activity may play an important role in inducing and/or maintaining elevated blood pressure levels in essential hypertension (282–284). Because of a lack of the direct measurement of sympathetic activity in humans, the determination of plasma norepinephrine concentrations has been taken as an index for sympathetic activity. Although there are numerous comparative studies in hypertensive and normotensive controls, it is still a matter of controversy whether or not plasma norepinephrine levels are elevated in patients with essential hypertension (for a recent review, see Reference 285).

To study changes in the properties of adrenoceptors, the targets for catecholamines, in hypertension, several experimental animal models have been used. The results, however, are rather conflicting: β-adrenoceptor number was found to be reduced in heart and vessels of spontaneously (SHR) (286,287), DOCA-salt (288,289), and renal hypertensive rats (290), but it was unchanged in heart of stroke-prone SHR (291) and in the kidney of DOCA-salt (289) and renal hypertensive rats (290). On the other hand, an increased number of β-adrenoceptors was found in the kidney of stroke-prone SHR (292) and in adipocytes of SHR (293). The reason for these divergent results (no changes, increases or decreases, in β-adrenoceptor number) is not known at present. However, there seems to be some evidence that in established experimental hypertension the responsiveness of cardiac β adrenoceptors is reduced (114).

To study β-adrenoceptor alterations in human essential hypertension, several groups have recently investigated β_2 adrenoceptors in circulating lymphocytes. There seems to be a general agreement that the number of β_2 adrenoceptors in circulating lymphocytes of patients with borderline or established essential hypertension is increased compared with normotensive controls (193,196,294–298), although two negative results have also been reported (299,300). In addition, a highly significant positive correlation between lymphocyte β_2-adrenoceptor density and mean arterial blood pressure was found (193,196,298). This increased β_2-adrenoceptor density was accompanied by an increased lymphocyte cyclic AMP response to isoproterenol stimulation (298,301). An increased β-adrenoceptor density associated with increased adenylate activation by β-adrenoceptor agonists has been recently found in the right atrium of essential hypertensive patients (302). Accordingly, in essential hypertension an increased sympathetic activity does not *decrease*, but rather *increases* β-adrenoceptor density; this is compatible with the view that in essential hypertension the physiological regulation of β adrenoceptors seems to be altered. This hypothesis has been recently supported by the observation that in essential hypertensive patients acute stimulation of the sympathetic activity by dynamic exercise led to a markedly lower increase in lymphocyte β_2-adrenoceptor density than in normotensive controls (see also Pathophysiological and Drug-induced Alterations: Desensitization) (197, 301). Interestingly, normalization of blood pressure in these patients with antihypertensives that do not directly influence β adrenoceptors [diuretics (303) and nifedipine or hydergine (304)] was accompanied by a reduction in lymphocyte β_2-adrenoceptor density to values not significantly different from those in age-matched normotensive controls. In addition, after normalization of blood pressure and lymphocyte β_2-adrenoceptor density, dynamic exercise produced in these patients, in contrast to the effects in untreated patients, significant increases in lymphocyte β_2-adrenoceptor density (304), indicating that the impairment of the acute regulation of β adrenoceptors seems to be corrected. These results support the view that in essential hypertension the increase in lymphocyte β_2-adrenoceptor density is directly related to the elevated blood pressure. Future experiments must show whether the increased β_2-adrenoceptor density is a cause or a consequence of the elevation in blood pressure in essential hypertension.

Chronic Renal Insufficiency

There is clinical evidence for a reduced sympathetic activity in patients on maintenance hemodialysis (305,306). Reduced baroreceptor sensitivity (307), abnormal responses to the Valsalva maneuver (308), dialysis-induced hypotension (309,310), elevated plasma norepinephrine levels (311-313), and blunted responsiveness to norepinephrine infusions (309) have been reported. In animal models of experimental uremia, an attenuated β-adrenoceptor responsiveness (assessed as heart rate response to isoproterenol) has been observed (314,315); this may be caused by a reduced cardiac β-adrenoceptor density and/or impaired adenylate cyclase activity (316). In patients on maintenance hemodialysis, the number of β_2 adrenoceptors in lymphocytes was found to be not different from that in healthy controls; on the other hand, lymphocyte cyclic AMP production in response to isoproterenol *and* NaF was diminished (317,318). In addition, in these patients the effects of isoproterenol infusion on systolic and diastolic blood pressure, as well as on heart rate, were significantly less than in age-matched healthy controls (318,319). Thus, in chronic uremia the number of β adrenoceptors is obviously not changed, but the β-adrenoceptor function is decreased. The reduced β-adrenoceptor responsiveness seems to be due to a post receptor defect, presumably to an impaired activity of the adenylate cyclase (318). Furthermore, in hypotensive patients on maintenance hemodialysis acute stimulation of the sympathetic activity by dynamic exercise (see Pathophysiological and Drug-induced Alterations: Desensitization) produced significantly lower increases in lymphocyte β_2-adrenoceptor density and lymphocyte cyclic AMP responses to isoproterenol than in healthy controls (320), indicating that in

Table 6 β-Adrenoceptor Determination in Various Diseases

Disease	Tissue and Species	β-Adrenoceptor changes	References
Diabetes mellitus			
Insulin dependent	Human leukocytes	No change	335
	Human lymphocytes	No change	Brodde, unpublished observations
Streptozotocin induced in rats	Rat ventricular myocardium	Decrease	336
	Rat skeletal muscle	Decrease	337
Ischemic heart disease	Canine ventricular myocardium	Increase	338
Hypoxia	Rat ventricular myocardium	Decrease	339
Depression	Human lymphocytes	Decrease	340, 341
	Human lymphocytes	No change	342
Leukemia	Human lymphocytes	Decrease	343-345
Atopic eczema	Human lymphocytes	No change	346, 347
Cystic fibrosis	Human lymphocytes	Decrease	348-350
Ewing's sarcoma		—[a]	351

[a]In contrast to other childhood cancers (osteosarcoma, neuroblastoma, lymphoma, and rhabdomyosarcoma) a high density of β adrenoceptors has been demonstrated in membrane preparations from human Ewing's sarcoma.

chronic uremia not only long-term, but also acute regulation of β adrenoceptors is impaired. Such an impaired regulation of β adrenoceptors—the target of endogenous catecholamines— may contribute to the disturbances of the sympathetic activity in chronic uremia leading to arterial hypotension.

Other Diseases

In a few further disease states β-adrenoceptor alterations have been determined; they are listed in Table 6.

FUTURE PERSPECTIVES

It is only 10 years ago that the first successful in vitro binding assay for β adrenoceptors was carried out (98). During the past decade a tremendous progress in the characterization of the β-adrenoceptor-adenylate cyclase system has been made. It is now widely accepted that β-adrenoceptors, rather than being static entities, are dynamically regulated by a wide variety of drugs, hormones, and physiological and pathological conditions. Very recently, β adrenoceptors have been purified to homogeneity (352-355), and photoaffinity techniques have been developed to label them (356-360). Thus, in the future, it would be possible to determine exactly the molecular structure of the three components of the β-adrenoceptor-adenylate cyclase system (the receptor, the G protein, and the catalytic unit of the adenylate cyclase). Precise knowledge of the molecular structure of the β-adrenoceptor system, however, would enable determination of how the three components are altered in disease- or drug-induced changes in β-adrenoceptor function. Furthermore, techniques have been developed to reconstitute purified components of the system (361). Such an approach would be helpful in identifying the component of the system responsible for changes in β-adrenoceptor responsiveness to β-adrenergic stimulation. Understanding of the molecular mechanisms involved in β-adrenoceptor regulation and the possibility of localizing disease-induced defects in the system may lead to the development of new therapeutic strategies and/or new drugs that will avoid treatment-induced changes in β-adrenoceptor function.

Another approach to characterize the β-adrenoceptor-adenylate cyclase system is the development of monoclonal antibodies against β_1 and β_2 adrenoceptors. The first successful experiments in this field have been reported by Venter and coworkers (362,363). The use of monoclonal antibodies would provide new insights into the properties of β_1 and β_2 adrenoceptors in different organs and would offer the possibility to investigate whether the function of one β-adrenoceptor subtype can be antagonized organ selectively without affecting that in other organs.

Finally, the specific cellular localization of β-adrenoceptor subtypes present in a single organ is still an open question. In several presumably homogeneous populations of cells, only one β-adrenoceptor subtype has been identified (113). Lau et al. (323) have presented evidence that in primary cultures of neonatal rat cardiac cells, β_1 adrenoceptors are localized on myoblasts but β_2 adrenoceptors are localized on fibroblasts. However, Homburger et al. (364) have demonstrated the coexistence of β_1 and β_2 adrenoceptors in C_6 glioma cells, which also strongly supports the view that *one* cell type can possess both β-adrenoceptor subtypes.

REFERENCES

1. H. H. Dale, J. Physiol. (Lond.) *34*, 163 (1906).
2. H. H. Dale, J. Physiol. (Lond.) *46*, 291 (1913).
3. R. P. Ahlquist, Am. J. Physiol. *153*, 586 (1948).
4. R. F. Furchgott in Handbook of Experimental Pharmacology, Vol. 33, Catecholamines (H. Blaschko and E. Muscholl, eds.), Springer, Berlin, 1972, pp. 283-335.
5. F. Murad, Y.-M. Chi, T. W. Rall, and E. W. Sutherland, J. Biol. Chem. *237*, 1233 (1962).
6. E. W. Sutherland, I. Oye, and R. W. Butcher, Recent Prog. Horm. Res. *21*, 623 (1965).
7. E. W. Sutherland, G. A. Robison, and R. W. Butcher, Circulation *37*, 279 (1968).
8. H. J. Schümann, M. Endoh, and O.-E. Brodde, Naunyn-Schmiedebergs Arch. Pharmacol. *289*, 291 (1975).
9. G. A. Robison, R. W. Butcher, and E. W. Sutherland, Cyclic AMP, Academic Press, New York, 1971.
10. M. Endoh, O.-E. Brodde, and H. J. Schümann, Naunyn-Schmiedebergs Arch. Pharmacol. *295*, 109 (1976).
11. K. H. Jakobs, W. Saur, and G. Schultz, J. Cyclic Nucleotide Res. *2*, 381 (1976).
12. K. H. Jakobs, K. Aktories, and G. Schultz, Adv. Cyclic Nucleotide Res. *17*, 135 (1984).
13. R. A. Salvador, K. I. Colville, S. A. April, and J. J. Burns, J. Pharmacol. Exp. Ther. *144*, 172 (1964).
14. J. J. Burns, R. A. Salvador, and L. Lemberger, Ann. N.Y. Acad. Sci. *139*, 833 (1967).
15. B. Levy, J. Pharmacol. Exp. Ther. *151*, 413 (1966).
16. N. C. Moran, Ann. N.Y. Acad. Sci. *139*, 649 (1967).
17. B. Levy, Arch. Int. Pharmacodyn. *170*, 418 (1967).
18. A. M. Lands, A. Arnold, J. P. McAuliff, F. P. Luduena, and T. G. Brown, Nature (Lond.) *214*, 597 (1967).
19. A. M. Lands, F. P. Luduena, and H. J. Buzzo, Life Sci. *6*, 2241 (1967).
20. A. Arnold and J. P. McAuliff, Experientia *24*, 436 (1968).
21. A. Arnold, J. P. McAuliff, D. F. Colella, W. V. O'Connor, and T. G. Brown, Jr., Arch. Int. Pharmacodyn. Ther. *176*, 451 (1968).

22. A. Arnold and W. H. Selberis, Experientia *24*, 1010 (1968).
23. A. Arnold and J. P. McAuliff, Arch. Int. Pharmacodyn. Ther. *179*, 381 (1969).
24. A. M. Lands, F. P. Luduena, and H. J. Buzzo, Life Sci. *8*, 373 (1969).
25. R. F. Furchgott, Ann. N.Y. Acad. Sci. *139*, 553 (1967).
26. M. J. Daly and G. P. Levy, in Trends in Autonomic Pharmacology, Vol. I (S. Kalsner, ed.), Urban and Scharzenberg, Baltimore, 1979, pp. 347-385.
27. E. J. Ariëns and A. M. Simonis, in Beta-Adrenoceptor Blocking Agents (P. R. Saxena and P. Forsyth, eds.), North Holland, Amsterdam, 1976, pp. 3-27.
28. E. Carlsson and A. Hedberg, Acta Physiol. Scand. (Suppl.) 440, 47 (1976).
29. E. J. Ariëns and A. M. Simonis, Biochem. Pharmacol. *32*, 1539 (1983).
30. E. Adler-Graschinsky and S. Z. Langer, Br. J. Pharmacol. *53*, 43 (1975).
31. S. Z. Langer, Pharmacol. Rev. *32*, 337 (1981).
32. H. Majewski, J. Auton. Pharmacol. *3*, 47 (1983).
33. M. J. Rand and H. Majewski, Clin. Exp. Hypertens. *A6*, 347 (1984).
34. L. Stjärne and J. Brundin, Acta Physiol. Scand. *94* , 139 (1975).
35. L. Stjärne and J. Brundin, Acta Physiol. Scand. *97*, 88 (1976).
36. C. Dahlöf, Acta Physiol. Scand. (Suppl.) *500*, 1 (1981).
37. T. C. Westfall, M. J. Peach, and V. Tittermary, Eur. J. Pharmacol. *58*, 67 (1979).
38. P. Hedqvist and A. Moawad, Acta Physiol. Scand. *95*, 494 (1975).
39. M. J. Stevens, R. E. Rittinghausen, R. L. Medcalf, and R. F. W. Moulds, Eur. J. Pharmacol. *83*, 263 (1982).
40. R. F. W. Moulds and M. J. Stevens, Gen. Pharmacol. *14*, 81 (1983).
41. T. J. Verbeuren, R. R. Lorenz, L. L. Aarhus, J. T. Shepherd, and P. M. Vanhoutte, J. Auton. Nerv. System. *8*, 261 (1983).
42. R. Hill, A. K. Keenan, M. Nelligan, and E. O'Malley, Br. J. Pharmacol. (Suppl.) *80*, 620p (1983).
43. M. Göthert and F. Hentrich, Br. J. Pharmacol. *85*, 933 (1985).
44. F. Hentrich, M. Göthert, and D. Greschuchna, Naunyn-Schmiedebergs Arch. Pharmacol. *330*, 245 (1985).
45. H. Lemoine, A. Ehle, and A. J. Kaumann, Naunyn-Schmiedebergs Arch. Pharmacol. *331*, 40 (1985).
46. R. R. Tuttle and J. Mills, Circ. Res. *36*, 185 (1975).
47. E. H. Sonnenblick, W. H. Frishman, and T. H. Lejemtel, N. Engl. J. Med. *300*, 17 (1979).
48. R. R. Ruffolo, Jr., K. Messick, and J. S. Horng, Naunyn-Schmiedebergs Arch. Pharmacol. *326*, 317 (1984).
49. R. R. Ruffolo, Jr., T. A. Spradlin, G. D. Pollock, J. E. Waddell, and P. J. Murphy, J. Pharmacol. Exp. Ther. *219*, 447 (1981).
50. T. P. Kenakin, J. Pharmacol. Exp. Ther. *216*, 210 (1981).
51. E. Carlsson, C. G. Dahlöf, A. Hedberg, H. Persson, and B. Tangstrand, Naunyn-Schmiedebergs Arch. Pharmacol. *300*, 101 (1977).
52. O. Rönn, C. Graffner, G. Johnsson, L. Jordö, P. Lundborg, and J. Wikstrand, Eur. J. Clin. Pharmacol. *5*, 9 (1979).

53. T. P. Kenakin and D. Beek, J. Pharmacol. Exp. Ther. *213*, 406 (1980).
54. N. Rohm, J. Wagner, and H. J. Schümann, Naunyn-Schmiedebergs Arch. Pharmacol. *315*, 85 (1980).
55. A. Hedberg, H. Mattsson, and E. Carlsson, J. Pharm. Pharmacol. *32*, 660 (1980).
56. T. P. Kenakin and D. Beek, J. Pharmacol. Exp. Ther. *220*, 77 (1982).
57. G. H. Apperly, G. M. Drew, and A. T. Sullivan, Eur. J. Pharmacol. *81*, 659 (1982).
58. N. Cook, A. Richardson, and D. B. Barnett, Eur. J. Pharmacol. *98*, 407 (1984).
59. T. P. Kenakin, J. Cardiovasc. Pharmacol. *7*, 208 (1985).
60. A. Nuttal and H. M. Snow, Br. J. Pharmacol. *77*, 381 (1982).
61. E. Malta, M. A. Mian, and C. Raper, Br. J. Pharmacol. *85*, 179 (1985).
62. D. Iakovidis, E. Malta, G. A. McPherson, and C. Raper, Br. J. Pharmacol. *68*, 677 (1980).
63. G. A. McPherson, E. Malta, P. Molenaar, and C. Raper, Br. J. Pharmacol. *82*, 897 (1984).
64. D. Dunlop and R. G. Shanks, Br. J. Pharmacol. *32*, 201 (1968).
65. M. A. van Baak, H. A. J. Struyker Boudier, and J. F. M. Smits, Clin. Exp. Hypertens. *A7*, 1 (1985).
66. A. Scriabine, Annu. Rev. Pharmacol. Toxicol. *19*, 269 (1979).
67. A. M. Strosberg, Arch. Int. Pharmacodyn. Ther. *222*, 200 (1976).
68. B. Ablad, E. Carlsson, and L. Ek, Life Sci. *12*, 107 (1973).
69. A. M. Barrett, J. Carter, J. D. Fitzgerald, R. Hull, and D. LeCount, Br. J. Pharmacol. *48*, 340P (1973).
70. A. M. Barrett, Postgrad. Med. J. *53*(Suppl. 3), 58 (1977).
71. J.-P. Boudot, I. Cavero, S. Fenard, F. Lefevre-Borg, P. Maroury, and A. G. Roach, Br. J. Pharmacol. *66*, 445 P (1979).
72. R. D. Smith and P. S. Wolf, in New Drugs Annual: Cardiovascular Drugs, Vol. 2 (A. Scriabine, ed.), Raven Press, New York, 1984, pp. 19-35.
73. H.-J. Schliep and J. Harting, J. Cardiovasc. Pharmacol. *6*, 1156 (1984).
74. X. L. Wang, M. Brinkmann, and O.-E. Brodde, Eur. J. Pharmacol. *114*, 157 (1985).
75. P. Pratesi and E. Grana, Adv. Drug Res. *2*, 127 (1965).
76. S. R. O'Donnell, Eur. J. Pharmacol. *19*, 371 (1972).
77. L. G. Letts, P. A. Easson, D.M. Temple, B. V. Lap, C. H. Lim, A. J. Orr, S. J. Pasaribu, S. N. Quessy, and L. R. Williams, Arch. Int. Pharmacodyn. Ther. *230*, 42 (1977).
78. J. L. Imbs, F. Miesch, J. Schwartz, J. Velly, G. Leclerc, A. Mann, and C. G. Wermuth, Br. J. Pharmacol. *60*, 357 (1977).
79. E. Holmberg, A.-B. Jeppson, C. J. Lamm, and B. Waldeck, Acta Pharmacol. Toxicol. (Copenh.) *46*, 150 (1980).
80. J. A. Nathanson, Br. J. Pharmacol. *73*, 97 (1981).
81. A. J. Bilski, S. Dorries, J. D. Fitzgerald, R. Jessup, H. Tucker, and J. L. Wale, Br. J. Pharmacol. *69*, 292 P (1980).
82. A. J. Bilski, S. E. Halliday, J. D. Fitzgerald, and J. L. Wale, J. Cardiovasc. Pharmacol. *5*, 430 (1983).

83. J. M. O. Arnold, P. C. O'Connor, J. G. Riddell, D. W. G. Harron, R. G. Shanks, and D. G. McDevitt, Br. J. Clin. Pharmacol. *19*, 619 (1985).
84. R. T. Brittain, J. B. Farmer, D. Jack, L. E. Martin, and W. T. Simpson, Nature (Lond.) *219*, 862 (1968).
85. V. A. Cullum, J. B. Farmer, D. Jack, and G. P. Levy, Br. J. Pharmacol. *35*, 141 (1969).
86. K. W. Dungan, Y. W. Cho, A. W. Gomoll, D. M. Aviado, and P. M. Lish, J. Pharmacol. Exp. Ther. *164*, 290 (1968).
87. J. Bergmann, H. Persson, and K. Wetterlin, Experientia *25*, 899 (1969).
88. S. R. O'Donnell, Eur. J. Pharmacol. *12*, 35 (1970).
89. K. Turnheim and O. Kraupp, Eur. J. Pharmacol. *15*, 231 (1971).
90. M. C. E. Gwee, M. W. Nott, C. Raper, and I. W. Rodger, Br. J. Pharmacol. *46*, 375 (1972).
91. G. H. Apperley, M. J. Daly, and G. P. Levy, Br. J. Pharmacol. *57*, 235 (1976).
92. Y. Yabuuchi, S. Yamashita, and S. S. Tei, J. Pharmacol. Exp. Ther. *202*, 326 (1977).
93. S. Yoshizaki, K. Tanimura, S. Tamada, Y. Yabuuchi, and K. Nakagawa, J. Med. Chem. *19*, 1138 (1976).
94. B. Levy, J. Pharmacol. Exp. Ther. *156*, 452 (1967).
95. R. J. Lefkowitz and E. Haber, Proc. Natl. Acad. Sci. USA *68*, 1773 (1971).
96. P. Cuatrecasas, G. P. E. Tell, V. Sica, I. Parikh, and K.-J. Chang, Nature *247*, 92 (1974).
97. R. J. Lefkowitz, L. E. Limbird, C. Mukherjee, and M. G. Caron, Biochim. Biophys. Acta *457*, 1 (1976).
98. R. J. Lefkowitz, C. Mukherjee, M. Coverstone, and M. G. Caron, Biochim. Biophys. Acta *60*, 703 (1974).
99. G. D. Aurbach, S. A. Fedak, C. J. Woodard, J. S. Palmer, D. Hauser, and T. Troxler, Science *186*, 1223 (1974).
100. G. Engel, D. Hoyer, R. Berthold, and H. Wagner, Naunyn-Schmiedebergs Arch. Pharmacol. *317*, 277 (1981).
101. K. Barovsky and G. Brooker, J. Cyclic Nucleotide Res. *6*, 297 (1980).
102. M. E. Maguire, P. M. Van Arsdale, and A. G. Gilman, Mol. Pharmacol. *12*, 335 (1976).
103. R. J. Lefkowitz, D. Mullikin, and M. G. Caron, J. Biol. Chem. *251*, 4686 (1976).
104. R. S. Kent, A. De Lean, and R. J. Lefkowitz, Mol. Pharmacol. *17*, 14 (1980).
105. K. E. J. Dickonson and S. R. Nahorski, J. Rec. Res. *3*, 123 (1983).
106. R. J. Lefkowitz, M. G. Caron, T. Michel, and J. M. Stadel, Fed. Proc. *41*, 2664 (1982).
107. J. M. Stadel, A. De Lean, and R. J. Lefkowitz, J. Biol. Chem. *255*, 1436 (1980).
108. D. Cassel and Z. Selinger, Proc. Natl. Acad. Sci. USA *74*, 3307 (1977).
109. D. Cassel and T. Pfeuffer, Proc. Natl. Acad. Sci. USA *75*, 2669 (1978).

110. D. M. Gill and R. C. Meren, Proc. Natl. Acad. Sci. USA *75*, 3050 (1978).
111. E. Carlsson, B. Ablad, A. Brandstrom, and B. Carlsson, Life Sci. *11*, 953 (1972).
112. P. B. Molinoff, Drugs *28*(Suppl. 2), 1 (1984).
113. K. P. Minneman, R. N. Pittman, and P. B. Molinoff, Annu. Rev. Neurosci. *4*, 419 (1981).
114. G. L. Stiles, M. G. Caron, and R. J. Lefkowitz, Physiol. Rev. *64*, 661 (1984).
115. D. B. Barnett, E. L. Rugg, and S. R. Nahorski, Nature *273*, 166 (1978).
116. E. L. Rugg, D. B. Barnett, and S. R. Nahorski, Mol. Pharmacol. *14*, 996 (1978).
117. K. P. Minneman, L. R. Hegstrand, and P. B. Molinoff, Mol. Pharmacol. *16*, 34 (1979).
118. A. De Lean, A. A. Hancock, and R. J. Lefkowitz, Mol. Pharmacol. *21*, 5 (1982).
119. A. J. Kaumann and H. Lemoine, Naunyn-Schmiedebergs Arch. Pharmacol. *331*, 27 (1985).
120. L. J. Bryan, J. J. Cole, S. R. O'Donnell, and J. C. Wanstall, J. Pharmacol. Exp. Ther. *216*, 395 (1981).
121. T. P. Kenakin and D. Beek, J. Pharmacol. Exp. Ther. *229*, 340 (1984).
122. C. Wilson and C. Lincoln, J. Cardiovasc. Pharmacol. *6*, 1216 (1984).
123. E. N. Juberg, K. P. Minneman, and P.-W. Abel, Naunyn-Schmiedebergs Arch. Pharmacol. *330*, 193 (1985).
124. A. J. Kaumann, Naunyn-Schmiedebergs Arch. Pharmacol. *332*, 406 (1986).
125. A. C. Dreyer and J. Offermeier, Pharmacol. Res. Commun. *7*, 151 (1975).
126. S. R. O'Donnell and J. C. Wanstall, Naunyn-Schmiedebergs Arch. Pharmacol. *308*, 183 (1979).
127. K. J. Broadley and M. J. Hawthorn, Br. J. Pharmacol (Suppl.) *78*, 136p (1983).
128. B. I. Costin, S. R. O'Donnell, and J. C. Wanstall, J. Pharm. Pharmacol. *35*, 752 (1983).
129. B. Ablad, B. Carlsson, E. Carlsson, C. Dahlöf, L. Ek, and F. Hultberg, Adv. Cardiol. *12*, 290 (1974).
130. J. Bonelli, Int. J. Clin. Pharmacol. *16*, 313 (1978).
131. O.-E. Brodde, K. Karad, H.-R. Zerkowski, N. Rohm, and J. C. Reidemeister, Naunyn-Schmiedebergs Arch. Pharmacol. (Suppl.) *321*, R242 (1982).
132. O.-E. Brodde, K. Karad, H.-R. Zerkowski, N. Rohm, and J. C. Reidemeister, Circ. Res. *53*, 752-758 (1983).
133. G. L. Stiles, S. Taylor, and R. J. Lefkowitz, Life Sci. *33*, 467 (1983).
134. A. Heitz, J. Schwartz, and J. Velly, Br. J. Pharmacol. *80*, 711 (1983).
135. S. Golf, R. Lovstad, and V. Hansson, Cardiovasc. Res. *19*, 636 (1985).
136. P. Robberecht, M. Delhaye, G. Taton, P. De Neef, M. Waelbroeck, J. M. De Smet, J. L. Leclerc, P. Chatelain, and J. Christophe, Mol. Pharmacol. *24*, 169 (1983).
137. A. Hedberg, F. Kempf, M. E. Josephson, and P. B. Molinoff, J. Pharmacol. Exp. Ther. *234*, 561 (1985).

138. K. E. J. Dickinson and S. R. Nahorski, Life Sci. *29*, 2527 (1981).
139. O.-E. Brodde, N. O'Hara, H.-R. Zerkowski, and N. Rohm, J. Cardiovasc. Pharmacol. *6*, 1184 (1984).
140. E. Gille, H. Lemoine, B. Ehle, and A. J. Kaumann, Naunyn-Schmiedebergs Arch. Pharmacol. *331*, 60 (1985).
141. M. Waelbroeck, G. Taton, M. Delhaye, P. Chatelain, J. C. Camus, R. Pochet, J. L. Leclerc, J. M. De Smet, P. Robberecht, and J. Christophe, Mol. Pharmacol. *24*, 174 (1983).
142. H.-R. Zerkowski, K. Ikezono, N. Rohm, J. C. Reidemeister, and O.-E. Brodde, Naunyn-Schmiedebergs Arch. Pharmacol., *332*, 142 (1986).
143. J. A. Ask, G. Stene-Larsen, K. B. Helle, and F. Resch, Acta Physiol. Scand. *123*, 81 (1985).
144. J. E. Brown, A. A. McLeod, and D. G. Shand, Clin. Pharmacol. Ther. *33*, 424 (1983).
145. O.-E. Brodde, K. D. Bock, N. Rohm, H.-R. Zerkowski, and J. C. Reidemeister, Verh. Dtsch. Ges. Inn. Med. *89*, 613 (1983).
146. O.-E. Brodde, ISI ATlas Sci.: Pharmacology *1*, 107 (1987).
147. G. Engel, Postgrad. Med. J. *57*(Suppl. 1) 77 (1981).
148. O.-E. Brodde, K. Karad, H.-R. Zerkowski, N. Rohm, and J. C. Reidemeister, Br. J. Pharmacol. (Suppl.) *78*, 72p (1983).
149. J. Zaagsma, P. J. C. M. van der Heijden, M. W. G. van der Schaar, and C. M. C. Bank, J. Rec. Res. *3*, 89 (1983).
150. S. Kakiuchi and T. W. Rall, Mol. Pharmacol. *4*, 367 (1968).
151. D.-M. Chuang and E. Costa, Proc. Natl. Acad. Sci. USA *76*, 3024 (1979).
152. D.-M. Chuang, Biochem. Biophys. Res. Commun. *105*, 1466 (1982).
153. J. M. Stadel, B. Strulovici, P. Nambi, T. N. Lavin, M. M. Briggs, M. G. Caron, and R. J. Lefkowitz, J. Biol. Chem. *258*, 3032 (1983).
154. B. Strulovici, J. M. Stadel, and R. J. Lefkowitz, J. Biol. Chem. *258*, 6410 (1983).
155. T. K. Harden, C. U. Cotton, G. L. Waldo, J. K. Lutton, and J. P. Perkins, Science *210*, 441 (1980).
156. R. C. Doss, J. P. Perkins, and T. K. Harden, J. Biol. Chem. *256*, 12281 (1981).
157. Y.-F. Su, T. K. Harden, and J. P. Perkins, J. Biol. Chem. *255*, 7410 (1980).
158. C. Hertel and M. Staehelin, J. Cell Biol. *97*, 1538 (1983).
159. C. Hertel, P. Müller, M. Portenier, and M. Staehelin, Biochem. J. *216*, 669 (1983).
160. R. H. Strasser, G. L. Stiles, and R. J. Lefkowitz, Endocrinology *115*, 1392 (1984).
161. T. K. Harden, Pharmacol. Rev. *35*, 5 (1983).
162. C. Hertel and J. P. Perkins, Mol. Cell. Endocrinol. *37*, 245 (1984).
163. R. J. Lefkowitz, M. G. Caron, and G. L. Stiles, N. Engl. J. Med. *310*, 1570 (1984).
164. Y.-F. Su, L. Cubeddu-Ximenez, and J. P. Perkins, J. Cyclic Nucleotide Res. *2*, 257 (1976).

165. Y.-F. Su, G. L. Johnson, L. Cubeddu-Ximenez, B. H. Leichtling, R. Ortmann, and J. P. Perkins, J. Cyclic Nucleotide Res. *2*, 271 (1976).
166. P. Mallorga, J.-F. Tallman, R. C. Henneberry, F. Hirata, W. T. Strittmatter, and J. Axelrod, Proc. Natl. Acad. Sci. USA *77*, 1341 (1980).
167. D. J. Kelleher, J. E. Pessin, A. E. Ruoho, and G. L. Johnson, Proc. Natl. Acad. Sci. USA *81*, 4316 (1984).
168. J. M. Stadel, P. Nambi, R. G. L. Shorr, D. F. Sawyer, M. G. Caron, and R. J. Lefkowitz, Proc. Natl. Acad. Sci. USA *80*, 3173 (1983).
169. D. R. Sibley, J. R. Peters, P. Nambi, M. G. Caron, and R. J. Lefkowitz, J. Biol. Chem. *259*, 9742 (1984).
170. L. T. Williams, R. Snyderman, and R. J. Lefkowitz, J. Clin. Invest. *57*, 149 (1976).
171. S. P. Galant, S. Underwood, L. Duriseti, and P. A. Insel, J. Lab. Clin. Med. *92*, 613 (1978).
172. O.-E. Brodde, G. Engel, D. Hoyer, K. D. Bock, and F. Weber, Life Sci. *29*, 2189 (1981).
173. H. R. Bourne and K. L. Melmon, J. Pharmacol. Exp. Ther. *178*, 1 (1971).
174. O.-E. Brodde, A. E.Daul, N. O'Hara, and A. M. Khalifa, J. Cardiovasc. Pharmacol. 7(Suppl. 6), S162 (1985).
175. O.-E. Brodde, R. Kretsch, K. Ikezono, H.-R. Zerkowski, and J. C. Reidemeister, Science *231*, 1584 (1986).
176. O.-E. Brodde, J. J. Beckeringh, and M. C. Michel, Trends Pharmacol Sci. *8*, 403 (1987).
177. M. E. Conolly and J. K. Greenacre, J. Clin. Invest. *58*, 1307 (1976).
178. H. G. Morris, S. A. Rusnake, J. C. Selner, K. Barzens, and P. J. Barnes, J.
178. H. G. Morris, S. A. Rusnake, J. C. Selner, K. Barzens, and P. J. Barnes, J. Cyclic Nucleotide Res. *3*, 439 (1977).
179. J. K. Greenacre, P. Schofield, and M. E. Conolly, Br. J. Clin. Pharmacol. *5*, 199 (1978).
180. S. P. Galant, L. Duriseti, S. Underwood, and P. A. Insel, N. Engl. J. Med. *299*, 933 (1978).
181. T. P. Lee, Res. Commun. Chem. Pathol. Pharmacol. *22*, 233 (1978).
182. S. P. Galant, L. Duriseti, S. Underwood, S. Allred, and P. A. Insel, J. Clin. Invest. *65*, 577 (1980).
183. D. P. Tashkin, M. E. Conolly, R. I. Deutsch, K. K. P. Hui, M. Littner, P. Scarpace, and I. Abrass, Am. Rev. Respir. Dis. *125*, 185 (1982).
184. W. Van den Berg, J. G. Leferink, J. K. Fokkens, J. Kreukniet, R. A. A. Maes, and P. L. B. Bruynzeel, J. Allergy Clin. Immunol. *69*, 410 (1982).
185. Y. Sano, G. Wastt, and R. G. Townley, J. Allergy Clin. Immunol. *72*, 495 (1983).
186. R. D. Aarons, A. S. Nies, J. G. Gerber, and P. B. Molinoff, J. Pharmacol. Exp. Ther. *224*, 1 (1983).
187. O.-E. Brodde, M. Brinkmann, R. Schemuth, N. O'Hara, and A. Daul, J. Clin. Invest. *76*, 1096 (1985).
188. A. O. Davies and R. J. Lefkowitz, J. Clin. Invest. *71*, 565 (1983).

189. K. K. P. Hui, M. E. Conolly, and D. P. Tashkin, Clin. Pharmacol. Ther. *32*, 566 (1982).
190. W. M. Samuelson and A. O. Davies, Am. Rev. Respir. Dis. *130*, 1023 (1984).
191. J. F. Krall, M. Connelly, and M. L. Tuck, J. Pharmacol. Exp. Ther. *214*, 554 (1980).
192. J. F. Tohmeh and P. E. Cryer, J. Clin. Invest. *65*, 836 (1980).
193. O.-E. Brodde, A. Prywarra, A. Daul, M. Anlauf, and K. D. Bock, J. Cardiovasc. Pharmacol. *6*, 678 (1984).
194. J. Butler, J. G. Kelly, K. O'Malley, and F. Pidgeon, J. Physiol. (Lond.) *344*, 113 (1983).
195. O.-E. Brodde, A. Daul, and N. O'Hara, Naunyn-Schmiedebergs Arch. Pharmacol. *325*, 190 (1984).
196. M. Middeke, J. Remien, and H. Holzgreve, J. Hypertens. *2*, 261 (1984).
197. O.-E. Brodde, A. E. Daul, N. O'Hara, X. L. Wang, and K. D. Bock, J. Hypertens. *3*(Suppl. 3), S149 (1985).
198. Y. Roan and S. P. Galant, Pediatr. Res. *16*, 591 (1982).
199. D. D. Schocken and G. S. Roth, Nature *267*, 856 (1977).
200. R. Landmann, H. Bittiger, and F. R. Bühler, Life Sci. *29*, 1761 (1981).
201. I. B. Abrass and P. J. Scarpace, J. Gerontol. *36*, 298 (1981).
202. V. Doyle, K. O'Malley, and J. G. Kelly, J. Cardiovasc. Pharmacol. *4*, 738 (1982).
203. N. O'Hara, A. E. Daul, R. Fesel, U. Siekmann, and O.-E. Brodde, Mech. Ageing Dev. *31*, 115 (1985).
204. R. D. Feldman, L. E. Limbird, J. Nadeau, D. Robertson, and A. J. J. Wood, N. Engl. J. Med. *310*, 815 (1984).
205. I. B. Abrass and P. J. Scarpace, J. Clin. Endocrinol. Metab. *55*, 1026 (1982).
206. J. R. Sporn, T. K. Harden, B. B. Wolfe, and P. B. Molinoff, Nature *194*, 624 (1976).
207. H. Kajiyama, K. Obara, Y. Nomura, and T. Segawa, Eur. J. Pharmacol. *77*, 75 (1982).
208. D. B. Bylund, L. R. Forte, D. W. Morgan, and J. R. Martinez, J. Pharmacol. Exp. Ther. *218*, 134 (1981).
209. T. E. Tenner, A. Mukherjee, and R. Kellyhester, Eur. J. Pharmacol. *77*, 61 (1982).
210. G. Glaubiger, B. S. Tsai, R. J. Lefkowitz, B. Weiss, and E. M. Johnson, Jr., Nature *273*, 240 (1978).
211. K. K. P. Hui and M. E. Conolly, N. Engl. J. Med. *304*, 1473 (1981).
212. G. Jennings, A. Bobik, and M. Esler, N. Engl. J. Med. *305*, 1019 (1981).
213. B. Davies, D. Sudera, C. Mathias, R. Bannister, and P. Sever, N. Engl. J. Med. *305*, 1017 (1981).
214. A. V. Chobanian, C. P. Tifft, H. Sackel, and A. Pitruzella, Clin. Exp. Hypertens, *A4*, 793 (1982).
215. D. G. McDevitt, Drugs *17*, 267 (1979).
216. B. N. C. Prichard, B. Tomlinson, R. J. Walden, and P. Bhattacharjee, J. Cardiovasc. Pharmacol. *5*(Suppl. 1), S56 (1983).

217. H. Boudoulas, R. P. Lewis, R. E. Kates, and G. Dalamangas, Ann. Intern. Med. *87*, 433 (1977).
218. G. Glaubiger and R. J. Lefkowitz, Biochem. Biophys. Res. Commun. *78*, 720 (1977).
219. R. D. Aarons and P. B. Molinoff, J. Pharmacol. Exp. Ther. *221*, 439 (1982).
220. R. D. Aarons, A. S. Nies, J. Gal, L. R. Hegstrand, and P. B. Molinoff, J. Clin. Invest. *65*, 949 (1980).
221. A. J. J. Wood, R. Feldman, and J. Nadeau, Clin. Exp. Hypertens. *A4*, 807 (1982).
222. C. R. Jones, K. F. White, and J. L. Reid, Br. J. Clin. Pharmacol. (Suppl.) *19*, 534P (1985).
223. O.-E. Brodde, A. Daul, N. Stuka, N. O'Hara, and U. Borchard, Naunyn-Schmiedebergs Arch. Pharmacol. *328*, 417 (1985).
224. Y. Giudicelli, D. Lacasa, B. Agli, and A. Leneveu, Eur. J. Clin. Pharmacol. *26*, 7 (1984).
225. P. B. Molinoff and R. D. Aarons, J. Cardiovasc. Pharmacol. *5*(Suppl. 1), S63 (1983).
226. O.-E. Brodde, R. Schemuth, M. Brinkmann, X. L. Wang, A. Daul, and U. Borchard, Naunyn-Schmiedebergs Arch. Pharmacol, *333*, 130 (1986).
227. M. Middeke, J. Remien, and H. Holzgreve, Verh. Dtsch. Ges. Inn. Med. *90*, 752 (1984).
228. E. Szecsi, S. Kohlschütter, W. Schiess, and E. Lang, Br. J. Clin. Pharmacol. *13*(Suppl. 2), 353S (1982).
229. R. J. Walden, P. Bhattacharjee, B. Tomlinson, J. Cashin, B. R. Graham, and B. N. C. Prichard, Br. J. Clin. Pharmacol. *13*(Suppl. 2), 359S (1982).
230. A. J. Kaumann, Naunyn-Schmiedebergs Arch. Pharmacol. *273*, 134 (1972).
231. J. C. Besse and A. D. Bass, J. Pharmacol. Exp. Ther. *154*, 224 (1966).
232. S. Kalsner, Br. J. Pharmacol. *36*, 582 (1969).
233. J. H. Exton, N. Friedmann, E. H.-A. Wong, J. P. Brineaux, J. D. Corbin, and C. R. Park, J. Biol. Chem. *247*, 3579 (1972).
234. A. O. Davies and R. J. Lefkowitz, Annu. Rev. Physiol. *46*, 119 (1984).
235. K. Mano, A. Akbarzadeh, and R. G. Townley, Life Sci. *25*, 1925 (1979).
236. J. B. Cheng, A. Goldfien, P. L. Ballard, and J. M. Roberts, Endocrinology *107*, 1646 (1980).
237. C. M. Fraser and J. C. Venter, Biochem. Biophys. Res. Commun. *94*, 390 (1980).
238. S. J. Foster and T. K. Harden, Biochem. Pharmacol. *29*, 2151 (1980).
239. A. O. Davies and R. J. Lefkowitz, J. Clin. Endocrinol. Metab. *51*, 599 (1980).
240. A. O. Davies and R. J. Lefkowitz, J. Clin. Endocrinol. Metab. *53*, 703 (1980).
241. A. Szentivanyi, J. Allergy *42*, 203 (1968).
242. S. T. Holgate, C. J. Baldwin, and A. E. Tattersfield, Lancet *2*, 375 (1977).
243. R. Ellul-Micallef and F. F. Fenech, Lancet *2*, 1269 (1975).

244. W. C. Stephan, T. W. Chick, B. P. Avner, and J. W. Jenne, J. Allergy Clin. Immunol. *65*, 105 (1980).
245. C. Davis and M. E. Conolly, Br. J. Clin. Pharmacol. *10*, 417 (1980).
246. J. P. Bilezikian and J. N. Loeb, Endocrine Rev. *4*, 378 (1983).
247. G. L. Stiles and R. J. Lefkowitz, Life Sci. *28*, 2529 (1981).
248. G. L. Stiles, J. M. Stadel, A. De Lean, and R. J. Lefkowitz, J. Clin. Invest. *68*, 1450 (1981).
249. L. T. Williams, R. J. Lefkowitz, A. M. Watanabe, D. R. Hathaway, and H. R. Besch, Jr., J. Biol. Chem. *252*, 2787 (1977).
250. J. S. Tsai and A. Chen, Nature *275*, 133 (1978).
251. J. Tse, J. R. Powell, C. A. Baste, R. E. Priest, and J. F. Kuo, Endocrinology *105*, 246 (1979).
252. J. Tse, R. W. Wrenn, and J. F. Kuo, Endocrinology *107*, 6 (1980).
253. O.-E. Brodde, H. J. Schümann, and J. Wagner, Mol. Pharmacol. *17*, 180 (1980).
254. C. C. Malbon, S. Y. Li, and J. N. Fain, J. Biol. Chem. *253*, 8820 (1978).
255. C. C. Malbon, J. Biol. Chem. *255*, 8692 (1980).
256. C. C. Malbon and M. L. Greenberg, J. Clin. Invest. *69*, 414 (1982).
257. R. S. Williams, C. E. Guthrow, and R. J. Lefkowitz, J. Clin. Endocrinol. Metab. *48*, 503 (1979).
258. K. K. P. Hui, R. N. Wolfe, and M. E. Conolly, Clin. Pharmacol. Ther. *32*, 161 (1982).
259. A. M. Ginsberg, W. E. Clutter, S. D. Shah, and P. E. Cryer, J. Clin. Invest. *67*, 1785 (1981).
260. R. G. G. Andersson, O. R. Nilsson, and J. F. Kuo, J. Clin. Endocrinol. Metab. *56*, 42 (1983).
261. G. Gross and I. Lues, Naunyn-Schmiedebergs Arch. Pharmacol. *329*, 427 (1985).
262. M. D. Miller and J. M. Marshall, Am. J. Physiol. *209*, 859 (1965).
263. J. M. Roberts, P. A. Insel, and A. Goldfien, Mol. Pharmacol. *20*, 52 (1981).
264. D. N. Hayashida, R. Leung, A. Goldfien, and J. M. Roberts, Am. J. Obstet. Gynecol. *142*, 389 (1982).
265. G. Berg, R. G. G. Andersson, and G. Ryden, Am. J. Obstet. Gynecol. *151*, 392 (1985).
266. C. A. Chidsey, E. Braunwald, and A. G. Morrow, Am. J. Med. *39*, 442 (1965).
267. J. A. Thomas and B. H. Marks, Am. J. Cardiol. *41*, 233 (1978).
268. G. S. Francis, S. R. Goldsmith, S. M. Ziesche, and J. N. Cohn, Am. J. Cardiol. *49*, 1152 (1982).
269. C. Kawai, Y. Yui, T. Hoshino, S. Sasayama, and A. Matsumori, J. Am. Coll. Cardiol. *2*, 834 (1983).
270. M. R. Bristow, Int. J. Cardiol. *5*, 648 (1984).
271. M. R. Bristow, N. E. Kantrowitz, R. Ginsburg, and M. B. Fowler, J. Mol. Cell. Cardiol. *17*(Suppl. 2), 41 (1985).
272. W. S. Colucci, R. W. Alexander, G. H. Williams, R. E. Rude, B. L. Holman,

M. A. Konstam, J. Wynne, G. H. Mudge, and E. Braunwald, N. Engl. J. Med. *305*, 185 (1981).
273. M. R. Bristow, R. Ginsburg, W. Minobe, R. S. Cubicciotti, W. S. Sageman, K. Lurie, M. E. Billingham, D. C. Harrison, and E. B. Stinson, N. Engl. J. Med. *307*, 205 (1982).
274. R. Ginsburg, M. R. Bristow, M. E. Billingham, E. B. Stinson, J. S. Schroeder, and D.C. Harrison, Am. Heart J. *106*, 535 (1983).
275. M. D. Snavely, H. J. Motulsky, E. Moustafa, L. C. Mahan, and P. A. Insel, Circ. Res. *51*, 504 (1982).
276. G. Tsujimoto, W. M. Manger, and B. B. Hoffman, Endocrinology *114*, 1272 (1984).
277. M. R. Bristow, R. Ginsburg, V. Umans, M. Fowler, W. Minobe, R. Rasmussen, P. Zera, R. Menlove, P. Shah, S. Jamieson, and E. B. Stinson, Cir. Res. *59*, 297 (1986).
278. O.-E. Brodde, S. Schüler, R. Kretsch, M. Brinkmann, H. G. Borst, R. Hetzer, J. C. Reidemeister, H. Warnecke, and H.-R. Zerkowski, J. Cardiovasc. Pharmacol. *8*, 1235 (1986).
279. M. Irmer, H. Wollschläger, and H. Just, Klin, Wochenschr. *59*, 639 (1981).
280. B. Sharma and J. F. Goodwin, Circulation *58*, 449 (1978).
281. P. D. V. Bourdillon, J. R. Dawson, R. A. Foale, A. D. Timmis, P. A. Poole-Wilson, and G. C. Sutton, Br. Heart J. *43*, 206 (1980).
282. B. Folkow, Physiol. Rev. *62*, 347 (1982).
283. J. De Champlain, Can. J. Physiol. Pharmacol. *56*, 341 (1978).
284. M. A. Weber and J. I. M. Drayer, Mineral Electrolyte Metab. *7*, 57 (1982).
285. D. S. Goldstein, Hypertension *5*, 86 (1983).
286. C. Limas and C. J. Limas, Biochem. Biophys. Res. Commun. *83*, 710 (1978).
287. C. J. Limas and C. Limas, Biochim. Biophys. Acta *582*, 533 (1979).
288. E. A. Woodcock, C. A. Olsson, and C. I. Johnston, Biochem. Pharmacol. *29*, 1465 (1980).
289. S. Yamada, H. I. Yamamura, and W. R. Roeske, Life Sci. *27*, 2405 (1980).
290. E. A. Woodcock and C. I. Johnston, Hypertension *2*, 156 (1980).
291. A. Mukherjee, R. M. Graham, A. I. Sagalowsky, W. Pettinger, and K. E. McCoy, J. Mol. Cell. Cardiol. *12*, 1263 (1980).
292. S. Yamada, T. Ishima, T. Tomita, M. Hayashi, and E. Hayashi, Eur. J. Pharmacol. *83*, 149 (1982).
293. G. M. Boriskina and Y. V. Postnov, Experientia *38*, 263 (1982).
294. O.-E. Brodde, A. Prywarra, M.Anlauf, A. Daul, and K. D. Bock, Hypertens. *1*(Suppl. 2), 263 (1983).
295. D. J. Fitzgerald, V. Doyle, E. T. O'Brien, J. G. Kelly, and K. O'Malley, J. Hypertens. *1*(Suppl. 2), 260 (1983).
296. R. Landmann, E. Bürgisser, and F. R. Bühler, J. Rec. Res. *3*, 71 (1983).
297. G. Bruschi, G. Orlandini, C. Pavarani, M. Spaggiari, L. Tacinelli, and A. Cavatorta, IRCS Med. Sci. *12*, 461 (1984).
298. O.-E. Brodde, N. Stuka, V. Demuth, R. Fesel, J. Bergerhausen, A. Daul, and K. D. Bock, Clin. Exp. Hypertens. *A7*, 1135 (1985).

299. M. S. Kafka, C. R. Lake, H.-G. Gullner, J. F. Tallman, F. C. Bartter, and T. Fujita, Clin. Exp. Hypertens. *1*, 613 (1979).
300. R. D. Feldman, L. E. Limbird, J. Nadeau, D. Robertson, and A. J. J. Wood, J. Clin. Invest. *73*, 648 (1984).
301. M. Middeke, S. Kirzinger, J. Remien, B. Krahl, and H. Holzgreve, Clin. Exp. Hypertens. *A8*, 135 (1986).
302. R. Hill, A. K. Keenan, and M. Neligan, Br. J. Pharmacol. (Suppl.)*87*, 106p (1986).
303. M. Middeke, J. Remien, S. Kirzinger, and H. Holzgreve, Eur. J. Pharmacol. *109*, 401 (1985).
304. O.-E. Brodde, M. C. Michel, and J. J. Beckeringh, J. Cardiovasc. Pharmacol. *10* (Suppl. 4), S50 (1987).
305. J. J. Lilley, J. Golden, and R. A. Stone, J. Clin Invest. *57*, 1190 (1976).
306. P. Zucchelli, A. Sturani, A. Zuccala, A. Santoro, E. Degli Esposti, and C. Chiarini, Contrib. Nephrol. *45*, 69 (1985).
307. T. G. Pickering, B. Gribbin, and D. O. Oliver, Clin. Sci. *43*, 645 (1972).
308. D. J. Ewing and R. Winney, Nephron *15*, 424 (1975).
309. A. Botey, J. Gaya, J. Montoliu, A. Torras, F. Rivera, J. Lopez-Pedret, and L. Revert, Proc. Eur. Dial. Transplant. Assoc. *18*, 586 (1981).
310. E. S. Kersh, S. J. Kronfield, A. Unger, R. W. Popper, S. Cantor, and K. Cohn, N. Engl. J. Med. *290*, 650 (1974).
311. H. M. Brecht, W. Ernst, and K. M. Koch, Proc. Eur. Dial. Transplant. Assoc. *12*, 281 (1975).
312. W. L. Henrich, F. H. Katz, P. B. Molinoff, and R. W. Schrier, Kidney Int. *12*, 279 (1977).
313. P. Grützmacher, W. Ernst, and W. Schoeppe, Nieren Hochdruckkrankheiten *11*, 122 (1982).
314. T. Drüeke, F. Lhoste, S. Larno, J. Zingraff, N. K. Man, P. Jungers, J. R. Boissier, and J. Crosnier, Proc. Eur. Dial. Transplant. Assoc. *13*, 464 (1976).
315. J. F. E. Mann, M. Hausen, A. Kutter, R. Sudhoff, and E. Ritz, Proc. Eur. Dial. Transplant. Assoc. *19*, 788 (1982).
316. J. F. E. Mann, M. Hausen, K.-H. Jakobs, A. Kutter, W. Nagel, W. Rascher, M. Schick, R. Sudhoff, and E.Ritz, Contrib. Nephrol. *41*, 108 (1984).
317. A. E. Daul, A. M. Khalifa, N. Graben, and O.-E. Brodde, Proc. Eur. Dial. Transplant. Assoc. *21*, 178 (1984).
318. O.-E. Brodde and A. Daul, Contrib. Nephrol. *41*, 99 (1984).
319. A. Ulman, T. Drüeke, J. Zingraff, and J. Crosnier, Clin. Nephrol. *7*, 58 (1977).
320. O.-E. Brodde and A. Daul, Contrib. Nephrol, *50*, 28 (1986).
321. A. A. Hancock, A. L. De Lean, and R. J. Lefkowitz, Mol. Pharmacol. *16*, 1 (1979).
322. T. Vago, M. Bevilacqua, R. Dagani, R. Meroni, G. Frigeni, C. Santoli, and G. Norbiato, Biochem. Biophys. Res. Commun. *121*, 346 (1984).
323. Y. H. Lau, R. B. Robinson, M. Rosen, and J. P. Bilezikian, Circ. Res. *47*, 41 (1980).
324. K. Dickinson, A. Richardson, and S. R. Nahorski, Mol. Pharmacol. *19*, 194 (1981).

325. S. R. Nahorski, D. B. Barnett, D. R. Howlett, and E. L. Rugg, Naunyn-Schmiedebergs Arch. Pharmacol. *307*, 227 (1979).
326. R. J. Summers, J. A. Stephenson, and M. J. Kuhar, J. Pharmacol. Exp. Ther. *232*, 561 (1985).
327. O.-E. Brodde, Biochem. Pharmacol. *31*, 1743 (1982).
328. G. Engel, R. Maurer, K. Perrot, and B. P. Richardson, Naunyn-Schmiedebergs Arch. Pharmacol. *328*, 354 (1985).
329. A. Hedberg, K. P. Minneman, and P. B. Molinoff, J. Pharmacol. Exp. Ther. *212*, 503 (1980).
330. H. Carswell and S. R. Nahorski, Br. J. Pharmacol. *79*, 965 (1983).
331. O.-E. Brodde, F.-J. Leifert, and H.-J. Krehl, J. Cardiovasc. Pharmacol. *4*, 34 (1982).
332. O.-E. Brodde, F. Kuhlhoff, J. Arroyo, and A. Prywarra, Naunyn-Schmiedebergs Arch. Pharmacol. *322*, 20 (1983).
333. A. S. Manalan, H. R. Besch, and A. M. Watanabe, Circ. Res. *49*, 326 (1981).
334. J. J. Moore and J. A. Whitsett, Placenta (Suppl. 3), 103 (1981).
335. P. Serusclat, S. G. Rosen, E. B. Smith, S. D. Shah, W. E. Clutter, and P. E. Cryer, Diabetes *32*, 825 (1983).
336. J. J. Savarese and B. A. Berkowitz, Life Sci. *25*, 2075 (1979).
337. E. N. Olson, D. A. Kelley, and P. B. Smith, Exp. Neurol. *73*, 154 (1981).
338. A. Mukherjee, T. M. Wong, L. M. Buja, R. J. Lefkowitz, and J. T. Willerson, J. Clin. Invest. *64*, 1423 (1979).
339. N. F. Voelkel, L. Hegstrand, J. T. Reeves, I. F. McMurty, and P. B. Molinoff, J. Appl. Physiol. *50*, 363 (1981).
340. J. Extein, J. Tallman, C. C. Smith, and F. K. Goodwin, Psychiatry Res. *1*, 191 (1979).
341. G. N. Pandey, M. W. Dyksen, D. L. Garver, and J. M. Davis, Am. J. Psychiatry *136*, 675 (1979).
342. J. Zohar, J. Bannet, D. Drummer, R. Fisch, R. P. Epstein, and R. H. Belmaker, Biol. Psychiatry *18*, 553 (1983).
343. J. R. Sheppard, R. Gormus, and C. F. Moldow, Nature *269*, 693 (1977).
344. J. M. Bidart, P. Motte, C. Bohuon, and D. Bellet, Leuk. Res. *5*, 443 (1981).
345. E. Paietta and J. D. Schwarzmeier, Eur. J. Clin. Invest. *13*, 339 (1983).
346. S. P. Galant, S. Underwood, S. Allred, and J. M. Hanifin, J. Invest. Dermatol. *72*, 330 (1979).
347. A. E. Ruoho, J. L. DeClerque, and W. W. Busse, J. Allergy Clin. Immunol. *66*, 46 (1980).
348. P. B. Davis, M. Braunstein, and C. Jay, Pediatr. Res. *12*, 703 (1978).
349. P. B. Davis, J. R. Shelhamer, and M. Kaliner, N. Engl. J. Med. *302*, 1453 (1980).
350. S. P. Galant, L. Norton, J. Herbst, and C. Wood, J. Clin. Invest. *68*, 253 (1981).
351. J. A. Whitsett, J. Burdsall, L. Workman, B. Hollinger, and J. Neely, JNCI *71*, 779 (1983).

352. G. Vauquelin, P. Geynet, J. Hanoune, and A. D. Strosberg, Proc. Natl. Acad. Sci. USA *74*, 3710 (1977).
353. O. Durieu-Trautmann, C. Delavier-Klutchko, G. Vauquelin, and A. D. Strosberg, J. Supramol. Struct. *13*, 411 (1980).
354. R. G. L. Shorr, R. J. Lefkowitz, and M. G. Caron, J. Biol. Chem. *256*, 5820 (1981).
355. R. G. L. Shorr, M. W . Strohsacker, T. N. Lavin, R. J. Lefkowitz, and M. G. Caron, J. Biol. Chem. *257*, 12341 (1982).
356. S. M. Wrenn and C. J. Homcy, Proc. Natl. Acad. Sci. USA *77*, 4449 (1980).
357. A. Rashidbaigi and A. E. Ruoho, Proc. Natl. Acad. Sci. USA *78*, 1609 (1981).
358. W. Burgermeister, M. Hekman, and E. J. M. Helmreich, J. Biol. Chem. *257*, 5306 (1982).
359. T. N. Lavin, P. Nambi, S. L. Heald, P. W. Jeffs, R. J. Lefkowitz, and M. G. Caron, J. Biol. Chem. *257*, 12332 (1982).
360. G. L. Stiles, R. H. Strasser, T. N. Lavin, L. R. Jones, M. G. Caron, and R. J. Lefkowitz, J. Biol. Chem. *258*, 8443 (1983).
361. R. J. Lefkowitz, R. A. Cerione, J. Codina, L. Birnbaumer, and M. G. Caron, J. Membr. Biol. *87*, 1 (1985).
362. C. M. Fraser, R. Greguski, B. Eddy, and J. C. Venter, J. Cell Biochem. *21*, 219 (1983).
363. J. C. Venter and C. M. Fraser, Fed. Proc. *42*, 273 (1983).
364. V. Homburger, M. Lucas, E. Rosenbaum, G. Vassent, and J. Bockaert, Mol. Pharmacol. *20,* 463 (1981).

9

Central Serotonin Receptors

Richard A. Glennon / Medical College of Virginia, Virginia Commonwealth University, Richmond, Virginia

Serotonin (5-hydroxytryptamine; 5-HT) has been the object of considerable research for over 30 years; now, with the recent identification of central 5-HT binding sites, has come a renewed interest in this neurotransmitter. The purpose of this review is to describe some of the progress that has been made with respect to the identification of these binding sites and agents that interact therewith; particular emphasis is placed on site-selective 5-HT agonists and antagonists. The general pharmacology of serotonin is not discussed; the interested reader is directed to reviews by Fuller (1,2) for an overview of 5-HT biosynthesis and metabolism and for a discussion of agents that have an effect on serotonergic mechanisms (including direct and indirect acting agonists, antagonists, synthesis inhibitors, and uptake inhibitors). Hamon et al. (3) have also reviewed 5-HT and 5-HT receptors. This chapter is limited to studies involving central 5-HT binding sites; excellent reviews of noncentral 5-HT receptors (including commentaries on the potential relationships between these receptors and central 5-HT sites) have been presented by Gershon (4), Fozard (5), and Bradley et al. (6).

HISTORICAL SURVEY

In 1972, Farrow and van Vunakis (7) identified 5-HT binding sites in rat brain homogenates using an equilibrium dialysis technique. Bennett and Aghajanian (8) 2 years later described their work on [^{3}H] LSD binding sites using a rapid filtration technique. Evidence that these sites may constitute 5-HT sites was provided to some extent by Bennett and Snyder (9), who employed [^{3}H] 5-HT as a radioligand. However, in 1976, Fillion et al. (10,11) suggested that sites labeled by [^{3}H] 5-HT and [^{3}H] LSD were dissimilar. That same year, Bennett and Snyder (12) noted similar results and suggested that 5-HT sites may exist in interconvertible agonist and antagonist conformations. They further suggested that 5-HT interacts only with the agonist conformation of serotonin sites, whereas LSD, a mixed agonist-antagonist, interacts with both conformations. This explained the observations that known agonists displayed a higher affinity for [^{3}H] 5-HT-labeled sites than for [^{3}H] LSD-labeled sites, whereas the opposite was generally the case for serotonin antagonists. These and other early studies using tritiated 5-HT and LSD as ligands have been reviewed by Fillion (13).

In 1978, Leysen et al. (14) and Creese and Snyder (15) independently reported that the tritiated neuroleptic agent spiperone (spiroperidol) labels certain populations of central 5-HT sites. This subsequently led Peroutka and Snyder (16) to suggest the existence of two major types of 5-HT binding sites. 5-HT$_1$ sites were defined as those that were labeled by [^{3}H] 5-HT, whereas 5-HT$_2$ sites (frontal cortex) were those sites labeled by [^{3}H] spiperone. Additional sites have been described; these are addressed in the following section.

SEROTONIN BINDING SITES AND RADIOLIGANDS

The localization and characterization of the two major populations of serotonin binding sites (i.e., 5-HT$_1$ and 5-HT$_2$) and the methodologies employed to investigate these sites have been reviewed by Leysen with Tollenaere (17-19) and by Hamon et al. (3). Table 1 lists 5-HT sites that have been reported to date; these sites are described here in more detail. Structures of radioligands that are currently used to label 5-HT binding sites are shown in Figure 1.

5-HT$_1$ Sites

In rat brain, 5-HT$_1$ sites predominate in the hippocampus, striatum, and frontal cortex; the most commonly employed ligand used for labeling these sites is [^{3}H] 5-HT. 5-HT$_1$ sites appear to represent a heterogeneous population of sites. Pedigo et al. (20) found that butaclamol and spiperone could distinguish between two types of 5-HT$_1$ sites; these sites were called 5-HT$_{1A}$ and 5-HT$_{1B}$ sites. 5-HT$_{1A}$ sites are sensitive to spiperone, whereas 5-HT$_{1B}$ sites display a 100- to 3000-fold lower affinity for spiperone (20,21). Serotonin itself binds nearly equally well to both subpopulations of sites. More recently, tritiated 8-hydroxy-(*N,N*-di-*n*-propylamino)tetralin (8-OH-DPAT) (22,23) and [^{125}I] (–)-iodocyanopindolol (24) have been used to selectively label 5-HT$_{1A}$ and 5-HT$_{1B}$ sites, respectively.

Table 1 Serotonin Binding Sites Described in the Literature

Site	Comments	References
Group A	*Sites that have been relatively well studied*	
5-HT_1	Labeled with $[^3H]$ 5-HT; heterogeneous (may consist of 5-HT_{1A}, 5-HT_{1B}, and 5-HT_{1C} subpopulations)	16
5-HT_{1A}	Labeled with $[^3H]$ 8-OH DPAT; SPIP-sensitive 5-HT_1 sites	20
5-HT_{1B}	Labeled with $[^{125}I]$ iodocyanopindolol; SPIP-insensitive 5-HT_1 sites	20,24
5-HT_2	Labeled with tritiated SPIP, KET, MES, DOB, and $[^{125}I]$ 2-iodo LSD	16,17,25 26,27
Group B	*Recently described or less well defined sites*	
5-HT_{1C}	Identified using $[^3H]$ MES; labeled with $[^{125}I]$ 2-I-LSD	31,32
5-HT_{2A}	$[^3H]$ SPIP-labeled 5-HT_2 sites with high affinity for methysergide	33,34
5-HT_{2B}	$[^3H]$ SPIP-labeled 5-HT_2 sites with low affinity for methysergide	33,34
5-HT_3	Identified (striatum) using $[^3H]$ 8-OH-DPAT	23
5-HT_3	Distinguished from 5-HT_1 and 5-HT_2 sites by antibodies from blood of an autistic patient	35
5-HT_3	Identified in (rat cortex) synaptosomal membrane preparations	36

5-HT_2 Sites

The greatest density of 5-HT_2 sites occurs in the cortex and, in particular, in the frontal cortex. $[^3H]$ spiperone (SPIP) has been the most widely used ligand for labeling 5-HT_2 sites, but $[^3H]$ ketanserin (KET) is rapidly gaining popularity. Another two ligands that have seen some application are $[^3H]$ mesulergine (MES) (25) and $[^{125}I]$ 2-iodo LSD (26). The most recently described ligand for 5-HT_2 sites is $[^3H]$ 1-(2,5-dimethoxy-4-bromophenyl)-2-aminopropane (DOB) (27); this ligand is unique in that, unlike SPIP and KET, for example, DOB is a 5-HT_2 agonist (28).

Miscellaneous Sites

There is a general recognition of the existence of two major populations of serotonin binding sites (i.e., 5-HT_1 and 5-HT_2 sites) and a growing acceptance of distinct 5-HT_{1A} and 5-HT_{1B} subpopulations. Although the 5-HT_1 and 5-HT_2 no-

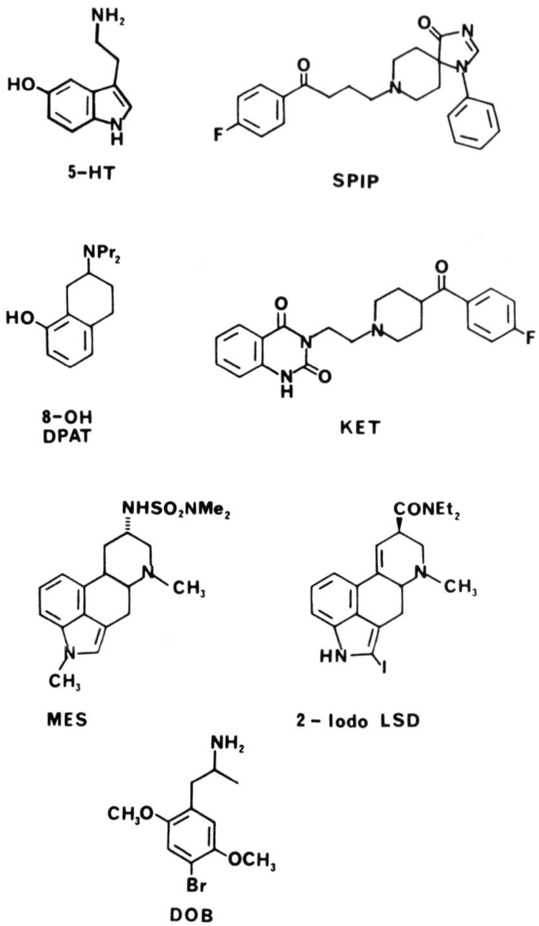

Figure 1 Structure of agents currently used as radioligands for studying serotonin binding sites. 5-HT, serotonin; SPIP, spiperone; 8-OH-DPAT; 8-hydroxy-2-(di-*n*-propylamino)tetralin; KET, ketanserin; MES, mesulergine; 2-Iodo-LSD, 2-iodolysergic acid diethylamide; DOB, 1-(2,5-dimethoxy-4-bromophenyl)-2-aminopropane.

menclature has been the most popular, some authors have referred to these binding sites as S_1 and S_2 sites. Because of possible confusion with electrophysiologically defined S_1, S_2, and S_3 serotonin receptors (29), S terminology is not used here. 5-HT_X nomenclature has also been used for peripheral serotonin receptors. 5-HT_1 sites display characteristics similar to those of peripheral "5-HT_1-like" receptors; there is also evidence for the existence of several subpopulations of these receptors (6). Peripheral 5-HT_2 receptors (i.e., the classic D receptor of Gaddum) are similar to 5-HT_2 sites (6).

Several other central 5-HT sites have been described; information on these sites is either very recent or is rather limited. Whether these sites represent distinct 5-HT receptors remains to be determined. Nevertheless, a brief discussion of these sites is included for the sake of completeness. 5-HT_{1C} serotonin sites were initially identified by autoradiography (30). These sites, which may also be present in rat and human cortex, have now been labeled in procine cortex homogenates with [^{3}H]MES (31,32). Although most studies have concluded that 5-HT_2 sites are homogeneous, there has been a suggestion of the existence of two subpopulations of sites (i.e., 5-HT_{2A} and 5-HT_{2B} sites) in mouse and human cortex (33,34), however, detailed follow-up studies substantiating these claims have been lacking. Several other types of serotonin sites have been described. For example, Hamon et al. (23) reported that whereas tritiated 8-OH-DPAT labeled 5-HT_{1A} sites in hippocampal homogenates, it labels sites in striatal homogenates that possess characteristics that are distinctly different from those of 5-HT_{1A} sites. These sites have been called 5-HT_3 sites. This same term (i.e., 5-HT_3 sites) has been used by Todd and Ciarenello (35) to describe sites they have identified in bovine and human cortex and by Robaut et al. (36) to describe a population of sites in crude membrane preparations or fractions enriched with synaptosomal membranes obtained from rat brain cortex. To date, there is no evidence to suggest that any of these 5-HT_3 sites correspond to peripheral 5-HT_3 receptors [i.e., the M receptors of Gaddum (6)]. Unfortunately, too little is known about this group of sites to warrant further discussion at this time.

Other tritiated ligands have been employed in an attempt to label serotonin binding sites, such as methiothepin, mianserin, metergoline, lisuride, and quipazine. For the most part, these agents appear to label more than one population of sites and have not met with the same success as those previously mentioned. See Hamon et al. (3) for a review of these ligands.

SELECTIVITY AND STRUCTURE-ACTIVITY RELATIONSHIPS

Members of various drug classes have been shown to interact at serotonin binding sites; at this time, the significance of these interactions is unknown. For example, β-carboline derivatives interact at both 5-HT_1 and 5-HT_2 sites; their affinity for these sites is rather low (i.e., in the micromolar range) (37,38), and all derivatives appear to possess a greater affinity for [^{3}H]tryptamine-labeled tryptamine binding sites (38,39). Benzodiazepine derivatives (e.g., chlordiazepoxide, diazepam, oxazepam, clonazepam, bromazepam, and lorazepam), including the triazolobenzodiazepines (e.g., midazolam), display little affinity for 5-HT_2 or [^{3}H] SPIP-labeled 5-HT_2 sites (40). However, serotonin may play a role in the mechanism of action of certain nonbenzodiazepine anxiolytics. For example, busipirone and the structurally related TVX Q 7821 (ipsapirone or isapirone) display a significant affinity for 5-HT_1 sites (41,42). These and related agents, unlike benzodiazepines, also demonstrate high affinity and selectivity for [^{3}H] 8-OH-DPAT-labeled 5-HT_{1A} sites (Table 2) (22,43). Conversely, [^{3}H] TVX Q 7821 labels serotonin binding sites (44). The role of 5-HT in the mechanism of action of nonbenzodiazepine anxiolytics promises to be an exciting area of research in the years to come.

The binding of neuroleptic agents to serotonin sites has been investigated. The butyrophenone derivative spiperone displays a high affinity for 5-HT_2 sites and, as already mentioned, tritiated spiperone is used to label 5-HT_2 sites. A problem associated with the use of this agent is that it also possesses an affinity for (and can label) 5-HT_{1A} sites (20). In general, neuroleptic agents display a higher affinity for 5-HT_2 sites than for 5-HT_1 sites. Nevertheless, their affinity for dopamine sites is usually higher than their affinity for 5-HT sites (although this is somewhat related to the chemical class to which the neuroleptic agent belongs). Table 3 lists some binding data for several examples of neuroleptic agents. In vivo binding studies have also been conducted; for example, see Ortmann et al. (48). Serotonin has long been speculated to play a role in the mechanism of action of antidepressants; for reviews of this topic see Briley (49) and Ogren (50). Most antidepressants display a low affinity for 5-HT_1 sites but possess a significant affinity for 5-HT_2 sites (Table 3). What little information is available from whole-animal studies suggests that certain of these

Table 2 Binding Profile of the Anxiolytic Agents Buspirone and TVX Q 7821

Binding site[a]	IC_{50} (nM)	
	Buspirone	TVX Q 7821
5-HT_{1A}	24	9.5
5-HT_{1B}	26,000	26,000
5-HT_2	2,100	10,000
Dopamine	380	3,000
Histamine	4,700	4,700
Benzodiazepine	>100,000	>100,000

[a]Includes dopamine (D_2), histamine (H_1), and tritiated flunitrazepam-labeled benzodiazepine (BDZ) sites.
Source: From Reference 43.

Table 3 Binding Profiles of Representative Neuroleptic Agents and Antidepressants[a]

Neuroleptics and antidepressants	5-HT$_1$ IC$_{50}$ (45,50)	5-HT$_1$ K$_i$ (52)	5-HT$_2$ K$_i$ (SPIP) (45,51)	5-HT$_2$ K$_i$ (KET) (46,43)	H$_1$ K$_i$ (52)	D$_2$ K$_i$ (47,52)	IMIP[b] IC$_{50}$ (53)
Neuroleptics							
Chlorpromazine	4,000	–	60	3	–	10	–
Chlorprothixene	1,000	–	10	–	–	4	–
Haloperidol	1,000	–	100	22	–	1	–
Pimozide	>20,000	–	–	6	–	1	–
Clozapine	1,000	–	50	3	–	120	–
Antidepressants							
Chlorimipramine	21,000	–	–	10	–	–	23
Imipramine	24,600	10,000	245	37	18	610	33
Amitryptilline	1,520	1,700	13	4	3	290	39
Nortryptilline	1,000	640	40	–	27	800	57
Desipramine	16,100	9,500	540	78	240	980	147
Doxepin	1,800	1,800	20	–	1	380	–
Mianserin	1,200	860	2	1	3	2,200	–

[a]Binding data (nM) are approximate. 5-HT$_2$ binding studies employed either tritiated spiperone (SPIP) or tritiated ketanserin (KET).
[b]See the Index for further discussion of [^{3}H] imipramine binding sites.
Binding data are given as IC$_{50}$ values for the neuroleptics and as K$_i$ values for the antidepressants.
Reference numbers are listed in parentheses. If two references are given, the first refers to neuroleptics and the second to antidepressants. H = histamine; D = dopamine.

agents behave as 5-HT$_2$ antagonists. Furthermore, there is evidence that chronic antidepressant treatment produces adaptive changes in postsynaptic 5-HT receptors, which appears to result in sub- or supersensitivity development depending upon the system being examined (50).

Numerous ergot alkaloids possess a very high affinity for serotonin binding sites (54); unfortunately, most of these agents display relatively little selectivity for one site versus another (see Table 4). Adrenergic agents also interact with serotonin binding sites; for example, [^{3}H] WB4101, an α-adrenergic label, is capable of labeling 5-HT$_{1A}$ sites (55). β-Adrenergic antagonists and hallucinogenic agents are discussed in subsequent sections.

At one time it was thought that 5-HT$_1$ sites and 5-HT$_2$ sites might constitute 5-HT agonist and antagonist sites, respectively; this type of thinking may have been an outgrowth of the "interconvertible conformation" hypothesis. However, current evidence does not support this concept. Although site-specific

Table 4 Binding Profiles of Selected Ergot Alkaloids

Alkaloids	5-HT_1	5-HT_2	α_1	α_2	D_2	H_1
			IC_{50} values (nM)[a]			
Ergotamine	9	4	22	3.5	6.5	>10,000
Dihydro-α-ergocriptine	40	17	11	6	2.5	–
Dihydro-β-ergocriptine	48	20	16	6	4.5	–
Dihydroergocristine	35	27	12	8	3	–
Bromocriptine	200	76	33	29	13	–
α-Ergocryptine	16	19	21	8	8.5	–
α-Ergocryptinin	600	812	1150	350	225	–

[a]Tritiated spiperone was used to label 5-HT_2 sites. α_1 and α_2 refer to adrenergic binding sites; D = dopamine; H = histamine.
Source: From Reference 54.

agents have yet to be developed, several site-selective agents are now available. The remainder of this section deals with the binding properties of agents that have been classified, on the basis of pharmacological studies, to behave as 5-HT agonists or antagonists. It should be borne in mind that certain of these agents can produce effects other than those that may be considered serotonergic, may act as mixed agonist-antagonists, and/or may act as agonists in some systems and as antagonists in others.

5-HT_1 AGONISTS

5-HT binds with high affinity (K_i = 1-6 nM) to 5-HT_1 sites. Surprisingly, very few structure-activity relationship (SAR) studies involving 5-HT or indolealkylamine analogs have been reported. (See Figure 2 for structures of indolealkylamines.) Engel et al. (56) showed that 5-HT has a 20-fold and 200-fold greater affinity for 5-HT_1 sites than 4-hydroxytryptamine (4-HT) and 6-hydroxytryptamine (6-HT), respectively. A similar rank order of potencies was obtained for a series of methoxy *N,N*-dimethyltryptamine (i.e., DMT) derivatives (i.e., 5-OMe > 4-OMe > 6-OMe > 7-OMe) (57). In general, N-monomethylation and O-methylation of 5-HT have relatively little effect on affinity; N,N-dialkylation and α-alkylation decrease affinity (Tables 5 and 6). With the recent discovery of subpopulations of 5-HT_1 binding sites, the available data should be interpreted with caution; nevertheless, these data can give an indication of the overall affinity of various indolealkylamines for 5-HT_1 sites. For example, 5-carboxamidotryptamine binds at 5-HT_1 sites with an affinity comparable to that of 5-HT (56); this agent appears to act as a serotonin agonist, but its subsite selectivity

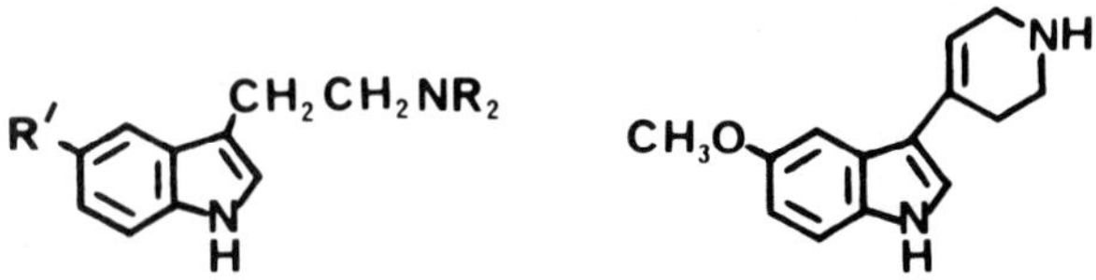

(a) (b)

Figure 2 Indolealkylamine derivatives. (a) Serotonin, R = H, R′ = OH; 5-methoxytryptamine (5-OMe-tryptamine), R = H, R′ = OCH_3; 5-carboxamidotryptamine, R = H, R′ = $CONH_2$; *N,N*-dimethyltryptamine (DMT), R = CH_3, R′ = H; bufotenine, R = CH_3, R′ = OH; 5-methoxy-DMT (5-OMe-DMT), R = CH_3, R′ = OCH_3. (b) RU 24969.

Table 5 Comparative 5-HT_1 and 5-HT_2 Binding Data for Several Indolealkylamines

		K_i (or IC_{50} *)(nM)[a]				
Agent		(56)	(57)*		(58)*	(59)*
5-HT_1 binding						
5-HT		3	3.6		2.6	6.1
5-OMe-Tryptamine		8	8.8		4.9	6.2
5-OH-*N*-Methyl tryptamine		5	5.0		–	8.1
Bufotenine		40	–		10	–
Tryptamine		68	176		87	76
5-HT_2 binding						
Tritiated ligand[a]	SPIP (45)	SPIP (59)	KET (46)	KET (60)	MES (25)	DOB (27)
5-HT	1000	1550	296	562	682	6
5-OMe-tryptamine	–	1800	742	686	3100	14
Tryptamine	5500	2800	1482	2894	–	–

[a]References are listed in parentheses.

Table 6 Further 5-HT_1 and 5-HT_2 Binding Data for Several Indolealkylamines[a]

Agent	K_i (nM) 5-HT_1	5-HT_2
5-OMe-Tryptamine	2	300
(+)-α-methyltryptamine	1,260	3,100
(−)-α-methyltryptamine	1,000	5,000
5-OMe-α-methyltryptamine	40	320
(+)-5-OMe-α-methyltryptamine	130	310
N,N-dimethyltryptamine (DMT)	175	1,200
1-*N,N*-trimethyltryptamine (1-TMT)	1,800	400
4-OMe-*N,N*-dimethyltryptamine (4-OMe-DMT)	50	1,300
Bufotenine (5-OH-DMT)	–[b]	480
5-OMe-*N,N*-dimethyltryptamine (5-OMe-DMT)	10	600
6-OMe-*N.N*-dimethyltryptamine (6-OMe-DMT)	800	7,300
6-*N,N*-trimethyltryptamine (6-TMT)	2,300	2,500
7-OMe-*N,N*-dimethyltryptamine (7-OMe-DMT)	6,600	5,400
7-OH-*N.N*-dimethyltryptamine (7-OH-DMT)	>10,000	>10,000
7-Bromo-*N,N*-dimethyltryptamine	220	170
5-OMe-7-*N,N*-trimethyltryptamine	510	360
5-OMe-*N,N*-diethyltryptamine (5-OMe-DET)	70	1,120

[a][^{3}H]5-HT and [^{3}H]KET were used to label 5-HT_1 and 5-HT_2 sites, respectively.
[b]Not determined.
Source: Binding data from Titeler, Lyon, and Glennon (unpublished data).

is still open to question (61-64). In general, detailed SAR studies for the binding of indolealkylamines to each of the 5-HT_1 subpopulations have yet to be reported. Likewise, although indolealkylamines bind at 5-HT_2 sites, relatively few agents have been examined (see below). An indolealkylamine for which a binding profile is available is 5-methoxy-*N,N*-dimethyltryptamine (5-OMe-DMT); Tricklebank et al. (65) reported the following (approximate) IC_{50} values: 5-HT_1 = 50 nM, 5-HT_{1A} = 25 nM, 5-HT_{1B} = 90 nM, and 5-HT_2 = 1250 nM. Nevertheless, even here there is some disagreement in that Sills et al. (21) reported a 50-fold selectivity of 5-OMe-DMT for 5-HT_{1A} over 5-HT_{1B} sites.

In 1979, Maj and coworkers (66) reported that the antidepressant trazodone behaves as a 5-HT antagonist at low doses but that at higher doses, particularly after a latency period, it acted as an agonist. They speculated that1-(3-chlorophenyl)piperazine (mCPP), a potential metabolite of trazodone, may be responsible for the agonist effects. Indeed, mCPP was identified later that year as a metabolite of trazodone (67). In fact, mCPP and related phenylpiperazines have been subsequently found to be metabolities of a variety of aryl-piperazine-containing centrally acting agents (68,69), and mCPP acts as a 5-HT agonist (70-72). Trazodone binds with a relatively high affinity (IC_{50} = 35 nM) and 50-fold selectivity to 5-HT_2 binding sites (73); however, its metabolite mCPP possesses a significant affinity for 5-HT_1 binding sites (74). The SAR of phenylpiperazines for 5-HT_1 binding has been reported: N(4)-methylation, replacement of the N(1) nitrogen with carbon, and ring opening, expansion, or methylation of the piperazine ring reduced affinity (75). Replacement of the 3-chloro group of mCPP by a trifluoromethyl group resulted in the most potent member of the series: 1-(3-trifluoromethylphenyl)piperazine (TFMPP) (74,75). (See Figure 3 for structures of arylpiperazines.) Although TFMPP is still one of the more

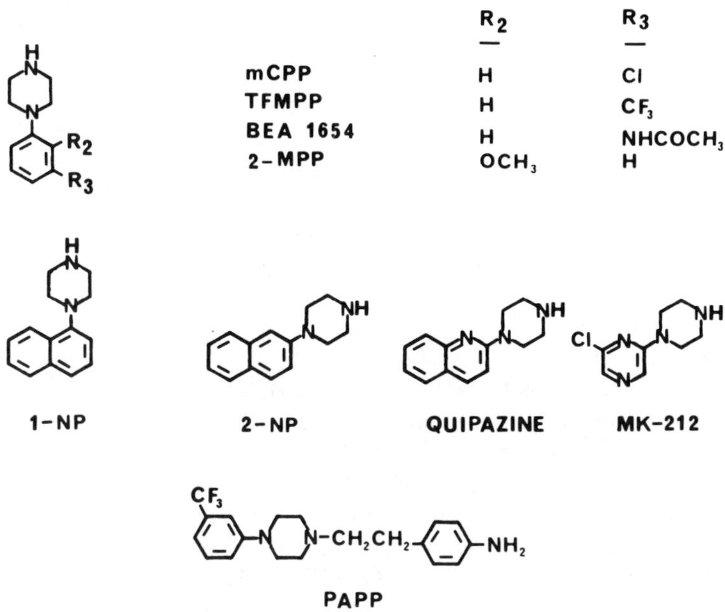

Figure 3 Structures of arylpiperazines that have activity as serotonin agonists and/or antagonists.

Table 7 Binding Selectivity of Several Arylpiperazine Derivatives[a]

Agent	K_i (nM) 5-HT_1	5-HT_2	5-HT_1 selectivity	Reference
Phenylpiperazine	134	2080	15	83
TFMPP	20	160	8	77
	90	590	6	79
	23	375	16	59
	116	298	3	80
mCPP	130	325	3	79
	60	260	4	59
Quipazine	1600	235	0.2	17
	200	1300	6	59
	230	230	1	83
MK-212	1400	6300	4	59
	3630	6000	2	80
2-MPP	35	3500	100	78

[a]Values may be approximate. 5-HT_2 studies used tritiated spiperone except in References 77 and 83, in which tritiated ketanserin was employed.

potent phenylpiperazine derivatives at 5-HT_1 sites, its selectivity for 5-HT_1 versus 5-HT_2 sites is only 10-fold; mCPP appears to be slightly less selective (see Table 7). Several related phenylpiperazines have recently been examined. 1-(3-Acetylaminophenyl)piperazine (BEA 1654) displays an affinity for 5-HT_1 sites comparable to that of TFMPP; although this agent also possesses a greater selectivity than TFMPP for 5-HT_1 versus 5-HT_2 sites, it is reportedly only a partial agonist (64,76). 1-(2-Methoxypheny)piperazine (2-MPP) possesses a 100-fold selectivity for 5-HT_1 over 5-HT_2 binding sites (77). mCPP and TFMPP possess a 3- to 70-fold selectivity for 5-HT_{1B} versus 5-HT_{1A} sites (21,78,81); comparable studies have not been conducted with 2-MPP. 2,3,4,4*a*,5,6-Hexahydro-9-(trifluoromethyl)-1H-pyrazino-[1,2-*a*] quinoline, a conformationally restricted analog of TFMPP, is approximately twice as potent as, but no more selective than, TFMPP at 5-HT_1 sites (80). CGS 12066B [i.e., 7-trifluoro-methyl-4-(4-methyl-1-piperazinyl)pyrrolo-[1,2-*a*] -quinoxaline] is a new, rather selective, 5-HT_{1B} agonist; its affinity (IC_{50} values in parentheses) for various neurotransmitter binding sites, such as 5-HT_{1A} (1160 nM), 5-HT_2 (6480 nM), dopamine D_1 (>5000 nM), dopamine D_2 (>5000 nM), α adrenergic (>6000), and α adrenergic (>1000 nM), is much less than its affinity for 5-HT_{1B} sites (IC_{50} = 47 nM) (78).

Table 8 Binding of Selected Phenylpiperazines

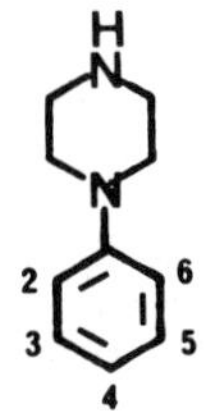

Agent or	Substituent	K_i (nM) 5-HT_1	5-HT_2 (KET)
	H	134	2,080
2-MPP	2-OMe	35	3,500
	2-OH	150	17,200
TFMPP	3-CF_3	20	160
	2,5-di-OMe	1,035	8,430
	2,5-di-OMe-4-Br	820	2,430
1-NP		5	18
2-NP		265	70
Quipazine		230	230

Source: Data from References 77 and 83.

Phenylpiperazines are commonly regarded as 5-HT_1 (and, more specifically, as 5-HT_{1B}) agonists; however, the situation seems to be much more complex than this. Benz fusion (b face) of phenylpiperazine, that is, 1-(1-naphthyl)piperazine (1-NP), enhances its affinity by about 30-fold for 5-HT_1 sites and about 100-fold for 5-HT_2 sites (Table 8) (82,83); 1-NP still possesses a three- to four-fold selectivity for 5-HT_1 sites. 1-[3-(Bis-trifluoromethyl)methylphenyl] piperazine (BTMP) is more potent than TFMPP at 5-HT_1 sites but is less potent at 5-HT_2 sites (82); as such, this agent possess a greater than 100-fold selectivity for 5-HT_1 sites. Both 1-NP and BTMP behave as 5-HT antagonists in the periphery (82,84), but preliminary evidence suggests that centrally BTMP may act as an agonist (85) and that 1-NP acts as a 5-HT_1 agonist (83) but as a 5-HT_2 antagonist (83,84). Benz fusion at the c face of phenylpiperazine, to afford 1-(2-naphthyl)piperazine (2-NP), reverses its selectivity (83). 2-NP is half as potent as phenylpiperazine at 5-HT_1 sites but 30 times more potent at 5-HT_2

sites; as such, this agent displays a fourfold selectivity for 5-HT_2 sites (83) and seems to behave as an agonist. Substitution at the N(4) nitrogen of TFMPP also produces some interesting results. The N(4)-(3-hydroxyphenethyl) analog LY-154 930 displays a 10-fold selectivity for 5-HT_1 sites and behaves as a peripheral 5-HT antagonist (82). A closely related analog, N(4)-(4-aminophenethyl)TFMPP (LY-165 163; PAPP), is selective for 5-HT_{1A} sites (81). mCPP and TFMPP discriminated between subsets of 5-HT_{1B} binding sites; PAPP is 100-fold more potent at 5-HT_{1A} sites than at 5-HT_{1B} sites and did not appear to discriminate between subsets of 5-HT_{1B} sites (81). [^{3}H] PAPP has been used as a radioligand to label 5-HT_{1A} sites (81). Phenylpiperazines (arylpiperazines), depending on their substituion pattern (and/or on the nature of the aryl group), can behave as 5-HT_{1A}, 5-HT_{1B}, or 5-HT_2 agonists. Furthermore, 1-NP possesses 5-HT_2 antagonist properties (83,84), and recently, Sanders-Bush (86) has provided evidence that certain phenylpiperazines, including mCPP and TFMPP, may also act as central 5-HT_2 antagonists. Other piperazines with activity as 5-HT_2 antagonists include a series of quinazolindione-substituted derivatives (e.g., TR 2515) (87,88). Thus, the terms "phenylpiperazines" and "5-HT_{1B} agonists" should no longer be considered synonymous.

Several structurally related arylpiperazines, including quipazine and 6-chloro-2-(1-pyrazinyl)pyrazine (MK-212), are among the oldest and best known piperazine derivatives (Figure 3). Quipazine acts as a mixed agonist-antagonist although its central effects are primarily of an agonist nature (89-91). Very few SAR data are available on quipazine analogs; removal of the N(4) nitrogen or replacement of the N(4) nitrogen by a methylene group reduces affinity for 5-HT_1 and 5-HT_2 sites by several orders of magnitude (83). MK-212 is a 5-HT agonist; however, its affinity for 5-HT_1 or 5-HT_2 sites is not particularly noteworthy (Table 7). Electrophysiological data suggest that MK-212 may be capable of discriminating between subsets of 5-HT receptors (92), but the appropriate binding studies have not yet been conducted. Certain arylpiperazines release endogenous stores of serotonin from hypothalamic slices (93). Although it has been suggested that this releasing property is unlikely to be singularly responsible for the in vivo effects of mCPP and TFMPP (because they do not alter 5-HT metabolism in a manner that is characteristic of other 5-HT releasers) (81), quipazine and MK-212 are even less potent than mCPP and TFMPP in this regard (93). Using a peripheral receptor preparation, it was demonstrated that MK-212 possesses a moderate affinity but a high efficacy for serotonin receptors (93); this may be one way to explain its activity as a serotonin agonist.

Certain tetrahydropyridylindole derivatives are potent 5-HT agonists (58); among the most potent of these is 5-methoxy-3-(1,2,3,6-tetrahydro-4-pyridyl)-1H-indole (RU-24969) (Figure 2) (58, 95). RU-24969 possesses a high affinity (IC_{50} = 6.8 nM) for 5-HT_1 sites (58) and displays a 500-fold selectivity for 5-HT_1 (IC_{50} = approximately 10 nM) versus 5-HT_2 sites (96). In addition,

RU-24969 possesses a 3- to 60-fold selectivity for 5-HT_{1B} versus 5-HT_{1A} sites; K_i values of 3 and 200 nM (21) and IC_{50} values of 2.4 and 81 nM (97) have been reported for the binding of RU-24969 to 5-HT_{1B} and 5-HT_{1A} binding sites, respectively. RU-24969 is much less potent at 5-HT_{1C} sites than at 5-HT_{1B} sites (30). With respect to binding properties, none of the other members of this series has been studied as well as RU-24969.

One of the most site-selective agents developed thus far is 8-hydroxy-2-(di-*n*-propylamino)tetralin (8-OH-DPAT). The serotonergic properties of 8-OH-DPAT were first reported by Arvidsson et al. (98), in 1981; a brief review of this agent appeared the following year (99). Binding data were not reported for this agent until 1983; Mason et al. (100) found that 8-OH-DPAT was a 5-HT_1-selective agent but that competition curves resulted in a very shallow Hill slope. That same year, Middlemiss and Fozard (101) demonstrated that 8-OH-DPAT was selective for 5-HT_{1A} sites. These findings have been supported by subsequent studies (Table 9 and 10). It may be added that 8-OH-DPAT also interacts at 5-HT_3 sites, although its affinity is only one-tenth that at 5-HT_{1A} sites (22).

Tritiated 8-OH-DPAT has been used as a radioligand to label 5-HT_{1A} sites (22,23). Some selected binding data are shown in Table 10. RU-24969 appears to possess an appreciable affinity for these 8-OH-DPAT-labeled sites, but it has been suggested that the low-affinity site for RU-24969 is a 5-HT_{1A} site (87). In keeping with the suggestion of Sills et al. (21) that the 5-HT_1 binding component of quipazine is primarily of a 5-HT_{1B} nature, quipazine possesses a low affinity for 8-OH-DPAT-labeled 5-HT_{1A} sites (Table 10). Binding data are also available for several indolealkylamine analogs; however, their SAR are rather unusual. Clearly, the 5-hydroxyl group of 5-HT is important for binding; its removal results in a 500-fold decrease in potency. *N,N*-dimethylserotonin (bufotenine) is one-third as potent as 5-HT itself; however, whereas O-methylation of

Table 9 Binding Profile of 8-OH-DPAT

IC_{50} (K_i*) (nM)					
5-HT_1	5-HT_{1A}	5-HT_{1B}	5-HT_{1C}	5-HT_2	References
70	–	–	–	6,000	Mason et al. (100)
190	6.5	3,800	–	>10,000	Middlemiss and Fozard (101)
26	1.85	6,880	–	47,000	Hamon et al. (103)
–	2	63,000	7,900	6,300	Hoyer (24)*
–	–	–	5,000	–	Pazos et al. (31)*
160	–	–	–	5,350	Titeler and Lyon (104)*

Table 10 Binding Characteristics of Selected Serotonergic Agents at Various 5-HT Binding Sites

	K_i values (nM)[a]						
Ligand	$5\text{-}HT_1$ [^{3}H]5-HT (28,78,110,111)	$5\text{-}HT_{1A}$ [^{3}H]8-OH-DPAT (165)	$5\text{-}HT_{1B}$ [^{125}I]ICYP (165)	$5\text{-}HT_{1C}$[b] [^{3}H]MES (165)	$5\text{-}HT_2$ [^{3}H]KET (165)	$5\text{-}HT_2$ [^{3}H]KET (78,83,111)	$5\text{-}HT_2$ [^{3}H]DOB (27)[c]
5-HT	2	3	25	35	2,950	560	6
O-Me-5-HT	2	–	400	–	–	690	14
5-OMe-DMT	50	8	350	–	–	600	15
		(166)	(166)			(166)	
8-OH-DPAT	160	2	63,000	7,250	7,100	5,500	700
TFMPP	20	175	25	-	–	160	–
		(166)	(166)				
RU-24969	2	5	4	400	1,000	780	–
(–)-DOB	4,200	–	–	–	–	60	0.4
ICYP	–	4	0.3	9,800	–	–	–
Ipsapirone	–	10	26,000	–	–	10,000	–
		(43)	(43)			(43)	
(–)-Pindolol	100	20	80	54,000	29,500	200,000	–
	(114)					(114)	
(+)-Pindolol	–	1,100[c]	4,700	65,000	–	82,600	
						(114)	

(–)-Propranolol	300 (114)	115	50	1,400	3,500	5,900 (114)[d]	–
Cinanserin	3,500 (124)	–	10,000	200	4	4	4
Ketanserin	Inactive	1,900	1,900	100	1	0.4 (46)	1
Mesulergine	–	300	13,000	1.5	4	2 (25)	–
Metergoline	20 (124)	6	25	0.5	1	0.3 (46)	–
Mianserin	1,100 (124)	–	4,700	10	10	1.4 (46)	–
Pirenperone	Inactive	1,700	6,600	60	1	0.3 (46)	–
Quipazine	230	3,300	320	200	645	230	17
Spiperone	160 (124)	40	4,800	1,150	2	0.5 (46)	1

[a]Binding data are approximate, and assays employed rat cortex unless otherwise noted. References are listed in parentheses.
[b]Pig choroid plexus.
[c]Also Titeler et al., unpublished data.
[d][^{3}H] spiperone was used as the radioligand.
[e]Pig cortex.

bufotenine reduces 5-HT_{1A} potency by half, O-methylation of 5-HT reduces potency by 40-fold (22,102). Almost no data are available on the SAR for the binding of DPAT analogs. An interesting finding is that ring opening of the saturated ring of 8-OH-DPAT, to afford 2-hydroxy-*N,N*-di-*n*-propyl phenethylamine, reduces potency by at least two orders of magnitude (105). Recently, Emerit et al. (106) have investigated 8-methoxy-2′-chloro-2-(di-*n*-propylamino)-tetralin as a potential irreversible 5-HT_{1A} ligand.

5-HT_2 Agonists

It has been fairly well established that serotonin is somehow involved in the mechanism of action of certain hallucinogenic agents (for reviews, see References 107 and 108). For this reason, it was not surprising to find that (+)-LSD displays a high affinity for 5-HT binding sites; however, Peroutka and Snyder (16) later reported that LSD was nearly equipotent at 5-HT_1 and at 5-HT_2 sites. Glennon and coworkers (109) found that the behavioral effects of the hallucinogen 1-(2,5-dimethoxy-4-methylphenyl)-2-aminopropane (DOM) could be antagonized by the 5-HT_2 antagonist ketanserin; this focused attention on a potential 5-HT_2 mechanism. The binding characteristics of a series of phenylisopropylamine (i.e., DOM) analogs was subsequently examined and it was found that several such agents displayed a relatively high affinity and selectivity for [^{3}H]ketanserin-labeled 5-HT_2 sites (28,111). Of the more potent and selective agents two are 1-(2,5-dimethoxy-4-X-phenyl)-2-aminopropane, where X = Br and I (i.e., DOB and DOI, respectively) (Table 11). DOB was selected as a starting point for an SAR study; each of the aromatic substituents was systematically removed or rerranged, and the terminal amine was modified (110). Each of these changes served only to decrease potency and/or selectivity. Ultimately, [^{3}H]DOB was prepared and examined as an agonist ligand for 5-HT_2 sites (27). Although these studies are still in progress, it appears that 5-HT possesses a much greater affinity for [^{3}H]DOB-labeled 5-HT_2 sites (K_i = 6 nM) than for [^{3}H]KET-labeled sites (K_i = approximately 500 nM). The affinity of the R(--) isomer of DOB for tritiated DOB sites is quite high (K_i = approximately 0.4 nM) (Table 10); as such, it possesses about a 5000-fold selectivity for these sites (27,110). Thus, these phenylisopropylamine derivatives appear to constitute the first class of 5-HT_2-selective agonists.

Other serotonin agonists interact at 5-HT_2 sites. For example, certain indolealkylamines (for example, see References 16, 17, and 28) and arylpiperazines (Table 8) possess significant affinities for these sites but suffer from an apparent lack of sensitivity. Although few tryptamine derivatives have been examined (Table 6), an interesting SAR is beginning to emerge. The presence of an α-methyl group has relatively little effect on affinity when racemates are examined, but it seems that the S(+) isomers of these types of agents possess a

Table 11 Binding Data for Selected Phenylisopropylamine Derivatives[a]

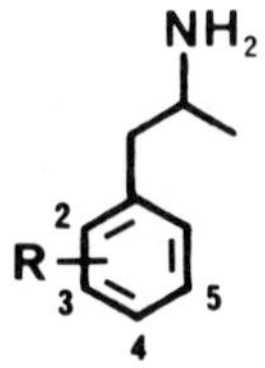

Agent	R_2	R_3	R_4	R_5	K_i (nM) 5-HT_1	5-HT_2
Amphetamine	H	H	H	H	7,600	43,000
OMA	OMe	H	H	H	3,500	8,130
MMA	H	OMe	H	H	2,660	7,850
PMA	H	H	OMe	H	80,000	33,600
3,4-DMA	H	OMe	OMe	H	64,600	43,300
2,5-DMA	OMe	H	H	OMe	1,020	5,200
2,4,5-TMA	OMe	H	OMe	OMe	46,800	1,650
MEM	OMe	H	OEt	OMe	35,500	2,220
DON	OMe	H	NO_2	OMe	14,100	300
DOM	OMe	H	Me	OMe	2,900	100
DOPR	OMe	H	nPr	OMe	3,170	69
DOB	OMe	H	Br	OMe	3,340	63
DOI	OMe	H	I	OMe	2,240	19
R(–)DOI	OMe	H	I	OMe	2,290	10

[a]Agents were evaluated as racemates except where noted.
Source: Data from References, 28, 110, and 111.

slightly greater affinity than their R(–) enantiomers. Alkyl substitution on the terminal amine (comparing 5-OMe-tryptamine, 5-OMe-DMT, and 5-OMe-DET; Table 6) decreases affinity for 5-HT_2 sites. Polar groups are not well tolerated at the indole 6 and, in particular, at the indole 7 position, although alkyl and bromo groups at the 7 position enhance affinity (Table 6). Although these agents are being discussed in this section, there is no evidence to suggest that they all behave as 5-HT agonists.

5-HT_1 Anatagonists

Classic serotonin antagonists generally possess a much higher affinity for 5-HT_2 sites than for 5-HT_1 sites (see Table 12); antagonists with high affinity and selectivity for 5-HT_1 sites remain elusive. Certain agents, such as metergoline and methiothepin, are not very selective for 5-HT_2 sites (17) and, on occasion, have been referred to as "nonselective" serotonin antagonists. Spiperone has been used in pharmacological studies as a 5-HT_{1A} antagonist, but this agent does possess a high affinity for 5-HT_2 and dopamine binding sites.

In 1977, Middlemiss and coworkers (113) reported that certain β adrenergic antagonists bind in a stereoselective manner to $[^3H]$5-HT-labeled serotonin sites. Subsequently, Nahorski and Willcocks (114) demonstrated that these agents were selective for 5-HT_1 sites; (–)-pindolol, for example, displays greater than

Table 12 Binding Site Selectivity for Serotonin Antagonists[a]

Antagonists	5-HT_2 K_i (nM)	Selectivity over 5-HT_1	D_2	α_1	α_2	H_1
Metergoline	0.3	50	75	125	1200	3500
Methysergide	0.9	100	200	2500	2500	>1000
Spiperone	0.5	300	0.3	20	>1000	>1000
Cyproheptadine	0.45	1600	65	200	1700	6
Mesulergine[b]		>1000	10	300	40	400
Pizotifen	0.3	5000	325	400	1600	6
Pirenperone	0.3	Intact[c]	50	25	10	50
Ketanserin	0.4	Intact	500	20	>1000	20
Ritanserin	0.3	Intact	75	100	160	40

[a]Values are approximate. K_i values for 5-HT_2 binding were obtained using tritiated ketanserin. Selectivity: the K_i value (IC_{50} in the case of mesulergine) at a particular site divided by the K_i (or IC_{50} in the case of mesulergine) at 5-HT_2 sites.
[b]Mesulergine: IC_{50} = 15 nM at SPIP-labeled 5-HT_2 (relative to 3 nM for ketanserin).
[c]Intact = inactive at 1 μM.
Source: Data derived from References 19, 25, 26, and 112.

Table 13 Binding of Nonselective β-Adrenergic Antagonists[a]

	5-HT$_1$		5-HT$_2$[b]	
Antagonists	(113)	(114)	(115)[c]	(114)
(±)-Propranolol	600	–	2,000[d]	–
(–)-Propranolol	210	295	1,700	5,900
(+)-Propranolol	12,000	13,300	13,000	16,100
(±)-Alprenolol	300	–	7,900[d]	–
(–)-Alprenolol	–	260	900	11,900
(+)-Alprenolol	–	21,100	–	24,500
(±)-Pindolol	790	–	–	–
(–)-Pindolol	–	99	–	205,500
(+)-Pindolol	–	29,600	–	82,600

[a]IC_{50} (nM). Data are approximate.
[b]Tritiated spiperone was employed as the radioligand.
[c]Reference 115, except where indicated.
[d]Data from Leysen (45).

a 2000-fold selectivity for 5-HT$_1$ versus 5-HT$_2$ sites (Table 13). Selective β_1- and β_2-adrenergic antagonists possess a lower affinity for 5-HT$_1$ sites than do the nonselective β-adrenergic antagonists (115). β-Adrenergic agonists are not without effect on serotonergic mechanisms (e.g., References 116 and 117), but their binding properties have not been investigated in detail. More recently, Middlemiss reinvestigated the binding of propranolol and found that the (–)-isomer (5-HT$_1$ IC_{50} = 200 nM) was about 40 times more potent at 5-HT$_1$ than at 5-HT$_2$ sites (118). The (+)-isomer was less potent and displayed only a fourfold selectivity for 5-HT$_1$ (IC_{50} = 4400 nM) sites. 5-HT$_{1B}$ binding data were also presented; (–)-propranolol was only half as potent at 5-HT$_{1B}$ sites (IC_{50} = 500 nM) as it was at 5-HT$_1$ sites; (+)-propranolol was essentially equiactive at both sites (118). At this point, it is still unclear whether these β-adrenergic antagonists display selectivity for 5-HT$_{1A}$ versus 5-HT$_{1B}$ sites (see also Table 10) or, for that matter, whether they act as antagonists at both sites. On the basis of various pharmacological studies, several groups of investigators seem to be of the opinion that these agents may be 5-HT$_{1A}$ antagonists (119-121); however, Doods et al. (97) have concluded that pindolol binds preferentially at a non-5-HT$_{1A}$ site. Nevertheless, these results provide needed support to the notion that it may eventually be possible to develop site-selective 5-HT$_1$ antagonists. More recently, isamoltane (1-[2-(1-pyrrolyl)-phenoxy]-3-isopropylamino-2-propanol) has been reported as a selective 5-HT$_{1B}$ receptor antagonist (167).

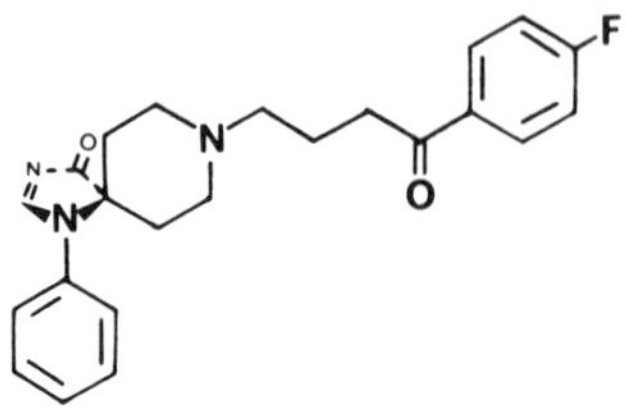

SPIP

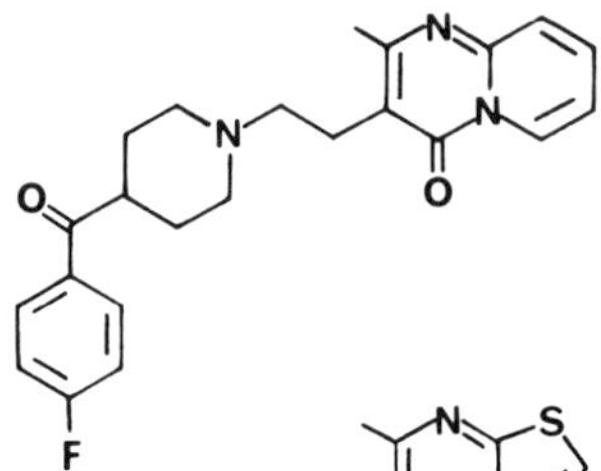

PIR

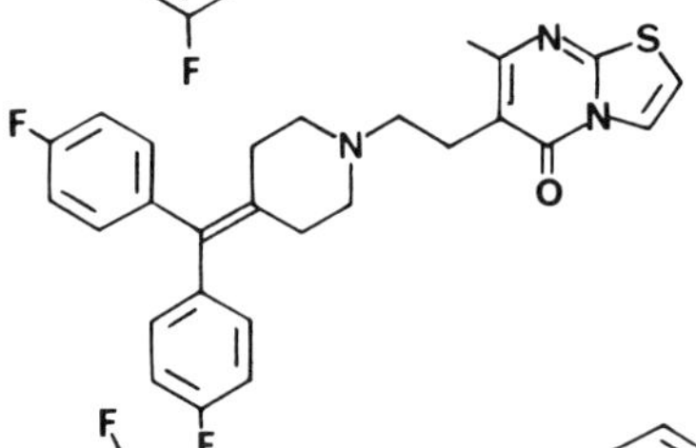

RIT

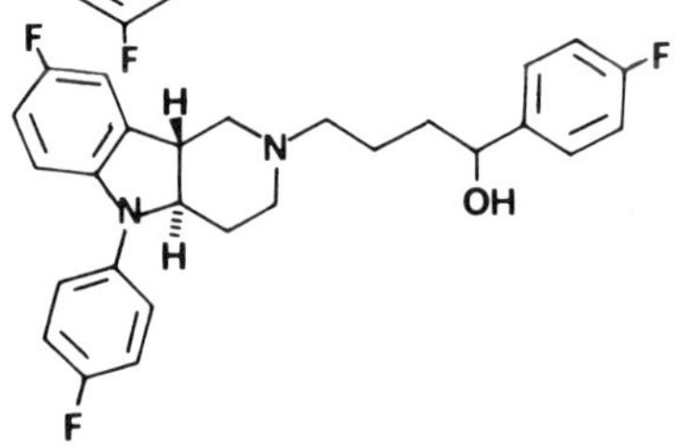

CP 52,215

Figure 4 Agents with activity as serotonin antagonists. SPIP, spiperone; PIR, pirenperone; RIT, ritanserin; CP 52,215, dihydroflutroline. See Figure 1 for the structure of ketanserin.

5-HT$_2$ Antagonists

As mentioned in the previous section and as shown in Table 12, classic serotonin antagonists bind with high affinity and selectivity to 5-HT$_2$ sites. (See Figures 4 and 5 for structures.) One of the first agents identified as a 5-HT$_2$-selective antagonist was spiperone. Janssen Pharmaceutica has recently developed a series of agents (e.g., ketanserin, pirenperone, and ritanserin) that is highly selective for 5-HT$_2$ versus 5-HT$_1$ sites. It is probably the introduction of these agents (ketanserin in particular) that has been one of the driving forces in the study of site-selective serotonergic agents. Unfortunately, as a class, there is a problem

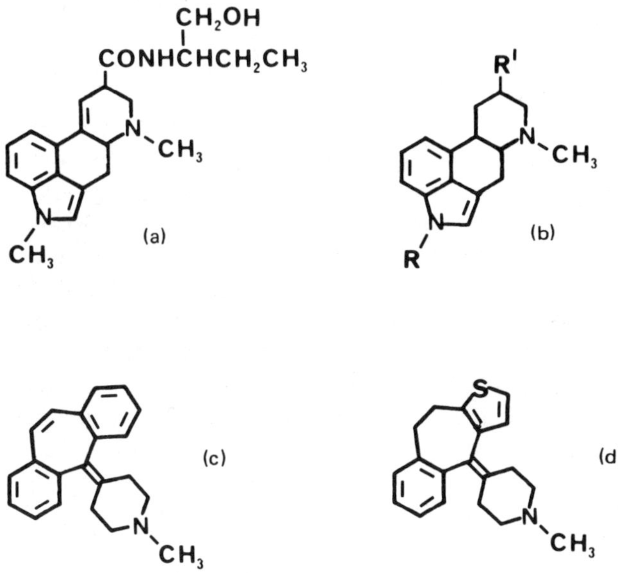

Figure 5 Structures of ergoline and tricyclic serotonin antagonists. (a) Methysergide; (b) metergoline ($R = CH_3$, R' $CH_2NHCOOCH_2C_6H_5$), LY 53857 [$R = iC_3H_7$, $R' = COOCH(CH_3)CH(OH)CH_3$]; (c) cyproheptadine; (d) pizotyline (pizotifen, BC-105). See Figure 1 for the structure of mesulergine.

associated with these 5-HT_2 antagonists in that many of them tend to display affinities for other neurotransmitter binding sites. Spiperone, for example, is more potent at dopamine sites than at 5-HT_2 sites; in addition, this agent interacts at 5-HT_{1A} sites. Pizotifen, like spiperone, possesses a significant affinity for 5-HT_{1A} sites (21,122) (although this affinity is less its affinity for 5-HT_2 sites). Ketanserin and pirenperone possess high affinities for adrenergic and histaminergic sites (Table 12), and Titeler and coworkers, for example, have used [^{3}H]KET to label adrenergic sites in porcine neurointermediate lobe homogenates (123). Nevertheless, these agents have proven to be extremely valuable tools for studying 5-HT_2-site interactions. Certain ergoline derivatives, although less selective for 5-HT_2 versus 5-HT_1 sites, possess significantly lower affinities for other neurotransmitter binding sites (Table 12). However, recent findings suggest that ergolines, such as methysergide and mesulergine (and LSD), also bind to 5-HT_{1C} sites (31,32).

Ketanserin (R 41468) and pirenperone (R 47465) are effective serotonin antagonists with high affinity for 5-HT_2 sites (46,124). Unfortunately, relatively little has been published in the way of SAR for this class of agents. Replacement of the quinazoline 2-oxo group of ketanserin with a sulfur atom seems to have little effect on its binding profile, as does methylation of the pyrido ring of pirenperone (125). Although it is somewhat structurally related to ketanserin and pirenperone, ritanserin (R 55667) probably represents a new generation of 5-HT_2 antagonists. Ritanserin is more potent than ketanserin in binding to dopamine D_2 sites and α-adrenergic sites but is less potent than ketanserin at histamine H_1 sites and α-adrenergic sites (112). There is good agreement between the in vitro and in vivo interactions of ritanserin at 5-HT_2 and histaminergic sites; however, the in vitro binding of ritanserin to D_2 and α sites was not evident under in vivo conditions (112). Leysen et al. (112) suggested that the true binding affinity of ritanserin to 5-HT_2 sites is difficult to assess because of the noncompetitive nature of the interaction (which has been ascribed to a very slow dissociation rate). The results of behavioral studies involving rodents supports the contention that ritanserin is a 5-HT_2 antagonist (126-128); ritanserin appears to be at least an order of magnitude less potent than pirenperone (126, 127).

LY 53857 is an ergoline analog that binds somewhat selectively to 5-HT_2 (IC_{50} = 24 nM) versus 5-HT_1 (IC_{50} = 350 nM) sites (84,129). Unlike ketanserin and pirenperone, this agent displays a much lower (300,000-fold) affinity for α-adrenergic receptors than for serotonin sites (129). Several other agents, such as tetrahydrotrazodone (THT) and dihydroflutroline, although not necessarily selective for serotonin, possess significant affinities for 5-HT_2 sites (see discussion in Reference 127). The behavioral effects of the purported 5-HT_2 agonist DOM have been antagonized by LY 53857, THT, and dihydroflutroline (127). Williams et al. (130) have described a cyproheptadine analog, L-646,462, that binds nearly equally well at dopamine and 5-HT_2 sites (K_i = approximately 25 nM) but seems to be greater than 100 times more potent as a peripheral 5-HT antagonist than as a central 5-HT antagonist. Xylamidine is another agent with a high affinity for 5-HT_2 sites (K_i = 1.5 nM) (46) but is relatively ineffective as a central 5-HT_2 antagonist when administered systemically. To date, there are no reported 5-HT_2-specific antagonists.

FUNCTION AND REGULATION

Effect of Guanine Nucleotides and Divalent Cations

Several groups of investigators have suggested that the binding of serotonergic agents to 5-HT binding sites may involve a multiple state interaction. That is, binding sites may exist in a high-affinity state when associated with a guanine nucleotide regulatory protein (G protein) and in a low-affinity state when dissociated from the G protein. The addition of guanine nucleotides, such as guanosine-5′-triphosphate (GTP) or the less readily hydrolyzable guanyl-5′-imidodiphosphate (GppNHp), can apparently convert the high-affinity state to a low-affinity state, whereas the addition of certain divalent cations can convert the low-affinity state to a high-affinity state. Thus, the affinity of 5-HT agonists is reduced in the presence of guanine nucleotides and is enhanced in the presence of divalent cations. Antagonist binding, which is presumably to low-affinity states, is not significantly altered in the presence of guanine nucleotides. Antagonists generally result in competition curves with Hill coefficients approximating unity, whereas agonists produce more shallow competition curves. The addition of guanine nucleotides has no effect on the Hill coefficients when antagonists are examined but resultes in increases in the Hill coefficients when agonists are employed. The effects have been demonstrated for 5-HT_1-site interactions by, for example, Peroutka et al. (131), Hamon et al. (102), and Sills et al. (132). Initially, GTP was found to be without effect on the binding of serotonin to [^{3}H] SPIP-labeled 5-HT_2 sites (131). However, Titeler and coworkers (27,60,111,133,134) and others (102,135) have now demonstrated this effect for [^{3}H] KET-labeled, [^{3}H] SPIP-labeled, and [^{3}H] DOB-labeled 5-HT_2 sites with various agonists and antagonists. Thus, there is evidence for the involvement of guanine nucleotide regulatory protein both for 5-HT_1 and 5-HT_2 binding. Similar studies have been conducted using [^{3}H] 8-OH-DPAT as the radioligand. Apparently, hippocampal 5-HT_1 sites (i.e., 5-HT_{1A} sites) exhibit decreased affinity for agonists in the presence of GTP; however, this does not appear to be the case with striatal 5-HT_1 sites (i.e., "5-HT_3" sites) (22). The binding of 5-HT to the 8-OH-DPAT -insensitive "5-HT_3" sites of Robaut et al. (36) was also unaffected by the addition of GTP.

Transducing Mechanism

Guanine nucleotides regulate the sensitivity of adenylate cyclase to various neurotransmitters. Therefore, the results provided in the previous section suggest an association between 5-HT binding sites and adenylate cyclase activity. However, Hamon et al. (102) have found that neither the modulation by GTP nor that due to divalent cations seems to depend on the possible association of 5-HT_1 sites with adenylate cyclase. For example, TFMPP and RU 24969, the binding of which is altered by GTP, do not interact with cyclase (102,136). Barbaccia et al. (137) have reported that 5-HT-stimulated adenylate cyclase is not linked to 5-HT_2 sites. The apparent affinity of the 5-HT receptor coupled to serotonin sensitive adenylate cyclase corresponds neither to that of 5-HT_1 or 5-HT_2 sites (3). This receptor may be distinct from these sites or may be more closely related to subpopulations of these sites. See Hamon et al. (3) and Fillion (13) for more detailed discussions.

Another possible biochemical effector involves increases in phosphoinositide (PI) turnover. Several groups have now provided evidence that the 5-HT_2

antagonist ketanserin can block 5-HT-stimulated PI turnover (138-141). Furthermore, 5-HT, 5-fluorotryptamine, and MK-212 stimulated PI turnover with potencies that resembled their potencies at 5-HT_2, but not 5-HT_1, binding sites (140); quipazine (141,142), but not 8-OH-DPAT (140), TFMPP, mCPP, or 1-NP (142), was active in this regard. In fact, the last three compounds blocked the effect of 5-HT (141). Conn and Sanders-Bush (140) have found that 5-HT-induced hydrolysis of phosphoinositides is not exclusively 5-HT_2-mediated in subcortical regions and have now obtained data suggesting that 5-HT_{1C} sites are also linked to PI hydrolysis (86,143).

Autoreceptors

Serotonin appears to regulate its own release through the activation of autoreceptors; for recent commentaries on this concept, see Gothert (144) and Moret (145). In 1982, Martin and Sanders-Bush sought to characterize these autoreceptors by examining the relative potencies of various serotonin agonists and antagonists (59). Their results demonstrated that 5-HT autoreceptors and 5-HT_1 binding sites display similar pharmacological characteristics and that there was a correlation between the effects of 5-HT agonists on K^+-evoked release of tritiated 5-HT from prelabeled synaptosomal preparations of rat hypothalamus and [^{3}H] 5-HT binding. Another two conclusions reached by these investigators was (1) that quipazine behaves as an antagonist at autoreceptors, and (2) that 5-HT autoreceptors located on serotonergic nerve terminals may be pharmacologically distinct from autoreceptors located on serotonergic cell bodies (59,146). Schlicker and Gothert had also reported that quipazine behaves as an antagonist (147). After the discovery of subpopulations of 5-HT_1 binding sites, Engel et al. (56) suggested that the potencies of 5-HT agonists at 5-HT autoreceptors correlated with their affinities for the low-affinity component of [^{3}H] 5-HT binding. Gozlan et al. (22) showed that [^{3}H] 8-OH-DPAT labels sites in the striatum (i.e., "5-HT_3 sites") that exhibit characteristics usually associated with presynaptic autoreceptors. However, Middlemiss (148) found that 8-OH-DPAT was without agonist or antagonist activity at 5-HT autoreceptors; in fact, subsequent studies led Middlemiss to conclude that 5-HT autoreceptors are most closely associated with the characteristics of 5-HT_{1B} binding sites (102,118). Verge et al. (149) have reported that 5-HT autoreceptors located on serotonergic cell bodies and/or dendrites do not correspond to 5-HT_{1B} sites but are most likely of the 5-HT_{1A} type; however, 5-HT autoreceptors controlling presynaptic 5-HT release correspond to neither 5-HT_{1A} nor 5-HT_{1B} sites. Thus, the situation is still somewhat less than resolved.

Functional Significance

Serotonin has been implicated in, for example, sleep disorders, certain aspects of sexual function, thermoregulation, depression, anxiety, appetite control, nociception, memory, and neuroendocrine disorders (e.g., see Reference 1). However, the exact role played by serotonin is, for the most part, unknown.

5-HT_{1A} agonists produce hypothermia, whereas 5-HT_2 agonists appear to be responsible for hyperthermia (150). 5-HT_{1A} agonists also seem to be involved in sexual function (99,151,152) and elicit certain aspects of the "serotonin syndrome in rodents, including tremor, forepaw treading, and head weaving, but not head-twitch behavior (which seems to be 5-HT_2 related) (119,150, 153). Certain agents with a high affinity for 5-HT_{1A} sites (e.g., buspirone, gepirone, and TVX Q 7821) constitute a new mechanistic class of anxiolytic agents that lack some of the undesirable side effects associated with the benzodiazepines (for example, see References 44, 154, and 155). On the other hand, several 5-HT_2 antagonist (most notably ritanserin) are also therapeutically effective in the treatment of certain anxiety disorders (125,156).

5-HT_{1B} sites may be involved in the serotonin syndrome (119,153) and may be functionally related to 5-HT autoreceptors (see Autoreceptors). The pharmacological significance of 5-HT_{1B} sites is poorly understood, and to date, evidence for the existence of 5-HT_{1B} sites in human brain is controversial. Relatively little is known about 5-HT_{1C} sites, although preliminary findings suggest an association with PI turnover (143). Because choroid plexus is rich in 5-HT_{1C} sites and because the main physiological role of the choroid plexus is the control of the volume and composition of cerebrospinal fluid, 5-HT_{1C} sites may play a role in the regulation of cerebrospinal fluid and/or in cerebral circulation (32).

Many 5-HT_2 agonists are hallucinogenic in humans (28), and the discriminative stimulus properties of DOM and DOI may be mediated through a 5-HT_2-related mechanism (28,108,109). Many 5-HT_2 antagonists possess activity as antidepressants (see Selectivity and Structure-Activity Relationships), although the 5-HT_2 (partial?) agonist quipazine also appears to possess an antidepressant profile (157). Janssen (125) has reviewed the potential therapeutic applications of 5-HT_2 antagonists, and Arvidsson et al. (150) have reviewed the behavioral effects produced by various serotonin agonists and antagonists. Glennon and coworkers (109,158,159), using a drug discrimination paradigm, have trained rats to discriminate site-selective serotonin agonists from saline. That is, groups of rats were trained to discriminate either a 5-HT_{1A} agonist (8-OH-DPAT) or a 5-HT_{1B} agonist (TFMPP) and one of several 5-HT_2 agonists [e.g., DOI and R(−)-DOB] from saline in a two-lever operant procedure. Tests of stimulus generalization revealed that animals trained to discriminate one of these selective agents recognized only 5-HT agonists that were selective for the same site (or occasionally, depending on the training drug, certain nonselective serotonin agonists, such as fenfluramine) but did not recognize serotonin agonists that were selective for one of the other sites. Unfortunately, very few clinical studies have been conducted with serotonin agonists. Quipazine, for example, produces significant gastrointestinal side effects (160), whereas many other agents (e.g.,

LSD, 5-OMe-DMT, DOM, and DOI) are hallucinogenic or are otherwise behaviorally active. mCPP has recently been evaluated in humans and a preliminary report has appeared (161); the authors state that their findings lend support to existing evidence that suggests a role for 5-HT in the regulation of prolactin, cortisol, body temperature, and mood. A number of studies indicate the involvement of serotonin in learning and memory; Altman et al. (162), for example, have found that alaproclate and zimeldine facilitated memory retrieval in rodents in a dose-dependent manner.

In addition to these effects of 5-HT_2 antagonists, a tremendous amount of work has been done with these agents in the cardiovascular area. Ketanserin and several related 5-HT_2 antagonists are effective, for example, in the treatment of hypertension. However, there is considerable controversy about whether these effects are due to 5-HT_2 antagonism or to α-adrenergic antagonism; the interested reader is directed to several recent treatments of this topic (163,164).

None of the agents mentioned here are site specific; as a consequence, it is difficult to draw conclusions regarding the relationships between a particular binding or receptor site and a given pharmacological response. This is particularly true for in vivo studies in which interactions may be very complex and involve more than one type of binding site (or only select populations of binding sites to which an agent has distributional access).

SPECULATION AND FUTURE DIRECTION

Remarkable progress has been made over the last several years with respect to serotonin research. Yet, much remains to be done. A number of serotonin binding sites have now been identified (Table 1), but some of these require further characterization (and, perhaps, in some cases, verification). Additional sites may remain to be identified. The effector or transducing mechanisms coupled to each of these sites needs clarification, as does the concept of multiple state binding.

Whether these central serotonin binding sites represent serotonin receptors has yet to be answered. Initially, several pharmacological roles were ascribed to 5-HT_2 sites, but certain investigators argued against a role for 5-HT_1 sites; however, now that subpopulations of 5-HT_1 sites have been identified it seems that these sites may indeed be pharmacologically relevant. Recently, Leysen et al. (112) have raised some interesting questions regarding in vitro and in vivo binding; it seems that (with respect to ritanserin, for example) an in vivo binding profile can differ significantly from an in vitro profile, and it may be difficult to determine real binding affinities for agents that dissociate slowly from binding sites (i.e., where equilibrium is not reached with a short incubation period). This may make it difficult to obtain correlations between binding affinities and pharmacological activities for a series of agents.

Some progress has been made with respect to the development of site-selective agents (Table 10). Several 5-HT_1 agonists have been identified; 8-OH-

DPAT appears to be a selective 5-HT_{1A} agonist, and certain piperazine derivatives may be 5-HT_{1B} selective, but a 5-HT_{1C}-selective agonist has not yet been developed. There is an urgent need for 5-HT_1-selective antagonists; certain ergolines can block 5-HT_1 sites but possess a higher affinity for 5-HT_2 sites. Spiperone, though possessing selectivity for 5-HT_{1A} versus 5-HT_{1B} sites, also possesses a high affinity for 5-HT_2 sites. β-adrenergic antagonists may constitute a useful group of 5-HT_1 antagonists; however, their subsite selectivity remains to be determined, and, obviously, these agents are not selective for serotonin. It has been recently proposed that certain phenylisopropylamine derivatives (such as DOB and DOI) may constitute the first class of 5-HT_2-selective agonists; additional work needs to be done in this area. 5-HT_2-selective antagonists abound; unfortunately, although these agents are selective for 5-HT_2 versus 5-HT_1, they are not necessarily selective for serotonin.

Very few new agents have been developed since the 5-HT_1 and 5-HT_2 concept was proposed (and even these have been serendipitous discoveries). This can probably be attributed to the novelty of this field of research (i.e., only 6 years have passed since the terms 5-HT_1 and 5-HT_2 were coined), and to the lack of complete characterization of these sites (and/or, conversely, to a lack of comprehensive structure-activity relationships of agents that interact with these sites). Much new information about these binding sites may become available if new site-selective (and, perhaps, site-specific) agents were available.

NOTE ADDED IN PROOF

A tremendous number of papers on serotonin have appeared in the literature since this chapter was originally written. Although limitations of space and time prevent inclusion of much of this newer work, two recent review articles by Fozard (167) and Glennon (168) might serve as a useful update. The behavioral pharmacology of serotonin has also been reviewed (169). Several points, however, are worthy of mention at this time. Four populations of 5-HT_1 sites are now acknowledged: 5-HT_{1A}, 5-HT_{1B}, 5-HT_{1C}, and 5-HT_{1D}; no agonists or antagonists selective for 5-HT_{1C} or 5-HT_{1D} sites have yet been reported. With regard to the 5-HT_3 sites in Table 1, those labeled by [^{3}H] 8-OH DPAT in the striatum are now considered to be presynaptic 5-HT_{1A} sites; the other two types of 5-HT_3 sites (as well as the 5-HT_{2A} and 5-HT_{2B} sites) listed in Table 1 remain to be substantiated. The currently accepted 5-HT_3 nomenclature (6) has been reserved for peripheral 5-HT receptors that were once referred to as M-receptors. Agents such as MDL-72222 and GR38032F are considered to be 5-HT_3-selective antagonists; a central 5-HT_3 binding site has recently been reported on the basis of radioligand binding studies. Perhaps one of the most exciting areas of serotonin research is the relationship between the second-generation anxiolytics (e.g., buspirone, gepirone, and ipsapirone) and 5-HT_{1A} mechanisms. Although these anxiolytics do not appear to be acting via a benzodiazepine mechanism, and whereas they may constitute 5-HT_{1A} mixed agonist-antagonists, their exact mechanism of action remains unknown (see Refs. 170 and 171 for reviews). Finally, a considerable amount of work has now been done on the transducing (effector) mechanisms that are coupled to each of the central 5-HT sites; this work has been reviewed by Conn and Sanders-Bush (172) and by Roth and Chuang (173).

REFERENCES

1. R. W. Fuller, in Biology of Serotonergic Transmission (N. N. Osborne, ed.), John Wiley and Sons, New York, 1982, p. 221.
2. R. W. Fuller, in Neuropharmacology of Serotonin (A.R. Green, ed.), Oxford University Press, Oxford, 1985, p. 1.
3. M. Hamon, S. Bourgoin, S. El Mestikawy, and C. Goetz, in Handbook of Neurochemistry, Vol. 6, Receptors in the Nervous System (A. Lajtha, ed.), Plenum Press, New York, 1982, p. 107.
4. M. Gershon, Annu. Rev. Neurosci. *4*, 227 (1981).
5. J. Fozard, Neuropharmacology *23*, 1437 (1984).
6. P. B. Bradley, G. Engel, W. Feniuk, J. R. Fozard, P. P. A. Humphrey, D. N. Middlemiss, E. J. Mylecharane, B. P. Richardson, and P. R. Saxena, Neuropharmacol. *25*, 563 (1986).
7. J. T. Farrow and H. van Vunakis, Nature *237*, 164 (1972).
8. J. L. Bennett and G. K. Aghajanian, Life Sci. *15*, 1935 (1974).
9. J. P. Bennett and S. H. Snyder, Brain Res. *94*, 523 (1975).
10. G. Fillion, M. P. Fillion, J. Jacob, and J. C. Rousselle, Br. J. Pharmacol. *58*, 425p (1976).
11. G. M. Fillion, J. C. Rousselle, M. P. Fillion, D. M. Beaudoin, M. R. Goiny, J. D. Deniau, and J. Jacob, Mol. Pharmacol. *14*, 50 (1978).
12. J. P. Bennett and S. H. Snyder, Mol. Pharmacol. *12*, 373 (1976).
13. G. Fillion, in Handbook of Psychopharmacology, Vol. 17 (L. L. Iversen, S. D. Iversen, and S. H. Snyder, eds.), Plenum Press, New York, 1983, p. 139.
14. J. Leysen, C. J. E. Niemegeers, J. P. Tollenaere, and P. M. Laduron, Nature *272*, 168 (1978).
15. I. Creese and S. H. Snyder, Eur. J. Pharmacol. *49*, 201 (1978).
16. S. Peroutka and S. H. Snyder, Mol. Pharmacol. *16*, 687 (1979).
17. J. E. Leysen and J. P. Tollenaere, Annu. Rep. Med. Chem. *17*, 1 (1982).
18. J. E. Leysen, Neuropharmacology *23*, 247 (1984).
19. J. E. Leysen, in Neuropharmacology of Serotonin (A. R. Green, ed.), Oxford University Press, Oxford, 1985, p. 79.
20. N. W. Pedigo, H. I. Yamamura, and D. L. Nelson, J. Neurochem. *6*, 22 (1981).
21. M. A. Sills, B. B. Wolfe, and A. Frazer, J. Pharmacol. Exp. Ther. *231*, 480 (1984).
22. H. Gozlan, S. El Mestikawy, L. Pichat, J. Glowinski, and M. Hamon, Nature *305*, 140 (1983).

23. M. Hamon, H. Gozlan, M. D. Hall, S. El Mestikawy, M. B. Emerit, and L. Pichat, in Regulation of Neurotransmitter Function: Basic and Clinical Aspects (P. Vizi and J. Magyar, eds.), Elsevier, Amsterdam, 1984, p. 141.
24. D. Hoyer, G. Engel, and H. O. Kalkman, Eur. J. Pharmacol. *118*, 1 (1985).
25. A. Closse, Life Sci. *32*, 2485 (1983).
26. M. J. Kadan, A. M. Krohn, M. J. Evans, R. L. Waltz, and P. R. J. Hartig, Neurochemistry *43*, 601 (1984).
27. M. Titeler, K. Herrick, R. A. Lyon, J. D. McKenney, and R. A. Glennon, Eur. J. Pharmacol. *117*, 145 (1985).
28. R. A. Glennon, M. Titeler, and J. D. McKenney, Life Sci. *35*, 2505 (1984).
29. G. K. Aghajanian, in Serotonin Neurotransmission and Behavior (B. L. Jacobs and A. Gelperin, eds.), MIT Press, Cambridge, Massachusetts, 1981, p. 156.
30. R. Cortes, J. M. Palacios, and A. Pazos, Br. J. Pharmacol. *82*, 202P (1984).
31. A. Pazos, D. Hoyer, and J. M. Palacios, Eur. J. Pharmacol. *106*, 531 (1985).
32. A. Pazos, D. Hoyer, and J. M. Palacios, Eur. J. Pharmacol. *106*, 539 (1985).
33. D. G. Morgan, J. Marcusson, and C. E. Finch, Soc. Neurosci. Abstr. *574*, 9 (1983).
34. J. Marcusson, D. G. Morgan, B. Winblad, and C. E. Finch, Soc. Neurosci. Abstr. *9*, 927 (1983).
35. R. D. Todd and R. D. Ciaranello, Proc. Natl. Acad. Sci. USA *82*, 612 (1985).
36. C. Robaut, M. P. Fillion, S. Dufois, C. Fayplle-Bauguen, J.-C. Rousselle, G. Gillet, S. Benkirane, and G. Fillion, Brain Res. *346*, 250 (1985).
37. H. Rommelspacher, G. Bruning, R. Susilo, M. Nick, and R. Hill, Eur. J. Pharmacol. *109*, 363 (1985).
38. P. L. Wood, C. Pilapil, F. LaFaille, N. P. V. Nair, and R. A. Glennon, Arch. Int. Pharmacodyn. *268*, 194 (1984).
39. C. S. Cascio and K. J. Kellar, Neuropharmacology *21*, 1219 (1982).
40. A. Fulton and G. D. Burrows, Prog. Neuropsychopharmacol. Biol. Psychiatr. *8*, 33 (1984).
41. T. Glaser and J. Traver, Eur. J. Pharmacol. *88*, 137 (1983).
42. L. A. Riblet, A. S. Eison, M. S. Eison, D. P. Taylor, D. L. Temple, and C. P. Vander Maelen, Psychopathology *17*, 69 (1984).
43. S. J. Peroutka, Biol. Psychiatry. *20*, 971 (1985).
44. J. Traber, M. A. Davies, W. U. Dompert, T. Glaser, T. Schuurman, and P. R. Seidel, Brain Res. *12*, 741 (1984).
45. J. E. Leysen, J. Physiol (Paris) *77*, 351 (1981).
46. J. E. Leysen, C. J. E. Niemegeers, J. M. Van Neuten, and P. J. Laduron, Mol. Pharmacol. *21*, 301 (1982).
47. D. R. Burt, I. Creese, and S. H. Snyder, Mol. Pharmacol. *12*, 800 (1976).
48. R. Ortmann, S. Bischoff, E. Radeke, O. Buech, and A. Delini-Stula, Naunyn-Schmiedebergs Arch. Pharmacol. *321*, 265 (1982).
49. M. Briley, in Neuropharmacology of Serotonin (A. R. Green, ed.), Oxford University Press, Oxford, 1985, p. 50.
50. S. O. Ogren, K. Fuxe, and L. Agnati, Pharmacopsychiatry *18*, 209 (1985).
51. H. Hall and S. O. Ogren, Eur. J. Pharmacol. *70*, 393 (1981).

52. S. Peroutka and S. H. Snyder, in Antidepressants: Neurochemical, Behavioral, and Clinical Perspectives, (S. J. Enna, J. B. Malick, and E. Richelson, eds.), Raven Press, New York, 1981, p. 75.
53. M. Sette, M. S. Briley, and S. Z. Langer, J. Neurochem. *40*, 622 (1983).
54. A. Closse, W. Frick, A. Dravid, G. Bolliger, D. Hauser, A. Sauter, and H.-J. Tobler, Naunyn-Schmiedebergs Arch. Pharmacol. *327*, 95 (1985).
55. A. B. Norman, G. Battaglia, A. L. Morrow, and I. Creese, Eur. J. Pharmacol. *106*, 461 (1985).
56. G. Engel, M. Gothert, E. Muller-Schweintzer, E. Schlicker, L. Sistonen, and P. A. Stadler, Naunyn-Schmiedebergs Arch. Pharmacol. *324*, 116 (1983).
57. D. L. Nelson and R. A. Glennon, unpublished data.
58. P. Hunt and C. Oberlander, in Serotonin: Current Aspects of Neurochemistry and Function (B. Haber, S. Gabay, M. R. Issidorides, and S. G. A. Alivasatos, eds.), Plenum Press, New York, 1979, p. 547.
59. L. L. Martin and E. Sanders-Bush, Naunyn-Schmiedebergs Arch. Pharmacol. *321*, 165 (1982).
60. G. Battaglia, M. Shannon, and M. Titeler, J. Neurochem. *43*, 1213 (1984).
61. W. Feniuk, P. P. A. Humphrey, and A. D. Watts, Br. J. Pharmacol. *82*, 209p (1984).
62. A. Shenker, S. Maayani, H. Weinstein, and J. P. Green, Eur. J. Pharmacol. *109*, 427 (1985).
63. S. Beck, W. P. Clarke, and J. Goldfarb, Eur. J. Pharmacol. *116*, 195 (1985).
64. P. R. Saxena and P. D. Verdouw, Br. J. Pharmacol. *84*, 533 (1985).
65. M. D. Tricklebank, C. Forler, D. N. Middlemiss, and J. R. Fozard, Eur. J. Pharmacol. *117*, 15 (1985).
66. J. Maj, W. Palider, and A. Rawlow, J. Neural Trans. *44*, 237 (1979).
67. M. Melzacka, J. Boksa, and J. Maj, J. Pharm. Pharmacol. *31*, 855 (1979).
68. V. Cioli, C. Corradino, D. Piccinelli, M. G. Rocchi, and P. Valeri, Pharmacol. Res. Commun. *16*, 85 (1984).
69. S. Caccia, M. H. Fong, S. Garattini, and A. Notarnicola, Biochem. Pharmacol. *34*, 393 (1985).
70. R. Samanin, T. Mennini, A. Ferraris, C. Bendotti, F. Borsini, and S. Garattini, Naunyn-Schmiedebergs Arch. Pharmacol. *308*, 159 (1979).
71. A. Rokosz-Pelc, L. Antikiewicz-Michaluk, and J. Vetulani, J. Pharm. Pharmacol. *32*, 220 (1980).
72. S. Caccia, M. Ballabio, R. Samanin, M. G. Zanini, and S. Garattini, J. Pharm. Pharmacol. *33*, 477 (1981).
73. L. A. Riblet and D. P. Taylor, Clin. Psychopharmacol. *1*, 17S (1981).
74. R. W. Fuller, H. W. Snoddy, N. R. Mason, and B. B. Molloy, Eur. J. Pharmacol. *52*, 11 (1978).
75. R. W. Fuller, N. R. Mason, and B. B. Molloy, Biochem. Pharmacol. *29*, 833 (1980).
76. H. M. Jennewein, J. Mierau, P. R. Saxena, and P. D. Verdouw, Br. J. Pharmacol. *82*, 208p (1984).
77. R. A. Lyon, M. Titeler, J. C. McKenney, P. Magee, and R. A. Glennon, J. Med. Chem. *29*, 630 (1986).

78. R. F. Neale, S. L. Fallon, W. C. Boyar, J. W. F. Wasley, L. L. Martin, B. Glaeser, C. M. Sinton, and M. Williams, Eur. J. Pharmacol. *136*, 1 (1987).
79. N. R. Mason, personal communication, 1981.
80. J. R. Huff, S. W. King, W. S. Saari, J. P. Springer, G. E. Martin, and M. Williams, J. Med. Chem. *28*, 945 (1985).
81. K. B. Asarch, R. W. Ransom, and J. C. Shih, Life Sci. *36*, 1265 (1985); J. C. Shih, personal communication, 1985.
82. M. Cohen and J. Wittenauer, J. Pharmacol. Exp. Ther. *233*, 75 (1985).
83. R. A. Glennon, R. M. Slusher, R. A. Lyon, M. Titeler, and J. D. McKenney, J. Med. Chem. *29*, 2375 (1986).
84. M. L. Cohen, R. W. Fuller, and K. D. Kurz, Hypertension *5*, 676 (1983).
85. R. Fuller, personal communication.
86. E. Sanders-Bush, personal communication.
87. E. Hong, R. Rion, and G. Rojas, Proc. West. Pharmacol. Soc. *27*, 1 (1984).
88. L. A. Barragan, J. A. Galindo-Morales, and E. Hong, Proc. West. Pharmacol. Soc. *27*, 329 (1984).
89. E. Hong, L. Sancilio, R. Vargas, and E. G. Pardo, Eur. J. Pharmacol. *6*, 274 (1969).
90. R. Rodriguez and E. G. Pardo, Psychopharmacologia *21*, 89 (1971).
91. M. J. R. Lansdown, H. L. Nash, P. R. Preston, D. I. Wallis, and R. G. Williams, Br. J. Pharmacol. *68*, 525 (1980).
92. G. G. Yarbrough, D. K. Singh, and D. Pettibone, J. Neuropharmacol. *23*, 1271 (1984).
93. D. J. Pettibone and M. Williams, Biochem. Pharmacol. *33*, 1531 (1984).
94. B. V. Clineschmidt, D. R. Reiss, D. J. Pettibone, and J. L. Robinson, J. Pharmacol. Exp. Ther. *235*, 696 (1986).
95. C. Euvrard and J. R. Boissier, Eur. J. Pharmacol. *63*, 65 (1980).
96. D. N. Middlemiss, J. Pharm. Pharmacol. *37*, 434 (1985).
97. H. N. Doods, H. O. Kalkman, A. DeJonge, M. J. M. C. Thoolen, B. Wilffert, P. B. M. W. M. Timmermans, and P. A. Van Zwieten, Eur. J. Pharmacol. *112*, 363 (1985).
98. L.-E. Arvidsson, U. Hacksell, J. L. G. Nilsson, S. Hjorth, A. Carlsson, P. Linberg, D. Sanchez, and H. Wikstrom, J. Med. Chem. *24*, 921 (1981).
99. S. Hjorth, P. Lindberg, D. Sanchez, H. Wikstrom, L.-E. Arvidsson, U. Hacksell, and J. L. G. Nilsson, J. Neural Trans. *55*, 169 (1982).
100. N. Mason, R. D. Marsh, K. W. Perry, H. D. Snoddy, and R. W. Fuller, Soc. Neurosci. Abstr. *9*, 1153 (1983).
101. D. N. Middlemiss and J. R. Fozard, Eur. J. Pharmacol. *90*, 151 (1983).
102. M. Hamon, C. Goetz, and H. Gozlan, in CNS Receptors—From Molecular Pharmacology to Behavior (P. Mandel and F. V. DeFeudis, eds.), Raven Press, New York, 1983, p. 349.
103. M. Hamon, S. Bourgoin, H. Gozlan, M. D. Hall, C. Goetz, F. Artaud, and A. S. Horn, Eur. J. Pharmacol. *100*, 262 (1984).
104. M. Titeler and R. A. Lyon, personal communication. 1985.
105. A. E. Hauck, D. N. Middlemiss, and R. A. Glennon, unpublished data.
106. M. B. Emerit, H. Gozlan, M. D. Hall, M. Hamon, and A. Marquet, Biochem. Pharmacol. *34*, 883 (1985).

107. B. L. Jacobs, in Psychopharmacology 1. Part 1: Preclinical Psychopharmacology (D. G. Grahame-Smith and D. J. Cowen, eds.), Excerpta Medica, Amsterdam, 1982, p. 344.
108. R. A. Glennon, in Neuropharmacology of Serotonin (A. R. Green, ed.), Oxford University Press, Oxford, 1985, p. 253.
109. R. A. Glennon, R. Young, and J. A. Rosecrans, Eur. J. Pharmacol. *91*, 189 (1983).
110. R. A. Glennon, J. D. McKenney, R. A. Lyon, and M. Titeler, J. Med. Chem. *29*, 194 (1986).
111. M. Shannon, G. Battaglia, R. A. Glennon, and M. Titeler, Eur. J. Pharmacol. *102*, 23 (1984).
112. J. E. Leysen, W. Gommeren, P. Van Gompel, J. Wynants, P. F. M. Janssen, and P. M. Laduron, Mol. Pharmacol. *27*, 600 (1985).
113. D. N. Middlemiss, L. Blakeborough, and S. R. Leather, Nature *267*, 289 (1977).
114. S. R. Nahorski and A. L. Willcocks, Br. J. Pharmacol. *78*, 107p, (1983).
115. A. R. Green, P. Johnson, and V. L. Nimgaonkar, Neuropharmacology *22*, 657 (1983).
116. P. C. Waldmeier, Naunyn-Schmiedebergs Arch. Pharmacol. *317*, 115 (1981).
117. A. R. Green, P. J. Cowen, V. L. Nimgaonkar, and D. G. Grahame-Smith, in Frontiers in Biochemical and Pharmacological Research in Depression (E. Usdin, ed.), Raven Press, New York, 1984, p. 285.
118. D. N. Middlemiss, Eur. J. Pharmacol. *101*, 289 (1984).
119. A. R. Green and D. J. Heal, in Neuropharmacology of Serotonin (A. R. Green, ed.), Oxford University Press, Oxford, 1985, p. 326.
120. M. Tricklebank, C. Forler, and J. Fozard, Eur. J. Pharmacol. *106*, 271 (1985).
121. M. D. Tricklebank, Br. J. Pharmacol. *81*, 26P (1984).
122. S. J. Peroutka, Brain Res. *344*, 167 (1985).
123. M. Battaglia, M. Shannon, B. Borgundvaag, and M. Titeler, J. Receptor Res. *4*, 713 (1984).
124. J. E. Leysen, F. Awouters, L. Kennis, P. M. Laduron, J. Vandenberk, and P. A. J. Janssen, Life Sci. *28*, 1015 (1981).
125. P. A. J. Janssen, J. Cardiovasc. Pharmacol. *7*(Suppl. 7), S2 (1985).
126. F. C. Colpaert, T. F. Meert, C. J. E. Niemegeers, and P. A. J. Janssen, Psychopharmacology *86*, 45 (1985).
127. R. A. Glennon and A. E. Hauck, Pharmacol. Biochem. Behav. *23*, 937 (1985).
128. G. M. Goodwin and A. R. Green, Br. J. Pharmacol. *84*, 743 (1985).
129. R. W. Fuller and S. H. Snoddy, Endocrinol. *105*, 923 (1979).
130. M. Williams, G. E. Martin, D. C. Remy, M. Hichens, A. W. Mangel, D. A. Taylor, G. G. Yarbrough, R. J. Bendesky, S. W. King, J. L. Robinson, J. A. Totaro, and B. V. Clineschmidt, J. Pharmacol. Exp. Ther. *229*, 775 (1984).
131. S. J. Peroutka, R. M. Lebovitz, and S. H. Snyder, Science *212*, 827 (1981).
132. M. A. Sillis, B. B. Wolfe, and A. Frazer, Mol. Pharmacol. *26*, 10 (1984).
133. G. Battaglia, M. Shannon, and M. Titeler, Life Sci. *32*, 2597 (1983).
134. M. Titeler, G. Battaglia, and M. Shannon, J. Receptor Res. *4*, 705 (1984).

135. D. A. Kendall and S. R. Nahorski, J. Pharmacol. Exp. Ther. *227*, 429 (1983).
136. M. Hamon, S. Bourgoin, A. Enjalbert, J. Bockaert, F. Hery, J. P. Ternaux, and J. Glowinski, Naunyn-Schmiedebergs Arch. Pharmacol. *294*, 99 (1976).
137. M. L. Barbaccia, N. Brunello, and E. Costa, J. Neurochem. *40*, 1671 (1983).
138. P. J. Conn and E. Sanders-Bush, Soc. Neurosci. Abstr. *10*, 223 (1984).
139. P. J. Conn and E. Sanders-Bush, Neuropharmacology *8*, 993 (1984).
140. P. J. Conn and E. Sanders-Bush, J. Pharmacol. Exp. Ther. *234*, 195 (1985).
141. D. A. Kendall and S. R. Nahorski, Br. J. Pharmacol. *82*, 206p (1984).
142. P. J. Conn and E. Sanders-Bush, Life Sci. *38*, 663 (1986).
143. E. Sanders-Bush, P. J. Conn, B. J. Hoffman, and P. R. Hartig, Soc. Neurosci. Abstr. *12*, 576 (1986).
144. M. Gothert, Trends Pharmacol. Sci. *3*, 437 (1982).
145. C. Moret, in Neuropharmacology of Serotonin (A. R. Green, ed.), Oxford University Press, Oxford, 1985, p. 21.
146. L. L. Martin and E. Sanders-Bush, Neuropharmacology *21*, 445 (1982).
147. E. Schlicker and M. Gothert, Naunyn-Schmiedebergs Arch. Pharmacol. *317*, 204 (1981).
148. D. N. Middlemiss, Naunyn-Schmiedebergs Arch. Pharmacol. *327*, 18 (1894).
149. D. Verge, G. Daval, A. Patey, H. Gozlan, S. El Mestikawy, and M. Hamon, Eur. J. Pharmacol. *113*, 463 (1985).
150. L.-E. Arvidsson, U. Hacksell, and R. A. Glennon, in Progress In Drug Research (E. Jucker, ed.), Birkhauser Verlag, Basel, 1986, p. 365.
151. S. Ahlenius, K. Larson, L. Svensson, S. Hjorth, A. Carlsson, P. Lindberg, H. Wikstrom, D. Sanchez, L.-E. Arvidsson, U. Hacksell, and J. L. G. Nilsson, Pharmacol. Biochem. Behav. *15*, 785 (1981).
152. S. Ahlenius and K. Larsson, Psychopharmacology *83*, 330 (1984).
153. I. Lucki, M. S, Nobler, and A. Frazer, J. Pharmacol. Exp. Ther. *228*, 133 (1984).
154. A. L. Johnston and S. E. File, Pharmacol. Biochem. Behav. *24*, 1467 (1986).
155. C. R. Gardner, Pharmacol. Biochem. Behav. *24*, 1479 (1986).
156. D. L. S. Ceulemans, M.-L. J. A. Hoppenbrouwers, Y. G. Gelders, and A. J. M. Reyntjens, Pharmacopsychiatry *18*, 303 (1985).
157. R. Rodriguez and E. G. Pardo, Psychopharmacologia *21*, 89 (1971).
158. R. A. Glennon and J. D. McKenney, Pharmacologist *27*, 194 (1985).
159. R. A. Glennon, Psychopharmacology *89*, S42 (1986).
160. E. A. Parati, P. Zanardi, D. Cocchi, T. Caraceni, and E. E. Muller, J. Neural Trans. *47*, 273 (1980).
161. E. A. Mueller, D. L. Murphy, T. Sunderland, and J. Jones, Psychopharmacol. Bull. *21*, 701 (1985).
162. H. J. Altman, D. A. Nordy, and S. O. Ogren, Psychopharmacology *84*, 496 (1984).
163. Proceedings of the 10th Scientific Meeting of the International Society of Hypertension, J. Cardiovasc. Pharmacol. *7*(Suppl. 7) (1985).

164. P. M. Vanhoutte, Serotonin and the Cardiovascular System, Raven Press, New York, 1958.
165. D. Hoyer, G. Engel, and H. O. Kalkman, Eur. J. Pharmacol. *118*, 13 (1985).
166. R. A. Lyon, M. Titeler, and R. A. Glennon, Soc. Neurosci. Abstr. *12*, 311 (1986).
167. J. R. Fozard, Trends Pharmacol. Sci. *8*, 501 (1987).
168. R. A. Glennon, J. Med. Chem. *30*, 1 (1987).
169. R. A. Glennon and I. Lucki, in *Serotonin* (E. Sanders-Bush, ed.). Humana Press, Clifton, New Jersey, in press.
170. J. Traber and T. Glaser, Trends Pharmacol. Sci. *8*, 432 (1987).
171. R. Young and R. A. Glennon, in *Serotonin Agonists as Psychoactive Drugs* (G. Gedulsky and R. Rech, eds.), NPP Press, Ann Arbor, in press.
172. P. J. Conn and E. Sanders-Bush, Psychopharmacol. *92*, 267 (1987).
173. B. L. Roth and D. Chuang, Life Sci. *41*, 1051 (1987).
174. P. C. Waldmeier, M. Williams, P. A. Baumann, S. Bischoff, M. A. Sills, and R. F. Neale, Naunyn-Schmiedebergs Arch. Pharmacol., in press.

10

Histamine Receptors

Armin Buschauer and Walter Schunack / Free University of Berlin, Berlin, Federal Republic of Germany

Jean-Michel Arrang, Monique Garbarg, and Jean-Charles Schwartz / Paul Broca Center of INSERM, Paris, France

J. Michael Young / University of Cambridge, Cambridge, England

HISTORICAL SURVEY

H_1 and H_2 Receptors

Since the chemical synthesis of histamine by Windaus and Vogt in 1907 (1) and the classic researches of Sir Henry Dale and his colleagues starting in 1910 (2,3) on its physiological effects, a huge literature has accumulated on the chemistry, natural occurrence and distribution, and involvement of histamine in a wide variety of actions and organs (4,5).

Barger and Dale (6) isolated histamine from ergot and showed it to be identical with a specimen of a substance prepared by Ackermann (7) by bacterial decarboxylation of histidine. The occurrence of histamine in intestinal mucosa was found in 1911 (8). Abel and Kubota showed that histamine was present as a normal constituent of many tissues (9).

The search for compounds antagonizing the pathophysiological effects of histamine led to the discovery of the first "antihistamines" in 1937 by Bovet and Staub (10), followed by the development of numerous more active and therapeutically used drugs in the succeeding decades, such as mepyramine and diphenhydramine. These compounds proved to be competitive antagonists of histamine in diverse organ preparations, such as the isolated ileum and trachea of the guinea pig and the perfused guinea pig lung (11). What soon became evident was that certain effects of histamine could not be antagonized by any of the large number of available antihistamines. The most prominent of these actions were gastric acid secretion and cardiac chronotropism. Ash and Schild recognized the implication in those exceptions and postulated a diversity of receptors for histamine action (12). They defined those receptors involved in the mepyramine- and diphenhydramine-sensitive histamine response as H_1 receptors. However, the question of whether the "non-H_1" receptors constituted single or multiple subclasses could not be answered. The chance to distinguish H_1 and non-H_1 responses to histamine by specifically acting antagonists, conversely, led to investigation of numerous compounds in different tissues in order to estimate their agonistic activity and to find some criterion of receptor specificity.

At the end of the 1960s, a group of researchers around J. W. Black at Smith, Kline and French (SK&F) embarked on breaking the hidden code of the histamine non-H_1 receptor. In 1972, burimamide, the first antagonist specifically acting at non-H_1 receptors, and some selective agonists were described (13). The development of the first histamine H_2-receptor antagonist opened a new era in the history of histamine. Burimamide was found to competitively antagonize several of the actions of histamine unaffected by the classic antihistamines, for example, the positive chronotropic effect at the isolated guinea pig atrium, the stimulation of gastric acid secretion, and the inhibition of electrically evoked rat uterus contraction. Thus, the existence of a second subclass of histamine receptors (H_2) was firmly established. Afterward a series of more active rather selective agonists and antagonists of these receptors was synthesized that acted as pharmacological tools for the study of histamine effects and led to the discovery that H_2 receptors were much more widely distributed than previously suspected.

Chemical modification of burimamide led to metiamide (14) and cimetidine (15,16), which became the first H_2 antagonist to be marketed for the treatment of peptic ulcers.

In the histamine H_2-receptor population, evidence for and against heterogeneity is reported. It is based on different degrees of potency and also of efficacy among the selective H_2-receptor agonists, on differences in the potency of H_2 antagonists, and on the inability of H_1 and H_2 antagonists to block some of the effects of histamine. All the data available so far suggest that, although it is premature to speak about H_2-receptor subtypes, further investigations are needed to check whether H_2 receptors represent homogeneous populations of histamine receptors (17,18).

H_3 Receptors

There are two main conditions that must be fulfilled together to ensure a reliable identification of a new receptor subclass for a given chemical messenger. The first is that a biologic response selectively mediated by this putative receptor should be identified. The second is that drugs should be available to reasonably differentiate the pharmacology of this response to that of previously identified receptor subclasses. The development of an in vitro model for measurement of histamine release from its neuronal stores in brain and the availability of a large variety of histamine agonists and antagonists with established potencies at H_1 and H_2 receptors paved the way in 1983 for the identification of H_3 receptors (19).

Starting in the 1970s, a large body of experimental data progressively established that histamine plays a neurotransmitter role in the brain (20-24). It is synthesized at a high rate in and released from a population of long neurons with a disposition resembling that of other monoaminergic neurons to act upon target cells via postsynaptic H_1 and H_2 receptors. In the case of other monoamines or even amino acid neurotransmitters, it was well known that their various actions in the central or peripheral nervous systems are mediated not only by postsynaptic but also by presynaptic receptors, which sometimes display a distinct pharmacology (25,26). Such autoreceptors had not been demonstrated to regulate neuronal histamine release, but in vitro models to study the release of the amine had been previously developed. Depolarization of slices from rat brain had been shown to release in a calcium-dependent manner endogenous histamine (27,28) or [^{3}H]histamine either taken up from the incubation medium (29) or synthesized from its precursor [^{3}H]amino acid (30). The latter model was selected, in spite of the technical difficulties introduced by the necessary separation of the [^{3}H]amine from large amounts of the [^{3}H]amino acid, because it was more likely to selectively reflect biochemical events occurring within histaminergic neurons. In fact, no active reuptake mechanism of histamine has been hitherto identified so that it is not established whether exogenous [^{3}H]histamine is able to selectively label the endogenous amine pool in histaminergic neurons.

Slices from cerebral cortex were used because in this cerebral region histamine synthesis occurs mainly if not solely, within terminals from extrinsic neurons (31-33). Depolarization of cortical slices by 30 mM K^+ (or veratridine) resulted in the release of about 15% of the tissue [^{3}H]histamine in the absence of any agent. However when nonradioactive histamine, N^α-methylhistamine or N^α,N^α-dimethylhistamine, was added to the medium, a dose-dependent inhibition of the release was observed (19). This inhibition was maximally 60%, rapidly reversible by washing the tissues, and occurred for low histamine concentrations, that is, lower by about one to two orders of magnitude than those required to activate either H_1 or H_2 receptors. In addition, the autoinhibitory response was completely antagonized in an apparently competitive manner by such compounds as burimamide or impromidine (the former, an H_2-receptor antagonist; the latter an H_2-receptor agonist). However, the model was reliable enough to allow a precise determination of the potencies of available histaminergic agents, and both the relative potencies of agonists and apparent dissociation constants of antagonists were found to clearly differ from corresponding values either at H_1 and H_2 receptors (see Selectivity and Structure-Activity Relationships). For example, 2-methylhistamine, an agonist at H_1 (or H_2) receptors was inactive, and either mepyramine or tiotidine were at least 1000 times less potent as antagonists than at H_1 and H_2 receptors, respectively. In summary, the autoinhibition of histamine release in brain slices seemed to display all the characteristics of a receptor-mediated event, and its pharmacological analysis showed that neither H_1 nor H_2 receptors were involved. Hence Arrang et al. proposed to designate this novel subclass of histamine receptors the H_3 receptors (19).

RADIOLIGAND BINDING STUDIES

Histamine H_1 Receptors

Radioligands

[^{3}H]Mepyramine (synonymous with [^{3}H]pyrilamine), the first [^{3}H]ligand introduced for H_1-receptor labeling (34), is still the most suitable ligand for most purposes. The original choice of mepyramine for labeling was based on its well-proven pharmacological specificity (probably the most common stumbling block with potential radioligands), its high affinity for the H_1 receptor (K_D = 0.5 nM in the guinea pig), and the relative ease with which one atom of tritium could be introduced into the molecule (Figure 1) (35). The evidence that the high-affinity binding of [^{3}H]mepyramine is to H_1 receptors rests primarily on the excellent correlation between the affinity constants of antagonists, derived from inhibition of [^{3}H]mepyramine binding, and the affinities obtained from inhibition of a functional response to histamine (usually contraction of the guinea pig ileum). This correlation has been demonstrated for both peripheral tissues of the guinea pig, notably homogenates of intestinal smooth muscle (34,36,37), heart (36), lung (38), and urinary bladder (39), and for homogenates of guinea pig brain (40,41).

Although [^{3}H]mepyramine has proved to be a satisfactory ligand in a wide range of tissues, problems can be encountered with its use. Binding to secondary, nonreceptor sites of medium affinity is the most common complication (37,42), especially at higher concentrations of [^{3}H]mepyramine (43), and the difficulties that can arise in certain tissues in separating the H_1-receptor component from the high level of nonspecific binding are well illustrated in studies on guinea pig (38) and human (44) lung. The problem of binding to secondary sites may be exacerbated in such species as the rat, in which the affinity of [^{3}H]mepyramine for the H_1 receptor is lower than in the guinea pig (41,42,45). It was, in part, in an attempt to overcome difficulties of this kind by using a ligand with greater potency at the H_1 receptor that led to an investigation of the properties of [^{3}H]doxepin (46).

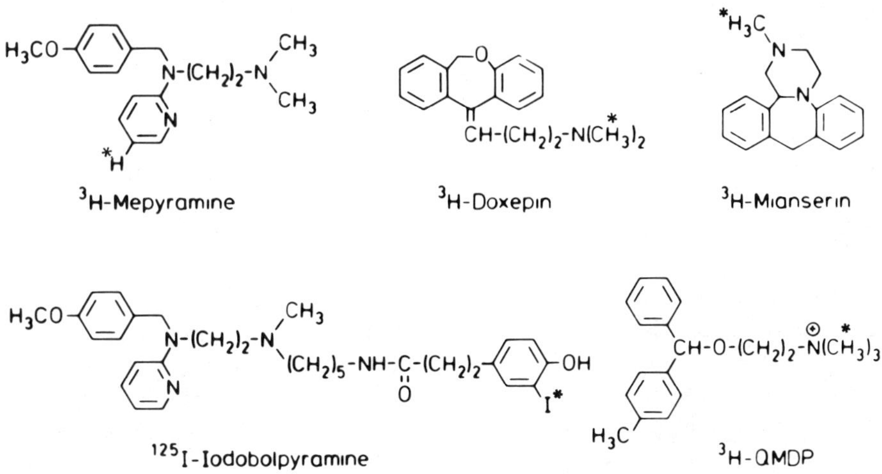

Figure 1 Structure of radioligands (asterisk indicates the position of the radiolabel).

Doxepin, which is used clinically as an antidepressant, has a very high affinity for the H_1 receptor, K_D = 0.03-0.05 nM (45,47), and consequently a significant receptor occupancy can be achieved at very low concentrations of the [^{3}H] derivative, at which nonspecific binding might be expected to be small. Further, doxepin has practically the same affinity for the H_1 receptor in guinea pig and rat brain (41,45,46) and has been labeled to high specific activity (80 Ci/mmol). However, doxepin has a high affinity for a number of receptor systems, and the binding of the radiolabeled compound to preparations of guinea pig, rat, and human brain is complex (46,48-50). It is not clear that the binding is adequately described by a two-site model (49), and this can lead to problems in identifying the H_1-receptor component. Further, in common with all radioligands of very high affinity that are used at very low concentration, the amount of ligand bound by the tissue may be a significant fraction of the total amount of [^{3}H] ligand added, which can introduce substantial errors in the determination of K_D (51).

[^{3}H] Mianserin, another putative antidepressant, has also been used to label histamine H_1 receptors (52). However, the affinity of [^{3}H] mianserin for the serotonin (5-HT) 5-HT_2 receptor is practically the same as that for the H_1 receptor, K_D = 3 nM in rat frontal cortex (52), so that both receptors are labeled unless binding to one is prevented by the addition of an appropriate antagonist. In guinea pig cerebellar homogenates, where the density of 5-HT_2 sites is low, the binding of [^{3}H] mianserin is almost wholly to H_1 receptors (52). Interestingly, it has been reported that the stereoisomers of mianserin have a similar potency in inhibiting the binding of [^{3}H] mepyramine to rat cerebral cortical membranes (53), whereas, in contrast, the S(+)enantiomer is 10 times more effective than the R(—)enantiomer in inhibiting the binding of [^{3}H] mianserin to 5-HT_2 sites in the same tissue (54).

For routine purposes [^{3}H] mepyramine is the ligand of choice for the H_1 receptor, but for specific purposes an alternative ligand may be advantageous. A very recent addition to the range of H_1-receptor ligands that promises to fill just such a role is [^{125}I] iodobolpyramine (55). The high specific activity of this compound, in combination with the high affinity of iodobolpyramine for the H_1 receptor, approximately 0.15 nM, means that sensitive assays can be performed with very small tissue samples. The particular utility of the [^{125}I] compound in autoradiographic studies is discussed below. The extent of the applicability of [^{125}I] iodobolpyramine has yet to be assessed, but attempts to use [^{125}I] iodobolpyramine to label rat brain membranes were unsuccessful, even though the parent compound, bolpyramine, has only a fourfold lower affinity in rat brain (55).

An interesting feature of the interaction of [^{125}I] iodobolpyramine with the H_1 receptor is the very slow rate of association and dissociation. In this it is reminiscent of astemizole, a highly lipophilic H_1 antagonist, which dissociates from the receptor so slowly as to be effectively irreversible (56). The binding of [^{3}H] astemizole to the H_1 receptor has been described, but the problems associated with its use make it unlikely that it will be of any practical use (56).

The synthesis of a quaternary ligand for the H_1 receptor, [^{3}H] *N*-4-dimethyldiphenhydramine ([^{3}H] QMDP), has also been reported recently (57). The binding properties of this ligand remain to be fully evaluated, although it has proven useful in kinetic studies. The possession of a permanent charge could make it useful for binding studies using intact cells; in certain cases the uptake of [^{3}H] mepyramine into the cells has been reported to lead to spurious antagonist-displaceable binding (58).

Receptor Heterogeneity and Species Variation

In the foregoing discussion the tacit assumption has been made that within a given species there is a single homogeneous population of H_1 receptors. There is as yet no substantial evidence to suggest that this assumption is not justified, although the history of other receptor systems would urge caution. In particular, there is no good indication from the binding properties of antagonists of any difference between central and peripheral H_1 receptors. A report that mequitazine, one of the newer nonsedative H_1 antagonists, had a lower affinity (K_D = 280 nM) for H_1 receptors in guinea pig brain homogenates than in guinea pig ileum or lung (59,60) has not been confirmed in other laboratories. The affinity of mequitazine deduced from the inhibition of the binding of

[^{3}H] mepyramine to guinea pig cerebellum, cerebral cortex, and hippocampus (K_D = 0.6, 0.6, and 1.1 nM, respectively) was similar to the value, 1.0 nM, derived from the inhibition of the histamine-induced contraction of the guinea pig ileum (61,62). Mequitazine was not greatly less potent, K_D = 6 nM, as an antagonist of [^{3}H] doxepin binding to human brain membranes (50).

Evidence that mequitazine does not achieve a high occupancy of H_1 receptors in vivo has been obtained using a technique for measuring [^{3}H] mepyramine binding to H_1 receptors in the brain of the living mouse following intravenous injection (63-65). It has been possible to demonstrate saturable binding of [^{3}H] mepyramine, a regional distribution of receptor sites comparable with that observed in studies on homogenates labeled in vitro, and inhibition by H_1 antagonists coadministered with [^{3}H] mepyramine. In studies of this kind mequitazine is a much weaker inhibitor of [^{3}H] mepyramine binding than its in vitro affinity in mouse brain, K_D = 6 nM, would suggest (63,64), and it seems likely that it is unable to cross the blood-brain barrier.

In making comparisons between central and peripheral receptors by means of binding experiments on tissue homogenates it must be borne in mind that most tissues are heterogeneous with respect to the types of cell present. This applies particularly to homogenates of brain regions, where histamine H_1 and H_2 receptors may be localized on blood vessels, a "peripheral tissue," as well as on neural elements. Direct measurements of [^{3}H] mepyramine binding to microvessel preparations isolated from bovine cerebral cortex show that the K_D values derived for mepyramine and other H_1 antagonists are closely similar to those measured on cortical gray matter (66,67). However, the percentage of H_1 receptors associated with the microvessels was estimated to be less than 1% of the total number of H_1-binding sites in bovine cerebral cortex.

The heterogeneity of H_1-receptor sites in rat brain has been suggested as a possible reason for an apparent difference in the number of sites labeled by [^{3}H] doxepin (high-affinity component of binding) and [^{3}H] mepyramine (48), although as the authors noted, not all the data were consistent with this explanation. A study of doxepin inhibition of [^{3}H] mepyramine binding failed to give any evidence of a preferential binding of doxepin to a subset of the sites labeled by [^{3}H] mepyramine (45), although it would be difficult to detect such a subset if it were only a small percentage of the total number of [^{3}H] mepyramine binding sites.

Although the evidence for differences in the characteristics of antagonist binding to H_1 receptors within a single species is not good, the evidence for differences between species is strong (41,42,45,55,68). The difference between H_1 receptors in guinea pig and rat brain, as reflected in antagonist affinities, is retained on solubilization, implying that the difference is a property of the receptor protein and is independent of the membrane environment (49).

H_1-Receptor Localization

Species variation seems to be a characteristic of every aspect of the histamine system. This is strikingly illustrated by the differences in the distribution of H_1 receptors in brain regions. In guinea pig brain the highest density of H_1 receptors is found in the cerebellum (40,41,69), a property that has made this tissue particularly suitable for studies of H_1-receptor binding. In contrast, in the rat (41,42) and human (41,50) brain the cerebellum is among the regions with the lowest H_1-receptor density. The hypothalamus is labeled to a moderate level in guinea pig, rat, and human brain, but in the cat it has twice the density found in any other region (70). In monkey brain (68) and in human brain (41,50), the greatest density is found in cortical areas.

The high levels of [^{3}H] mepyramine binding sites in guinea pig cerebellum, which possesses only low levels of other markers for the histamine system, was initially puzzling and raised the question of whether the density of [^{3}H] mepyramine binding sites was a reliable indication of the number of *functional* H_1 receptors. The good correlation between the density of [^{3}H] mepyramine binding sites and the magnitude of histamine-induced inositol phospholipid breakdown in regions of guinea pig brain (71) has gone some way toward answering this question, as has the demonstration that iontophoretically applied histamine depresses the firing of the great majority of Purkinje cells in guinea pig cerebellum, but to a considerably lesser extent in the rat (72). Similarly, the notable insensitivity of rat small intestine to histamine is reflected in the very low level of high-affinity [^{3}H] mepyramine binding (42).

Measurement of [^{3}H] mepyramine binding to homogenates of dissected brain areas gives only limited information on the distribution of the binding within a region. A more satisfactory approach (73) has been to label cryostat sections of tissues on microscope slides at 4°C, followed by autoradiography. Using this technique the detailed distribution of [^{3}H] mepyramine binding sites has been described in regions of guinea pig (74,75) and rat (76) brain. An even clearer demonstration of the distribution in guinea pig brain has been provided by the use of [^{125}I] iodobolpyramine (Figure 2) (55). This autoradiograph required only 2 days' exposure compared with 4-6 weeks with [^{3}H] mepyramine. The heavy labeling in the molecular layer of the cerebellum, in the thalamus, and in the nucleus accumbens was particularly apparent.

[^{3}H] Mepyramine labeling in the molecular layer of guinea pig cerebellum is greatly depressed after local injection of kainic acid (75). This implies that the H_1 receptors are localized on intrinsic neurones, probably on Purkinje cell dendrites, but the lack of any significant effect of kainic acid on [^{3}H] mepyramine binding in rat striatum and hippocampus (77) points to binding sites on the terminals of extrinsic neurons or glial cells in these tissues.

The autoradiographic technique has also been used to demonstrate the presence of H_1 receptors in the spinal cord (78,79) and sensory ganglion (78). In-

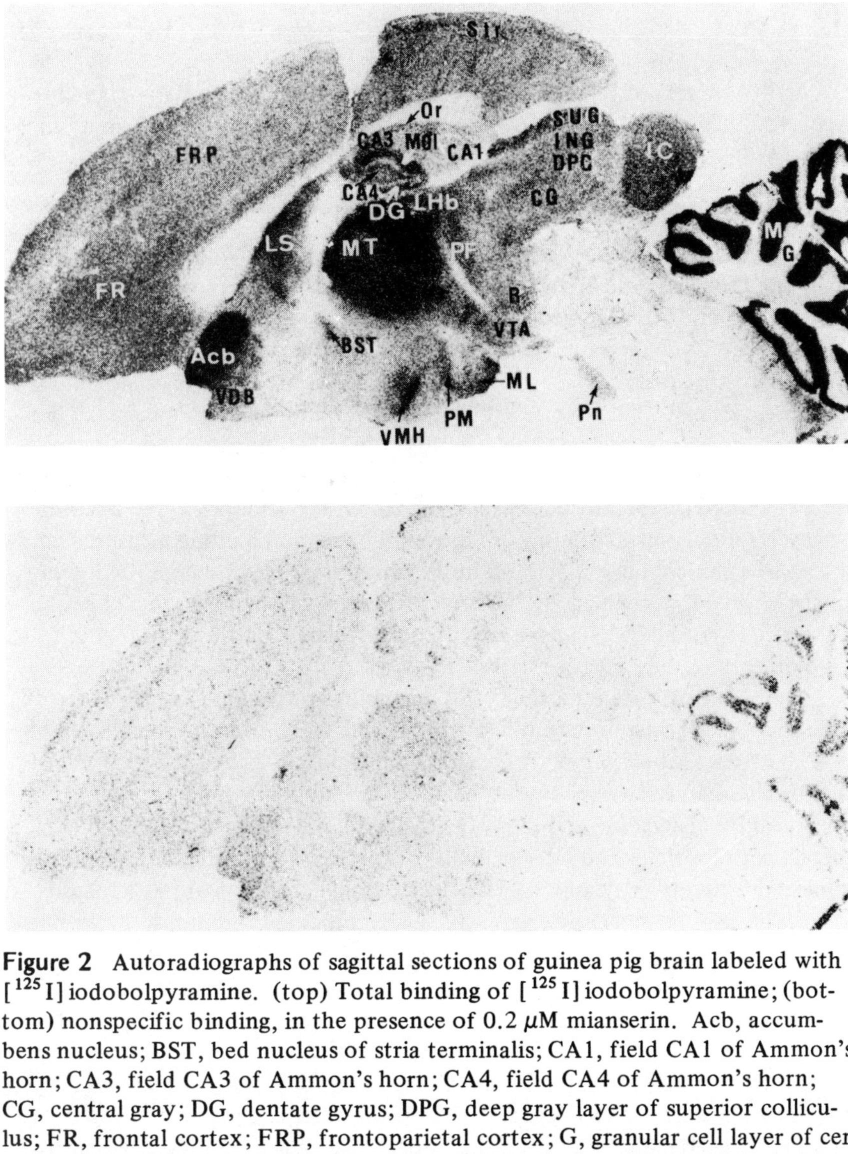

Figure 2 Autoradiographs of sagittal sections of guinea pig brain labeled with [^{125}I] iodobolpyramine. (top) Total binding of [^{125}I] iodobolpyramine; (bottom) nonspecific binding, in the presence of 0.2 μM mianserin. Acb, accumbens nucleus; BST, bed nucleus of stria terminalis; CA1, field CA1 of Ammon's horn; CA3, field CA3 of Ammon's horn; CA4, field CA4 of Ammon's horn; CG, central gray; DG, dentate gyrus; DPG, deep gray layer of superior colliculus; FR, frontal cortex; FRP, frontoparietal cortex; G, granular cell layer of cerebellum; IC, inferior colliculus; ING, intermediate gray layer of superior colliculus; LHb, lateral habenular nucleus, LS, lateral septal nucleus; M, molecular layer of cerebellum; ML, medial mamillary nucleus, lateral part; Mol, molecular layer of dentate gyrus; MT, medial thalamic nucleus; Or, oriens layer of hippocampus; PF, parafascicular thalamic nucleus; PM, premamillary nucleus; Pn, pontine nuclei; R, red nucleus; str, striate cortex; SUG, superficial gray layer of superior colliculus; VDB, nucleus of the vertical limb of the diagonal band; VMH, ventromedial hypothalamic nucleus; VTA, ventral tegmental area. (From Reference 55.)

terestingly, with regard to the roles of histamine and substance P in cutaneous allergy (80), histamine H_1 receptors are present on some substance P-containing cells in dorsal root ganglia of the rhesus monkey (81).

Characteristics of H_1 Ligand-Receptor Interactions

There is no indication that the interaction between [^{3}H]mepyramine and [^{3}H]-QMDP and the histamine H_1 receptor at physiological temperatures is anything other than a simple one-step equilibrium between antagonist and receptor (57, 62). However, both the rate constants for complex formation K_1 and dissociation K_{-1} decrease markedly as the temperature is lowered, so that at 4°C K_{-1} becomes very small. The kinetics of [^{3}H]QMDP binding at the lower temperature remain consistent with a simple equilibrium (57), but it is not certain that this is true for [^{3}H]mepyramine (62).

The marked temperature sensitivity of the rate constants may be a feature of many H_1 antagonists, and incomplete equilibration in binding measurements can lead to significant errors in inhibition constants. This problem of equilibration has long been recognized (82) but is seldom heeded. The anomalous value for the affinity of mequitazine already discussed can be partly explained in this way (Young, unpublished observations).

The binding of histamine H_1 agonists in guinea pig brain, as mirrored by the inhibition of [^{3}H]mepyramine binding in 50 mM Tris buffer, is sensitive to ions and GTP, whereas the binding of antagonists is reported to be little affected (49, 83). Sodium and, to a lesser extent, lithium inhibit agonist binding, whereas it is enhanced by manganese and, less effectively, by magnesium. GTP, and its stable analog GppNHp, and GDP cause a twofold decrease in the affinity of histamine and 2-pyridylethylamine. The effect is small, but it is of particular interest in the light of recent evidence that a GTP binding protein may be involved in coupling receptors for "calcium mobilizing" agonists with the membrane-bound phospholipase C, which leads to polyphosphoinositide breakdown (see Functions) (84). The effects of GTP is lost on solubilization of the receptor, but the effect of sodium and manganese is retained (49).

Pretreatment of guinea pig cerebellar membranes with the thiol alkylating agent *N*-ethylmaleimide (NEM) causes a leftward shift of the curve of histamine inhibition of [^{3}H]mepyramine binding, with some decrease in the Hill coefficient (85). The shift is less with partial agonists and undetectable with antagonists. The effect on histamine binding is retained in the solubilized receptor. However, the effect of NEM is not prevented by the concurrent presence of H_1 antagonists of histamine and its action is thus, presumably, at a site remote from the agonist binding site. Exactly how NEM produces its effect will probably only be properly understood when more is known of the structure of the receptor and the way it interacts in the membrane with regulatory and/or effector units. Progress in this direction has been limited.

Solubilization of H_1-receptor preparation with digitonin has no marked effect on the binding properties of [^{3}H] mepyramine (86), [^{3}H] doxepin (49), or [^{3}H] mianserin (87), although there appears to be a decrease in the affinity of antagonists. A fourfold purification of the solubilized receptor has been achieved by affinity chromatography on a wheat germ agglutinin column, indicating that the H_1 receptor is a glycoprotein (88). Target size inactivation analysis of lyophilized membranes from bovine and human cerebral cortex gives a molecular weight of 160,000 daltons (89). The higher value of 430,000 daltons obtained from sedimentation analysis of a digitonin-solubilized preparation from rat brain very probably reflects the large amounts of bound digitonin (49).

Histamine H_2 Receptors

Compared with the wealth of binding studies on the H_1 receptor, information on the histamine H_2 receptor is very limited. Early progress was limited by the lack of an antagonist with a sufficiently high affinity for the receptor, and attempts to use [^{3}H] cimetidine and [^{3}H] ranitidine both failed. [^{3}H] cimetidine binds with high affinity to an "imidazole" site and [^{3}H] ranitidine to a separate nonreceptor site (90-92). The first convincing demonstration of receptor-selective binding came with a study of the binding of [^{3}H] tiotidine to homogenates of guinea pig cerebral cortex (93). The affinity constants deduced for a range of H_2 antagonists of various structural types were in excellent agreement with affinities derived from inhibition of histamine-induced adenylate cyclase activity in guinea pig gastric mucosal homogenates or the positive chronotropic effect in guinea pig right atrium (92,93). In spite of this success, [^{3}H] tiotidine does not appear to be of wide applicability as a radiolabel for the H_2 receptor. Even in guinea pig cerebral cortical homogenates the nonspecific binding at 2 nM [^{3}H] tiotidine accounts for 50-60% of the total binding (93), and attempts to use the ligand in rat brain (94) or guinea pig gastric mucosa or right atrium have been unsuccessful (92,93). However, by a suitable choice of experimental conditions it has been possible to detect H_2-receptor binding in homogenates of guinea pig lung parenchyma (95) and the human HL-60 myeloid leukemia cell line (96), although in both systems the level of nonspecific binding was high.

In the light of the difficulties it is not surprising that information on the properties and distribution of H_2-receptor binding is sparse. Of four regions of guinea pig brain examined, binding of [^{3}H] tiotidine was highest in the striatum, 212 pmol/g protein, moderate in cerebral cortex and hippocampus, 141 and 97 pmol/g protein; and undetectable in cerebellum and pons medulla (97). These values are similar to H_1-receptor densities in guinea pig brain (40,97), but the low levels of H_2 receptors in cerebellum is in marked contrast to the relatively high levels, approximately 250 pmol/g protein, in this region in the guinea pig.

Furter progress with [^{3}H] antagonist binding will probably have to await the synthesis of a compound with more suitable properties, including an affinity an order of magnitude greater than that of [^{3}H] tiotidine. However, a development of particular interest has been the demonstration that H_2 receptors in rat brain homogenates can be labeled selectively with [^{3}H] histamine (94,98). There is an excellent correlation between IC_{50} values for H_2-antagonist inhibition of [^{3}H] histamine binding and dissociation constants from inhibition of tissue responses. However, the most striking feature of a very detailed study is the complexity of the properties of agonists and antagonists as inhibitors of [^{3}H] histamine binding (94,98,99).

The inhibition curves of antagonists had Hill coefficients that varied from 1.0 to 0.6, and consequently some curves fit well to a one-site model, but for others a two-site model was necessary. However, for the latter group the apparent proportions of high- and low-affinity sites varied between antagonists, although all antagonists achieved a maximum level of inhibition virtually identical to that given by unlabeled histamine. The correlation with dissociation constants from inhibition of physiological responses was better for the IC_{50} of the inhibition curve than for the dissociation constants of either the low- or high-affinity site. Most surprisingly, the interaction between tiotidine (Hill coefficient 0.97) or cimetidine (0.59) and [^{3}H] histamine was not competitive. Increasing the concentration of [^{3}H] histamine had only a minimal effect on the shape or position of the tiotidine and cimetidine inhibition curves. Moreover, the H_2 antagonists appear to be noncompetitive with respect to each other. In contrast, in digitonin-solubilized preparations all H_2 antagonists examined had Hill coefficients of 1.0 and were competitive with [^{3}H] histamine. However, the affinities now bore no relationship to the physiological constants.

Binding data for H_2 agonists were similarly complex, all giving biphasic inhibition curves of [^{3}H] histamine binding, which became simple hyperbolas in solubilized preparations (94,99). However, the affinities in solution correlate well with the lower affinity site in the membrane-bound state and suggest that [^{3}H] histamine is labeling the same site in both preparations. This conclusion is further strengthened by the similarity of the dissociation constant for histamine in the two preparations, 4 and 5 nM (100).

Indirect evidence that the binding of [^{3}H] histamine is to the H_2 receptor is provided by the reciprocal actions of GTP or its analogs and Mg^{2+} ions on histamine affinity (100,101). Such an effect of GTP on agonist binding is a characteristic of adenylate cyclase-coupled systems. The reduction in histamine binding induced by GTP is retained in the solubilized preparation (102). This effect of guanine nucleotides makes it likely that the site of [^{3}H] histamine binding is the same as that observed in an earlier study (101), although the characteristics of H_2-antagonist inhibition differed quite markedly. How the complex binding of [^{3}H] histamine is to be related to H_2 structure or function is not clear, particularly since the EC_{50} for H_2-receptor-mediated tissue responses to histamine is usually well above the nanomolar range. H_2 antagonists may act as allosteric inhibitors of histamine binding to the membrane-bound receptor, but receptor structure or organization is somehow altered on solubilization. One comfort to be taken from this bewildering behavior is that complex binding patterns must contain more information on receptor structure and function than very simple ones. Whether the secrets of [^{3}H]histamine binding will be easily unlocked remains to be seen.

SELECTIVITY AND STRUCTURE-ACTIVITY RELATIONSHIPS

H_1-Receptor Agonists

Physicochemical properties of histamine [1] and relations to the biologic activity have been summarized in numerous reviews (Table 1) (103,104). There is evidence that the biologically active form of histamine at both H_1 and H_2 receptors is the N^{τ}H-tautomer [1a] (105) of the monocation (Figure 3). A series of heterocyclic ethylamines, such as 2-(2-pyridyl)ethylamine [2] and 2-(2-thiazolyl)ethylamine [4], which cannot tautomerize, are also H_1 agonists (4,104, 106). Thus, the 1,3-prototropic tautomerism of the imidazole ring appears not to be functionally involved in H_1-receptor stimulation. The common structural

Table 1 Structures and Activities of Selective H_1-Receptor Agonists at the Isolated Guinea Pig Ileum and Atrium

		H_1 (ileum)		H_2 (atrium)		
H_1-Agonists	No.	Relative activity	Intrinsic activity	Relative activity	Intrinsic activity	Reference
Histamine	1	100	1	100	1	
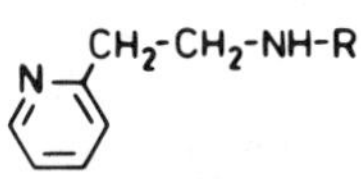 R = H	2	5.6	1	[2.5]	0.5	103
R = CH_3	3	8		[1.5]	0.4	103
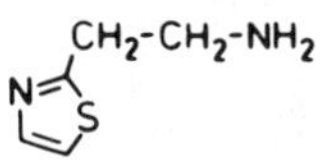	4	11.2,(26)	1	[2.2]	0.6	109,(103)
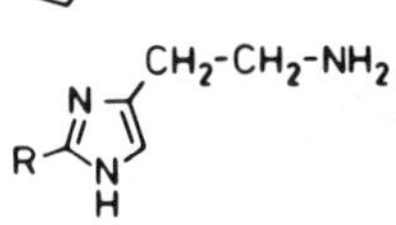 R = CH_3	5	14.1,(16.5)	1	4.4	0.9	108,(103)
R = 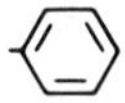	6	19.1	1	–	0.4	108
R = 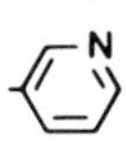	7	11	1	–	0.3	108

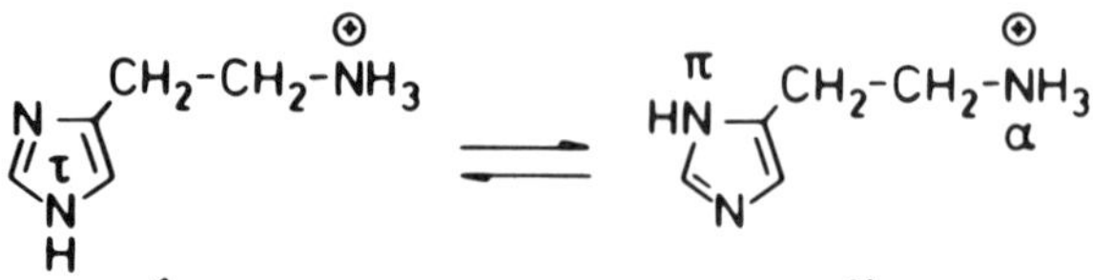

Figure 3 Tautomeric forms of the histamine monocation.

feature of active heterocyclic histamine analogs is an aminoethyl side chain ortho to a pyridine-type nitrogen, comparable to the N^{τ}H-tautomer of histamine. Thus, the imidazole NH group is regarded to have no special significance in H_1 in contrast to H_2 activity.

Compounds [2] and [4] are only weak but are selective H_1 agonists. A derivative of [2], which is monomethylated at the side chain nitrogen (betahistine [3]), has about 8% of the activity of histamine at the guinea pig ileum and is therapeutically used in the treatment of Menière's disease (107). Many attempts have been made to synthesize selectively acting histamine agonists by modifying the parent molecule. Introduction of substituents at the imidazole ring 2 or 5 position decreases potency at both H_1 and H_2 receptors but leads to selective agonists (13). 2-Methylhistamine [5] was found to have about 14% (108) of the activity of histamine at the guinea pig ileum but only approximately 4% (108) at the atrium, whereas 5-methylhistamine selectively stimulates H_2 receptors. Increasing the size of the aliphatic substituent leads to a gradual decrease of receptor affinity and intrinsic activity, especially in the series of 5-substituted histamine analogs (108,109). The reduction of efficacy induced by substituents at ring 2 position is less pronounced than at the 5 position and, additionally, in contrast to the latter, aromatic substituents are also tolerated without a decrease in intrinsic activity. For example, 2-phenylhistamine [6] has about 20% of the potency of histamine and full intrinsic activity at the guinea pig ileum but is only a weak partial agonist, achieving 40% of the maximal response at the guinea pig right atrium. Similarly, 2-(3-pyridyl)histamine [7] is a selective H_1 agonist with 11% of the potency of histamine at the H_1 receptor (108,110).

It was proposed that the dual activity of histamine at H_1 and H_2 receptors depends on distinct "essential conformations" of the histamine monocation, which can adopt two preferred conformations, namely, trans and gauche conformers (104,111).

Based on studies in a series of methyl-substituted histamines, the trans conformation with coplanarity of the ring, as well as the side-chain carbon and nitrogen atoms, was suggested to be essentially involved at some stages during the interaction at the H_1-receptor site (104). Various conformationally restricted or rigid analogs of histamine have been investigated for H_1-agonistic activity, but all these compounds showed only weak action in the range of 1% relative to histamine, frequently not achieving maximal response (104,109). Similar results, without correlation to the absolute configuration, were found in a series of chiral analogs of histamine and thiazolylethylamine [4] (109). None of all compounds currently known is more potent than histamine in H_1-receptor systems.

H_1-Receptor Antagonists

Histamine H_1 antagonists are widely used for the treatment of diverse allergic disorders (112,113) and under some special clinical conditions (114,115). If one takes into account some newer aspects of the development of antihistamines, the currently available compounds may be divided into three groups, although this may be considered an arbitrary act:

1. The early or classic antihistamines and related compounds
2. Drugs with combined H_1-antihistaminic and "antianaphylactic" activity
3. Newer more selective H_1 antagonists that are devoid of sedative side effects

The chemistry, structure-activity relationships, and pharmacology of H_1 antagonists are reviewed in numerous papers (104,112,116,117) [in particular the compounds of the first group (118-120)]. Therefore, the volume of this contribution is restricted to some fundamental aspects.

Most of the early antihistamines reveal a certain structural resemblance to histamine as they may be considered substituted ethylamines corresponding to the general formula in Figure 4.

In contrast to histamine, these agents are generally lipophilic compounds due to aryl, heteroaryl, or arylmethyl groups (Ar^1, Ar^2-$(CH_2)_n$; n = 0, 1) attached at a group or an atom X. Representative examples for chemical groups according to that formula are ethylenediamines, such as mepyramine [8], ethanolamines, such as diphenhydramine [9], and propylamines, such as chlorpheniramine [10], where X represents either N, CH–O, or CH (Table 2). Closely related are propenamines and butenamines (n = 1), such as triprolidine [11]

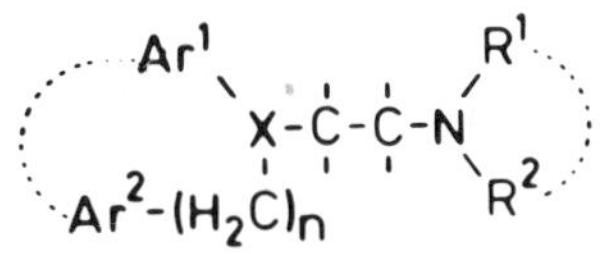

Figure 4 General formula of the early antihistamines.

Table 2 Structure and Antihistaminic Activity of Some Classic H_1 Antagonists at the Isolated Guinea Pig Ileum

No.	Compound	Structure	pA_2	Reference
8	Mepyramine	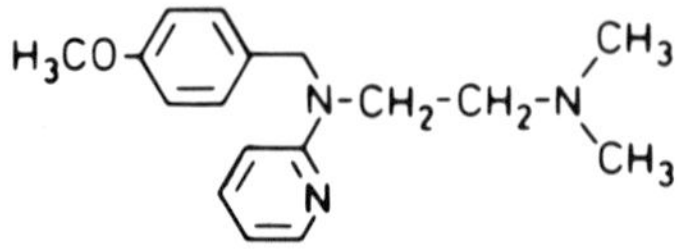	9.36 9.0	104 112
9	Diphenhydramine	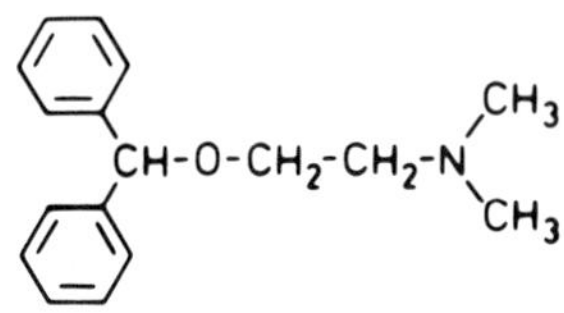	8.14 7.7-7.8	104 112
10	Chlorpheniramine	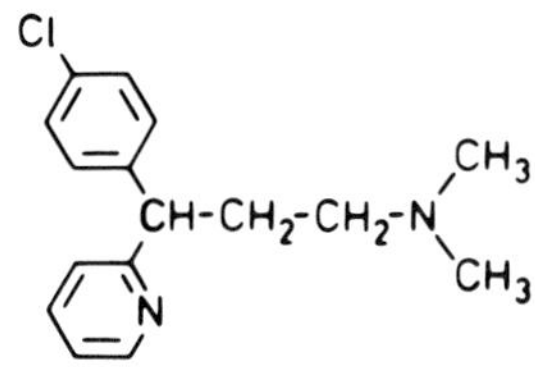	9.04	104
11	Triprolidine	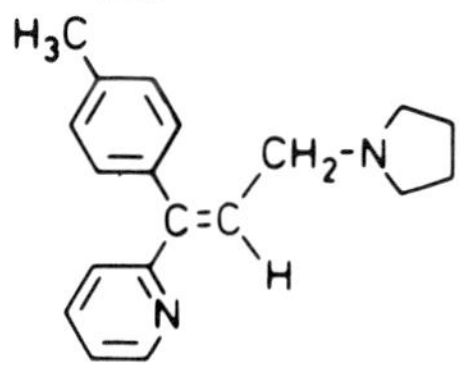	9.94	104
12	Promethazine	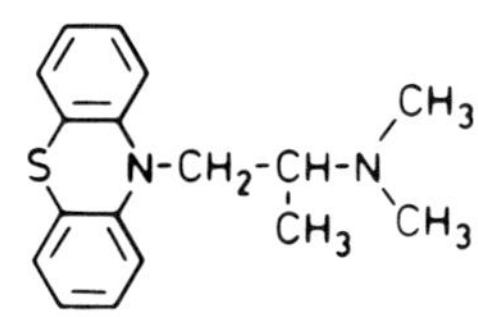	8.93	104
13	Clemastine	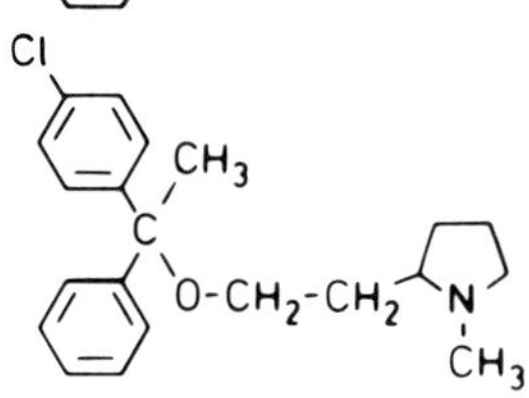	9.45 9.8	104 112
14	Dimetindene	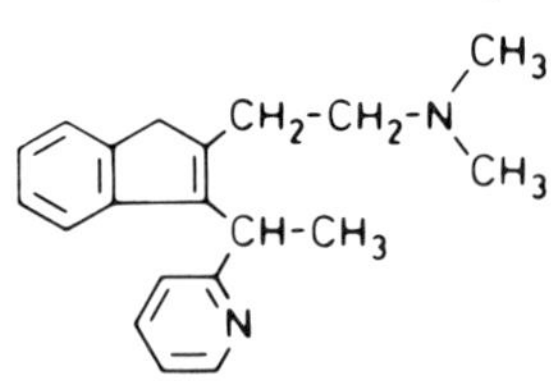	9.05	112

and pyrrobutamine, where X and the neighboring C atom form together a C=C double bond, and the piperazines, such as cyclizine, where X represents CH–N and the two nitrogen atoms are bridged to form a ring (104,119). Furthermore, Ar^1 and Ar^2 may be connected to form together with X a tricyclic ring system like phenothiazine, for example, promethazine [12]. Another common feature of the classic H_1 antagonists, apart from lipophilicity, is that they provide a strongly basic amino group, generally pyrrolidino or dimethylamino, which is almost quantitatively protonated under physiological conditions. The cationic form is considered the biologically active one, because the corresponding quaternary ammonium salts also act as H_1 antagonists. The ammonium side chain is thought to compete with the protonated amino group of histamine at the receptor, whereas the aromatic rings provide additional binding to accessory lipophilic regions (104). However, the application of QSAR analysis revealed no simple correlation between lipophilicity and antihistaminic activity (120,121). Steric effects of substituents also appear to be important, and probably only one of the aromatic rings contributes via hydrophobic binding. Generally, the antagonistic potency is substantially increased by introduction of substituents, preferably in the para position of one of the aromatic rings. For example, chlorpheniramine [10] is about one order of magnitude more active than the parent compound pheniramine. Some structures are considerably sensitive to stereochemistry, particularly if an asymmetric center is not close to the side-chain nitrogen but to the aromatic rings. For example, in the pheniramine series the S (+)-enantiomers are the more active. The antagonistic activity of triprolidine resides with the E isomer. As the fine detail of chemical structure may affect antihistaminic potency, it is evident that specific molecular interactions are involved in the binding of H_1 antagonists at the receptor.

Beyond the structural classes just mentioned, a number of H_1 antihistaminics are closely related (e.g., clemastine [13]), as well as compounds with low chemical resemblance to the general formula in Figure 4, for example dimetindene [14].

The mechanism of action of mepyramine and diphenhydramine at H_1 receptors was defined as competitive. Nevertheless, most of the antihistaminic drugs, including newer compounds, demonstrate a dual mode of action in different in vitro preparations, consisting of a competitive and a noncompetitive component (112). In some cases, the dose ratios for competitive and noncompetitive activities differ, by only a factor of 10 or less. For example, clemastine was found to have a pA_2 of 9.8 and a pD_2' of 8.6 at the guinea pig ileum. Dimetindene revealed pA_2-pD_2' values of 9.05;6.3 in the same preparation and 9.12:7.99 at the aorta of the rabbit (112). The early antihistamines and related compounds vary in potency and receptor specificity as well. Consequently, some side effects are more or less pronounced in comparison of the different chemical classes, owing to varying degrees of affinity to binding sites of other neurotransmitters, such as catecholamines, acetylcholine, dopamine, and serotonin.

Conversely, H_1-antagonistic activity is also found as a side effect within other groups of drugs, such as antipsychotics, tricyclic antidepressants, anticholinergics, and serotonin antagonists. Ketotifene [15] (122) has been developed from the latter and is chemically closely related to cyproheptadine and pizotifene. Compound [15] may be considered a representative example for a group of newer compounds that combine H_1-antihistaminic activity with antagonism of other mediators and/or a pronounced mast cell stabilizing effect. In anesthetized guinea pigs, intravenous injection of ketotifene antagonizes histamine-induced bronchoconstriction with a potency 20 times that of clemastine. Moreover, [15] is an antagonist of other constrictor agents, such as acetylcholine, serotonin, bradykinin, and slow-reacting substance, and also an inhibitor of the release of myotonic mediators. As a consequence of these properties, [15] may prevent bronchospasm produced by mediator release, leukotrienes in particular, and reduce bronchial hyperreactivity in asthmatic patients, whereas classic antihistamines are ineffective in the treatment of bronchial asthma (122). Azatadine (123), an analog of cyproheptadine, azelastine (124), and oxatomide [16] (125) are further examples of antihistamines with "antiallergic activity" (Table 3). However, it is difficult to estimate the relative contribution of the inhibition of mediator release versus antagonism of mediators to antiallergic properties and clinical efficacy. Likewise, these drugs also demonstrate antagonistic activity

Table 3 Structures of Some H_1 Antagonists with Additional "Antiallergic" Properties

No.	Compound	Structure
15	Ketotifene	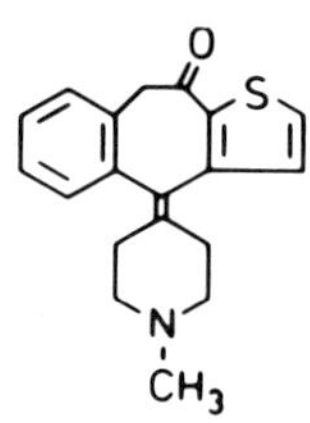
16	Oxatomide	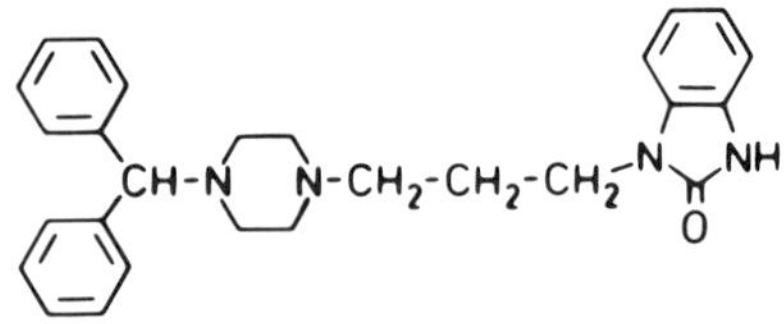

at other receptors. For example, azatadine binds relatively strongly at muscarinic receptors, somewhat less markedly at serotonin and very little at dopamine receptors, whereas oxatomide is reported to have relatively high affinity to serotonin and dopamine receptors but no activity at muscarinic receptors (116).

Sedation is the most common side effect of all the compounds just mentioned; this side effect is generally disadvantageous in the treatment of allergic disorders. In the recent past, several selective nonsedating antihistamines have been developed, such as terfenadine [17] (126,127) and astemizole [18] (128, 129), representing a worthwhile improvement in side-effect profile on traditional H_1 antagonists (Table 4). The lack of central effects in therapeutical dosage appears to be rather a consequence of pharmacokinetics than of different affinity to cerebral and peripheral H_1 receptors [see mequitazine (59)], as these compounds are not capable of penetrating the blood-brain barrier.

A further essential advantage of these peripherally acting antihistamines consists in the prevention of central interactions with other drugs. Terfenadine shows a slow onset of effect and ensuing slow dissociation from histamine receptors in the isolated guninea pig ileum. Astemizole dissociates slowly from H_1 receptors in vitro, too, and is an exceptionally long-acting compound in vivo. Compound [18] is a very potent selective H_1 antagonist, with negligible binding to serotonin receptors, and is devoid of affinity to muscarine and dopamine receptors.

Other newer H_1-receptor antagonists with a low incidence of sedative effects are acrivastine (BM 825C) (130), an acrylic acid derivative of triprolidine, loratadine (SCH 29851) (131), an analog of azatadine (AHR-11325) (132), and

Table 4 Structures of Some "Nonsedating" H_1 Antagonists

No.	Compound	Structure
17	Terfenadine	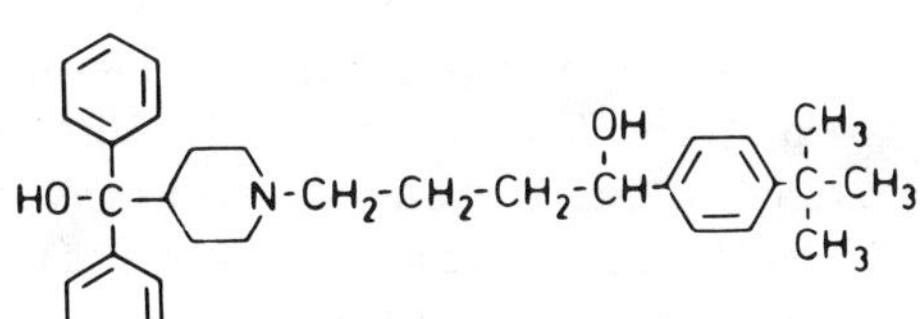
18	Astemizole	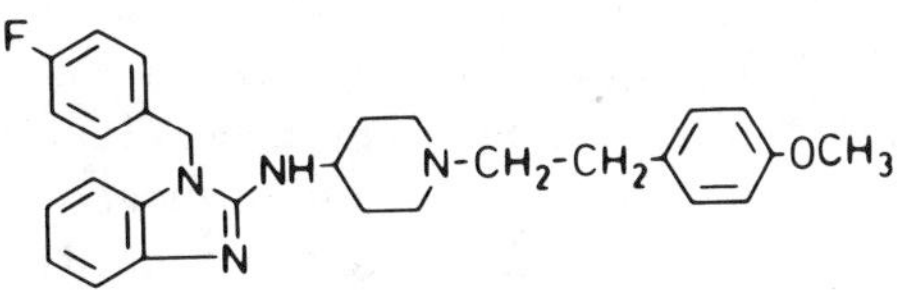

Table 5 Structure and Antihistaminic Activity of SK&F 93319 and Temelastine at the Isolated Guinea Pig Ileum and Atrium

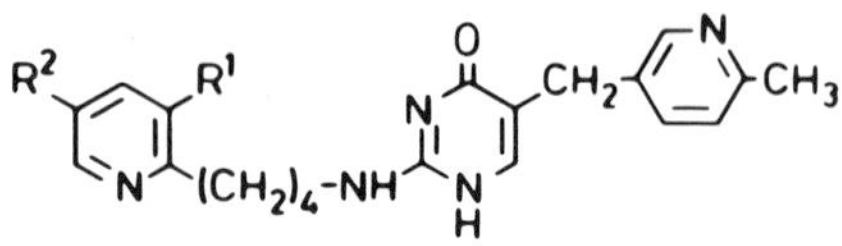

No.	Compound	R^1	R^2	pA_2(ileum)	pA_2(atrium)
20	SK&F 93319	OCH_3	H	7.77	7.49
19	Temelastine	CH_3	Br	9.55	∿ 5.9 (noncompetitive)

temelastine (SK&F 93944) [19] (117,133). The last agent is a potent H_1 antagonist with negligible penetration into the central nervous system (Table 5). The compound has been developed from a series of aminopyrimidones with combined H_2- and H_1-antagonistic activities, such as SK&F 93319 [20] (134).

The same pyrimidone moiety is also a structural part of some other potent H_2 antagonists, such as lupitidine. Replacement of the methoxy group of [20] by a methyl group and introduction of a bromine atom in the 5 position of the pyridine ring enhance the H_1-antagonistic activity about 100-fold, whereas the H_2-antagonistic potency is substantially decreased. Nevertheless, compounds with pharmacodynamic properties, like [20], combining H_1 and H_2 antagonism, may also find increasing interest, as blockade of both receptors plays an important role under some clinical conditions (114,115).

H_2-Receptor Agonists

Functional chemical requirements of histamine as an agonist at H_2 receptors are considered the protonated side-chain nitrogen of the monocation, the $N^{\tau}H$-tautomer of the imidazole ring, and the ability of the imidazole ring to undergo 1,3-prototropic tautomerism (104). Histamine is about one order of magnitude less active in such H_2-receptor systems as the isolated guinea pig atrium ($pD_2 \cong$ 6.0) than in such H_1 receptor systems as the isolated guinea pig ileum ($pD_2 \cong$ 6.9). The introduction of substituents at C-5 of the imidazole ring leads to selective H_2 agonists (13,109). 5-Methylhistamine [21] was found to have about 71% (135) of the activity of histamine at the guinea pig right atrium but only 0.9% (135) at the guinea pig ileum (Table 6).

Lengthening the alkyl substituent or introduction of electron-withdrawing groups leads to gradual decrease of the H_2-agonistic potency (104,109,135). Compounds with bulky substituents, like butyl, alicyclic, or aromatic groups, are inactive. An interesting attempt has been made to correlate substituent

Table 6 Structure and Activity of Some Selective H_2 Agonists at the Isolated Guinea Pig Atrium (H_2) and Ileum (H_1)

Structure	Compound		No.	Relative activity (histamine = 100)		Reference
				H_2	H_1	
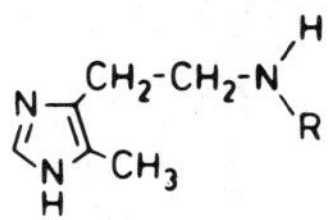	R = H		21	71 (43)	0.9 (0.23)	135,(13)
	R = CH_3		22	95 (36)	1.3 (0.16)	135,(103)
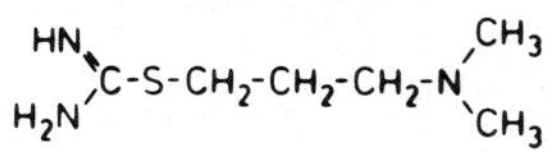		Dimaprit	23	71	<10^{-4}	103
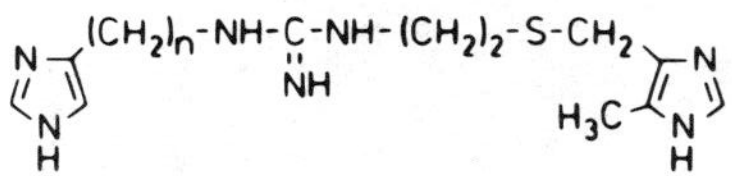	n = 3	Impromidine	24	4810	<10^{-3} (pA_2 = 5.47)	144,104
	n = 2		25	13 (80%max.)		109
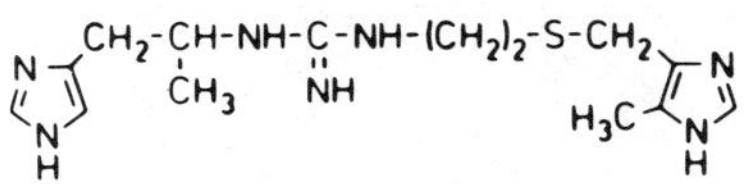	R-(−)	Sopromidine	26	741	[1.7] (20%max.)	146,109
	S-(+)		27	0 (pA_2 = 5.63)	0	146,109

parameters of the 5 substituents with "corrected" pD_2 values, that is, pD_2 values determined in vitro were converted with respect to the actual mole fraction of the "active" N^{τ}H-tautomer (136). The results indicate that both electronic and steric parameters of the 5 substituent affect H_2 affinity, analogously found in the imidazole series of H_2 antagonists (136). Comparable to 5-methylhistamine, N^{α}-5-dimethylhistamine [22] is also a selective H_2 agonist (see Table 6). Increasing the substituent at the side-chain nitrogen leads to reduction or loss of activity (135). The 1,3-prototropic tautomerism of the imidazole ring appears to be important at H_2 receptors, although there is some doubt owing to several unexplained results. For example, 2-(2-aminoethyl)-1-methylimidazole (N^{π}-methylisohistamine) (137) is a weak, slightly selective H_2 agonist (1.4% activity, relative to histamine) with an activity ratio of H_2 to H_1 of about 7:1, achieving 82% (atrium) and 87% (ileum) of the maximal response (138), whereas isohistamine (139), although able to tautomerize, is purely a weak H_1 agonist without any activity at H_2 receptors. It has been suggested that the histamine monocation may act statically in hydrogen bonding or dynamically as a proton-transfer agent at the receptor (104). Therefore attention was drawn to alternative 1,3-tautomeric amidine systems.

Dimaprit [23] (140,141) is a highly selective H_2 agonist with an aminoalkyl chain similar to that of N^{α},N^{α}-dimethylhistamine and an isothiourea group that may be considered a bioisostere of imidazole in this particular case (104).

The agonistic activity of dimaprit varies in different preparations. At the guinea pig atrium, 71%, and at the isolated rat uterus, 17.5% of the potency of histamine were found. Several investigations have been carried out to enhance the activity of dimaprit by structural modifications (104,142,143), such as replacing the dimethylamino group by other basic structures, quaternization, lengthening or shortening the carbon chain, substituting the amidine nitrogen, or synthesis of bis(isothiourea) analogs, but none of these efforts were successful. An important increase in activity was reached by the highly selective impromidine [24] (144). Impromidine consists of a homohistamine moiety [3-(4-imidazolyl)propylamine], which is integrated in a strongly basic guanidine system and additionally connected with a structural part known from the H_2-antagonist cimetidine.

Impromidine is about 48 times as potent as histamine in the isolated guinea pig atrium and approximately 17 times more active in stimulation of gastric acid secretion (rat). On the rat uterus preparation it is 9 times more potent than histamine but only achieves 80% of maximal response.

Although the unsubstituted 3-(4-imidazolyl)propylguanidine (SK&F 91486) is only a partial agonist at H_2 receptors (104) this structural part is probably responsible for the agonistic character of impromidine, whereas the cimetidine part contributes to receptor affinity by providing additional binding (104,145). The protonated guanidino group and the imidazole ring may have the same function as the side-chain ammonium group and the imidazole ring of the histamine monocation. At least two nitrogen atoms of the guanidine system appear to be required, namely, the unsubstituted one and that bearing the imidazolylpropyl group, as replacement by sulfur mutually leads to loss of agonistic activity (145). Replacement of the homohistamine part by the isomeric 2-(3-aminopropyl)imidazole yields an H_2 antagonist approximately as active as cimetidine (109). The alternative incorporation of histamine leads to a substantial decrease in activity ([25] Table 6).

Extensive investigations have been carried out to clarify the stereochemical requirements of histamine agonists (for a review, see Reference 109). Chiral histamine and impromidine analogs have been synthesized and tested in H_1- and H_2-receptor systems. In a series of histamine analogs with branched or ring-incorporated side chain, no degree of stereoselectivity was found at the guinea pig ileum and only small differences in activity were found at the guinea pig atrium. Replacement of the homohistamine group of impromidine by the side chain branched α-methylhistamine, however, resulted in an unique difference in the activity of the enantiomers (see Table 6). The R–(–)–enantiomer [26] sopromidine (109,146) has 7.4 times the activity of histamine at the atrium, but the S(+)-enantiomer [27] shows no efficacy but is an H_2 antagonist with about 24% of the activity of metiamide. Thus, sopromidine illustrates that a high degree of stereoselectivity is characteristic of the H_2 receptor. Sopromidine stimulates acid secretion (mouse) to the same extent as histamine. Moreover, sopromidine is a highly selective H_2 agonist, achieving only 20% of histamine's maximal response at the guinea pig ileum. Recent stereochemical modifications of impromidine and analogous compounds consist in chain branching of the "cimetidine moiety" (see Table 7). Introduction of a methyl substituent at the carbon atom α to the aromatic ring gave enantiomers, that is, [28,29], of substantially reduced H_2-agonistic activity (108). Alkyl branching of the ethylene group gave compounds with pronounced H_2-agonistic potency, [30-35] (148), in particular if a methyl group was introduced adjacent to the guanidine nitrogen. The R-configurated [32] was found to be more potent than impromidine and to exceed the activity of the S enantiomer [33] by sevenfold (148).

The "cimetidine moiety" of impromidine may be replaced in a wide range without loss of efficacy. Incorporation of structural parts known from other H_2-antagonist series, such as guanidinothiazoles and dimethylaminomethylfuranes, also leads to compounds with impromidinelike activity ([36-39], Table 8) (149-151). The alternative introduction of an H_2-unspecific phenyl ring leads only to a three to sixfold reduction in activity [41] (150). In contrast, replacement of the 5-methyl-4-imidazolyl group of impromidine by the isomeric 1-methyl-2-imidazolyl substituent [40], which is a structural part of the H_2 antagonist "isocimetidine" (152), leads to a considerable decrease in activity. Taken together, R should be an aromatic substituent to achieve acceptable H_2-agonistic potency (150). Aliphatic groups give much less active agonists.

Table 7 Structure and Agonistic Activity of Some Chiral Derivatives of Impromidine with Branched Methylthioethyl Chain at the Isolated Guinea Pig Atrium

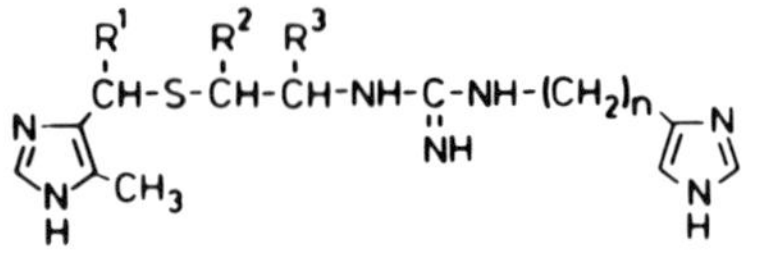

No.	n	R^1	R^2	R^3	Configuration	Relative activity Histamine = 100	Intrinsic activity Histamine = 1	Activity ratio R : S	Reference
28	2	CH_3	H	H	R	2	0.7		147
29	2	CH_3	H	H	S	3	0.9		
30	3	H	CH_3	H	R	1860	1	∿ 4:1	148
31	3	H	CH_3	H	S	430	1		
32	3	H	H	CH_3	R	10700	1	∿ 7:1	148
33	3	H	H	CH_3	S	1480	1		
34	3	H	H	C_2H_5	R	1660	1	∿ 3:1	148
35	3	H	H	C_2H_5	S	540	0.85		

The connecting chain may be devoid of a sulfur atom or replaced by an alicyclic ring (149). For example an analog of [39] with three-membered carbon chain is about 30 times as potent as histamine at the guinea pig atrium (151).

Selective histamine H_2 agonists have proved to be very useful pharmacological tools for the investigation of histamine receptors. A clinical use has not been established except for the application of betazole [2-(3-pyrazolyl)ethylamine], a very weak agonist with some selectivity toward H_2 receptors, as a diagnostic agent for testing gastric acid secretion capacity. Nevertheless, in the recent past, Baumann et al. (153-155) have found that treatment with H_2 agonists may be a new therapeutic approach to catecholamine-insensitive heart failure. The divergent potencies of certain H_2 agonists in different H_2-receptor systems, such as rat uterus, gastric acid secretion, and guinea pig atrium, although unexplained at present, suggest that it may be possible to separate therapeutically desired and disadvantageous effects (17,18).

H_2-Receptor Antagonists

In the well-documented search for a selective antagonist at "non-H_1 receptors," the structure of histamine served as a chemical starting point (see References 103, 104, 156, and 157). Consequently, the first H_2 antagonists developed were

Table 8 Structure and Agonistic Activity of Some Impromidine Analogs at the Isolated Guinea Pig Atrium

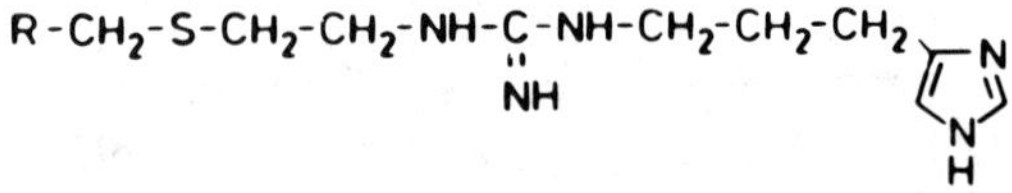

No.	R	Relative activity (histamine = 1)	Intrinsic activity (histamine = 1)	Reference
36	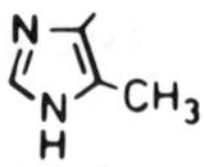 Impromidine	48	1	144
37	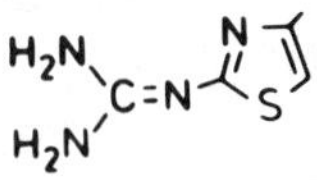	∿ 50	1	149,150
38	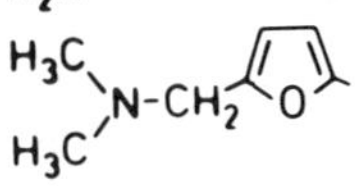	∿ 40	0.9	150
39	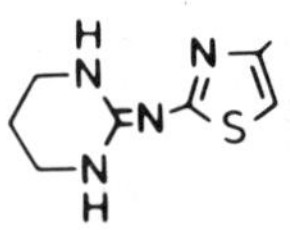	22	1	151
40	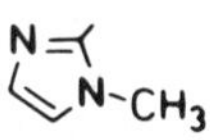	1.3	1	151
41	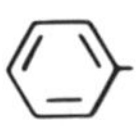	∿ 15	1	150

imidazole derivatives. Burimamide (13) was the first H_2 antagonist to be examined in humans, but it was not considered sufficiently potent for therapeutic use. Chemical modification of burimamide led to metiamide (14) and cimetidine (15,16); the latter was first marketed in 1976 for the treatment of conditions associated with gastric hyperacidity. The guanidinothiazole tiotidine (158) and the basically substituted furan derivative ranitidine (159) became known 3 years later as potent H_2-receptor antagonists. Thus, in contrast to previous views it was demonstrated that a basic imidazole ring is not required for H_2 affinity but can be successfully replaced by a dimethylaminomethylfuran or guanidinothiazole group. Whereas tiotidine had to be withdrawn owing to the discovery of stomach lesions in chronically dosed rats (160), ranitidine became the second compound launched onto the market. Afterwards, numerous chemically different H_2 antagonists were synthesized, bearing only little or no structural resemblance to the original chemical lead histamine. Despite considerable diversity, the actually known H_2 antagonists can be divided into two main structural classes (161):

1. Compounds containing a flexible connecting chain
2. Compounds with diaryl structures

H_2-Antagonists Containing a Flexible Chain

The common structural features of this group of compounds are an aromatic ring system, a flexible connecting chain, and a polar (uncharged at physiological pH) planar group (see Figure 5). As the polar moiety has been modified in a

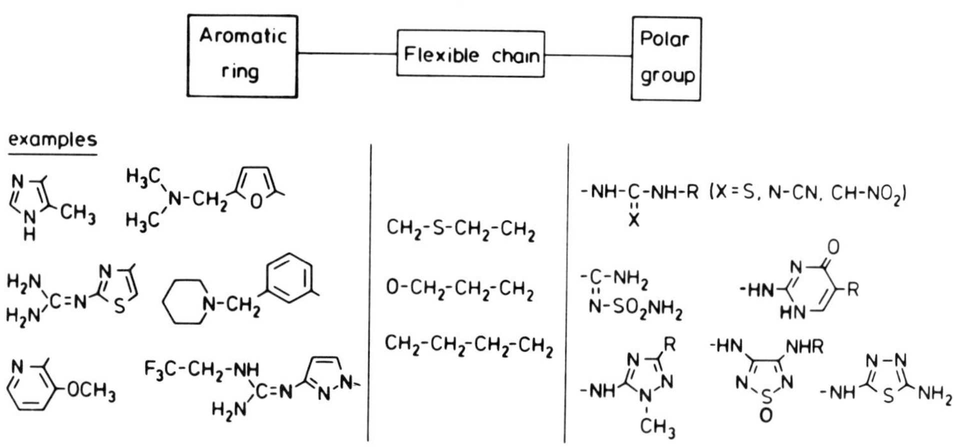

Figure 5 Structural features of H_2-receptor antagonists containing a flexible connecting chain.

large scope, the H_2 antagonists are usually divided into groups according to the aromatic rings. Four main series comprise the currently most important compounds, namely, the imidazole, aminomethylfuran, guanidinothiazole, and aminomethylphenoxy series. Metiamide [42] was the first H_2-antagonist clinically effective in inhibiting gastric acid secretion but had to be abandoned when several cases of granulocytopenia were uncovered (Table 9). Replacement of the thiourea by the bioisosteric cyanoguanidine group (162) led to cimetidine [43]. The cyanoguanidine group in cimetidine has been replaced by numerous alternative uncharged systems with varying lipophilicity and dipole moment orientation (163,164), such as 2-nitro-1,1-ethenediamine, 2-amino-3-nitropyrrol, or 2-amino-4-pyrimidone (isocytosine). The isocytosine system leads to compounds of enhanced H_2-antagonistic activity if the pyrimidine ring is substituted at the 5 position with lipophilic groups in particular (165,166), such as methylenedioxybenzyl substituent as in oxmetidine [44] (167).

Oxmetidine is a noncompetitive antagonist about 16 times as potent as cimetidine in vitro and 2-3 times as potent in inhibiting gastric acid secretion after oral administration. The compound was withdrawn owing to toxic side effects on the liver (168). Recently, a possible binding site model was proposed for H_2-receptor antagonists with substituents at C-5 of the isocytosine group based on calculations of the energies of interaction (169).

A number of attempts have been made to enhance the activity of cimetidine by structural modifications. Etintidine, an analog of cimetidine with a propargyl instead of a methyl substituent at the cyanoguanidine moiety, was found to be about two to three times more active than cimetidine in vitro (170). Modifications to the side chain, like chain branching or replacing the sulfur atom by alternative heteroatom linkages, substantially decrease H_2-antagonistic potency (171-175). Introduction of neutral or basic substituents at the imidazole ring 2 position gave compounds with considerably reduced H_2 affinity (e.g., [45,46], Table 9) (176-178). Histamine is thought to act at H_2 receptors in the "1,4-tautomeric" form of the monocation. The antagonists of the imidazole series should mimic the imidazole part of histamine and bind to the receptor in a similar manner. Therefore, in the development of metiamide and cimetidine, special attention was drawn on optimizing ring pK_a and increasing the percentage of the "active tautomer" in equilibrium. Retrospective structure-activity studies in a wider series of metiamide and histamine analogs revealed that receptor affinity may also be affected by electronic and steric effects of the ring substituents (136). A substituent of optimal size at the 5 position may induce a favorable conformation of the side chain. The pyridine-type nitrogen of cimetidine and related compounds appears to be more important for the H_2-receptor affinity than the 1,3-tautomeric system. Structural isomers of cimetidine (e.g., [48]) and analogous pyrimidones containing a 1-methyl-2-imidazolyl group instead of the normal imidazole moiety are about as potent as the respective reference

Table 9 Structure and Activity of Some H_2 Antagonists of the Imidazole Series (A) and Some Derivatives and Isomeric Analogs (B), at the Isolated Guinea Pig Atrium

A

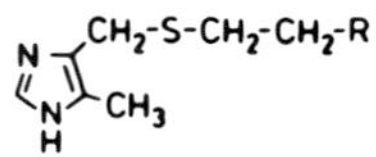

No.	Compound	R	pA_2	Ref.
42	Metiamide	NH-C(=S)-NHCH$_3$	6.04 6.25	14 177
43	Cimetidine	NH-C(=N-CN)-NHCH$_3$	6.10 6.40	15 220
44	Oxmetidine	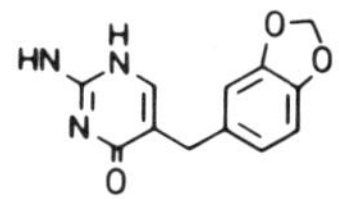	non-comp [7.1]	167 188

B

R-CH$_2$-S-CH$_2$-CH$_2$-NH-C(=X)-NHCH$_3$

No.	R	X	pA_2	Ref.
45	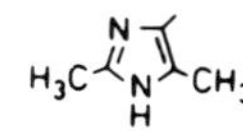	S	4.98	177
46	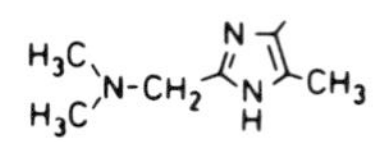	N-CN	5.35	178
47	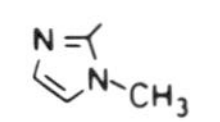	S	6.08	179
48		N-CN	6.3	152

compounds in vitro (152). The corresponding isomer, that is, [47], of metiamide was found to be approximately three times more active than the ring N-unsubstituted parent molecule in the isolated guinea pig atrium (179). Moreover, H_2-antagonistic activity is also found in other nitrogen-containing heterocyclic series (163,179), for example, isothiazoles, thiazoles, and pyridines, generally [except the oxazole derivative, CRC 1970 (180)] if the side chain is attached adjacent to the heterocyclic pyridine-type nitrogen. Substitution in the ortho position to the side chain may enhance H_2-antagonistic activity, not in the imidazole series alone; for example, SK&F 93319 ([20], Table 5) (134), a 3-methoxypyridine connected by a four-membered carbon chain with an isocytosine moiety, is about 25 times as potent as cimetidine in vitro.

Within the aminoalkylfuran series of H_2 antagonists (181,182), the optimal substitution pattern proved to be the 2,5-disubstituted heterocycle. Highest potency was found if the tertiary amine was a dimethylaminomethyl group. In contrast to the imidazole series, compounds with a nitroethenediamine moiety, like ranitidine [49] (159), are more active than the analogous cyanoguanidines (Table 10). Ranitidine is about 4-6 times more potent than cimetidine in vitro and 4-10 times more potent in vivo as an inhibitor of histamine-induced gastric acid secretion, depending on the route of administration (183), but is also a "short-acting" compound. Lupitidine [50] (184) is a very potent H_2-receptor antagonist with extended duration of action that contains the same 5-substituted pyrimidone moiety as [20]. The trials were suspended because lupitidine was found to induce mucosal hyperplasia and dysplasia in rats at chronic high dosage (185). Combination of the dimethylaminomethylfuran guanidinothiazole, or piperidinomethylphenoxy moiety with the 3,4-diamino-1,2,5-thiadiazole-1-oxide system gave potent H_2 antagonists with varying degrees of duration of action (186-188). BMY 25271 [51] is a short-acting antagonist 30-50 times more active than cimetidine in inhibiting gastric acid secretion in dogs after oral administration. Further development of the diaminothiadiazole oxide moiety yielded a series of 5-substituted 3,4-diaminoisothiazole-1,1-dioxides. L-643 728 [52] (189) demonstrates unsurmountable antagonism in vitro and was found to be orally somewhat less potent than ranitidine as inhibitor of gastric acid secretion. As shown by nizatidine [53] (190,191), the furan ring of ranitidine can be replaced by a 2,4-substituted thiazole ring. Nizatidine is reported to be 5-18 times as potent as cimetidine in vitro.

The guanidinothiazole moiety of tiotidine [54] (158) and related compounds is a group with exceptionally high H_2-receptor affinity and probably may not be considered equivalent to the dimethylaminomethylthiazole group (Table 11). Simple 4-methyl-2-guanidinothiazole was found to be an H_2-antagonist with even higher potency than burimamide in vitro (149). The guanidinothiazole group has been incorporated into many H_2 antagonists, such as pyrimidones (192), thiadiazole oxide derivatives (e.g., BL 6341 A, [55]),

Table 10 Structure and Activity of H_2 Antagonists of the Furan Series and Related Compounds at the Guinea Pig Atrium

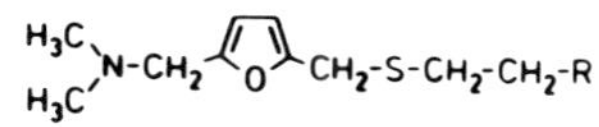

No.	Compound	R	pA_2	Ref.	No.	Compound	R	pA_2	Ref.
49	Ranitidine	$NH-C(=CH-NO_2)-NHCH_3$	7.2 6.8	159 188	51	BMY 25271	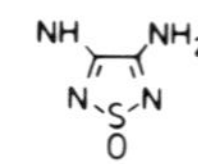	7.6 8.2 7.7	187 186 188
50	Lupitidine	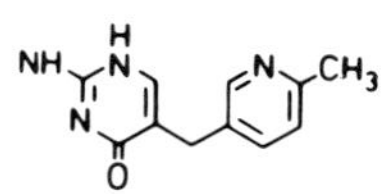	7.8	184	52	L-643728	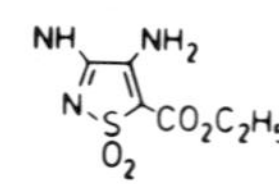	—	189
53	Nizatidine	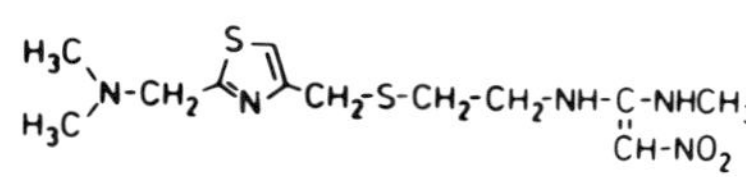						7.1 (rat uterus)	191

Table 11 Structure and Activity of H_2 Antagonists of the Guanidinothiazole Series and Related Drugs at the Guinea Pig Atrium

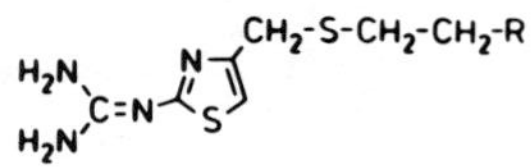

No.	Compound	R	pA_2	Ref.	No.	Compound	R	pA_2	Ref.
54	Tiotidine	NH-C(=N-CN)-NHCH$_3$	7.8	158	57		C(=N-CONH$_2$)-NH$_2$	7.1	195
55	BL 6341A	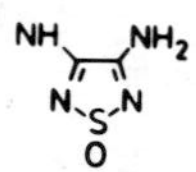	7.6 8.0 8.2	187 188 186	58	Famotidine	C(=N-SO$_2$NH$_2$)-NH$_2$	7.8 8.0	179 188
56	Tuvatidine (HUK 978)	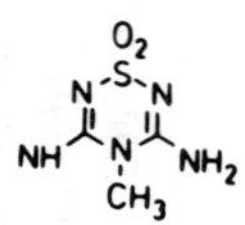		194	59	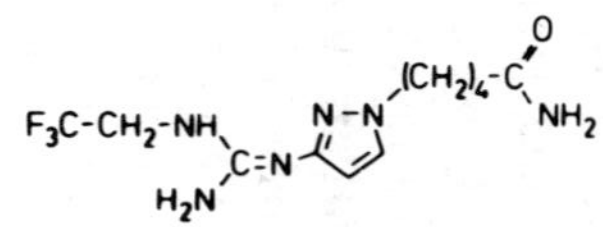 Grafitidine (ICI 162846)			198,179

1,2,4,6-thiatriazine-1,1-dioxides (193,194) (e.g., tuvatidine [56]), hydantoins (193), N-cyanamidines (195), N-carbamoylamidines (195), or sulfamoylamidines like famotidine [58] (196,197). Famotidine was launched onto the market in Japan in 1985. It is reported to be about 50-70 times as potent as cimetidine in the isolated guinea pig atrium and 40-50 times more potent as inhibitor of gastric acid secretion in dog and rat (196,197), respectively, with a 1.3 times longer duration of action. Recently, a trifluorethylguanidinopyrazole with a valeramide side chain, grafitidine (ICI 162 846, [59]) (179,198), became known as an H_2 antagonist. Compound [59] may be considered an analog to drugs of the guanidinothiazole series, where the polar group may also lack the usual chain NH function (see [57,58]).

The discovery of the aminomethylphenoxypropyl group as an alternative structural part with high affinity for H_2 receptors has considerably enlarged the scope of modifications to the polar moiety. Highest potency was generally found in compounds with piperidinomethyl substituent.

The triazole derivatives lamtidine [60] and loxtidine [61] are very potent, extremely long-acting H_2 antagonists (Table 12) (199-202). Lamtidine is reported to reduce gastric acid secretion in the dog more than 50% 24 hr after oral administration of 0.1 mg/kg. Compounds with thiadiazole oxide (L 643 441, [62]) (203) or squaramide moiety (e.g., BMY 25 368, [67]) (188) are described to have qualitatively similar pharmacological profiles. Numerous substances of this series produce unsurmountable antagonism in the isolated guinea pig right atrium and are difficult to remove from the tissue by washing (188,202, 203). Potency, tightness of binding at the H_2 receptor, and duration of action considerably vary depending on the structure of the basic meta-phenylene substituent and the polar moiety as well (188,204).

Unfortunately, it now seems that unsurmountable H_2-receptor blocking drugs with an extremely extended duration of action may induce mucosal changes. The trials with loxtidine were suspended after the discovery of gastric carcinoid tumors in the rat (205,206).

In the search for further structural alternatives of the "polar group," besides cyanoguanidines, nitroethenediamines, and others (207,208), many novel "urea equivalents" have been incorporated into antagonist molecules (179, 209), for example, thieno- and benzoisothiazole-1,1-dioxides (210,211) (e.g., WY-45087, [66]), isomeric thiadiazole diamines (188,212,213) [63,64] and oxadiazole diamines (214,215) [65], diaminopyrimidines (216) [68], or amides like the nicotineamide-N-oxide derivative SR-58042 (217) and TZU 0460 (roxatidine [69]) (218). The latter is reported to be a competitive antagonist about as potent as ranitidine in vitro and in vivo, with a somewhat longer duration of action. Although only one nitrogen atom of the polar moiety appears to be required, compounds [60-64] are very potent H_2 antagonists, supporting the idea that ring-incorporated guanidine or diamidine systems

Table 12 Structure and Activity of the H_2 Antagonists of the Piperidinomethylphenoxy Series at the Guinea Pig Atrium

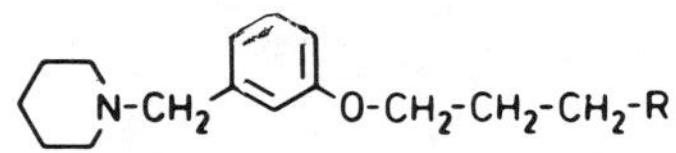

No.	Compound	R		pA_2	Ref.
60	Lamtidine	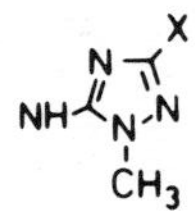	X = NH_2	–	200
61	Loxtidine		X = CH_2OH	–	201
62	L 643441	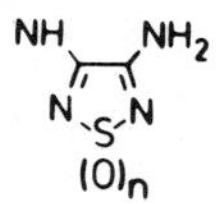	n = 1	[7.8]	186
63	BMY 25405		n = 0	8.5	188
64	TAS	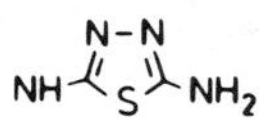		7.3 8.2	179 212
65		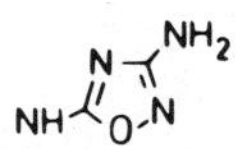		7.6	215
66	Wy 45086	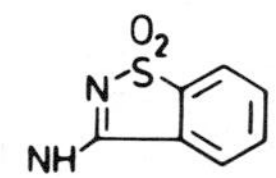		8.1	210
67	BMY 25368	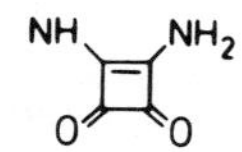		[7.9]	188
68		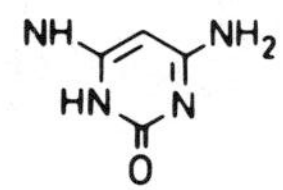		7.7	216
69	TZU 0460	NH-C(=O)-CH_2-O-C(=O)-CH_3		7.0	179

Table 13 Structure and H_2-Antagonistic Activity of Some N,N′-Substituted Thioureas, Cyanoguanidines, and Nitroethenediamines at the Guinea Pig Atrium

$$R^1\text{-}CH_2\text{-}S\text{-}CH_2\text{-}CH_2\text{-}NH\text{-}\underset{\underset{X}{\parallel}}{C}\text{-}NH\text{-}CH_2\text{-}CH_2\text{-}S\text{-}CH_2\text{-}R^2$$

No.	R^1	R^2	X	pA_2	Ref.
70	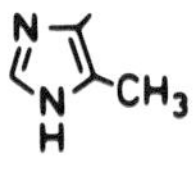	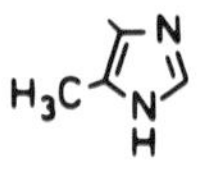	S	7.1	219
71			N-CN	7.1	219
72			$CH\text{-}NO_2$	6.6	219
73	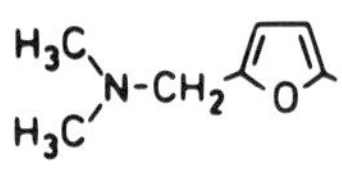	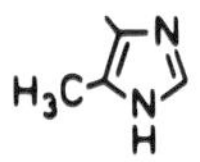	N-CN	7.4	220
74			$CH\text{-}NO_2$	8.3	220
75		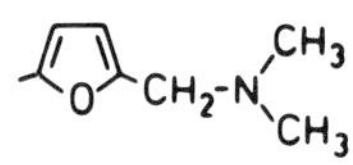	N-CN	7.3	220
76			$CH\text{-}NO_2$	8.4	220
77	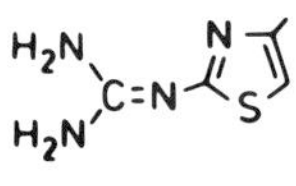	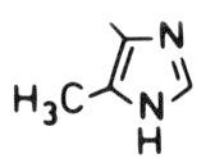	N-CN	8.1	192
78		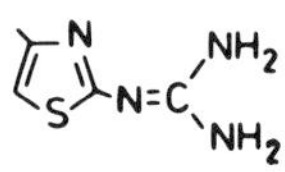	N-CN	8.4	192

with attenuated basicity particularly contribute high receptor affinity. Another approach to obtain H_2 antagonists of increased potency consists in doubling key portions of antagonist molecules of the single structural classes mentioned, for example, by synthesis of N,N′-substituted thioureas, cyanoguanidines, or nitroethenediamines (Table 13) (192,219,220). The potency-enhancing effect is dependent on the nature of both the aromatic rings and the polar groups. Thus, [76] is about 100 times as potent as cimetidine in vitro and approximately 60 times more active in inhibition of the histamine-induced acid secretion (rat) after IV application.

This approach has been extended to connect two heteroarylalkylguanidine residues by a carbon chain, obtaining symmetrical bis-guanidinoalkanes. Here an optimum was found in a compound with eight-membered carbon chain, and it was suggested that receptor binding sites may have closely repeating units (179).

H_2-Antagonists with Diaryl Structures

Several investigations have been carried out to replace the characteristic connecting chain of the early H_2 antagonists by conformationally restricted systems. The most potent compounds were found in the guanidinothiazole series. For example, the replacement of the "normal" methylthioethyl chain of tiotidine by a metaphenylene ring leads to an almost 10-fold increase in H_2-receptor affinity (ICI 127032, [79] (Table 14) (149). Stereochemical factors seem to be more important than electronic effects of the benzene ring, as the cis-1,3-cyclopentyl and cis-1,3-cyclohexyl analogs are potent H_2-blocking compounds, too ($pA_2 \cong 8.0$) (149). Analogous modifications in the furan or imidazole series proved to be unsuccessful. However, interestingly, H_2-antagonistic activity was found in a series of imidazolylphenylformamidines (221-223), although the amidine group is protonated under physiological conditions and, consequently, is different from the usually required neutral polar structures. In contrast to the phenylene analogs of tiotidine in the imidazolylphenylformamidine series, H_2-antagonistic activity resides with the para isomers. For example, mifentidine [80] (221) is about 30 times more potent than cimetidine in vitro. Crucial features responsible for activity have been identified: (1) the positively charged formamidino group, (2) an uncharged tautomeric imidazole ring, and (3) a critical distance between charged and the uncharged parts determined by a para-substituted phenyl ring (223). Considering the physicochemical properties

Table 14 Structure and H_2-Antagonistic Activity of Some Diaryl Derivatives at the Guinea Pig Atrium

No.	Compound	Structure	pA_2	Ref.
79	ICI 127 032	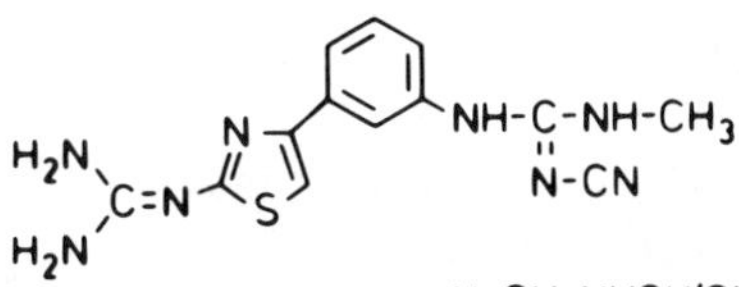	8.7	149
80	Mifentidine	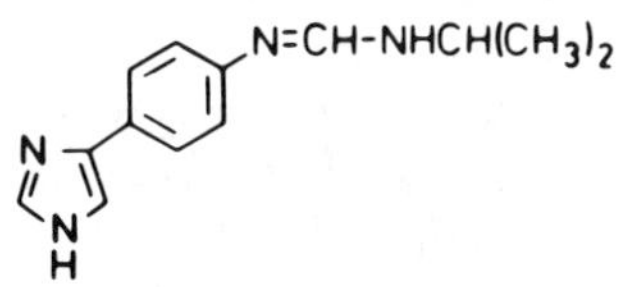	7.6	221
81		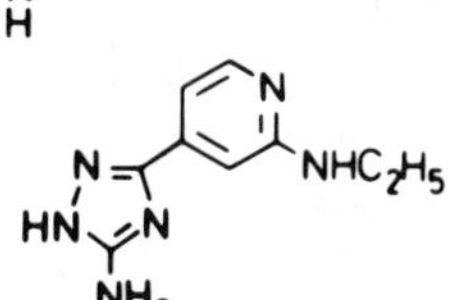	6.6	225

of mifentidine and the analogous guanidinothiazolylphenylformamidine derivative DA 4643 (224), both types of antagonists may interact with the receptor at the same site, binding at closely similar molecular structures.

Another type of diaryl structure with H_2-antagonistic properties has been developed by replacing the key portions of the histamine molecule by bioisosteres, namely, such pyridyltriazoles as [81] (225,226). Compound [81] is more potent than cimetidine in vitro. A QSAR study on a subset of analogs closely related to [81] showed that gastric acid antisecretory activity increased with decreasing lipophilicity (226).

H_3-Receptor Agonists and Antagonists

A variety of compounds, originally synthesized to interact with either H_1 or H_2 receptors, have been useful to start defining the chemical selectivity of H_3 receptors. Most of them were tested on the in vitro [^{3}H] histamine release model, but there is little doubt that receptors also controlling histamine synthesis in brain display a very similar pharmacology.

Regarding *agonists*, it appears that small modifications of the histamine molecule have even more dramatic effects on its recognition by H_3 receptors than by either H_1 or H_2 receptors. This can be regarded as consistent with the higher apparent potency of histamine as reflected by its lower EC_{50} value at H_3 receptors than at H_1 or H_2 receptors: the amine interaction with the H_3 receptor may involve additional bondings than with the two other receptor subclasses.

The integrity of the imidazole ring seems much more important for recognition by the H_3 than by H_1 or H_2 receptors, as shown by comparing various ring-methylated histamine derivatives. Indeed, histamine methylation into N^{τ}-methylhistamine, the metabolite formed in vivo under the action of histamine-*N*-methyltransferase, results in a complete loss of activity at the three receptor subclasses. However, whereas either 2-methylhistamine or 4-methylhistamine retains full agonist activity and gains selectivity at H_1 or H_2 receptors, only partially losing potency (13), both compounds are completely inactive at H_3 receptors with potencies relative to histamine of less than 0.08 and 0.008%, respectively (19).

Replacement of the imidazole ring with other heterocycles while keeping the ethylamine side chain, as in 2-thiazolylethylamine and 2-pyridylethylamine (or beta histine, its N-methyl derivative), leads to compounds retaining significant agonist activity at H_2 and particularly at H_1 receptors (104), whereas these modifications lead to a complete loss of agonist activity at H_3 receptors, some of these compounds, like betahistine, behaving as weak antagonists (19,227).

"Opening" the imidazole ring (as well as modifying the side chain) of histamine results in dimaprit, a compound selectively retaining full agonist activity at H_2 receptors (141), leads to a complete loss of agonist activity at H_3 receptors

Table 15 Comparison of Potencies of N^{α}-Substituted Histamine Derivatives on the Guinea Pig Ileum Contraction (H_1 Receptor), Atrial Rate (H_2 Receptor), and Inhibition of Rat Cortical [^3H] Histamine Release (H_3 Receptor)

Agents		Activity on biological systems		
		Ileum contraction (H_1-receptor)	Atrium rate (H_2-receptor)	Brain histamine release (H_3-receptor)
Histamine	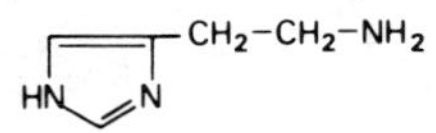	100	100	100
N^{α}-methyl-histamine	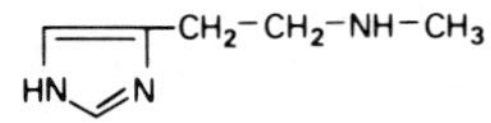	81	185	270
N^{α},N^{α}-dimethyl-histamine	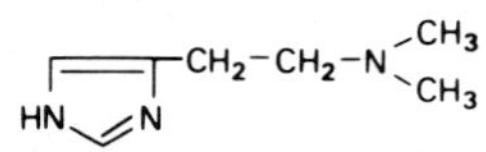	44	51	170
N^{α}-ethyl-histamine	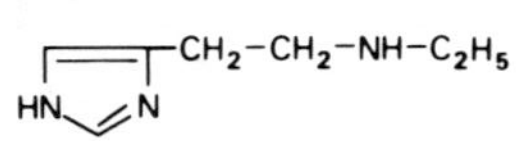	7	21	56
N^{α}-propyl-histamine	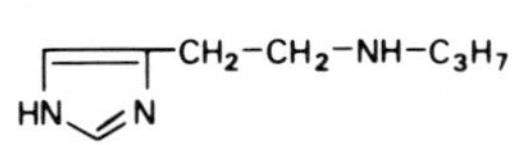	0.2	0	< 0.6
4-(2-piperidino-ethyl)-imidazole	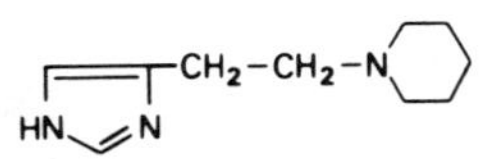			< 0.03

(19). It can be remarked that impromidine and sopromidine, two other selective and highly potent H_2-receptor agonists in which several structural modifications were introduced, also completely lose their agonist activity at the H_3 autoreceptor.

In contrast, modifications introduced at the level of the aminoethyl side chain appear to be rather well tolerated by the H_3 receptor, sometimes more

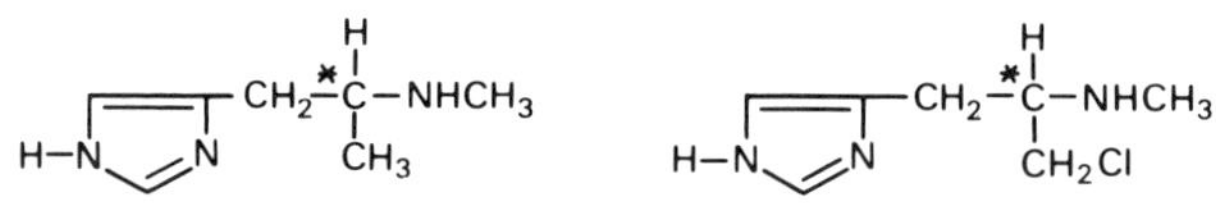

I = R(+)α, N$^{\alpha}$-DIMETHYL HISTAMINE

III = S(+) N$^{\alpha}$-METHYL-α-CHLOROMETHYLHISTAMINE

II = S(−)α, N$^{\alpha}$-DIMETHYL HISTAMINE

IV = R(−) N$^{\alpha}$-METHYL-α-CHLOROMETHYLHISTAMINE

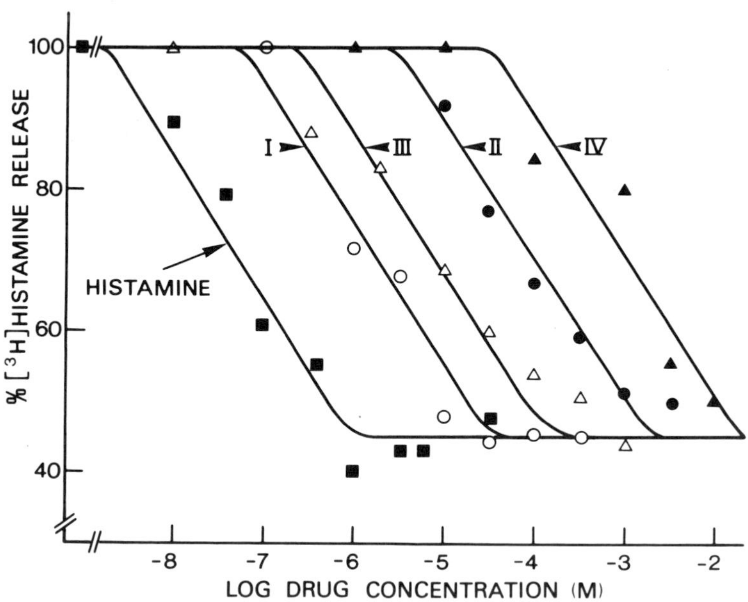

Figure 6 Effect of chiral histamine derivatives on the K^+-evoked release of [^{3}H]histamine from slices of rat cerebral cortex.

than by the H_1 or H_2 receptors. This is shown in the series of N^α-alkylhistamine derivatives in Table 15 and Figure 6. The inhibitory activity of histamine at the autoreceptor is fully mimicked by its N^α-methyl or N^α,N^α-dimethyl derivatives, which appear even more potent than histamine, whereas the reverse is true at H_1 or H_2 receptors. Extending the length of the N^α substituents progressively leads to a loss of agonist activity at the three receptor subclasses, although this is less marked in the case of H_3 receptors for which N^α-ethylhistamine still retains more than half its histamine potency. Modification of the side chain of N^α-methylhistamine in the α position by methyl or chloromethyl residues allows it to retain some agonist activity at H_3 receptors and introduces a chiral center, thus providing an interesting probe to assess the stereoselectivity of these receptors. Both series of compounds display a high difference in activity of the two enantiomers (two orders of magnitude), much more marked than and the inverse of that shown at H_2 receptors (228). In contrast, H_1 receptors are not differentially activated by the two enantiomers (229).

Regarding *antagonists*, structure-activity relationships can only be proposed at an initial stage. However, the importance of (an unsubstituted) imidazole ring for recognition by the H_3 autoreceptor seems again illustrated by several examples. The H_1-receptor antagonists tested, which lack an imidazole ring, were inactive, whereas several H_2-receptor antagonists, particularly those with an unsubstituted imidazole ring, were quite potent, sometimes more than at H_2 receptors. The best example is burimamide, the first H_2-receptor antagonist described (13), which was approximately 100-fold more potent at H_3 than at H_2 receptors. A similarly high antagonist potency ($K_i \cong 0.1$ μM) was found for impromidine, in which, interestingly, one of the two imidazole rings is unsubstituted. In contrast, metiamide and cimetidine, in which the imidazole ring is methylated, display a much lower affinity than burimamide, and replacement of the imidazole ring by other heterocyclic rings in ranitidine and tiotidine also leads to a loss of affinity toward H_3 autoreceptors.

It may be important to underscore that metiamide displays similar affinity constants at H_2 and H_3 receptors: since in some previous studies histamine-induced responses were often categorized as mediated by H_2 receptors simply on the basis of their blockade by this agent, this may merit reinvestigation with agents more clearly discriminating H_2 and H_3 receptors.

Finally, it should be mentioned that H_3 receptors have so far been evidenced only in brain, where they are not likely to be reached in vivo by either agonists or antagonists available at present (which all appear rather hydrophilic molecules). Since all these agents are also far from being selective, the development of new agents is required to precisely define the functions of histamine H_3 receptors.

FUNCTION AND REGULATION

The responses to histamine mediated by the three classes of receptors are, indeed, extremely different from one tissue to the other. It is outside the scope of the present chapter to describe these various responses, and for a detailed account the reader may refer to recent review volumes (230,231). However, it may be emphasized that similar biochemical mechanisms are presumably involved in the various responses mediated by each receptor subclass. Some light has recently been thrown on these mechanisms, particularly in the central nervous system.

Histamine H_1 Receptor and Calcium Translocation

Stimulation of H_1 receptors elicits a large variety of responses in numerous organs. These include contraction of smooth muscles in the bronchial, gastrointestinal, genitourinary, or vascular systems, exocrine gland secretion, glycogenolysis in brain, and excitatory responses of neurons in the peripheral and central nervous system. It may appear difficult to understand how such diverse final effects follow the stimulation of a single receptor subclass, but some common characteristics emerge when they are considered in detail. First, all these responses require intact cells to be observed, which is not the case for H_2-receptor-mediated responses. Second, all H_1-receptor-mediated responses seem to be linked in some manner or other with Ca^{2+} ion movements. This was clear for a long time regarding the contraction of smooth muscles elicited by histamine as well as other types of stimulant compounds. This was more recently evidenced for the glycogenolytic response to histamine in brain slices (232). This response is reduced in the absence of Ca^{2+}. Activation of phosphorylase is unlikely to be related to activation of Ca^{2+}-dependent protein kinase C since phosphorylase kinase is a poor substrate for protein kinase C (233). More probably, the calmodulin subunit of phosphorylase kinase is modulated by the Ca^{2+} ions released by inositol-1,4,5-triphosphate (234,235). Another well-characterized H_1-receptor-mediated effect in intact cells that is dependent upon Ca^{2+} availability is the stimulation of cyclic GMP formation, which can be observed in such diverse preparations as a cloned endothelial cell line (236), bovine tracheal strips (237), cultured neuroblastoma cells (238), and cervical ganglia (239).

Finally, in slices from guinea pig brain, histamine stimulates cyclic AMP accumulation via activation of both H_1 and H_2 receptors. Although the response mediated by H_2 receptors is not modified in the absence of Ca^{2+}, that involving H_1 receptors is reduced (240).

Taken togetheer, these various observations led to the proposal that all H_1-receptor-mediated responses involved either directly or indirectly a Ca^{2+} translocation process (241).

Histamine H_1 Receptors and the Phosphatidylinositol Cycle

Michell has advanced the hypothesis that receptor-mediated changes in membrane phosphatidylinositol turnover are linked in a causal manner with Ca^{2+} gating processes in a variety of biologic systems (242). It is now well established that activators of this turnover act by increasing the catalytic activity of phospholipase C present in the plasma membrane. This process generates two distinct intracellular signals (Figure 7). The first is inositol-1,4,5-triphosphate, which mobilizes Ca^{2+} from its intracellular stores, such as in the endoplasmic reticulum. The second is diacylglycerol, which activates protein kinase C and thereby promotes the phosphorylation of intracellular regulatory proteins.

The first indication that H_1 receptors were coupled with the phosphatidylinositol cycle came in 1975-1976 (243,244). In the brain an H_1-receptor-mediated activation of phospholipid turnover was first shown in 1982 (245).

More recently, a direct demonstration that the step catalyzed by stimulation of H_1 receptors in brain slices is the breakdown of phosphatidylinositol 4,5-biphosphate was recently provided. In slices from rat or guinea pig brain preincubated with [^{3}H] inositol in the presence of 5-10 mM Li^+ (to inhibit inositol 1-phosphatase), histamine markedly stimulates the accumulation of [^{3}H] inositol 1-phosphate (71,246,247). The response shows an absolute requirement of extracellular Ca^{2+} and is selectively mediated by H_1 receptors (248,249). In addition, the response amplitude in several brain regions seems related with H_1-receptor density and, in contrast with the glycogenolytic re-

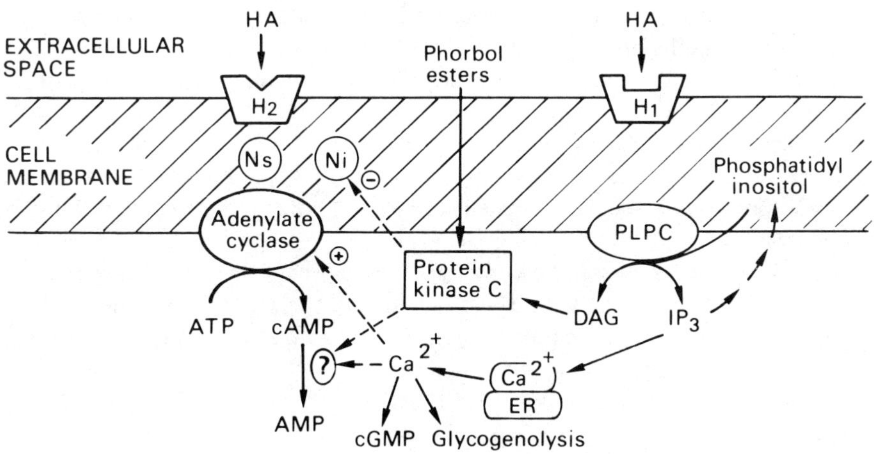

Figure 7 Intracellular responses mediated by H_1 and H_2 receptors in nervous tissues.

sponse, there is no obvious receptor reserve. However, some characteristics of the response, such as the effects of partial agonists differing in various areas of guinea pig brain, suggest it may be more complex than purely activation of the inositol cycle (250). Hence, although these various studies do not entirely clarify the relationship of calcium gating to inositol phospholipid metabolism, they lend further support to the proposition that all H_1-receptor-mediated responses are secondary to changes in intracellular concentrations of Ca^{2+}.

It is not yet clear whether all final physiological responses mediated by H_1 receptors involve the two intracellular signals generated by the phosphatidylinositol cycle. At least this appears to be the case for stimulation of cyclic AMP accumulation in brain slices: the potentiating effect of H_1-receptor stimulation is only partially mimicked by phorbol esters (251,252), which are known to activate protein kinase C in the same way as diacylglycerol. This suggests that inositol-1,4,5-triphosphate also participates. The precise level(s) at which the two intracellular signals generated by the phosphatidylinositol cycle interact with the system generating the cyclic AMP signal is (are) not known (see Figure 7).

Histamine H_2 Receptors and Adenylate Cyclase

Stimulation of H_2 receptors by histamine also elicits a large variety of biologic responses, depending on the tissue considered. The most widely studied because of its therapeutic importance is the proton secretory effect at the level of parietal cells from gastric glands (253). However, many other different effects mediated by H_2 receptors are well characterized in peripheral tissues (254-257). These include the positive chronotropic action in the auricular muscle, inotropic action in the ventricular muscle, the lipolytic effect in fat cells, the cytolytic activity of T lymphocytes, the inhibition of histamine release from basophils, the lysosomal enzyme release from neutrophils, and increased contractility of uterus.

In brain, several electrophysiological responses have been described. The best characterized of these consists in a potentiation of various excitatory signals resulting from a decrease of calcium-activated potassium conductance in the hippocampus (258).

Does a single biochemical mechanism underlie these various effects? The best candidate seems to be activation of adenylate cyclase, although the links between cyclic AMP and the physiological responses are not always clear and their existence remains debated in some cases.

Coupling of H_2 receptors with adenylate cyclase was first characterized on a cell-free preparation from guinea pig brain (259). The enzyme is activated by histamine with an EC_{50} value of 10 μM, which increases its maximal reaction velocity without altering the affinity for its substrate Mg-ATP. A similar response is found in membranes from the ventricular muscle or gastric mucosa. In addition, increases in cyclic AMP levels were demonstrated in a number of unbroken cell preparations from tissues in which physiological responses mediated by H_2 receptors had been previously described. These include fat cells, blood cells, uterus, and others (254).

It is widely believed that in many systems the link between cyclic AMP and physiological responses is constituted by phosphorylation mechanisms catalyzed by cyclic AMP-dependent protein kinases. However, this was apparently not so far demonstrated in the case of histamine. Furthermore, the role of the nucleotide in responses like acid secretion by gastric parietal cells is debated (260). Nevertheless, in many tissues the various physiological responses mediated by H_2 receptors are potentiated by phosphodiesterase inhibitors (which protect the intracellular nucleotide from degradation) and mimicked by cell-penetrating derivatives of the cyclic nucleotide. This strongly suggests that the latter is, indeed, the second messenger of histamine in these systems.

Histamine H_3 Receptors and Control of Histamine Synthesis and Release

The designation of receptors responsible for the control of histamine release in brain as H_3, that is, a novel subclass, stems from their pharmacology being distinct from that of the H_1 and H_2 subclasses. In fact, H_3 receptors also differ from the other subclasses by their localization and functions. In brain, histamine appears to act as a neurotransmitter for a small number of highly divergent neurons controlling states of sleep and wakefulness, energy metabolism, vascular permeability, and pituitary hormone secretion (20-24,261). Whereas the action of the amine on target cells is mediated by H_1 and H_2 receptors, various studies indicate that H_3 receptors are located on the histaminergic neurons themselves where they control both the biosynthesis and the release of the neurotransmitter.

In brain slices, the autoinhibitory action of histamine on its own release was evident in the presence of tetrodotoxin, an agent that blocks potential generation: this tends to exclude the participation of interneurons or nearby axon terminals possibly impinging on histaminergic terminals (262). In addition, the autoinhibitory effect was shown to persist in striatal slices from rats having previously received a local injection of kainate. This neurotoxin selectively ablates perikarya and thereby abolishes the action of histamine on target cells, whereas it does not affect histaminergic neurons themselves (263). Moreover, the autoinhibitory effect of histamine on its own release, mediated by pharmacologically identified H_3 receptors, was still evident on a synaptosomal preparation, that is, isolated nerve endings (262).

H_3 autoreceptors also appear to modulate histamine biosynthesis in histaminergic nerve terminals. It was shown 10 years ago that in brain slices, $[^3H]$-histamine formation from its $[^3H]$ amino acid precursor is significantly increased when the preparation is depolarized (30). This augmented synthesis is largely prevented in the presence of exogenous histamine acting through typical H_3 receptors (264). The mechanism(s) by which these two autoinhibitory effects occur are not entirely clarified but they could involve a modulation of Ca^{2+} ions influxes consequent to the depolarization. A distinct feature of H_3 receptors is their high sensitivity to histamine since the responses they mediate are triggered by the amine at concentrations much lower (by about two orders of magnitude) than those required to activate H_1 or H_2 postsynaptic receptors. This suggests that the "brake" that H_3 receptors impose onto histaminergic synapses in brain is a tonic one and therefore that H_3-receptor antagonism will, indeed, result in a facilitation of histaminergic transmission. Such an effect may, at least partially, explain some central effects of a drug such as betahistine, utilized in the treatment of Menière's syndrome, that has been shown to improve mental functions in elderly patients, perhaps through increases in regional cerebral blood flow.

Hitherto, histamine H_3 receptors were only evidenced on histaminergic neurons in brain, but this does not mean that this constitutes their sole localization, as other classes of autoreceptors such as the adrenergic α_2 receptors, are known to mediate not only presynaptic but also postsynaptic effects.

SPECULATIONS AND FUTURE DIRECTIONS

Drugs Acting at Peripheral Histamine Receptors

At present, therapeutically used drugs acting at histamine receptors are almost exclusively antagonists and there is no lack of potent compounds. Increasing receptor specificity and avoidance of central side effects are predominant objects in drug design of H_1-receptor antagonists. Nevertheless, it appears to be worthwhile to attempt to improve additional "antianaphylactic" properties of antihistamines, such as mast cell stabilizing effects, in order to increase their clinical efficacy in the prevention of severe allergic disorders. H_2 antagonists continue to play a dominant role in the treatment of peptic ulcers. A shift in emphasis has taken place, however, in therapy as well as in drug design, with regard to potency and duration of action, as total suppression of acid secretion appears to be not desirable.

Histamine agonists have been investigated in experimental pharmacology, predominantly to characterize histamine receptors and histamine antagonists; however, one is inclined to deny the therapeutic applicability owing to the deleterious pathophysiological effects of histamine. The currrently known H_1-receptor agonists are all weakly active compounds, and it will be a worthwhile and important task to develop highly active H_1 agonists. As has been demonstrated by H_2 agonists, potent, highly selective drugs are needed, not only for experimental use but also for new therapeutic approaches. H_2-receptor agonists may become a valuable tool for the treatment of heart failure as well as catecholamine-insensitive cardiomyopathy (153-155), and further indications are conceivable. Impromidine may not be considered as an optimal drug. There is some hope, however, that the incidence of disadvantageous effects, such as stimulation of heart rate and gastric acid secretion, as well as the arrhythmogenic effect, limiting the application of impromidine, may be considerably reduced by novel compounds (151), which are currently under investigation. Besides some "tissue selectivity," many of these new H_2 agonists, demonstrate a novel pharmacological profile as they combine high H_2-agonist activity with H_1-antagonistic potency in the range of commercially available antihistamines. The potential therapeutic value of such a dual mode of action and the optimal relationship of both effects have to be assessed.

Histamine Receptors in Brain

There are several reasons to believe that the three subclasses of histamine receptors in brain constitute important targets for new classes of CNS-acting drugs that are likely to be developed in the forthcoming years.

Histaminergic neurons are probably few in the brain (about 1000 seem present in rat brain), which suggests that modification of histaminergic neurotransmission may result in rather discrete, well-defined physiological consequences. Simple biologic assay systems for agonists and antagonists of histamine at the three receptor subclasses are available, and intensive chemical synthesis work in the field of histamine pharmacology during the last decade has allowed the definition of useful structure-activity relationships. In spite of this, a single class of centrally acting histaminergic agents, the sedative H_1 antihistamines, are currently used in therapeutics. Theoretically, new classes could be constituted by selective H_1, H_2, or H_3 agonists, as well as H_2 and H_3 antagonists able to cross the blood-brain barrier. Whereas H_1 and H_2 agonists seem likely to have important drawbacks, that is, the propensity to stimulate peripheral histamine receptors, for instance in smooth muscles or gastric glands, this may not be the case for the other drug classes.

It is still difficult to predict the effects of these various compounds in physiology and, even more, in therapeutics, but several hypotheses can be formulated. It is likely that facilitation of histaminergic neurotransmission (via stimulation of postsynaptic or blockade of presynaptic receptors) may have analeptic activity and promote energy metabolism in the CNS. Conversely, inhibition of histaminergic transmission in the CNS, via blockade of H_2 receptors or stimulation of H_3 receptors, may have sedating, anticonvulsant effects and diminish microvessel permeability. It is hoped that such new drugs will soon become available to assess the validity of these speculations.

REFERENCES

1. A. Windaus and W. Vogt, Ber. Dtsch. Chem. Ges. *40*, 3691 (1907).
2. G. Barger and H. H. Dale, J. Physiol. (Lond.) *40*, 38 (1910).
3. H. H. Dale and P. P. Laidlaw, J. Physiol. (Lond.) *41*, 318 (1910).
4. M. Rocha e Silva (ed.), Histamine and Anti-Histaminics, Handbook of Experimental Pharmacology, Vol. 18, Part 1, Springer-Verlag, Berlin, 1966.
5. M. Rocha e Silva (ed.), Histamine II and Anti-Histaminics, Handbook of Experimental Pharmacology, Vol. 18, Part 2, Springer-Verlag, Berlin, 1978.
6. G. Barger and H. H. Dale, J. Chem. Soc. *97*, 2592 (1910).
7. D. Ackermann, Hoppe-Seylers Z. Physiol. Chem. *65*, 504 (1910).
8. G. Barger and H. H. Dale, J. Physiol. (Lond.) *41*, 499 (1911).
9. J. J. Abel and S. Kubota, J. Pharmacol. Exp. Ther. *13*, 243 (1919).
10. D. Bovet and A.-M. Staub, C. R. Soc. Biol. (Paris) *124*, 547 (1937).
11. O. Arunlakshana and H. O. Schild, Br. J. Pharmacol. Chemother. *14*, 48 (1959).
12. A. S. F. Ash and H. O. Schild, Br. J. Pharmacol. Chemother. *27*, 427 (1966).
13. J. W. Black, W. A. M. Duncan, G. J. Durant, C. R. Ganellin, and M. E. Parsons, Nature (Lond.) *236*, 385 (1972).
14. J. W. Black, W. A. M. Duncan, J. C. Emmett, C. R. Ganellin, T. Hesselbo, M. E. Parsons, and J. H. Wyllie, Agents Actions *3*, 133 (1973).
15. R. W. Brimblecombe, W. A. M. Duncan, G. J. Durant, C. R. Ganellin, M. E. Parsons, and J. W. Black, Br. J. Pharmacol. *53*, 435P (1975).
16. R. W. Brimblecombe, W. A. M. Duncan, G. J. Durant, J. C. Emmett, C. R. Ganellin, and M. E. Parsons, J. Int. Med. Res. *3*, 86 (1975).
17. G. Bertaccini and G. Coruzzi, Ital. J. Gastroenterol. *15*, 51 (1983).
18. G. Bertaccini and G. Coruzzi, Pharmacology *23*, 1 (1981).
19. J. M. Arrang, M. Garbarg, and J.-C. Schwartz, Nature *302*, 832 (1983).
20. J.-C. Schwartz, Life Sci. *17*, 503 (1975).
21. J.-C. Schwartz, Annu. Rev. Pharmacol. Toxicol. *17*, 325 (1977).
22. J.-C. Schwartz, H. Pollard, and T. T. Quach, J. Neurochem. *35*, 26 (1980).
23. J.-C. Schwartz, M. Garbarg, and H. Pollard, in The Handbook of Physiology. The Nervous System IV (F. E. Bloom, W. B. Mountcastle and S. R. Geiger, eds.), American Physiological Society, Bethesda, 1986, p. 257.
24. L. B. Hough and J. P. Green, in Handbook of Neurochemistry (A. Lajtha, ed), Vol. 6. Plenum Press, New York, 1984, p. 145.
25. S. Z. Langer, Br. J. Pharmacol. *60*, 481 (1977).
26. K. Starke, Annu. Rev. Pharmacol. Toxicol. *21*, 7 (1981).
27. C. Atack and A. Carlsson, J. Pharm. Pharmacol. *24*, 990 (1972).
28. K. M. Taylor and S. H. Snyder, J. Neurochem. *21*, 1215 (1973).
29. N. Subramanian and A. H. Mulder, Eur. J. Pharmacol. *35*, 203 (1976).
30. M. Verdiere, C. Rose, and J.-C. Schwartz, Eur. J. Pharmacol. *34*, 157 (1975).
31. M. Baudry, M. P. Martres, and J.-C. Schwartz, J. Neurochem. *21*, 1301 (1973).
32. M. Garbarg, G. Barbin, J. Feger, and J.-C. Schwartz, Science *186*, 833 (1974).
33. G. Barbin, J. C. Hirsch, M. Garbarg, and J.-C. Schwartz, Brain Res. *92*, 170 (1975).

34. S. J. Hill, J. M. Young, and D. H. Marrian, Nature *270*, 361 (1977).
35. D. H. Marrian, S. J. Hill, J. K. M. Sanders, and J. M. Young, J. Pharm. Pharmacol. *30*, 660 (1978).
36. R. S. L. Chang, V. T. Tran, and S. H. Snyder, J. Pharmacol. Exp. Ther. *209*, 437 (1979).
37. S. J. Hill and J. M. Young, Mol. Pharmacol. *19*, 379 (1981).
38. H. Carswell and S. R. Nahorski, Eur. J. Pharmacol. *81*, 301 (1982).
39. N. Kondo, K. Taniyama, and C. Tanaka, Eur. J. Pharmacol. *114*, 89 (1985).
40. S. J. Hill, P. C. Emson, and J. M. Young, J. Neurochem. *31*, 997 (1978).
41. R. S. L. Chang, V. T. Tran, and S. H. Snyder, J. Neurochem. *32*, 1653 (1979).
42. S. J. Hill and J. M. Young, Br. J. Pharmacol. *68*, 687 (1980).
43. A. J. Hadfield, N. R. Robinson, and S. J. Hill, Biochem. Pharmacology *32*, 2449 (1983).
44. T. B. Casale, D. Rodbard, and M. Kaliner, Biochem. Pharmacol. *34*, 3285 (1985).
45. J. Aceves, S. Moriscal, K. E. Morrison, and J. M. Young, Br. J. Pharmacol. *84*, 417 (1985).
46. V. T. Tran, R. Lebovitz, L. Toll, and S. H. Snyder, Eur. J. Pharmacol. *70*, 501 (1981).
47. J. Figge, P. Leonard, and E. Richelson, Eur. J. Pharmacol. *58*, 479 (1979).
48. J. E. Taylor and E. Richelson, Eur. J. Pharmacol. *78*, 279 (1982).
49. L. Toll and S. H. Snyder, J. Biol. Chem. *257*, 13593 (1982).
50. S. Kanba and E. Richelson, Brain Res. *304*, 1 (1984).
51. J. E. Taylor and E. Richelson, Eur. J. Pharmacol. *67*, 41 (1980).
52. S. J. Peroutka and S. H. Snyder, J. Pharmacol. Exp. Ther. *216*, 142 (1981).
53. R. M. Pinder and A. M. L. van Delft, Br. J. Clin. Pharmacol. *15*, 2695 (1983).
54. A. M. Johnson, D. R. Nelson, J. M. Nicholass, P. E. Richards, and D. R. Thomas, Br. J. Pharmacol. *80*, 651P (1983).
55. M. Korner, M. L. Bouthenet, C. R. Ganellin, M. Garbarg, C. Gros, R. J. Ife, N. Sales, and J.-C. Schwartz, Eur. J. Pharmacol. *120*, 151 (1986).
56. P. M. Laduron, P. F. M. Janssen, W. Gommeren, and J. E. Leysen, Mol. Pharmacol. *21*, 294 (1982).
57. J. M. Treherne and J. M. Young, Br. J. Pharmacol. *87*, 72P (1986).
58. J.-M. Maloteaux, A. Gossuin, C. Waterkeyn, and P.-M. Laudron, Biochem. Pharmacol. *32*, 2543 (1983).
59. A. Uzan, G. Le Fur, and C. Malgouris, J. Pharm. Pharmacol. *31*, 701 (1979).
60. G. Le Fur, C. Malgouris, and A. Uzan, Life Sci. *29*, 547 (1981).
61. R. M. Wallace, Ph.D. Thesis, University of Cambridge, 1983.
62. R. M. Wallace and J. M. Young, Mol. Pharmacol. *23*, 60 (1983).
63. T. T. Quach, A. M. Duchemin, C. Rose, and J.-C. Schwartz, Eur. J. Pharmacol. *60*, 391 (1979).
64. T. T. Quach, A. M. Duchemin, C. Rose, and J.-C. Schwartz, Neurosci. Lett. *17*, 49 (1980).
65. D. Diffley, V. T. Tran, and S. H. Snyder, Eur. J. Pharmacol. *64*, 177 (1980).

66. S. J. Peroutka, M. A. Moskowitz, J. F. Reinhard, and S. H. Snyder, Science *208*, 610 (1980).
67. A. J. Culvenor and B. Jarrott, Neurochem. Int. *3*, 365 (1981).
68. B. Bielkiewicz and D. A. Cook, Can. J. Physiol. Pharmacol. *63*, 756 (1985).
69. V. T. Tran, R. S. L. Chang, and S. H. Snyder, Proc. Natl. Acad. Sci. USA *75*, 6290 (1978).
70. J. E. Taylor, T. L. Yaksh, and E. Richelson, Brain Res. *243*, 391 (1982).
71. P. R. Daum, C. P. Downes, and J. M. Young, Eur. J. Pharmacol. *87*, 497 (1983).
72. J. Marwaha, B. J. Hoffer, and F. Freedman, Exp. Neurol. *74*, 285 (1981).
73. W. S. Young, III, and M. J. Kuhar, Brain Res. *179*, 255 (1979).
74. J. M. Palacios, W. S. Young, III, and M. J. Kuhar, Eur. J. Pharmacol. *58*, 295 (1979).
75. J. M. Palacios, J. K. Wamsley, and M. J. Kuhar, Brain Res. *214*, 155 (1981).
76. J. M. Palacios, J. K. Wamsley, and M. J. Kuhar, Neuroscience *6*, 15 (1981).
77. R. S. L. Chang, V. T. Tran, and S. H. Snyder, Brain Res. *190*, 95 (1980).
78. M. Ninkovic, S. P. Hunt, and J. R. W. Gleave, Brain Res. *241*, 197 (1982).
79. V. S. Seybold, Brain Res. *342*, 291 (1985).
80. F. Lembeck, Trends Neurochem. Sci. *6*, 106 (1983).
81. M. Ninkovic and S. P. Hunt, Neurosci. Lett. *53*, 133 (1985).
82. D. Rodbard, H. J. Ruder, J. Vaitukaitis, and H. S. Jacobs, J. Clin. Endocrinol. Metab. *33*, 343 (1971).
83. R. S. L. Chang and S. H. Snyder, J. Neurochem. *34*, 916 (1980).
84. S. Cockcroft and B. D. Gomperts, Nature *314*, 534 (1985).
85. E. Yeramian, M. Garbarg, and J.-C. Schwartz, Mol. Pharmacol. *28*, 155 (1985).
86. M. Gavish, R. S. L. Chang, and S. H. Snyder, Life Sci. *25*, 783 (1979).
87. B. Chan and B. K. Madras, Eur. J. Pharmacol. *83*, 1 (1982).
88. M. Garbarg, E. Yeramian, M. Korner, and J.-C. Schwartz, in Frontiers in Histamine Research (C. R. Ganellin and J.-C. Schwartz, eds.), Pergamon Press, Oxford, 1985, p. 19.
89. T. Kuno, N. Kubo, and C. Tanaka, Biochem. Biophys. Res. Commun. *129*, 639 (1985).
90. I. R. Smith, M. T. Cleverley, C. R. Ganellin, and K. M. Metters, Agents Actions *10*, 422 (1980).
91. D. R. Bristow, J. R. More, J. R. Mearn, and L. E. Martin, Br. J. Pharmacol. *72*, 548P (1980).
92. T. J. Rising and D. B. Norris, in Frontiers in Histamine Research (C. R. Ganellin and J.-C. Schwartz, eds.), Pergamon Press, Oxford, 1985. p. 61.
93. G. A. Gajtkowski, D. B. Norris, T. J. Rising, and T. P. Wood, Nature *304*, 65 (1983).
94. J. W. Wells, D. L. Cybulsky, M. Kandel, S. I. Kandel, and G. H. Steinberg, in Frontiers in Histamine Research (C. R. Ganellin and J.-C. Schwartz, eds.), Pergamon Press, Oxford, 1985, p. 69.
95. J. C. Foreman, D. B. Norris, T. J. Rising, and S. E. Webber, Br. J. Pharmacol. *86*, 475 (1985).

96. C. L. Johnson and D. G. Sawutz, in Frontiers in Histamine Research (C. R. Ganellin and J.-C. Schwartz, eds.), Pergamon Press, Oxford, 1985, p. 79.
97. D. B. Norris, G. A. Gajtkowski, and T. J. Rising, Agents Actions *14*, 543 (1984).
98. G. H. Steinberg, M. Kandel, S. I. Kandel, and J. W. Wells, Biochemistry *24*, 6107 (1985).
99. G. H. Steinberg, M. Kandel, S. I. Kandel, and J. W. Wells, Biochemistry *24*, 6115 (1985).
100. G. H. Steinberg, J. G. Eppel, M. Kandel, S. I. Kandel, and J. W. Wells, Biochemistry *24*, 6095 (1985).
101. G. Barbin, J. M. Palacios, E. Rodergas, J.-C. Schwartz, and M. Garbarg, Mol. Pharmacol. *18*, 1 (1980).
102. D. L. Cybulski, S. I. Kandel, J. W. Wells, and A. G. Gornall, Eur. J. Pharmacol. *72*, 407 (1981).
103. C. R. Ganellin, in Histamine Receptors (T. O. Yellin, ed.), Spectrum Publications, New York, 1979, p. 377.
104. C. R. Ganellin, in Pharmacology of Histamine Receptors (C. R. Ganellin and M. E. Parsons, eds.), Wright PSG, London, 1982, p. 10.
105. J. W. Black and C. R. Ganellin, Experientia *30*, 111 (1974).
106. G. J. Durant, C. R. Ganellin, and M. E. Parsons, J. Med. Chem. *18*, 905 (1975).
107. T. J. Wilmot and G. N. Menon, J. Laryngol. Otol. *90*, 833 (1976).
108. M. Hepp, P. Dziuron, and W. Schunack, Arch. Pharm. (Weinheim) *312*, 637 (1979).
109. W. Schunack, A. Buschauer, S. Büyüktimkin, P. Dziuron, S. Elz, G. Gerhard, E. Lebenstedt, H.-G. Lennartz, S. Schwarz, M. Spitzhoff, and R. Steffens, in VIIIth International Symposium on Medicinal Chemistry Vol. 2 (R. Dahlbom and J. L. G. Nilsson, eds.), Swedish Pharmaceutical Press, Stockholm, 1985, p. 169.
110. P. Dziuron and W. Schunack, Eur. J. Med. Chem. Chim. Ther. *10*, 129 (1975).
111. L. B. Kier, J. Med. Chem. *11*, 441 (1968).
112. D. Reinhardt and U. Borchardt, Klin. Wochenschr. *60*, 983 (1982).
113. J. H. Musser, A. F. Kreft, and A. J. Lewis, Annu. Rep. Med. Chem. *20*, 71 (1985).
114. W. Lorenz and A. Doenicke, N. Engl. Reg. Allergy Proc. *6*, 37 (1985).
115. W. Lorenz and A. Doenicke, N. Engl. Reg. Allergy Proc. *6*, 174 (1985).
116. G. Vanden Bussche, Pharm. Weekbl. [Sci.] *118*, 495 (1983).
117. G. J. Durant, C. R. Ganellin, R. Griffiths, C. A. Harvey, D. A. A. Owen, and G. S. Sach, in Frontiers in Histamine Research (C. R. Ganellin and J.-C. Schwartz, eds.), Pergamon Press, Oxford, 1985, p. 3.
118. A. F. Harms, W. Hespe, W. T. Nauta, R. F. Rekker, H. Timmerman, and J. de Vries, in Drug Design, Vol. VI (E. J. Ariens, ed.), Academic Press, New York, 1975, p. 1.

119. A. F. Casy, in Handbook of Experimental Pharmacology, Vol. 18, Part 2, (M. Rocha e Silva, ed.), Springer-Verlag, Berlin, 1978, p. 175.
120. W. T. Nauta and R. F. Rekker, in Handbook of Experimental Pharmacology, Vol. 18, Part 2 (M. Rocha e Silva, ed.), Springer-Verlag, Berlin, 1978, p. 215.
121. R. F. Rekker, in Strategy in Drug Research (J. A. Keverling Buisman, ed.), Elsevier, Amsterdam, 1982, p. 315.
122. L. P. Craps and U. M. Ney, Respiration *45*, 411 (1984).
123. S. Tozzi, F. E. Roth, and I. L. A. Tabachnick, Agents Actions *4*, 264 (1974).
124. K. Tasaka and M. Akagi, Arzneimittelforsch. *29*, 488 (1979).
125. D. M. Richards, R. N. Brogden, R. C. Heel, T. M. Speight, and G. S. Avery, Drugs *27*, 210 (1984).
126. E. M. Sorkin and R. C. Heel, Drugs *29*, 34 (1985).
127. C. Rose, T. T. Quach, C. Llorens, and J.-C. Schwartz, Arzneimittelforsch. *32*, 1171 (1982).
128. D. M. Richards, R. N. Brogden, R.C. Heel, T. M. Speight, and G. S. Avery, Drugs *28*, 38 (1984).
129. F. Janssens, J. Torremans, M. Janssen, R. A. Stokbroekx, M. Luyckx, and P. A. J. Janssen, J. Med. Chem. *28*, 1943 (1985).
130. A. F. Cohen, M. Hamilton, C. Burke, J. Findlay, and A. W. Peck, Br. J. Clin. Pharmacol. *17*, 647P (1984).
131. A. Barnett, L. C. Iorio, W. Kreutner, S. Tozzi, H. S. Ahn, and A. Gulbenkian, Agents Actions *14*, 590 (1984).
132. C. A. Leonard, C. B. Jackson, D. J. Stephens, A. G. Proakis, and R. S. Alphin, Pharmacologist *26*, 221 (1984).
133. G. J. Durant, C. R. Ganellin. R. Griffiths, C. A. Harvey, R. J. Ife, D. A. A. Owen, M. E. Parsons, and G. S. Sach, Br. J. Pharmacol. *82*, 232P (1984).
134. R. C. Blakemore, T. H. Brown, D. G. Cooper, G. J. Durant, C. R. Ganellin, R. J. Ife, M. E. Parsons, A. C. Rasmussen, and G. S. Sach, Br. J. Pharmacol. *80*, 437P (1983).
135. H.-G. Lennartz, M. Hepp, and W. Schunack, Eur. J. Med. Chem. Chim. Ther. *13*, 229 (1978).
136. G. J. Durant, T. H. Brown, J. C. Emmett, C. R. Ganellin, H. D. Prain, and R. C. Young, in The Chemical Regulation of Biological Mechanisms (A. M. Creighton and S. Turner, eds.), The Royal Society of Chemistry, Special Publication No. 42, London, 1982, p. 27.
137. A. Buschauer and W. Schunack, Arch. Pharm. (Weinheim) *316*, 891 (1983).
138. A. Buschauer, 2-Substituierte Imidazole als Bausteine neuer H_2-Antagonisten, Ph.D. thesis, Mainz, 1984.
139. A. Buschauer, K. Wegner, and W. Schunack, Arch. Pharm. (Weinheim) *317*, 9 (1984).
140. M. E. Parsons, D. A. A. Owen, C. R. Ganellin, and G. J. Durant, Agents Actions 7, 31 (1977).

141. G. J. Durant, C. R. Ganellin, and M. E. Parsons, Agents Actions *7*, 39 (1977).
142. G. J. Sterk, H. van der Goot, and H. Timmerman, 14th meeting of the European Histamine Research Society, Aachen, 1985, abstract 3.1.
143. G. J. Sterk, H. van der Goot, and H. Timmerman, Eur. J. Med. Chem. Chim. Ther. *19*, 545 (1984).
144. G. J. Durant, W. A. M. Duncan, C. R. Ganellin, M. E. Parsons, R. C. Blakemore, and A. C. Rasmussen, Nature *276*, 403 (1978).
145. G. J. Durant, C. R. Ganellin, D. W. Hills, P. D. Miles, M. E. Parsons, E. S. Pepper, and G. R. White, J. Med. Chem. *28*, 1414 (1985).
146. W. Schunack and G. Gerhard, Naunyn-Schmiedebergs Arch. Pharmacol. *319*, R56 (1982).
147. S. Büyüktimkin and W. Schunack, Pharm. Z. Sci. Ed. *128*, 1239 (1983).
148. S. Elz, Ph.D. thesis, University of Mainz, 1986.
149. D. J. Gilman, D. F. Jones, K. Oldham, J. M. Wardleworth, and T. O. Yellin in The Chemical Regulation of Biological Mechanisms (A. M. Creighton and S. Turner, eds.), The Royal Society of Chemistry, Special Publication No. 42, London, 1982, p. 58.
150. G. J. Sterk, H. van der Goot, and H. Timmerman, Frontiers in Histamine Research International Symposium, Jouy-en-Josas (near Paris), France, July 25-27, 1984, abstract 72.
151. A. Buschauer, H. Schickaneder, W. Schunack, S. Elz, I. Szelenyi, G. Baumann, and K.-H. Ahrens, European Patent Appl. EP 0 199845 (1986).
152. A. Buschauer and W. Schunack, Arzneimittelforsch. *35*, 1223 (1985).
153. G. Baumann, S. B. Felix, G. Riess, U. Loher, L. Ludwig, and H. Blömer, J. Cardiovasc. Pharmacol. *4*, 542 (1982).
154. G. Baumann, D. Mercader, U. Busch, S. B. Felix, U. Loher, L. Ludwig, H. Sebening, C. D. Heidecke, S. Hagl, F. Sebening, and H. Blömer, J. Cardiovasc. Pharmacol. *5*, 618 (1983).
155. G. Baumann, B. Permanetter, and A. Wirtzfeld, Pharmacol. Ther. *24*, 165 (1984).
156. G. J. Durant, in Recent Advances in Receptor Chemistry (F. Gualtieri, M. Giannella, and C. Melchiorre, eds.), Elsevier/North-Holland, Amsterdam, 1979, p. 245.
157. C. R. Ganellin, J. Med. Chem. *24*, 913 (1981).
158. T. O. Yellin, S. H. Buck, D. J. Gilman, D. F. Jones, and J. M. Wardleworth, Life Sci. *25*, 2001 (1979).
159. J. Bradshaw, R. T. Brittain, J. W. Clitherow, M. J. Daly, D. Jack, B. J. Price, and R. Stables, Br. J. Pharmacol. *66*, 464P (1979).
160. C. S. Street, R. E. Cimprich, and J. L. Robertson, Scand. J. Gastroenterol. *19*, Suppl. 101, 109 (1984).
161. C. R. Ganellin, in Frontiers in Histamine Research (C. R. Ganellin and J.-C. Schwartz, eds.), Pergamon Press, Oxford, 1985, p. 47.
162. G. J. Durant, J. C. Emmett, C. R. Ganellin, P. D. Miles, M. E. Parsons, H. D. Prain, and G. R. White, J. Med. Chem. *20*, 901 (1977).
163. T. H. Brown and R. C. Young, Drugs Fut. *10*, 51 (1985).
164. R. C. Young, G. J. Durant, J. C. Emmett, C. R. Ganellin, M. J. Graham,

R. C. Mitchell, H. D. Prain, and M. L. Roantree, J. Med. Chem. *29*, 44 (1986).
165. J.-P. Spengler, K. Wegner, and W. Schunack, Agents Actions *14*, 566 (1984).
166. J.-P. Spengler and W. Schunack, Arch. Pharm. (Weinheim) *317*, 425 (1984).
167. R. C. Blakemore, T. H. Brown, G. J. Durant, J. C. Emmett, C. R. Ganellin, M. E. Parsons, and A. C. Rasmussen, Br. J. Pharmacol. *70*, 105P (1980).
168. G. F. Rush, M. Ripple, and R. Chenergy, J. Pharmacol. Exp. Ther *233*, 741 (1985).
169. H.-D. Höltje, P. Baranowski, J.-P. Spengler, and W. Schunack, Arch. Pharm. (Weinheim) *318*, 542 (1985).
170. R. L. Cavanagh, J. J. Usakewicz, and J. P. Buyniski, J. Pharmacol. Exp. Ther. *224*, 171 (1983).
171. K. Wegner, E. Fritschi, and W. Schunack, Arch. Pharm. (Weinheim) *311*, 98 (1978).
172. A. Borchers, N. Büyüktimkin, and W. Schunack, Sci. Pharm. *51*, 3 (1983).
173. J. Herke and W. Schunack, Arch. Pharm. (Weinheim) *312*, 35 (1979).
174. J. Herke and W. Schunack, Arch. Pharm. (Weinheim) *315*, 641 (1982).
175. J. Herke and W. Schunack, Eur. J. Med. Chem. Chim. Ther. *14*, 203 (1979).
176. P. Dziuron and W. Schunack, Arch. Pharm. (Weinheim) *309*, 583 (1976).
177. W. Schunack, H. Engler, and E. Fritschi, Arzneimittelforsch. *29*, 595 (1979).
178. A. Buschauer, K. Wegner, and W. Schunack, Eur. J. Med. Chem. Chim. Ther. *17*, 505 (1982).
179. C. R. Ganellin, G. J. Durant, J. C. Emmett, D. W. Hills, R. J. Ife, P. D. Miles, and M. E. Parsons, in VIIIth International Symposium on Medicinal Chemistry, Vol. 2. (R. Dahlbom and J. L. G. Nilsson, eds.), Swedish Pharmaceutical Press, Stockholm, 1985, p. 153.
180. G. Comisso, R. Toso, G. Gratton, F. Kajfez, and V. Sunjic, Acta Pharm. Jugoslav. *29*, 125 (1979).
181. J. Bradshaw, M. E. Butcher, J. W. Clitherow, M. D. Dowle, R. Hayes, D. B. Judd, J. M. McKinnon, and B. J. Price, in The Chemical Regulation of Biological Mechanisms (A. M. Creighton and S. Turner, eds.), The Royal Society of Chemistry, Special Publication No. 42, London, 1982, p. 45.
182. M. J. Daly and B. J. Price, in Progress in Medicinal Chemistry, Vol. 20 (G. P. Ellis and G. B. West, eds.), Elsevier, Amsterdam, 1983, p. 337.
183. M. J. Daly, Drugs Today *18*, 615 (1982).
184. R. C. Blakemore, T. H. Brown, G. J. Durant, C. R. Ganellin, M. E. Parsons, A. C. Rasmussen, and D. A. Rawlings, Br. J. Pharmacol. *74*, 200P (1981).
185. G. R. Betton and G. K. Salmon, Scand. J. Gastroenterol. *19*(Suppl. 101), 103 (1984).
186. W. C. Lumma, P. A. Anderson, J. J. Baldwin, W. A. Bolhofer, C. N. Ha-

becker, J. M. Hirshfield, A. M. Pietruszkiewicz, W. C. Randall, M. L. Torchiana, S. F. Britcher, B. V. Clineschmidt, G. H. Denny, R. Hirschmann, J. M. Hoffman, B. T. Phillips, and K. B. Streeter, J. Med. Chem. *25*, 207 (1982)
187. A. A. Algieri, G. M. Luke, R. T. Stanbridge, M. Brown, R. A. Partyka, and R. R. Crenshaw, J. Med. Chem. *25*, 210 (1982).
188. J. P. Buyniski, R. L. Cavanagh, A. W. Pricio, A. A. Algieri, and R. R. Crenshaw, in Highlights in Receptor Chemistry (C. Melchiorre and M. Giannella, eds.), Elsevier, Amsterdam, 1984, p. 195.
189. J. J. Baldwin, W. A. Bolhofer, S. F. Britcher, B. V. Clineschmidt, N. P. Gould, C. N. Habecker, J. M. Hoffman, W. C. Lumma, R. G. Pendleton, K. M. Strohmaier, and M. L. Torchiana, in VIIIth International Symposium on Medicinal Chemistry, Vol. 2 (R. Dahlbom and J. L. G. Nilsson, eds.), Swedish Pharmaceutical Press, Stockholm, 1985, p. 204.
190. D. C. Evans, R. R. Ruffolo, M. W. Warrick, and T. M. Lin, Fed. Proc. *43*, 1074 (1984).
191. T. M. Lin, D. C. Evans, M. W. Warrick, R. P. Pioch, and R. R. Ruffolo, Gastroenterology *84*, 1231 (1983).
192. K-A. Trumm and W. Schunack, Arzneimittelforsch. *33*, 188 (1983).
193. D. B. Norris, G. A. Gajtkowski, T. P. Wood, and T. J. Rising, Agents Actions *16*, 170 (1985).
194. J. D. Michael, J. D. Coombes, S. J. Cousins, D. B. Norris, T. J. Rising, B. C. Ross, and A. Steward, 190th National Meeting of the American Chemical Society, Division of Medicinal Chemistry, Chicago, September 8-13, 1985, abstract 33.
195. I. Yanagisawa, Y. Hirata, and Y. Ishii, J. Med. Chem. *27*, 849 (1984).
196. T. Takagi, M. Takeda, and H. Maeno, Arch. Int. Pharmacodyn. *256*, 49 (1982).
197. M. Takeda, T. Takagi, Y. Yashima, and H. Maeno, Arzneimittelforsch. *32*, 734 (1982).
198. T. O. Yellin and D. J. Gilman, European Patent Appl. 0 060 094 (1982).
199. R. T. Brittain, M. J. Daly, J. M. Humphray, and R. Stables, Br. J. Pharmacol. *76*, 195P (1982).
200. R. Stables, M. J. Daly, and J. M. Humphray, Agents Actions *13*, 166 (1983).
201. R. Stables and J. M. Humphray, 12th Meeting European Histamine Research Society, Brighton, 1983, abstract 012.
202. R. T. Brittain and D. Jack, J. Clin. Gastroenterol. *5*(Suppl. 1), 71 (1983).
203. M. L. Torchiana, R. G. Pendleton, P. G. Cook, C.A. Hanson, and B. V. Clineschmidt, J. Pharmacol. Exp. Ther. *224*, 514 (1983).
204. D. E. Bays and B. J. Price, in VIIIth International Symposium on Medicinal Chemistry, Vol. 2 (R. Dahlbom and J. L. G. Nilsson, eds.), Swedish Pharmaceutical Press, Stockholm, 1985, p. 183.
205. D. Poynter, C. R. Pick, R. A. Harcourt, S. A. M. Selway, G. Ainge, I. W. Harman, N. W. Spurling, P. A. Fluck, and J. L. Cook, Gut *26*, 1284 (1985).

206. R. T. Brittain, D. Jack, J. J. Reeves, and R. Stables, Br. J. Pharmacol. *85*, 843 (1985).
207. A. Buschauer, S. Postius, I. Szelenyi, and W. Schunack, Arzneimittel-Forsch. *35*, 1025 (1985).
208. A. Buschauer, I. Krämer, and W. Schunack, Arch. Pharm. (Weinheim) *319*, 434 (1986).
209. G. L. Garay and J. M. Muchowski, Annu. Rep. Med. Chem. *20*, 93 (1985).
210. S. T. Nielsen, P. Dove, G. Palumbo, A. Sandor, C. B. Buonato, G. Schiehser, A. Santilli, and D. P. Strike, Fed. Proc. *43*, 1074 (1984).
211. P. Dove, A. Sandor, G. J. Palumbo, C. B. Buonato, G. Schiehser, A. Santilli, and D. P. Strike, Gastroenterology *86*, 1195 (1984).
212. K. Wegner, Arch. Pharm. (Weinheim) *319*, 85 (1986).
213. H. Toyofuku, Y. Tsuriya, T. Kuroda, H. Aoki, and H. Nagasawa (Wakamoto Pharmaceutical Co., Ltd.), German Patent DE 3426533 (1985).
214. G. Sorba, R. Calvino, A. Defilippi, A. Gasco, and M. Orsetti, Eur. J. Med. Chem. Chim. Ther. *20*, 571 (1985).
215. I. Krämer and W. Schunack, Arch. Pharm. (Weinheim) *318*, 888 (1985).
216. A. Buschauer, Pharm. Z. Sci. Ed. *130*, 2067 (1985).
217. Drugs Fut. 10, 128 (1985).
218. M. Tarutani, H. Sakuma, K. Shiratsuchi, and M. Mieda, Arzneimittelforsch. *35*, 703 (1985).
219. A. Borchers, H. Engler, I. Szelenyi, and W. Schunack, Arzneimittelforsch. *32*, 1509 (1982).
220. A. Borchers, S. Postius, I. Szelenyi, and W. Schunack, Arzneimittelforsch. *34*, 751 (1984).
221. A. Donetti, E. Cereda, E. Bellora, A. Gallazzi, C. Bazzano, P. Vanoni, P. Del Soldato, R. Micheletti, F. Pagani, and A. Giachetti, J. Med. Chem. *27*, 380 (1984).
222. F. Pagani, R. Micheletti, A. Brambilla, A. Schiavone, E. Montagna, C. Ciprandi, and A. Giachetti, Arzneimittelforsch. *35*, 451 (1985).
223. E. Cereda, E. Bellora, R. Micheletti, F. Pagani, A. Giachetti, and A. Donetti, in VIIIth International Symposium on Medicinal Chemistry, Vol. 2 (R. Dahlbom and J. L. G. Nilsson, eds.), Swedish Pharmaceutical Press, Stockholm, 1985, p. 196.
224. G. Bietti, R. Micheletti, F. Pagani, A. Giachetti, A. Donetti, and G. Trummlitz, in VIIIth International Symposium on Medicinal Chemistry, Vol. 2 (R. Dahlbom and J. C. G. Nilsson, eds.), Swedish Pharmaceutical Press, Stockholm, 1985, p. 211.
225. C. A. Lipinski, J. Med. Chem. *26*, 1 (1983).
226. C. A. Lipinski, J. L. LaMattina, and L. A. Hohnke, J. Med. Chem. *28*, 1628 (1985).
227. J.-M. Arrang, M. Garbarg, T. T. Quach, M. D. Trung Tuong, E. Yeramian, and J.-C. Schwartz, Eur. J. Pharmacol. *111*, 73 (1985).
228. J.-M. Arrang, J.-C. Schwartz, and W. Schunack, Eur. J. Pharmacol. *117*, 109 (1985).

229. W. Schunack, S. Schwarz, G. Gerhard, S. Büyüktimkin, and S. Elz, in Frontiers in Histamine Research (C. R. Ganellin and J.-C. Schwartz, eds.), Pergamon Press, Oxford, 1985, p. 39.
230. C. R. Ganellin and M. E. Parsons (eds.), Pharmacology of Histamine Receptors, Wright PSG, London, 1982.
231. C. R. Ganellin and J.-C. Schwartz (eds.), Frontiers in Histamine Research, Pergamon Press, Oxford, 1985.
232. T. T. Quach, A. M. Duchemin, C. Rose, and J.-C. Schwartz, Mol. Pharmacol. *17*, 301 (1980).
233. Y. Takai, V. Kikkawa, K. Kaibuchi, and Y. Nishizuka, in Advances in Cyclic Nucleotide and Protein Phosphorylation Research, Vol. 18 (P. Greengard and G. A. Robison, eds.), Raven Press, New York, 1984, p. 119.
234. P. Cohen, A. Burchell, J. C. Foulkes, P. T. W. Cohen, T. C. Vanaman, and A. C. Mairn, FEBS Lett. *92*, 287 (1978).
235. D. Verveken, P. van Veldhofen, C. Proost, H. Carton, and H. De Wulf, J. Neurochem. *38*, 1286 (1982).
236. V. Buonassisi and J. C. Venter, Proc. Natl. Acad. Sci. USA *73*, 1612 (1976).
237. S. Katsuki, and F. Murad, Mol. Pharmacol. *13*, 330 (1977).
238. E. Richelson, Science *201*, 69 (1978).
239. R. E. Study and P. Greengard, J. Pharmacol. Exp. Ther. *207*, 767 (1978).
240. J.-C. Schwartz, G. Barbin, A. M. Duchemin, M. Garbarg, J. M. Palacios, T. T. Quach, and C. Rose, in Receptors for Neurotransmitters and Peptide Hormones (G. Pepeu, M. J. Kuhar, and S. J. Enna, eds.), Raven Press, New York, 1980, p. 169.
241. J.-C. Schwartz, Life Sci. *25*, 895 (1979).
242. R. H. Michell, Biochim. Biophys. Acta *415*, 81 (1975).
243. R. O. Friedel and S. M. Schanberg, J. Neurochem. *24*, 819 (1975).
244. S. S. Jafferji and R. M. Michell, Biochem. Pharmacol. *25*, 1429 (1976).
245. M. J. Berridge, J. P. Downes, and M. R. Hanley, Biochem. J. *206*, 587 (1982).
246. E. Brown, D. A. Kendall, and S. R. Nahorski, J. Neurochem. *42*, 1379 (1984).
247. J. Donaldson and S. J. Hill, Br. J. Pharmacol. *85*, 499 (1985).
248. D. A. Kendall and S. R. Nahorski, J. Neurochem. *42*, 1388 (1984).
249. H. Carswell, P. R. Daum, and J. M. Young, in Frontiers in Histamine Research (C. R. Ganellin and J.-C. Schwartz, eds.), Pergamon Press, Oxford, 1985, p. 27.
250. P. R. Daum, C. P. Downes, and J. M. Young, J. Neurochem. *43*, 25 (1984).
251. J.-C. Schwartz, J.-M. Arrang, M. Garbarg, and W. Schunack, in Innovative Approaches in Drug Research (A. F. Harms, ed.), Elsevier, Amsterdam 1986, p. 73.
252. J.-C. Schwartz, J.-M. Arrang, M. Garbarg, and M. Korner, J. Exp. Biol., *124*, 203 (1986).

253. C. F. Code, in Pharmacology of Histamine Receptors (C. R. Ganellin and M. E. Parsons, eds.), Wright PSG, London, 1982, p. 217.
254. C. L. Johnson, in Pharmacology of Histamine Receptors (C. R. Ganellin and M. E. Parsons, eds.), Wright PSG, London, 1982, p. 146.
255. R. Levi, D. A. A. Owen, and J. Trzeciakowski, in Pharmacology of Histamine Receptors (C. R. Ganellin and M. E. Parsons, eds.), Wright PSG, London, 1982, p. 236.
256. M. E. Parsons, in Pharmacology of Histamine Receptors (C. R. Ganellin and M. E. Parsons, eds.), Wright PGS, London, 1982, p. 323.
257. M. Plaut and L. M. Lichtenstein, in Pharmacology of Histamine Receptors (C. R. Ganellin and M. E. Parsons, eds.), Wright PSG, London, 1982, p. 392.
258. M. L. Haas, in Frontiers in Histamine Research (C. R. Ganellin and J.-C. Schwartz, eds.), Pergamon Press, Oxford, 1985, p. 215.
259. L. R. Hegstrand, P. D. Kanof, and P. Greengard, Nature *260*, 163 (1976).
260. K.-Fr. Sewing and W. Beil, in Frontiers in Histamine Research (C. R. Ganellin and J.-C. Schwartz, eds.), Pergamon Press, Oxford, 1985, p. 275.
261. J.-C. Schwartz, G. Barbin, A. M. Duchemin, M. Garbarg, C. Llorens, H. Pollard, T. T. Quach, and C. Rose, in Pharmacology of Histamine Receptors (C. R. Ganellin and M. E. Parsons, eds.), Wright PSG, London, 1982, p. 351.
262. J.-M. Arrang, M. Garbarg, and J.-C. Schwartz, Neuroscience *15*, 553 (1985).
263. M. Garbarg, G. Barbin, J. M. Palacios, and J.-C. Schwartz, Brain Res. *150*, 638 (1978).
264. J.-M. Arrang, M. Garbarg, and J.-C. Schwartz, in Frontiers in Histamine Research (C. R. Ganellin and J.-C. Schwartz, eds.), Pergamon Press, Oxford, 1985, p. 143.

11
GABA Receptors

Povl Krogsgaard-Larsen / The Royal Danish School of Pharmacy, Copenhagen, Denmark

The neutral amino acid γ-aminobutyric acid (GABA, [1]) (Figure 1) is present in every region of the mammalian central nervous system (CNS). Although many aspects concerning the functions of GABA are still incompletely understood, it is now generally accepted that GABA has a neurotransmitter function and that it fulfills the main criteria established for the identification of an inhibitory neurotransmitter (1-5).

GABA is involved in the control of many physiological mechanisms, including the regulation of the secretion of prolactin (6,7) and other hormones, including the growth hormone (7,8). Furthermore, GABA plays a role in the regulation of cardiovascular functions, such as blood pressure and heart rate (9-13) and is involved in the sensation of pain (14-18) and anxiety (19-21). There are strong indications that activation of the GABA system modifies feeding (22) and aggressive behavior (23) in animals.

The growing interest in the pharmacology of GABA has been stimulated by the findings that GABA is apparently involved in the development of certain neurological and psychiatric diseases, such as epilepsy, Huntington's chorea, Parkinson's disease, and, perhaps, schizophrenia and Alzheimer's disease. Thus, analyses of brain samples from sites near seizure foci in epileptic patients and in animals made epileptic have revealed severe impairments of the GABA system (24,25). Low levels of (*S*)-glutamate-1-carboxylyase (GAD), the GABA-synthesizing enzyme, and reduced GABA uptake capacity indicate that decreased activity and probably partial degeneration of GABA neurons are factors of importance in epileptic phenomena (26,27). Low levels of GABA and GAD have been measured in postmortem brain tissue from choreic patients (28,29). In Parkinson's disease there is an imbalance between the GABA and dopamine

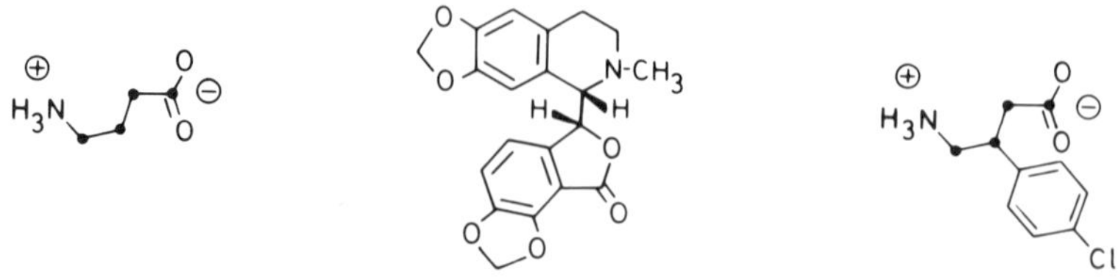

1 2 3

Figure 1 The structures of GABA [1], the GABA-A antagonist bicucilline [2], and the GABA-B agonist baclofen [3].

systems, and in some brain areas of parkinsonian patients, especially the substantia nigra, GAD activity and GABA receptor density are below normal levels (30). Decreased GABA activities in certain regions of brains from dying schizophrenic patients suggest that GABA is involved in the pathophysiology of schizophrenia (31-33). The precise role of GABA in schizophrenia and, perhaps, in Alzheimer's disease (34), is far from clear, however. There is some evidence of a GABA-ergic contribution to the symptoms in depressed patients (33) and to the action of antidepressant drugs (35), and remarkably low plasma GABA levels have been measured in alcoholics (33).

In the development of strategies for pharmacological intervention in the GABA neurotransmission process, the degree of degeneration of GABA neurons must be taken into consideration (36). In stages of the diseases in which GABA neurons are still functioning, but at an abnormally low level, presynaptic mechanisms including, the uptake systems and GABA: 2-oxoglutarate aminotransferase (GABA-T) as well as the postsynaptic receptors, are potential pharmacological sites of attack. In extensive neuronal degeneration, only the postsynaptic GABA receptors survive, and deprived of their neurotransmitter, such receptors apparently develop hypersensitivity to GABA agonists (37). In the advanced stages of epilepsy or Huntington's chorea, in which GABA neurons in some brain areas are extensively degenerated, GABA agonist therapies may be particularly pertinent. It must be emphasized, however, that hypersensitive GABA receptors may not be functioning optimally, and furthermore, it is normally very difficult to estimate the degree of degeneration of GABA neurons in patients suffering from the diseases concerned. In any case, the most direct, although perhaps not the most flexible, approach (38) to stimulation of GABA-mediated neurotransmission appears to be activation of the postsynaptic GABA receptors. Consequently, design and development of specific GABA agonists suitable for pharmacological and clinical studies have been, and continue to be, an active research field (36,39,40), and GABA agonists have been studied in a variety of animal models and in different groups of patients.

CLASSIFICATION

Electrophysiological and receptor binding studies have been extensively used to detect GABA receptors and to identify subtypes of these receptors (41,42). The electrophysiological demonstration of a specific neuronal response to GABA (43) and, subsequently, the discovery of selective antagonism of this action of GABA by the alkaloid bicuculline [2] (Figure 1) (44) represent milestones in this research field. GABA receptors are now divided into two main classes: GABA-A receptors, which are sensitive to bicuculline, and bicuculline-insensitive GABA receptors, which, like the GABA-A receptors, are heterogeneous and include the GABA-B receptors.

Electrophysiological studies indicate that GABA-A receptors are coupled to a chloride ion channel (1,2). Activation of the GABA receptor results in a net influx or efflux of chloride ions, depending on the prevailing concentration gradient (45). GABA-A receptors include postsynaptic (axosomatic or axodendritic) (Figure 2) as well as presynaptic (axoaxonic) (Figure 2) receptors, which have similar characteristics (46,47). GABA-A receptors include extrasynaptic receptors, which apparently have agonist specificities different from those of synaptic GABA receptors (46,48,49). Postsynaptic GABA-A receptors regulate the passage of chloride ions such that receptor activation causes hyperpolariza-

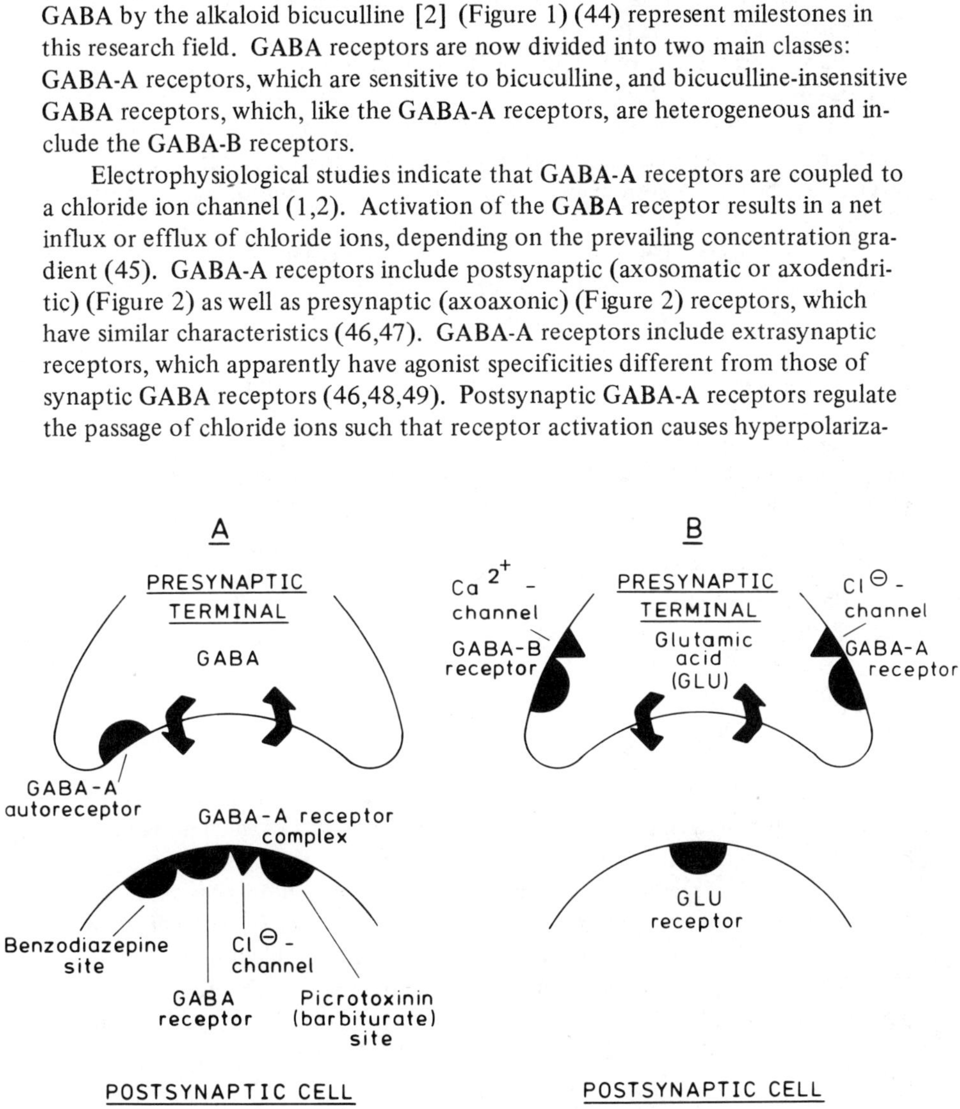

Figure 2 A schematic illustration of the different types of GABA receptors.

tion of the cell membrane and, thus, decreased sensitivity of the neuron to excitatory input. In contrast, activation of presynaptic GABA-A receptors normally leads to a net efflux of chloride ions, causing partial depolarization (46,50). This phenomenon is generally referred to as presynaptic inhibition, since the evoked release of transmitter from the partially depolarized axon is diminished, as exemplified in Figure 2 for a nerve terminal releasing the exitatory neurotransmitter glutamic acid (GLU).

The synaptic release of GABA may be regulated by a GABA-A autoreceptor (Figure 2). This type of receptor receives its input from the transmitter released from the axon terminal, on which the autoreceptor is located, to influence further transmitter release (51,52). Receptors of this particular class have been described for the GABA system exclusively on the basis of neurochemical studies (53-57).

There is some evidence supporting the view that GABA-A receptors can be further subdivided. Thus, the agonist profile (44,58,59) and sensitivity to antagonists (60-62) of GABA-A receptors in the mammalian spinal cord and cerebral cortex appear to be dissimilar.

As mentioned, not all GABA receptors are sensitive to bicuculline [2]. Within the group of [2]-insensitive GABA receptors the GABA-B receptors have been most extensively studied. These receptors are activated by GABA and, in contrast to GABA-A receptors, by baclofen [3] (Figure 1). Although very little is known about the physiological role of the GABA-B receptors, they seem to regulate the release of some, but not all, neurotransmitters. Accumulating evidence indicates that GABA-B receptors are predominantly located presynaptically and that they affect neurotransmitter release via regulation of a calcium channel (Figure 2) (63,64).

Bicuculline-Sensitive GABA Receptors

Postsynaptic Receptor Complex

Our present knowledge of the structure of the postsynaptic GABA receptor complex is summarized (65,66) in Figure 3. The chloride channel is regulated by the GABA-A receptor, consisting of more than one binding site.

The GABA-A receptor sites have been the subject of numerous studies. The pioneer work in this field was performed by Peck et al. (67), Zukin et al. (68), and Enna and Snyder (69), who demonstrated saturable binding of $[^3H]$GABA to brain synaptic membranes. The presence of two kinetically distinct types of GABA binding sites was demonstrated in thoroughly washed or detergent-treated synaptic membranes (70,71). Subsequently, radioactive muscimol [4] (Figure 3) was introduced as a versatile ligand for GABA-A receptor binding studies (72-75), and more recently, the specific GABA-A agonists isoguvacine [5] (76), piperidine-4-sulfonic acid (P4S [6]) (77), and 4,5,6,7-

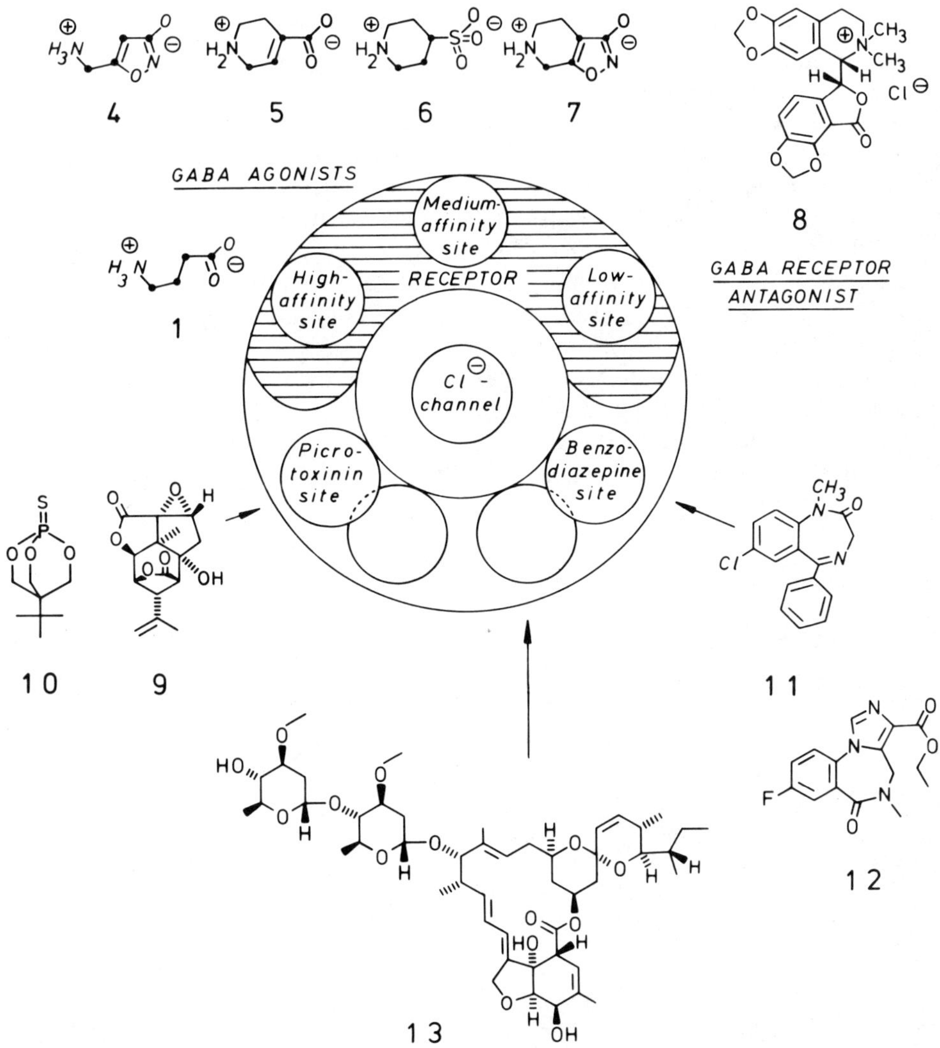

Figure 3 A schematic illustration of the postsynaptic GABA receptor complex, the structures of some GABA-A agonists and antagonists, and the sites of action of a number of drugs.

tetrahydroisoxazolo[5,4-*c*]pyridin-3-ol (THIP [7]) (Figure 3) (78) have been labeled and used for GABA-A receptor binding studies. These binding studies have been carried out under different experimental conditions, and consequently, more or less dissimilar binding characteristics have been reported by different research groups (67-86). Nevertheless, the pharmacological profiles for displacement of labeled GABA or GABA analogs compare favorably with data obtained in more intact systems using electrophysiological techniques (5,41,42,87,88). In addition to labeled GABA or GABA agonists, the tritiated GABA antagonist bicuculline methochloride (BMC [8]) (Figure 3) has been used for studies of GABA receptor sites in vitro (89-91). Like the GABA agonist THIP (Figure 4) (78,92), BMC (91,93) was shown to interact preferentially with low-affinity GABA receptor sites.

Computer-aided nonlinear regression analyses of binding data are, in most cases, consistent with the presence of two or three (78,87,94) GABA receptor binding sites that are not interconvertible under the assay conditions (94). An analysis of the binding data for [^{3}H]GABA and [^{3}H]THIP [7] (78) is il-

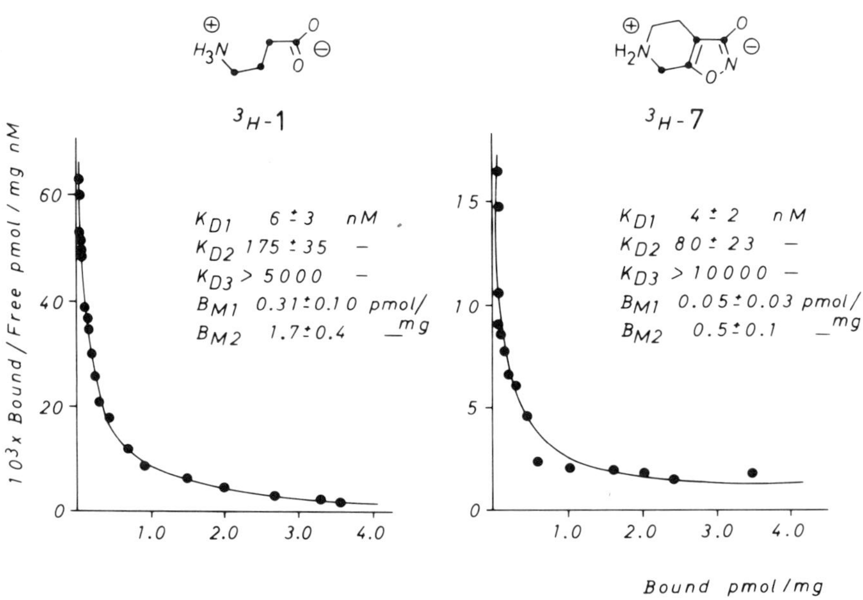

Figure 4 Scatchard plots of the specific binding of radioactive GABA [1] and THIP [7] to rat brain membranes. The binding parameters are derived from computer-fitted nonlinear regression analyses of the data (From Reference 78).

lustrated in Figure 4. The affinity of the low-affinity binding site for both these ligands, as well as for [^{3}H]P4S [6], was too low for satisfactory characterization, however (78).

The physiological relevance of these multiple binding sites is unknown. In general, low-affinity receptor sites are likely to correspond to functional receptors (95). In the case of GABA this view may be supported by the observations that the coupling between the GABA receptor and the benzodiazepine (BZD) site (Figure 3) seems to involve low-affinity GABA receptor sites (94). It has been proposed that the different affinities may reflect anatomic rather than functional differences. Thus, high-affinity GABA binding sites were proposed to relate to postsynaptic GABA-A receptors, whereas low-affinity sites may reflect presynaptic GABA-A receptors (Figure 2) (96,97).

The GABA receptor function appears to be modulated by various additional units that can be detected in vitro as distinct binding sites for the BZD (98,99), such as diazepam [11] or the BZD antagonist Ro 15-1788 [12], and for picrotoxinin [9] (65) or the cage convulsant *tert*-butyl bicyclophosphothionate (TBPS [10]) (100-103). There is some evidence of heterogeneity for both these binding sites (41,42) and for the existence of a distinct binding site at the postsynaptic GABA receptor complex for the avermectins, including ivermectin and avermectin B_{1a} [13] (Figure 3) (104-109).

The physiological role of the additional sites ("pharmacological receptors") at the GABA receptor complex is unknown, but the intimate contact and allosteric interactions between these sites as detected in vitro may reflect certain aspects of the dynamic properties of the GABA receptors (65,94).

There is strong evidence that the picrotoxinin binding sites represent the pharmacological receptors for the barbiturates (65,100,103,110,111). Not only the sedative, hypnotic, and anticonvulsant barbiturates, such as amobarbital and pentobarbital, interact with this site, but also the convulsant barbiturates, such as 5-ethyl-5-(1,3-dimethylbutyl)barbituric acid, are inhibitors of the binding of radioactive dihydropicrotoxinin (65,110) and TBPS (100,102,103).

The clinically used anxiolytic BZDs, such as [11] (Figure 3) and flunitrazepam, bind very tightly to the particular site at the GABA receptor complex, and these compounds and the BZD antagonists, such as [12], have been shown to affect the GABA neurotransmission in vivo and under a variety of conditions (42,112).

The interaction in vitro of GABA agonists with the GABA receptor-BZD receptor complex (Figure 3) has been extensively studied (88,113-115). These studies have disclosed striking differences between the effects on BZD binding of different "structural classes" of GABA agonists and an appreciable effect of temperature and, in particular, chloride ions, on the degree of GABA agonist-induced stimulation of BZD binding (Figure 5) (114-116). Whereas there is generally a positive correlation between the potencies of these GABA analogs

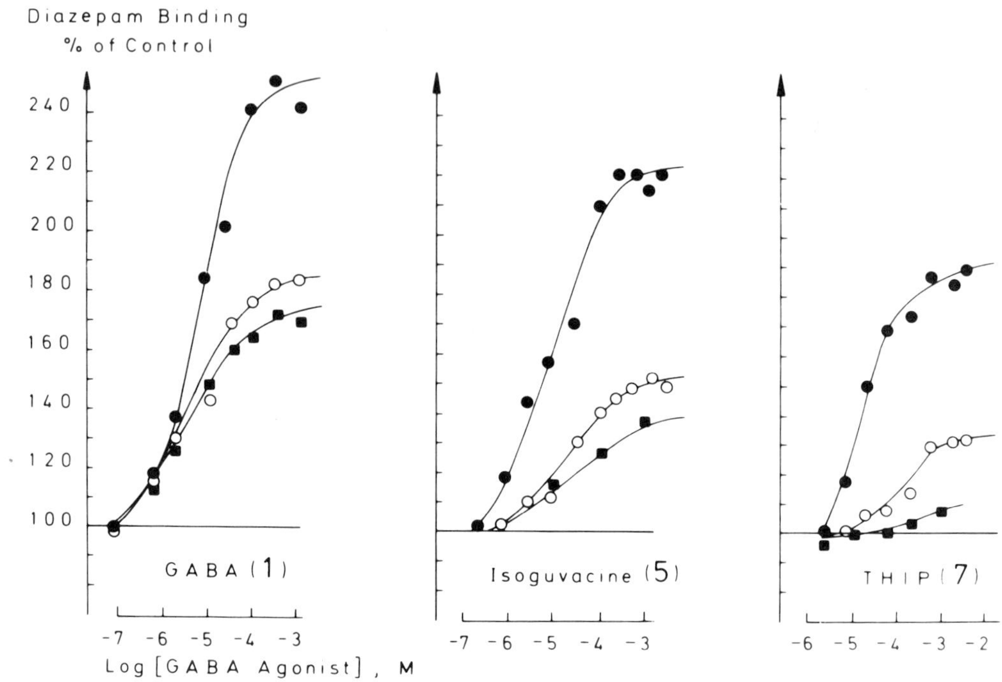

Figure 5 Effects of GABA [1], isoguvacine [5], and THIP [7] on the binding of radioactive diazepam [11] at 4°C in the absence of chloride (■) and at 30°C in the absence (○) or presence (●) of 150 mM sodium chloride.

in vivo and as inhibitors of GABA binding (36,40,88), there is no simple correlation between the potencies of GABA agonists and their ability to stimulate BZD binding in vitro. For the latter effect of GABA agonists, three structural parameters appear to be of major importance: (1) the structure of the acid moiety of the GABA agonist, (2) the degree of substitution at the basic nitrogen atom, and (3) the conformational mobility of the entire molecule. These aspects are exemplified in Figure 5 for GABA and the GABA-A agonists isoguvacine [5] and THIP [7] (Figure 3).

The pharmacological importance of these dissimilar effects of GABA agonists on BZD binding in vitro is unclear. The effects of [5] and [7], for example, are reminiscent of partial GABA agonist activities, but so far electrophysiological studies have not disclosed partial GABA agonist profiles of these very potent GABA-A agonists (114,117,118).

Presynaptic Receptors

GABA nerve terminals impinge on 1a primary afferent fibers, where GABA decreases the evoked release of sensory transmitter(s), and a presynaptic depolarizing inhibitory process mediated by GABA may be confined to terminals of such primary afferent fibers (119). The lack of direct electrophysiological evidence for different pharmacological profiles of these presynaptic GABA receptors (Figure 2) and the GABA-A receptors in the postsynaptic receptor complex has hampered progress in establishing GABA as the transmitter at presynaptic sites (46). It has recently been shown, however, that the terminal axonal arborizations in the spinal cord of neurons in the red nucleus have GABA receptors different from those on the cell bodies, as illustrated by the lack of sensitivity to BMC [8] (Figure 3) of the depressant actions of GABA as well as the GABA-A agonist P4S [8] at receptors on cell bodies (119).

There is circumstantial neurochemical evidence for the existence of presynaptically located GABA-A receptors in the brain. Thus, GABA enhances the potassium-evoked release of transmitter from cerebellar cortex slices in a bicuculline [2]-sensitive manner (120). A similar effect of GABA on slices of occipital cortex was, however, not sensitive to [2] (121). Similarly, GABA is capable of facilitating the potassium-evoked release of dopamine from slices of corpus striatum, but again, antagonism by [2] is not always apparent (122-125). Facilitation by GABA of the potassium-stimulated release of GLU (see Figure 2) from striatal tissue has been reported (126). This effect appears to be mediated by GABA receptors sensitive to [8] but showing a profile of agonist sensitivity different from that characterizing the GABA-A receptors in the postsynaptic receptor complex (Figure 3). The physiological significance of this observation, which has interesting therapeutic prospects, is not clear, however (63). On the other hand, the evoked release of GLU from cultured cerebellar granule cells has been shown to be under inhibitory control by GABA receptors sensitive to [8] (127). These receptors showed an agonist profile typical for GABA-A receptors, [4] and [7] (Figure 3) being active, whereas no effect of baclofen could be demonstrated. Interestingly, these effects of GABA-A agonists proved to be mediated by low- rather than high-affinity GABA receptors (see earlier discussion and Figure 3) (127).

Autoreceptors

Although the existence of GABA autoreceptors (Figure 2) has not yet been supported by studies employing electrophysiological techniques, neurochemical data are consistent with the existence of such receptors (53-56,128). Studies on the effects of GABA itself on the depolarization-induced release of labeled GABA from preloaded slices or synaptosomes prepared from brain tissue are hampered by homoexchange phenomena. Thus, the uptake of GABA applied to the exterior surface of the tissue preparations produces an apparent efflux of the labeled transmitter owing to 1:1 exchange.

The most convincing pieces of evidence for the existence of GABA autoreceptors are derived from in vitro studies on GABA release using such GABA agonists as [5], [6], and [7] (Figure 3), which are not substrates for the GABA uptake mechanisms (129-132). These specific GABA-A agonists, as well as a number of GABA analogs, such as [4], that interact selectively with GABA-A receptors, inhibit in a dose-dependent manner the evoked release of GABA from brain tissue preparations (133,134). The sensitivity of the receptors mediating this phenomenon to GABA agonists and GABA antagonists, such as [2] and [8], appears to be very similar to that of the postsynaptic GABA-A receptors, and furthermore, the GABA-B agonist baclofen [3] is inactive at the receptors concerned (135). On the other hand, the GABA homolog δ-aminolevulinic acid [14] (Figure 6) is apparently a relatively selective agonist at GABA-A autoreceptors (55,136), and in contrast to postsynaptic GABA-A receptors, autoreceptors do not seem to be coupled to BZD receptor sites (Figure 2) (137).

Bicuculline-Insensitive GABA Receptors

Baclofen-Sensitive GABA Receptors

Curtis et al. (138) and Davies and Watkins (139) observed that the depressant effects of baclofen [3] (Figure 1) on mammalian central neurons could not be blocked by bicuculline [2] or BMC [8]. Subsequently, it was demonstrated that neither [2] nor picrotoxinin [9] blocked the depressant action of GABA on the evoked release of norepinephrine from sympathetic nerve terminals (140). In this preparation, the GABA-A agonists [5] and [6] (Figure 3) were inactive and [4] was much weaker than GABA, whereas the particular effect of GABA was mimicked by [3] (141). Similarly, GABA and [3] have been shown

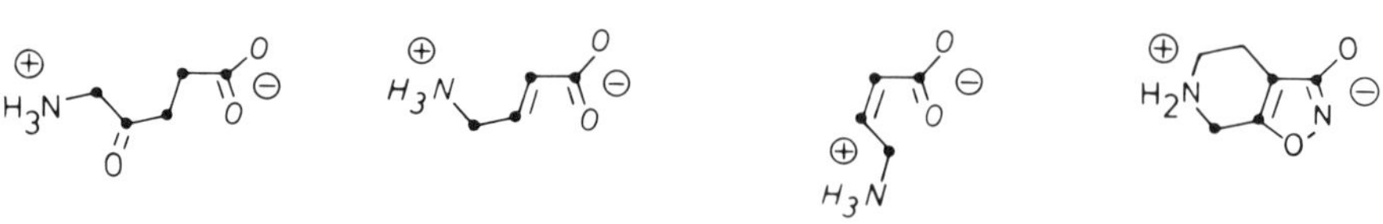

14 15 16 7

Figure 6 The structures of some analogs of GABA with different mechanisms of action.

to decrease neurotransmitter release in the CNS by activation of receptors at which [2] and [8] are not antagonists (142). On the basis of these observations, the existence of a subclass of GABA receptors, GABA-B receptors, distinctly different from the "classic" GABA-A receptors, was proposed (142). The GABA-B receptor sites can be readily demonstrated in brain synaptic membranes using radiolabeled binding techniques with either [^{3}H] baclofen or [^{3}H] GABA, in the latter case under conditions in which GABA-A receptor sites are blocked by [5] (143-145). Since these original observations, GABA-B receptors have been detected and described in a variety of tissue preparations of central or peripheral origin (146).

In addition to the different pharmacological profiles of the GABA-A and GABA-B receptors, major differences between these receptor systems have been disclosed (41,42). Although pre- as well as postsynaptic GABA-A receptors are coupled to chloride channels, GABA-B receptors are coupled to calcium channels (Figure 2), and in contrast to GABA-A receptors, GABA-B receptors appear to be associated with a second messenger system. Thus, GABA as well as *R*(–)-[3], but not *S*(+)-[3], are capable of potentiating the response to stimulation of adenylate cyclase by catecholamines, whereas neither compound has any significant effect on basal adenylate cyclase activity in brain slices or membranes (147,148). On the other hand, GABA and *R*(–)-[3], but not *S*(+)-[3], were shown to inhibit forskolin-induced cAMP production (147,148). These effects may to some extent reflect GABA-B receptor mechanisms.

There are other major differences between the GABA-A and GABA-B receptors. Although treatment of synaptic membranes by detergents increases the binding capacity of GABA-A receptor sites (70), a decrease in GABA-B binding capacity results, possibly reflecting that GABA-B binding macromolecules are more readily denaturated or solubilized (144-146). An association between a large population of postsynaptic GABA-A receptors and BZD receptor sites is well documented (Figure 3) (41,42), but it seems unlikely that GABA-B receptors are linked to BZD receptor sites (146). Finally, the regional distributions of GABA-A and GABA-B receptors in the CNS are different. In the cerebellum, GABA-B sites are confined to the molecular layer, whereas GABA-A sites predominate in the granule cell layer (149-151). In the spinal cord, the largest density of GABA-B sites is detected in the dorsal horn laminae I, II, and III, whereas the GABA-B sites seem to be more uniformly distributed throughout the dorsal and ventral horns (146).

Baclofen-Insensitive Receptors

Some GABA analogs, including *cis*-4-aminobut-2-enoic acid [16] (Figure 6) (152), some muscimol analogs (153), and *cis*-2-(aminomethyl)cyclopropanecarboxylic acid (CAMP) (154), depress the firing of central neurons in a manner similar to that of [3] (138,139). The depressant effects of these compounds are not reversed by [2] or [8], but in contrast to [3], and the nonspecific GABA-A/B agonist *trans*-4-aminobut-2-enoic acid [15], they do not interact with GABA-B sites in binding experiments (59,155).

Among these compounds, [16] has attracted particular interest. The cis configuration of [16] prompted Andrews and Johnston (156) to propose that this compound interacts with a class of GABA receptors that, in contrast to the GABA-A receptors, bind GABA in a folded conformation. This idea was rekindled by the observation that [16] and CAMP proved to be more potent as inhibitors of bicuculline [2]- and isoguvacine [5]-insensitive GABA binding than of the binding of *R*(–)-baclofen [3] (157), suggesting the existence of a class of GABA receptors different from GABA-A as well as GABA-B receptors (157). Recent receptor binding studies have so far failed to support this proposal, and besides the structural similarity between [16] and GABA and the very weak effect of [16] on GABA-A receptor binding (158), there actually is very little evidence that the depressant action of [16] is mediated by GABA receptors.

The potent GABA-A agonist THIP [7] (Figures 3 and 6) (117,118) has been subject to extensive animal behavioral studies (159,160). Quite surprisingly, not all the pharmacological effects of [7] can be reversed by [2]. Thus, neither the effects of [7] on the electroencephalographic (EEG) patterns in animals (161), nor the very potent nonopioid analgesic effects of [7] (16-18,162,163), nor its effects on sexual behavior in the male rat (164) are sensitive to [2]. Since [7] does not affect GABA-B receptor binding (145), these observations seem to indicate that [7], in addition to its GABA-A agonist activity, interacts with a distinct subtype of GABA receptors. A relationship between these unclarified aspects of the mechanism of action of [7] and those of [16] cannot at present be excluded.

STRUCTURE-ACTIVITY RELATIONSHIPS

By systematic alterations of the structural parameters of GABA, including bioisosteric replacement of the carboxyl and amino groups of GABA by a variety of acid and basic groups, a very broad spectrum of analogs has been developed (36,39,40,160). A necessary but certainly not sufficient condition for potent and specific GABA agonist activity is the presence of a negatively and a positively charged group. A considerable degree of conformational flexibility is a trait of the molecule of GABA (Figure 7). This molecular characteristic is apparently essential for the synaptic activity of GABA but inherently makes GABA itself unsuitable for studies of its active conformations at different synaptic recognition sites. An approach to these problems involves the design of GABA analogs of more rigid structure.

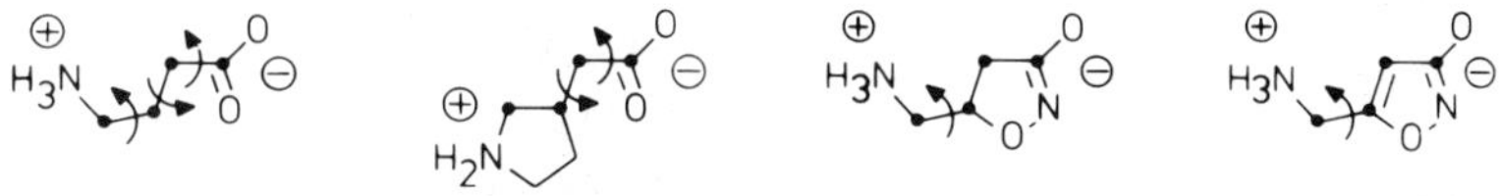

	1	17	18	4
AG	— — —	— — —	— — — —	— — — —
A	0.03	0.3	0.008	0.006
UPT-N	15	75	1500	2500
UPT-G	35	20	4000	2000

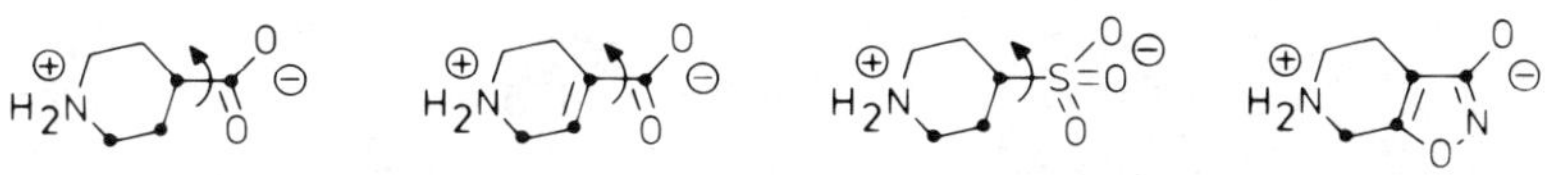

	19	5	6	7
AG	— — —	— — — —	— — — —	— — —(—)
A	0.3	0.04	0.03	0.1
UPT-N	>5000	>5000	>5000	>5000
UPT-G	>5000	>5000	>5000	>5000

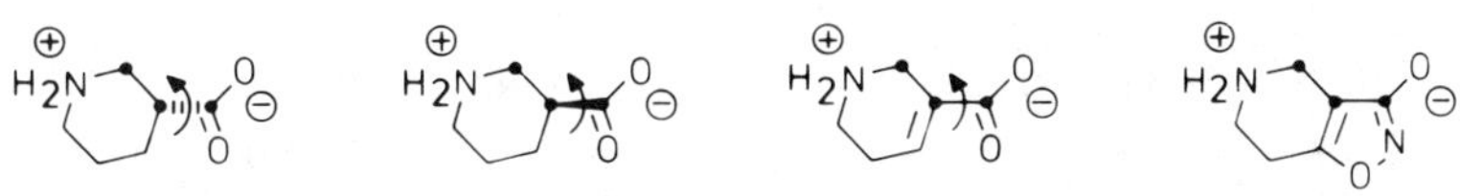

	(R)-(–)-20	(S)-(+)-20	21	22
AG	0	0	0	0
A	>100	>100	>100	80
UPT-N	70	500	100	5000
UPT-G	30	1000	25	300

Figure 7 Structure, conformational flexibility, and biologic and in vitro activity of GABA [1] and a number of GABA analogs. AG: BMC [8]-sensitive GABA agonist (GABA-A agonist) activity relative to that of GABA (— — —) as determined microelectrophoretically (44). A: Inhibition of GABA-A receptor binding (IC_{50}, μM) using rat brain membranes (78). UPT-N and UPT-G: Inhibition of neuronal (N) and glial (G) GABA uptake using brain slices and cultured astrocytes, respectively, at GABA concentrations of 1 μM (IC_{50}, μM) (167).

GABA-A Agonists

Structural and Conformational Factors

Like [15] (Figure 6) (152,158), homo-β-proline [17] (114,165) interacts with GABA-A receptors as well as with GABA uptake, whereas dihydromuscimol [18] (118) and muscimol [4] (153) are quite selective GABA-A agonists. Still, certain degrees of rotational freedom allow these compounds to adopt conformations recognizable by the GABA transport carriers (Figure 7). Such conformations are outside the range of possibilities for isonipecotic acid [19] (131), isoguvacine [5] (117,118), and THIP [7] (117,118), however, all of which are specific GABA-A agonists.

The very narrow and distinctly different ranges of active conformations of GABA with respect to different synaptic mechanisms are illustrated by the observations that *R*(–)- and *S*(+)-nipecotic acid [20] (166,167), guvacine [21] (131,168), and 4,5,6,7-tetrahydroisoxazolo[4,5-*c*]pyridin-3-ol (THPO [22]) (167,169) inhibit GABA uptake without significantly affecting GABA receptors (Figure 7). The different profiles of these compounds with respect to neuronal and glial GABA uptake seem to indicate that GABA binds to these transport mechanisms in different conformations (129). Compound [22] is the most selective inhibitor of glial GABA uptake (129,170), and in contrast to [20] (171,172) and [21] (129), [22] is not a substrate for GABA transport mechanisms (173).

The specific GABA-A agonist THIP [7] is capable of penetrating the blood-brain barrier (BBB) (174,175), it is well tolerated by various animal species and the human (159,160), and it is active after oral administration (159,160). These aspects have focused much pharmacological and therapeutic interest on [7] (see later section) and prompted the synthesis of various structurally related compounds as exemplified in Figure 8. Even minor alterations of the structure of

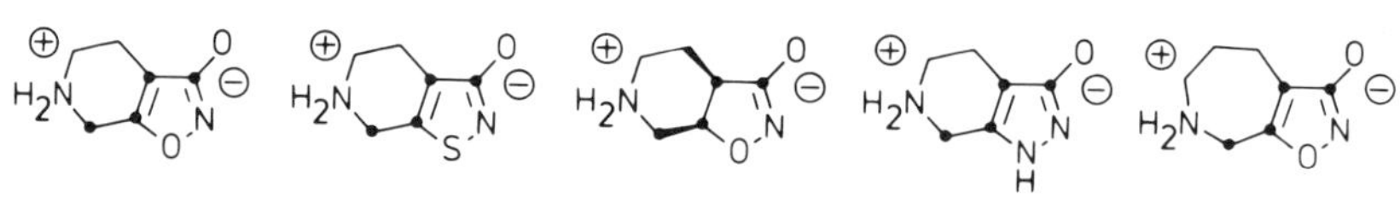

	7	23	24	25	26
AG	– – –(–)	–	N.t.	N.t.	N.t.
A	0.1	40	>100	>100	>100

Figure 8 Structure and biologic and in vitro activity of the GABA-A agonist THIP [7] and a number of structurally related compounds (see legend for Figure 7).

[7] result in pronounced or complete loss of activity. Although thio-THIP [23] is a weak GABA-A agonist (176), dihydro-THIP [24] (177), aza-THIP [25] (178), and the ring homolog of [7], 5,6,7,8-tetrahydro-4*H*-isoxazolo[5,4-*c*]-azepine-3-ol (THIA [26]) (179) show no detectable affinity for the GABA-A receptor sites. These structure-activity studies illustrate that very strict structural constraints are imposed on agonists at the GABA-A receptors.

Electronic and Stereochemical Factors

All GABA analogs with specific actions at GABA synaptic mechanisms are zwitterions. It is normally difficult to map out precisely the electronic structures of these zwitterions, including the degree of delocalization of their charges in vitro, and it is impossible to determine the exact electronic structures of GABA agonists during their interaction with the GABA receptors. Nevertheless, it is necessary to consider the approximate charged structures of GABA analogs in structure-activity studies on GABA agonists (40,118,180).

The very high potency of [18] as a GABA-A agonist (118) is consistent with its structural similarity with GABA, and the much lower activity of isomuscimol [28] is readily understandable on the basis of its very different charged structure (Figure 9) (118). The electronic structure of the sulfonate group of 3-aminopropanesulfonic acid (3APS [27]) is different from those of the negatively charged moieties of GABA and [18], which may explain why the mechanism of action of [27] and other aminosulfonic acid GABA agonists, such as [6] (Figure 3), is apparently somewhat different from that of GABA, [18], [4] and analogs of [4] (40,114).

As illustrated in Figure 9, the degree of charge delocalization of the imidazole and guanidino groups of imidazole-4-acetic acid (IAA [29]) and β-guanidinopropionic acid [32], respectively, are different from that of the amino group of GABA (40). Similarly, the positive charges of 2-aminothiazoline-4-acetic acid [30] (181) and *cis*-3-[(aminoiminomethyl)thio] propenoic acid [31] (157) are more delocalized than that of GABA. The relationship between structure and GABA-A receptor affinity of these compounds seems to indicate that a high degree of delocalization of the positive charges is tantamount to reduced affinity.

Stereostructure-activity studies have disclosed different effects of the optical isomers of chiral GABA analogs (36,40,167,182), as exemplified in Figure 10. The degree of stereoselectivity depends, however, on the structure of the chiral GABA analogs, being a function of the conformational flexibility of the compounds. Thus, the *S*(–) and *R*(+) forms of the flexible GABA analog 4-methyl-GABA [33] are equally effective at GABA-A receptor sites, and both interact with the neuronal as well as the glial GABA uptake systems in vitro (36). Introduction of double bonds into these molecules to give the *S*(–) and *R*(+) forms of *trans*-4-aminopent-2-enoic acid [34] has quite dramatic effects

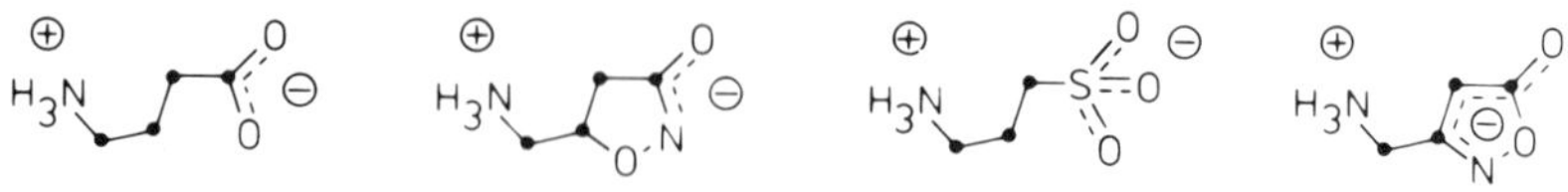

	1	18	27	28
A	0.03	0.008	0.08	30

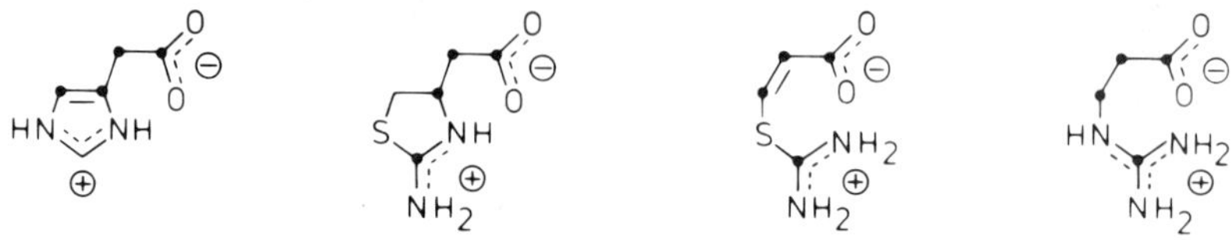

	29	30	31	32
A	0.1	0.5*	*ca.* 0.07	4

Figure 9 Structure, approximate charge delocalization, and GABA-A receptor affinity (A) of GABA [1] and a number of GABA analogs (see legend for Figure 7). The IC_{50} marked with an asterisk was determined under conditions in which the IC_{50} values for GABA [1] and isoguvacine [5] were 0.1 and 0.49 μM, respectively (From Reference 181).

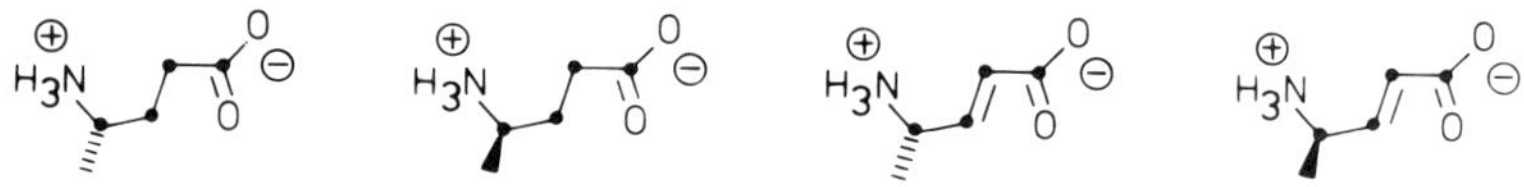

	(*S*)-(−)-	(*R*)-(+)-	(*S*)-(−)-	(*R*)-(+)-
	33		34	
A	5	5	4	>100
UPT-N	750	200	>5000	160
UPT-G	1000	120	>5000	500

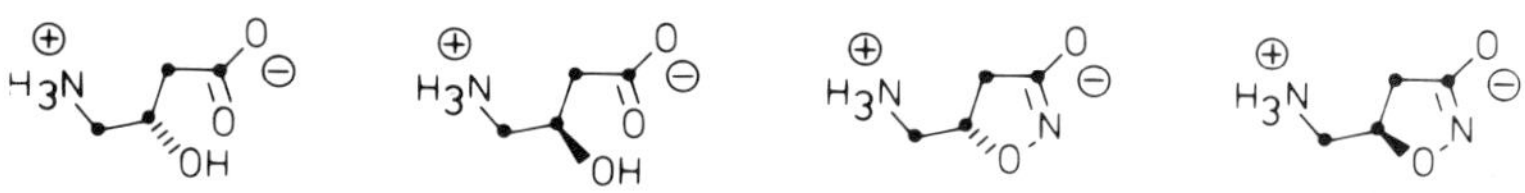

	(*S*)-(+)-	(*R*)-(−)-	(*S*)-(+)-	(*R*)-(−)-
	35		18	
A	0.4	1	0.004	0.3
UPT-N	130	400	>5000	800
UPT-G	300	800	>5000	2000

Figure 10 Structure and in vitro activity of a number of chiral GABA analogs with known absolute configuration (see legend for Figure 7).

(167,183). Although the (*S*) form of this conformationally restriced GABA analog specifically and quite effectively interacts with GABA-A receptors in vivo and in vitro (184), *R*(+)-[34] interacts with the GABA uptake systems without affecting GABA-A receptor sites (Figure 10) (167,184).

Similarly, the *S*(+) and *R*(–) isomers of the conformationally flexible GABA analog 3-hydroxy-4-aminobutyric acid [35] show a low degree of stereoselectivity as GABA-A agonists (Figure 11) and as GABA uptake inhibitors, whereas the less flexible bioisosteres of these optical isomers, *S*(+)- and *R*(–)-dihydromuscimol [18], respectively, have very different pharmacological profiles. *S*(+)-[18] is a very potent GABA-A agonist (Figure 11) showing the highest affinity for the GABA-A receptor sites so far described and no affinity for GABA uptake, but *R*(–)-[18] is a much less active GABA-A agonist (Figure 11) and inhibits GABA uptake (Figure 10) (182).

These stereostructure-activity relationships, which support earlier findings of different active conformations of GABA at GABA-A receptors and GABA uptake systems (36,88), are of importance for the design of new drugs with specific actions at synapses, at which GABA is the transmitter.

GABA-A Antagonists

As illustrated in Figure 3, the classic GABA antagonists 1*S*,9*R*(+)-bicuculline [2] (Figure 12) and picrotoxinin [9] have different sites of action at the postsynaptic GABA-A receptor complex. Both [2] and BMC [8] appear to interact directly with the GABA-A receptor (93,96,97), possibly with an "antagonist conformational state" of the receptor (89). It has recently been demonstrated that [8] selectively binds to low-affinity sites of the GABA-A receptors (91,93). In agreement with the stereoselectivity of the activation of the GABA-A receptors in the CNS (Figure 10), these receptors are blocked in a stereoselective manner (89,185). Thus, 1*R*,9*S*(--)-[2] has very weak GABA antagonist properties, and it does not significantly affect the binding of radioactive [8]. Although narcotine [36] has the same absolute stereochemistry as [2] (Figure 12), it is 10 times weaker than [2] as a GABA-A antagonist (185). The methoxy group on the tetrahydroisoquinoline ring of [36] may force the molecule to adopt a conformation different from that in which [2] binds to the GABA-A receptor. The (1*R*,9*S*) and (1*S*,9*S*) isomers of [36] have negligible GABA-A antagonist properties, and neither (1*R*,9*R*)-hydrastine [37] nor the (1*R*,9*S*) or (1*S*,9*S*) isomers of [37], none of which has the same absolute configuration as [2] (Figure 12), have significant GABA-A antagonist properties (185,186).

In recent years a number of GABA-A antagonists structurally unrelated to [2] have been developed. Pitrazepin [38] blocks the effects of synaptically released GABA and also antagonizes the actions of exogenous GABA, but not those of baclofen [3] (187). However, in contrast to [2[, [38] is an inhibitor

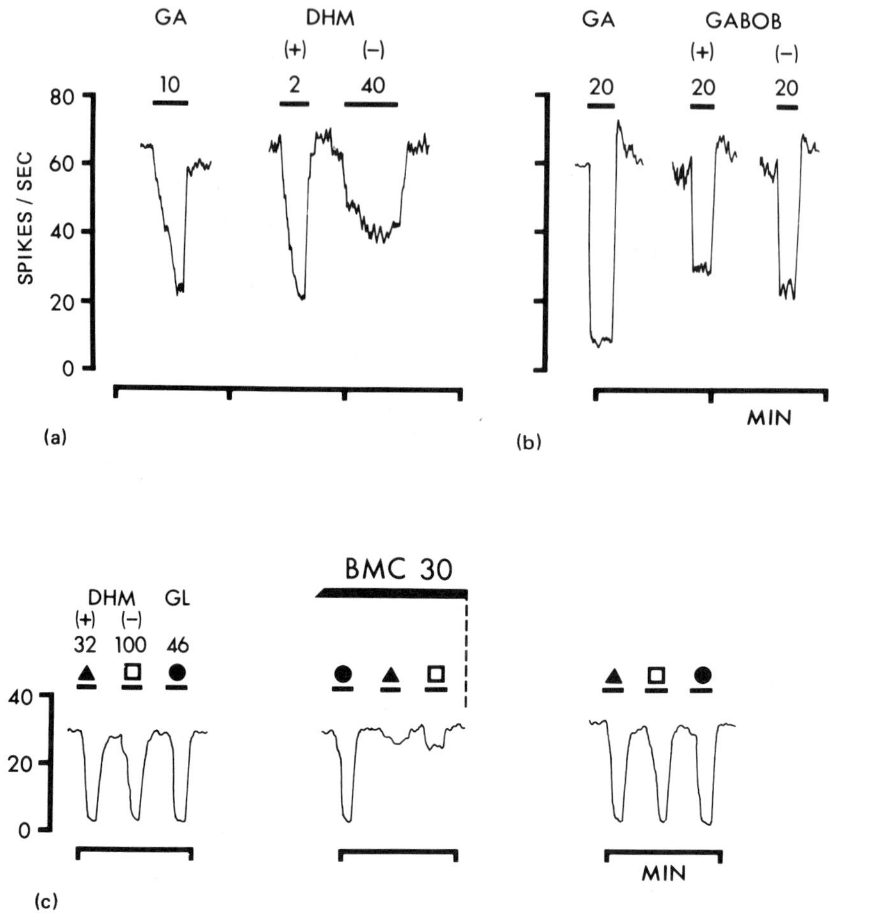

Figure 11 Comparison of the inhibitory effects of GABA (GA, [1]), glycine (GL), (*S*)-(+)-DHM [18], (*R*)-(−)-DHM [18], (*S*)-(+)-GABOB[35], and (*R*)-(−)-GABOB [35] on the firing of two spinal interneurons (a and c) and a Renshaw cell (b) maintained with continuously ejected (*RS*)-homocysteic acid (a, 19 nA; b, 8 nA; c, 13 nA). The amino acids were ejected for the times and with the currents (nA) indicated by the horizontal bars (for details, see Reference 182). (c) Before, during, and 2 min after the microelectrophoretic ejection of BMC [8] for a total time of 2.5 min, ceasing at the vertical broken line. Ordinates: firing rate, spikes per second. Abscissas: time, minutes. (Reproduced with permission from Reference 182.)

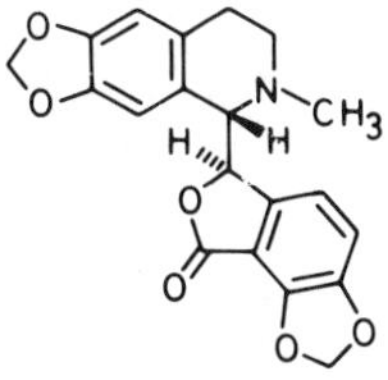

2

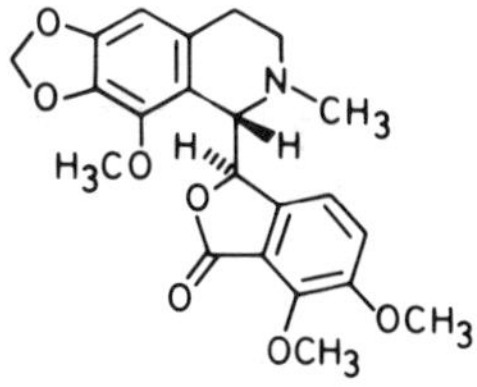

36

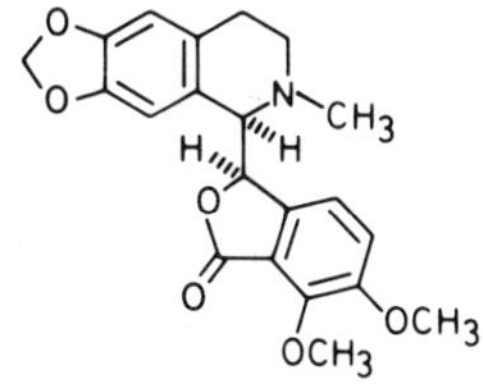

37

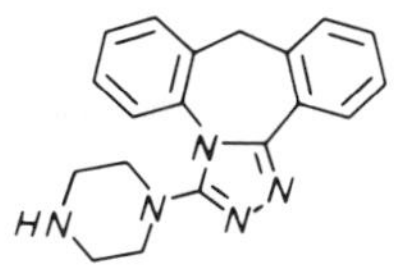

38

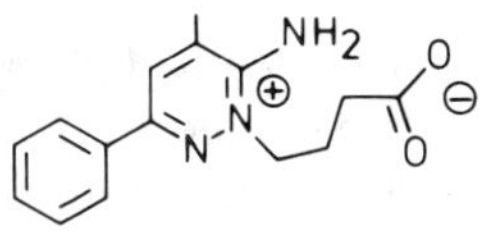

39

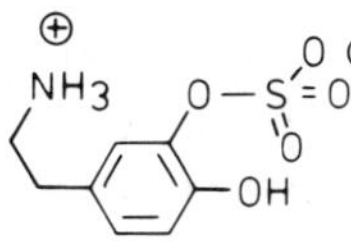

40

Figure 12 The structures of bicuculline [2], some structurally related alkaloids, and the GABA-A antagonists pitrazepin [38], SR 95103 [39], and dopamine-3-*O*-sulfate [40].

of the binding of radioactive flunitrazepam, and [38] is substantially more potent than [2] as an inhibitor of the binding of muscimol [4] (187). The aminopyridazine analog of GABA, SR 95103 [39], is a very potent GABA-A antagonist (188). In contrast to [38], compound [39] does not affect BZD binding, but like [2] it antagonizes the GABA-induced enhancement of BZD binding (188). In the cat spinal cord, [39] blocks the inhibitory effects of glycine as well as the effects of GABA (D. R. Curtis, unpublished). Dopamine-3-*O*-sulfate [40], as well as the 4-*O*-isomer, both metabolites of dopamine, have been shown to possess GABA-A antagonist profiles (189). These compounds appear to be the first examples of endogenous antagonists for the GABA-A receptors.

GABA-B Agonists

The therapeutic effects of baclofen [3] on certain types of spasticity (190) have prompted the synthesis of a variety of structurally related compounds, as exemplified in Figure 13. The pharmacological profiles of these compounds have been assessed on the basis of microelectrophoretic experiments on single neurons and pharmacological studies on reflexes in the cat (191). Alterations of the amino acid as well as the aromatic moiety of [3] have resulted in substantial or complete loss of pharmacological effects. Replacement of the chloro atom of [3] by a trifluoromethyl group to give [41] led to complete loss of activity.

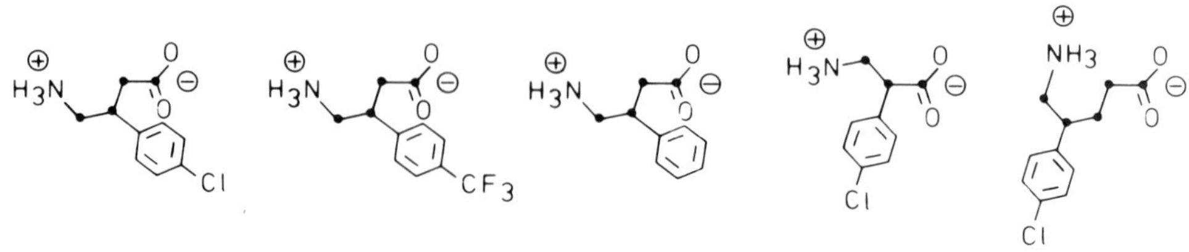

3 41 42 43 44

Figure 13 The structures of the GABA-B agonist baclofen [3] and some structurally related compounds.

Compound [42], containing an unsubstituted phenyl group, was much weaker than [3] in terms of reflex inhibiting properties and somewhat weaker as a depressant of the spontaneous firing of single neurons (191). Shortening or extension of the amino acid backbone of [3] also gave weaker compounds [43] and [44] showing no significant effects on reflexes, and both compounds were weaker than [3] as neuronal depressants (191). The unsaturated analogs of [3], *cis*- and *trans*-4-amino-3-(4-chlorophenyl)but-2-enoic acid have been synthesized, but no pharmacological data were reported (192).

A number of analogs of [3] have been tested as inhibitors of the binding of radioactive [3] or GABA to GABA-B receptor sites (145). All the analogs tested were much less active than unlabeled [3] and GABA ([42], being some two orders of magnitude weaker than [3]).

A number of aliphatic and heterocyclic GABA analogs (Figure 14) have been shown to interact with GABA-B receptor sites, but none of these com-

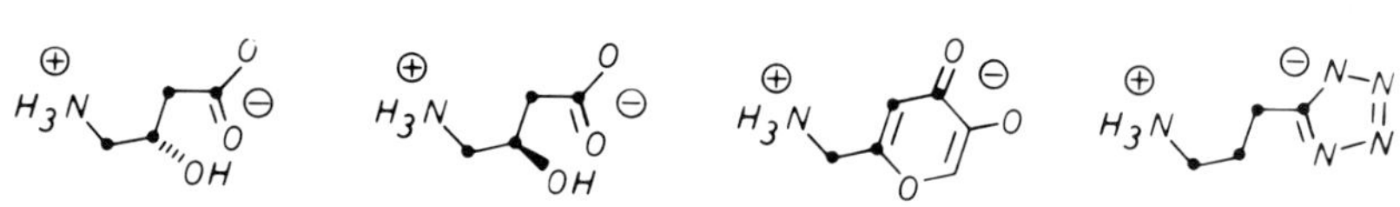

(S)-(+)-35 (R)-(−)-35 45 46

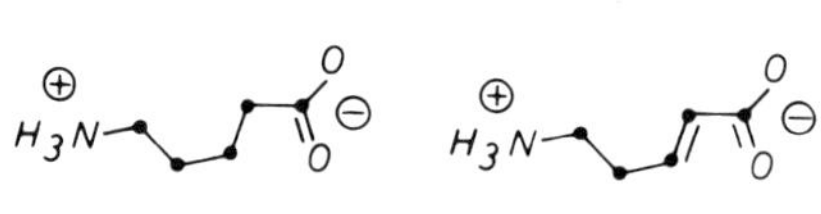

47 48

Figure 14 The structures of some GABA analogs.

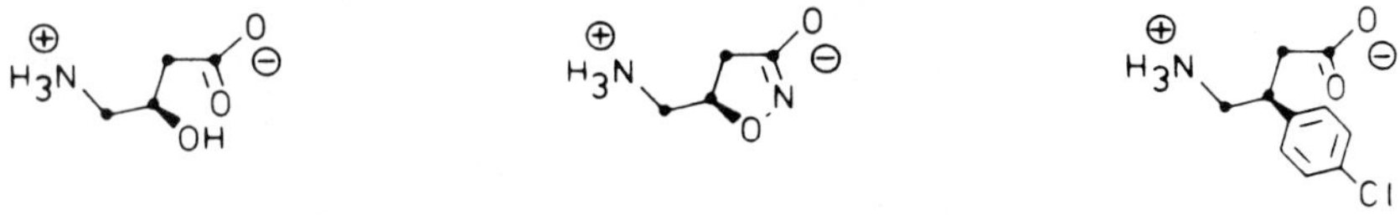

	(R)-(−)-35	(R)-(−)-18	(S)-(+)-3
A	1	0.3	>100
B	0.4	5	15
UPT-S	67	70	>300

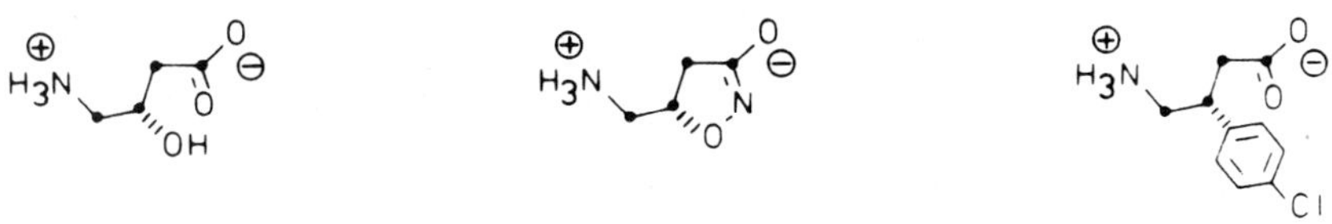

	(S)-(+)-35	(S)-(+)-18	(R)-(−)-3
A	0.4	0.004	>100
B	2.5	13	0.08
UPT-S	30	>300	>300

Figure 15 Structure and in vitro activity of some chiral GABA analogs with known absolute configuration. B: Inhibition of GABA-B receptor binding to rat brain membranes (IC_{50}, μM) using a procedure slightly different from that published (145). UPT-S: Inhibition of synaptosomal GABA uptake at a GABA concentration of 50 nM (IC_{50}, μM) (114). (See legend for Figure 7.)

pound exhibits specific GABA-B agonist or antagonist properties. The *S*(+) and *R*(−) isomers of [35] interact with GABA-B receptor sites as well as GABA uptake and GABA-A sites (Figure 15), and kojic amine [45] is a moderately potent agonist at GABA-A (193,194) as well as GABA-B receptors (195). 5-(3-Aminopropyl)tetrazole [46] has recently been shown to interact weakly with GABA-A and GABA-B receptor sites (196).

5-Aminopentanoic acid (DAVA [47]) has a very weak antagonist profile at peripheral GABA-B receptors (197), it binds weakly to central GABA-B receptor sites (Figure 16) (198); and it interacts more or less effectively with all

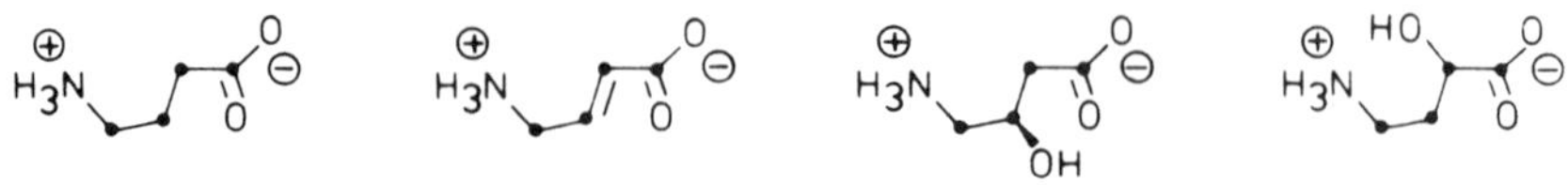

	1	15	(R)-(−)-35	49
	PARENT AMINO ACIDS			
A	0.03	0.1	1	15
B	0.03	2	0.4	>100

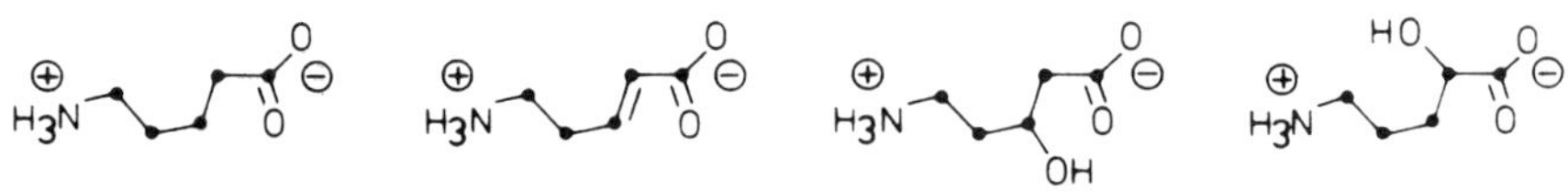

	47	48	50	51
	HOMOLOGUES			
A	4	65	15	50
B	10	>100	>100	23

Figure 16 Structure and in vitro activity of GABA [1], some GABA analogs, and their homologs (see legends for Figures 7 and 15.)

recognition sites at synapses containing GABA-A receptors (39,183). The unsaturated analog of [47], 5-aminopent-2-enoic acid [48], does not interact significantly with GABA-B, GABA-A, or GABA uptake sites (Figure 16) (183,198).

Comparative stereostructure-activity studies using the enantiomers of [35], [18], and [3] have disclosed that GABA-A and GABA-B receptor sites exhibit opposite stereoselectivity (Figure 15) (198). Thus, whereas *S*(+)-[35] is more potent than *R*(–)-[35] as an inhibitor of GABA-A binding, GABA-B sites show the opposite stereoselectivity, *R*(–)-[35] showing about one-fifth the affinity of *R*(–)-[3] for GABA-B sites. Similarly, *R*(–)-[18] is a more potent inhibitor of GABA-B binding than *S*(+)-[18], which is a powerful GABA-A agonist (Figure 11) and the most potent inhibitor of GABA-A binding so far described (Figure 10) (182).

S(+)-[18], evidently, almost perfectly mimics the active conformation of GABA at GABA-A receptors, whereas *R*(–)-[35] and (*R*)-(–)-[3] seem to reflect the conformation adopted by GABA during its binding to GABA-B receptor sites. In this connection it is interesting that *R*(–)-[35] and *R*(–)-[3] actually have opposite stereochemical orientations of the hydroxy and 4-chlorophenyl groups, respectively (Figure 15). This observation suggests strongly that the polar hydroxy group in the former GABA analog interacts with a structural element of the GABA-B receptor site different from that binding the lipophilic 4-chlorophenyl group of *R*(–)-[3] during its interaction with GABA-B receptor sites. Interestingly, 3-methoxy-4-aminobutyric acid, in which the hydroxy group of [35] has been replaced by the much less polar methoxy group, has no significant affinity for GABA-B receptor sites (198).

A prerequisite for a satisfactory mapping of the pharmacology of the GABA-B receptors is the development of specific GABA-B antagonists. The weak antagonist profile of DAVA [47] at certain GABA-B receptors (197) has prompted syntheses and in vitro studies of the homologs of a number of GABA analogs, including [15], [35], and [49] (Figure 16) (198). These homologs are typically much weaker than the parent compounds as inhibitors of GABA-B, as well as GABA-A, receptor binding, with one exception, namely, 2-hydroxy-5-aminopentanoic acid (2-OH-DAVA [51]). The corresponding parent amino acid [49] does not affect GABA-B binding, but [51] is only slightly less active than [47] at GABA-B sites (198). The mechanisms underlying this relatively weak GABA-B affinity of [51] have not been studied yet.

PHARMACOKINETIC ASPECTS OF GABA AGONISTS

The ability of neutral amino acids, for which active transport processes do not exist, to penetrate the lipophilic BBB is a function of two parameters: (1) the lipophilic character of the carbon backbone and substituents of the amino acid, and (2) the ratio between the concentrations of ionized and un-ionized molecules in solution (I/U ratio, ionization constant) (174,176). A high degree of lipophilicity and a low I/U ratio (the latter a function of the difference between the pK_a values of the amino acids) are tantamount to an increased ability of the compounds to penetrate the BBB (Figure 17).

The highly lipophilic character of the 4-chlorophenyl group allows [3] to enter the CNS after peripheral administration to animals and humans (199). GABA itself and most GABA analogs, including the specific GABA uptake inhibitor *R*(–)-[20] (Figure 17), are very polar compounds that in addition have high I/U ratios, and such compounds do not easily penetrate the BBB. THIP [7] and muscimol [4] are also very polar compounds, but the low I/U ratios for these GABA analogs allow penetration of the BBB (174-176). The I/U ratio for the selective glial GABA uptake inhibitor THPO [22] (Figure 7) is higher than those of [7] or [4] and whereas [22] appears to penetrate the BBB in young chickens, it does not easily enter the brain after systemic administration in adult mice (Figure 17) (200). In contrast to [4] (201-203), [7], [22], or [20] does not modify GABA-T activity in vitro (118,204). THIP [7] is well tolerated

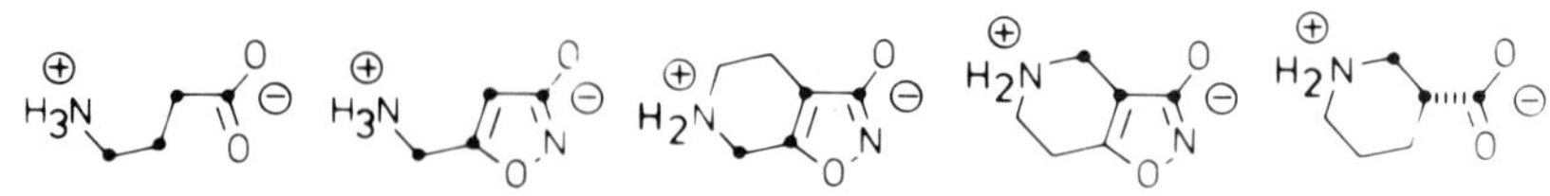

	1	4	7	22	(R)-(-)-20
AG	— — —	— — — —	— — —	0	0
A	0.03	0.006	0.1	80	>100
UPT-S	2	240	>500	160	4
GABA-T	1.92	1.27	No effect	No effect	No effect
pKA	4.0 : 10.7	4.8 : 8.4	4.4 : 8.5	4.3 : 9.1	3.9 : 10.3
I/U	800 000	900	1500	2500	250 000
BBB	No	Yes	Yes	Yes/No	No

Figure 17 Structure, biologic and in vitro activity, and pharmacokinetic properties of GABA [1] and some GABA analogs. GABA-T: Effects on GABA-T activity (K_m, mM) (118,203); I/U: I/U ratio (174,176); BBB: ability to penetrate the blood-brain barrier (174,175). (See legends for Figures 7 and 15.)

by various animal species and humans, it is active after oral administration, and it is excreted unchanged and, to some extent, in a conjugated form in animal and human urine (159,205).

Progabide [52] is a lipophilic derivative of GABA, or rather GABA-amide [53] (Figure 18). The spectroscopic data for [52] (206) indicate that the imino

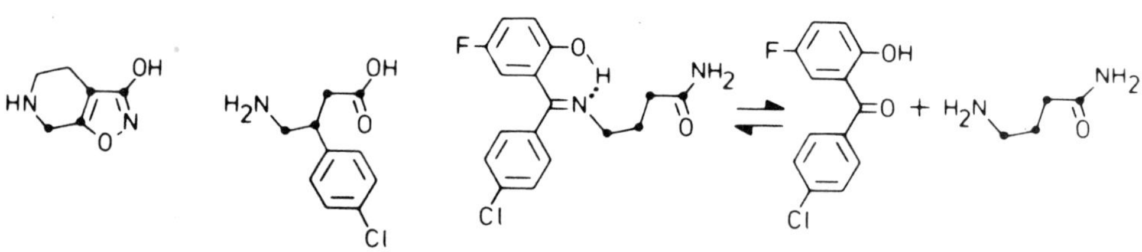

7 3 52 53

Figure 18 The structures of THIP [7], baclofen [3], and progabide [52] and the proposed constituents of an aqueous solution of [52].

group in this molecule is stabilized by a hydrogen bond, and in aqueous solution [52] is probably in equilibrium with the hydrolysis products [53] and the substituted benzophenone shown in Figure 18. In vivo, [52] appears to be converted into the corresponding carboxylic acid (SL 75.102), which is an active metabolite, whereas [53] is hydrolyzed to give GABA (207,208). These aspects make a detailed analysis of the mechanisms underlying the pharmacological and clinical effects of [52] difficult (209).

PHARMACOLOGICAL AND CLINICAL STUDIES OF GABA AGONISTS

The GABA-A agonist THIP [7], the GABA-B agonist baclofen [3], and the GABA agonist and GABA prodrug progabide [52] (Figure 18) have been and continue to be subjects of a wide variety of pharmacological and clinical studies. The anticonvulsant and nonopioid analgesic effects of [7] (159,160), the effects on spasticity of [3] (199,210), and the anticonvulsant actions and effects on psychiatric disorders of [52] (211,212) are of particular interest and have been reviewed. In this chapter only a few aspects of GABA agonist-induced analgesia and anticonvulsant effects are discussed.

The antinociceptive effects of [4], [7], [3], and morphine have been compared in a variety of test models. In general, the relative potencies of these analgesics are strongly dependent on the test system used (16,213). Furthermore, analgesia induced by [3] as well as by the GABA-A agonists [4] and [7] are generally insensitive to bicuculline [2] (16-18,162,163,213). As mentioned earlier, these observations have led to the suggestion that this effect may be mediated by a distinct subtype of GABA receptors. The insensitivity of analgesia induced by [3], [4], or [7] to naloxone indicates that these effects are not mediated by opiate receptors (16,18,162), in agreement with the observation that [7], in contrast to morphine, does not cause respiratory depression in animals or humans (213,214).

The potentiation of the analgesic effect of [7] by cholinergics and the antagonism of this effect by atropine reflects an as yet unclarified functional interaction between GABA and acetylcholine neurons in the CNS rather than a direct action of [7] on muscarinic receptors (18,162). Compound [7] is a clinically active analgesic comparable in potency to morphine (214,215), but the apparent analgesic effect of [3] in spastic patients may be the result of a reduction of the muscle spasms rather than a direct analgesic effect (216). It has been demonstrated that GABA and [3], in contrast to [4], potentiate the evoked release of met-enkephalin (217). This effect was insensitive to [2] but, quite surprisingly, antagonized by picrotoxinin (217).

At rather high doses, progabide [52] has a broad spectrum of anticonvulsant activities in experimental models and in epileptic patients (211). The structural similarity between [52] (Figure 18) and the BZD, such as diazepam [11]

(Figure 3), may explain some of the pharmacological effects of [52]. In view of its complex pharmacokinetics (see previous section), it is difficult to analyze in detail the mechanism of action of [52]. THIP [7] has anticonvulsant effects in a variety of animals (159,218-220). However, [7] was not an anticonvulsant in baboons with photosensitive epilepsy, and at relatively high doses actually induced epilepti-form activity (220). Similar effects have been observed for [4] (221), and it is interesting to note that in a particular strain of mice, [7] suppressed myoclonic jerks induced by [4] (222). In a single-blind controlled trial of [7] in epileptic patients, in which [7] was added to concomitant antiepileptic treatment, [7] did not show significant effects, although a trend was observed for lower seizure frequency, notably during a period of submaximal doses of [7] (223).

Like the GABA-A agonists [4] and [7], the GABA-B agonist [3] has anticonvulsant properties (199), and the changes in EEG patterns in rats induced by [4], [7], [3], and [52] have been studied (161,224). The electrographic epileptiform activities induced by [4] and [7] were similar and were somewhat different from the EEG abnormalities provoked by [52] (224). On the other hand, the patterns of pharmacological interactions of [7] or [52] with a number of BZD are very similar (225). Quite surprisingly, the EEG abnormalities induced by [7] were shown to be similar to, although not identical with, those induced by [3] (161). Involvement of the excitatory amino acid neurotransmitters, such as GLU (Figure 2b), in these effects of [7] seems unlikely, and furthermore, disinhibitory GABA neuronal mechanisms do not seem to underlie the effects of [7] on the EEG patterns in animals (161,224). As mentioned earlier, it is noteworthy that these effects of the GABA-A agonist [7] could not be reversed by [2] (161).

It can be concluded that many aspects of the pharmacology of the GABA receptors are far from being fully elucidated. A number of potent agents with specific effects related to the function of GABA as a neurotransmitter are now available for pharmacological and, in some cases, clinical studies. It is to be anticipated that the therapeutic prospects of pharmacological manipulation of the GABA receptors will be further elucidated in the near future.

ACKNOWLEDGMENTS

This work was supported by grants from the Danish Medical Research Council. The secretarial and technical assistance of Mrs. B. Hare and Mr. S. Stilling is gratefully acknowledged.

REFERENCES

1. D. R. Curtis and G. A. R. Johnston, Ergeb. Physiol. *69*, 97 (1974).
2. K. Krnjević, Physiol. Rev. *54*, 418 (1974).

3. E. Roberts, T. N. Chase, and D. B. Tower (eds.), GABA in Nervous System Function, Raven Press, New York, 1976.
4. F. Fonnum (ed.), Amino Acids as Chemical Transmitters, Plenum Press, New York, 1978.
5. P. Krogsgaard-Larsen, J. Scheel-Krüger, and H. Kofod (eds.), GABA-Neurotransmitters. Pharmacochemical, Biochemical and Pharmacological Aspects, Munksgaard, Copenhagen, 1979.
6. L. Grandison and A. Guidotti, Endocrinology *105*, 754 (1979).
7. E. E. Müller, D. Cocchi, V. Locatelli, J. A. Apud, M. L. Tappaz, C. Masotto, A. Novelli, and G. Racagni, in The GABA Receptors (S. J. Enna, ed.), Humana Press, Clifton, New Jersey, 1983, p. 257.
8. S. J. Enna, Biochem. Pharmacol. *30*, 907 (1981).
9. M. J. Antonaccio, L. Kerwin, and D. G. Taylor, Neuropharmacology *17*, 597 (1978).
10. M. J. Antonaccio, L. Kerwin, and D. G. Taylor, Neuropharmacology *17*, 783 (1978).
11. J. A. DiMicco, K. Gale, B. Hamilton, and R. A. Gillis, Science *204*, 1106 (1979).
12. F. V. DeFeudis, Neurochem. Int. *3*, 113 (1981).
13. J. A. DiMicco, N. L. Alsip, and J. H. Wible, Neuropharmacology *23*, 819 (1984).
14. I. K. Ho, H. H. Lok, and E. L. Way, Life Sci. *18*, 1111 (1976).
15. E. V. DeFeudis, Prog. Neurobiol. *19*, 1 (1982).
16. R. C. Hill, R. Maurer, H.-H. Buescher, and D. Roemer, Eur. J. Pharmacol. *69*, 221 (1981).
17. T. Andree, D. A. Kendall, and S. J. Enna, Life Sci. *32*, 2265 (1983).
18. A. Grognet, F. Hertz, and F. V. DeFeudis, Gen. Pharmacol. *14*, 585 (1983).
19. J. F. Tallman, S. M. Paul, P. Skolnick, and D. W. Gallager, Science *207*, 274 (1980).
20. R. Hoehn-Saric, Psychopharmacology *80*, 338 (1983).
21. J. A. Gray, J. Mellanby, and C. Buckland, Neuropharmacology *23*, 827 (1984).
22. N. Blavet, F. V. DeFeudis, and F. Clostre, Psychopharmacology *76*, 75 (1982).
23. P. Mandel, L. Ciesielski, M. Maitre, S. Simler, G. Mack, and E. Kempf, in GABA-Biochemistry and CNS Functions (P. Mandel and F. V. DeFeudis, eds.), Plenum Press, New York, 1979, p. 475.
24. R. G. Fariello, P. L. Morselli, K. G. Lloyd, L. F. Quesney, and J. Engel (eds.), Neurotransmitters, Seizures, and Epilepsy II, Raven Press, New York, 1984.
25. G. Nistico, P. L. Morselli, K. G. Lloyd, R. G. Fariello, and J. Engel (eds.), Neurotransmitters, Seizures, and Epilepsy III, Raven Press, New York, 1986 (in press).
26. C. E. Ribak, A. B. Harris, L. Anderson, J. E. Vaughn, and E. Roberts, Science *205*, 211 (1979).

27. S. M. Ross and C. R. Craig, J. Neurochem. *36*, 1006 (1981).
28. T. N. Chase, N. S. Wexler, and A. Barbeau, (eds.), Huntington's Disease, Raven Press, New York, 1979.
29. S. J. Enna, E. D. Bird, J. P. Bennett, D. B. Bylund, H. I. Yamamura, L. L. Iversen, and S. H. Snyder, N. Engl. J. Med. *294*, 1305 (1976).
30. K. G. Lloyd, L. Shemen, and O. Hornykiewicz, Brain Res. *127*, 269 (1977).
31. J. Stevens, K. Wilson, and W. Foote, Psychopharmacologia *39*, 105 (1974).
32. D. P. Van Kammen, Am. J. Psychiatry *134*, 138 (1977).
33. F. Petty and J. A. Coffman, Neuropharmacology *23*, 859 (1984).
34. P. Davies, Brain Res. *171*, 319 (1979).
35. K. G. Lloyd and A. Pilc, Neuropharmacology *23*, 841 (1984).
36. P. Krogsgaard-Larsen, J. Med. Chem. *24*, 1377 (1981).
37. B. L. Waszczak, C. Hume, and J. R. Walters, Life Sci. *28*, 2411 (1981).
38. R. G. Fariello and M. K. Ticku, Life Sci. *33*, 1629 (1983).
39. R. D. Allan and G. A. R. Johnston, Medicinal Res. Rev. *3*, 91 (1983).
40. P. Krogsgaard-Larsen, P. Jacobsen, and E. Falch, in The GABA Receptors (S. J. Enna, ed.), Humana Press, Clifton, New Jersey, 1983, p. 149.
41. S. J. Enna (ed.), The GABA Receptors, Humana Press, Clifton, New Jersey, 1983.
42. N. G. Bowery (ed.), Actions and Interactions of GABA and Benzodiazepines, Raven Press, New York (1984).
43. D. R. Curtis and J. C. Watkins, Pharmacol. Rev. *17*, 347 (1965).
44. D. R. Curtis, A. W. Duggan, D. Felix, and G. A. R. Johnston, Brain Res. *32*, 69 (1971).
45. S. J. Enna and J. P. Gallagher, Int. Rev. Neurobiol. *24*, 181 (1983).
46. D. R. Curtis, in Amino Acids as Chemical Transmitters (F. Fonnum, ed.), Plenum Press, New York, 1978, p. 55.
47. D. R. Curtis, D. Lodge, J. C. Bornstein, M. J. Peet, and J. D. Leah, Exp. Brain Res. *48*, 387 (1982).
48. D. A. Brown, Trends Neurosci. *2*, 271 (1979).
49. B. E. Alger and R. A. Nicoll, J. Physiol. (Lond.) *328*, 125 (1982).
50. M. A. Simmonds, in Actions and Interactions of GABA and Benzodiazepines (N. G. Bowery, ed.), Raven Press, New York, 1984, p. 27.
51. S. Z. Langer, Br. J. Pharmacol. *60*, 481 (1977).
52. K. Starke, Annu. Rev. Pharmacol. Toxicol. *21*, 7 (1981).
53. P. R. Mitchell and I. L. Martin, Nature *274*, 904 (1978).
54. S. R. Snodgrass, Nature *274*, 392 (1978).
55. M. J. W. Brennan and R. C. Cantrill, Nature *280*, 514 (1979).
56. S. Arbilla, L. Kamal, and S. Z. Langer, Eur. J. Pharmacol. *57*, 211 (1979).
57. M. J. W. Brennan, R. C. Cantrill, M. Oldfield, and P. Krogsgaard-Larsen, Mol. Pharmacol. *19*, 27 (1981).
58. D. R Curtis, A. W. Duggan, D. Felix, G. A. R. Johnston, and H. McLennan, Brain Res. *33*, 57 (1971).
59. P. Krogsgaard-Larsen, in Metabolic Relationship Between Glutamine, Glutamate and GABA in the CNS (L. Hertz, E. Kvamme, E. G. McGeer, and A. Schousboe, eds.), Alan. R. Liss, New York, 1983, p. 537.

60. D. W. Straughan, M. J. Neal, M. A. Simmonds, G. G. S. Collins, and R. G. Hill, Nature *233*, 352 (1971).
61. J. Arnt and P. Krogsgaard-Larsen, Brain Res. *177*, 395 (1979).
62. P. Krogsgaard-Larsen, H. Hjeds, D. R. Curtis, J. D. Leah, and M. J. Peet, J. Neurochem. *39*, 1319 (1982).
63. N. G. Bowery, in The GABA Receptors (S. J. Enna, ed.), Humana Press, Clifton, New Jersey, 1983, p. 177.
64. K. Kato, M. Goto, and H. Fukuda, Gen. Pharmacol. *13*, 445 (1982).
65. R. W. Olsen, J. Neurochem. *37*, 1 (1981).
66. J. F. Tallman, in The GABA Receptors (S. J. Enna, ed.), Humana Press, Clifton, New Jersey, 1983, p. 93.
67. E. J. Peck, J. M. Schaeffer, and J. H. Clark, Biochem. Biophys. Res. Commun. *52*, 394 (1973).
68. S. R. Zukin, A. B. Young, and S. H. Snyder, Proc. Natl. Acad. Sci. USA *71*, 4802 (1974).
69. S. J. Enna and S. H. Snyder, Brain Res. *100*, 81 (1975).
70. S. J. Enna and S. H. Snyder, Mol. Pharmacol. *13*, 442 (1977).
71. D. V. Greenlee, P. C. Van Ness, and R. W. Olsen, Life Sci. *22*, 1653 (1978).
72. M. J. Leach and J. A. Wilson, Eur. J. Pharmacol. *48*, 329 (1978).
73. K. Beaumont, W. S. Chilton, H. I. Yamamura, and S. J. Enna, Brain Res. *148*, 153 (1978).
74. M. Williams and E. A. Risley, J. Neurochem. *32,* 713 (1979).
75. Y. J. Wang, P. Salvaterra, and E. Roberts, Biochem. Pharmacol. *28*, 1123 (1979).
76. A. M. Morin and C. Wasterlain, Life Sci. *26*, 1239 (1980).
77. P. Krogsgaard-Larsen, A. Snowman, S. C. Lummis, and R. W. Olsen, J. Neurochem. *37*, 401 (1981).
78. E. Falch and P. Krogsgaard-Larsen, J. Neurochem. *38*, 1123 (1982).
79. D. J. Wong and J. S. Horng, Life Sci. *20*, 445 (1977).
80. A. Guidotti, G. Toffano, and E. Costa, Nature *275*, 553 (1978).
81. T. H. Chiu and H. C. Rosenberg, Eur. J. Pharmacol. *58*, 335 (1979).
82. C. Napias, M. O. Bergman, P. C. Van Ness, D. V. Greenlee, and R. W. Olsen, Life Sci. *27*, 1001 (1980).
83. C. C. Jordan, A. I. Matus, W. Piotrowski, and D. Wilkinson, J. Neurochem. *39*, 52 (1982).
84. M. R. Quinn and R. H. Cagan, J. Neurochem *39*, 1381 (1982).
85. F. V. DeFeudis, J. Neurosci. Res. *10*, 107 (1983).
86. J. Kardos, K. Maderspach,and M. Simonyi, Neurochem. Int. *7*, 737 (1985).
87. R. W. Olsen, M. O. Bergman, P. C. Van Ness, S. C. Lummis, A. E. Watkins, C. Napias, and D. V. Greenlee, Mol. Pharmacol. *19*, 217 (1981).
88. P. Krogsgaard-Larsen and E. Falch, Mol. Cell. Biochem. *38*, 129 (1981).
89. H. Möhler and T. Okada, Nature *267*, 65 (1977).
90. H. Möhler and T. Okada, Mol. Pharmacol. *14*, 256 (1978).
91. R. W. Olsen and A. M. Snowman, J. Neurochem. *41*, 1653 (1983).
92. J. H. Skerritt and G. A. R. Johnston, Neurosci. Lett. *38*, 315 (1983).
93. R. W. Olsen, E. W. Snowhill, and J. K. Wamsley, Eur. J. Pharmacol. *99*, 247 (1984).

94. R. W. Olsen, in The GABA Receptors (S. J. Enna, ed.), Humana Press, Clifton, New Jersey, 1983, p. 63.
95. N. J. M. Birdsall, A. S. V. Burgen, and E. C. Hulme, in Cholinergic Mechanisms and Psychopharmacology (D. J. Jenden, ed.), Plenum Press, New York, 1977, p. 25.
96. R. C. Frere, R. L. Macdonald, and A. B. Young, Brain Res. *244*, 145 (1982).
97. L. M. Nowak, A. B. Young, and R. L. Macdonald, Brain Res. *244*, 155 (1982).
98. R. F. Squires and C. Braestrup, Nature *266*, 732 (1977).
99. H. Möhler and T. Okada, Science *198*, 849 (1977).
100. R. F. Squires, J. E. Casida, M. Richardson, and E. Saederup, Mol. Pharmacol. *23*, 326 (1983).
101. J. Seifert and J. E. Casida, J. Neurochem. *44*, 110 (1985).
102. R. Ramanjaneyula and M. K. Ticku, J. Neurochem. *42*, 221 (1984).
103. G. Maksay and M. K. Ticku, J. Neurochem. *44*, 480 (1985).
104. M. Williams and G. G. Yarbrough, Eur. J. Pharmacol. *56*, 273 (1979).
105. P. Supavilai and M. Karobath, J. Neurochem. *36*, 798 (1981).
106. S. S. Pong and C. C. Wang, J. Neurochem. *38*, 375 (1982).
107. M. Williams and E. A. Risley, Eur. J. Pharmacol. *77*, 307 (1982).
108. C. C. Wang, J. Med. Chem. *27*, 1 (1984).
109. M. Williams and E. A. Risley, J. Neurochem. *42*, 745 (1984).
110. R. W. Olsen, M. K. Ticku, D. Greenlee, and P. Van Ness, in GABA-Neurotransmitters. Pharmacochemical, Biochemical and Pharmacological Aspects (P. Krogsgaard-Larsen, J. Scheel-Krüger, and H. Kofod, eds.), Munksgaard, Copenhagen, 1979, p. 165.
111. G. A. R. Johnston, in The GABA Receptors (S. J. Enna, ed.), Humana Press, Clifton, New Jersey, 1983, p. 107.
112. J. F. Tallman and D. W. Gallager, Annu. Rev. Neurosci. *8*, 21 (1985).
113. C. Braestrup, M. Nielsen, P. Krogsgaard-Larsen, and E. Falch, Nature *280*, 331 (1979).
114. E. Falch, P. Jacobsen, P. Krogsgaard-Larsen, and D. R. Curtis, J. Neurochem. *44*, 68 (1985).
115. E. H. F. Wong and L. L. Iversen, J. Neurochem. *44*, 1162 (1985).
116. P. Supavilai and M. Karobath, Neurosci. Lett. *19*, 337 (1980).
117. P. Krogsgaard-Larsen, G. A. R. Johnston, D. Lodge, and D. R. Curtis, Nature *268*, 53 (1977).
118. P. Krogsgaard-Larsen, H. Hjeds, D. R. Curtis, D. Lodge, and G. A. R. Johnston, J. Neurochem. *32*, 1717 (1979).
119. D. R. Curtis and R. Malik, Proc. R. Soc. Lond. *B223*, 25 (1984).
120. N. G. Bowery, A. Doble, D. R. Hill, A. L. Hudson, and M. J. Turnbull, Br. J. Pharmacol. *70*, 77P (1980).
121. S. Arbilla and S. Z. Langer, Naunyn-Schmiedebergs Arch. Pharmacol. *306*, 161 (1979).
122. J. C. Stoof and A. H. Mulder, Eur. J. Pharmacol. *46*, 177 (1977).
123. M. F. Giorguieff, M. L. Kemel, J. Glowinski, and M. J. Besson, Brain Res. *139*, 115 (1978).

124. M. S. Starr, Eur. J. Pharmacol. *53*, 215 (1979).
125. C. Ennis and B.Cox, Eur. J. Pharmacol. *70*, 417 (1981).
126. R. Mitchell, Biochem. Pharmacol. *31*, 2684 (1982).
127. E. Meier, J. Drejer, and A. Schousboe, J. Neurochem. *43*, 1737 (1984).
128. M. J. W. Brennan and R. C. Cantrill, J. Neurochem. *33*, 721 (1979).
129. P. Krogsgaard-Larsen, Mol. Cell. Biochem. *31*, 105 (1980).
130. P. Krogsgaard-Larsen, E. Falch, A. Schousboe, D. R. Curtis, and D. Lodge, J. Neurochem. *34*, 756 (1980).
131. P. Krogsgaard-Larsen and G. A. R. Johnston, J. Neurochem. *30*, 1377 (1978).
132. A. Schousboe, O. M. Larsson, and P. Krogsgaard-Larsen, Neurochem. Int. *7*, 505 (1985).
133. M. J. W. Brennan, R. C. Cantrill, and P. Krogsgaard-Larsen, in GABA and Benzodiazepine Receptors (E. Costa, G. Di Chiara, and G. L. Gessa, eds.), Raven Press, New York, 1980, p. 157.
134. M. J. W. Brennan and R. C. Cantrill, J. Neurochem. *31*, 1339 (1978).
135. R. O. Lockerbie and P. R. Gordon-Weeks, Neurosci. Lett. *55*, 273 (1985).
136. M. J. W. Brennan, R. C. Cantrill, and S. Kramer, Int. J. Biochem. *12*, 833 (1980).
137. M. J. W. Brennan, J. Neurochem. *38*, 264 (1982).
138. D. R. Curtis, C. J. A. Game, G. A. R. Johnston, and R. M. McCulloch, Brain Res. *70*, 493 (1974).
139. J. Davies and J. C. Watkins, Brain Res. *70*, 501 (1974).
140. N. G. Bowery and A. L. Hudson, Br. J. Pharmacol. *66*, 108P (1979).
141. N. G. Bowery, A. Doble, D. R. Hill, A. L. Hudson, J. S. Shaw, M. J. Turnbull, and R. Warrington, Eur. J. Pharmacol. *71*, 53 (1981).
142. N. G. Bowery, D. R. Hill, A. L. Hudson, A. Doble, D. N. Middlemiss, J. Shaw, and M. Turnbull, Nature *283*, 92 (1980).
143. D. R. Hill and N. G. Bowery, Nature *290*, 149 (1981).
144. N. G. Bowery, D. R. Hill, and A. L. Hudson, Br. J. Pharmacol. *78*, 191 (1983).
145. N. G. Bowery, D. R. Hill, and A. L. Hudson, in Problems in GABA Research from Brain to Bacteria (Y. Okada and E. Roberts, eds.), Excerpta Medica, Amsterdam, 1982, p. 302.
146. N. G. Bowery, D. R. Hill, A. L. Hudson, G. W. Price, M. J. Turnbull, and G. P. Wilkin, in Actions and Interactions of GABA and Benzodiazepines (N. G. Bowery, ed.), Raven Press, New York, 1984, p. 81.
147. S. J. Enna and E. W. Karbon, Neuropharmacology *23*, 821 (1984).
148. D. R. Hill, N. G. Bowery, and A. L. Hudson, J. Neurochem. *42*, 652 (1984).
149. G. P. Wilkin, A. L. Hudson, D. R. Hill, and N. G. Bowery, Nature *274*, 584 (1981).
150. J. M. Palacios, W. Scott Young, and M. J. Kuhar, Proc. Natl. Acad. Sci. USA *77*, 670 (1980).
151. J. M. Palacios, J. K. Wamsley, and M. J. Kuhar, Brain Res. *222*, 285 (1981).

152. G. A. R. Johnston, D. R. Curtis, P. M. Beart, C. J. A. Game, R. M. McCulloch, and B. Twitchin, J. Neurochem. *24*, 157 (1975).
153. P. Krogsgaard-Larsen, G. A. R. Johnston, D. R. Curtis, C. J. A. Game, and R. M. McCulloch, J. Neurochem. *25*, 803 (1975).
154. R. D. Allan, D. R. Curtis, P. M. Headley, G. A. R. Johnston, D. Lodge, and B. Twitchin, J. Neurochem. *34*, 652 (1980).
155. C. A. Drew, G. A. R. Johnston, and R. P. Weatherby, Neurosci. Lett. *52*, 317 (1984).
156. P. R. Andrews and G. A. R. Johnston, Biochem. Pharmacol. *28*, 2697 (1979).
157. G. A. R. Johnston and R. D. Allan, Neuropharmacology *23*, 831 (1984).
158. G. A. R. Johnston, R. D. Allan, S. M. E. Kennedy, and B. Twitchin, in GABA-Neurotransmitters. Pharmacochemical, Biochemical and Pharmacological Aspects (P. Krogsgaard-Larsen, J. Scheel-Krüger, and H. Kofod, eds.), Munksgaard, Copenhagen, 1979, p. 149.
159. P. Krogsgaard-Larsen, E. Falch, and V. Christensen, Drugs Future *9*, 597 (1984).
160. P. Krogsgaard-Larsen, E. Falch, and H. Hjeds, Prog. Med. Chem. *22*, 67 (1985).
161. G. T. Golden and R. G. Fariello, in Neurotransmitters, Seizures and Epilepsy II (R. G. Fariello, P. L. Morselli, K. G. Lloyd, L. F. Quesney, and J. Engel, eds.), Raven Press, New York, 1984, p. 237.
162. D. A. Kendall, M. Browner, and S. J. Enna, J. Pharmacol. Exp. Ther. *220*, 482 (1982).
163. S. H. Zorn and S. J. Enna, Brain Res. *338*, 380 (1985).
164. A. Agmo and R. Paredes, Eur. J. Pharmacol. *112*, 371 (1985).
165. O. M. Larsson, P. Thorbek, P. Krogsgaard-Larsen, and A. Schousboe, J. Neurochem. *37*, 1509 (1981).
166. G. A. R Johnston, P. Krogsgaard-Larsen, and A. L. Stephanson, J. Neurochem. *26*, 1029 (1976).
167. A. Schousboe, P. Thorbek, L. Hertz, and P. Krogsgaard-Larsen, J. Neurochem. *33*, 181 (1979).
168. G. A. R. Johnston, P. Krogsgaard-Larsen, and A. L. Stephanson, Nature *258*, 627 (1975).
169. A. Schousboe, O. M. Larsson, L. Hertz, and P. Krogsgaard-Larsen, Drug Dev. Res. *1*, 115 (1981).
170. A. Schousboe, O. M. Larsson, J. D. Wood, and P. Krogsgaard-Larsen, Epilepsia *24*, 531 (1983).
171. G. A. R. Johnston, A. L. Stephanson, and B. Twitchin, J. Neurochem. *26*, 83 (1976).
172. O. M. Larsson, P. Krogsgaard-Larsen, and A. Schousboe, J. Neurochem. *34*, 970 (1980).
173. N. Seiler, S. Sarhan, P. Krogsgaard-Larsen, H. Hjeds, and A. Schousboe, Gen. Pharmacol. *16*, 509 (1985).
174. P. Krogsgaard-Larsen, B. Schultz, H. Mikkelsen, T. Aaes-Jørgensen, and K. P. Bøgesø, in Amino Acid Neurotransmitters (F. V. DeFeudis and P. Mandel, eds.), Raven Press, New York, 1981, p. 69.

175. F. Moroni, M. C. Forchetti, P. Krogsgaard-Larsen, and A. Guidotti, J. Pharm. Pharmacol. *34*, 676 (1982).
176. P. Krogsgaard-Larsen, H. Mikkelsen, P. Jacobsen, E. Falch, D. R. Curtis, M. J. Peet, and J. D. Leah, J. Med. Chem. *26*, 895 (1983).
177. R. Nordmann, P. Graff, R. Maurer, and B. H. Gähwiler, J. Med. Chem. *28*, 1109 (1985).
178. P. Krogsgaard-Larsen and T. R. Christiansen, Eur. J. Med. Chem. *14*, 157 (1979).
179. L. Brehm, P. Krogsgaard-Larsen, K. Schaumburg, J. S. Johansen, E. Falch, and D. R. Curtis, J. Med. Chem. *29*, 224 (1986).
180. D. R. Armstrong, R. J. Breckenridge, and C. J. Suckling, J. Theor. Biol. *97*, 267 (1982).
181. D. R. Bristow, M. M. Campbell, L. L. Iversen, J. A. Kemp, G. Marshall, K. J. Watling, and E. H. F. Wong, unpublished data, 1985.
182. P. Krogsgaard-Larsen, L. Nielsen, E. Falch, and D. R. Curtis, J. Med. Chem. *28*, 1612 (1985).
183. T. Honoré, H. Hjeds, P. Krogsgaard-Larsen, and T. R. Christiansen, Eur. J. Med. Chem. *13*, 429 (1978).
184. P. Krogsgaard-Larsen, E. Falch, M. J. Peet, J. D. Leah, and D. R. Curtis, in CNS Receptors–from Molecular Pharmacology to Behaviour (P. Mandel and F. V. DeFeudis, eds.), Raven Press, New York, 1983, p. 1.
185. S. J. Enna, J. F. Collins, and S. H. Snyder, Brain Res. *124*, 185 (1977).
186. N. G. Bowery, J. F. Collins, G. Cryer, T. D. Inch, and N. J. McLaughlin, in GABA Biochemistry and CNS Functions (P. Mandel and F. V. DeFeudis, eds.), Plenum Press, New York, 1979, p. 339.
187. B. H. Gähwiler, R. Maurer, and H. J. Wüthrich, Neurosci. Lett. *45*, 311 (1984).
188. J.-P. Chambon, P. Feltz, M. Heaulme, S. Restle, R. Schlichter, K. Biziere, and C. G. Wermuth, Proc. Natl. Acad. Sci. USA *82*, 1832 (1985).
189. N. T. Buu, J. Duhaime, and O. Kuchel, Life Sci. *35*, 1083 (1984).
190. D. Burke, C. J. Andrews, and L. Knowles, J. Neurol. Sci. *14*, 199 (1971).
191. H.-R. Olpe, A. Glatt, and W. Bencze, Brain Res. Bull. *5*(Suppl. 2), 507 (1980).
192. R. D. Allan and H. Tran, Aust. J. Chem. *34*, 2641 (1981).
193. G. G. Yarbrough, M. Williams, and D. R. Haubrich, Arch. Int. Pharmacodyn. *241*, 266 (1979).
194. J. G. Atkinson, Y. Girard, J. Rokach, C. S. Rooney, C. S. McFarlane, A. Rackham, and N. N. Share, J. Med. Chem. *22*, 99 (1979).
195. S. J. Enna (ed.), in The GABA Receptors, The Humana Press, Clifton, New Jersey, 1983, p. 1.
196. G. Schlewer, C. G. Wermuth, and J. P. Chambon, Eur. J. Med. Chem. *19*, 181 (1984).
197. M. Muhyaddin, P. J. Roberts, and G. N. Woodruff, Br. J. Pharmacol. *77*, 163 (1982).
198. E. Falch, A. Hedegaard, L. Nielsen, B. R. Jensen, H. Hjeds, and P. Krogsgaard-Larsen, J. Neurochem. *47*, 898 (1986).

199. A. Delini-Stula, in GABA-Neurotransmitters. Pharmacochemical, Biochemical and Pharmacological Aspects (P. Krogsgaard-Larsen, J. Scheel-Krüger, and H. Kofod, eds.), Munksgaard, Copenhagen, 1979, p. 482.
200. J. D. Wood, D. D. Johnston, P. Krogsgaard-Larsen, and A. Schousboe, Neuropharmacology *22*, 139 (1983).
201. M. Baraldi, L. Grandison, and A. Guidotti, Neuropharmacology *18*, 57 (1979).
202. A. Maggi and S. J. Enna, Neuropharmacology *18*, 361 (1979).
203. L. J. Fowler, D. H. Lovell, and R.A. John, J. Neurochem. *41*, 1751 (1983).
204. P. Krogsgaard-Larsen and G. A. R. Johnston, J. Neurochem. *25*, 797 (1975).
205. B. Schultz, T. Aaes-Jørgensen, K. P. Bøgesø, and A. Jørgensen, Acta Pharmacol. Toxicol. (Copenh.) *49*, 116 (1981).
206. J.-P. Kaplan, B. M. Raizon, M. Desarmenien, P. Feltz, P. M. Headley, P. Worms, K. G. Lloyd, and G. Bartholini, J. Med. Chem. *23*, 702 (1980).
207. G. Bartholini, B. Scatton, B. Zivkovic, and K. G. Lloyd, in GABA-Neurotransmitters. Pharmacochemical, Biochemical and Pharmacological Aspects (P. Krogsgaard-Larsen, J. Scheel-Krüger, and H. Kofod, eds.), Munksgaard, Copenhagen, 1979, p. 326.
208. P. Worms, H. Depoortere, A. Durand, P. L. Morselli, K. G. Lloyd, and G. Bartholini, J. Pharmacol. Exp. Ther. *220*, 660 (1982).
209. K. G. Lloyd, S. Arbilla, K. Beaumont, M. Briley, G. De Montis, B. Scatton, S. Z. Langer, and G. Bartholini, J. Pharmacol. Exp. Ther. *220*, 672 (1982).
210. S. S. Oja, P. Kontro, and P. Lähdesmäki, Prog. Pharmacol. *1*, 1 (1977).
211. P. L. Morselli and L. Bossi, in Neurotransmitters, Seizures, and Epilepsy II (R. G. Fariello, P. L. Morselli, K. G. Lloyd, L. F. Quesney, and J. Engel, eds.), Raven Press, New York, 1984, p. 253.
212. K. G. Lloyd, P. L. Morselli, H. Depoortere, V. Fournier, B. Zivkoviz, B. Scatton, C. Broekkamp, P. Worms, and G. Bartholini, Pharmacol. Biochem. Behav. *18*, 957 (1983).
213. M. D. Hynes, J. D. Leander, R. C. A. Frederickson, P. P. K. Ho, D. W. Johnson, and R. A. Archer, Drug Dev. Res. *4*, 405 (1984).
214. T. Lindeburg, S. Foelsgaard, H. Sillesen, E. Jacobsen, and H. Kehlet, Acta Anaesthesiol. Scand. *27*, 10 (1983).
215. M. Kjaer and H. Nielsen, Br. J. Clin. Pharmacol. *16*, 477 (1983).
216. C. F. Terrence, D. M. Potter, and G. H. Fromm, Clin. Neuropharmacol. *6*, 241 (1983).
217. J. Sawynok and F. S. Labella, Eur. J. Pharmacol. *70*, 103 (1981).
218. W. Löscher, Neuropharmacology *21*, 803 (1982).
219. W. Löscher, H.-H. Frey, R. Reiche, and D. Schultz, J. Pharmacol. Exp. Ther. *226*, 839 (1983).
220. B. Meldrum and R. Horton, Eur. J. Pharmacol. *61*, 231 (1980).
221. T. A. Pedley, R. W. Horton, and B. S. Meldrum, Epilepsia *20*, 409 (1979).
222. M. K. Menon, Neuropharmacology *20*, 1183 (1981).

223. H. R. Petersen, I. Jensen, and M. Dam, Acta Neurol. Scand. *67*, 114 (1983).
224. R.G. Fariello, G. T. Golden, P. F. Reyes, G. W. Alexander, and R. J. Schwartzman, in Neurotransmitters, Seizures, and Epilepsy II (R. G. Fariello, P. L. Morselli, K. G. Lloyd, L. F. Quesney, and J. Engel, eds.), Raven Press, New York, 1984, p. 245.
225. A. Y. Chweh, E. A. Swinyard, H. H. Wolf, and H. J. Kupferberg, Life Sci. *36*, 737 (1985).

12

Benzodiazepine Receptors and Tissue Function

Michael Williams / CIBA-GEIGY Corporation, Summit, New Jersey

Richard W. Olsen / University of California School of Medicine, Los Angeles, California

HISTORICAL PERSPECTIVE

The benzodiazepine anxiolytics, Librium (chlodiazepoxide) and Valium (diazepam) (Figure 1) are among the most widely used psychotropic agents in the world today (1). Yet these compounds were only discovered to have antianxiety activity in 1958 (2) and were introduced as drugs 2 years later, in 1960.

In addition to their antianxiety (anxiolytic) activity, the benzodiazepines (BZ) are also potent anticonvulsant, sedative, and muscle relaxant agents (3). These compounds are also able to potentiate the sedative actions of alcohol and can cause dependence.

Efforts to determine the mechanism of action of the benzodiazepines were initiated simultaneously with their introduction into the clinic and mainly involved studies of their electrophysiological actions (4,5). It was well established by the early 1970s that the BZ could potentiate the effects of the inhibitory amino acid neurotransmitter, γ-aminobutyric acid (GABA), an effect that was thought to involve central serotonergic mechanisms (6). Newer anxiolytics, such as buspirone (Figure 2) (7), have refocused attention on the involvement of serotonin in anxiety.

A major step forward, in 1977, was the identification by two groups, one in Switzerland (8) and another in Denmark (9), using radiolabeled diazepam, of distinct receptors for the BZ in mammalian tissues. This discovery presaged a tremendous effort in basic research to establish further the mechanism of the anxiolytic action of the BZ in the hope that newer compounds would be discovered that would be anxioselective, that is, anxiolytic without the complicating side effects associated with the use of the classic BZ. Some 10 years later,

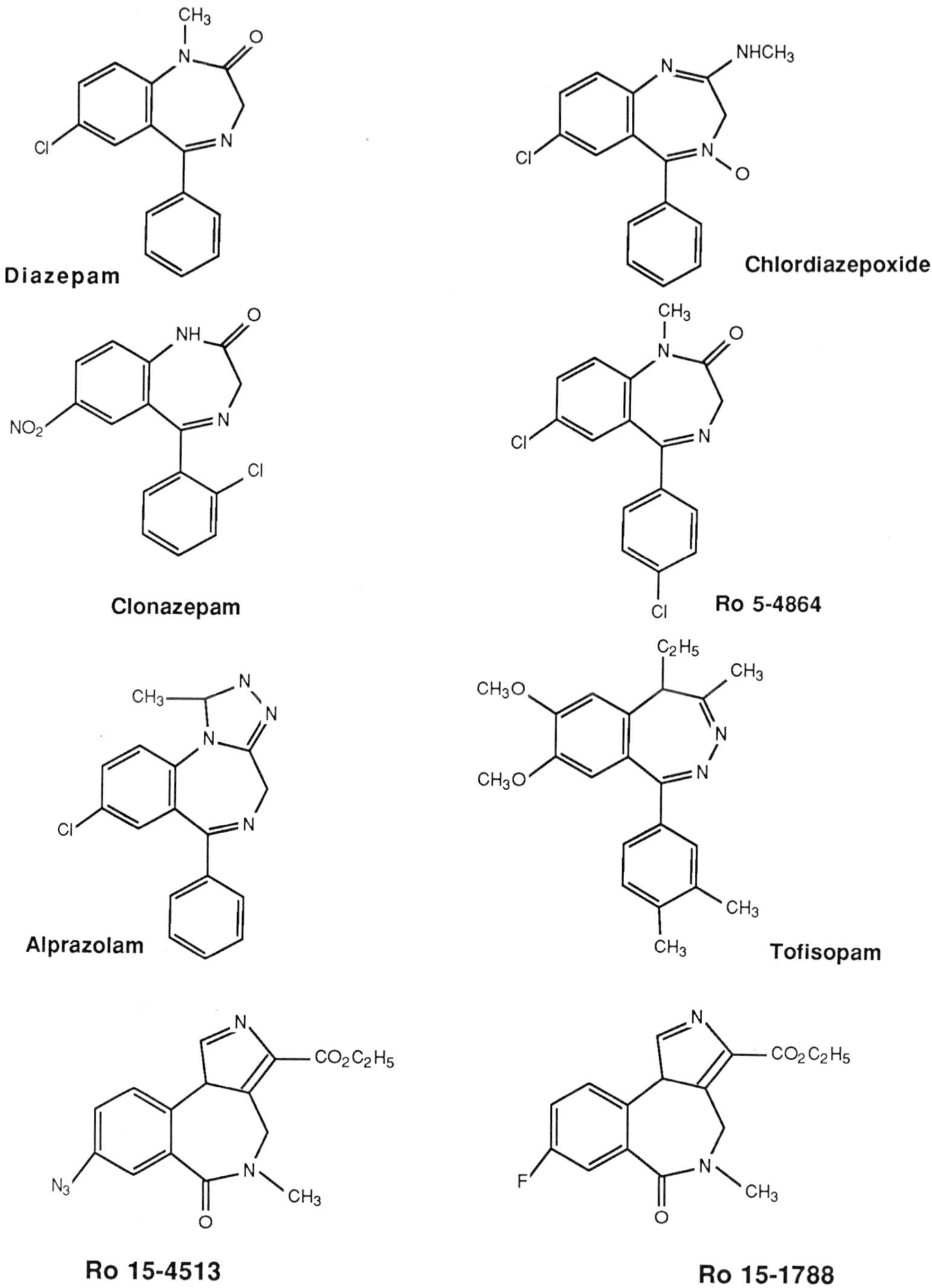

Figure 1 Structures of classic BZ receptor ligands.

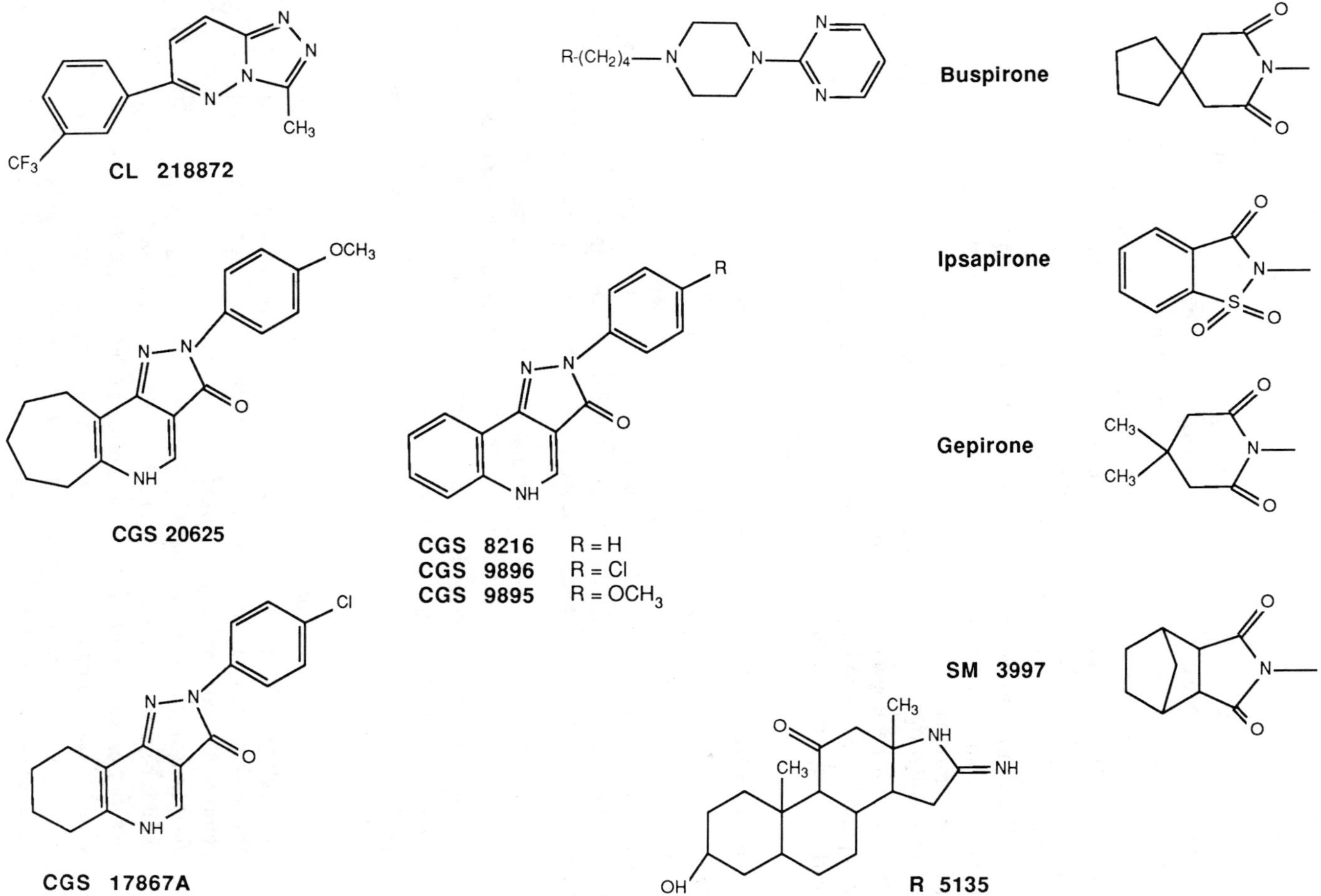

Figure 2 Atypical BZ receptor ligands and 5-HT-1A agonists.

this research effort has yet to yield such an entity. It has, however, provided an enormous body of literature that has in many aspects changed the traditional concepts of how drugs work (10,11). In addition, the basic aspects of BZ-related research has become bifurcated. On the one hand, newer antianxiety agents, such as buspirone and ipsapirone (Figure 2), produce their effects independently of the BZ receptor (7,12); the BZ pharmacophore itself appears to be a generalized ligand for peptide receptors (13,14).

In addition to the identification of central-type BZ receptors, studies with radiolabeled BZ also led to the identification of the peripheral-type BZ receptor (15), which has been implicated in the regulation of calcium levels in heart (16) and mitochondria (17).

The identification of a specific receptor for the BZ has also led to an intensive search for endogenous ligands for this recognition site. Although many such compounds have been identified, none has as yet been unequivocally identified as the "enkephalin" of the BZ receptor.

CONCEPTS IN BENZODIAZEPINE RECEPTOR ACTIVATION

Receptor theory is usually concerned with three classes of agent that interact with the receptor. These, which can be referred to as ligands, are agonists, partial agonists, and antagonists (18). However, newly discovered compounds, including the BZ, Ro 15-1788 (Figure 1) (19) and the pyrazoloquinoline, CGS 8216 (Figure 2) (20), have led to a new concept, that of *inverse agonist* (21).

1. By definition (18), an *agonist* is a ligand that produced a maximal response due to its efficacy, the latter an index of the ability of a compound to evoke a biologic response. This is a concept distinct from that of affinity, which is a measure of the ability of a ligand to bind to its receptor.
2. A *partial agonist* is an agonist which reduced efficacy, that is, it is a ligand that although potentially having high affinity for a given receptor is unable to evoke a maximal response in terms of biologic activity.
3. By definition (18), an *antagonist* is a ligand that has zero efficacy. That is, the ligand is able to bind but is unable to elicit a biologic response. From this definition, a partial agonist is also a partial antagonist and can also be referred to as an agonist-antagonist.
4. *Inverse agonist*, a term first used by Braestrup et al. (21) in relation to BZ receptor-related entities, is a ligand that binds to a given receptor and has negative efficacy; that is, it produces effects opposite and distinct from those of either a partial or full agonist. An inverse agonist can, but need not always, have antagonistic actions for the agonist properties associated with receptor activation.

CLASSIFICATION OF BENZODIAZEPINE RECEPTORS

Central BZ Receptors

The binding of [^{3}H] diazepam to rat brain membranes allowed the characterization of a high-affinity binding site (8,9) that was sensitive to the BZ, clonazepam (Figure 1), and through which diazepam was thought to produce its antianxiety effects. Diazepam, as discussed in further detail below, also bound to sites in peripheral tissues, especially the kidney, which led to the description of the "kidney-type" or peripheral BZ receptor (17). This differed from the central receptor inasmuch as binding of the ligand was not especially sensitive to clonazepam but was displaced by Ro 5-4864 (Figure 1), a BZ that has low affinity for the central receptor and has no antianxiety activity but may be proconvulsant (22).

Based on the diverse pharmacological properties of the BZ, considerable effort was expended in an attempt to identify central receptor subtypes, the logic being that one type may mediate the anxiolytic effects of these compounds and other subtypes may be responsible for their side effects. Based on kinetic and thermodynamic analyses and the actions of a novel new chemical entity (NCE), CL 218,872 (Figure 2), that had antianxiety actions but no apparent sedative effects (23), it was proposed (15,24) that two BZ receptor subtypes were present in brain: the BZ-1 receptor, localized to the cerebellum, activation of which mediated the antianxiety actions of diazepam and the BZ-2 receptor, enriched in hippocampal tissue, through which the side effects of BZ-type compounds were mediated. In addition to CL 218,872 being subsequently found to have side effects in preclinical models similar to those of the BZ (25), the logic of their being anxiety-related function in a portion of the brain, the cerebellum, that is involved in motor coordination is somewhat unlikely. Subsequently, BZ receptors in cerebellum have been found to be labeled with a selective hypnotic, zolpidem (Figure 3) (26), further weakening this hypothesis. In addition, the whole issue of receptor subtypes as opposed to receptor states has been questioned (27). The attractiveness of a hypothesis involving distinct receptor subtypes that mediate the different actions of a therapeutic agent has continued to focus interest on defining such entities. The results with zolpidem (26) provide strength to the possibility, although perhaps not in the original context on which the BZ-1/BZ-2 receptor concept was based. In addition to the possibility that BZ-1 receptors mediate sedation, it has also been suggested, again on the basis of studies with zolpidem, that the BZ-2 receptor may mediate the myorelaxant actions of the BZ (28). Using behavioral discrimination models for the nonsedating pyrazoloquinoline, CGS 9896 (Figure 2), and for zolpidem (Figure 2), however, an alternative functional receptor nomenclature has been proposed (29). The type A receptor at which CGS 9896 acts as an agonist is proposed to mediate the anxiolytic actions of BZ-

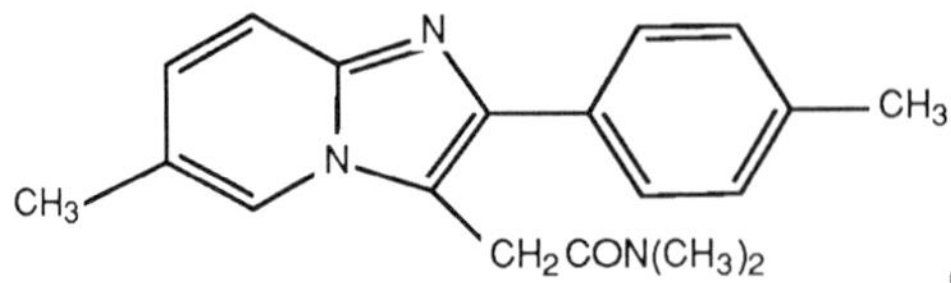

Zolpidem

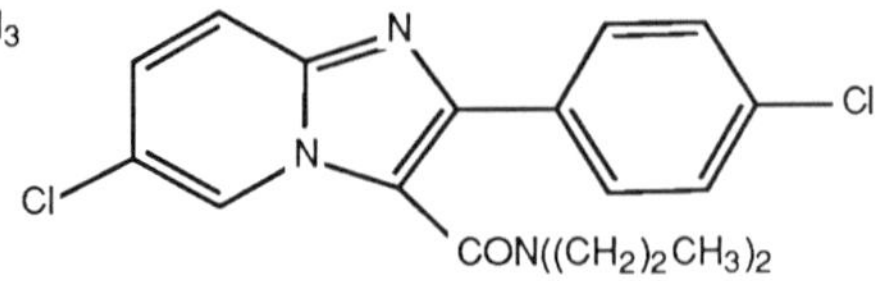

Alpidem

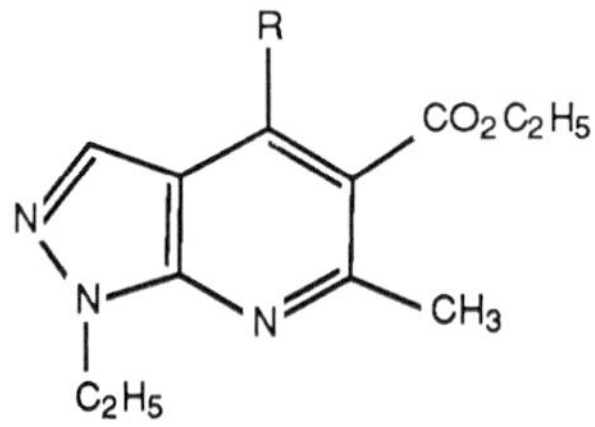

Etazolate: R = NHN=C(CH_3)$_2$

Cartazolate: R = NH (CH_2)$_3$$CH_3$

Tracazolate: R = NH C_3H_7

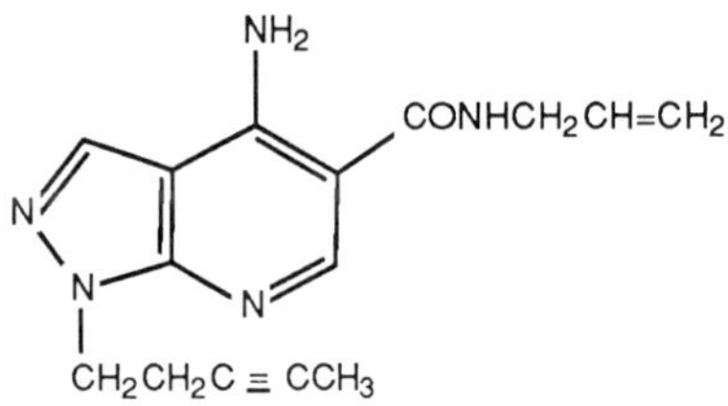

ICI 190, 622

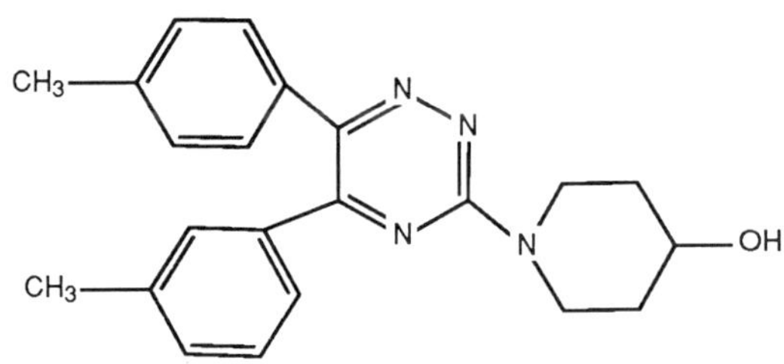

LY 81067

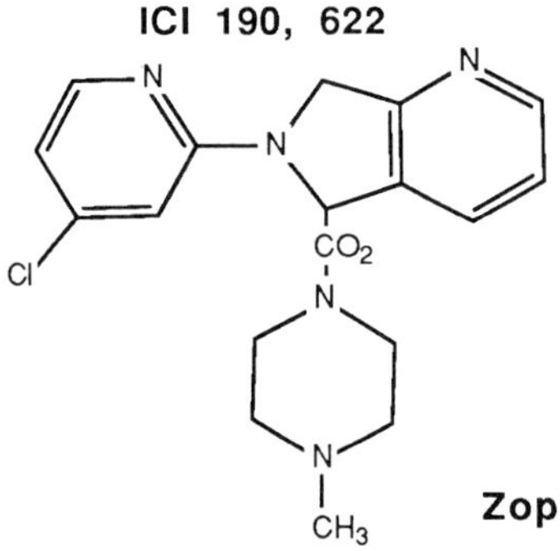

Zopiclone

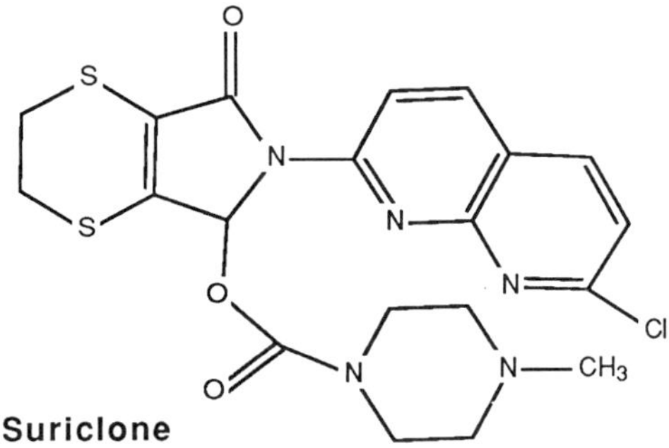

Suriclone

Figure 3 Atypical BZ receptor ligands.

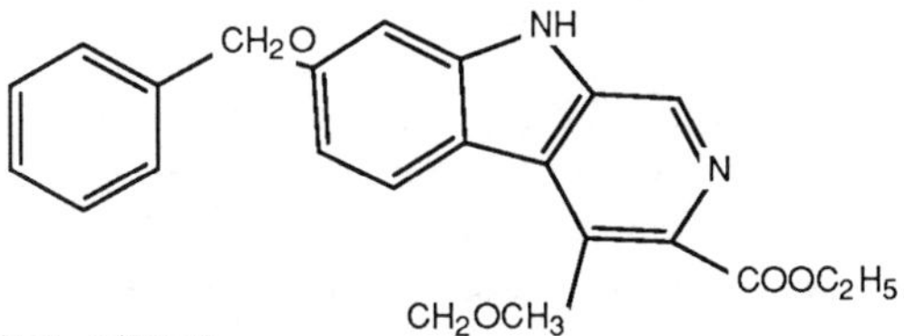

ZK 93423

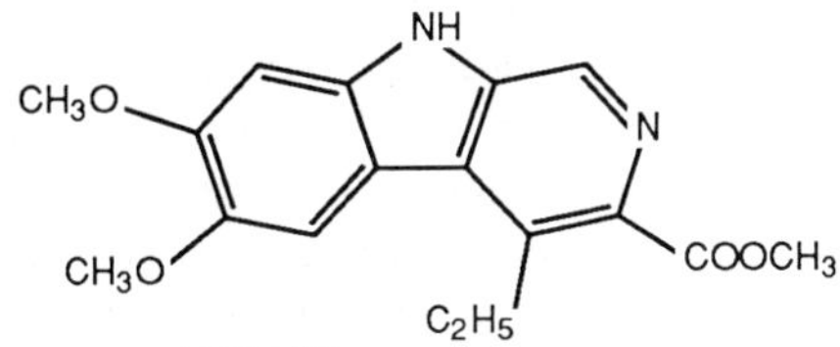

DMCM

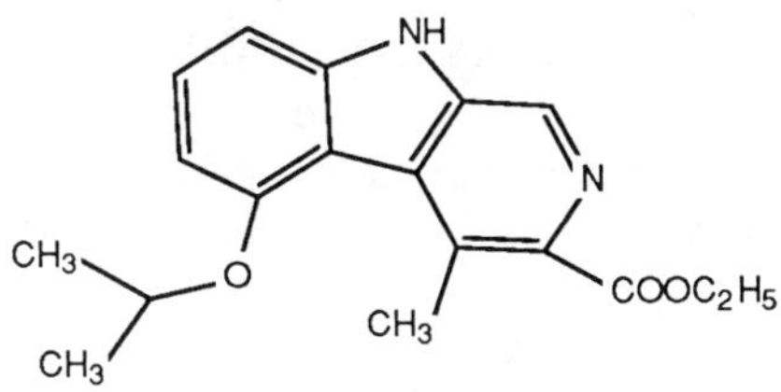

ZK 93426

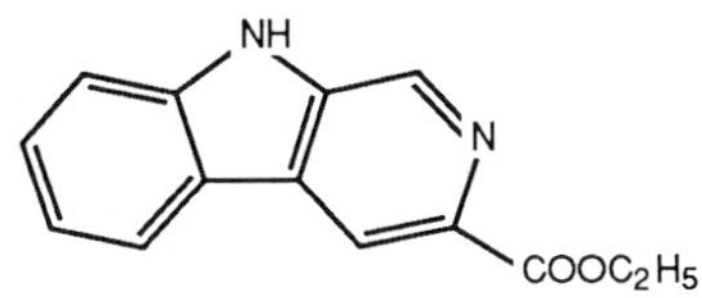

Beta-CCE

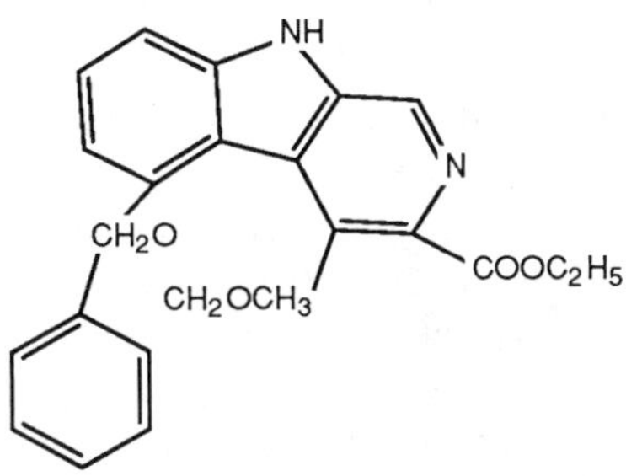

ZK 91296

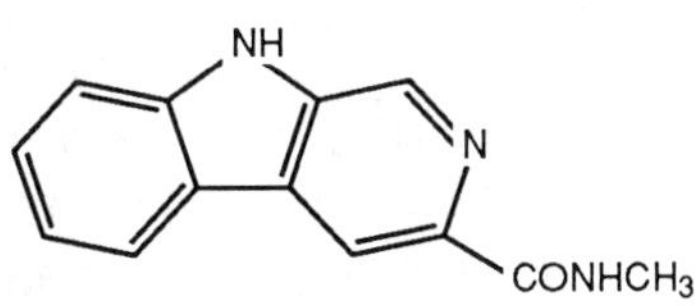

FG 7142

Figure 3 (Continued)

type ligands. The type S receptor is that through which the sedative actions of such componds are mediated. Zolpidem is an agonist at this site. The use of a functional (behavioral) as distinct from a biochemical (binding) nomenclature may better serve to define the issue of BZ receptor subtypes.

The ability to measure direct interactions with the central BZ receptor in vitro provided the means to randomly evaluate new compounds for their ability to displace [^{3}H] diazepam and also to better understand the molecular aspects of receptor function. In addition to CL 218,872, a number of other entities have been identified through screening.

Potential Endogenous Ligands

Using the radioligand binding assay for the central BZ receptor, many individuals both in academia and industry began to screen novel natural entities for activity at the receptor. Of these, several proteins, including BZ-competitive factor 1 (30), GABA-modulin (31), a peptide of 10 kD termed endozepine (32), nepenthin (33), tribulin (34), thyroid hormone (35), and diazepam binding inhibitor (DBI) (36) have been described. In addition, several purine nucleosides (37) as well as thromboxane A_2 (38), have been reported to have affinity for the central BZ receptor. Interest in many of these entities has been limited, only DBI and its constituent peptides, ODN (octadecaneuropeptide) and ON (octaneuropeptide), still fostering research efforts (39,40). The peptide appears to be an antagonist of the BZ receptor, being proconvulsant (35). The gene for DBI has been cloned (41). However, there is some controversy about the biologic activity of this compound, many more groups having been unable to detect activity than those who do. Another class of compound with activity at the central BZ receptor is that of the β-carbolines (Figure 3). These naturally occurring tricylics, originally isolated from human urine (42) and bovine brain (43), have potent activity at the BZ receptor (10^{-9} M), with a wide spectrum of activity, and certain members of this chemical series have been developed as clinical candidates (44) as well as made into adducts with the 1,4-BZ nucleus (45). Although the β-carbolines appear throughout nature, many exist that do not interact with the BZ receptor conplex. Another class of compounds with activity at the BZ receptor are a series of steroid hormones (46) that may be of interest within the more global context of the role of stress in the regulation of immune function (47). R-5135 (Figure 2) is a GABA receptor antagonist that can modulate the BZ-receptor complex. A series of steroids lacking overt metabolic function has also been identified as BZ receptor modulators (48). The possibility has been raised that, given the widespread use of the BZ, some of these "natural" ligands may in fact be BZ contamination, especially in those instances in which human brain was used as source material. In a study (49) using brains that were preserved since the 1940s, long before any BZ were publicly available, BZ-like material was, however, identified, reducing the likelihood that many of these observations are artifactual.

Non-BZ Receptor Ligands

In addition to CL 218,872, several other NCE with diverse chemical structures have been identified with potent activity at the central BZ receptor.

CGS 9896 and related compounds CGS 9896 (Figure 2) (50) and the related pyrazoloquinolines, CGS 9895 (51) and CGS 17867A (52), and the pyrazolopyridine, CGS 20625 (53), have high affinity for the central BZ receptor and are all active in preclinical animal models predictive of anxiolytics.

In contrast to diazepam, none of these compounds has significant muscle relaxant, sedative, or alcohol-potentiating activity and may thus be termed anxioselective, although this facet of the activity of this class of compounds, like that of CL 218,872, is not universally accepted (54). However, in early clinical trials CGS 9896 was found to be efficacious and completely devoid of the sedative and alcohol-potentiating actions associated with the BZ. CGS 8216, the chloro analog of CGS 9896 (20), was one of the first inverse agonists described. A related thienyl analog, S-135 (Figure 2), has also been reported to be an inverse agonist (55).

Etazolate, Cartazolate, and Related Compounds A series of pyrazolopyridines (Figure 3), etazolate (SQ 20,009), cartazolate (SQ 65,396), and tracazolate (ICI 136,753), were found to *increase* the binding of BZ to the central receptor and have been reported to be active in the rat conflict models predictive of antianxiety action (56,57). A theory relating to how such compounds may have antianxiety activity has been proposed (11), and similar types of compounds differing in their chemical structure, such as the diaryltriazine, LY 81067 (Figure 3), have similar preclinical profiles (58). ICI 190,662 (Figure 3) is the first of a series of pyrazolopyridines with direct agonist activity at the central BZ receptor (59).

Zopiclone and Suriclone The cyclopyrrolones, zopiclone and suriclone (Figure 3), have high affinity for the central BZ receptor and have animal activity indicative of hypnotics and anxiolytics, respectively (60,61). Their interaction with the BZ receptor is thought to differ from that of other ligands, consistent with a distinct binding site, unique for this chemical structure (62), eliciting a decrease in the apparent number of BZ sites without affecting receptor affinity. In addition (63), the cyclopyrrolones accelerate radioligand dissociation, suggesting an allosteric interaction with BZ site per se.

PK 8165 and PK 9084 The phenylquinolines PK 8165 and PK 9084 (Figure 4) have high affinity for the BZ receptor (64) and were reported to enhance punished response without sedative properties. The antianxiety actions of this class of compound have been questioned (65). In addition, they have been found to potentiate the hypnotic actions of the BZ (66), to reduce rat locomotor activity (67), and to cause sedation in clinical trials (68).

EMD 28422 EMD 28422 (Figure 4) is a novel, lipophilic, N^6-substituted adenosine analog that has been reported to enhance BZ in vitro (69). Its activity as an antianxiety agent in vivo is, however, controversial (70). It may be noted, however, that there is a considerable body of data linking adenosine to the anxiolytics (71), although this may be a reflection of the sedative actions of these compounds rather than their antianxiety effects.

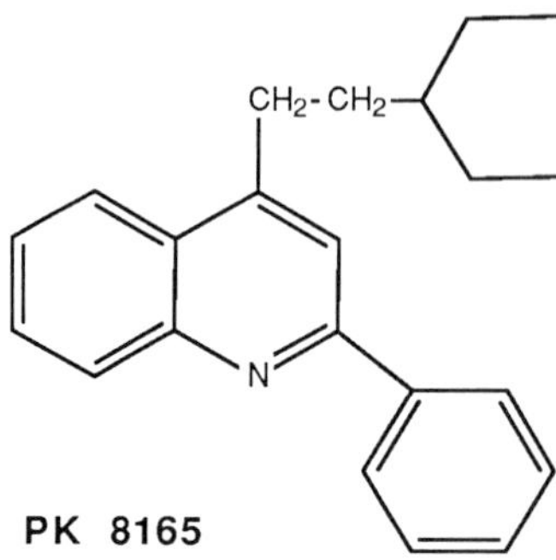

PK 8165

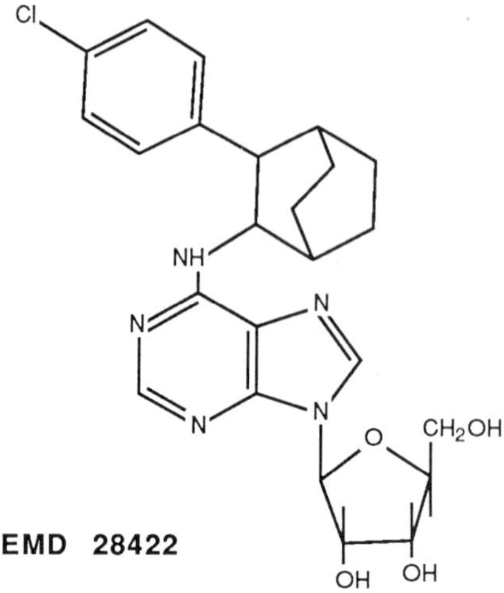

EMD 28422

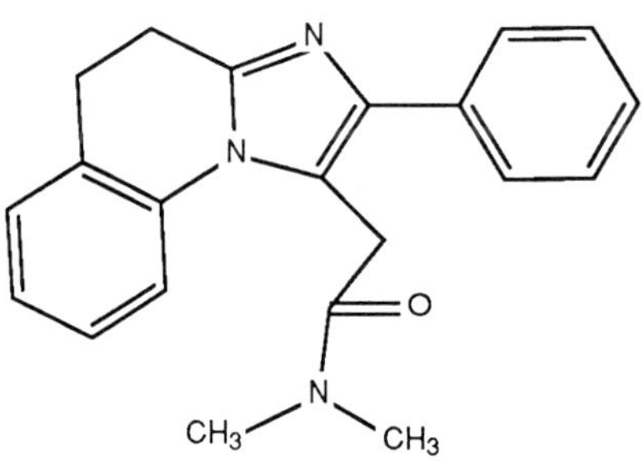

Imidazoquinoline (Ludwig)

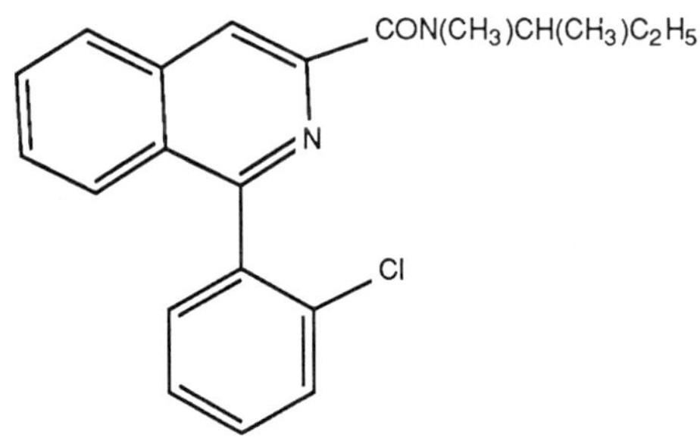

PK 11195

Figure 4 Atypical BZ receptor ligands.

The β-Carbolines The β-carbolines (Figure 13) have been the most consistently active of the endogenous factors that are candidates as ligands for the BZ receptor. Although the initial identification of the ethyl-β-carboline-3-carobxylate (β-CCE) (42) in human urine is considered an extraction artifact (15,71), more recent work (43) using bovine brain as the source material has removed this argument. β-CCE was initially found to enhance anxiety in several different test paradigms (53). β-CCE has also been reported to be proconvulsant, and various β-carbolines have been shown to have a spectrum of biologic activity at the BZ-receptor complex from full agonist [ZK 93423 (72)] to antagonist [ZK 93426 (73)] to inverse agonist [DMCM (74)], although it may be noted that ZK 91296 has been reported to have anxiogenic activity (75). FG 7142 was the first β-carboline reported to be anxiogenic in the human (76).

Ro 15-1788 and Ro 15-4503 These two BZ (Figure 1) are inverse agonist-antagonist entities. Ro 15-1788 was the first inverse agonist reported (19) and is currently in clinical use to reverse BZ-induced anesthesia (77). Ro 15-4503 is another BZ antagonist that in addition to being proconvulsant (78) has the unique property of antagonizing the sedative actions of alcohol (79). The possibility that such an entity could be used to develop a drug to treat chronic alcoholism has been clouded by the ethical issues surrounding the additional use of any such compound to mask the central nervous system (CNS) effects related to recreational alcohol consumption.

Alpidem and Zolpidem The imidazopyridine zolpidem (Figure 3), as already indicated (27), is a potent sedative-hypnotic with high affinity for the proposed BZ-1 receptor. A related entity, alpidem, has been reported to be an anxiolytic with weak myorelaxant and sedative effects both in preclinical models and in human volunteers (80). Despite a different pharmacological profile to that of zolpidem, alpidem appears to interact selectively with the BZ-1 receptor (81), adding further confusion to the classification of BZ receptor subtypes.

BZ Agonists Despite the plethora of new chemical entiries with high affinity for the central BZ receptor, BZ still maintain a consistent place in the marketplace. The triazolobenzodiazepine, alprazolam (Figure 1), has an antidepressant component in addition to its anxiolytic profile (82) that serves to enhance its clinical attractiveness. The 3,4-BZ, tofisopam (Figure 1), has been reported to be clinically active although it is inactive in preclinical antianxiety models (53). A novel series of imidazoquinolines (Figure 4) (83) that modulate BZ receptor function have been recently reported.

The Peripheral or "Kidney-Type" Benzodiazepine Receptor

Interest in the peripheral BZ receptor remained mostly academic for a long period of time. Binding assays using Ro-5-4864 (17) and an "antagonist," PK 11195 (Figure 4) (84), identified a discrete distribution of peripheral BZ receptor sites in the zona glomerulosa of the adrenal gland, testicular Leydig cells, the olfactory nerves of the brain, and the kidney, indicating some unique functional role for such receptors. Peripheral sites have also reported to be present in the central nervous system (85). Endogenous ligands for the peripheral BZ receptor have been reported from mammalian tissue sources (86-90). Ligands selective for the peripheral BZ receptor have a variety of pharmacological actions, including modulation of immune function (47), cardiac action potential generation (16,91), stimulation or inhibition of cell proliferation (92-94), gene expression (95), and platelet function (96,97), as well as modulation of convulsive threshold (22,53).

Peripheral BZ Receptors in Guinea Pig Heart

Based on studies showing that Ro 5-4864 shortened and reduced contractile force in intact guinea pig hearts (16,91), it was suggested that the peripheral BZ receptor may be involved in the regulation of calcium flux (100). Previous work in guinea pig ileum (98), had suggested a similar role. Using PK 11195 in addition to Ro 5-4864 (91,99), it was suggested that peripheral BZ ligands may have potential use in the treatment of angina and cardiac ischemia, acting as negative chronotropic agents. The type of peripheral BZ receptor involved in the proposed modulation of calcium actions in cardiac tissue has been questioned (101). Despite the findings that calcium entry blockers can interact with peripheral BZ recognition sites (102) and the peripheral BZ ligands, Ro 5-4864 and PK 11195, can inhibit calcium uptake into guinea pig synaptosomes (103), the possibility that the peripheral BZ receptor is linked to voltage-sensitive calcium channels has yet to be unequivocally demonstrated (103).

Peripheral BZ Receptors in Mitochondria

Peripheral-type BZ receptors have been identified at the subcellular level in the outer mitochondrial membrane (17,104) and have been shown to be associated with a voltage-sensitive anion channel designated as porin. PK 11195 binding is inhibited by the porphyrins, the most active of which is protoporphyrin IX (105). The presence of porphyrins in cytochromes, such as hemoglobin, catalase, and various peroxidases, may explain a functional role for the mitochondrial peripheral BZ receptor in cellular metabolic processes. The relationship of such effects to the discrete pharmacological events modulated by ligands for the peripheral BZ receptor remains to be determined, however.

THE CENTRAL BENZODIAZEPINE-RECEPTOR COMPLEX

The BZ receptor is part of a large macromolecular complex (Figure 5) first proposed in 1980 (10) on the basis of observations that the BZ not only enhance the postsynaptic response to GABA but that BZ receptor binding in vitro in brain homogenates is allosterically enhanced by $GABA_A$ receptor agonists (105). Furthermore, BZ (106) and GABA (107,108) binding in vitro can be allosterically enhanced by sedative-hypnotic barbiturates and related agents (57,109). Such enhancement is chloride dependent and picrotoxin sensitive (110). Thus, distinct binding sites for GABA, BZ, barbiturates, picrotoxinlike convulsants, and chloride ions appear to be tightly associated (3,10,11,111, 112), as indicated in Figure 6. Representative compounds active at the four different recognition sites on the BZ-receptor complex are indicated in the figure and discussed in other sections.

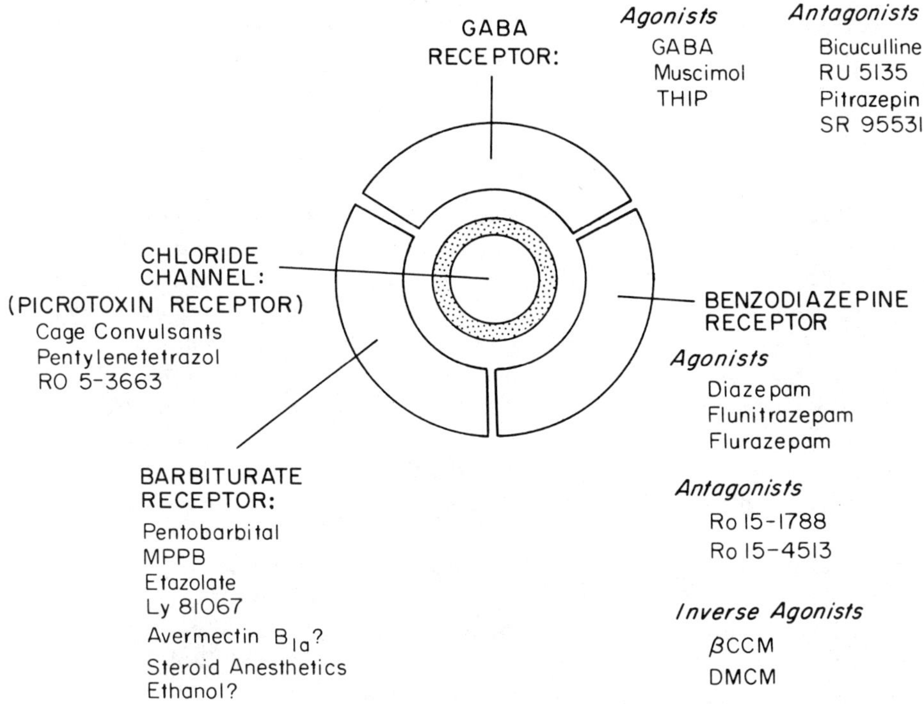

Figure 5 Model of the BZ-GABA receptor complex. Four different and separate drug receptors are hypothesized, and examples of ligands for each site are shown. No implications about protein subunit structure are intended.

The BZ receptor, as part of the GABA receptor-chloride ion channel complex, is an important site of action for many centrally active drugs. Thus CNS depressants and excitatory agents with potential clinical utility and showing a great variety of chemical structures share this common target site in the nervous system, the BZ-GABA-receptor complex. In addition, a number of other entities, including steroids (48) and the macrolide anthelmintic, avermectin, also interact with the complex (113).

Further evidence for the existence of this complex has been obtained by the solubilization of receptor binding activities from brain membranes using mild detergents, which has resulted in the copurification of the BZ receptor protein with GABA binding, convulsant binding, and barbiturate modulation intact (114–117). The definition of a receptor complex with allosterically, drug-related associated sites similar to that described for the nicotinic cholinergic receptor (118) and for the more recently described NMDA-PCP-receptor complex (119) has markedly changed the focus in terms of delineating ligands that modulate receptor function. In addition, the relatively recent discovery of the BZ-receptor complex has raised closer parallels between receptors and the more classic, multiply regulated enzyme complexes (18).

The GABA Shift

The discovery of a receptor binding site and the classification of the various types of ligand that interact with it requires validation in vivo in terms of a behavioral response. Thus the classification of the various agonists, antagonists, and inverse agonists already described has required their testing in appropriate animal models. Once classes of such compounds were established, the ability to predict in vivo activity using in vitro techniques became attractive. Based on the ability of the various classes of compound to bind to the central BZ receptor site, the effect of GABA on the affinity of the ligand for the BZ site, the ability of the ligand to increase [^{35}S]TBPS binding to the chloride channel and the interaction of the ligand with etazolate-modulated GABA radioligand binding, it has been possible to provide a set of criteria that allow the prediction of in vivo pharmacological activity (21,120). These criteria are outlined in Table 1. It may be noted, however, that the designations, especially for the GABA shift (21), and the ability of the amino acid to increase or decrease ligand affinity, appear to be predictive only within broad classes (121). Thus a GABA shift of 0.7 or less is reasonably predictive of an inverse agonist and a shift of 2 or greater indicates

Table 1 GABA Shift Parameters[a]

Compound type binding	Effect of GABA on BZ receptor affinity (GABA shift)	Effect on TBPS binding	Effect on etazolate-stimulated muscimol
Full agonist	Increased (>2)	+ 30-40%	+ 25%
Partial agonist	Unchanged or increased (0.8-1.5)	∿ 20%	+ 10-12%
Agonist-antagonist	Unchanged or decreased (0.8-1.2)	∿ 20%	+ 10-12%
Antagonist	Unchanged or decreased (0.8-1.2)	−10 to + 5%	0 to + 5%
Inverse agonist	Decreased (<0.7)	−20 to 25%	−10 to 15%

[a]GABA shift = IC_{50} in absence of GABA/IC_{50} in presence of GABA. TBPS = *t*-butylbicyclophosphothionate.
Source: Adapted from References 11 and 120 and unpublished data.

agonist activity. Between the two values, any mixture of the three primary activities can exist and needs in vivo characterization.

MOLECULAR ASPECTS OF BENZODIAZEPINE RECEPTORS

The BZ binding protein was identified as a peptide subunit of 51 kD on SDS gel electrophoresis by photoaffinity labeling with [^{3}H] flunitrazepam (122). Thus the receptor complex must contain at least this peptide, probably in more than one copy per oligomer, and possibly other subunits. The BZ-GABA-receptor complex is an ion channel receptor, analogous in many respects to the nicotinic acetylcholine receptor, which is made up of five subunits comprising four separate peptides with a stoichiometry of $\alpha_2\beta\gamma\delta$ (123). Each of the five subunits contributes one wall to the ion channel. The voltage-regulated sodium channel is a single peptide, but it has a size of over 200 kD and contains four repeats of a 50 kD domain, each contributing one of the four walls of the channel (118). Other receptors that do not form channels themselves but regulate target cell activity by coupling through GTP binding proteins, such as the muscarinic acetylcholine receptor (124) and the β-adrenergic receptor (125), appear to be monomeric single-subunit proteins.

The BZ binding protein has been solubilized from mammalian brain membrane homogenates with mild detergents and shown to copurify with both GABA receptor binding activity and picrotoxin ([^{35}S] TBPS) binding activity (114-117,126,127). The molecular weight of the "native" protein (in detergent) was estimated by sucrose gradient-gel filtration columnchromatography at about 350 kD (114), but a molecuar weight of 220 kD has also been reported (117, 127). The latter molecular weight was also consistent with irradiation target size estimations of 220 kD for both GABA and BZ binding activities. Purification of the protein on a BZ affinity column to a specific activity of about 3000 pmol GABA binding (1000 pmol BZ binding) per milligram protein led to the detection of two peptide bands on SDS-PAGE, one at 53 kD (α) and the other at 56 kD (β) (117). The α subunit could be photoaffinity labeled with [^{3}H] flunitrazepam, and assuming a molecular weight of 220 kD, a structure of $\alpha_2\beta_2$ ($2 \times 53 + 2 \times 57 = 220$ kD) was proposed (117).

The BZ receptor protein purified in Triton X-100 or deoxycholate detergent contained high-affinity GABA binding activity but lacked convulsant binding and barbiturate modulation (11,117,126). These other activities could be retained, however, by using the detergent CHAPS for solubilization and purification (128-130). When affinity chromatography was carried out in CHAPS detergent, the same purity (two bands on SDS-PAGE) was obtained for a receptor preparation showing [^{35}S] TBPS binding activity (129). Furthermore, when samples purified in other detergents, showing no "convulsant-barbiturate" activity, were returned to CHAPS detergent, the "convulsant-barbiturate" activity, returned (130), further indicating that no additional proteins are needed for that activity and that all activities reside on the same macromolecular complex (131-133).

The 220 kD molecular weight model has, however, been questioned by data from irradiation target analysis (134) in which a size for GABA and BZ binding activities of about 50 kD, but 137 kD for [^{35}S]TBPS binding and about 350 kD for allosteric interactions between the binding sites was found, suggesting oligomeric subunit interactions. The irradiation target size for CHAPS-solubilized [^{35}S]TBPS binding activity was also of the order of 300-500 kD (130). The difference between these values (300-500 kD) and 220 kD could be due to the presence of either additional subunits or more than two copies each of the 53 kD (α) and 56 kD (β) subunits. Four peptide bands were obtained on SDS-PAGE with an affinity-purified BZ receptor complex (126), two of which were previously reported (117) plus bands at 31 and 47 kD. These latter two bands do not appear necessary for chloride channel function as discussed below (132). Other work (133,134) has independently established that the GABA binding site is carried by the 56-57 kD (β) subunit as shown by photoaffinity labeling with [^{3}H]muscimol. Binding sites for barbiturates and convulsants have not yet been identified.

Evidence that the α and β subunits may be the only integral peptides of the receptor complex came from the first development of monoclonal antibodies to the partially purified protein (135). A family of monoclonal antibodies was developed that could precipitate BZ-GABA binding activity. These recognized either one or the other of the 51-53 kD and 55-57 kD subunits, or neither, on Western blot analysis, and no other peptides (135). The β-subunit antibody is designated bd-17; that for the α subunit is termed bd-24 (136). The same two peptides were found in all brain regions. Final evidence that these two peptides are necessary and sufficient for GABA-activated and drug-modulated chloride channel function has been provided by cloning (135). A homology between BZ-receptor complex subunits in bovine and human, but not rat or cat brain, has been reported (136). At the cellular level, BZ subunit immunoreactivity to the two BZ antibodies indicated that recognition sites were present on both pre- and postsynaptic axodendritic and axosomatic synapses (136).

GABA-BZ receptor-chloride channels, as well as other important proteins, can be expressed in such cells as frog oocytes that do not normally express these proteins by injecting crude native messenger RNA (mRNA) from tissues like mammalian brain that do express such proteins (137). The activities can then be assayed electrophysiologically. Complementary DNA (cDNA) can be synthesized that corresponds to this brain mRNA, cut into small pieces, and inserted into suitable vectors, such as the coliphage λ for cloning. Clones that contain cDNA coding for the BZ-GABA receptor can then be identified by several approaches. For example, specific antibodies can be used to identify clones that synthesize antigenic protein in an expression library, such as *Escherichia coli*.

One approach (132) has been to obtain partial protein sequences for receptor subunits by fragmentation of the purified protein and fractionation of the peptides. Oligonucleotide probes were synthesized based on the genetic code corresponding to the peptide sequences, and the corresponding cDNA clones in phage λ were identified by hybridization (double-helix annealing). The large cDNA clones thus obtained were sequenced, and an open reading frame coding for protein sequences in the purified peptides was found. The cDNA sequences gave the deduced peptide sequences for both α and β subunits. These were proven to be the correct cDNAs by engineering the corresponding mRNAs and injecting them into oocytes and measuring the appearance of chloride channels regulated by GABA and BZ (132). In addition to studying BZ receptor characteristics at the transcriptional and translational levels, posttranscriptional modification by such processes as glycosylation and phosphorylation (138) may be anticipated to provide a clearer understanding of the molecular basis of the potential differences in subunit composition and possible receptor heterogeneity.

Using the techniques of molecular biology will thus provide more definitive information about receptor protein structure and function that will increase knowledge about the structure of the drug binding sites and other functional domains. The two GABA-BZ receptor subunits described to date show great homology with each other and also considerable homology with the nicotinic acetylcholine receptor (118), as well as similar hydropathicity profiles (132). This suggests that there may be a family of GABA receptor genes and proteins, as well as gene "superfamily" for all ion channel receptors. Cloning also allows the determination of the genomic DNA sequence and clues to the regulation of expression of the gene, including developmental aspects and possible alterations in human disease. Such information will be of use in understanding the mechanism of receptor activation and will consequently aid in the process of rational drug design.

STRUCTURE-ACTIVITY RELATIONSHIPS

The numerous BZ synthesized as part of Hoffman-La Roche's interest in this class of chemical compound, together with the novel chemical entities identified using radioligand binding assays, have led to a large body of data for the determination of the structure-activity relationship of the central BZ receptor. Crystallographic studies of Ro 15-1788, CGS 8216, and a series of β-carbolines suggested a receptor model with seven conformationally mobile binding points (139). A more extensive study (140) has led to the delineation of a general stereochemical model of the BZ receptor comprising a diffuse yet substantial planar region characterized by carbonylic-iminic hydrogen bonding and an ability of ligands with different pharmacological properties, that is, agonist-antagonist, to push the proposed receptor macromolecule into a conformation mediating these various activities. Using a technique termed "combinatorial optimization," the steric requirements of Ro 15-1788 and ZK 93426 have been compared (71) and found to be very similar in nature. Minor substitutions in active BZ receptor ligands can cause dramatic changes in biologic activity. The removal of a chloride group from the agonist CGS 9896 produces the inverse agonist CGS 8216 (20,50;52). Similarly, displacing the phenyl of the triazolopyridazine CL 218,872 (Figure 2) from the 5 to 7 position as represented by SR 95195 has been reported (141) to cause a shift from agonist to inverse agonist activity, although it may be noted that these two structures, like CGS 8216 and CGS 9896, differ in the presence of a substitution on the phenyl ring. Thus it may be questioned whether the removal of the orthotrifluoromethyl group rather than reorientation of the phenyl ring is responsible for the change in biologic activity in the triazolopyridazine series. As with many such receptor models, the transition from hypothetic minimum energy conformation and assumptions about hydrogen bonding and hydrophobic regions has yet to result in the synthesis of the ideal compounds.

PHARMACOLOGICAL ROLES FOR THE BENZODIAZEPINE PHARMACOPHORE

The major focus of this chapter has been on the potential anxiolytic activity of BZ receptor ligands. Thus the sedative and muscle relaxant actions of this drug class have not been covered in any great detail. Although the anticonvulsant and sedative actions of the BZ involve the BZ-GABA-receptor complex, as do the sedative actions of the barbiturates and the anticonvulsant actions of compounds that work through the picrotoxinin recognition site, there are effective compounds in these therapeutic areas that can produce effects somewhat similar to those of the BZ via mechanisms that do not appear to involve the BZ-receptor complex. In terms of antianxiety drugs, these are covered here. The muscle relaxant meprobamate and the anticonvulsant carbamazepine do not appear to produce their effects via the BZ receptor (11). Conversely, the BZ are considered primarily in terms of the effects that they produce by modulating the BZ-receptor complex, but some unexpected activities associated with the BZ chemical nucleus have been discovered in recent years.

Studies on the binding of thyrotropin releasing hormone (TRH) to rat amygdala tissues identified chlordiazepoxide (142) and midazolam (143) as high-affinity ligands for the TRH receptor. Although the physiological significance of this binding site is unclear (143), this observation was the first that BZ may interact with receptors for peptidic neuromodulators. The actions of platelet aggregation factor (PAF) were also found to be selectively blocked by a BZ, in this case the "crowned" BZ, alprazolam (96). Subsequent studies (97) showed that other BZ ligands had selective activity in modifying neutrophil

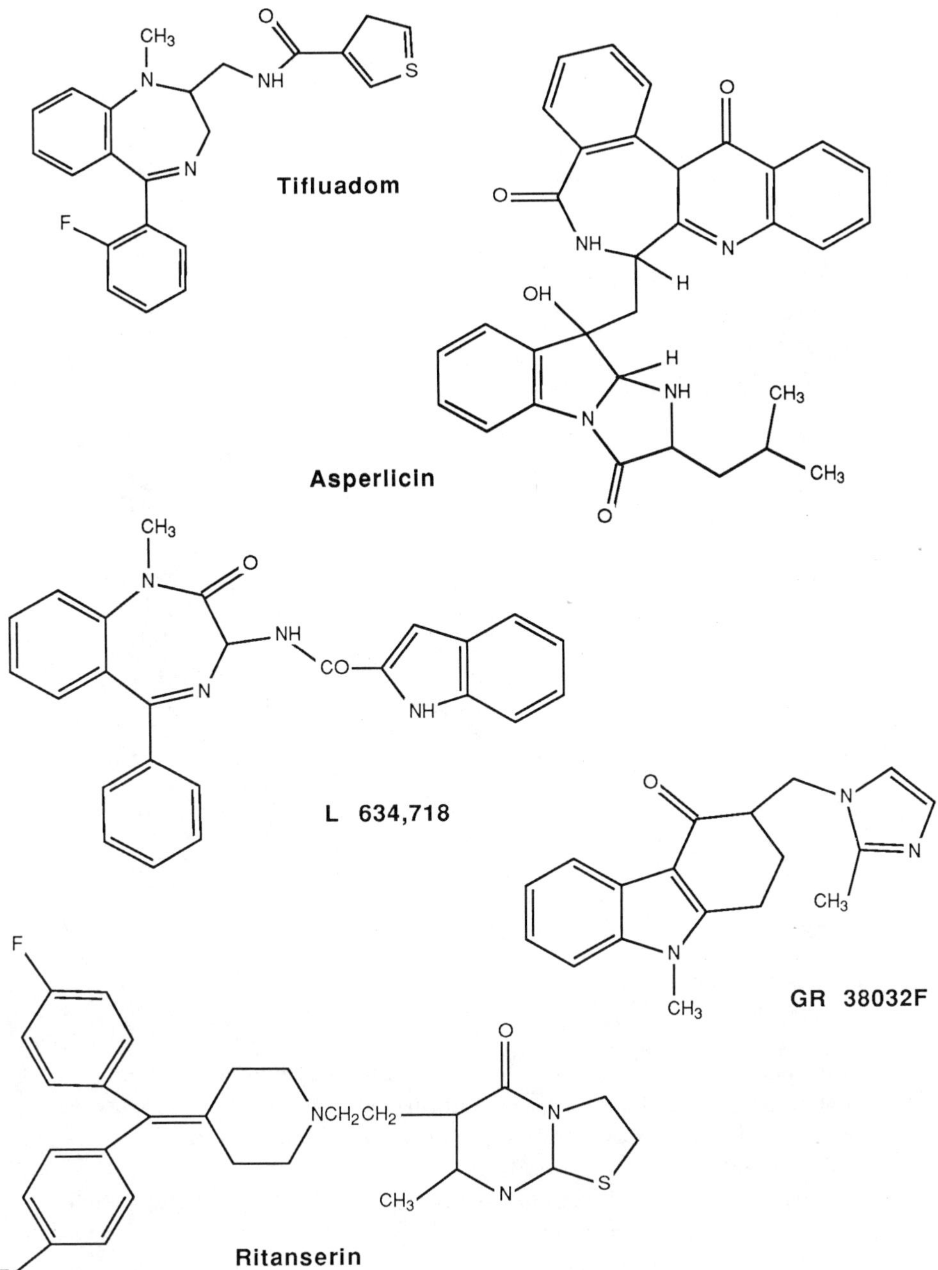

Figure 6 Novel antianxiety agents and BZ receptor ligands.

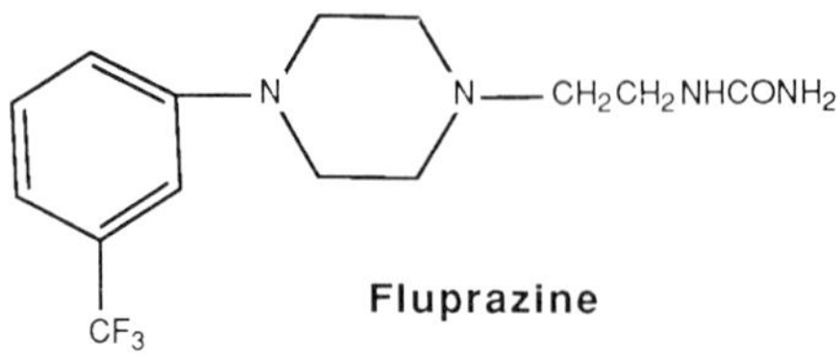

Fluprazine

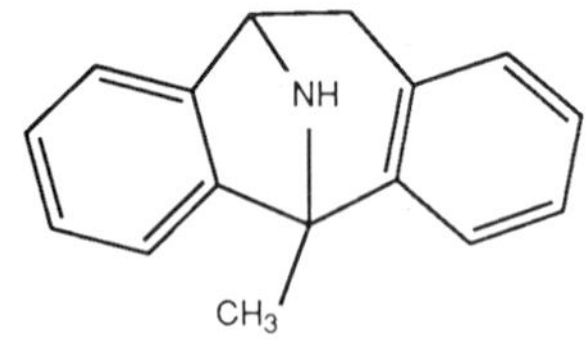

MK 801

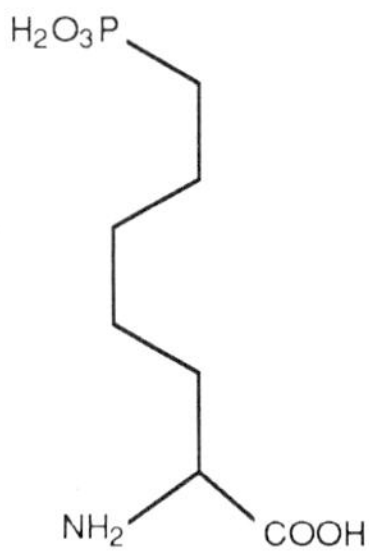

AP7

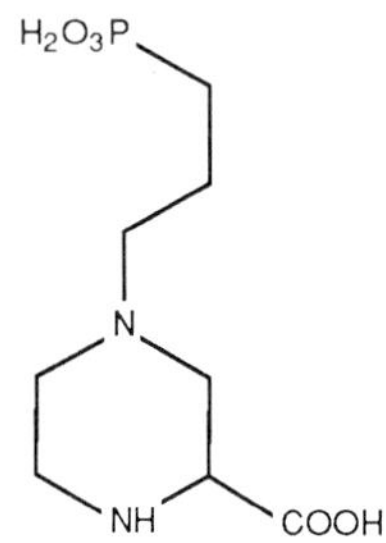

CPP

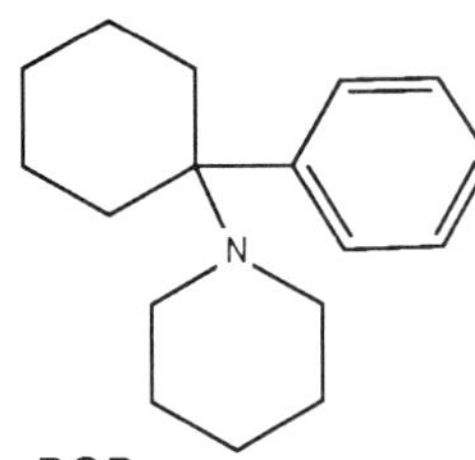

PCP

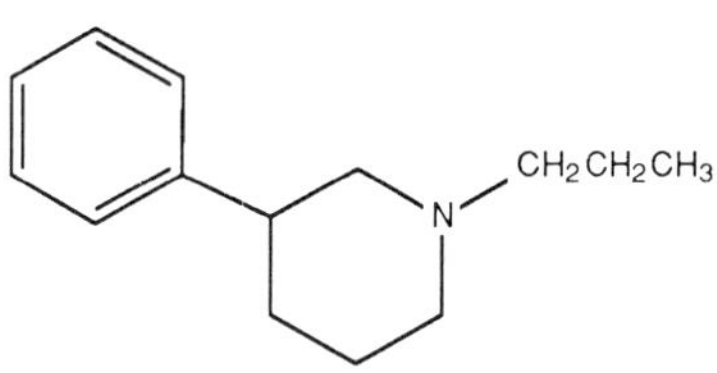

3-PPP

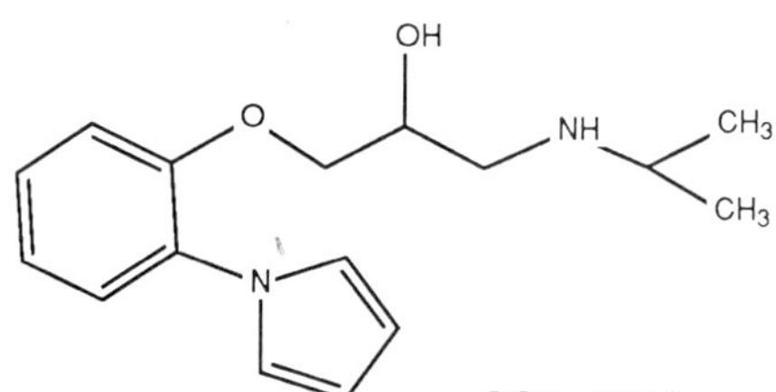

CGP 361A

Figure 6 (Continued)

function. The "crowned" BZ, tifluadom (Figure 6), was found by random screening to be an effective κ-opiate-receptor ligand (144) with analgesic activity and an affinity for the κ receptor of approximately 60 nM. This BZ also interacts with the central TRH receptor (145).

Ro 5-4864 and diazepam, but not the centrally selective BZ clonazepam, can stimulate chemotaxis in human monocytes via interaction with discrete receptor sites (47). This effect can be blocked, like the effects of Ro 5-4864 on guinea pig heart function (92), by PK 11195. Coupled with evidence that BZ can promote differentiation and inhibit cell proliferation (93-95), this has suggested

(47) that BZ with activity at the peripheral site may have growth-controlling effects. In another series of random screens using a binding assay for the peripheral cholecystokinin (CCK) receptor, a nonpeptidic lead (146), asperlicin (Figure 6), led to the synthesis of the high-affinity (IC_{50} = 500 pM) peripheral CCK ligand, L 634,718 (14). Finally, the actions of atrial natriuretic factor (ANF) can be antagonized in vivo by a number of BZ receptor ligands, including alprazolam, diazepam, and PK 11195 (147), although the mechanistic basis for this effect is unclear.

These seemingly fortuitous findings taken together with the possibility, as yet unproven, that the endogeneous ligand for the central BZ receptor is peptidic in nature, have led to the proposal (13,14) that the 1,4-benzazepine nucleus is a pharmacophore for many peptidic receptors. In one scenario, this property has been ascribed to the ability of the BZ to bind to an Ariens-type "accessory" binding site (14). Given the many thousands of BZ synthesized to date plus the availability of NCE that interact with the BZ receptor yet are different from the BZ in chemical structure, it may be anticipated that the search for nonpeptidic ligands for the many classes of peptide receptors thus identified will focus on the BZ. The challenge will be, however, to produce ligands like asperlicin that are selective for their given type of receptor and will not, in replacing the peptides as potential therapeutic agents, provide agents efficacious in any number of disease states but that have associated CNS side effects.

ANTIANXIETY AGENTS OF THE FUTURE

Although this topic is strictly outside the framework of this chapter and the receptors in question are the topic of other chapters, it is worthwhile mentioning briefly evidence to support the possibility that agents interacting with receptors other than the BZ may be clinically efficacious antianxiety agents.

Serotonin

As noted in the introduction, early studies indicated that the monoamine serotonin (5-HT) may be an effector agent in the genesis of anxiety. The serendipitous finding that the putative atypical dopamine receptor antagonist buspirone was a clinically effective anxiolytic (7) devoid of the side effect associated with the BZ led to the subsequent finding that this compound was a partial agonist of the 5-HT-1A receptor aided in the discovery of other 5-HT-1A receptor ligands with antianxiety activity, including gepirone (7,16), ipsapirone (12), and SM 3997 (Figure 2) (148). Antagonists of the 5-HT-2 receptor have also been shown to have antianxiety activity in animal models. Such compounds include ritanserin (149) and fluprazine (Figure 6) (150). The novel 5-HT-3 receptor antagonist, odanserin (GR 38032F; Figure 6), has also been reported to have antianxiety activity in preclinical models (151). In addition, this compound appears to have antipsychotic and antiemetic actions (152).

N-Methyl-D-Aspartate Receptors

In 1982, Squires, suggested that antagonists of excitatory amino acids, such as *N*-methyl-D-aspartate (NMDA) may be equivalent in terms of their pharmacological activities to agonists of the inhibitory amino acid receptor (11). Since diazepam is a potentiator of the actions of the inhibitor amino acid GABA, it appeared plausible and has in fact been shown that competitive NMDA receptor antagonists, such as AP7 and CPP (Figure 6) (119,153), as well as the noncompetitive antagonist MK 801 (Figure 6) (154), have anxiolytic properties in preclinical models. Such compounds do, however, have limiting side effects, such as muscle relaxation and an enhancement of locomotor activity similar to that seen with the dissociative anesthetic, phencyclidine (Figure 6) (155,156).

Other Neuroeffectors

Initial studies on the mehanism of action of buspirone provided a considerable body of information to suggest that this novel non-BZ anxiolytic produced its pharmacological effects via interactions with central dopamine (DA) receptor systems (157). As indicated, the compound rather appears to be a partial 5-HT-1A agonist (7). More recently, however, the DA autoreceptor agonist, 3-PPP (Figure 6), has been reported to have antianxiety activity (158), as has the β-adrenoceptor blocker, CGP 361A (Figure 6), which is equiactive with diazepam and more efficacious than propranolol as an antianxiety agent (159). Thus other monoamine systems may be potentially involved in the etiology of anxiety.

Other work has shown anxiogenic actions of corticotropin releasing factor (160,161), and adrenocorticotrophic hormone has long been proposed to have anxiogenic properties (162).

REFERENCES

1. D. E. Goodwin, *Anxiety*, Oxford University Press, New York, 1985.
2. L. H. Sternbach, Prog. Drug Res. *22*, 229-266 (1978).
3. W. Haefley, E. Kyburz, M. Gerecke, and H. Mohler, Adv. Drug Res. *14*, 166 (1985).
4. R. F. Schmidt, M. E. Vogel, and M. Zimmermann, Naunyn-Schmiedebergs Arch. Pharmacol. *258*, 69 (1967).
5. P. Polc, H. Mohler, and W. Haefley, Naunyn-Schmiedebergs Arch. Pharmacol. *284*, 319 (1974).
6. E. Costa and A. Guidotti, Annu. Rev. Pharmacol. Toxicol. *19*, 531 (1979).

7. M. J. Eison, D.A. Taylor, and L. A. Riblet, in *Drug Discovery and Development* (M. Williams and J. B. Malick, eds.), Humana Press, Clifton, New Jersey, 1987, p. 387.
8. H. Mohler and T. Okada, Science *198*, 849 (1977).
9. R. F. Squires and C. Braestrup, Nature *266*, 732 (1977).
10. F. Leeb-Lundberg and R. W. Olsen, in *Psychopharmacology and Biochemistry of Neurotransmitter Receptors* (H. I. Yamamura, R. W. Olsen, and E. Usdin, eds.), Elsevier, New York, 1980, p. 593.
11. M. Williams, Prog. Neuro-Psychopharmacol. Biol. Psychiatr. *8*, 209 (1984).
12. J. Traber, M. A. Davies, W. V. Pompert, T. Glaser, T. Schuurman, and P. R. Seidel, Brain Res. Bull. *12*, 741 (1984).
13. M. Williams, J. W. F. Wasley, and B. A. Petrack Chimi Oggi, September 11, (1987).
14. B. E. Evans, K. E. Rittle, M. G. Bock, R. M. DiPardo, R. M. Freidinger, W. L. Whittier, N. P. Gould, G. F. Lundell, C. E. Hominick, D. F. Veber, P. S. Anderson, R. S. L. Chang, V. J. Lotti, D. J. Cerino, T. B. Chen, P. J. King, K. A. Kunkel, J. P. Springer, and J. Hirshfield, J. Med. Chem. *30*, 1229 (1987).
15. R. F. Squires, Handbook Neurochem. *6*, 261 (1984).
16. M. Mestre, T. Carriot, C. Berlin, A. Uzan, C. Renault, M.-C. Dubreucq, C. Gueremy, and G. Le Fur, Life Sci. *35*, 953 (1984).
17. R. R. H. Anholt, Trends Pharmacol. Sci. *7*, 506 (1986).
18. M. Williams and S. J. Enna, Annu. Rep. Med. Chem. *21*, 211 (1986).
19. W. Hunkeler, H. Mohler, L. Pieri, P. Polc, E. P. Bonetti, R. Cumin, R. Schaffner, and W. Haefley, Nature *290*, 514 (1980).
20. A. J. Czernik, B. Petrack, H. J. Kalinsky, S. Psychoyos, W. D. Cash, C. Tsai, R. K. Rinehart, F. R. Granat, R. A. Lovell, D. E. Brundish, and R. Wade, Life Sci. *30*, 363 (1982).
21. C. Braestrup, R. Schmiechen, G. Neff, M. Neilsen, and E. N. Petersen, Science *216*, 1241 (1982).
22. S. E. File and S. Pellow, Brain Res. *278*, 370 (1983).
23. R. F. Squires, D. I. Benson, C. Braestrup, J. Coupet, C. A. Klepner, V. Myers, and B. Beer, Pharmacol. Biochem. Behav. *10*, 825 (1979).
24. A. S. Lippa, J. Coupet, E. N. Greenblatt, C. A. Klepner, and B. Beer, Pharmacol. Biochem. Behav. *11*, 99 (1979).
25. N. R. Oakley, B. J. Jones, and D. W. Straughan, Neuropharmacology *23*, 797 (1984).
26. S. Arbilla, J. Allen, A. Wick, and S. Z. Langer, Eur. J. Pharmacol. *130*, 257 (1986).
27. I. L. Martin, C. L. Brown, and A. Doble, Life Sci. *32*, 1925 (1983).
28. R. Niddam, A. Dubois, B. Scatton, S. Arbilla, and S. Z. Langer, J. Neurochem. *49*, 890 (1987).
29. S. J. Cooper, T. C. Kirkham, and L. B. Estall, Trends Pharmacol. Sci. *8*, 180 (1987).
30. G. D. Collelo, D. M. Hockenbery, H. B. Bosmann, S. Fuchs, and K. Folkers, Proc. Natl. Acad. Sci. USA *75*, 6319 (1978).

31. G. Toffano, A. Guidotti, and E. Costa, Proc. Natl. Acad. Sci. USA *75*, 4024 (1978).
32. M. Shoyab, L. E. Gentry, H. Marquardt, and G. J. Todaro, J. Biol. Chem. *261*, 11968 (1986).
33. J. H. Woolf and J. C, Nixon, Biochemistry *20*, 4263 (1981).
34. M. Sandler, V. Glover, A. Clow, and I. Armano, in *Pharmacology of the Benzodiazepines* S. M. Paul, P. Skolnick, J. F. Tallman, D. Greenblatt, and E. Usdin, eds.), MacMillan, London, 1982, p. 583.
35. S. Nagy and A. Lathja, J. Neurochem. *40*, 414 (1983).
36. E. Costa and A. Guidotti, Biochem. Pharmacol. *34*, 3399 (1985).
37. T. Asano and S. Spector, Proc. Natl. Acad. Sci. USA *76*, 977 (1979).
38. A. I. Ally, M. S. Manku, D. F. Horrobin, R. A. Karmali, R. O. Morgan, and M. Marazyn, Neurosci. Lett. *7*, 31 (1978).
39. P. Ferrero, M. R. Santi, B. Conti-Tronconi, E. Costa, and A. Guidotti, Proc. Natl. Acad. Sci. USA *83*, 827 (1986).
40. M. Kavaliers and M. Hirst, Brain Res. *383*, 343 (1986).
41. P. W. Gray, Neuropharmacology *26*, 863 (1987).
42. C. Braestrup, M. Neilsen, and C. E. Olsen, Proc. Natl. Acad. Sci. USA *77*, 2288 (1980).
43. C. Pena, J. H. Medina, M. L. Nova, A. C. Paladini, and E. Robertis, Proc. Natl. Acad. Sci. USA *83*, 4952 (1986).
44. L. H. Jensen, E. N. Petersen, and C. Braestrup, Life Sci. *33*, 393 (1983).
45. R. H. Dodd, C. Oannes, M.-C. Potier, L. Prado de Carvalho, J. Rossier, and P. Potier, J. Med. Chem. *30*, 1248 (1987).
46. J. Deckert and P. J. Marangos, Life Sci. *39*, 675 (1986).
47. M. R. Ruff, C. B. Pert, R. J. Weber, L. M. Wahl, S. M. Wahl, and S. M. Paul, Science *229*, 1281 (1985).
48. M. D. Majewska, N. L. Harrison, R. D. Schwartz, J. L. Barker, and S. M. Paul, Science *232*, 1004 (1986).
49. A. L. De Blas and P. Friedrick, Soc. Neurosci. Abstr. *12*, 185 (1986).
50. D. A. Bennett and B. A. Petrack, Drug Dev. Res. *4*, 75 (1985).
51. D. A. Bennett, C. L. Amrick, D. E. Wilson, P. S. Bernard, N.Yokoyama, and J. M. Liebman, Drug. Dev. Res. *6*, 313 (1985).
52. D. A. Bennett, C. L. Amrick, D. E. Wilson, C. A. Boast, P. Loo, P. S. Bernard, M. Schmutz, S. C. Gerhardt, A. Braunwalder, K. Klebs, N. Yokoyama, and J. M. Liebman, Drug Dev. Res. *11*, 219 (1987).
53. M. Williams, D. A. Bennett, P. S. Loo, A. F. Braunwalder, C. L. Amrick, D. E. Wilson, T. N. Thompson, M. Schmutz, N. Yokoyama, and J. W. F. Wasley, J. Pharmacol. Exp. Ther. submitted (1988).
54. S. E. File, Neuropharmacology *26*, 877 (1987).
55. S. Takada, H. Shindo, T. Sasatani, A. Matsushita, M. Eigyo, K. Kawasaki, and S. Murate, J. Med. Chem. *30*, 454 (1987).
56. B. Beer, C. A. Klepner, A. S. Lippa, and R. F. Squires, Pharmacol. Biochem. Behav. *9*, 849 (1978).
57. M. Williams and E. A. Risley, Life Sci. *24*, 833 (19790.

58. D. T. Wong, F. P. Bymaster, and W. B. Lacefield, Drug Dev. Res. *3*, 433 (1983).
59. T. M. Bare, J. F. Resch, and J. B. Patel, Annu. Rep. Med. Chem. *22*, 11 (1987).
60. J. C. Blanchard, A. Boileau, C. Garret, and L. Julou, Life Sci. *24*, 2417 (1979).
61. J. C. Blanchard and L. Julou, J. Neurochem. *40*, 601 (1983).
62. R. R. Trifeletti and S. H. Snyder, Mol. Pharmacol. *26*, 458 (1984).
63. S. H. Snyder, in *Fast and Slow Chemical Signalling in the Nervous System* (L. L. Iversen and E. C. Goodman, eds.), Oxford University Press, Oxford, 1986, p. 106.
64. G. Le Fur, J. Mizoule, M. C. Burgevin, O. Ferri, M. Heaulme, A. Gauthier, C. Gueremy, and A. Uzan, Life Sci. *28*, 1439 (1981).
65. S. Pellow and S. E. File, Pharmacol. Biochem. Behav. *24*, 525 (1986).
66. J. Mizoule, J. Rataud, M. Uzan, M. Mazaider, M. Daniel, A. Gauthier, C. Ollat, C. Gueremy, C. Renault, M.-C. Dubroeucq and G. Le Fur, Arch. Int. Pharmacodyn. Ther. *271*, 189 (1984).
67. S. E. File and S. Pellow, Neurosci. Lett. *50*, 197 (1984).
68. R. von Frenkell, M. Ansseau, and D. Bonnett, Int. Clin. Psychopharmacol. *1*, 24 (1986).
69. P. Skolnick, K.-L. Lock, B. Paugh, P. Marangos, R. Windsor, and S. M. Paul, Pharmacol. Biochem. Behav. *12*, 685 (1980).
70. H. Lal, S. Gherezghhiher, and J. Carney, Drug. Dev. Res. *3*, 75 (1983).
71a. M. Williams, Annu. Rev. Pharmacol. Toxicol. *27*, 315 (1987).
71b. I. L. Martin, Neuropharmacology *26*, 957 (1987).
72. T. Klockgether, I. Pardowitz, M. Schwarz, K.-H. Sontag, and L. Turski, Br. J. Pharmacol. *86*, 357 (1985).
73. C. Braestrup, T. Honore, M. NIelson, E. N. Petersen, and L. H. Jensen, Biochem. Pharmacol. *33*, 859 (1984).
74. C. Braestrup and M. Neilsen, J. Neurochem. *37*, 333 (1981).
75. E. N. Petersen, L. F. Jensen, T. Honore, C. Braestrup, J. W. Kehr, D. W. Stephens, H. Wachtel, D. Seidelman, and R. Schmiechen, Psychopharmacology *83*, 240 (1984).
76. R. Dorow, R. Horowski, G. Paschelke, M. Amin, and C. Braestrup, Lancet *2*, 98 (1983).
77. J. Wolff, P. Carl, T. G. Caluser, and B. Mikeelsen, Anesthesia *41*, 1001 (1986).
78. R. G. Lister and D. J. Nutt, Trends Neurosci. *10*, 223 (1987).
79. P. D. Suzdak, J. R. Glowa, J. N. Crawley, R. D. Schwartz, P. Skolnick, and S. M. Paul, Science *234*, 1234 (1986).
80. B. Saletu, J. Grunberger, and L. Linzmayer, Int. Clin. Psychopharmacol. *1*, 145 (1986).
81. S. Arbilla, S. Tan, P. George, J. Allen, A. Wick, and S. Z. Langer, Br. J. Pharmacol. *92*, 538p (1987).
82. J. Ludwig, French Patent 86000834 (1987).
83. H. H. Theibot, L. Doare, A. J. Peuch, and P. Simon, Eur. J. Pharmacol. *84*, 103 (1982).

84. G. Le Fur, M. L. Perrier, N. Vaucher, F. Imbault, A. Flamier, J. Benavides, A. Uzan, C. Renault, M.-C. Dubroeucq, and C. Gueremy, Life Sci. *32*, 1839 (1983).
85. R. R. H. Anholt, E. B. DeSousa, M. L. Oster-Granite, and S. H. Snyder, J. Pharmacol. Exp. Ther. *233*, 517 (1985).
86. M. Gavish and F. Fares, Eur. J. Pharmacol. *107*, 283 (1985).
87. T. Nishikawa, E. C. Cantor, and S. Spector, Fed. Proc. *44*, 1830 (1985).
88. C. R. Mantione, B. A. Weisman, M. E. Goldman, S. M. Paul, and P. Skolnick, FEBS Lett. *176*, 69 (1984).
89. K. Beaumont, A. K. Cheung, M. L. Geller, and D. D. Fanestil, Life Sci. *33*, 175 (1985).
90. J. D. Elsworth, D. Deward, V. Glover, B. L. Goodwin, A. Clow, and M. Sandler, J. Neural Trans. *67*, 45 (1986).
91. M. Mestre, G. Bouetard, A. C. Uzan, C. Renault, M.-C. Dubroeucq, and G. Le Fur, Eur. J. Pharmacol. *112*, 257 (1985).
92. J. K. T. Wang, J. I. Morgan, and S. Spector, Proc. Natl. Acad. Sci. USA *81*, 753 (1984).
93. J. K. T. Wang, J. I. Morgan, and S. Spector, Proc. Natl. Acad. Sci. USA *81*, 3770 (1984).
94. F. Zavala, J. Haumont, and M. Lenfant, Eur. J. Pharmacol. *103*, 561 (1985).
95. T. Curran and J. I. Morgan, Science *229*, 1265 (1985).
96. E. Kornecki, Y. H. Ehrlich, and R. H. Lenox, Science *226*, 1454 (1984).
97. J. Cassals-Stenzel, J. Franke, T. Friedrich, and J. Lichey, Br. J. Pharmacol. *91*, 799 (1987).
98. J. P. Hullihan, S. Spector, T. Taniguchi, and J. K. T. Wang, Br. J. Pharmacol. *78*, 321 (1983).
99. M. Mestre, T. Carriott, G. Neliak, A. Uzan, C. Renault, M.-C. Dubrouecq, G. Gueremy, A. Doble, and G. Le Fur, Life Sci. *39*, 329 (1986).
100. E. H. Cantor, A. Kennessey, G. Semenuck, and S. Spector, Proc. Natl. Acad. Sci. USA *81*, 1549 (1984).
101. M. Holck and W. Osterrieder, Eur. J. Pharmacol. *118*, 293 (1985).
102. D. Rampe and D. J. Triggle, Eur. J. Pharmacol. *134*, 189 (1987).
103. R. R. H. Anholt, P. L. Pedersen, E. B. DeSousa, and S. H. Snyder, J. Biol. Chem. *261*, 576 (1986).
104. A. Verma, J. S. Nye, and S. H. Snyder, Proc. Natl. Acad. Sci. USA *84*, 2256 (1987).
105. J. F. Tallman, J. W. Thomas, and D. W. Gallager, Nature *274*, 383 (1978).
106. F. Leeb-Lundberg, A. Snowman, and R. W. Olsen, Proc. Natl. Acad. Sci. USA *77*, 7468 (1980).
107. M. Willow and G. A. R. Johnston, J. Neurosci. *1*, 364 (1981).
108. R. W. Olsen and A. M. Snowman, J. Neurosci. *2*, 1812 (1982).
109. F. Leeb-Lundberg, A. Snowman, and R. W. Olsen, J. Neurosci. *1*, 471 (1981).
110. P. Supavilai and M. Karobath, Eur. J. Pharmacol. *60*, 111 (1979).
111. R. W. Olsen, J. Neurochem. *37*, 1 (1981).

112. R. W. Olsen, Annu. Rev. Pharmacol. Toxicol. *22*, 245 (1982).
113. M. Williams and E. A. Risley, J. Neurochem. *42*, 745 (1984).
114. F. A. Stephenson, A. E. Watkins, and R. W. Olsen, Eur. J. Biochem. *123*, 291 (1982).
115. M. Gavish and S. H. Snyder, Proc. Natl. Acad. Sci. USA *78*, 1939 (1981).
116. T. Sanao, Y. Yamada, and N. Ogaswara, J. Neurochem. *40*, 209 (1983).
117. E. Sigel, F. A. Stephenson, C. Mamalaki, and E. A. Barnard, J. Biol. Chem. *258*, 6595 (1983).
118. M. Noda, S. Shimizu, T. Tanabe, T. Takai, T. Kayano, T. Ikeda, H. Takahashi, H. Nakayama, Y. Kanaoka, N. Minamino, K. Kangawa, H. Matsuo, M. Raftery, T. Hirose, S. Inayama, H. Hayashida, T. Miyata, and S. Numa, Nature *312*, 121 (1984).
119. J. Lehmann, J. Schneider, and M. Williams, Annu. Rep. Med. Chem. *22*, 31 (1987).
120. P. L. Wood, P. Loo, A. Braunwalder, N. Yokoyama, and D. L. Cheney, J. Pharmacol. Exp. Ther. *231*, 572 (1984).
121. D. A. Bennett, Pharmacol. Biochem. Behav. *41*, 241 (1987).
122. H. Mohler, M. K. Battersby, and J. G. Richards, Proc. Natl. Acad. Sci. USA *77*, 1666 (1980).
123. J.-P. Changeux, A. Devillers-Thiery, and P. Chemouilli, Science *225*, 1335 (1984).
124. T. Kubo, K. Fukada, A. Mikami, A. Maeda, H. Takahasi, M. Mishina, T. Haga, K. Haga, A. Ichiyama, K. Kanagawa, M. Lijima, H. Matsuo, T. Hirose, and S. Numa, Nature *232*, 411 (1986).
125. R. A. F. Dixon, B. K. Kobilka, D. J. Strader, J. L. Benovic, H. G. Dohlman, T. Frielle, M. A. Bolanowski, C. D. Bennett, E. Rands, R. E. Diehl, R. A. Mumford, E. E. Slater, I. S. Sigal, M. G. Varon, R. J. Lefkowitz, and C.D. Strader, Nature *321*, 75 (1986).
126. G. B. Stauber, R. W. Ransom, A. I. Dilber, and R. W. Olsen, Eur. J. Biochem. *167*, 125 (1987).
127. L.-R. Chang and E. A. Barnard, J. Neurochem. *39*, 1507 (1982).
128. F. A. Stephenson and R. W. Olsen, J. Neurochem. *39*, 1579 (1982).
129. E. Sigel and E. A. Barnard, J. Biol. Chem. *259*, 7219 (1984).
130. R. G. King, M. Nielsen, G. B. Stauber, and R. W. Olsen, Eur. J. Biochem. *169*, 555 (1987).
131. M. Nielsen, T. Honore, and C. Braestrup, Biochem. Pharmacol. *34*, 3633 (1985).
132. P. R. Schofield, M. G. Darlison, N. Fujita, D. R. Burt, F. A. Stephenson, H. Rodriguez, L. M. Rhee, J. Ramachandran, V. Reale, T. A. Glencourse, P. H. Seeburg, and E. A. Barnard, Nature *328*, 221 (1987).
133. L. Deng, R. W. Ransom, and R. W. Olsen, Biochem. Biophys. Res. Commun. *138*, 1308 (1986).
134. S. O. Casalotti, F. A. Stephenson, and E. A. Barnard, J. Biol. Chem. *261*, 15013 (1986).

135. P. Haring, C. Stahli, P. Schoch, T. Tackacs, T. Staehelin, and H. Mohler, Proc. Natl. Acad. Sci. USA *82*, 4837 (1985).
136. J. G. Richards, P. Shoch, P. Haring, B. Tackacs, and H. Mohler, J. Neurosci. *7*, 1866 (1987).
137. T. G. Smart, L. Constanti, G. Bilbe, D. A. Brown, and E. A. Barnard, Neurosci. Lett. *40*, 55 (1983).
138. P. M. Sweetnam and J. W. Tallman, Mol. Pharmacol. *29*, 299 (1986).
139. P. W. Codding and A. K. S. Muir, Mol. Pharmacol. *28*, 178 (1985).
140. P. A. Borea, G. Gilli, V. Bertolasi, and V. Ferretti, Mol. Pharmacol. *31*, 334 (1987).
141. K. Bizere, J. Bouruignon, J. P. Chambon, M. Healume, A. Perio, S. Tebib, and G. G. Wermuth, Br. J. Pharmacol. *90*, 183 (1987).
142. N. A. Sharif and D. R. Burt, J. Neurochem. *43*, 742 (1984).
143. R. K. Rinehart, B. Barbaz, S. Iyengar, F. Ambrose, D. J. Steel, R. F. Neale, B. Petrack, H. Bittiger, P. L. Wood, and M. Williams, J. Pharmacol. Exp. Ther. *238*, 178 (1986).
144. D. Romer, H. Buscher, R. C. Hill, R. Maurer, T. J. Pletcher, H. Zueger, W. Benson, E. Finner, W. Milkowsi, and P. W. Thies, Nature *298*, 759 (1982).
145. S. Das and H. N. Bhargava, Neuropharmacology *26*, 1141 (1987).
146. R. S. L. Chang, V. J. Lotti, R. L. Monaghan, J. Birnbaum, E. O. Stapley, M. A. Goetz, G. Albers-Schonberg, A. A. Patchett, T. M. Liesch. O. D. Hensens, and J. P. Springer, Science *230*, 177 (1985).
147. R. L. Webb, R. Della Puca, R. D. Ghai, M. B. Zimmerman, P. Loo, A. Braunwalder, R. K. Rinehart, R. D. Robson, and M. Williams, J. Cardiovascu. Pharmacol. submitted (1988).
148. Drugs of the Future *11*, 949 (1986).
149. J. A. Barone, R. H. Bierman, J. W. Cornish, A. Hsuan, N. D. Drake, and J. L. Colaizzi, Drug Intell. Clin. Pharmacol. *20*, 770 (1986).
150. K. G. Lloyd, H. Depoortere, B. Scatton, H. Schoemaker, B. Zinkovic, P. Manouri, S. Z. Langer, P. L. Morselli, and G. Bartholini, Pharmacol. Biochem. Behav. *23*, 645 (1985).
151. B. J. Jones, N. J. Oakley, and M. B. Tyers, Br. J. Pharmacol. *90*, 88 p. (1987).
152. B. A. Costall, R. B. Naylor, and M. B. Tyers, Br. J. Pharmacol. *90*, 85 p. (1987).
153. D. N. Stephens, B. S. Meldrum, R. Weidmann, C. Schneider, and M. Grutzner, Psychopharmacology *90*, 166 (1986).
154. B. V. Clineschmidt, M. Williams, J. J. Witoslawski, P. R. Bunting, E. A. Risley, and J. A. Totaro, Drug Dev. Res. *2*, 147 (1982).
155. W. Koek, J. H. Woods, and P. Ornstein, Life Sci. *39*, 973 (1986).
156. W. Koek, J. H. Woods, and P. Ornstein, Psychopharmacology *91*, 297 (1987).
157. H. C. Stanton, D. P. Taylor, and L. A. Riblet, in *Neurobiology of the Nucleus Accumbens* (R. B. Chronister and J. F. DeFrance, eds.), Haer Institute, Brunswick, Maine, 1982, p. 316.

158. S. Hjorth, A. Carlsson, and J. A. Engel, Psychopharmacology *92*, 371 (1987).
159. V. M. Puttenkammer, H. J. Gaertner, A. Mahal, J. May, U. Binz, and H. Heimann, Arzneimittel. Forsch. *37*, 721 (1987).
160. K. T. Britton, G. Lee, R. Dana, S. C. Risch, and G. F. Koob, Life Sci. *39*, 1281 (1986).
161. A. J. Dunn and S. E. File, Horm. Behav. in press (1987).
162. D. De Weid, Life Sci. *20*, 195 (1977).

13

Receptors for Excitatory Amino Acids: Ligand Binding and Functional Assays

John W. Ferkany / Nova Pharmaceutical Corporation, Baltimore, Maryland

HISTORICAL PERSPECTIVE

Evidence is substantial that the acidic, dicarboxylic amino acids L-glutamate (L-GLU) and L-aspartate (L-ASP) function as mediators of chemical synaptic transmission in the vertebrate central nervous system (CNS) and at the neuromuscular junction of nonvertebrate species (1-6). More limited biochemical, pharmacological, and physiological investigations suggest that other amino acids or amino acid-like compounds, including cysteic acid, cysteine sulfinic acid (CSA), homocysteic acid (HCA) (7-12), the tryptophan metabolite quinolinic acid (QUIN) (13-17), and the dipeptide *N*-acetyl-L-aspartyl-L-glutamate (NAAG) (18-31) (Figure 1), are neuroexcitants that serve in a neuromodulatory or neurotransmitter capacity. It is probable that the physiologic significance of small acidic compounds as excitatory neurotransmitters rivals that of GABA as an inhibitory agent in brain.

The existence of L-GLU (and other amino acids) in nervous tissue has been established for some time, and transport mechanisms for L-GLU were described as early as 1949 (32). However, the concept of amino acids as putative neurotransmitters and modulators rather than components of intermediary metabolism was slow to gain acceptance. Studies initiated by Hayashi (33) and subsequently confirmed by Curtis et al. (34-36) revealed that iontophoretic application of L-GLU, L-ASP, and HCA effected a rapid depolarization of nervous tissue. Biochemical investigations followed, and in rapid order the criteria necessary to establish L-GLU and L-ASP as transmitter candidates were fulfilled.

L-Glutamate and L-ASP are transported in brain slices and synaptosomes by a high-affinity, Na^+-dependent mechanism (37-39), and release of both amino

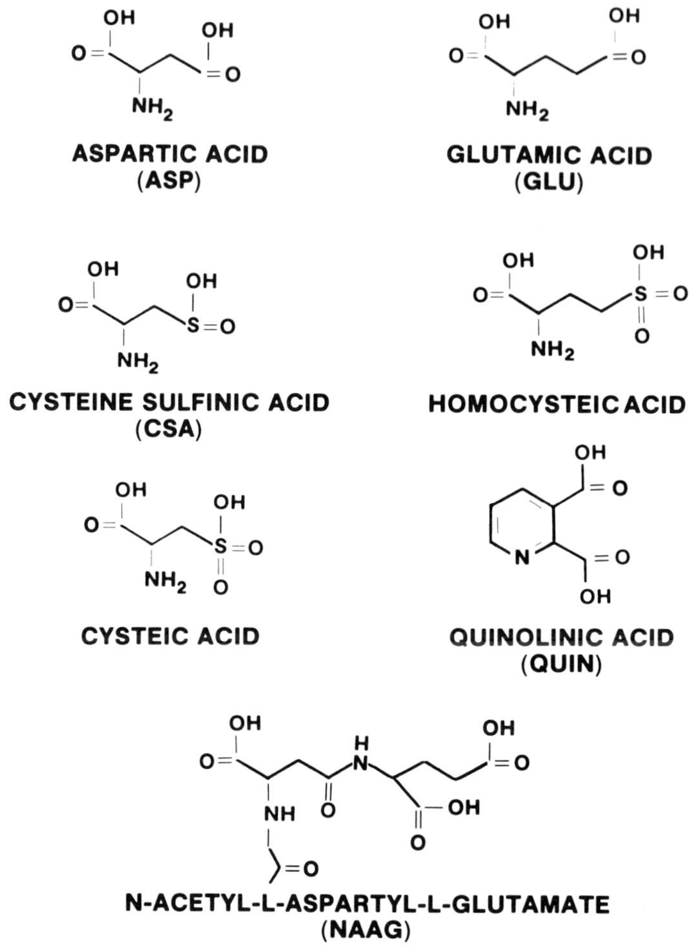

Figure 1 Structure of excitatory amino acids found in mammalian brain. All the compounds are found in mammalian brain to varying degrees and are potent neuroexcitants.

acids can be evoked by elevated extracellular K^+ or electrical stimulation (40-42). Stimulus-evoked release is Ca^{2+}-dependent, suggesting that releasable fractions are coupled to neuronal pools (40-44). In accord with participation in CNS metabolic activity, differences in the regional distribution of L-GLU and L-ASP are not as marked as for non-amino acid neurotransmitter and modulators; however, lesion or ablation of putative excitatory amino acid (EAA) pathways is accompanied by a decline in the uptake or evoked release of EAA in distant loci innervated by the interrupted tract (45-49). Antibodies specific for the enzyme glutaminase have been utilized with other approaches to identify L-GLU containing neurons in cerebellum, cerebral cortex, and retina (50-54).

Electrophysiologic, biochemical, and pharmacological studies provide evidence for multiple classes of EAA receptors in brain (1-6). Among the initial compounds unequivocally identified to interact at a subpopulation of EAA receptors were the pyrrolidines, domoic and kainic acids (KA) (Figure 2). Isolated

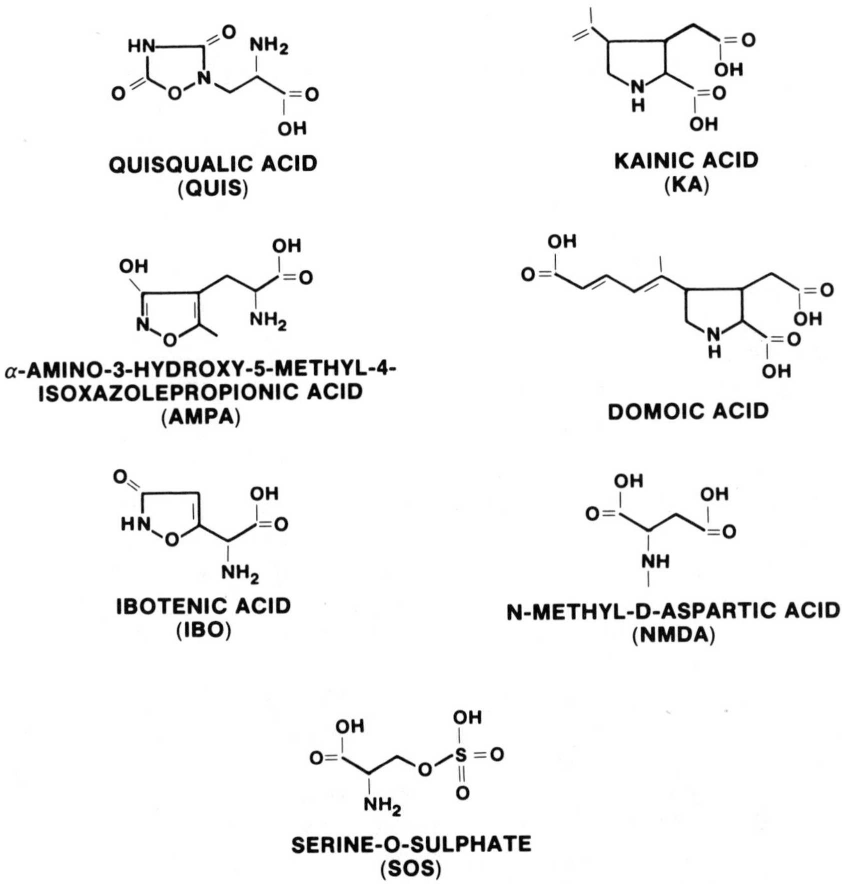

Figure 2 Structures of excitatory amino acid agonist analogs. The compounds shown are not endogenous to brain but are potent neuroexcitants and/or neurotoxins. The actions of the compounds are thought to be mediated by interactions by excitatory amino acid receptors.

during a search for the active anthelmintic principle of *Digenea simplex* (56), KA and domoic acid are potent neuroexcitotoxins (56). Other synthetic or naturally occurring products thought to interact with EAA receptors include quisqualic acid (QUIS), a cyclic analong of L-GLU, the isoxazole neurotoxin ibotenic acid (IBO), the synthetic analog of L-ASP, *N*-methyl-D-aspartic acid (NMDA), and the QUIS-type receptor agonist, AMPA (α-amino-3-hydroxy-5-methyl-4-isoxazolepropionic acid; Figure 2) (1-6,57).

Drawing from work with prototypical agonists and antagonists and excluding Na^+-dependent transport sites, at least four receptors for EAA have been described (3,4). These include

1. Receptors activated by QUIS and AMPA and preferentially antagonized by glutamic acid diethyl ester (GDEE) (Figure 3)
2. Receptors responsive to NMDA and antagonized by α-amino adipic acid (ADA), D-(–)-2-amino-5-phosphonopentanoic acid (DAP5), and D-(–)-2-amino-7-phosphonheptanoic acid (DAP7) (Figure 3) and noncompetitively by Mg^{2+}
3. Receptors activated by KA and domoic acid for which no highly selective antagonist has been discovered and
4. Receptors antagonized by L-(+)-2-amino-4-phosphonobutyric acid (LAP4); the endogenous transmitter(s) for L-(+)-AP4 receptors have not been identified with certainty but are unlikely to be L-GLU or L-ASP (58-60)

Beyond these fundamental receptor categories, it has become increasingly apparent that subdivisions may exist within each receptor category. For example, receptors for NMDA appear to be pharmacologically distinct in different regions of brain (17,61,62), and one subpopulation of NMDA receptors may be allosterically linked to the site of action of the dissociative anesthetics and certain of the benz(F)isoquinolines (63-71). The novel tricyclic anticonvulsant MK801 may effect its primary action via an allosteric modulation of NMDA receptors (72,73). Thus, it has been suggested that at least some EAA receptors are composed of supramolecular complexes analogous to the GABA-benzodiazepine-barbiturate effector proteins (67,71).

In this chapter, EAA receptors are explored from the perspectives of ligand binding and in vitro functional assays. Those interested in the electrophysiology of EAA neurotransmission, a more comprehensive discussion of the neurochemistry of EAA, or a description of EAA neurotransmission in invertebrate species are referred to any of several excellent reviews (1-6,17). Details of the theoretical and practical considerations of assay methods are found elsewhere in this volume.

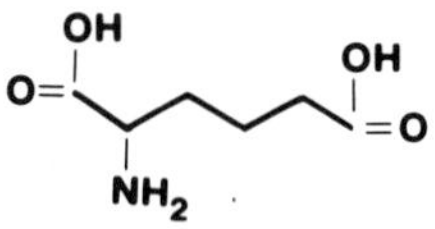

α-AMINO ADIPIC ACID
(ADA)

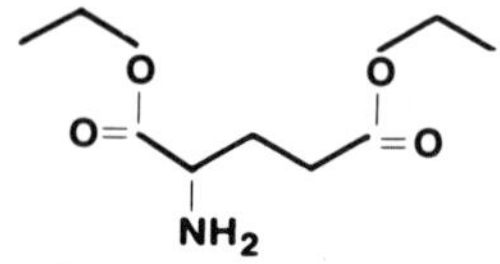

GLUTAMIC ACID DIETHYL ESTER
(GDEE)

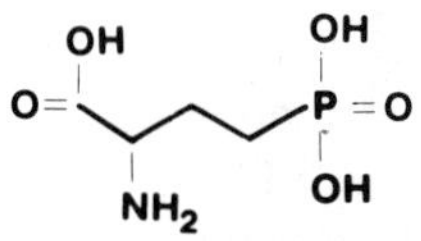

2-AMINO-4-PHOSPHONO
BUTYRIC ACID
(AP4)

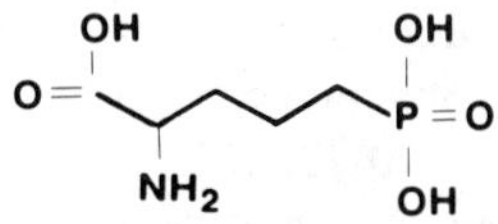

2-AMINO-5-PHOSPHONO
PENTANOIC ACID
(AP5)

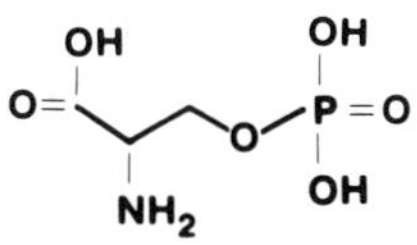

SERINE-O-PHOSPHATE
(SOP)

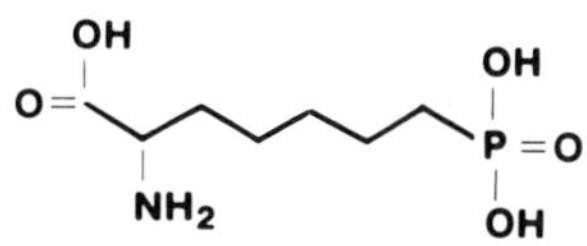

2-AMINO-7-PHOSPHONO
HEPTANOIC ACID
(AP7)

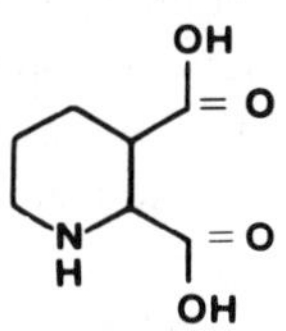

CIS-2,3-PIPERIDINE
DICARBOXYLIC ACID

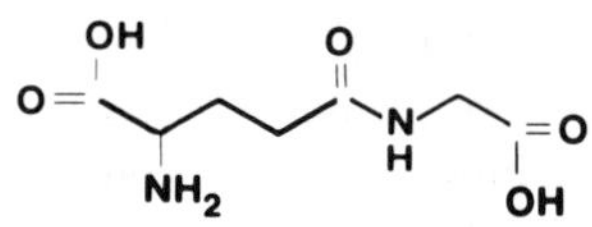

γ-D-GLUTAMYL GLYCINE

Figure 3 Structures of prototypical excitatory amino acid receptor antagonists. The compounds shown interact with varying degrees of specificity with NMDA-, QUIS, KA-, and L-(+)-AP4-preferring receptors in brain. For instance, ADA is the prototypical NMDA antagonist, whereas GDEE is most useful in defining QUIS receptors. AP5 and AP7 are highly selective for NMDA receptors. The remainder of the compounds antagonize EAA-mediated responses with a range of selectivities and potencies, although L-(±)-AP4 has been used to identify a unique receptor category.

LIGAND BINDING ASSAYS

Tissue Homogenates

The methodological aspects of EAA binding assays are similar in most respects to those of assays for other neurotransmitter receptors. Tissue preparations reported to be useful to study receptor characteristics include whole particulate fractions of brain (24,72-79), crude microsomal preparations (24,80-97), purified synaptic junctional membranes (98-103), and postsynaptic densities (97,97, 104-106). Without exception, the CNS has been the source of tissue and binding of EAA to peripheral tissues is virtually absent. Contingent upon the assay, either fresh (81-87,90,92,93,98-100) or previously frozen (24,74-77,79,91,94-97,104-108) tissue preparations are used. Ligands currently available are universally labeled with tritium to high specific activities (20-70 Ci/mmol).

In contrast to binding assays that utilize vacuum filtration to terminate the reaction, the EAA ligand-receptor complex is frequently separated using high-speed macro- or microcentrifugation at reduced temperatures. This latter aspect of EAA assays reflects the relatively low binding affinity (K_D = 0.1-10 μM) of most EAA ligands. Similarly, those unfamiliar with EAA assays should not be surprised when the receptor number (B_{max}) achieves units of picomoles per milligram protein, although this contrasts sharply with assays for other neurotransmitters in which receptors are present in densities of femtomoles per milligram protein. The high apparent density of EAA receptors in brain may reflect the widespread distribution and importance of EAA neurotransmission in normal function. Alternatively, the observation may be an artifact of nonselective interactions of previously used EAA ligands with multiple recognition sites in brain (e.g., receptors, enzymes, and transport sites).

As is apparent from Tables 1 and 2 and the discussion that follows, compounds defined by electrophysiologic criteria to interact with a single EAA receptor in vivo or in vitro frequently recognize more than one radiolabeled binding site in vitro. Conversely, compounds acting in vivo at one or more EAA receptor(s) may demonstrate little affinity for tritium-labeled sites in vitro. These caveats notwithstanding, radioligand binding assays for EAA remain useful, and at the practical level, these methods provide a rapid procedure to screen for novel compounds having affinity for EAA recognition sites. When applied to tissues obtained postmortem, differences in the patterns of EAA binding can yield useful information on the pathophysiology of CNS disorders. Further, recently introduced ligands appear to be more selective in recognizing single receptor populations and provide information on the pharmacological characteristics and localization of receptors in brain. As more selective ligands become available, or as assay procedures are refined to make available ligands more specific, it should become increasingly possible to examine the molecular and pharmacological characteristics of EAA receptors in vitro.

Ligands

Ligands useful for EAA binding assays that are available from commercial sources include tritium-labeled L-GLU, L-ASP, QUIN, NAAG, KA, AMPA, Dl-(±)-AP4, and DL-(±)-AP7. Specific binding of DL-(±)-AP5 to brain membranes has been described (90). The novel NMDA receptor [^{3}H] CPP [3-(2-carboxypiperazine-4-yl)propyl-1-phosphonic acid] and [^{3}H]CGS 19755 (256) are commercially available.

Presented in the following sections are brief descriptions of EAA receptor binding assays. The sections are grouped according to receptor subtype and are intended to provide a general introduction to the area. Readers interested in details of particular assays are urged to consult the references cited. A summary of the pharmacological characteristics of EAA binding assays has been compiled (Tables 1 and 2).

NMDA Receptor Assays

Reports describing the specific binding of [^{3}H] QUIN to brain membranes have not appeared in the literature, although this may be due to the recent commercial introduction of the ligand. [^{3}H] NMDA has been available for some time, but attempts to detect specific binding to crude particulate fractions of brain have been largely unsuccessful. [^{3}H] NMDA binds to purified postsynaptic densities (PSD) (105); however, the interaction is not robust and only 25-30% of the binding is specific.

Explanations for the failure to detect specific binding of [^{3}H] NMDA to brain are not immediately apparent, although it has been suggested (4) that a low density of NMDA receptors and/or a low affinity of the ligand for the binding site may preclude detection of binding using current technologies. Alternatively, methods of tissue preparations may alter the conformation of the receptor or remove necessary regulatory factors. The detection of NMDA recognition sites in PSD using [^{3}H] L-GLU as a ligand (see below) make the latter explanation unlikely.

Ferkany and Coyle (109) (and unpublished observations), as well as Manallack and Beart (95), have reported specific and saturable binding of the NMDA receptor antagonist [^{3}H] DL-(±)-AP7 to crude microsomal fractions of rat forebrain (K_D = 3.0 μM). Of the compounds that inhibit binding, QUIS, serine-*O*-sulfate (SOS), and IBO are potent (K_i = 0.6-3.0 μM), followed in order by L-GLU = ADA > CSA = HCA = DL-(±)-AP4 > DL-(±)-AP5 (Table 1). Whereas NMDA is ineffective to inhibit binding when tested at 1000 μM, the NMDA agonist QUIN and various QUIN analogs are weak (K_i = 100-300 μM) competitive inhibitors at the binding site (110). Surprisingly, L-(+)-stereoisomers of the α-amino-Ω-phosphonocarboxylic acids are more potent at displacing [^{3}H]-

Table 1 Potencies of Various Compounds to Inhibit Specific [^{3}H] Ligand Binding to EAA Recognition Sites on Brain Membranes In Vitro[a]

	Ligand (μM)							
	L-GLU[b]							
Compound	A	B	AP5	AP7	L-GLU[c]	AP4	NAAG	L-ASP
Agonists								
NMDA	7.2	>100	11	>100	>100	>100	>100	>100
QUIS	1.1	89	250	0.6	0.1-0.6	0.4	0.47	>100
L-GLU	0.2	0.1-0.4	0.9	5.6	0.2-1.0	0.8-1.7	0.46	2.5
IBO	7.1	>100	–	3	0.4-3.1	0.5-10	0.53	10->100
CSA	–	3.4	13	23	4.7-7.1	7.1	2.37	47-125
HCA	–	2-59	4	14	1.6	6.3	1.34	–
AMPA	–	–	–	–	28	25	>100	–
L-ASP	2.1	0.8-1.8	11	>1000	2.1-28	5.5-7.0	3.0	2.7
D-GLU	9.7	3.9-244	–	>1000	2.6-15	7-18	5.0	>100
D-ASP	1.6	>270	–	>1000	27-50	15-56	>100	>100
SOS	–	–	25	1.1	–	<<100	0.12	–
QUIN	91	–	350	110	–	>100	>100	–
KA	29	>100	>100	>1000	>100	>100	>100	>100

Antagonists								
DL-(+)-AP4	3300	100	260	17	1.2-4.6	3.6	3.86	–
D-(–)-AP4	–	–	–	–	75	10-30	–	–
L-(+)-AP4	–	–	–	–	5	1.5-2.2	–	–
DL-(±)-AP5	1.0	7	1.7	21	4.4	6-40	6.0	–
D-(–)-AP5	–	–	0.6	–	14	126	–	–
L-(+)-AP5	–	–	55	–	16	20	–	–
DL-(±)-AP7	4.6	–	2.4	3.6	13	19-32	11.4	360
DL-(±)-ADA	–	30	–	–	1.3	3.6	–	–
D-(–)-ADA	–	>100	–	5.8	0.7-2.4	16	–	–
L-(+)-ADA	–	–	–	–	3.4	3	–	–
cis-PDA	–	–	78	>1000	–	–	>100	–
DGG		–	40	>1000	–	–	>100	–
GDEE	–	–	–	>1000	–	–	>100	–
CPP	–	–	6.5	–	–	–	–	–

[a]The numbers represent the potency or the range of potencies (IC_{50} or K_i) of various compounds in inhibiting specific binding of the indicated ligand to brain membranes in vitro. Notations and references are as follows: A, postsynaptic densities (104,105); B, crude microsomal fraction or synaptic plasma membranes (177); AP5 (92); AP7 (109); AP4 (93,100); NAAG (25); L-ASP (90,91,102); CPP (122).

[b][^{3}H]GLU minus calcium chloride.

[c]L-GLU plus Ca^{2+}/Cl^- or Cl^- (79,83,85,103).

Table 2 Potencies of Excitatory Amino Acids to Inhibit [^{3}H]Kainate, [^{3}H]AMPA, or [^{3}H]CPP Binding to Brain[a]

Compound	[^{3}H]Kainate (μM)			[^{3}H]AMPA (μM)			[^{3}H]CPP (μM)
	A	B	C	D	E	F	
Agonists							
Domoic acid	–	–	0.013	–	–	–	–
Kainic acid	0.002	0.021	0.02	6	22	>5000	>100
AMPA	–	–	–	0.3	0.004[b] 0.105[c]	12	>100
CPP	–	–	–	–	–	–	0.17
QUIS	0.02	0.26	0.61	0.3	0.04	32	19
L-GLU	0.06	0.72	1.50	1.3	1.3	43	0.15
NMDA	>100	>100	>100	>100	–	–	4.1
IBO	3.4	36	9.4	>100	–	500	2.6
HCA	71	87	–	–	–	–	–
L-ASP	>100	>100	>100	>100	–	700	1.5
D-ASP	>100	>100	>100	>100	–	1000	1.1

NMLA	–	–	–	–	–	–	29
D-GLU	–	–	–	>100	110	>1000	7.1
Antagonists							
DL-(±)-AP3	–	–	–	–	–	–	>100
DL-(±)-AP4	–	–	100	–	–	>1000	>100
D-(–)-AP5	–	–	>100	–	–	>1000	0.18
L-(+)-AP5	–	–	>100	–	–	>1000	8.2
DL-(±)-AP6	–	–	–	–	–	–	13
DL-(±)-AP7	–	–	>100	–	–	>1000	1.1
ADA	–	–	–	>100	–	–	–
GDEE	>100	>100	>100	>100	–	>1000	>100
GAMS	–	–	–	–	–	–	>100

[a]Numbers shown are the potency (IC_{50} or K_i) of various compounds in inhibiting specific binding of [^{3}H]kainic acid, [^{3}H]AMPA, or [^{3}H]-CPP to brain membranes. Sources are as follows: A, forebrain, high-affinity site (75); B, forebrain, low-affinity site (75); C, cerebellum (74); D, forebrain (87); E, forebrain plus 100 mM KSC (89); F, forebrain, pH 5.5 (88); CPP (108).
[b]High-affinity site.
[c]Low-affinity site.

AP7 in vitro, whereas in physiologic assays, the D-(−) enantiomers are clearly the more potent antagonists of NMDA-evoked responses (112-116).

Initial expectations that [^{3}H] AP7 would provide a sensitive ligand to detect NMDA receptors in vitro do not appear fully justified by available evidence. Although the site labeled by AP7 exhibits characteristics associated with EAA receptors, classification of this recognition site into the current EAA receptor scheme must await further evaluation.

Binding sites with more convincing characteristics of NMDA-type receptors have been identified using the lower homolog of AP7, [^{3}H] DL-(±)-AP5 (92), as well as [^{3}H] L-GLU (96,104,105). Specific binding of [^{3}H] DL-(±)-AP5 to rat forebrain membranes is potently inhibited by D-(−)-AP5, L-GLU, and DL-(±)-AP7. Additionally, SOS, HCA, CSA (NMDA receptor preferring agonists), and NMDA are moderately potent inhibitors, with K_i between 4 and 25 μM. In contrast, both QUIS and KA are 500- to 1500-fold less potent in displacing binding when compared with D-(−)-AP5 (Table 1). Should tritium-labeled DL-(±)-AP5 become commercially available, it may represent a useful ligand to establish an NMDA receptor binding assay.

Using [^{3}H] L-GLU, Fagg and coworkers (97,104,105) have provided evidence for a binding site on postsynaptic densities that can be equated with NMDA receptors. The two populations of recognition sites detected can be discriminated according to the potency of QUIS or NMDA to inhibit binding. In the presence of QUIS (10 μM) to mask "QUIS-preferring receptors," NMDA-sensitive binding represents 58% of the total receptor population. The ligand-receptor interaction can be inhibited by NMDA, L-GLU, IBO, DL-(±)-AP5, DL-(±)-AP7, and D- and L-ASP (K_i = 1-10 μM; Table 1). The NMDA agonist QUIN is also an effective displacer (K_i = 100 μM), and intriguingly, the dipeptide phenylalanyl-L-glutamate (97,104,117) is equipotent to NMDA. More recently (67), NMDA-sensitive binding of [^{3}H] L-GLU to less pure synaptic junctional membranes has been reported (118). However, binding is markedly time dependent, and with prolonged incubations (>20 min), the NMDA-sensitive component decreases substantially, suggesting the presence of a labile recognition site. Whether the latter system represents a useful method to characterize the NMDA receptor must await additional verification.

When examined in the presence of 100 μM NMDA to mask NMDA-sensitive binding sites, residual [^{3}H] L-GLU binding to PSD is inhibited by QUIS and KA (K_i = 1 and 29 μM, respectively). Both the QUIS- and NMDA-sensitive components of the reaction are resistant to prior freezing of the tissue, thereby differentiating them from the Ca^{2+}/Cl^--dependent [^{3}H] L-GLU binding site. The localization of the NMDA-sensitive site to PSD suggests a possible correspondence to receptors involved in synaptic transmission.

Very recently, Watkins et al. have identified 3-(2-carboxypiperazine-4-yl)-propyl-1-phosphonic acid (Figure 4) as a potent and selective NMDA antagonist

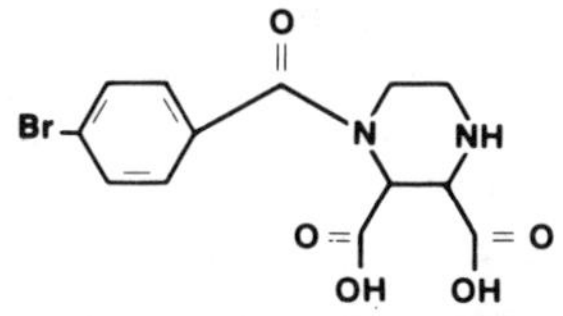

1-(p-bromobenzoyl)-PIPERAZINE-2, 3 - DICARBOXYLIC ACID

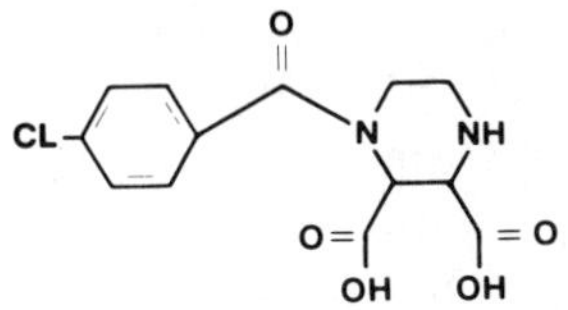

1-(p-chlorobenzoyl)-PIPERAZINE-2, 3 - DICARBOXYLIC ACID

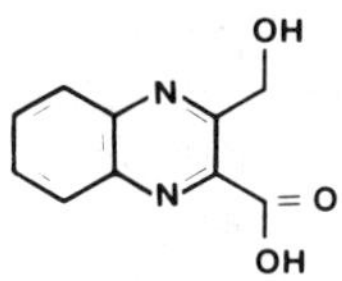

3-HYDROXY-2-QUINOXALINE CARBOXYLIC ACID

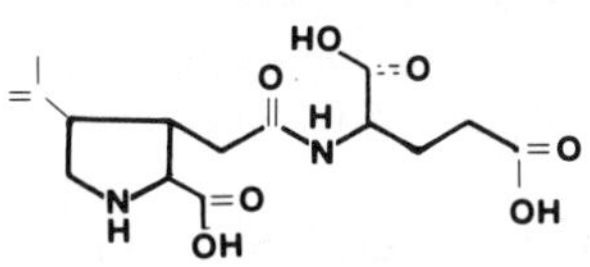

γ-KAINYL-L-GLUTAMIC ACID

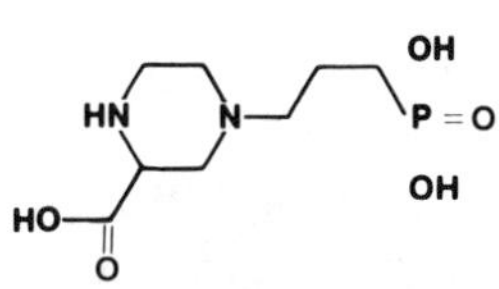

3-(2-CARBOXYPIPERAZIN-4-YL) PROPYL-1-PHOSPHONIC ACID (CPP)

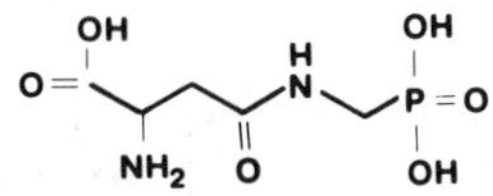

β-D-ASPARTYLAMINOMETHYL-PHOSPHONIC ACID

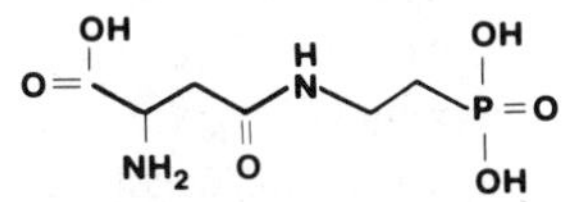

β-D-ASPARTYLAMINOETHYL-PHOSPHONIC ACID

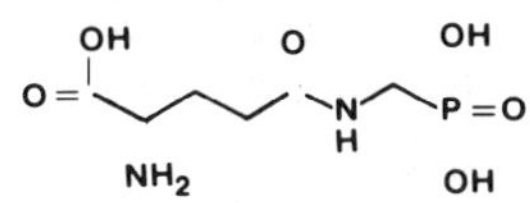

γ-D-GLUTAMYLAMINOMETHYL-PHOSPHONIC ACID

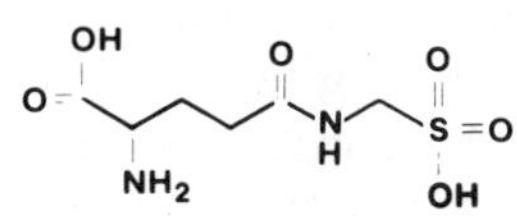

γ-D-GLUTAMYLAMINOMETHYL-SULPHONIC ACID (GAMS)

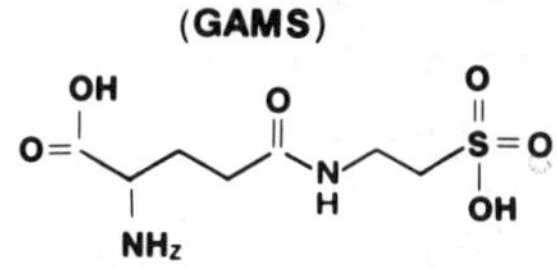

γ-D-GLUTAMYLTAURINE

Figure 4 Structures of novel excitatory amino acid receptor antagonists. The compounds shown have been reported to be EAA antagonists in vivo or in vitro and represent departures from the structures of the prototypical EAA antagonists shown in Figure 3.

(119-121). Using both in vitro and in vivo procedures, CPP appears at least an order of magnitude more potent than DL-(±)-AP7 as an NMDA antagonist. The compound blocks agonist responses in spinal neurons and in the hippocampal slice, with the order of potency against EAA being NMDA > HCA > L-ASP > L-GLU > KA > QUIS (120,121). At doses that do not affect Schaffer collateral synaptic transmission, CPP reversibly blocks the induction of long-term potentiation of Schaffer synaptic responses, although the compound has no effect on lateral perforant path responses or on the inhibition of these responses by AP4 (121). Intraperitoneal drug administration (ED_{50} = 10 mg/kg) effectively prevents audiogenic seizures in mice and seizures induced by NMDA, mercaptopropionic acid, and maximal electroschock (120,123). In binding assays, CPP is a potent and competitive inhibitor at the site labeled by [^{3}H] AP5 (K_i = 0.48 μM), and the distribution of CPP-sensitive [^{3}H] L-GLU binding determined by autoradiographic analysis is consistent with the identification of an NMDA receptor (120, 121).

Tritium-labeled CPP binding to Triton-X-100 treated crude synaptic membrane preparations is saturable and specific (107,108). Binding of the ligand is rapid (equilibrium $\leqslant$ 20 min; 23°C) and fully reversible following the addition of L-GLU. Saturation analysis reveals only a single population of recognition sites with an apparent K_D of 168 nM and an apparent B_{max} of 5.6 pmol/mg protein (108). Both the subcellular and regional distribution of [^{3}H] CPP binding are consistent with the identification of an NMDA receptor population. Binding is most enriched in synaptosomal subfractions and in the hippocampus, a region of brain known to have a high density of NMDA receptors. No specific binding is detected in peripheral tissues.

Most importantly, binding of [^{3}H] CPP to brain in inhibited in order of potency by CPP = L-GLU = D-(−)-AP5 > DL-(±)-AP7 = D-ASP = L-ASP $\geqslant$ IBO > NMDA (K_i = 4.0 μM) >> QUIS (K_i = 20 μM) = *cis*-2,3-piperidine dicarboxycyclic acid >> KA (Table 2). Binding is stereoselective, with L-AP5 and *N*-methyl-L-aspartate less potent than the D isomers in inhibiting the reaction. Furthermore, a variety of EAA agonists and antagonists, including AMPA, DL-(±)-AP4, γ-D-glutamylaminomethylsulfonic acid (GAMS; Figure 4), kynurenic acid, and glutamic acid diethyl ester (GDEE) are without activity when tested at concentrations of 100 μM. Likewise, such compounds as 2-chloroadenosine, dexoxadrol, γ-aminobutyric acid (GABA), muscimol, baclofen, glycine, and levoxadrol, which are known to interact with other receptors, are similarly inactive.

As noted, CPP has been reported to be a systemically actve anticonvulsant. In addition, the compound exhibits an anxiolytic action in rodents (122) and prevents ischemic brain damage in gerbils (124). Both properties are expected of an NMDA receptor antagonist. CPP may represent a novel compound to identify NMDA receptors in vitro, and if substantiated by additional investigations, the observations of high-affinity, NMDA-sensitive [^{3}H] CPP receptors would represent a significant advance in EAA receptor pharmacology. Unlike currently available techniques that require tedious membrane preparations, the assay described would permit the routine examination of NMDA receptors using simple tissue preparation techniques. More recently, [^{3}H]CGS 19755 has permitted the use of filtration instead of centrifugation as an isolation technique (246).

Quisqualate Receptor Assays

Unequivocal identification of a QUIS-preferring receptor has proven elusive using binding techniques. As discussed in the previous section, QUIS-sensitive [^{3}H] L-GLU binding sites have been identified in PSD. Whether these sites correspond to the physiologic QUIS receptor is not clear; for instance, KA interacts with the NMDA-insensitive binding component with a potency only 25-fold less than QUIS itself. Further, the prototypical QUIS antagonist GDEE does not recognize the QUIS-sensitive [^{3}H] L-GLU binding site in vitro.

Electrophysiologic investigations suggest that L-GLU is a partial agonist at multiple EAA receptor subtypes (1-6), an observation corresponding with the binding of [^{3}H] L-GLU to multiple receptors (81-86,98,99,102-106). Thus, there has been widespread interest to identify compounds selective for QUIS-preferring receptors that would be useful as radioligands in receptor binding assays. One such compound of interest is AMPA (Figure 2).

Initially synthesized as an analog of IBO, AMPA appears to be a potent and selective QUIS-receptor agonist antagonized by GDEE in vivo (125-127). The two populations of binding sites for [^{3}H] AMPA present in brain membranes (K_D = 0.009; 2.4 μM) are both sensitive to QUIS, L-GLU, and KA (Table 2) (87,128). Freezing the tissue prior to assay increases specific binding by 60%; however, this has been reported to eliminate the high-affinity component of the reaction (K_d = 0.4 μM). In the presence of potassium thiocyanate (KSCN), the affinity of [^{3}H] AMPA to bind to previously frozen Triton-X-100-treated brain membranes is enhanced eight fold, and two receptor populations are detected and the pharmacological profile of the ligand interaction is unaltered (88). Under all conditions, IBO and GDEE fail to inhibit binding at concentrations of 100 μM (87,88,128).

In a single report (89), specific binding of [^{3}H] AMPA to brain membranes was described using buffers adjusted to maintain a pH of 5.5, a condition claimed to optimize binding. At the reduced pH, a single population of receptors was detected, and AMPA, L-GLU, and QUIS inhibited binding with affinities of 12-55 μM. Notably, the affinity of KA to displace [^{3}H] AMPA was >5000 μM. It remains to be determined whether these conditions reveal a more "QUIS-preferring" receptor population labeled by [^{3}H] AMPA or are an artifact of the unusual assay conditions.

Sites labeled in PSD by [^{3}H] L-GLU or by [^{3}H] AMPA in crude particulate preparations of brain exhibit characteristics of QUIS-type receptors. However, the recognition sites interact with KA, albeit with moderate affinity, as well as with QUIS, and have no affinity for the prototypical QUIS antagonist GDEE. It would therefore be tenuous to classify either ligand or method as a pure QUIS-preferring receptor binding assay. Notably, however, QUIS is a potent inhibitor of specific [^{3}H]KA binding in vitro (see following section and Table 2), and in vivo, many antagonists of QUIS-mediated responses antagonize responses to KA to a similar extent. Thus, cross-reactivity of KA and QUIS in binding assays is not totally unexpected. As a potential route to unequivocally identify QUIS receptors in vitro, it would be of interest to examine the binding of [^{3}H]-AMPA to PSD since this preparation contains only a small population of receptors for KA (4,104).

Kainic Acid Receptor Assays

The most widely used ligand employed to label KA receptors is [^{3}H]KA, and this provides one of the most pharmacologically selective EAA binding assays (75,76,130). Unfortunately, the physiologic importance of KA receptors is unclear since an endogenous transmitter or modulator substance has not been identified that interacts with KA receptors in vivo. This notwithstanding, interest in KA remains substantial, and in animals, KA-induced lesions and epileptiform activity provide useful models for human neurological disorders (131-138). Alterations in [^{3}H]KA receptors have been reported to accompany human diseases (76,77,139).

Tritium-labeled KA binds with nanomolar affinities to two populations of receptors in mammalian forebrain, although only a single receptor is detected in cerebellum (76,77,139). Both high-affinity (K_D = 2-5 nm) and low-affinity (K_D = 20-40 nM) binding sites are pharmacologically identical, with the order of potency of inhibition being KA $>$ QUIS $\geqslant$ L-GLU $>>$ IBO (Table 2). Domoic acid (Figure 2), a naturally occurring structural analog of KA is the only compound identified as more potent than KA to inhibit binding, and the differences in potency are not great (74).

Structure-activity profiles of [^{3}H]KA binding are consistent with the identification of the receptor mediating the neuroexcitotoxic actions of KA in vivo (74,75,141). Modification of the isopropylene side chain (α-keto-KA) results in a 10- to 20-fold reduction of potency to inhibit specific binding and correlates well with the relative neurotoxicity of α-keto-KA in vivo. Reduction of the methylene side chain (dihydro-KA), esterification of the carboxyl moieties (KA-diethyl ester), or formation of α-allo-KA results in substantial reductions in neurotoxic activity in vivo and of receptor recognition in vitro. Excitatory amino acid analogs thought to interact with other receptor subtypes, including GDEE, NMDA, ADA, and the phosphono amino acids, are devoid of activity (74). In contrast to an initial report (142), it now appears unlikely that folic acid or folate derivatives interact with KA receptors in vivo or in vitro (143-146).

Potent and selective KA receptor antagonists have not been identified, and antagonists that do exist antagonize QUIS-mediated responses in brain. However, the amino terminal glutamate containing dipeptides GAMS and glutamyltaurine (GLU TAU; Figure 4) and two novel piperazine derivatives [1-(parachlorobenzyl)- and 1-(parabromobenzyl)piperazine-2,3-dicarboxylic acids] (Figure 4) have been reported to show some selectivity for KA-mediated responses (147-153). β-KA, in which the carboxyl group is trans to its normal position, has also been reported to be a KA antagonist (120,154). These latter compounds display anticonvulsant activity following intracerebroventricular (ICV) or intraperitoneal (IP) administration, suggesting a pure KA receptor antagonist may have therapeutic utility in epilepsy. Reports on the potency of novel KA antagonists to interact with [^{3}H] KA binding sites are lacking.

Ca^{2+}/Cl^--Dependent Receptor Assays

Using freshly prepared tissues, [^{3}H] L-GLU binds to a distinctive population of recognition sites on brain membranes (4,98,99,103). Specific binding is complex, and in the presence of Ca^{2+} and Cl^- the apparent receptor number is increased two- to threefold relative to assays lacking these ions (98). Whereas Ca^{2+} appears to play a permissive role, Cl^- ions are required to observe the increase in receptor number. Binding isotherms for [^{3}H] L-GLU are monophasic, suggesting a single receptor population; however, various compounds discriminate multiple receptor subtypes (103). The pharmacological profile for binding is not consistent with the identification of a QUIS-, KA-, or NMDA-preferring receptor, and a number of EAA agonists and antagonists, including QUIS, IBO, CSA, L-ASP, AP4, AP5, AP7, and ADA interact with the site at low to moderate micromolar concentrations (Table 1). Inactive compounds include NMDA, KA, and GDEE. Freezing the tissue prior to assay eliminates the influence(s) of ions and alters the pharmacological profile of the ligand-receptor interaction (Table 1) (99,102,103,106).

Use of [^{3}H] DL-(±)-AP4 provides a more direct method to examine the Ca^{2+}/Cl^--dependent EAA binding sites in brain since the interaction occurs at a single recognition site. As described for [^{3}H] L-GLU, binding of [^{3}H] DL-(±)-AP4 is optimal using freshly prepared tissue and is absent if Ca^{2+}/Cl^- is omitted from the assay (93,100). The pharmacological profile of [^{3}H] DL-(±)-AP4 binding is similar to Ca^{2+}/Cl^--dependent [^{3}H] L-GLU binding (Table 1). Of interest, Butcher et al. (94) have recently demonstrated that purine nucleotides decrease the affinity of [^{3}H] DL-(±)-AP4 for its recognition site, suggesting EAA receptors may be subject to regulation by nucleotides in vivo.

The correlation of Ca^{2+}/Cl^--dependent EAA binding sites to a physiologically important receptor is a matter of current dispute. Electrophysiologically,

L-(+)-AP4 is a potent antagonist of excitatory activity at synapses of the lateral olfactory tract (26), spinal cord (112,155), and the lateral perforant-dentate gyrus pathway and mossy fiber CA_3 synapses of the hippocampus (156-158). However, the identity of the transmitter antagonized by L-(+)-AP4 is unknown, although it is unlikely to be L-GLU (15,58-60,159,160). Further, it is not clear that the antagonist activity of L-(+)-AP4 is mediated by pre- or postsynaptic receptors, and it has been suggested that presynaptic inhibition of transmitter release may be involved (159). Ca^{2+}/Cl^--dependent [^{3}H] L-GLU receptors are present in purified synaptic plasma membranes, but not in PSD, suggesting little association with postsynaptic elements (104). In mice lesioned at birth with methylazoxymethanol acetate to reduce the number of cerebellar granule cells, specific DL-(±)-AP4 binding is reduced by 20-30% in the adult animal, indicating some receptors are located presynaptically on excitatory neurons in cerebellum (Ferkany, unpublished). The pharmacological correlation between [^{3}H] AP4 binding and inhibition of hippocampal synaptic activity has been argued to be good (161,162) or poor (163).

Pin et al. (164) have suggested that Ca^{2+}/Cl^--dependent [^{3}H] L-GLU binding sites are an artifact of tissue preparation representing the interaction of L-GLU with intracellular chloride-stimulated transport sites exposed when vesicular elements are formed in membrane homogenates. The hypothesis that [^{3}H]-DL-(±)-AP4 binding occurs to nonsynaptic elements (indeed, structures previously having an intracellular location) follows closely and is not unreasonable since EAA are found within the cell and intracellular transport mechanisms are expected. However, Cl^--stimulated [^{3}H] L-GLU binding to intact cells (PC-12) has been demonstrated in the absence of Na^+, a condition that precludes access of the ligand to intracellular sites. By default, the recognition sites are suggested to be located on the extracellular surface (165).

To summarize, in the presence of Ca^{2+}/Cl^- both [^{3}H] L-GLU and [^{3}H]-DL-(±)-AP4 identify a unique receptor population in vitro. Whether these sites correspond to the receptor mediating the actions of L-(+)-AP4 in vivo remains to be determined.

N-Acetyl-L-Aspartyl-L-Glutamate

N-acetyl-L-aspartyl-L-glutamate is an endogenous dipeptide isolated from brain that elicits DL-(±)-AP4-sensitive excitatory responses in terminal synapses of the lateral olfactory tract (LOT) (24-31). Direct measurement of NAAG or visualization of the peptide with specific antibodies indicates a marked regional distribution in brain and spinal cord (18,22,27-29,166), with exceptional levels of the peptide occurring in the mitral cells of the LOT (22). Notably, the actions of the LOT transmitter are antagonized by DL-(±)-AP4, making interesting the suggestion that NAAG may be the endogenous excitatory substance. Changes in the brain content of NAAG demonstrate a distinct ontogenetic profile (27) and a complex, membrane-bound enzymatic mechanism that cleaves the peptide and results in the sequestration of L-GLU has been described, providing an hypothesis to describe the inactivation of endogenous NAAG analogous to the mechanism that exists for acetylcholine (20,21,23,168). The enzyme responsible for metabolizing NAAG may represent a unique peptidase. Of great interest, the most potent inhibitor of enzyme activity is QUIS, leading to speculation that some actions of the latter compound may be indirect and due to the inhibition of enzyme activity in vivo. When injected into striatum, NAAG induces cortical seizures in rats; however, initiation of seizure activity is not coincident with the injection and may be secondary to metabolism of the parent compound (Blakley, personal communication).

Unlabeled NAAG is equipotent to QUIS in inhibiting Cl^--dependent specific binding of [^{3}H] L-GLU to crude homogenates of brain (Table 1) (24,31). Like other EAA analogs, NAAG displaces only a portion of the total sites labeled by [^{3}H] L-GLU, suggesting the peptide interacts with a subpopulation of L-GLU receptors (24,31). The possibility that the site involved represents a Cl^--dependent L-GLU transport site (see previous section) cannot be excluded based on current evidence.

Direct labeling of [^{3}H] NAAG recognition sites in brain has been reported to have a regional distribution and ontogeny distinct from sites labeled by [^{3}H] L-GLU (24,167). Specific binding of [^{3}H] NAAG to crude microsomal preparations of rat forebrain is inhibited most potently by L-cysteine, L-GLU, SOS, QUIS, and IBO (K_i = 0.12-0.5 μM). Among the phosphono amino acids, DL-(±)-AP4 is most potent to inhibit peptide binding (K_i = 3.9 μM), although DL-(±)-AP5 and DL-(±)-AP7 display substantial affinity for the receptor (Table 1).

Caution is warranted in interpreting the results obtained from binding studies employing [^{3}H] NAAG in the absence of peptidase inhibitors. Thus, recent work using tritiated NAAG has identified the metabolism of the compound under routine assay conditions (Ferkany, unpublished; Abreu and Blakely, personal communication). Since commercially available [^{3}H] NAAG is labeled on the glutamate residue, spurious [^{3}H] L-GLU binding was detected.

Sufficient evidence is now available to suggest that NAAG may function as a neurotransmitter or neuromodulator, possibly as the endogenous transmitter for the L-(+)-AP4-sensitive receptor (26). The identification of NAAG and other peptides that interact as modulators or antagonists at EAA receptors make it interesting to speculate that compounds other than amino acids may be the active mediators of excitatory synaptic transmission (97,117,147-150,168,169). Caution should be exercised, however, in utilizing [^{3}H] NAAG as a radiolabel since hydrolysis of the compound may occur rapidly in vivo or in vitro. Identification of a specific enzyme(s) that metabolizes NAAG, and specific inhibitors of the enzyme(s) should allow the development of a binding assay using this interesting compound.

L-Aspartate Receptor Assays

L-Aspartate interacts with a range of potencies in all established EAA binding assays except those using [^{3}H] KA. Although investigations using [^{3}H] L-ASP are less extensive than for other EAA, binding sites for [^{3}H] L-ASP have been detected in the CNS (90,91,101,102).

Specific binding of [^{3}H] L-ASP to brain (90,91) and spinal cord (101) is stereoselective, and D-ASP is 90-fold less potent to inhibit binding than the L isomer. Notably, L-GLU is a potent displacer, whereas QUIS, KA, GDEE, DL-(±)-AP4, and ADA are without appreciable activity (Table 1). Consistent with the hypothesis that receptors for [^{3}H] L-ASP are involved in neurotransmission, binding of the ligand is enriched ninefold in synaptic junctional membranes (102). Although L-GLU displaces L-ASP binding (K_i = 0.2-0.8 μM) when measured in the same tissue, the number of receptors for [^{3}H] L-GLU is two- to threefold greater than for L-ASP. Summarizing available pharmacological, biochemical, and kinetic data, Foster and Fagg (4) have suggested that [^{3}H] L-ASP and Ca^{2+}/Cl^--dependent [^{3}H] L-GLU binding occur at a similar or identical receptor. Whether this receptor corresponds to the L-GLU bindsite on PSD (104,105) is unclear but would seem unlikely as NMDA fails to displace specifically bound [^{3}H] L-ASP from synaptic junctional membranes.

Autoradiographic Analysis

The distribution of [^{3}H] L-GLU (110,170-172), [^{3}H] KA (173-177), [^{3}H]-AMPA (116,178), and [^{3}H] NAAG (167) receptors in brain has been investigated extensively using autoradiographic methods. Caveats stated previously for individual assays apply equally to autoradiographic analysis.

Limitations of space do not permit a detailed discussion of findings, but in general, the results support those gathered using other methods of analysis. For example, electrophysiologic investigations have revealed the stratum lucidum of the hippocampus to be exquisitely sensitive to KA, and this corresponds well with the high density of [^{3}H] KA binding sites localized autoradiographically to this region of brain (173,177). In contrast, areas of brain showing little sensitivity to KA have a low receptor density in autoradiographs. Quantitative analysis of [^{3}H] NAAG binding indicates a receptor density substantially less than for [^{3}H] L-GLU throughout brain (25,167).

Descriptions of the pharmacological profile of EAA receptors using autoradiographic approaches are less extensive than for standard binding assays. Most exciting, however, are methods describing the localization of QUIS-, KA-, and NMDA-preferring receptors (110,171,172). Under appropriate conditions,

[^{3}H] L-GLU binds to an apparently homogeneous population of receptors that can be discriminated into multiple binding sites by QUIS and NMDA. Receptors with a high affinity for QUIS can be further subdivided into QUIS- and KA-sensitive sites, whereas receptors with a low affinity for QUIS display a markedly higher affinity for NMDA. These results are similar to those reported by Fagg and coworkers (104,105) and permit the direct visualization of three (QUIS, KA, and NMDA preferring) of the four major subtypes of EAA receptors.

FUNCTIONAL ASSAYS

Radioligand binding assays are useful to examine the ontogeny, distribution, structure-activity profile, and biochemical characteristics of neurotransmitter receptors in normal and pathological tissues. However, discrimination of agonists from antagonists is not generally possible and the assays do not describe the functional status of the receptor. To explore the latter, methods have been developed to examine events occurring consequent to the drug-receptor interaction. Many of these "functional assays" involving second messenger systems are described in detail in Chapter 4. For this reason, the following discussion is abbreviated, representing an overview of procedures useful to examine EAA receptor pharmacology.

Cyclic GMP Formation

The action of L-GLU in stimulating the formation of cGMP in slices of rat cerebellum was described by Ferrendelli et al. in 1973 (179). Subsequent work confirmed this observation, which was extended to include KA and NMDA receptor agonists (166,180-186). General agreement exists that the order of potency of EAA to stimulate cGMP formation in slices of adult rat cerebellum is KA > NMDA >> L-GLU = L-ASP (180,181,183). The efficacy of KA in stimulating nucleotide production is greater than for other EAA agonists. GDEE is an effective antagonist against L-GLU, whereas ADA and Mg^{2+} antagonize responses to NMDA (192,183). The action of EAA to elicit an increase in cGMP in regions of brain other than cerebellum is less robust or absent, suggesting the possibility that EAA receptors in cerebellum are functionally distinct from those in other regions of brain (180). It should be noted, however, that early studies were performed in the absence of phosphodiesterase inhibitors, and the failure to detect EAA-mediated stimulation of cGMP formation may be due to differential rates of metabolism of the nucleotide among regions of brain.

When slices of adult rat cerebellum are examined, the cGMP response to EAA agonist yields two- to threefold increases. In contrast, when slices are prepared from immature rats (8-12 days postgestation), as much as a 200-fold increase in cGMP production is obtained (184). Although the large response is clearly more desirable from a practical perspective, the pharmacological profile of EAA-mediated cGMP formation is distinct from that observed in the adult animal. For instance, the order of potency of EAA to elicit the response becomes NMDA > L-ASP >≫ L-GLU > KA and KA becomes less efficacious than NMDA or L-GLU (182,183,185). In the immature cerebellum both KA- and NMDA-mediated responses are variably antagonized by ADA (182), whereas GDEE antagonizes the actions of L-GLU (182,186) or L-GLU and NMDA (185). DL-(±)-AP5 has been reported to be without effect on the action of any agonist (185), to antagonize the actions of NMDA (186), or to antagonize NMDA, KA, and IBO (187). The nonspecific EAA antagonist *cis*-2,3-piperidinedicarboxylic acid enhances cGMP formation apparently through an NMDA-type receptor in immature rat cerebellar slices, an effect that is antagonized by DL-(±)-AP7 (166).

Garthwaite has recently provided evidence that the low potency of L-GLU and L-ASP to stimulate cGMP formation may be secondary to the transport of these amino acids to intracellular spaces (186). Using dispersed suspensions of cerebellum (immature), the potency of L-GLU and L-ASP to stimulate cGMP formation increased 80- and 15-fold, respectively, compared with intact slices. Of interest, QUIS appears to be devoid of action to stimulate cGMP formation in any system.

Agonist-evoked stimulation of cGMP formation remains a useful functional assay to explore EAA receptor pharmacology. Notably, the assay appears most useful to detect compounds with agonist activities. Further work is required to document the utility of the system to discriminate specific EAA receptor antagonists, particularly when neonatal tissues are used. It would be of interest to extend available results to cell culture if lines could be identified that responded to EAA with an increase in nucleotide formation.

NMDA-Evoked Acetylcholine Release

NMDA evokes a dose-dependent release (EC_{50} = 50 μM) of radiolabeled acetylcholine (ACh) from striatal slices preloaded with either [^{14}C]- or [^{3}H]choline (188,189). The action is competitively antagonized by AP5 and AP7, and the D-(−)isomers are the more potent inhibitors. Compounds thought to interact with NMDA receptors in vivo, including IBO and QUIN, mimick the effects of NMDA (111,188-190). Likewise, L-GLU induces a release of ACh, an action sensitive to inhibition by Mg^{2+}, indicating the involvement of NMDA receptors (189). Both KA and QUIS are essentially devoid of activity in stimulating transmitter release.

NMDA-evoked ACh release provides a simple and reproducible functional assay to examine NMDA-preferring receptors in vitro and has been used with success to identify novel NMDA receptor antagonists. Intriguingly, the effects of NMDA are not limited to ACh and the agonist evokes an overflow of dopamine in dual-label experiments (Ferkany, unpublished; K. Johnson, personal communication).

KA-Evoked EAA Release

Superfusion of striatal, hippocampal, or cerebellar slices or hippocampal synaptosomes with KA evokes a dose-dependent release of endogenous L-GLU and L-ASP as determined by high-performance liquid chromatography (HPLC) or gas chromatographic-mass spectrometric (GC-MS) analysis (41,191,192). The action is not mimicked by NMDA or IBO or antagonized by DL-(±)-AP4 or DL-(±)-AP5 (41). KA does not evoke an increase over basal transmitter release when slices are preloaded with [^{3}H] L-GLU or [^{3}H] D-ASP (40). However, in the presence of electrical stimulation, KA evokes an additional overflow of [^{3}H] D-ASP from cortical slices (193), whereas L-GLU has been reported to decrease K^+-evoked release of [^{3}H] D-ASP from hippocampal slices in a GDEE-sensitive manner (194).

KA-evoked release of endogenous L-GLU and L-ASP provides a functional assay to selectively examine KA-preferring receptors in vitro. However, the method is tedious since it requires the analysis of individual superfusion samples using HPLC or GC-MS technologies.

Sodium Flux Assays

The properties of EAA receptors have been analyzed in brain slices using agonist-induced stimulation of [^{22}Na] efflux (195). Several amino acids, including L- and D-GLU and NMDA but not DL-HCA, produce progressively smaller responses upon successive applications.

D-(−)-ADA discriminates among EAA agonists, completely blocking responses to NMDA, D-GLU, and DL-HCA while partially antagonizing QUIS and KA and being without effect on L-GLU. The order of potency of several EAA to stimulate [$^{22}Na^+$] efflux correlates with their relative effects measured using electrophysiological techniques but does not correlate with CA^{2+}/Cl^--dependent [^{3}H] L-GLU binding. The similarities in the properties of EAA receptors evidenced with the [$^{22}Na^+$] efflux assay and with the electrophysiological approach in the in vitro hippocampal slice preparation indicates that the same receptors are likely to be sampled using the two techniques.

Phosphatidylinositol Turnover

EAA agonists (L-GLU, L-ASP, IBO, and QUIS) stimulate phosphatidylinositol (PI) turnover in slices of immature (6-8 day) rat hippocampus and cerebellum

(196-199) and in primary cultures of cerebellar granule cells (200) with EC_{50} in the low micromolar range. Maximal stimulation of PI turnover can exceed 14-fold that of baseline, providing an excellent signal-to-noise ratio. However, NMDA, KA, and QUIS have no effect, and with the exception of IBO, EAA agonists do not stimulate PI turnover in slices prepared from adult animals (197). In contrast, EAA agonists inhibit norepinephrine (NE)-stimulated PI turnover in adult animals. The observation that IBO but not NMDA or QUIS is an effective agonist in adult tissues lends support to the growing body of evidence that the actions of IBO are not mediated through an NMDA receptor mechanism.

The utility of EAA-mediated stimulation of PI turnover as a routine assay to examine the actions of agonists in vitro remains to be determined. Caveats mentioned earlier (see Cyclic GMP Formation) concerning the use of immature tissues apply to this assay as well. For example, the actions of L-GLU to stimulate PI turnover are antagonized by D-(−)-AP5 and DL-(±)-AP7, an observation inconsistent with the lack of efficacy for NMDA (197,199). Furthermore, the EAA antagonists DL-(±)-AP4 and phospho-*O*-serine mimic the action of agonists in adult tissues to antagonize NE-induced increases in PI turnover (197). The QUIS and KA antagonists GAMS and GLU TAU have no effect on QUIS or enhance agonist actions (199). In primary cultures of cerebellar granule cells, all EAA agonists (L-GLU, L-ASP, QUIS, KA, IBO, QUIN, and NMDA), activate PI turnover, limiting the usefulness of the assay to identify novel receptor subtype-specific agonists or antagonists (200).

REGULATORY INFLUENCES

As described in a previous section (see Quisqualate Receptor Assays), Ca^{2+} can have substantial effects on the characteristics of [^{3}H] L-GLU binding to brain. The divalent has been reported to increase [^{3}H] L-GLU receptor number through activation of a protease-mediated event that reveals normally occluded receptors (84). The effect of Ca^{2+} is temperature-dependent (suggesting active processes are involved), shows a marked ontogenetic profile, and is irreversible as extensive washing of membranes following exposure to Ca^{2+} does not result in a decline in receptor density (81,82). The effect of Ca^{2+} is reduced or absent in some regions of brain (82), and EAA receptors may be regulated at the molecular level by different processes throughout the CNS. Notably, however, it is by no means established that the actions of Ca^{2+} observed in vitro are of functional consequence in vivo since physiologic concentrations of Ca^{2+} elicit the response. These ion concentrations would normally be available in the extracellular fluids and presumably induce a maximum response at all times.

The effects of Cl^- to reveal [^{3}H] DL-(±)-AP4 binding sites on brain membranes is not limited to Cl^- and can be mimicked by NO^{3-} and Br^- in high concentrations (98,99). As with Ca^{2+}, it is not clear that the action of Cl^- to reveal EAA-like binding sites in vitro is of functional significance in the CNS. Mg^{2+} antagonizes the actions of NMDA, most likely by blockade of NMDA-activated ion channels (183,187,188,196,201-205). Since physiologic concentrations of Mg^{2+} are required to elicit receptor blockade, it has been suggested that NMDA receptors do not play a role in normal transsynaptic neurotransmission or are active only in the presence of a prior depolarization of the nerve terminal (206). Mg^{2+} has no effect on [^{3}H] L-GLU binding to NMDA receptors in vitro (104, 105,108). Other compounds that have been reported to influence the affinity or number of EAA receptors include KA (201), guanine and purine nucleotides (94,207,208), and a novel series of dipeptides represented by phenylalanyl-L-glutamate (97,117,169).

At the molecular level it has been widely assumed that the effects of EAA receptor activation are mediated by transmembrane ion fluxes. Direct linkage to second messenger systems (e.g., adenylate cyclase) has been difficult to demonstrate. For example, it is by no means clear that EAA agonist-induced increases in cGMP occur in regions of brain outside the cerebellum, making it difficult to suggest cGMP represents a common transducer. As discussed previously, it remains to be determined if EAA-induced increases in PI turnover are of physiologic significance.

Most exciting, however, has been the recent demonstration that NMDA receptors may represent a supramolecular complex analogous to the GABA-benzodiazepine-barbiturate effector proteins (66,67). Thus, the dissociative anesthetics ketamine and phencyclidine (PCP), some benz(f)isoquinolines, and benzomorphan opiates have been reported to interact with NMDA receptors in vitro and in vivo (63-65,68-70,209). The action of the latter compounds may represent an "inverse agonist" effect or an allosterically linked antagonist mechanism presumably associated with the NMDA-activated, Mg^{2+}-sensitive ion channel. A close association between NMDA and PCP receptors has been demonstrated (71), and in vitro, the binding of ligands specific for PCP receptors is markedly enhanced in the presence of L-GLU and is antagonized by Mg^{2+} (67; see also 247).

The linkage between NMDA and PCP receptors suggests that EAA antagonists may elicit PCP-like behavioral effects. In rodents, NMDA antagonists produce changes in behavior similar to those seen with PCP (70,210). However, it appears that the novel anticonvulsant MK801 interacts with the site linked to NMDA receptors (71-73), suggesting that therapeutically useful agents targeting the regulatory site can be developed. Furthermore, the clinically useful anethestic ketamine may act indirectly via NMDA receptors, making it interesting to speculate that direct-acting NMDA antagonists would be useful analgesic agents. In the latter case, the condition of surgical anesthesia would limit side-effect liabilities due to unwanted changes in the psychological state.

Whether receptors for KA and QUIS are similarly linked to neuroregulatory complexes remains to be determined.

STRUCTURE-ACTIVITY RELATIONSHIPS

As might be predicted from the simplicity of EAA, the structural requirements for compounds to have agonist or antagonist activity is not straightforward (1-6, 74,75,141,211). At the simple level, L-GLU and L-ASP differ by a single $-CH_2-$ grouping in chain length, although the compounds most likely interact with distinct receptors. In contrast, the higher homolog of L-GLU (ADA) is the prototypical NMDA antagonist. Substitution of phosphono moieties for Ω-terminal carboxyl groups of amino acids has been encouraging, leading to the development of several EAA antagonists. The most specific of these (AP5 and AP7) are most potent and selective to antagonize NMDA receptors. The closest analog of L-GLU, L-(+)-AP4, fails to antagonize L-GLU, whereas D-(−)-AP4 is a nonselective EAA receptor antagonist (155-158). DL-(±)-2-amino-6-phosphonohexanoic acid, the corresponding phosphono analog of ADA, is at best a weak antagonist of EAA (74,187), as is the case for the L-ASP analog, DL-(±)-2-amino-3-phosphonoproprionic acid (212,213). Modifications of the basic aminophosphonocarboxylic acid structure are known to result in the loss of antagonist activity (157-159). Substitution of sulfonyl grouping (serine-*O*-sulfate) for Ω-carboxyl moieties confers a QUIS and/or a NMDA type of agonist activity, whereas the corresponding phosphono-containing compound (serine-*O*-phosphate) is an NMDA antagonist (157,158,172,198).

Until recently, cyclic analogs of EAA have frequently proven to be agonists (147,214,215). Structure-activity requirements are complex, and although it was initially believed that KA interacted with L-GLU receptors, this now appears unlikely although L-GLU can be superimposed on the pyrrolidine. The difficulties in the rational design and synthesis of EAA agonists or antagonists is exemplified by AMPA, which is closely related to IBO but recognizes QUIS receptors in brain. QUIN, the unsaturated analog of *cis*-(±)-2,3-piperidine dicarboxylic acid, is a cyclic analog of NMDA with NMDA agonist properties (13-16,111,190). Of interest, the aromatic analog of QUIN (pthalic acid) exhibits NMDA-type agonist activity, suggesting that nitrogen moieties are not an absolute requirement for activity at EAA receptors (190). Similarly, the structure of the prototypical QUIS-preferring receptor antagonist (GDEE) indicates that free carboxylic acids are not necessary for receptor recognition. As mentioned previously, *cis*-(±)-2,3-piperidine dicarboxylic acid, the saturated analog of QUIN, is either a partial agonist (166) or a nonspecific EAA antagonist in a number of systems (147,214,215).

Recent success has been achieved in developing potent and selective cyclic EAA antagonists, and CPP has been introduced as a compound interacting with

NMDA receptors (107,108,119-129,153,210,216). Preliminary reports suggest that CPP may elicit CNS-mediated effects following systemic administration, making it useful in studying the function of NMDA receptors in the intact animal.

The relative failure of simple structural modifications of EAA to produce potent, selective, and specific EAA antagonists (particularly for QUIS and KA receptors) has led to the development of more novel chemical approaches. For example, Jones et al. (147) have synthesized and evaluated for antagonist activity a number of amino terminal aspartyl or glutamyl dipeptides. The most potent of these (β-D-aspartylaminomethylphosphonic acid, γ-D-glutamylamino-methylphosphonic acid, and β-D-aspartylaminoethylphosphonic acid; Figure 4) are as selective and potent as AP7 as antagonists for NMDA-mediated responses in brain. Substitution of sulfonyl for phosphono moieties on these same dipeptides (GAMS and GLU TAU) produced more potent inhibitors of QUIS- and KA-mediated neuroexcitation. Not only do the latter compounds provide lead structures to develop QUIS and/or KA antagonists but they demonstrate that terminal sulfonic acid moieties do not necessarily confer agonist properties to EAA analogs. Dipeptide analogs of KA have been synthesized, and one of these (γ-kainyl-L-glutamic acid; Figure 4) is anticonvulsant in mice (217). Interestingly, the anticonvulsant activity has been attributed to antagonism of NMDA-preferring receptors, demonstrating that simple modifications may result in EAA antagonists with spectra of receptor specificity distinct from that of the parent compound. Other modifications of KA, including β-KA, α-kainyl-glycine, and β-kainyl-glycine, have also been reported to be anticonvulsant in animal models, whereas kainylamino-methylphosphonic acid is devoid of biologic activity (154,217).

EAA receptor pharmacology is currently in an infant status, and most antagonists are close structural analogs of receptor agonists. Thus, the evaluation of compounds structurally unrelated to EAA for agonist or antagonist activity is not unwarranted. In this regard, several novel piperazine derivatives and a hydroxyquinoxaline carboxylic acid (Figure 4) have been reported to have EAA receptor-mediated anticonvulsant activities or to antagonize EAA-mediated events in vivo and in vitro (151-153,214,218). These initial results are encouraging to the search for novel EAA antagonists.

THERAPEUTIC APPLICATIONS

Substantial evidence suggests that EAA are involved in the etiology of various pathophysiological conditions affecting the CNS (7). Thus, QUIS (219), KA (129,131-133), IBO (220-222), and QUIN (13,14,17,223) have been used to produce animal models of human epilepsy. The anatomic and neurochemical lesions and deficiencies produced by EAA in animals have characteristics similar to those detected in the brains of patients dying with Huntington's disease (129, 133,135) and epilepsy (131,132,134,136–138). Administration of KA can produce a limbic structure lesion that mimics Ammon's horn sclerosis, an abnormality frequently found in temporal lobe epilepsy. Research on this particular disorder, which is resistant to currently available antiepileptics, has suggested that EAA are involved (224-226). In addition to Huntington's disease and epilepsy, EAA may contribute to Alzheimer's disease (14,139,227,228), the neuronal death following hypoglycemia (229), stroke, and other forms of cerebral ischemia (124,230), and to hereditary olivopontocerebellar atrophy (231, 232).

The conceptual link between EAA and human disease has produced a growing interest in identifying antagonists for clinical application. Although available compounds are limited in number, receptor specificity, and activity following systemic administration, preliminary results have been encouraging. Besides interfering with the neurotoxic and convulsant actions of NMDA, QUIN, and IBO, NMDA antagonists protect against audiogenically- (120,233), chemically- and electroshock-induced seizures in rodents (120,122,123,234,235) and prevent myoclonus in the photosensitive baboon (236). AP7, CPP, and CGS 19755 have been reported to markedly reduce ischemic brain damage in the carotid artery occlusion model of stroke (124,230,237,248), to attenuate insulin-induced hypoglycemic brain damage (229), and to participate in an anticonflict response in rodents (238,239). The less potent EAA antagonist, γ-D-glutamylglycine (Figure 3), protects cultured rat hippocampal neurons from degeneration under conditions of oxygen depletion and blocks the toxicity of exogenously applied L-GLU and L-ASP (240). The prototypical QUIS antagonist GDEE blocks homocysteine- and alcohol withdrawal-induced seizures in rodents (241,242). As mentioned earlier, other EAA antagonists, including GAMS, GLU TAU, β-KA, and kainyl-L-glutamate, have been reported to have anticonvulsant activity. Although none of these compounds penetrate the blood-brain barrier in sufficient quantities following systemic administration to be therapeutically useful, it seems logical given these encouraging results to continue to explore pharmacological means to antagonize the actions of EAA.

Whether EAA antagonists become clinically useful will ultimately depend on the side-effect liabilities of the compounds. The extent of EAA-mediated neurotransmission and neuromodulation in brain rivals that associated with the inhibitory GABA systems. At the conceptual level, it can certainly be argued that antagonists of EAA would result in massive consequences well beyond the desired therapeutic actions. Recent observations that receptors for PCP and NMDA are closely associated suggest that NMDA antagonists may induce adverse psychological responses (206). Prototypical NMDA antagonists have been reported to interfere with learning in rodents (243), an observation that is consistent with the suggestion that EAA receptors may be involved in the formation of memory substrates (244). The dogmatic view that EAA antagonists would lack receptor-mediated peripheral side effects has been challenged by the recent observation that agonists induce dose-dependent contractile responses in isolated preparations of the rat ileum (245) (also Ferkany, unpublished). Thus, future investigations may reveal a role for EAA neurotransmission in the local control of peripheral organs comparable to that emerging for GABA.

These caveats notwithstanding, the pursuit of EAA antagonists remains warranted if only to further understand the basic mechanisms underlying CNS activities. Furthermore, EAA neurotransmission is not monolithic and, instead, involves multiple receptor classifications, with receptor subtypes possibly existing within each category. The development of compounds interacting with subtypes of receptors may provide therapeutically useful agents with limited toxic potential. Whether these hopes are ultimately realized must await the development of a broad array of systemically active EAA antagonists.

SUMMARY

The pharmacology of EAA receptors is complex, and few available ligand binding or functional assays appear totally selective for a single receptor subtype. In part this may be attributed to the relative lack of potent and selective compounds for EAA receptors. However, recent efforts to develop receptor-specific agonists and antagonists have been encouraging, and it now appears possible to define the NMDA receptor by antagonist profile. The introduction of selective antagonists for KA- and QUIS-preferring receptors remains to be accomplished.

Few if any of the available compounds penetrate to brain in sufficient quantities following systemic administration to represent useful therapeutic agents. Given the conceptual link between EAA and some human neuropathologies, it is important to identify compounds exhibiting receptor selectivity and CNS activity following systemic administration. Compounds with these properties may be useful as anticonvulsants or for the prophylactic, symptomatic, or therapeutic treatment of human neurodegenerative disorders.

NOTE ADDED IN PROOF

During the period between the preparation and publication of this chapter, important advances have taken place in the field of excitatory amino acid receptor pharmacology. Chief among these has been rapidly expanding understanding of the linkage between phencyclidine (PCP) and NMDA receptors (Trends Neurosci. *10*, 7, 1987). Reports have also appeared indicating that glycine can modulate activity at the NMDA/PCP receptor complex (J. W. Johnson and P. Ascher, Nature, *325*, 529, 1987) and it has been suggested that NMDA, KA, and QUIS may activate a single ion channel (C. E. Jahr and C. E. Stevens, Nature, *325*,

522, 1987). Growing evidence suggests that Cl-dependeng L-GLU "binding" is a sequestration phenomena mediated by an exchange system (M. Kessler et al., J. Neurochem. *48*, 1191, 1987). Putative antagonists for QUIS-type receptors have been announced (J. Drejer and T. Honore, Soc. Neurosci. Abstr. *13*, 758, 1987) and CGS 19755 has been identified as being more potent than CPP to antagonize the actions of NMDA.

These advances notwithstanding, the guidelines contained in the current chapter remain practical. Furthermore, it is gratifying that an area of neuroscience that drew limited attention in the past is now growing at such a rapid pace!

REFERENCES

1. J. Davies, R. H. Evans, A. A. Francis, A. J. Jones, and J. C. Watkins, in Neurotransmitters and Their Receptors (U. Littauer, I. Silman, V. Teichberg, and Z. Vogel, eds.), Wiley, London, 1980, p. 333.
2. J. T. Coyle, in Neurotransmitter Receptors (S. J. Enna and H. I. Yamamura, eds.), Chapman and Hall, London, 1980, p. 5.
3. J. C. Watkins and R. H. Evans, Annu. Rev. Pharmacol. Toxicol. *21*, 165 (1981).
4. A. C. Foster and G. E. Fagg, Br. Res. Rev. *7*, 103 (1984).
5. J. W. Ferkany and J. T. Coyle, in Neuromethods: Receptor Binding, Vol. 4 (A. A. Boulton, G. B. Baker, and P. Hrdina, eds.), Human Press, Clifton, New Jersey, 1986, p. 171.
6. J. Davies, R. H. Evans, A. A. Francis, and J. C. Watkins, J. Physiol. (Paris) *75*, 641 (1979).
7. J. T. Coyle, S. J. Bird, R. Gulley, J. V. Nadler, J. Nicklas, and J. W. Olney, Neurosci. Res. Prog. Bull. *19*, (1981).
8. M. Recasens, V. Varga, D. Nanopoulos, G. Saadour, G. Vincendos, and J. Binavides, Brain Res. *239*, 153 (1982).
9. B. Parsons and T. C. Rainbow, Brain Res. *294*, 193 (1984).
10. H. McLennan and D. Lodge, Brain Res. *169*, 83 (1979).
11. L. Fagni, M. Baudry, and G. Lynch, J. Neurosci. *3*, 1538 (1983).
12. M. Baudry, K. Kramer, L. Fagni, M. Recasens, and G. Lynch, Mol. Pharmacol. *24*, 222 (1983).
13. R. Schwarcz, W. O. Whetsell, and R. M. Mangano, Science *219*, 316 (1983).
14. A. C. Foster, J. F. Collins, and R. Schwarcz, Neuropharmacology *22*, 1331 (1983).
15. T. W. Stone and M. N. Perkins, Eur. J. Pharmacol. *72*, 411 (1981).
16. M. N. Perkins and T. W. Stone, J. Pharmacol. Exp. Ther. *226*, 551 (1983).
17. T. W. Stone and J. H. Connick, Neuroscience *15*, 597 (1985).
18. R. D. Blakely, L. Ory-Lavollee, and J. T. Coyle, Soc. Neurosci. Abstr. *12*, 62 (1986).
19. G. L. Westbrook and M. L. Mayer, Soc. Neurosci. Abstr. *12*, 62 (1986).
20. M. B. Robinson, L. Ory-Lavollee, D. C. Thompson, and J. T. Coyle, Soc. Neurosci. Abstr. *12*, 62 (1986).

21. R. D. Blakley, L. Ory-Lavollee, R. C. Thompson, and J. T. Coyle, J. Neurochem. *47*, 1013 (1986).
22. R. D. Blakely, L. Ory-Lavollee, R. Grzanan, K. J. Koller, and J. T. Coyle, Brain Res. *402*, 393 (1987).
23. M. B. Robinson, R. D. Blakley, R. Couto, and J. T. Coyle, J. Biol. Chem. *262*, 14498 (1987).
24. R. Zaczek, K. Koller, R. Cotter, D. Heller, and J. T. Coyle, Proc. Natl. Acad. Sci. USA *80*, 1116 (1983).
25. K. J. Koller and J. T. Coyle, Mol. Pharmacol. (in press).
26. J. M. French-Mullen, K. Koller, R. Zaczek, J. T. Coyle, N. Hori, and D. O. Carpenter, Proc. Natl. Acad. Sci. USA *82*, 3897 (1985).
27. K. Koller and J. T. Coyle, Dev. Brain Res. *317*, 137 (1984).
28. K. J. Koller, R. Zaczek, and J. T. Coyle, J. Neurochem. (in press).
29. C. B. Cangro, D. E. Garrison, P. A. Luongo, M. E. Truckenmiller, M. A. Namboodini, and J. H. Neak, Soc. Neurosci. Abstr. *11*, 108 (1985).
30. R. Blakley, L. Ory-Lavollee, R. Thompson, and J. T. Coyle, Soc. Neurosci. Abstr. *11*, 108 (1985).
31. K. J. Koller and J. T. Coyle, Eur. J. Pharmacol. *98*, 193 (1984).
32. J. R. Stern, L. V. Eggleston, R. Hems, and H. A. Krebs, Biochemistry *44*, 410 (1949).
33. T. Hayashi, Keio J. Med. *3*, 183 (1954).
34. D. R. Curtis, J. W. Phillis, and J. C. Watkins, Br. J. Pharmacol. *11*, 262 (1961).
35. D. R. Curtis and J. C. Watkins, J. Physiol. (Lond.) *166*, 1 (1983).
36. D. R. Curtis, A. W. Duggan, D. Felix, and G. A. R. Jonston, Brain Res. *41*, 283 (1972).
37. J. P. Bennett, W. J. Logan, and S. H. Snyder, Science *178*, 997 (1972).
38. W. J. Logan and S. H. Snyder, Brain Res. *42*, 413 (1972).
39. A. Schousboe, J. Drejer, and L. Hertz, in Glutamine and Glutamate in Mammals (E. Kvamme, ed.), CRC Press, Boca Raton, Florida, (in press).
40. J. W. Ferkany and J. T. Coyle, Brain Res. *278*, 279 (1983).
41. J. W. Ferkany and J. T. Coyle, J. Pharmacol. Exp. Ther. *225*, 399 (1983).
42. A. C. Hamberger, G. H. Chiang, E. S. Nylen, S. W. Scheff, and C. W. Cotman, Brain Res. *168*, 515 (1979).
43. S. J. Potashner, J. Neurochem. *34*, 177 (1978).
44. J. Reubi, Neuroscience *5*, 2145 (1980).
45. P. L. McGeer, E. G. McGeer, U. Scherer, and K. Singh, Brain Res. *128*, 369 (1977).
46. W. F. White, J. V. Nadler, A. Hamberger, C. W. Cotman, and J. Cummins, Nature *270*, 356 (1977).
47. J. Storm-Mathisen and M. W. Upsahl, Neurosci. Lett. *9*, 65 (1978).
48. R. Zaczek, J. C. Hedreen, and J. T. Coyle, Exp. Neurol. *65*, 145 (1979).
49. I. Walaas and F. Fonnum, Neuroscience *5*, 1691 (1980).
50. K. R. Magnuson, A. J. Beitz, A. A. Larson, J. E. Madl, and R. Altschuler, Soc. Neurosci. Abstr. *11*, 579 (1985).
51. A. J. Beitz, A. A. Larson, J. E. Madl, R. A. Altschuler, and M. M. Mullet, Soc. Neurosci. Abstr. *11*, 865 (1985).

52. G. J. Wiet, D. A. Godfrey, C. D. Ross, and J. D. Dunn, Soc. Neurosci. Abstr. *11*, 1048 (1985).
53. C. D. Ross, M. Bowers, and D. A. Godfrey, Soc. Neurosci. Abstr. *11*, 1216 (1985).
54. K. K. Skaggs, R. J. Wenthold, and R.A. Altschuler, Soc. Neurosci. Abstr. *9*, 1090 (1983).
55. T. Takemoto, in Kainic Acid as a Tool in Neurobiology (E. G. McGeer, J. W. Olney and P. L. McGeer, eds.), Raven Press, New York, 1978, p. 1.
56. H. Shinozaki, and S. Konish, Brain Res. *24*, 369 (1970).
57. J. C. Watkins, in Kainic Acid as a Tool in Neurobiology (E. G. McGeer, J. W. Olney, and P. L. McGeer, eds.), Raven Press, New York, 1978, p. 37.
58. T. H. Lanthorn, A. H. Ganong, and C. W. Cotman, Brain Res. *290*, 174 (1984).
59. L. Fugni, M. Baudry, and G. Lynch, Brain Res. *261*, 167 (1983).
60. N. Hori, C. R. Auker, D. J. Braitman, and D. O. Carpenter, Cell. Mol. Neurobiol. *1*, 115 (1981).
61. M. Perkins and T. W. Stone, J. Pharmacol. Exp. Ther. *80*, 263 (1983).
62. M. Perkins and T. W. Stone, Brain Res. *259*, 172 (1983).
63. S. C. Berry and D. Lodge, Biochem. Pharmacol. *33*, 3829 (1984).
64. N. A. Anis, S. C. Berry, N. R. Burton, and D. C. Lodge, Br. J. Pharmacol. *79*, 565 (1983).
65. L. D. Snell and K. M. Johnson, J. Pharmacol. Exp. Ther. *235*, 50 (1985).
66. S. McPherson, P. L. Wood, and J. Lehmann, Soc. Neurosci. Abstr. *11*, 824 (1985).
67. P. Loo, A. Braunwalder, J. Lehmann, and M. Williams, Eur. J. Pharmacol. *123*, 467 (1986).
68. L. M. Aanonsen and G. L. Wilcox, Neurosci. Lett. *67*, 191 (1986).
69. C. R. Honey, Z.Miljkovic, and J. F. MacDonald, Neurosci. Lett. *61*, 135 (1985).
70. L. D. Snell and K. M. Johnson, J. Pharmacol. Exp. Ther. *238*, 938 (1986).
71. W. F. Maragos, D. C. M. Chu, J. T. Greenamyre, J. B. Penney, and A. B. Young, Eur. J. Pharmacol. *123*, 173 (1986).
72. B. V. Clineschmidt, G. E. Martin, and P. R. Bunting, Drug Dev. Res. *2*, 123 (1982).
73. E. H. F. Wong, J. A. Kemp, T. Priestley, A. R. Knight, G. N. Woodruff, and L. L. Iversen, Proc. Natl. Acad. Sci. USA *83*, 7104 (1986).
74. J. T. Slevin, J. F. Collins, and J. T. Coyle, Brain Res. *265*, 169 (1983).
75. E. D. London and J. T. Coyle, Mol. Pharmacol. *15*, 492 (1979).
76. E. D. London, H. I. Yamamura, E. D. Bird, and J. T. Coyle, Biol. Psychiatry *16*, 155 (1981).
77. H. Henke, Neurosci. Lett. *14*, 247 (1979).
78. K. Biziere, H. Thompson, and J. T. Coyle, Brain Res. *183*, 421 (1980).
79. J. Slevin, J. Collins, K. Lindsley, and J. T. Coyle, Brain Res. *249*, 353 (1982).
80. T. Honore, J. Laurdisen, and P. Krogsgaard-Larsen, J. Neurochem. *36*, 1302 (1981).
81. M. Baudry, D. Arst, M. Oliver, and G. Lynch, Dev. Brain Res. *1*, 37 (1981).

82. M. Baudry, E. Smith, and G. Lynch, Mol. Pharmacol. *20*, 280 (1981).
83. M. Baudry and G. Lynch, J. Neurochem. *36*, 811 (1981).
84. M. Baudry and G. Lynch, Proc. Natl. Acad. Sci. USA *77*, 2298 (1980).
85. L. Werling, K. Doman, and J. V. Nadler, J. Neurochem. *41*, 586 (1983).
86. J. V. Nadler, A. Wang, and L. Werling, J. Neurochem. *44*, 1791 (1985).
87. E. Hosli, P. Krogsgaard-Larsen, and L. Hosli, J. Neurochem. *38*, 173 (1982).
88. D. Murphy, E. Snowhill, and M. Williams, Neurochem. Res. *12*, 775 (1987).
89. C. Tsai and J. Lehmann, Soc. Neurosci. Abstr. *11*, 109 (1985).
90. A. DiLauro, J. Meek, and E. Costa, J. Neurochem. *38*, 1261 (1982).
91. N. Sharif and P. J. Roberts, Brain Res. *211*, 293 (1981).
92. H. J. Olverman, A. W. Jones, and J. C. Watkins, Nature *307*, 460 (1984).
93. S. P. Butcher, J. F. Collins, and P. J. Roberts, Br. J. Pharmacol. *80*, 355 (1983).
94. S. P. Butcher, P. J. Roberts, and J. F. Collins, Biochem. Pharmacol. *35*, 991 (1986).
95. D. T. Manallack and P. M. Beart, Neurosci. Lett. *68*, 11 (1986).
96. K. Wu, R. Carlin, and P. Siekevitz, J. Neurochem. *46*, 831 (1986).
97. J. Baud and G. E. Fagg, Neurosci. Lett. *70*, 228 (1986).
98. E. Mena, G. E. Fagg, and C. W. Cotman, Brain Res. *243*, 378 (1982).
99. G. E. Fagg, A. C. Foster, E. E. Mena, and C. W. Cotman, J. Neurosci. *2*, 958 (1982).
100. D. T. Monaghan, M. C. McMills, A. R. Chamberlin, and C. W. Cotman, Brain Res. *278*, 137 (1983).
101. S. P. Butcher and P. J. Roberts, Biochem. Pharmacol. *33*, 2439 (1984).
102. A. C. Foster, E. E. Mena, G. E. Fagg, and C. W. Cotman, J. Neurosci. *1*, 620 (1981).
103. G. E. Fagg, A. C. Foster, E. E. Mena, and C. W. Cotman, Eur. J. Pharmacol. *88*, 105 (1983).
104. G. E. Fagg and A. Matus, Proc. Natl. Acad. Sci. USA *81*, 6876 (1985).
105. G. E. Fagg and J. Baud, Soc. Neurosci. Abstr. *11*, 109 (1985).
106. G. E. Fagg, E. E. Mena, D. T. Monaghan, and C. W. Cotman, Neurosci. Lett. *38*, 157 (1983).
107. D. E. Murphy, J. Schneider, J. Lehmann, and M. Williams, Soc. Neurosci. Abstr. *12*, 59 (1986).
108. D. E. Murphy, J. Schneider, C. Boehm, J. Lehmann, and M. Williams, J. Pharmacol. Exp. Ther. *240*, 778 (1987).
109. J. W. Ferkany and J. T. Coyle, Life Sci. *33*, 1295 (1983).
110. D. T. Monaghan, V. R. Holets, D. W. Toy, and C. W. Cotman, Nature *306*, 176 (1983).
111. J. Lehmann, P. Schaefer, J. Ferkany, and J. T. Coyle, Eur. J. Pharmacol. *96*, 111 (1983).
112. R. H. Evans, A. A. Francis, A. W. Jones, D. A. S. Smith, and J. C. Watkins, Br. J. Pharmacol. *75*, 65 (1982).
113. P. J. Roberts, G. A. Foster, N. A. Sharif, and J. F. Collins, Brain Res. *238*, 475 (1982).
114. D. M. Close, J. Neil, and A. D. Lofwy, Brain Res. *249*, 393 (1982).
115. N. L. Harris and M. A. Simmonds, Br. J. Pharmacol. *84*, 381 (1985).

116. K. Wu, R. Carlin, and P. Siekevitz, Soc. Neurosci. Abstr. *11*, 646 (1985).
117. J. Ferkany, R. Zaczek, A. Markl, and J. T. Coyle, Neurosci. Lett. *44*, 281 (1984).
118. J. B. Monahan, J. Michel, W. F. Hood, L. M. Pullan, and R. P. Compton, Soc. Neurosci. Abstr. *12*, 60 (1986).
119. J. C. Watkins and A. W. Jones, U.K. Patent Application 6B 2 157 685A HMSO London, 1985.
120. J. Davies, R. H. Evans, P. L. Herrling, A. W. Jones, H. J. Olverman, P. Pook, and J. C. Watkins, Brain Res. *382*, 169 (1986).
121. E. W. Harris, A. H. Ganon, D. T. McNaghan, J. C. Watkins, and C. W. Cotman, Brain Res. *382*, 174 (1986).
122. J. Lehmann, J. Schneider, S. McPherson, D. E. Murphy, P. Berhard, C. Tsai, D. A. Bennett, G. Pastor, D. J. Steel, C. Boehm, D. L. Cheney, J. M. Liebman, P. L. Wood, and M. Williams, J. Pharmacol. Exp. Ther. *240*, 737 (1987).
123. A. G. Chapman, B. S. Meldrum, N. Nanji, and J. C. Watkins, Soc. Neurosci. Abstr. *12*, 60 (1986).
124. S. C. Gerhardt, P. Bernard, G. Pastor, and C. A. Boast, Soc. Neurosci. Abstr. *12*, 59 (1986).
125. L. Hosli, E. Hosli, R. Lehmann, and P. Eng, Neurosci. Lett. *36*, 59 (1983).
126. P. Krogsgaard-Larsen, T. Honore, J. J. Hanson, D. R. Curtis, and D. Lodge, Nature *284*, 64 (1980).
127. A. King, A. Nistri, and C. Rovira, Neurosci. Lett. *55*, 77 (1985).
128. E. Hosli, P. Krogsgaard-Larsen, and L. Hosli, Brain Res. *268*, 177 (1983).
129. R. Schwarcz and J. T. Coyle, Brain Res. *127*, 235 (1977).
130. J. R. Simon, J. F. Contrera, and M. J. Kuhar, J. Neurochem. *26*, 141 (1976).
131. R. Schwarcz, R. Zaczek, and J. T. Coyle, Eur. J. Pharmacol. *50*, 209 (1978).
132. J. V. Nadler, Life Sci. *29*, 2031 (1981).
133. J. T. Coyle and R. Schwarcz, Nature *263*, 244 (1976).
134. R. Zaczek, J. Collins, and J. T. Coyle, Neurosci. Lett. *24*, 181 (1981).
135. J. T. Coyle, J. W. Ferkany, and R. Zaczek, Neurobehav. Toxicol. Teratol. *5*, 617 (1983).
136. E. W. Lothman, R. C. Collins, and J. R. Ferrendelli, Neurology *31*, 806 (1981).
137. M. Pisa, P. R. Sanberg, M. E. Corcoran, and H. C. Fibiger, Brain Res. *200*, 481 (1980).
138. E. W. Lothman and R. C. Collins, Brain Res. *218*, 299 (1981).
139. J. W. Geddes, D. T. Monaghan, C. W. Cotman, I. T. Loti, R. C. Kim, and H. C. Chui, Science *230*, 1179 (1985).
140. E. D. London, N. Kelman, and J. T. Coyle, Brain Res. *192*, 463 (1979).
141. J. T. Coyle, R. Zaczek, J. Slevin, and J. C. Collins, in Glutamate as a Neurotransmitter (G. DiChiara and G. L. Gessa, eds.), Raven Press, New York, 1981, p. 337.
142. A. Ruck, S. Kramer, J. Metz, and M. J. Brennan, Nature *287*, 852 (1980).

143. J. Ferkany, J. T. Slevin, R. Zaczek, and J. T. Coyle, Neurobehav. Toxicol. Teratol. *4*, 573 (1982).
144. P. J. Roberts, G. A. Foster, and E. M. Thomas, Nature *243*, 654 (1981).
145. I. G. Morgan and S. El-Lakany, Neurosci. Lett. *34*, 69 (1982).
146. A. J. Kehl, H. McLennan, and G. L. Collingridge, Neuroscience *11*, 111 (1984).
147. A. W. Jones, D. A. Smith, and J. C. Watkins, Neuroscience *13*, 573 (1984).
148. J. Davies and J. C. Watkins, Brain Res. *327*, 173 (1985).
149. M. J. Croucher, B. S. Meldrum, A. W. Jones, and J. C. Watkins, Brain Res. *322*, 111 (1984).
150. A. G. Chapman, G. P. Hart, B. S. Meldrum, L. Tursk, and J. C. Watkins, Neurosci. Lett. *52*, 325 (1985).
151. J. Davies, A. W. Jones, M. J. Sheardown, D. A. S. Smith, and J. C. Watkins, Neurosci. Lett. *52*, 79 (1984).
152. A. G. Chapman, G. P. Hart, B. S. Meldrum, L. Turski, and J. C. Watkins, Neurosci. Lett. *55*, 325 (1985).
153. A. H. Ganong, A. W. Jones, J. C. Watkins, and C. W. Cotman, J. Neurosci. *6*, 9330 (1986).
154. J. F. Collins, A. J. Dixon, G. Badman, G. DeSarro, A. G. Chapman, G. P. Hart, and B. S. Meldrum, Neurosci. Lett. *51*, 455 (1983).
155. J. Davies and J. C. Watkins, Brain Res. *235*, 378 (1982).
156. E. W. Harris and C. W. Cotman, Exp. Brain Res. *52*, 455 (1983).
157. R. K. Freund, S. L. Crooks, J. F. Koerner, and R. L. Johnson, Brain Res. *291*, 150 (1984).
158. J. F. Koerner, R. L. Johnson, R. K. Freund, M. B. Robinson, and S. L. Crooks, Brain Res. *272*, 299 (1983).
159. A. G. Ganong, M. N. Perkins, J. A. Flatman, and C. W. Cotman, Soc. Neurosci. Abstr. *11*, 1003 (1985).
160. M. B. Robinson, E. R. Whittemore, R. L. Marks, and J. F. Koerner, Brain Res. *381*, 187 (1986).
161. S. L. Crooks, M. B. Robinson, R. L. Johnson, and J. F. Koerner, Pharmacologist *27*, 664 (1985).
162. M. B. Robinson, S. L. Crooks, R. K. Freund, J. R. Koerner, and R. L. Johnson, Soc. Neurosci. Abstr. *11*, 104 (1985).
163. G. E. Fagg and T. H. Lanthorn, Br. J. Pharmacol. *86*, 743 (1985).
164. J. P. Pin, J. Bechaert, and M. Recasens, FEBS Lett. *175*, 31 (1984).
165. C. E. Chandler and E. E. Mena, Soc. Neurosci. Abstr. *11*, 824 (1985).
166. M. J. Leach, C. M. Marder, and H. M. Canning, Eur. J. Pharmacol. *121*, 173 (1986).
167. M. E. Abreu, R. D. Blakley, and J. T. Coyle, Soc. Neurosci. Abstr. *11*, 108 (1985).
168. R. D. Blakley, L. Ory-Lavollee, R. C. Thompson, and J. T. Coyle, J. Neurochem. *47*, 1013 (1986).
169. M. Kessler, M. Baudry, J. T. Cummins, S. Way, and G. Lynch, J. Neurosci *6*, 355 (1986).
170. S. Halpain, C. M Wieczorek, and T. C. Rainbow, J. Neurosci. *4*, 2247 (1984).

171. J. T. Greenamyre, A. B. Young, and J. B. Penney, J. Neurosci. *4*, 2133 (1984).
172. J. T. Greenamyre, J. M. Olson, J. B. Penney, and A. B. Young, J. Pharmacol. Exp. Ther. *233*, 254 (1985).
173. D. T. Monaghan and C. W. Cotman, Brain Res. *252*, 91 (1982).
174. H. Henke, A. Beaudet, and M. Cuenod, Brain Res. *219*, 95 (1981).
175. A. C Foster, E. E. Mena, D. T. Monaghan, and C. W. Cotman, Nature *289*, 73 (1981).
176. M. Berger and Y. Ben-Ari, Neurosci. Lett. *39*, 237 (1983).
177. J. R. Unnerstall and J. K. Wamsley, Eur. J. Pharmacol. *86*, 361 (1983).
178. D. T. Monaghan, D. Yao, H. J. Olverman, and C. W. Cotman, Soc. Neurosci. Abstr. *10*, 557 (1984).
179. J. A. Ferrindelli, M. M. Chang, and D. A. Kinscheri, J. Neurochem. *22*, 545 (1974).
180. M. J. Schmidt, J. J. Ryan, and B. B. Molloy, Brain Res. *112*, 113 (1976).
181. J. Ferkany, R. Zaczek, and J. T. Coyle, Nature *298*, 757 (1982).
182. J. Garthwaite, Neuroscience *7*, 2491 (1982).
183. J. Garthwaite and R. Balazs, in Glutamate as a Neurotransmitter (G. Di Chiara and L. Gessa, eds.), Raven Press, New York, 1981, p. 317.
184. J. Garthwaite and R. Balazs, Nature *275*, 328 (1978).
185. G. A. Foster and P. J. Roberts, Life Sci. *27*, 215 (1980).
186. J. Garthwaite, Br. J. Pharmacol. *85*, 297 (1985).
187. P. J. Roberts, G. A. Foster, N. A. Sharif, and J. F. Collins, Brain Res. *238*, 475 (1982).
188. B. Scatton and J. Lehmann, Nature *297*, 422 (1982).
189. J. Lehmann and B. Scatton, Brain Res. *252*, 77 (1982).
190. J. Lehmann, J. W. Ferkany, P. Schaefer, and J. T. Coyle, J. Pharmacol. Exp. Ther. *232*, 873 (1985).
191. J. Ferkany and J. T. Coyle, Brain Res. *278*, 279 (1983).
192. A. Poli, A. Contestabile, P. Migani, L. Rossi, C. Rondelli, M. Virgili, R. Bissoli, and O. Barnabei, J. Neurochem. *45*, 1677 (1985).
193. S. J. Potashner and D. Gerard, J. Neurochem. *40*, 1548 (1983).
194. G. J. McBean and P. J. Roberts, Nature *291*, 593 (1981).
195. M. Baudry, K. Kramer, L. Fagni, M. Recasens, and G. Lynch, Mol. Pharmacol. *24*, 227 (1983).
196. F. Nicoletti, M. Iadarola, J. T. Wroblewski, and E. Costa, Soc. Neurosci. Abstr. *11*, 822 (1985).
197. F. Nicolletti, M. J. Iadarola, J. T. Wroblewski, and E. Costa, Proc.Natl. Acad. Sci. USA *83*, 1931 (1986).
198. F. Nicolletti, J. T. Wroblewski, M. J. Iadarola, and E. Costa, Neuropharmacology *25*, 335 (1986).
199. P. A. Bauman and P. C. Waldmeier, VIth International Conference on Cyclic Nucleotides, Calcium and Protein Phosphoylation, Foundation for Advanced Education in the Sciences, Bethesda, Maryland, Raven Press, New York, 1986, p. 31.

200. F. Nicolletti, J. T. Wroblewski, A. Novelli, H. Alho, A. Giodotti, and E. Costa, J. Neurosci. *6*, 1905 (1986).
201. B. Ault, R. H. Evans, A. A. Francis, and J. C. Watkins, J. Physiol. (Lond.) *307*, 413 (1980).
202. J. Davies and J. C. Watkins, Brain Res. *139*, 364 (1977).
203. N. L. Harrison and M. A. Simmonds, Br. J. Pharmacol. *84*, 381 (1985).
204. E. J. Coan and G. L. Collingridge, Neurosci. Lett. *53*, 21 (1985).
205. M. J. Myer and G. L. Westbrook, J. Physiol. (Lond.) *361*, 65 (1985).
206. R. Dingledine, Trends Neurosci. *9*, 47 (1986).
207. K. Fuxe, L. F. Agnati, and M. F. Celani, Neurosci. Lett. *35*, 233 (1983).
208. N. A. Sharif and P. J. Roberts, Biochem. Pharmacol. *30*, 3019 (1981).
209. S. C. Berry, S. L. Dawkins, and D. C. Lodge, Br. J. Pharmacol. *83*, 179 (1984).
210. P. S. Bernard and D. A. Bennett, Soc. Neurosci. Abstr. *12*, 60 (1986).
211. P. Krogsgaard-Larsen and T. Honore, Trends Pharmacol. Sci. *4*, 31 (1983).
212. W. F. White, Brain Res. *164*, 177 (1979).
213. R. H. Evans, A. A. Francis, A. W. Jones, D. A. Smith, and J. C. Watkins, Br. J. Pharmacol. *75*, 65 (1982).
214. J. Davies, R. H. Evans, A. A. Francis, A. W. Jones, and J. C. Watkins, J. Neurochem. *36*, 1305 (1981).
215. S. Birley, J. F. Collins, M. N. Perkins, and T. W. Stone, Br. J. Pharmacol. *77*, 7 (1982).
216. T. W. Stone and J. F. Collins, Neuroscience *17*, 629 (1986).
217. O. Goldberg and V. I. Teichberg, J. Med. Chem. *28*, 1957 (1985).
218. U. Erez, H. Frenk, O. Goldberg, and V. I. Teichberg, Eur. J. Pharmacol. *110*, 31 (1985).
219. H. Fukuda, T. Tanaka, M. Kaijima, H. Nakai, and Y. Yonemasu, Neurosci. Lett. *59*, 53 (1985).
220. J. T. Coyle, in Handbook of Psychopharmacology (L. L. Iversen, S. D. Iversen, and S. H. Snyder, eds.), Plenum Press, New York, 1982, p. 237.
221. R. Schwarcz, A. C. Foster, E. D. French, W. O. Whetsell, and C. Kohler, Life Sci. *35*, 19 (1984).
222. C. Kohler and R. Schwarcz, Neuroscience *8*, 819 (1983).
223. R. Schwarcz, W. O. Whetsell, and A. C. Foster, in Excitotoxins (K. Fuxe, P. J. Roberts, and R. Schwarcz, eds.), Pergamon Press, Oxford, 1984, p. 122.
224. J. V. Nadler, B. W. Perry, and C. W. Cotman, Nature *271*, 676 (1978).
225. J. W. Olney, In Excitotoxins (K. Fuxe, P. Roberts, and R. Schwarcz, eds.), Plenum Press, New York, 1984, p. 82.
226. R. K. Freund, S. L. Crooks, J. E. Koerner, and R. L. Johnson, Brain Res. *291*, 150 (1984).
227. E. Roberts, in Strategies for the Development of an Effective Treatment for Senile Dementia (E. Crook and L. Gershon, eds.), Mark Power Associates, New Camarin, Connecticut, 1981, p. 247.
228. J. Greenamyre, J. Penney, A. Young, C. J. D'Amato, S. P. Hicks, and I. Shoulson, Science *227*, 1496 (1985).

229. T. Wielock, Science *230*, 681 (1985).
230. R. P. Simon, J. H. Swan, T. Griffiths, and B. S. Meldrum, Science *226*, 850 (1984).
231. J. T. Coyle, Trends Neurosci. *5*, 287 (1982).
232. A. Plaitakis, S. Berl, and M. Yarh, Science *216*, 193 (1982).
233. M. J. Croucher, J. F. Collins, and B. S. Meldrum, Science *216*, 899 (1982).
234. S. J. Czuczwar and B. S. Meldrum, Eur. J. Pharmacol. *83*, 335 (1982).
235. B. S. Meldrum, M. J. Croucher, S. J. Czuczwar, J. F. Collins, K. Curry, M. Joseph, and T. W. Stone, Neuroscience *9*, 925 (1983).
236. B. S. Meldrum, M. J. Croucher, G. Badman, and J. F. Collins, Neurosci. Lett. *39*, 101 (1983).
237. H. Hagberg, P. Andersson, S. Butcher, M. Sandberg, A. Lehmann, and A. Hamberger, Neurosci. Lett. *68*, 311 (1986).
238. D. A. Bennet, C. L. Corredi, and J. Lehmann, Soc. Neurosci. Abstr. *11*, 106 (1985).
239. A. Z. Zarkovsky, Pharmacol. Biochem. Behav. *24*, 1215 (1986).
240. S. Rothman, J. Neurosci. *4*, 1884 (1984).
241. W. J. Freed and E. K. Michaelis, Pharmacol. Biochem. Behav. *8*, 509 (1977).
242. W. J. Freed, Epilepsia *26*, 30 (1985).
243. R. G. M. Morris, E. Anderson, G. S. Lynch, and M. Baudry, Nature *319*, 774 (1986).
244. G. Lynch and M. Baudry, Science *224*, 1057 (1984).
245. F. Moroni, S. Luzzi, S. Franchi-Micheli, and L. Zilletti, Neurosci. Lett. *68*, 57 (1986).
246. D. E. Murphy, A. J. Hutchison, S. D. Hurt, M. Williams, and M. A. Sills. Br. J. Pharmacol. (in press).
247. P. S. Loo, A. Braunwalder, J. Lehmann, M. Williams, and M. A. Sills, Mol. Pharmacol. *32*, 820 (1987).
248. M. F. Jarvis, D. E. Murphy, M. Williams, S. C. Gehardt, and C. A. Boast, Synapse (submitted).

14

Cholinergic Receptors

Carl Kaiser and W. Janusz Rzeszotarski / Nova Pharmaceutical Corporation, Baltimore, Maryland

HISTORICAL PERSPECTIVES

Neurotransmission in the parasympathetic (cholinergic) part of the autonomic nervous system is mediated by acetylcholine (ACh) [1]. Acetylcholine and other compounds that produce similar pharmacological effects by substituting for ACh in interacting with a receptor are termed cholinergics or parasympathomimetics and have as their fundamental action the excitation or inhibition of postganglionic sympathetic nerves. Additional effects are also exerted on ganglia, the neuromuscular junction, and cells that receive little or no parasympathetic innervation but that possess cholinergic receptors. Uptake of choline (Ch) [2] and ACh biosynthesis (1-4) are altered in certain central nervous system (CNS) disorders (5-8). The storage (9-20), release (21-26) [in which Ca^{2+} plays a pivotal role (27-36)], and interaction of ACh with cholinergic receptors, in conjunction with synaptic inactivation, determine the performance of the cholinergic system and the quality of the message transmitted.

$CH_3COOCH_2CH_2N(CH_3)_3^+\ OH^-$ [1]

$HOCH_2CH_2N(CH_3)_3^+\ OH^-$ [2]

ACh was probably the first major neurotransmitter in the process of evolution, and it was also the first to be unequivocably identified. As a consequence, it has been very thoroughly investigated, especially with respect to its action on peripheral receptors. The mode of interaction with these receptors is complicated, not only by the conformational flexibility of the molecule, but also by the great diversity of structures that behave as mimics (agonists) or blockers (antagonists) at these receptors. Thus, it is important to review in perhaps greater detail than for other receptors the structure-activity relationship(s) (SAR) among agonists and antagonists for this neurotransmitter before proceeding to a consideration of radioligand binding studies. As a consequence, the format of this chapter varies somewhat from that of others in this book.

Clearly, the classes of chemicals and SAR within various classes of cholinergic agonists, as well as those of agents that can antagonize the effects of ACh and its simulators, are the source of much information concerning the nature of the cholinergic receptors with which they interact. More recently, a great deal of research has been directed toward subpopulations of ACh receptors and the actual isolation, purification, total identification, synthesis, and reconstitution of these receptors. In this chapter some of these studies are briefly reviewed in an attempt to better understand the cholinergic receptors and their binding sites so that a more rational approach may be undertaken to the design of new agents with the potential for stimulating or blocking these receptors, or their subtypes, in a selective, therapeutically useful manner.

MODULATION OF ACETYLCHOLINE BIOSYNTHESIS AND METABOLISM

Once released into the synapse, ACh behaves as a neurotransmitter that associates with a proteinaceous macromolecular receptor. The result of this interaction is a change in the action potential of the nerve resulting from movement of Na^+, K^+, and Cl^- in and out of the membrane. Clearly, this sequence must be terminated after the interaction of ACh with the receptor and subsequent dissociation with the agonist-receptor complex. Termination of the action of ACh is achieved by its hydrolysis into Ch [2] and acetic acid in a reaction catalyzed by the enzyme acetylcholinesterase (AChE). The great velocity of this hydrolysis is responsible for the discrete localized action of ACh. In fact, this reaction precludes the therapeutic use of ACh or the existence of a presynaptic reuptake mechanism. ACh is unable to pass the blood-brain barrier because of its cationic character and its rapid inactivation by AChE (37,38).

In each stage of the process from Ch uptake to the metabolism of ACh, means for chemical intervention have been determined. Substances that interfere with ACh biosynthesis and release clearly behave pharmacologically in a manner similar to that of anticholinergics. They reduce the availability of ACh and produce, for example, neuromuscular blockade. Conversely, inhibitors of ACh metabolism, that is, AChE inhibitors, prolong or imitate the action of the neurotransmitter.

Compounds that interfere with the Ch uptake mechanism, such as hemicholinium-3 (HC-3) [3] (39-42), cause a deficiency of the ACh precursor at the nerve terminal; its affinity for a carrier uptake system is more than 50 times greater than that of Ch itself. Other evidence suggests that HC-3 is transported by the uptake mechanism in place of Ch and that it also competes with Ch at the intracellular binding sites. In either event, the blockade of nervous transmission by HC-3 is a reversible process; it can be overcome by excess Ch. HC-3 exists in tautomeric forms. In aqueous solution it is a bis-hemiketal [3], from which its name is derived. In the solid state it is the keto tautomer, the bis-phenacylcholine [4] (43).

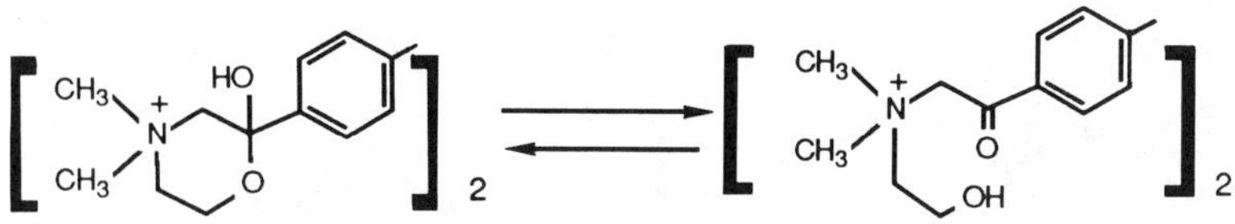

3 4

A bromoethylamine (mustard) derivative of [4], that is, [5], produces a substantially greater inhibition of Na^+-dependent, high-affinity Ch uptake. This compound, in contrast to HC-3, produces an irreversible inhibition, suggesting that it binds covalently to the uptake system in a biethyleniminium form of the bis-hemiketal [6] (44).

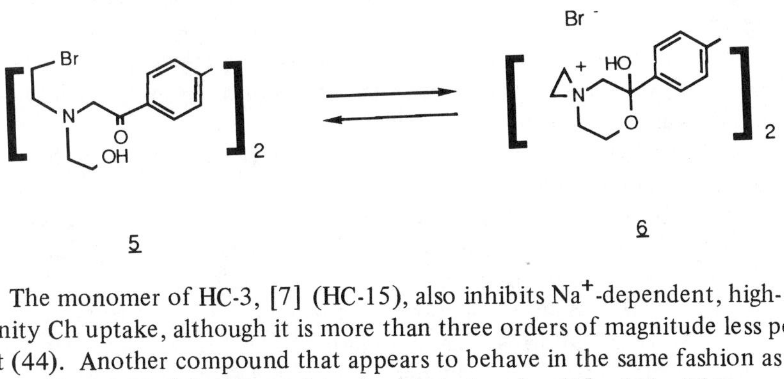

5 6

The monomer of HC-3, [7] (HC-15), also inhibits Na^+-dependent, high-affinity Ch uptake, although it is more than three orders of magnitude less potent (44). Another compound that appears to behave in the same fashion as HC-3 is WIN 4981 [8] (45), as does the triethyl analog [9] of Ch. This compound is readily acetylated in a reaction catalyzed by the enzyme choline acetyltransferase (ChAT) to produce acetyltriethylcholine, which is inactive at ACh receptors (46).

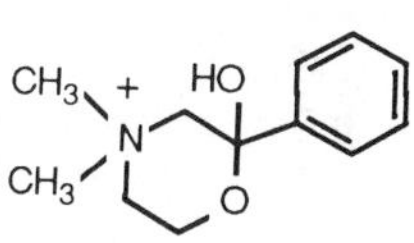

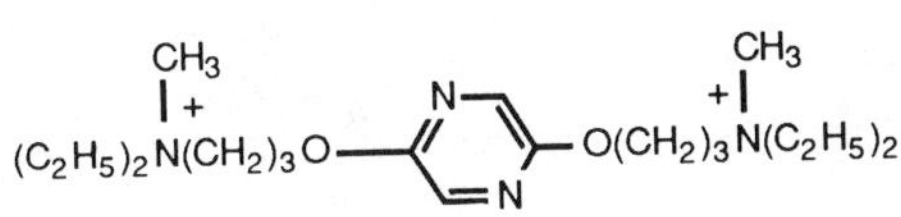

$(C_2H_5)_3\overset{+}{N}(CH_2)_2OH$

7 8 9

Although in vivo effective inhibitors of ACh biosynthesis have not yet been developed, a number of substances that are in vitro inhibitors of ChAT have been achieved. One of the more potent of these is trans-4-(1-naphthylvinyl)pyridine (4-NVP) [10] (47,48).

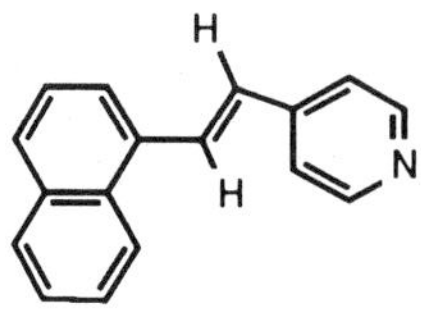

10

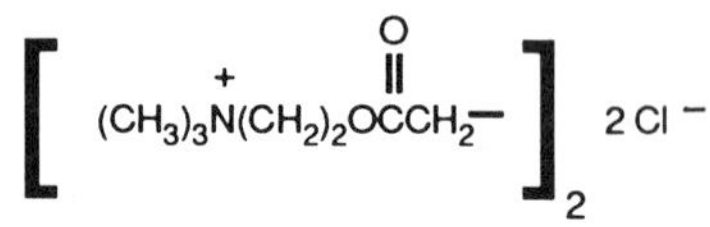

11

The synaptic release of ACh can be inhibited by Ca^{2+} complexing agents, such as the tetracycline antibiotics. In high doses they can produce a presynaptic neuromuscular block (33). Other compounds, such as suxamethonium chloride [11], its homologs, and other analogs, are short-acting neuromuscular blocking agents (49,50) that may owe their activity to inhibition of release of ACh from prejunctional sites (43). Another agent, the 2-aminocyclohexanol [12] (AH 5183), appears to act in the overall uptake-metabolism sequence. It may block the life cycle of the ACh vesicles by inhibiting the process in which the vesicles are refilled with ACh from the axoplasm (51,52), although some of its action is mediated postsynaptically.

AChE inhibitors have been described in detail in recent reviews (References 53 and 54, for example), and therefore are only outlined here. The inhibition of AChE leads to accumulation of released or leaked ACh with consequent, often lethal, effects. Inhibitors of AChE are usually divided into three classes: (1) fully reversible inhibitors, such as edrophonium [13], (2) carbamylating agents, such as the alkaloid physostigmine [14], and (3) phosphorylating agents that have provided a number of chemical warfare agents, such as soman [15], and insecticides, such as paraoxan [16] (55,56).

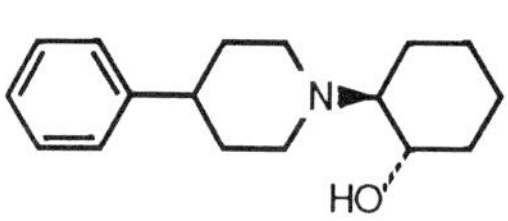

12

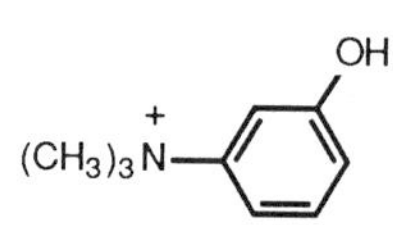

13

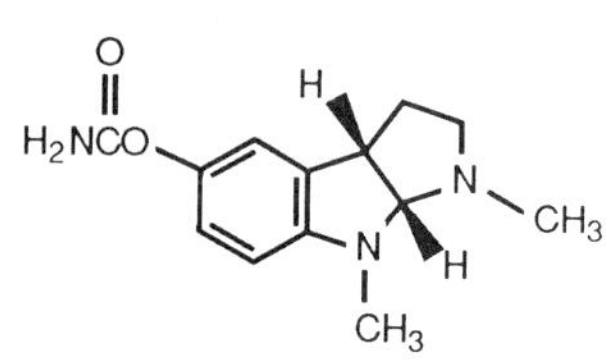

14

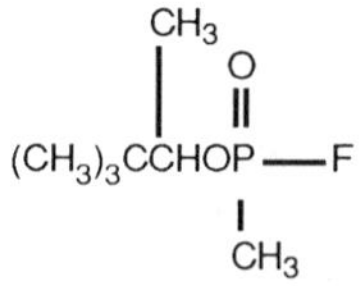

15

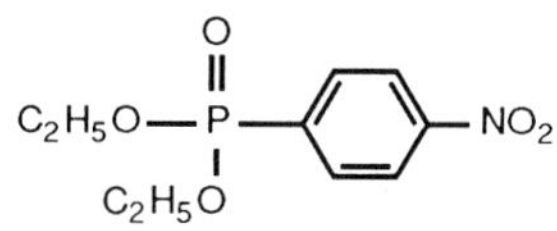

16

SUBTYPES OF CHOLINERGIC RECEPTORS AND SELECTIVE LIGANDS

The recognition of the heterogeneity of ACh receptors (AChR) began with the studies of Dale in 1914 (57). Low doses of ACh caused a reduction in blood pressure, and higher doses decreased heart rate. These effects were blocked by atropine but were not affected by vagal section. This suggested that the ACh-induced effects were produced peripherally. When larger doses of ACh were given to atropinized animals, blood pressure increased. This hypertensive effect was not influenced by destruction of the spinal cord, thus indicating that this action was also peripherally mediated. The latter effect was blocked by a dose of nicotine sufficient to paralyze all autonomic ganglia. The hypotensive, cardiac inhibitory effect of ACh was described (57) to be of a muscarinic type, being mimicked closely by the naturally occurring (+)–(2*S*,3*R*,5*S*)-muscarine [17] (58). The hypertensive effect was characterized as like (–)–(2*S*)-nicotine. The dual actions of ACh are now referred to as muscarinic and nicotinic and are believed to be produced by stimulation of muscarinic (mAChR) and nicotinic (nAChR) ACh receptors, respectively.

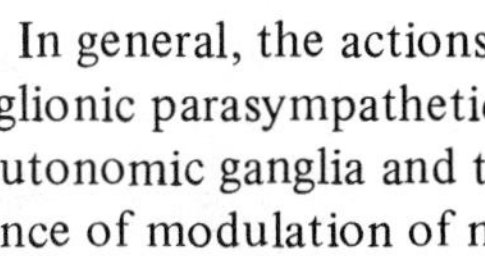

17

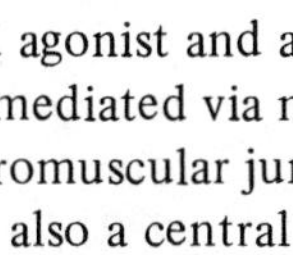

18

In general, the actions of ACh-related agonist and antagonist drugs at postganglionic parasympathetic terminals are mediated via mAChR whereas effects at autonomic ganglia and the skeletal neuromuscular junction are the consequence of modulation of nAChR. ACh is also a central neurotransmitter (59), and there is evidence for both m- (60) and nAChR (61,62) in the CNS. The principal effects of stimulating mAChR are bradycardia, increased flow of secretion from most secretory (salivary, bronchial, gastric, and sweat) glands, peripheral vasodilation, miosis, and increased contractions and/or peristaltic

Table 1 Biochemical Muscarinic Responses (63)

System	Effect
Adenylate cyclase	Inhibition
Guanylate cyclase	Activation
Regulation of phosphodiesterase	Activation
Polyphosphoinositide breakdown	Activation
Protein phosphorylation	Modulation
K^+ channels	Opened (conduction tissue of the heart)
K^+ channels (M channels)	Closed (synpathetic ganglia, hippocampus)
Ca^{2+} channels	Opened
Mobilization of intracellular Ca^{2+}	Activation

motility in the gastrointestinal tract, that is, effects similar to those produced by muscarine. Some of the biochemical responses to mAChR stimulation are summarized in Table 1 (63).

nAChR agonists stimulate and then block the end plates of autonomic and skeletal muscles. The effect of activation of subytpes of AChR in the CNS is less clear because most information in this area has derived from the study of nonselective anticholinergics (64-66). These compounds disturb memory and orientation. They also produce hallucinations, mania, and delirium.

SAR and comparison of the action of agents at m- and nAChR are further complicated by the probability that nicotine and many nAChR agonists also act indirectly through the release of endogenous ACh (67). In addition, as long as 30 years ago it was recognized that mAChR in the heart had properties different from those in the gastrointestinal tract; however, it was not until 1978 that substantial evidence for the existence of subpopulations of these receptors began to accumulate (68). Major evidence for these subtypes derived from the study of cholinergic regulation at the lower esophageal sphincter. Vagal stimulation and some mAChR agonists, such as McN-A-343 [19], produce relaxation of this sphincter; however, others, such as bethanechol [20], cause contraction. The effects of both mAChR agonists are blocked by atropine. Further studies indicated that this sphincter is innervated by intrinsic neurons that bear mAChR (termed M_1) that are different from those (termed M_2) that innervate smooth muscle cells in this sphincter. Stimulation of the M_1 receptors causes relaxation, whereas activation of M_2 receptors produces contraction. Similar complexities may underlie gastric acid secretion. Binding studies of the interaction of mAChR agonists with cerebral cortical membranes also indicate two subtypes of these receptors (69). Subsequently, binding studies with the tritiated M_1-selective AChR antagonist pirenzepine [21] (70) indicated that M_1 receptors predominate in the corpus striatum, cerebral cortex, hippocampus, and perhaps autonomic ganglia. Conversely, in the heart, cerebellum, and ileum the low-affinity M_2-AChR appear to have greater significance.

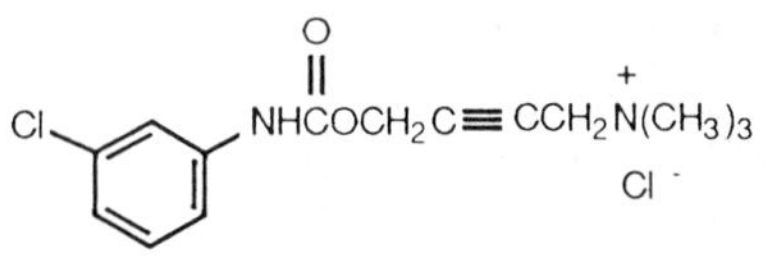

19

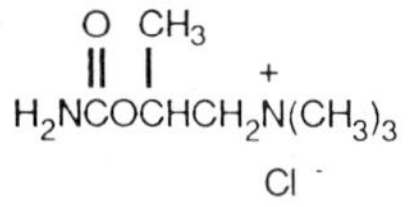

20

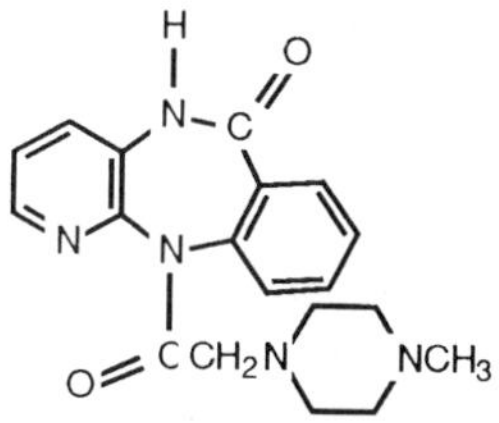

21

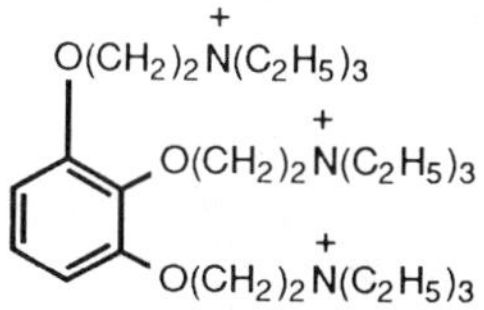

22

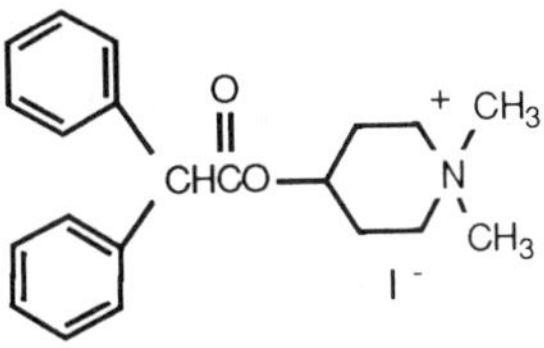

23

It is believed that the M_1-AChR may be important in the generation of phosphorylated derivatives of inositol and in the regulation of Ca^{2+} fluxes. The M_2 receptor is apparently linked to adenylate cyclase; receptor activation causes inhibition of the enzyme. The M_2 receptor may also be present on presynaptic neurons, where its stimulation inhibits the release of ACh. It also appears to interact with guanine nucleotide-binding regulatory proteins (68).

A number of agonists and antagonists has been produced and called "selective" for M_1- and M_2-AChR on the basis of binding studies or functional assays of mAChR coupled to second messenger systems, such as stimulation of phosphatidylinositol turnover, inhibition of adenylate cyclase, and potentiation of cAMP phosphodiesterase. At present, a substantial amount of confusion exists in defining the mAChR subtypes. Based on binding studies, subtypes of mAChR have recently been defined (71) as follows:

"The M_1 receptor is one which shows a high degree of sensitivity to pirenzepine [21] and which exhibits high affinity for the drug in binding studies. The M_2 receptor is one which shows a low sensitivity to pirenzepine and exhibits low binding. Whereas it appears that M_1 receptors are homogeneous, there are indications that M_2 receptors may be heterogeneous, and thus may need further definition." The latter qualification is discussed in greater detail subsequently.

The functional definition divides mAChR into excitatory M_1 and inhibitory M_2 subtypes. On this basis, the excitatory M_1-AChR is insensitive to *N*-ethylmaleimide (NEM), guanine nucleotides, ethylenediaminetetraacetic acid (EDTA), and high salt concentrations (72). Activation of these receptors mediates stimulation of phosphoinositol breakdown (73) and closes K^+ channels. The inhibitory M_2-AChR are insensitive to NEM, guanine nucleotides, and EDTA. Its inhibitory activity is expressed by guanosine triphosphate (GTP)-dependent inhibition of adenylate cyclase by muscarinic agonists and opening of K^+ channels (74-77). The neuromuscular blocking agent gallamine [22] has been reported to show selectivity for this receptor (78). Recently, studies with 4-DAMP methiodide [23] (79) suggest differences in M_2 receptors on ileum and atria (80). Studying the pharmacological and electrophysiological properties of other M_2 receptor antagonists, e.g., [24] (silahexacyclium) resulted in the suggestion that the M_2 receptors may consist of two subtypes, $M_{2\alpha}$ (cardiovascular/vascular type) and $M_{2\beta}$ (smooth muscle/glandular type) (81). Based on the actions of the mAChR antagonists 4-diphenylacetoxy-*N*-methylpiperidine (4-DAMP) [23], AF-DX 116 [25], and pirenzepine [21], another study suggested three populations of muscarinic receptors: M_1 (sympathetic ganglia), M_2 (cerebellum and heart), and M_3 (pons-medulla, cortex, and salivary glands) (82).

It has also been proposed that three subpopulations of agonist binding sites can coexist and interconvert (83,84). Thus, in the cerebral cortex one can expect agonist binding sites that are super-high (SH), high (H), and low (L) affinity. The proportions of SH, H, and L sites may vary in different regions of the brain or other tissues. When rat myocardial mAChR are studied under normal conditions, guanine nucleotides like GTP or GppNHp shift the predominant H sites to L sites. Magnesium and other divalent ions convert H sites to SH sites or potentiate the action of guanine nucleotides (85). The opposite is known for binding of antagonists where a medium of low ionic strength is used. In rat cerebral cortex magnesium ions behave differently, converting part of the L sites to H and SH sites.

These and other observations led to the conclusion (72) that the existence of three states for each mAChR is an artifact produced by the coupling of guanine nucleotide binding protein to the M_2-AChR. According to these findings, there are only two basic populations of mAChR that do not interconvert. The high-affinity M_2-AChR represents a complex of agonist, receptor, GTP binding protein, and bivalent cation (Mg^{2+} or Mn^{2+}). That complex disintegrates in the presence of NEM, which does not affect the high affinity M_1-AChR. The high-affinity M_1-AChR have an affinity that overlaps that of uncoupled low-affinity M_2-AChR, creating an artifact, that is, the appearance of three total states. Based on this hypothesis it is easy to understand why in tissues rich in M_2-AChR (cholinergic nerves, superior colliculus, brain stem, heart, and smooth muscle) all potent agonists are M_2 selective. The coupling of M_2-AChR to the GTP binding protein does not appear to be universal (85). Thus, the mAChR on canine gastric smooth muscle is regulated by guanine nucleotides. This supports the thesis of more than one subclass or state of that receptor.

McN-A-343 [19] binds to and activates the myocardial mAChR, but unlike the neuromuscular blocker gallamine [23] its binding is noncompetitive. These findings led to another hypothesis postulating the existence of a McN-A-343-gallamine binding site in myocardial tissue. This site is distinct from the ACh binding sites in both M_1- and M_2-AChR (86).

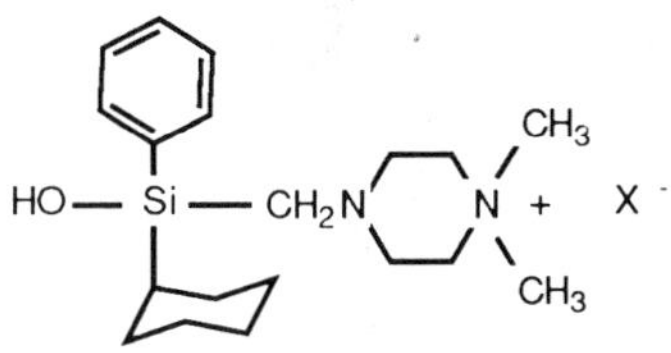

24

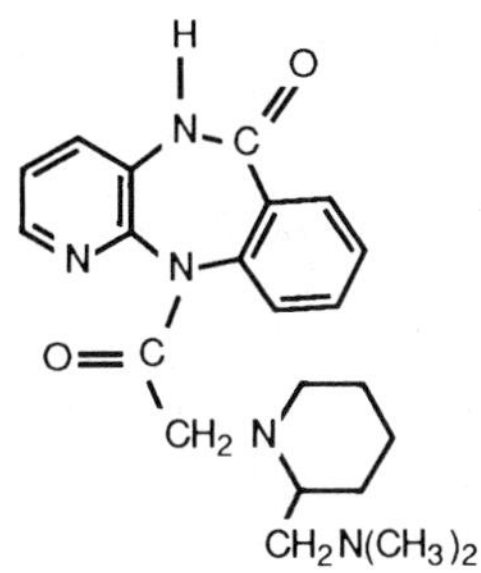

25

nAChR also appear to exist in more than one population. It has been suggested that there may be N_G-AChR located on autonomic ganglia, N_S-AChR located in striated muscle, and nAChR in the CNS (87). At present, the matter of ACh receptor subtypes remains unresolved.

STRUCTURE-ACTIVITY RELATIONSHIPS

The uncertainty concerning sublcasses of cholinergic receptors, coupled with enormous differences in investigative techniques utilized to study agonists and antagonists, makes the task difficult at best: comparing varying and contradictory reports from the past and present to arrive at meaningful SAR that will permit extrapolation to enhance our understanding of cholinergic receptors and their binding sites. Nonetheless, it is important to consider available SAR, especially in light of the receptor subpopulations. Such considerations offer the potential for the development of new, selective, and therapeutically useful drugs.

Cholinergic Agonists Derived from Acetylcholine

ACh itself has almost no therapeutic applications. This is because of its short duration of action, resulting from hydrolysis in the gastrointestinal tract following oral administration, and because of its rapid metabolism by both AChE and butyrylcholinesterase in the plasma. Its clinical use is restricted to topical application in the eye to produce a rapid, but brief, miosis (88). ACh acts at both m- and nAChR. Because of its inactivity on oral administration, rapid metabolic hydrolysis, and lack of specificity for subtypes of receptors, many studies have been directed toward the derivation of new, selective, longer acting agents that may be more effective on oral administration. The new cholinergics fall into two groups: (1) ACh and its synthetic relatives and (2) naturally occurring cholinomimetic alkaloids, notably pilocarpine, muscarine, and arecoline.

The many analogs, derivatives, and congeners of ACh [1] include those in which: (1) the quaternary ammonium group has been modified (53,88-91), (2) the acetyl group has been replaced by other moieties, (3) the alkylene bridge connecting the quaternary ammonium and acetyl or other group has been changed, and (4) the ester has been replaced by another kind of substituent. Unfortunately, as noted earlier, uniform pharmacological or biochemical data are unavailable and thus only qualitative SAR comparisons are possible.

Modifications of the quaternary ammonium group of ACh have taken two directions: replacement of the N-methyl groups and substitution of other atoms for the nitrogen (53). In general both kinds of modification lead to a decrease or loss of activity. An exception is [26], which has only modestly decreased activity; it is one-third to one-fifth as potent as ACh, and is markedly more effective than its uncyclized analog [27] (92,93).

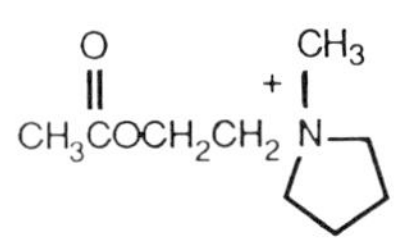

26

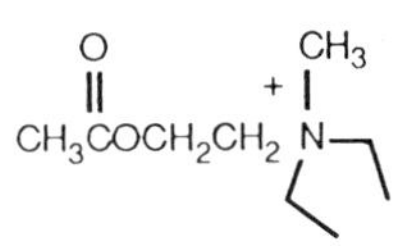

27

Also, replacement of one of the quaternary methyl groups of ACh and related compounds with a methoxyl sometimes enhances both muscarinic and nicotinic activity (93), but amine oxides have little cholinergic activity (95). The decrease or loss of activity noted upon replacement of the quaternary methyl groups with hydrogen or other alkyl groups has been rationalized by molecular orbital calculations (96), which suggest that extension of even a single methyl group decreases the charge density on the nitrogen (97); this may be significant in receptor binding and activation. Replacement of the trimethylammonium head with other cationic groups (91,98) or even a tert-butyl (99) moiety results in greatly diminished muscarinic activity.

A large number of ACh analogs, such as [29a-k], in which the acetyl group has been replaced with other acyl groups, has been studied (100-106).

$RCOOCH_2CH_2X^+(CH_3)_n$

[28] a, X = P; n = 3
b, X = As; n = 3
c, X = S; n = 2
d, X = C; n = 3

$RCOOCH_2CH_2N(CH_3)_3{}^+$

[29] a, R = H; b, R - C_2H_5; c, R = n-C_3H_7; d, R = i-C_3H_7; e, R = n-C_4H_9; f, R = $HOCH_2$; g, R = $CH_3CH(OH)$; h, R = H_2NCH_2; i, R = $CH_2{=}CH$; j, R = CH_3CO; k, R = $BrCH_2$; l, R = H_2N

$ROCH_2CH_2N(CH_3)_3{}^+$

[30] a, R = NO_2; b, R = $(CH_3)_2PO$; c, R = $CH_3C{=}S$; d, R = CH_3CH_2

$CH_3C({=}O)XCH_2CH_2N(CH_3)_3{}^+$

[31] a, X = S; b, X = Se; c, X = NH; d, X = CH_2

Ch itself possesses some of the cholinergic properties of ACh; however, acetylation of the hydroxyl functionality to give ACh affords optimum potency. Both lower [29a] (100) and higher [29b–e] (100,101) acyl homologs, as well as those bearing a polar or unsaturated group, such as [29f-k], are less potent than ACh and some are antagonists (100,102-106). Carbamate esters of Ch, such as [20] (bethanechol) and [29l] (carbachol), are generally more potent (107-109) and more toxic than ACh (110). Replacement of the acetyl group of ACh with various other groups, such as [30a-d] (111-114) and [31a-d] (106, 114-116), uniformly decreases but does not abolish cholinergic activity.

The length of the alkylene bridge connecting the acetoxy and quaternary nitrogen groups of ACh is very important. Thus, if the ethylene bridge is increased to propylene, as in acetyl-γ-homocholine [32a] (116) or butylene [32b] (117), cholinergic potency is markedly decreased. Branching of the ethylene bridge also has a pronounced effect on cholinergic properties. Thus, acetyl-β-methylcholine [33a] is a more potent, longer acting muscarinic agent than ACh, but its nicotinic effects are decreased (110). The increased duration of action of methacholine [33a] is attributed to its slower rate of hydrolysis (54), which apparently results from the poor affinity of the compound for the catalytic site of AChE. A similar prolonged activity is seen with the carbamate esters; for example, the β-methyl derivative bethanechol [20] is longer acting than the methylene-bridged compound carbachol [29l]. The muscarinic effects of these β-methyl derivatives resides almost entirely in the (+)-S isomer; the (−)-enantiomers are only 1/240 as potent as ACh (118,119). Introduction of other alkyl substituents into the α or β position of ACh to provide [33b-h], for example, uniformly decreases cholinergic activity and selectivity (54,120-123).

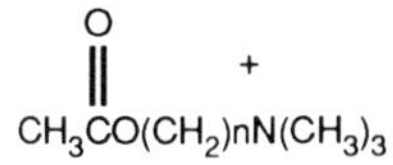

32a, n = 3; b, n = 4

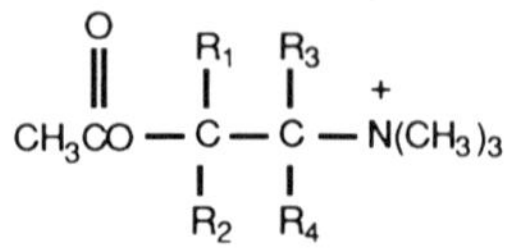

	R_1	R_2	R_3	R_4
33a	CH_3	H	H	H
b	H	H	H	CH_3
c	CH_3	H	CH_3	H
d	CH_3	H	H	CH_3
e	H	H	CH_3	CH_3
f	CH_3	CH_3	H	H
g	n-Pr	H	H	H
h	n-Bu	H	H	H

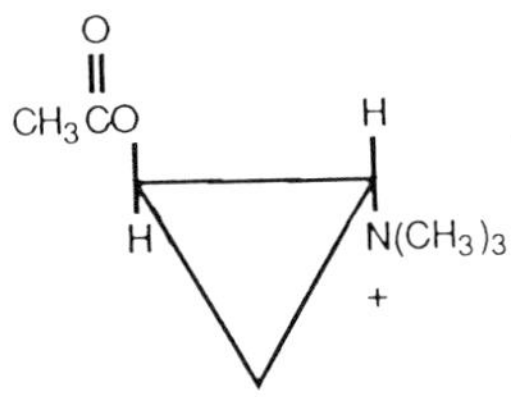

34

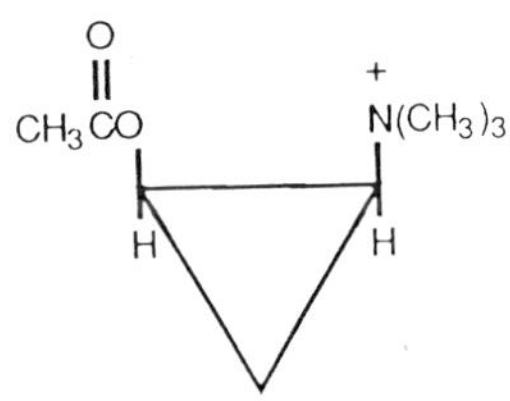

35

The ethylene bridge of ACh has been incorporated into ring systems in various ways. These more rigid compounds, although having the disadvantage of added bulk, provide information about the orientation of the ester and ammonium groups (or their imitators) during binding with the cholinergic receptors. The smallest amount of bulk is incorporated when the presumably important functionalities are joined by a cyclopropane ring. Both the E [34] and Z [35] isomers of 2-acetoxycyclopropyltrimethylammonium (ACTM) have been studied (124,125). The (+)-(E)-(1S,2S) enantiomer of [34] is equal to or more potent than ACh as a muscarinic, but it has only weak nicotinic activity. The (-)-(E)-(1R,2R) enantiomer of [34] is more than two orders of magnitude less effective than the dextrarotatory antipode, and the racemate [35] is practically inactive at mAChR, although it also displays very weak nicotinic effects. The racemic E-cyclobutyl analog [36] is much less potent than the racemic cyclopropyl homolog [34] (126). Some related cyclopropyl and cyclobutyl derivatives likewise cause only weak muscarinic but appreciable nicotinic actions (127,128). Cyclopentyl [37a] and cyclohexyl [37b] homologs have little or no cholinergic activity (67,129).

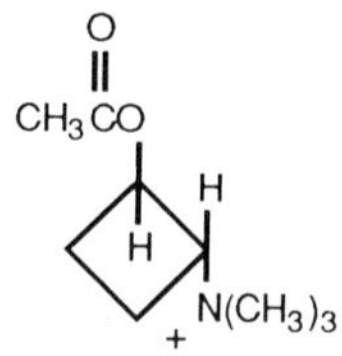

36

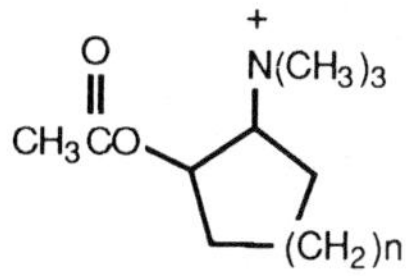

37a. n = 1; b. n = 2

Some other cyclized analogs of ACh are represented by [38] (130,131) and [39] (131). The piperidine [38a] is only weakly effective, whereas E- (but not Z-[38b]) is active at mAChR, where it is about 1/10 as potent at ACh. At nAChR the Z isomer has the greater potency (131). The morpholinone [39], a rigid conformational (quasi-ringlike) extreme of ACh, is only weakly effective at both m- and nAChR, casting some doubt on the proposal of Scheuler (131) that ACh may interact with these subtypes in different conformations. Some cyclohexyl-fused analogs of [37b] are represented by stereoisomers of the E-decalin [40a]. Among these isomers, the greatest muscarinic activity is displayed by the 2,3-E diaxial form; however, even this isomer is more than two orders of magnitude less potent than ACh (120). Among a series of ring-methylated E-decalins, the 2,3-dimethyl compound [40b] is the most potent muscarinic; however, it is only 1/50 as potent as ACh (133).

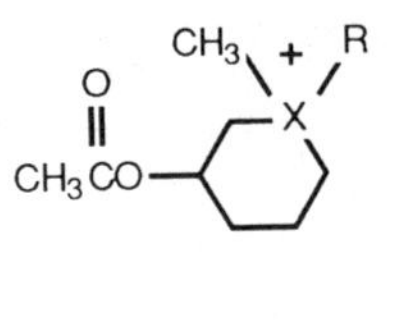

38a. X=N; R=CH_3

b. X=S; R=----

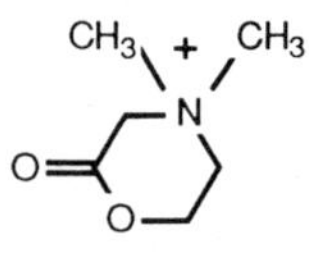

39

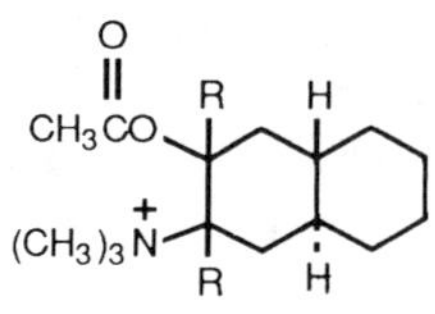

40a. R=H; b. R=CH_3

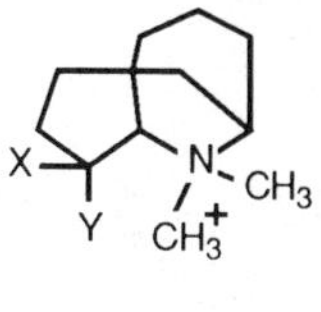

41a, X = CH_3CO_2; Y = H
41b X = H; Y = CH_3CO_2

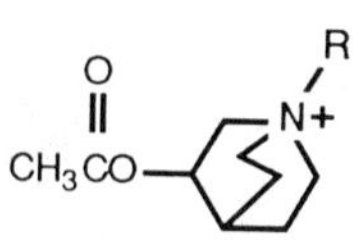

42a, R = H
42b, R = CH_3

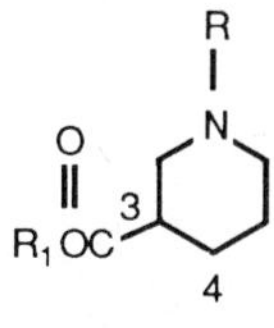

43a, R = R_1 = CH_3; 3,4 = =
44b, R = H; R_1 = C_2H_5; 3,4 = H_2

Some very rigid cyclized ACh derivatives are the quaternary tropyl acetates [41a] and [41b]. Both are very potent nicotinic agents; however, they are extremely weak muscarinics (134). Conversely, another derivative of ACh, 3-quinuclidinyl acetate (aceclidine) [42a], is a potent and selective muscarinic agent (135); however, quaternarization to give [42b], as well as various other modifications, decreases the potency (102,105,135). The chirality of aceclidine is very significant. The (+)-S enantiomer is much more potent than the (-)-R isomer (136). This observation is particularly significant as the 3-quinuclidinol-based anticholinergics require the opposite absolute configuration (137), an observation that can be rationalized on the basis of receptor models (see Nature of Cholinergic Receptors: Muscarinic Receptors).

Another cyclic compound, more distantly related to ACh is the alkaloid arecoline [43a], which is an agonist of both n- and mAChR (66). Its close analog ethyl nipecotate [43b] displays stereoselective cholinergic activity as measured by inhibition of GTP-induced stimulation of rat striatal cAMP. In this test, the R enantiomer is nearly 10-fold more potent than its S counterpart (138). A series of bioisosteric analogs of arecoline in which the ester is replaced by a 3-methoxyisoxazole group inhibits the binding of [^{3}H] propylbenzilyl choline mustard (PrBCM) to rat brain membranes, and these compounds penetrate the blood-brain barrier (139).

Although the only three cholinergics derived from ACh that are used therapeutically, methacholine, carbachol, and bethanechol, have an esteratic carbonyl group, this function is not imperative for activity. As noted earlier, even choline [2] and the ketone [31d] retain some muscarinic activity. The latter compound is about 1/12 as potent as ACh (114,140). Even ethers of Ch retain some, albeit reduced, activity (141).

Cholinergic Agonists Structurally Unrelated to Acetylcholine

The most obvious cholinergic agents that do not bear a direct structural relationship to ACh are the alkaloids muscarine [17] and nicotine [18]. Another alkaloid with cholinergic activity is pilocarpine [44]. As in the case of muscarine and nicotine, pilocarpine is stereoselective; it requires the Z-(3S,4R) absolute configuration (142). The E isomer (isopilocarpine) is nearly inactive.

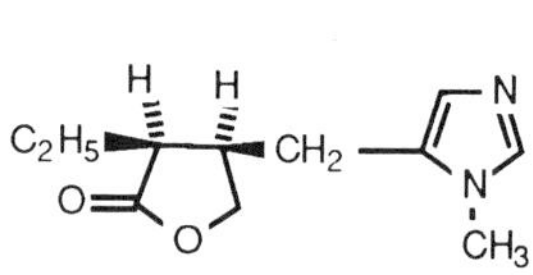

44

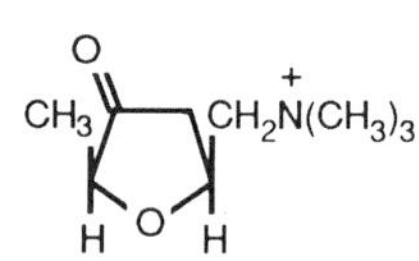

45

Muscarine [17] provides a semirigid molecule of known absolute stereochemistry and thus may be helpful in establishing the structural requirements for muscarinic activity and information about the nature of mAChR. The enantiomer of the naturally occurring (+)-(2S,3R,5S)-muscarine is practically inactive, as are its three diastereoisomers, epimuscarine, allomuscarine, and epiallomuscarine (143). The keto analog of muscarine, muscarone [45], has muscarinic activity, but it also has some nicotinic properties and has decreased stereoselectivity. 4,5-Dehydromuscarone [46] is about equipotent to the isomers of [45] (144). 4,5-Dehydromuscarine retains a high degree of activity, as does the aromatic congener methylfurtrethonium [47] (143). The dioxolane [48a] (F2268), in which the carboxylic acid group of ACh is replaced with an acetal group, is very stereospecific. The (+)-(Z)-(2S,4R) isomer, that is, the antipode with the same stereochemistry is that of naturally occurring muscarine, is more than two orders of magnitude more potent than ACh in guinea pig ileum. In general, structural alteration of [48a] markedly decreases potency (145,146); however, the oxathiolane methiodide [48d], with the same absolute stereochemistry, that is, (+)-(Z)-(2R,5R), is 10 times more potent than muscarine in a guinea pig ileum preparation. The sulfoxide [48c] and sulfone [48d] also have potent muscarinic actions, although different steric requirements are observed in various preparations, further suggesting subtypes of mAChR (147). The tetrahydrofuran [48e], as well as the reverse ether, have only slightly decreased muscarinic potency compared to [48a] (148). Even the cyclopentane derivatives [49a] and [49b] are highly potent muscarinics (149), thus contradicting proposals (144) that the ether oxygen of muscarine plays a pivotal role in receptor binding.

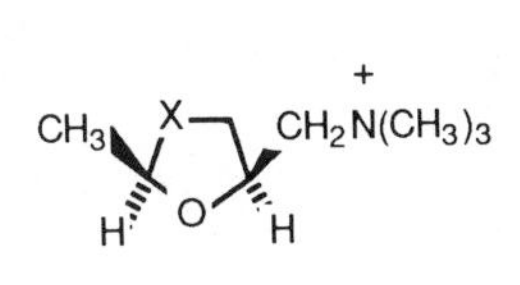

46

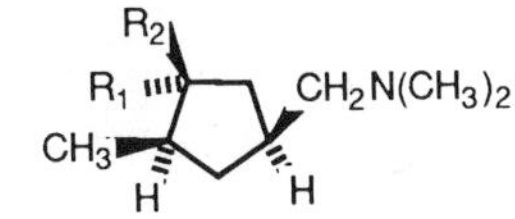

47

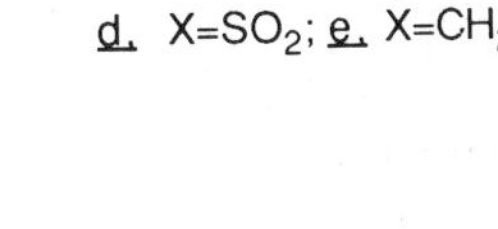

48a. X=O; b. X=S; c. X=SO; d. $X=SO_2$; e. $X=CH_2$

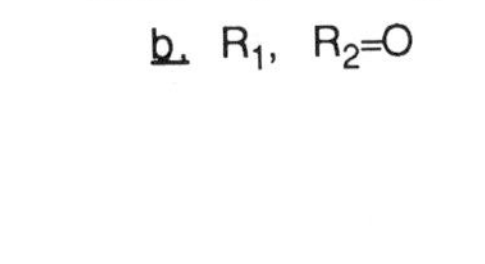

49a. R_1=OH; R_2=H

b. R_1, R_2=O

The growing interest in Alzheimer's disease has emphasized the need for centrally acting selective M_1-AChR agonists that in addition to an M_2-AChR selective antagonist, and perhaps an AChE inhibitor, could support failing cholinergic transmission in the brain. Several conformationally rigid compounds, spiro derivatives somewhat related to [47] and [48b], namely, AF102B [50] and RS86 [51] (150), produced beneficial effects in cognitive impairments (151).

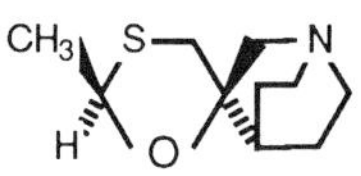

50

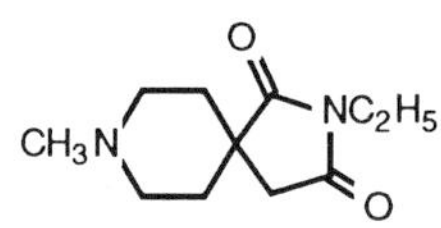

51

Tremorine [52] is metabolized to the lactam oxotremorine [53a], which is almost equipotent to ACh as a muscarinic agent (152); it appears selective for M_2-AChR, and it has been used as a tritiated ligand for these receptors. Introduction of a methyl group into the butynyl chain or pyrrolidine ring of [53a] provides [53b] (153) and [54a] (154), both of which are anticholinergics in which practically all of the activity resides in the R enantiomer. Conversely, replacement of the pyrrolidine ring of [54a] with a dimethylamino group affords [54b], which is a weak agonist. Quaternarization of [54b] gives [54c], a potent agonist resembling oxotremorine in activity (155). Again, the R enantiomer is more potent than its S counterpart. The mustard analog of oxotremorine, (BM 123) [54d], is an irreversible inhibitor of ACh (156). The allene analog [55] lacks cholinergic agonist or antagonist activity in mice (157).

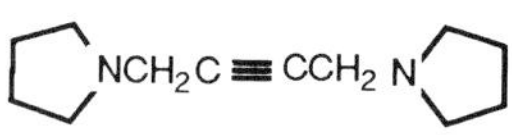

52

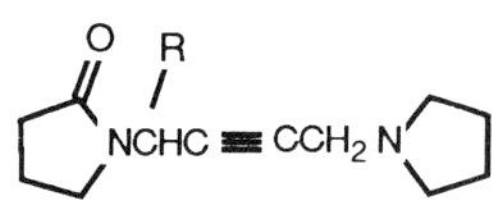

53a. R=H; b. R=CH_3

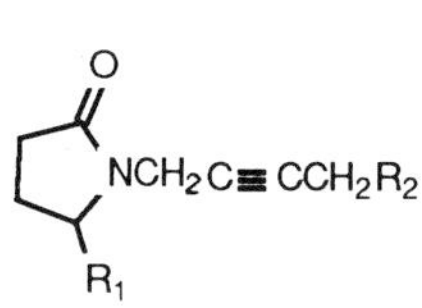

54a. R_1=CH_3; R_2=N  b. R_1=CH_3; R_2=$N(CH_3)_2$; c. R_1=CH_3; R_2 = $N(CH_3)_3^+$; d = R_1 = H, R_2 = $N(CH_3)CH_2CH_2Cl$

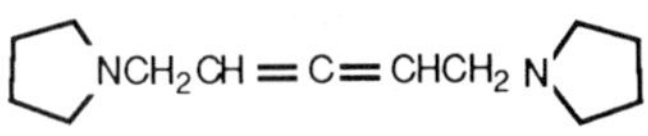

55

The first muscarinic agonist showing specificity toward M_1-AChR was the acetylenic "ganglionic stimulant" McN-A-343 [19] (158). Olefinic analogs of [19], the E-[56] and Z-[57] isomers, are weaker than McN-A-343, but retain an M_1-selective agonist profile in the pithed rat; the E isomer is less potent than its Z counterpart (153).

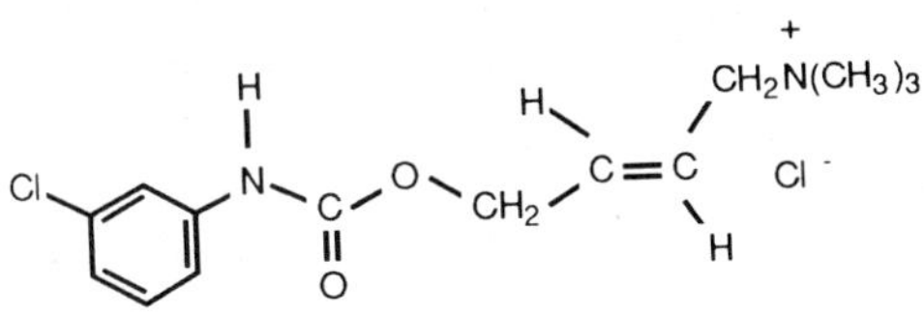

56

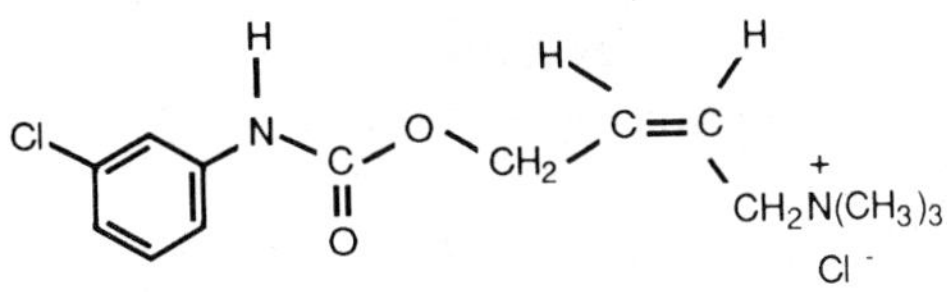

57

Cholinergic Receptor Antagonists

The observation that atropine [48] and scopolamine [59], two alkaloids isolated from *Solanaceus* plants, block the action of ACh at mAChR in various tissues to produce antispasmodic, antisecretory, and mydriatic effects has led to an extensive search for compounds in which these activities are separated. The search for a synthetic replacement of the natural muscarinics began over a hundred years ago (159). The enormous number of AChR antagonists prepared from the day Ladenburg (160) produced homatropine [60], the first semisynthetic analog of atropine, represents a "gold mine" that remains to be properly exploited. Most of these compounds have not been studied for cholinergic receptor binding properties.

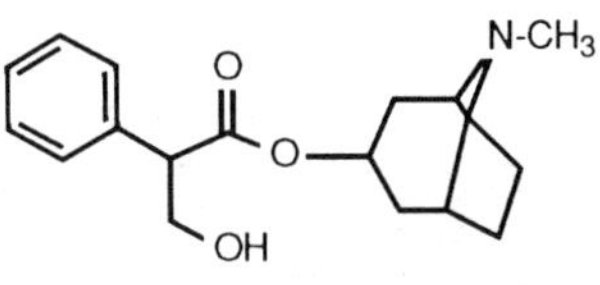

58

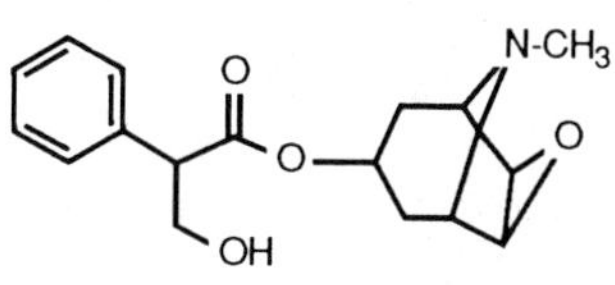

59

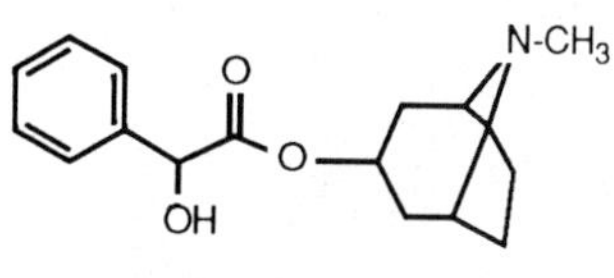

60

Muscarinic Blocking Agents

The pursuit of new selective anticholinergic agents has taken several paths (161): (1) dissection of atropine into simpler molecules, (2) molecular modification of cholinergic agonists, and (3) alteration of cholinergic drugs based on principles of bioisosterism. Some examples of the resulting kinds of structures with anticholinergic activity are the thioester thiphenamil [$Ph_2C(OH)COSCH_2CH_2N(C_2H_5)_2$] (162), the alkane tricyclamol [$Ph(c{-}C_6H_{11})C(OH)CH_2CH_2{-}c{-}N(CH_2)_4(CH_3)^+$] (163), the ether $Ph_2CHOCH_2CH_2N(CH_3)_2$, and the amides $Ph_2C(CONH_2)CH_2CH(CH_3)N(CH_3)_2$ (aminopentamide) (164) and $PhCH(CH_2OH)CON(C_2H_5)CH_2$–4-pyridine (mydriacyl) (165).

Gradual reduction of the size of the 3 α-tropanol part of the atropine molecule provides esters like those of 4-hydroxy-*N*-methylpiperidine [61] (166) or dialkylaminoethanol [62] (167). The influence on biologic activity of the character, size, and shape of the aminoalcohol parts of the esters has been most thoroughly investigated by Barlow and his colleagues (168-170). Alterations of the tropic acid portion of the molecule, such as [63-66], were also studied very thoroughly by several other groups (168,171-173). An obvious conclusion from these studies is that compounds carrying an α-disubstituted glycolic acid have greater receptor affinity than do other structural variants of tropic acid.

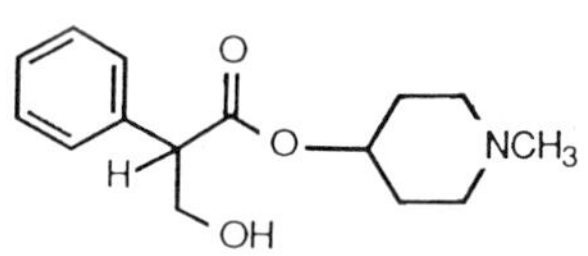

61

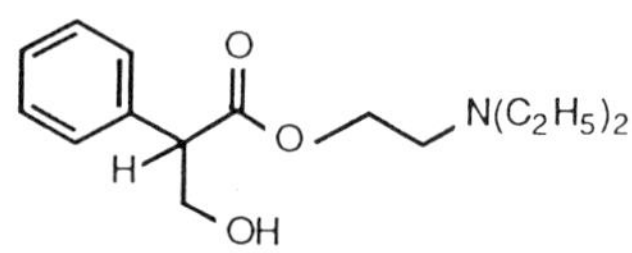

62

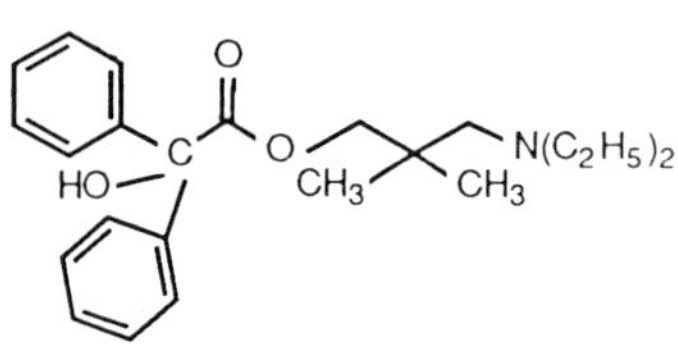

63

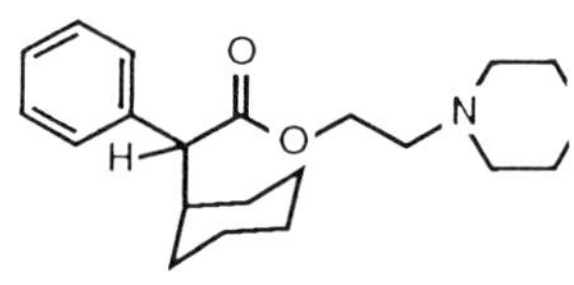

64

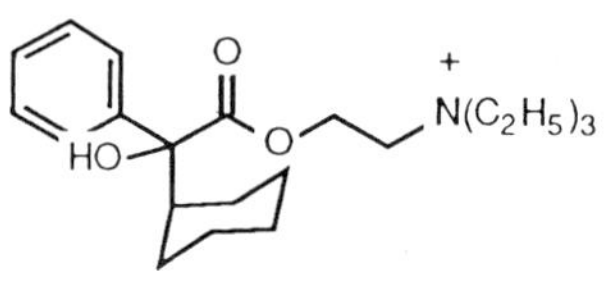

65

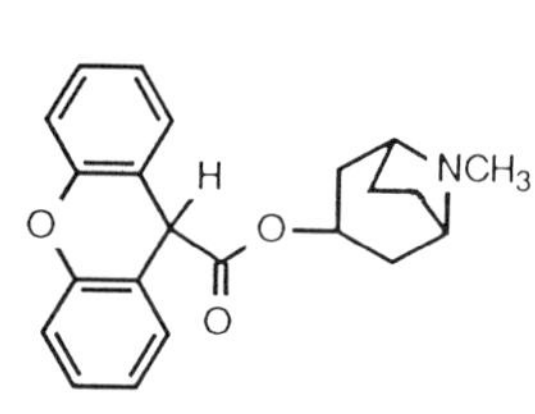

66

Alteration of the 3α-tropanol portion of atropine has also led to novel and potent anticholinergic compounds. Esterification of 3-quinuclidinol [67] (174, 175) with benzilic acid affords one of the most potent mAChR antagonists, 3-quinuclidinyl benzilate (QNB) [68] (176,177).

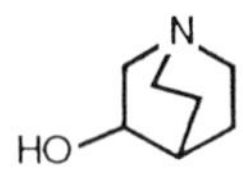

67

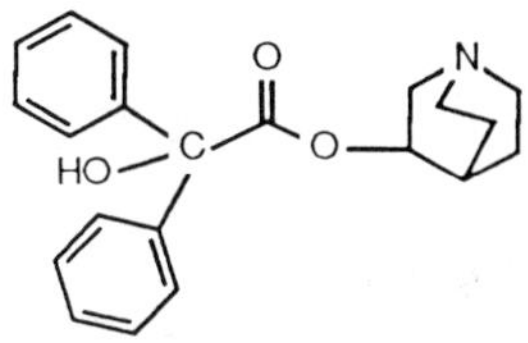

68

The investigation of QNB analogs to determine the structural requirements vesting high-affinity, receptor specificity, and a desired tissue distribution pattern, was initially met with limited success (178,179). mAChR-rich fractions from rat and dog ventrical muscle and rabbit cuadate putamen were used in these binding studies. The far greater affinity of esters of 3-quinuclidinol compared with those of an open-chain dialkylaminoalcohol like benactyzine [69] or monocyclic alcohols like 4-hydroxy-*N*-methylpiperidine [70] is striking.

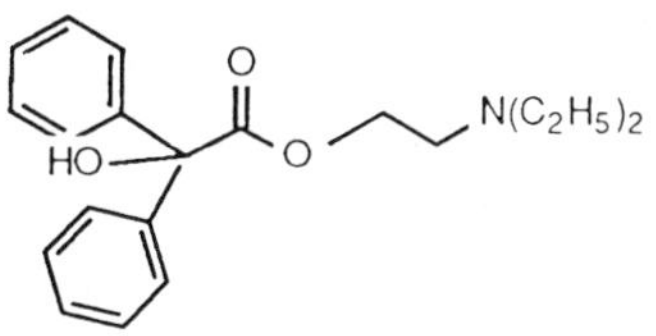

69

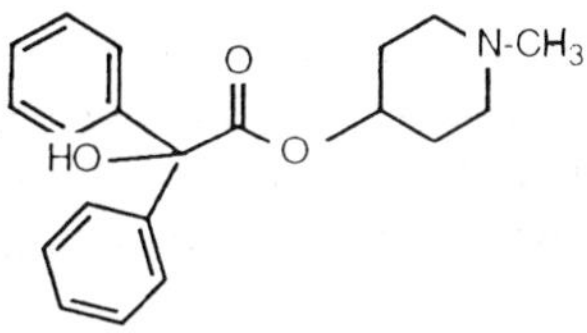

70

Halogen substitution of one of the phenyl rings of QNB has a variable influence on receptor affinity; however, a para-iodo substituent decreases affinity only modestly. Binding studies suggest the activity of the bulky, lipophilic para-iodo derivative may derive from interaction with an additional hydrophobic binding region (178).

Conformational studies utilizing x-ray diffractometric methods (180) or computer-assisted molecular modeling systems (181) suggest that the phenyl rings of QNB are orthogonal. The isolipophilic (178) xanthene analog [71] (QNX) of [68], in which both of the pheny rings must reside in nearly the same plane, is 25-fold less effective in mAChR binding studies. This indicates that the planar conformation is not favored by cardiac mAChR (178,182).

Replacement of one of the phenyl rings of **QNB** with the small and polar propargyl group [72a] leads to nearly a 175-fold decrease in binding. Conversely, in a receptor binding study, **QNB** and its cyclohexyl [72b], cyclopentyl [72c], and n-butyl [72d] analogs were almost equally effective, although the affinity of the cyclopentyl analog [72c] was greatest (178).

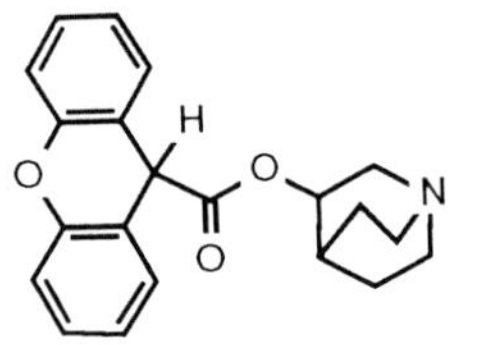

71

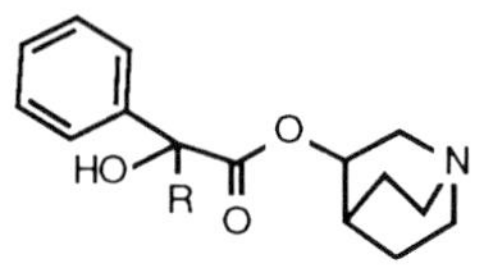

72a $R = CH_2C{\equiv}CH$; b. $R = c\text{-}C_6H_{11}$

c. $R = c\text{-}C_5H_9$; d. $R = n\text{-}C_4H_9$

Molecular modification of cholinergic agonists has likewise led to an abundance of AChR antagonists. In general, the principle employed here is to substitute the methyl portion of the acetyl group with large bulky substituents. It seems clear that among this kind of AChR antagonist substitution of the methyl group with cyclic systems by virtue of van der Waals interactions, π-π stacking, or hydrophobic bonding, combined with bonding of the amine (or ammonium) group, produces a strong drug-receptor complex that effectively blocks the approach of ACh to the receptor (65). In discussing modification of ACh, structural features are referred to as indicated in Figure 1.

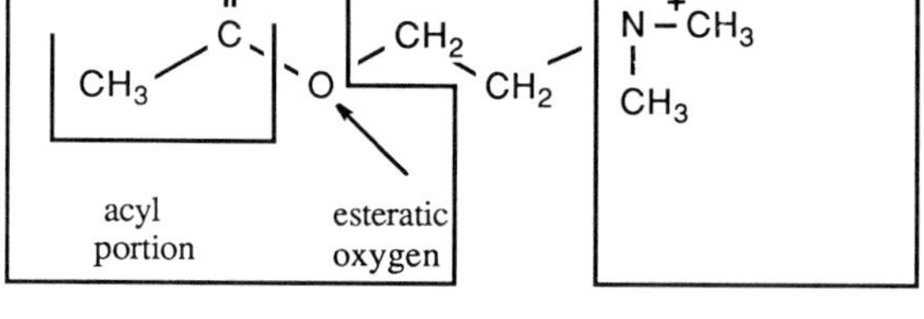

Figure 1 Terminology for portions of the acetylcholine (ACh) molecule.

The cationic head is an essential group in anticholinergic agents, as was noted for the related agonistic compounds. It is reasonable to assume that this part of the molecule is associated with a negatively charged field on the mAChR. It seems probable that the cationic head binds first with the receptor and that weaker physical and/or chemical interactions with the cyclic group(s) follow. The stability of the drug-receptor complex depends on the basicity of the amino function and steric factors. Maximal anticholinergic activity results when the methyl groups of β-dimethylaminoethyl benzilate methochloride are replaced by ethyl groups, but larger groups decrease or abolish activity (183). This suggests that in AChR antagonists the cationic portion should contain groups large enough to prevent close contact with the receptor and yet enable neutralization of the charge (184,185). The electrostatic forces in blocking agents may aid in orienting the molecule toward the receptor, and the nature of the substituents holds the cationic head in place. A methyl group is generally, although not always, optimum in quaternary derivatives (186,187). Apparently the substituents on the nitrogen atom actively participate in the receptor interaction. Thus, the carbon isostere 3,3-dimethylbutyl benzilate, $Ph_2C(OH)CO_2(CH_2)_2C(CH_3)_3$, like its ammonium counterpart retains anticholinergic activity (188).

Introduction of cyclic groups into the acyl portion of ACh reverses activity from agonist to antagonist. A phenyl group is the most common (159), but cyclohexyl, cyclopentyl, and related unsaturated groups, such as cyclohexenyl, cyclopentenyl, and even aromatic heterocycles, including the 2- and 3-thienyl groups, are also effective. In general, the most potent anticholinergics contain two cyclic substituents (189). Although many anticholinergic drugs contain only one cyclic system on the acyl moiety, in general this carbon also bears an aliphatic group, or more advantageously, a hydroxyl or hydroxymethyl group, as in atropine [58] and homatropine [59]. In fact, replacement of the acyl methyl group of many cholinergics with a dicyclic methanolic group, as illustrated in Table 2 [73-76], represents a general method for converting cholinergics into anticholinergics (64). Some of the generalities of cyclic substitution are illustrated in Table 3 (189-191).

Generalizations concerning the contributions of cyclic systems to anticholinergic activity are not reliable. Apparently, the effectiveness of these groups is dependent upon the overall molecular structure. In general, a flat cyclic group seems beneficial, although not essential (see Table 3), and as noted earlier it is advantageous if the two cyclic groups are not in the same plane. Thus, QNX [71] is only 1/25 as effective as QNB [68] in binding studies for M_2-AChR, although the two compounds are equipotent at the M_1 subclass (178). Likewise, the fluorene derivative [85] is a less potent anticholinergic than its diphenyl counterpart (189). It seems important that the substitutent be cyclic; the n-hexyl derivative [86] is ineffective, whereas its c-hexyl counterpart [83] is an effective anticholinergic substance (185).

Table 2 Anticholinergic Drugs Derived from Cholinergic Precursors

Cholinergic	Anticholinergic
$CH_3COOCH_2CH_2N(CH_3)_3^+$ 1	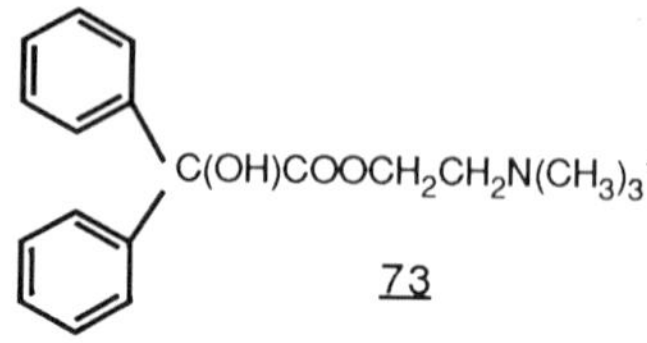 73
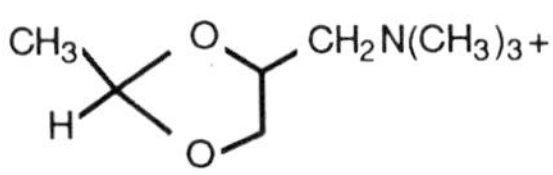 48	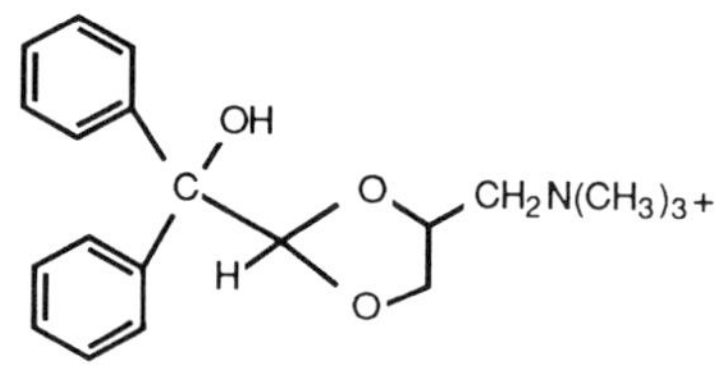 74
	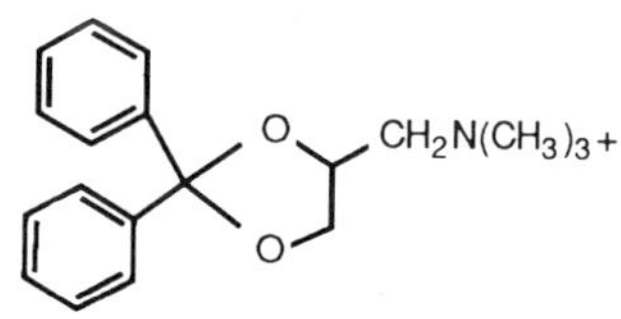 75
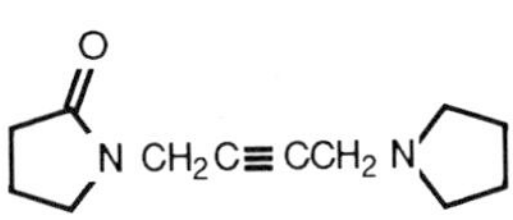 53a	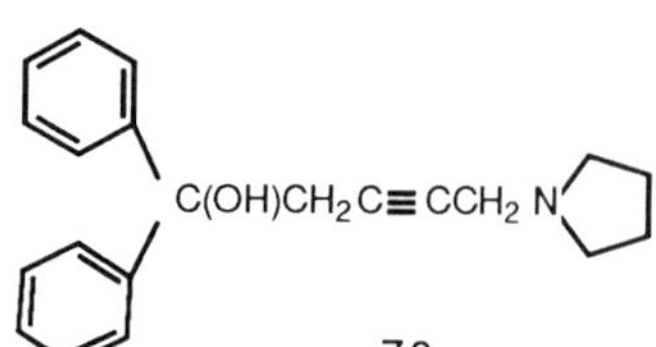 76

Source: From Reference 64.

Table 3 Effect of Cyclic Groups on Anticholinergic Activity (Rabbit Intestine)

$RCOOCH_2CH_2N(C_2H_5)_2$				
Group A, R =		Group B, R =		Activity ratio B/A
$PhCH_2$	[77]	Ph_2CH	[78]	6.7
$PhCH_2$	[77]	Ph_3C	[79]	0.7
CH_2 (OH)	[80]	PhCH(OH)	[81]	22.3
PhCH(OH)	[81]	$Ph_2C(OH)$	[82]	114.0
Ph_2CH	[78]	$PhC(c{-}C_6H_{11})(OH)$	[83]	3.3
Ph_2CH	[78]	$(c{-}C_6H_{11})_2CH$	[84]	10.0

Source: From References 189-191.

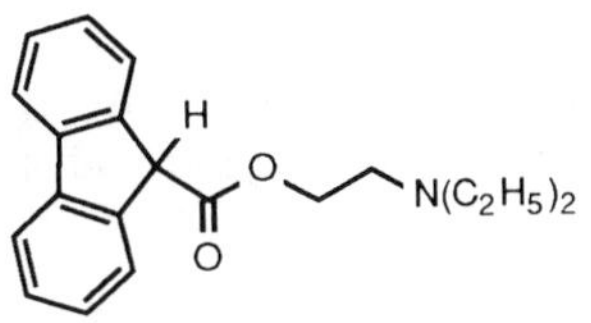

85

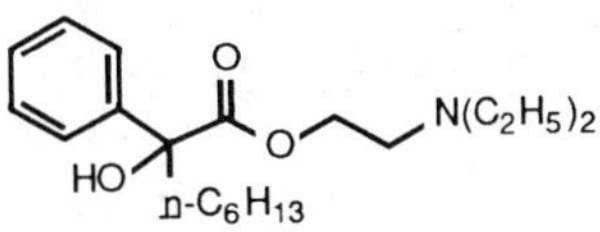

86

The nature of the alkyl chain (length, form, lateral branching, heteroatoms, and so on) connecting the cationic head and the ester end (Figure 1) of anticholinergics influences activity very significantly. The importance of the ester grouping, originally thought to be highly significant, is now in question. This is strikingly illustrated by the equal potency of [78] and its carbon isostere, $Ph_2CH(CH_2)_4N(C_2H_5)_2$ (159). Longer alkylene bridges decrease or abolish anticholinergic activity (192,193). This has led to the "rule of five": for anticholinergic activity the most important factor is that there be a five-atom separation between the bulky cyclic residue(s) and the amine terminus (98). In a series of anticholinergics in which the ester group is replaced by a carbon chain, such as $Ph_2C(OH)CH_2CH_2N(C_2H_5)_2$, for maximum activity there are three atoms in the main chain (159). In distinction to these observations, studies of a series of tetramines in atrial and ileum tissue indicates methoctramine [87a] to be the most selective M_2-AChR antagonist, whereas medotromine [87b] is the most potent (194,195).

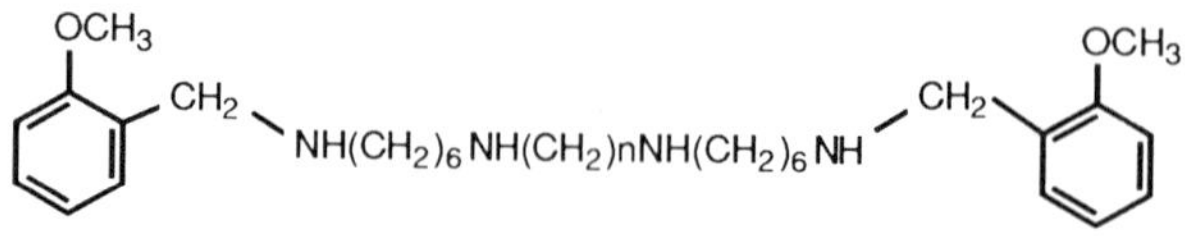

87a, n=8; b, n=12

ACh and related esters can exist in two conformational extremes: "extended" (Figure 2) and "quasi-ring" conformations (Figure 3).

The quasi-ring conformation of ACh closely simulates the structure of muscarine (196). Yet, if this were the only conformation of ACh to interact with mAChR, how could the potent muscarinic activity of the (+)-(E)-(1S,2S) enantiomer of ACTM [34] be rationalized? In this rigid system, which mimics the extended conformation of ACh, the distance between the cationic head and the ester is 5-7 Å. Thus, it appears that both the quasi ring and extended conformations of ACh antagonists interact at the same mAChR. At nAChR the same possibility, that the quasi-ring and extended conformatons of ACh may interact, seems plausible.

There is relatively little information about the influence of branching of the main chain on anticholinergic activity. Introduction of a methyl group α to the esteratic oxygen generally decreases potency (197). The presence of a hydroxyl group in anticholinergic compounds appears quite important, and most advantageously this group is α to the carboxyl (198,199). Thus, atropine [58] is about an order of magnitude more potent than homatropine [60].

As in the case of AChR agonists, a high degree of stereoselectivity is also observed for antagonists (200). The stereoselectivity of atropine, the racemic form of hyoscyamine, has long been recognized (201). Such pharmacological differences in enantiomers are particularly important because these antipodes have identical physical properties, and thus, barring differences in their metabolism, deviations in their pharmacological activity may reasonably be attributed to receptor-related events. In all examples of enantiomeric pairs in which the absolute configuration at an asymmetric benzylic carbon has been established, the relative spatial orientation of groups in the active enantiomer is identical, although the R and S nomenclature, which is a convention and defines only the

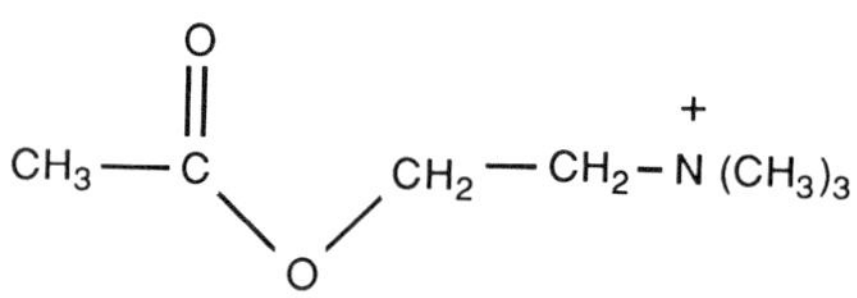

Figure 2 Extended conformation of ACh.

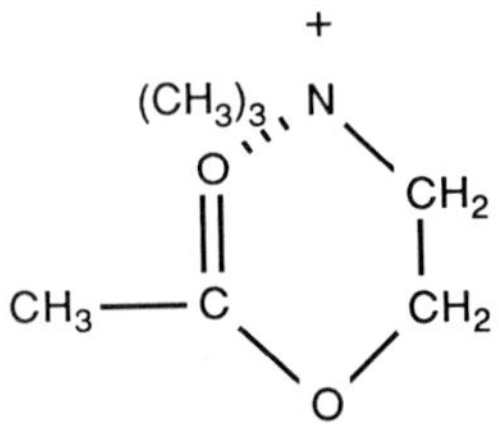

Figure 3 Quasi-ring conformation of ACh.

disposition of rule-defined groups (202), varies. In most cases in which the asymmetric center is not located at the benzylic position, enantiomeric potency ratios are relatively small (64,200). Some examples that demonstrate these concepts are illustrated by [88-96]. For more complete information it is suggested that earlier reviews (References 64 and 200, for example) be consulted.

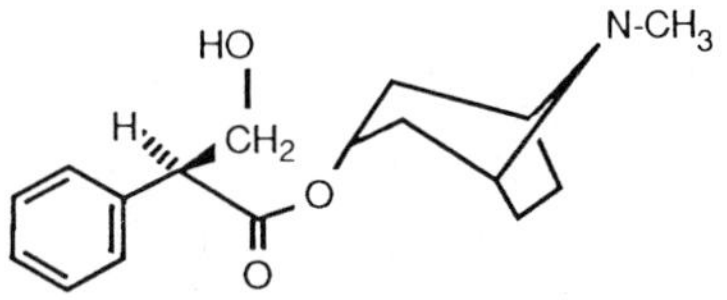

88 [(-)-S]

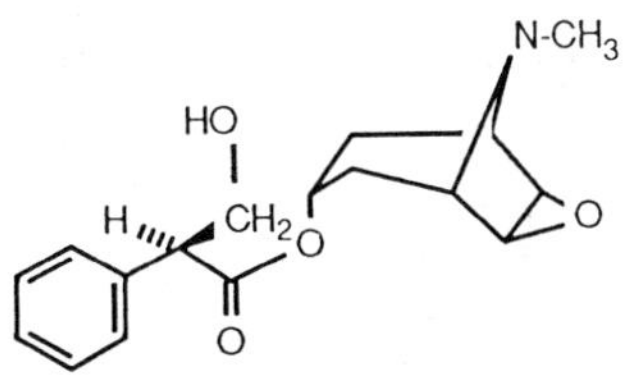

89 [(-)-S]

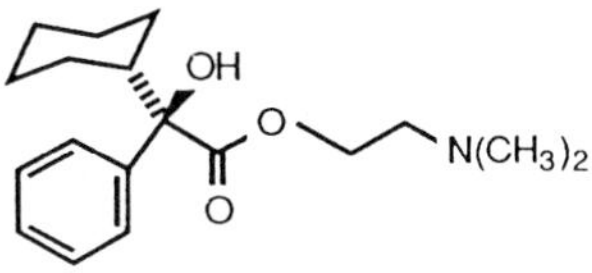

90 [(-)-R]

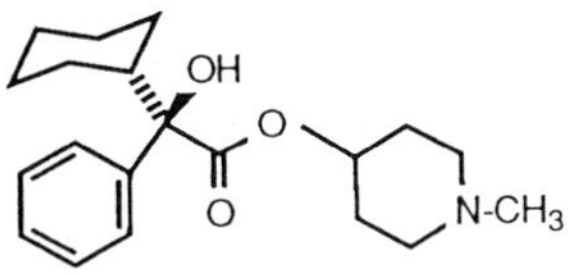

91 [(-)-R]

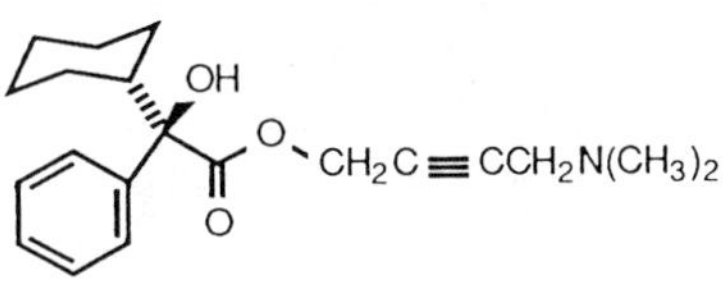

92 [(-)-R]

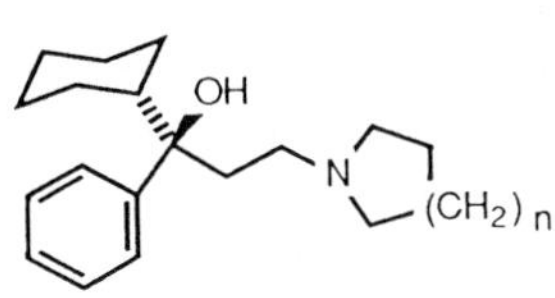

93, n = 1 (R)

94, n = 2 (R)

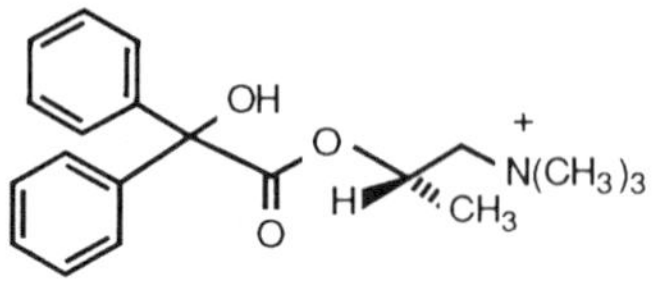

95 [(-)-S]

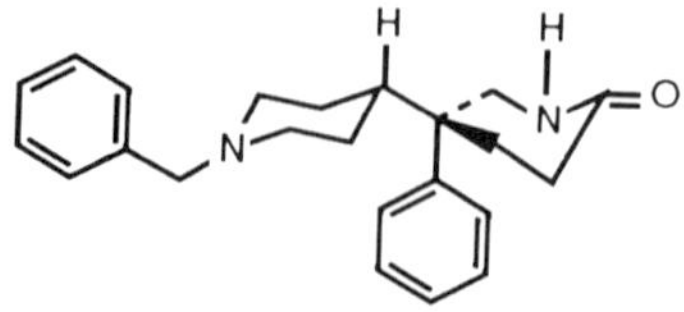

96 [(+)-S]

In a test for mydriasis production in mice, the S enantiomer of [88] is 200 times more potent than its R antipode; that is, the isomeric ratio is 200 (203). For hyosine [89] in a variety of tests the S/R potency ratio varies from 5 to 76 (204). In [90] the R/S potency ratio (note the change in configurational notation) ranges from >22 to 123 (205). For the 4-piperidinol [91] the R/S isomeric ratio varies from 16 to 272 (205). In the case of the acetylenic derivative stereoselectivity is not great; the R/S potency ratio is >3-10; however, its methyl quaternary salt shows greater enantioselectivity, >40 (206). The more potent enantiomer of procyclidine [93] has the absolute configuration S; its S/R isomeric ratio ranges from >18 (207) to 375 (208). Its piperidine homolog benzhexol [94] also demonstrates enantioselectivity, ranging from 4.8 (207) to 160 (208). The anticholinergic [95] is illustrative of an agent in which the asymmetric center is not in the benzylic position; it shows very little, if any, enantioselectivity (209). Another long-acting and potent anticholinergic is benzetimide [96] (dexetimide) (210). Structurally it seems diverse from the typical anticholinergics; however, certain structural similarities have been cited (211). In relaxing guinea pig ileum, this compound demonstrates remarkable enantioselectivity; the S enantiomer [96] is more than three orders of magnitude more effective than its R counterpart (210). The striking enantioselectivity of both muscarinic agonists and antagonists offers reassurance that the effects of these agents are receptor mediated.

Nicotinic Blocking Agents

The effects of ACh at the skeletal neuromuscular junction are mediated by nAChR. Drugs that block the action of ACh at nAChR induce neuromuscular blockade, which may also be the consequence of depolarization of the motor end plate. Such blockade was first noticed with curare, a poisonous extract from the bark of various trees of the *Strychnos* and *Chondodendron* families (212,213). It was a few years later that curare extracts were recognized to contain quaternary ammonium salts (214,215), and these quaternary salts, as well as other strongly cationic compounds (216), can cause this block. The ability of simple onium ions to block frog sartorius muscle is related to the charge density of the central ion (217). Thus, the electron-repelling influence of alkyl substituents Me > n – Pr > Et determines the potency order, that is,

$Me_4N^+ > (n\text{-}Pr)_4N^+ > Et_4N^+$. In addition to the charge density of onium compounds, other factors, such as secondary binding interactions and stereochemical factors, affect antagonist-receptor association. For example, in a series of alkyltrimethylammonium compounds, potency increases with increasing alkyl chain length (218). The quaternary ammonium compounds may bind to the anionic center of AChE, thus blocking the approach of ACh; that is, the compounds are AChE inhibitors. The relative potency of quaternary ammonium ions as inhibitors of AChE is Me_4N^+ (1), Et_4N^+ (5), $(n\text{-}Pr)_4\text{–}N^+$ (100), $(n\text{-}Bu)_4N^+$ (50), n-Pr$(Me)_3N^+$ (5), n-Bu$(Me)_3N^+$ (7.5), n-$C_5H_{11}(Me)_3N^+$ (10), and n-$C_7H_{15}(Me)_3$-N^+ (25) (219). In general, the greatest potency among asymmetric quaternary nitrogen compounds is observed when the larger substituent is oriented equatorially (220). Stereochemical requirements at the centers immediately adjacent to the quaternary nitrogen atom are significant but poorly understood (43), although it has been noted that interactions of agonists and antagonists can be rationalized by a model whereby the modulators interact with the receptor at the right and rear faces (221). Tetraethylammonium chloride and hexamethonium, once used in the treatment of hypertension, have been replaced by more potent and longer lasting quaternary ammonium salts, such as chlorisondamine [97] and trimethidium [98], but these are not well absorbed orally; the hindered tertiary amines mecamylamine [99], an agent that may distinguish nAChR subtypes (222), and pempidine [100] are well absorbed but are too short-acting and have side effects (223).

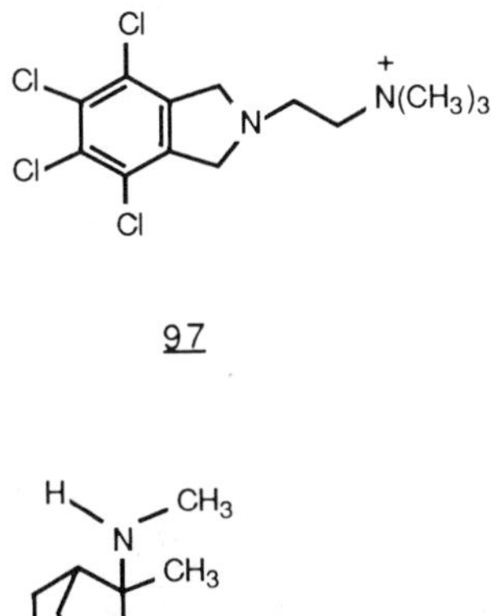

97

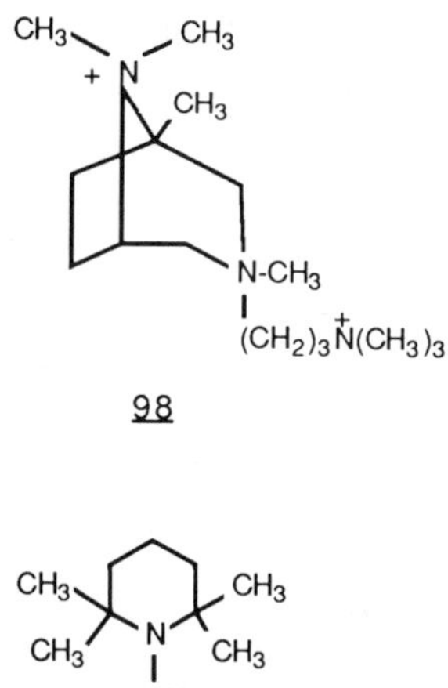

98

99

100

Bisquaternary salts are particularly potent neuromuscular blocking agents that act at nAChR. Although not itself a bisquaternary salt, (+)-tubocurarine [101] (224) led to the derivation of polymethylene bisquaternary ammonium salts with antinicotinic properties (225). The potency of these compounds varies with the length of the polymethylene chain. Separation of the ammonium heads by a pentamethylene bridge affords a compound [102a] with practically no activity; however, activity is observed with the hexamethylene-bridged compound hexamethonium [102b]. This effect gradually increases to a maximum in decamethonium [102c] and then decreases with further lengthening of the polymethylene bridge. Of prime importance to the activity of these compounds is that they be long and slender (226). This requirement indicates that the anionic receptor is recessed from the surface in some sort of a cavity. Access to the cleft for anionic and hydrophobic binding is essential for induction of a typical agonist response as a consequence of conformational changes in the receptor protein, ion channeling, and depolarization. Such factors as chain branching and increasing the N-alkyl group dimensions prohibit entry into the cleft to prevent ACh from reaching the receptors, and they therefore act as competitive antagonists (227). The function of the second quaternary center is likely to promote antagonist-receptor association at a subsidiary anionic site that forms an integral part of the receptor. This is in contrast to a one-point attachment theory (228) in which the receptor is suggested to be shielded against agonist approach by the bulk of the antagonist but is supported by the properties of monoquaternary steroids. These compounds are very weak neuromuscular blocking agents in contrast with the highly potent bisquaternary steroid pancuronium [103] (229).

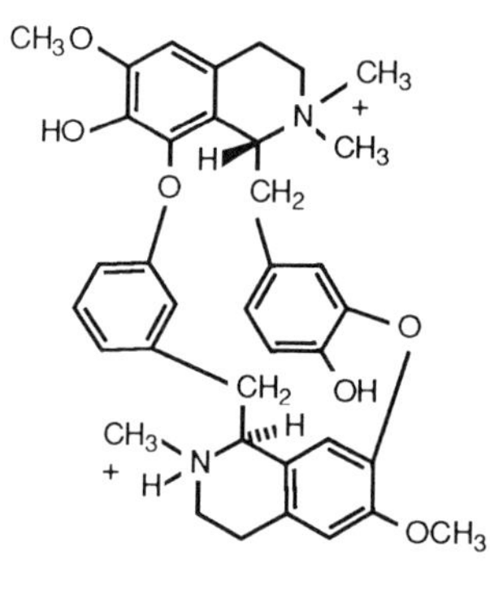

101

$(CH_3)_3\overset{+}{N}(CH_2)_n\overset{+}{N}(CH_3)_3$

102a, n = 5; b, n = 6; c, n = 10

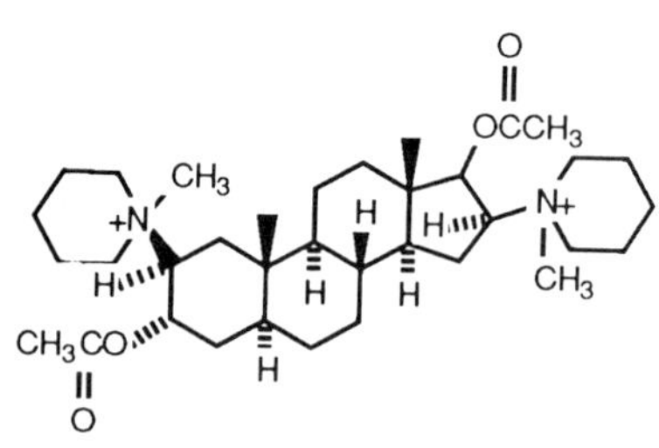

103

NATURE OF CHOLINERGIC RECEPTORS

Analysis of SAR of ACh agonists and antagonists, especially stereostructure-activity relationships, should provide information about the cholinergic receptors. A major difference between m- and nAChR agonists is in their stereochemical requirements. Thus, the natural (–)-(2S)-nicotine [18] is only slightly more potent than its (+)-(2R) antipode depending, of course, on the test system employed (230). Similarly, acetyl-α-methylcholine [33b], which is predominantly nicotine-like, shows little stereoselectivity; the R and S enantiomers are nearly equipotent (54). Both the E and Z isomers of the sulfonium compound [38b] are active at nAChR, but only the E isomer is effective at mAChR. Likewise, both the endo and exo isomers of the quaternary tropyl acetate are very potent nicotinic agents with relatively little muscarinic activity.

Conversely, many mAChR agonists uniformly display a high degree of stereoselectivity (200). Thus, the muscarinic effects of bethanechol [20] and carbachol [291] reside almost entirely in the enantiomers with the S stereochemistry. The potent muscarinic cyclopropyl analog of ACh, ACTM [34] also shows great stereoselectivity. Many other examples could be cited, including muscarine [17] itself and the "superpotent" dioxolane [48a]. Aceclidine [42a] is another example of a potent and selective muscarinic agent that makes specific stereochemical demands. In many of these compounds, equivalent centers in the stereoselective muscarinic agents can be superimposed (200,231).

Considering the high stereoselectivity of muscarinic agonists compared with the relatively nonselective behavior of nicotinic agonists, it is difficult to totally accept the hypothesis of Clothia and Pauling (232-234) that the conformation of ACh at both receptors is the same but that opposite sides of the neurotransmitter react with the two receptor subtypes. It is suggested that the "methyl side" activates mAChR whereas the "carbonyl side" activates nAChR. X-ray diffraction data and theoretical conformational analysis of muscarinic agonists (235-237) support only the first part of the hypothesis, that is, that the "methyl side" (Figure 4) of ACh will bind preferentially to mAChR. There is no proof

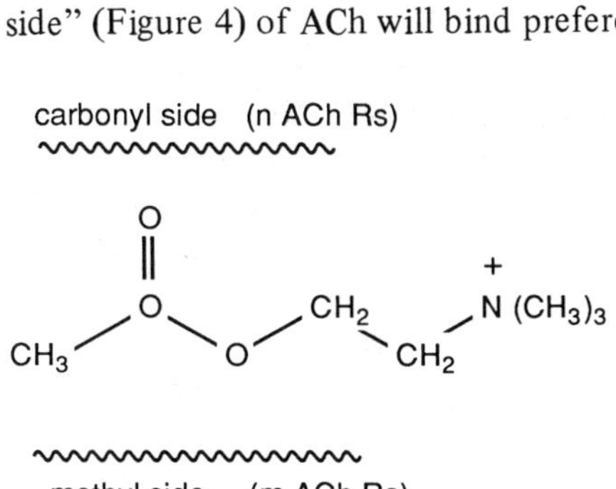

Figure 4 Proposed binding of ACh to m- and nAChR. (From Reference 234.)

for the suggestion that the "carbonyl side" will favor the nAChR. This has led to disfavor of the postulate. The lack of significant nicotinic properties in almost all rigid congeners of ACh has impeded definition of specific geometry with respect to nicotinic effects.

Muscarinic Receptors

The interaction of cholinergic agonists and antagonists with mAChR has been the subject of intense study. A major objective of this research has been to determine both the conformation of these molecules that interacts with the receptor and a definition of the nature and location of the critical binding sites. In addition to traditional SAR studies (54), a variety of techniques, such as radioligand labeling (238), nuclear magnetic resonance (NMR) spectroscopy (239,240), x-ray diffraction (241), and molecular orbital calculations (242,243), have been utilized in an effort to determine the receptor-bound conformation of muscarinic agonists and antagonists. Despite these intensive studies, the precise conformation of the flexibile mAChR ligands remains uncertain, although many models of the possible nature and location of critical binding sites have been proposed (54,238,244). A recently suggested model (245,246) is based on x-ray diffractometric data in combination with molecular orbital calculations. Several assumptions are made in this model. Initially, it is assumed that the cationic head of the agonist interacts directly with an anionic receptor site, most probably a carboxylate anion. The receptor oxygen probably lies on the threefold axis of the trimethylammonium group or is nearly colinear with the N–H bond, in the case of a protonated head to form a linear N–H–––O bond. The distance between the nitrogen of the onium head and the receptor oxygen, represented by point P in Figure 5, is calculated as 3 Å. Second, a second negative electrostatic potential region of the esteratic or ether oxygen that can bind to another site is assumed. This site is a positive residue on the receptor that may act as a hydrogen bond donor, as indicated in Figure 5; it is symbolized by the point Q. It is further assumed that as point P and Q are receptor sites and agonist binding is of high affinity and stereoselective, the distance between the two is invariant, with a maximum deviation of no more than 0.3 Å. Also, the absolute location of P and Q, the distance between binding sites, and the "interaction dihedral angle" between the planes NPO and POQ, ∢PNOQ (see Figure 5), should be narrowly defined. For a group of muscarinic agents this was calculated to vary between 100.5 and 117.0°. Finally, the model emphasizes the importance of a minimum conformational energy for known potent muscarinic agonists. This is estimated to be about 3-4 kcal/mol. The distance between point P and the acetyl methyl carbon (point C) is also important.

An appealing feature of this model is that it rationalizes the stereoselectivity of muscarinic agonists, such as (+)-(2S,3R,5S)-muscarine [17], (+)-(Z)-(2S,4R)-2-methyl-4-trimethylammoniummethyl-1,3-dioxolane [48a], (+)-ACTM (34),

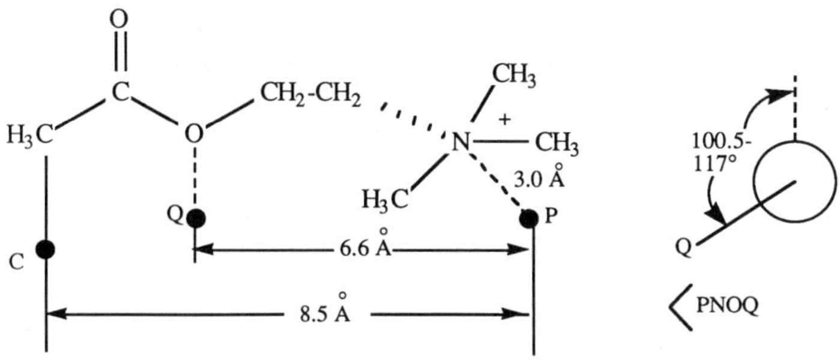

Figure 5 Proposed interaction of the nitrogen of acetylcholine with the carboxylate anion (P) of mAChR and of the esteratic oxygen of ACh with the receptor's electrophilic group Q, for example, a hydrogen bonding proton. On the right side of the figure the angle formed by the planes NPO and POQ, ∢PNOQ, is depicted in a Newman projection. (From References 245 and 246.)

(+)-(3S)-3-acetoxyquinuclidine [42a], and (+)-(S)-acetyl-β-methylcholine [33a]. Inactive isomers of these muscarinic agonists are unable to attain a geometry acceptable for interaction with the muscarinic pharmacophore unless they are forced into a conformation with a minimum conformational energy considerably greater than 3-4 kcal/mol. Another feature of the model is that it addresses the difficultly rationalized problem of the relationship between chain length and agonist potency. The results indicate that for high potency the distance P-C should be about 8.5 Å.

The ACh-like cholinergics and atropinelike anticholinergics, which contain similar pharmacophoric groups, have been assumed to bind with similar sites on mAChR. It has been suggested that accessory groups, primarily cyclic or hydroxyl moieties on the acetyl methyl group of ACh, provide additional contact(s) with the mAChR by chemical and physical interactions to strengthen overall binding and prevent access of ACh or other cholinergic agonists to the receptor (247). Although progress has been made in understanding these interactions, many aspects of anticholinergic drug-receptor associations are yet to be established. Considerable data and speculation on this topic have been presented (159,197,248,249). Clearly, important SAR, and especially differences between agonists and antagonists, must be considered in trying to explain the association of antimuscarinic drugs with mAChR, or possibly their subpopulations.

Although the cationic head is required for all mAChR binding agents, agonist and antagonist structural requirements are different. Whereas the small methyl group on the nitrogen is important for agonist activity, inhibitory actions result when larger groups are placed on the cationic portion. Perhaps they prevent close contact with the receptor and appropriate charge neutralization

(250). The cationic portions of blocking agents provide electrostatic forces that may orient the antagonist toward the receptor and hold other binding groups in place. The criticality of substitution of the acetyl methyl group has already been cited. The ester group, or even an ether oxygen, is not required for anticholinergic activity. The hydroxyl group in atropinelike anticholinergics may hydrogen bond with a site on mAChR that is rich in electrons. Also, the contribution of the hydroxyl group to the free energy of adsorption is about 2 kcal/mol. Perhaps the single most important structural feature in providing information about the mode of interaction of anticholinergics with mAChR is the absolute stereochemistry of the substituted acetoxy methyl group of these agents. In the anticholinergics derived from ACh, enantiomeric selectivity is marked and invariably demonstrates a consistent steric relationship of similar substituents (250,251). The classic view of binding of quaternary lachesine to the muscarinic receptor is shown in Figure 6, although many other possibilities have been advanced (250).

A recent study has employed a group of conformationally restricted analogs of benactyzine to determine the orientation of the side chain during receptor interaction (252,253). Results of this study are essentially in agreement with the model presented earlier. That is, as illustrated in Figure 7, rigid congeners simulating "conformation 1" lack anticholinergic activity, whereas those that resemble "conformation 2" have considerable activity.

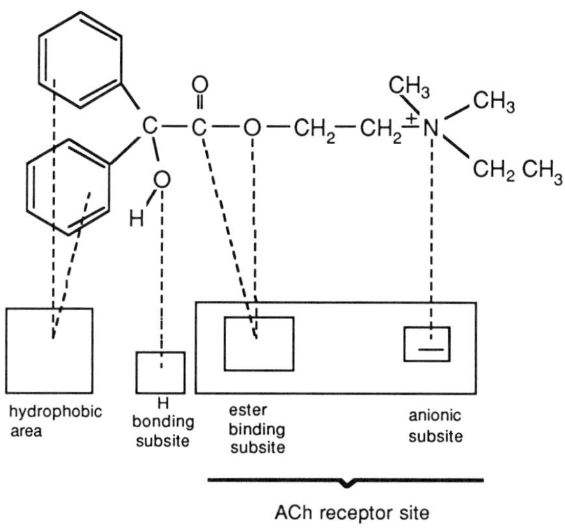

Figure 6 Classic depiction of binding of lachesine to the muscarinic receptor. (From Reference 250.)

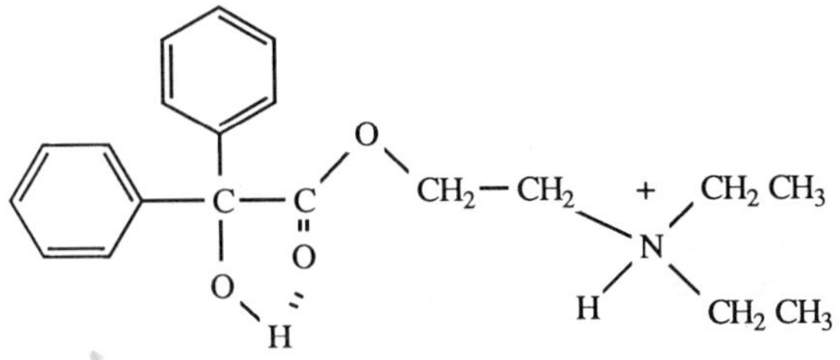

Conformation 1

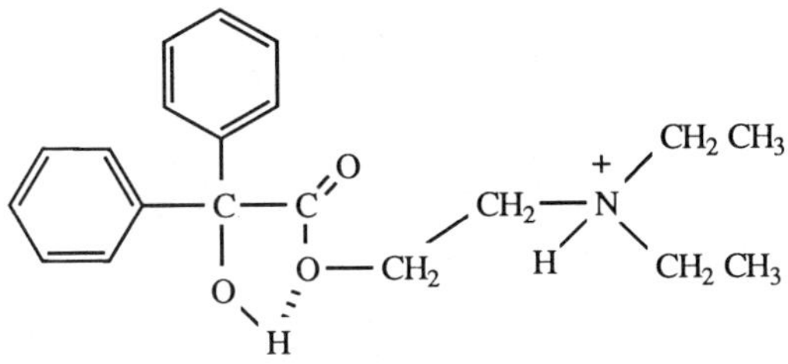

Conformation 2

Figure 7 Conformations 1 and 2 of benactyzine. (From Reference 252.)

Perhaps the anticholinergics interact with AChR in a conformation similar to that depicted for benactyzine in Figure 7, with the hydroxyl group bonding to a complementary nucleophilic site on the receptor as has been suggested earlier. If such hydrogen bonding is indeed important for receptor interaction, the SAR for anticholinergic tropic acid esters may be anticipated to differ from those of benzilic ester anticholinergics, such as benactyzine. This is because the cyclic groups or substituents on the benzylic carbon in these two classes of anticholinergics are displaced by one methylene unit. Clearly, the tropic acid esters are accommodated by the receptor, most likely in a manner that involves supplemental binding of the phenyl group with a complementarily located accessory binding site on the receptor. This notion is substantiated by the potent anticholinergic activity of atropine compared with that of its desphenyl counterpart. In addition to the phenyl group, an additionaal relatively small group, such as methyl, is tolerated by the mAChR. This is illustrated by the potent anticholinergic activity of QNA [104]. The R,R diastereomer of QNA has anticholinergic activity approximately equal to that of atropine (253). Conversely, larger groups, such as an α-phenyl substituent in tropic acid ester anticholinergics, unlike in the benzilic acid series, apparently are not advantageously oriented for binding with mAChR. Thus, as noted earlier, a second group on the benzylic carbon of atropine, as illustrated by [105], lowers anticholinergic activity (159). Obviously, the information available is insufficient to define the precise location or rotameric orientation of the cyclic groups on the benzylic carbon of these kinds of anticholinergic agents that is best suited for binding with a complementary accessory site on the mAChR. Apparently, some conformational liberties are permissible; however, advancement of models that define the location of such secondary binding sites is not possible at present.

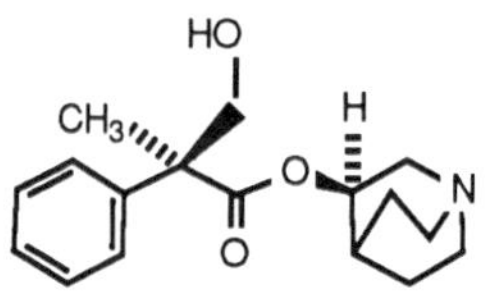

104

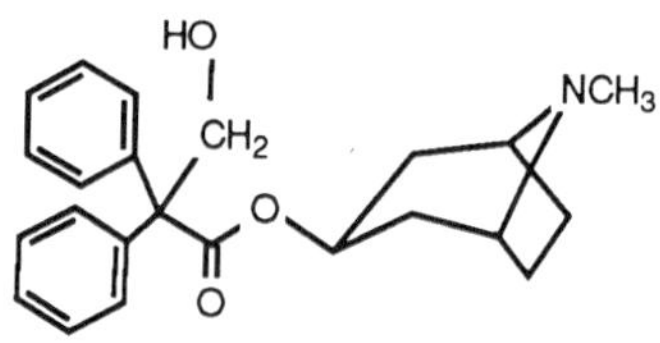

105

Advances in the study of mAChR, primarily as a consequence of their relatively low concentrations in various tissues , have been relatively slow. Earlier radioligand binding studies suggested that mAChR are discrete membrane components that recognize agonists and antagonists (254–256). mAChR have been successfully solubilized with a variety of detergents (257). More recently, these receptors have also been purified and analyzed (258).

Nicotinic Receptors

With nAChR the situation has been almost completely reversed from that of the mAChR. SAR studies (259-261) provide relatively little information other than that compounds with onium heads separated by six (hexamethonium) to ten (dexamethonium) methylene units are effective antagonists and stereochemical considerations are of much less significance than for mAChR. The nature of nAChR has been established mainly by study of nerve-muscle preparations and by studies on the electric tissue of electric eels, such as *Torpedo marmorata, Torpedo californica*, and *Electrophorus electricus*. As a result the nAChR is among the best known of all receptors. It is a pentamer having two α subunits and one each of β, γ, and δ (262), and it has an apparent molecular weight of 320,000 daltons (263). The primary structure of the five subunits of *T. californica* has been determined by cloning and sequencing of complementary DNA clones (262,264,265). A combination of molecular genetic, electrophysiological, and structural techniques is beginning to present a picture of the nAChR channel that is more refined and exact than could have been expected only a few years ago (262). The nAChR apparently has a great deal of structural similarity to AChE (266). The anionic center on both the receptor and enzyme has a high proportion of glutamate and aspartate residues (267), an observation of particular significance because the catalytic surface of AChE with two anionic centers is relatively well understood (54). The presence in close proximity to the receptor's anionic site(s) of an easily reducible disulfide group has also been discerned (268). Receptor activity is apparently associated with conformational changes in the receptor protein (269). Binding experiments with purified nAChR indicate that agonists promote a conformational alteration in the receptor, thus changing the low-affinity resting state to a high-affinity state (270).

Structure elucidation of α-bungarotoxin (α-BuTX) and various snake venoms has aided the understanding of the mechanism of action of the neuromuscular blocking antagonists of nAChR. These natural toxins possess at least one, and frequently two or more, sets of adjacent strongly basic amino acid units that are relevant to blocking potency (271). Each pair of adjacent basic amino acids can bridge across receptor subunits, in the same fashion as bisquaternary ammonium antagonists, to block the responses to ACh. A structural model for the ligand-receptor binding sites at the nAChR has been advanced (272).

CHOLINERGIC RECEPTOR BINDING STUDIES

Binding studies have begun to address the fundamental question of whether cholinergic agonists and antagonists occupy identical, similar, or different binding sites. Substantial differences in the stereochemical requirements for cholinergic agonists and antagonists are frequently observed (230,231). Thus, binding studies represent a more fundamental approach to the measurement of drug-receptor interaction; however, such studies with agonists may reflect only affinity and give no indication of intrinsic activity. Additionally, it must be emphasized that agonist and antagonist binding sites are at least partially distinct (231).

Paton and Rang (273) first studied the muscarinic receptor binding of [^{3}H]atropine [48] to guinea pig ileum. Later, many other research groups studied the interaction of brain preparations with cholinergic agonists or alkylating agents derived from muscarinic agonists as well as a number of atropine-like compounds. Snyder and his colleagues (274) utilized the potent reversible muscarinic receptor antagonist [^{3}H]QNB [68] to study mAChR in the CNS and periphery. They found that the regional distribution of this radioligand essentially paralleled that of other cholinergic markers, suggesting that most cholinergic synapses are muscarinic, particularly in the CNS. The inhibition of [^{3}H]QNB binding to homogenates of brain and guinea pig ileum by various muscarinic drugs correlated with their pharmacological potencies; however, nicotinic agents and noncholinergic drugs had almost no affinity (274). Subse-

Table 4 Some Ligands for Cholinergic Receptors

Compound	Receptor(s) (action)[a]	References
ACh [1]	N, M (Ag)	61, 293, 294
Muscarine [17]	M_1, M_2 (Ag)	58
Nicotine [18]	N_G, N_S (Ag)	61, 289–295
Carbachol (CCh) [291]	N, M (Ag)	36, 281
Hemicholinium-3 [3]	–	296, 297
QNB [68]	M_1, M_2 (An)	17, 285, 294, 295
Atropine [58]	M_1, M_2 (An)	258, 273, 285
N-Methylatropinium	M_1, M_2 (An)	285
N-Methylscopolamine (NMS) [59]	M_1, M_2 (An)	285
N-Propylcholine benzilate (PrBCh)	M_1, M_2 (An)	285
N-Methyl-4-piperidyl benzilate (4-NMPB)	M_1, M_2 (An)	36
Dexetimide [96]	M_1, M_2 (An)	258, 296, 297
Propylbenzilyl choline mustard (PBrCM)	M_1, M_2 (An)[c]	295
α-Bungarotoxin (α-BuTX)	N_G, N_S (An)[c]	61, 288-292, 295
Tubocurarine [101]	N_G, N_S (An)[c]	290, 292
Dexamethonium [102c]	N_G, N_S (An)	298
4-(*N*-Maleimido)benzyltrimethyl-ammonium (MBTA)	N_G, N_S (An)[c]	290
Pirenzepine (PZ) [21]	M_1 (An)	182, 282, 283
QNX [71]	M_1 (An)	284
McN-A-343 [19]	M_1 (Ag)	70, 279
Oxotremorine [53a]	M_2 (Ag)	68, 287, 296, 297
Methylfurtrethonium [47]	M_1, M_2 (Ag)	69
cis-Methyldioxolane (CD) [48a]	M_1, M_2 (Ag)	277
Gallamine [22]	N, M_2 (An)	279
4-DAMP [23]	M_2 (An)	278
Secoverine	M_2 (An)	279
Dihydro-β-erthyroidine (DBE)	N	304

[a]Ag = agonist, An = antagonist; reversible binding unless indicated otherwise.
[b]Inhibitor of high-affinity, Na^+-dependent Ch uptake.
[c]Irreversible inhibitor.

quently, many studies have revealed the presence of high-affinity binding sites for radioactively labeled muscarinic receptor antagonists in subcellular fractions from brain, smooth muscle, and intact tissues (238,275). Some of the radioligands and the receptors with which they interact are listed in Table 4 (276-298,304), although it should be emphasized that this compilation is only illustrative, not exhaustive.

Radioligands have also been employed to distinguish subtypes of muscarinic receptors. The concept of mAChR heterogeneity was first verified by measurements of agonist binding (276) using [^{3}H] ACh, [^{3}H] methylfurtethonium (69), and [^{3}H] cis-methyldioxolane (277). Deviations from simple mass action behavior seen in agonist-[^{3}H] antagonist competition experiments occurred as a result of receptor heterogeneity. The modulatory effects of guanine nucleotides, sulfhydryl reagents, and ions are useful in studying the significance of muscarinic receptor heterogeneity (276). More recent experiments with the muscarinic agonists McN-A-343 (279-280) and carbachol (281) and the nonclassic muscarinic antagonists, such as the M_2-AChR-selective 4-diphenylacetoxy-*N*-methylpiperidine (4-DAMP) [23] (278), and secoverine (279), as well as the M_2, nAChR antagonist gallamine [22] and the M_1-AChR-selective antagonist pirenzepine [21] (279), indicate structural differences in the ligand recognition sites from different tissues (282).

The use of selective tritiated ligands in quantitative autoradiography (182) and immunohistochemistry, for example methods using monoclonal antibodies (295), permit localization of mAChR. For example, [^{3}H] hemicholinium-3 [3] and [^{3}H] pirenzepine [21] in rat brain slices label high-affinity, Na^+-dependent Ch uptake sites and postsynaptic M_1-AChR, respectively. The absence of M_1-AChR binding sites in some anatomic regions rich in Ch uptake sites suggests that the postsynaptic AChR in these locations are likely to be nicotinic or of a different muscarinic class. [^{3}H] Oxotremorine [53a] in turn selectively labels the M_2-AChR (296,297). Thus, M_1-AChR appear to be densely located in cortex, hippocampus, and striatum, whereas M_2-AChR are located in cholinergic nerves, brain stem, heart, and peripheral tissues. It seems highly probable that ligands identified in the future will be useful in the differentiation and determination of the locality of various other subpopulations of cholinergic receptors.

The properties of both brain (249-258) and peripheral (259,260) nAChR have been studied by several groups. Apparently these sites in the CNS are similar to those in muscle, and their regional distribution in the brain has been determined. α-Bungarotoxin (299), which binds to all nAChR but not to AChE (300), is especially useful for the localization of these receptors. [^{131}I] α-BuTX has been used as a marker in the isolation of nicotinic receptors, as well as for determination of their number and distribution (298-300). The sites identified by this radioiodinated ligand are found on postsynaptic membranes in the periphery (287,288).

The strong and reproducible effects of nicotine on behavioral and other physiological responses suggest an important role for central nAChR. Studies indicate that [^{3}H] nicotine is a more effective ligand than [^{125}I] α-BuTX in the brain (61). Other studies suggest the presence of presynaptic nicotinic autoreceptors in rat hippocampus (293). After repeated administration of nicotine, an increase in nAChR was noted (294,301). Two explanations have been advanced: nicotine may interfere with the mechanism of action or release of ACh, or nicotine or a metabolite may antagonize receptors in the brain (302). Clearly, however, these studies establish with certainty the presence of central nAChR.

FUTURE DIRECTIONS

Cholinergic research will likely follow that of other receptor studies (303). Recent advances in technology will continue to be used to isolate, purify, characterize, and synthesize (by recombinant DNA techniques) subpopulations of cholinergic receptors. Indeed, a great deal of progress in this vein has already been ahcieved with nAChR. The interaction of these receptors with agonists and antagonists after they have been reconstituted in a manner that simulates their natural environment will then be examined by x-ray crystallographic, NMR, and other spectroscopic techniques to discern precise binding modes and sites. With this knowledge, quantitative SAR techniques, combined with computer-assisted molecular modeling, may be employed to rationally design new, specifically acting cholinergic agents. Until these methodologies can be perfected, classical pharmacological and binding studies, coupled with traditional screening and medicinal chemical approaches, will likely continue to dominate the search for more potent, selective, and therapeutically useful cholinergic agonists and antagonists devoid of toxicity and other unwanted side effects, their primary focus, at the present time, being related to the treatment of such diseases as Alzheimer's.

REFERENCES

1. B. Katz, Nerves, Muscle and Synapse, McGraw-Hill, New York, 1966.
2. R. S. Jope, Brain Res. Rev. *1*, 313 (1979).
3. C. J. Cavallito, H. S. Yun, J. C. Smith, and F. F. Foldes, J. Med. Chem. *12*, 134 (1969).
4. R. Birks and F. C. MacIntosh, Can. J. Biochem. Physiol. *39*, 787 (1961).
5. R. Strong, P. Hicks, L. Hsu, R. T. Bartus, and S. J. Enna, Neurobiol. Aging *1*, 59 (1980).
6. C. L. Murray and H. C. Fibiger, Neuroscience *14*, 1025 (1985).
7. D. L. Price, P. J. Whitehouse, R. G. Struble, A. W. Clark, J. T. Coyle, M. R. Delony, and J. C. Hedreen, Neurosci. Comment. *1*, 84 (1982).
8. M. N. Rossor, J. F. Rehfeld, P. C. Emerson, C. Q. Mountjoy, M. Roth, and L. L. Iversen, Life Sci. *29*, 405 (1981).

9. J. I. Hubbard, Physiol. Rev. *53*, 674 (1973).
10. M. Israel, J. Gautrun, and B. Lesbats, C. R. Sceances Acad. Sci. Ser. D (Paris) *266*, 273 (1968).
11. A. W. Clark, A. Mauro, H.E. Longenecker, and W. P. Hurlburt, Nature (Lond.) *225*, 703 (1970).
12. H. A.Lester, Sci. Am. *236*, 106 (1977).
13. J. I. Hubbard and D. F. Wilson, J. Physiol. (Lond.) *228*, 307 (1973).
14. L. T. Potter, J. Physiol. (Lond.) *206*, 145 (1970).
15. S. W. Kuffler and D. Yoshikami, J. Physiol. (Lond.) *206*, 145 (1970).
16. P. W. Ramwell, J. E. Shaw, and J. Kucharski, Science *149*, 1390 (1969).
17. E. D. P. DeRobertis and H. S. Bennett, J. Biophys. Biochem. Cytol. *1*, 47 (1955).
18. B. Katz, The Release of Neural Transmitter Substances, Liverpool University Press, Liverpool, 1969.
19. J. Musick and J. I. Hubbard, Nature (Lond.) *237*, 279 (1972).
20. W. Feldberg, Physiol. Rev. *25*, 596 (1948).
21. P. Fatt and B. Katz, J. Physiol. (Lond.) *115*, 320 (1951).
22. L. Tauc, Physiol. Rev. *62*, 857 (1982).
23. E. M. Kosower and R. Werman, Nature New Biol. *233*, 121 (1971).
24. F. A. Dodge, R. Miledi, and R. Rahamimoff, J. Physiol. (Lond.) *200*, 267 (1969).
25. U. Meiri and R. Rahamimoff, J. Physiol. (Lond.) *215*, 709 (1971).
26. J. I. Hubbard, S. F. Jones, and E. M. Landau, J. Physiol. (Lond.) *194*, 355; *196*, 75 (1968).
27. A. W. Clark, W. P. Hurlburt, and A. Mauro, J. Cell Biol. *52*, 1 (1972).
28. J. Heuser and R. Miledi, Proc. R. Soc. Lond. [Biol.] *179*, 247 (1971).
29. R. W. Abers and G. J. Koval, Biochim. Biophys. Acta *60*, 359 (1962).
30. Z. L. Blioch, I. M. Glagoleva, E. A. Liberman, and V. A. Nenashev, J. Physiol. (Lond.) *199*, 11 (1968).
31. L. L. Simpson, J. Pharm. Pharmacol. *20*, 889 (1968).
32. I. M. Glagoleva, E. A. Liberman, and Z. M. Khashaev, Biofizika *15*, 76 (1970).
33. C. B. Pittinger and R. Adamson, Annu. Rev. Pharmacol. *12*, 169 (1972).
34. E. S. Vizi, Release mechanisms of acetylcholine and the role of Na+-K+-activated ATPase, in Cholinergic Mechanisms (P. G. Waser, ed.), Raven Press, New York, 1975, pp. 199-211.
35. M. K. Ticku and D. J. Triggle, Gen. Pharmacol. *7*, 133 (1976).
36. M. Sokolovsky, M. Cohen-Armon, Y. Egozi, D. Gurwitz, Y. I. Henis, Y. Kloog, E. Moseona-Amir, and G. Schrieber, Trends Pharmacol. Sci. (Suppl.), *5*, 39 (1985).
37. K. Renkawek, M. R. Murray, M. Spatz, and I. Klatzo, Exp. Neurol. *50*, 194 (1976).
38. J. K. Marquis and E. B. Fishman, Trends Pharmacol. Sci. *6*, 387 (1985).
39. R. I. Birks and F. R. MacIntosh, Br. Med. Bull. *13*, 157 (1957).
40. J. P. Long and F. W. Schueler, J. Am. Pharm. Assoc. Sci. Ed. *43*, 79 (1954).
41. F. W. Schueler, J. Pharmacol. Exp. Ther. *115*, 127 (1955).

42. F. C. MacIntosh, R. I. Birks, and P. B. Sastry, Nature (Lond.) *178*, 1181 (1956).
43. J. B. Stenlake, in Burger's Medicinal Chemistry, 4th ed., Part III (M. E. Wolff, ed.), Wiley, New York, 1981, pp. 431-463.
44. L. I. Smart, J. Med. Chem. *26*, 104 (1983).
45. R. M. Gesler, A. B. Lasher, J. O. Hoppe, and E. A. Steck, J. Pharmacol. Exp. Ther. *125*, 323 (1959).
46. A. S. V. Burgen, G. Burke, and M. Desbarats-Schonbaum, Br. J. Pharmacol. Chemother. *11*, 308 (1956).
47. J. C. Smith, C. J. Cavallito, and F. F. Foldes, Biochem. Pharmacol. *16*, 2438 (1967).
48. H. L. White and C. J. Cavallito, Biochim. Biophys. Acta *206*, 343 (1970).
49. F. F. Foldes, T. S.. Mackoj, R. D. Hunt, P. G. McNail, and P. C. Carberry, J. Am. Med. Assoc. *150*, 1559 (1952).
50. R. Fusco, G. Palazzo, S. Chiavarelli, and D. Bovet, Gazz. Chim. Ital. *79*, 836 (1949).
51. R. I. Brittain, G. P. Levy, and M. B. Tyers, Eur. J. Pharmacol. *8*, 93 (1969).
52. I. G. Marshall, Br. J. Pharmacol. *38*, 502 (1970).
53. J. G. Cannon, in Burger's Medicinal Chemistry, 4th ed., Part III (M. E. Wolff, ed.), Wiley, New York, 1981, pp. 354-360.
54. A. H. Beckett, Ann. N.Y. Acad. Sci. *144*, 675 (1967).
55. A. G. Karczmar, Anticholinesterase Agents in International Encyclopedia of Pharmacology and Therapeutics, Vol. 1 (A. G. Karczmar, ed.), Pergammon Press, Oxford, 1970, Sec. 13, p. 1.
56. J. E. Gearien, in Medicinal Chemistry, 3rd ed., Part II (A. Burger, ed.), Wiley, New York, 1970, pp. 1307-1313.
57. H. H. Dale, J. Pharmacol. Exp. Ther. *6*, 147 (1914).
58. E. Hardegger and F. Lohse, Helv. Chim. Acta *40*, 2383 (1957).
59. P. B. Bradley, Int. Rev. Neurobiol. *11*, 1 (1968).
60. A. Bebbington and R. W. Brimblecome, Adv. Drug Res. *2*, 143 (1965).
61. B. J. Morley, G. R. Farley, and E. Javel, Trends Pharmacol. Sci. *3*, 225 (1983).
62. M. J. Marks and A. C. Collins, Pharmacol. Biochem. Behav. *22*, 283 (1985).
63. N. J. M. Birdsall and E. C. Hulme, Trends Pharmacol. Sci. *3*, 459 (1983).
64. T. D. Inch and R. W. Brimblecombe, Int. Rev. Neurobiol. *16*, 67 (1974).
65. B. V. Rama Sastry, in Burger's Medicinal Chemistry, 4th ed., Part III (M. E. Wolff, ed.), Wiley, New York, 1981, pp. 361-411.
66. N. Weiner, in The Pharmacological Basis of Therapeutics, 7th ed. (A. G. Gilman, L. S. Goodman, T. W. Rall, and F. Murad, eds.), Macmillan, New York, 1985, pp. 130-144.
67. A. F. Casy, Prog. Med. Chem. *11*, 1 (1975).
68. B. I. Hirshowitz, R. Hammer, A. Giachetti, J. J. Kevins, and R. R. Levine (eds.), Symposium (various authors). Subtypes of muscarinic receptors, Trends Pharmacol. Sci. (Suppl.) *4*, 1-103 (1985).
69. N. J. M. Birdsall, A. S. V. Burgen, and E. C. Hulme, Mol. Pharmacol. *14*, 723 (1978).

70. L. K. Choo, F. Mitchelson, and Y. M. Vong, Naunyn-Schmiedebergs Arch. Pharmacol. *328*, 430 (1985).
71. B. I. Hirschowitz, R. Hammer, A. Giachetti, J. Keirns, and R. R. Levine, Trends Pharmacol. Sci. (Suppl.) Preface (1985).
72. L. T. Potter, D. D. Flynn, H. E. Hanchett, D. L. Kalinoski, J. Luber-Narod, and D. C. Mash, Trends Pharmacol. Sci. (Suppl.) *4*, 22 (1984).
73. J. N. Fain, S.-H. Lin, I. Litosch, and M. Wallace, Life Sci. *32*, 2055 (1983).
74. M. Watson, W. R. Roeske, and H. I, Yamamura, Life Sci. *31*, 2019 (1982).
75. M. Watson, H. I. Yamamura, and W. R. Roeske, Life Sci. *32*, 3001 (1983).
76. K. H. Jakobs, K. Aktories, and G. Schultz, Naunyn-Schmiedebergs Arch. Pharmacol. *310*, 113 (1979).
77. D. W. Gil and B. B. Wolfe, J. Pharmacol. Exp. Ther. *232*, 608 (1985).
78. G. Jagadeesh and P. V. Sulakhe, Eur. J. Pharmacol. *109*, 311 (1985).
79. R. B. Barlow, and M. K. Shepherd, Br. J. Pharmacol. *85*, 427 (1985).
80. R. B. Barlow, International Symposium on Muscarinic Cholinergic Mechanisms (S. Cohen, J. Finberg, and M. Sokolovsky, eds.). The Tel-Aviv University Press, Tel-Aviv, Israel, November 30 to December 4, 1986.
81. G.Lambrecht, V. Moser, J. Riotte, M. Wagner, J. Wess, G. Gmelin, R. Tacke, H. Zilch, and E. Mutschler, Program and Abstracts, International Symposium Muscarinic Cholinergic Mechanisms, (S. Cohen, J. Finberg, and M. Sokolovsky, eds.), The Tel-Aviv University Press, Tel-Aviv, Israel, November 30 to December 4, 1986, p. 25.
82. K. J. van Charldorp, D. Davidesko, H. N. Doods, A. deJonge, and P. A. van Zweiten, Program and Abstracts, International Symposium on Muscarinic Cholinergic Mechanisms, (S. Cohen, J. Finberg, and M. Sokolovsky, eds.), The Tel-Aviv University Press, Tel-Aviv, Israel, November 30 to December 4, 1986, p. 37.
83. N. J. M. Birdsall, A. S. V. Burgen, and E. C. Hulme, Mol. Pharmacol. *15*, 725 (1978).
84. N. J. M. Birdsall, E. C. Hulme, and J. M. Stockton, Trends Pharmacol. Sci. Suppl. 4 (1984).
85. S. M. Collins, Subtypes of Muscarinic Receptors, The Second International Symposium, (R. R. Levine, N. J. M. Birdsall, A. Giachetti, R. Hammer, L. L. Iversen, D. J. Jenden, and R. A. North, eds.), Elsevier Science Publishers, New York, Abstract 31, 1985.
86. N. J. M. Birdsall, E. C. Hulme, J. C. Stockton, and E.F. Wong, in Mechanism of Drug Action, Academic Press, New York, 1983, p. 27.
87. N. Weiner and P. Taylor, in The Pharmacological Basis of Therapeutics, 7th ed. (A. G. Gilman, L. S. Goodman, T. W. Rall, and F. Murad, eds.), Macmillan, New York, 1985, p. 94.
88. A. B. Ruzzuti, Am. J. Ophthamol. *63*, 484 (1967).
89. R. L. Stehle, K. I. Melville, and F. K. Oldham, J. Pharmacol. Exp. Ther. *56*, 473 (1936).
90. P. Holton and H. R. Ing, Br. J. Pharmacol. *4*, 190 (1949).
91. H. R. Ing, P. Kordik, and D. P. H. Tudor Williams, Br. J. Pharmacol. *7*, 103 (1952).

92. H. Kibinger, A. Wagner, and R. Zerban, Naunyn-Schmiedebergs Arch. Pharmacol. Exp. Pathol. *295*, 81 (1976).
93. H. L. Friedman, Postganglionic Parasympathetic Stimulants (Muscarinic Drugs), in Medicinal Research Series, Vol. 1 (A. Burger, ed.), Marcel Dekker, New York, 1967, pp. 79-131.
94. L. L. Darko, J. G. Cannon, J. P. Long, and T. F. Burks, J. Med. Chem. *8*, 841 (1965).
95. J. G. Cannon, R. V. Smith, G. A. Fisher, J. P. Long, and F. W. Benz, J. Med. Chem. *14*, 66 (1971).
96. B. Pullman, P. Courriere, and J. L. Coubeils, Mol. Pharmacol. *7*, 397 (1971).
97. I. Kimura, I. Morishima, T. Yonezawa, and M. Kimura, Chem. Pharm. Bull. (Tokyo) *22*, 429 (1974).
98. H. R. King, Science *109*, 264 (1949).
99. J. Bannister and V. P. Whittaker, Nature (Lond.) *167*, 605 (1951).
100. H. C. Chang and J. H. Gaddum, J. Physiol. (Lond.) *79*, 255 (1933).
101. R. Hunt and R. de Taveau, Br. Med. J. 1788 (1906).
102. B. W. J. Ellenbrock, R. J. F. Nivard, J. M. Van Rossum, and E. J. Ariëns, J. Pharm. Pharmacol. *17*, 393 (1965).
103. B. V. Rama Sastry, C. C. Pfeiffer, and A. Lasslo, J. Pharmacol. Exp. Ther. *130*, 346 (1960).
104. D. Bovet and F. Bovet-Nitti, Medicaments du Systeme Nerveux Végatatif, S. Karger, Basel, 1948.
105. V. P. Whittaker, Biochem. Pharmacol. *1*, 342 (1959).
106. C. Y. Chiou, J. Pharm. Sci. *64*, 469 (1975).
107. J. M. Van Rossum and E. J. Ariëns, Arch. Int. Pharmacodyn. Ther. *118*, 418 (1959).
108. R. T. Major and H. T. Bonnet, U. S. Patent 2,347,367 (1945); Chem. Abstr. *39*, 4721 (1945).
109. A. Ercoli, Ann. Chim. Appl. *25*, 263 (1935).
110. G. B. Koelle, in The Pharmacological Basis of Therapeutics (L. S. Goodman and A. Gilman, eds.), 5th ed., Macmillan, New York, 1975, pp. 468, 471.
111. R. Hunt and R. R. Renshaw, J. Pharmacol. Exp. Ther. *25*, 315 (1925).
112. R. R.Renshaw and C. Y. Hopkins, J. Am. Chem. Soc. *51*, 953 (1929).
113. V. E. Gmiro, N. B. Brovtsyna, L. N. Poznyakova, A. F. Danilov, S. A. Shelkovnikov, and N. V. Khromov-Borisov, Dokl. Akad. Nauk, SSSR *265*, 488 (1982).
114. B. K. Beznasko, V. A. Dobrina, D. V. Joffe, O. I. Kolganova, N. Y. Lukomskaya, M. Y. Milhelson, L. L. Protas, and S. A. Shelkovnitorovnikov, Farmakol. Toxsikol. *43*, 44 (1980).
115. K. A. Scott and H. G. Mautner, Biochem. Pharmacol. *13*, 907 (1964).
116. A. Blankart, in Festschrift Emil C. Barell, 1936, p. 284; Chem. Abstr. *31*, 2675 (1937).
117. A. M. Lands and C. J. Cavallito, J. Pharmacol. Exp. Ther. *110*, 369 (1954).
118. M. M.-L. Chan and J. B. Robinson, J. Med. Chem. *17*, 1057 (1974).
119. B. W. Ellenbroek and J. M. Van Rossum, Arch. Int. Pharmacodyn. Ther. *125*, 216 (1960).

120. E. E. Smissman, W. L. Nelson, J. LaPidus, and J. L. Day, J. Med. Chem. *9*, 458 (1966).
121. G. H. Cocolas, E. C. Robinson, and W. L. Dewey, J. Med. Chem. *13*, 299 1970).
122. A. Simonart, Arch. Int. Pharmacodyn. Ther. *48*, 328 (1934).
123. R. Hunt and R. de M. Taveau, J. Pharmacol. Exp. Ther. *1*, 303 (1909).
124. P. D. Armstrong, J. G. Cannon, and J. P. Long, Nature (Lond.) *220*, 65 (1968).
125. C. Y. Chiou, J. P. Long, J. G. Cannon, and P. D. Armstrong, J. Pharmacol. Exp. Ther. *166*, 243 (1969).
126. J. G. Cannon, T. Lee, V. Sanaran, and J. P. Long, J. Med. Chem. *18*, 1027 (1975).
127. J. G. Cannon, A. B. Rege, T. L. Gruen, and J. P. Long, J. Med. Chem. *15*, 71 (1972).
128. J. G. Cannon, Y. Lin, and J. P. Long, J. Med. Chem. *16*, 27 (1973).
129. J. B. Kay, J. B. Robinson, B. Cox, and D. Polkonjak, J. Pharm. Pharmacol. *22*, 214 (1970).
130. E. Mutschler and G. Lambrecht, Trends Pharmacol. Sci. (Suppl.) *4*, 39 (1984).
131. F. W. Schueler, J. Am. Pharm. Assoc. Sci. Ed. *45*, 197 (1956).
132. G. Lambrecht, Eur. J. Med. Chem. *12*, 41 (1977).
133. E. E. Smissman and G. S. Chappell, J. Med. Chem. *16*, 23 (1969).
134. S. Archer, A. M. Lands, and T. R. Lewis, J. Med. Pharm. Chem. *5*, 423 (1962).
135. M. D. Mashovskii, Khim. Farmatsev. Zh. *3*, 3 (1967).
136. J. B. Robinson, B. Belleau, and B. Cox, J. Med. Chem. *12*, 848 (1969).
137. A. Meyerhöffer, J. Med. Chem. *15*, 994 (1972).
138. A. Giachetti, International Symposium on Muscarinic Cholinergic Mechanisms, (S. Cohen, J. Finberg, and M. Sokolovsky, eds.), The Tel-Aviv University Press, Tel-Aviv, Israel, November 30 to December 4, 1986.
139. P. Sauerberg, J.-J. Larsen, E. Falch, and P. Krogsgaard-Larsen, J. Med. Chem. *29*, 1004 (1986).
140. J. H. Welsh and R. Taub, J. Pharmacol. Exp. Ther. *103*, 62 (1951).
141. A. Simonart, J. Pharmacol. Exp. Ther. *46*, 157 (1932); *50*, 1 (1934).
142. A. Noordam, Rec. Trav. Chim. Fr. *98*, 467 (1979).
143. P. G. Waser, Pharmacol. Rev. *13*, 465 (1961).
144. A. H. Beckett, B. H. Warrington, R. Griffiths, E. S. Pepper, and K. Bowden, J. Pharm. Pharmacol. *28*, 728 (1976).
145. B. Belleau and J. Puranen, J. Med. Chem. *6*, 325 (1963).
146. E. Teodori, F. Gualtieri, P. Angeli, L. Brasili, M. Gianella, and M. Pigini, J. Med. Chem. *29*, 1610 (1986).
147. P. Angeli, L. Brasili, F. Gualtieri, and E. Teodori, Program and Abstracts, International Symposium on Muscarinic Cholinergic Mechanisms, (S. Cohen, J. Finberg and M. Sokolovsky, eds.), The Tel-Aviv University Press, Tel-Aviv, Israel, November 30 to December 4, 1986, p. 15.
148. C. Melchiorre, P. Angeli, M. Gianella, F. Gualtieri, M. Pigini, M. L. Cingolani, G. Gamba, L. Leone, P. Pigini, and L. Re, Eur. J. Med. Chem. *13*, 357 (1978).

149. R. S. Givens and D. R. Rademacher, J. Med. Chem. *17*, 457 (1974).
150. R. Spiegel, Psychiatry Res. *11*, 1 (1984).
151. A. Fisher, E. Heldman, R. Brandeis, I. Karton, Z. Pittel, S. Dachir, A. Levy, and F. Mizobe, Program and Abstracts, International Symposium on Muscarinic Cholinergic Mechanisms, (S. Cohen, J. Finberg, and M. Sokolovsky, eds.), The Tel-Aviv University Press, Tel-Aviv, Israel, November 30 to December 4, 1986, p. 12.
152. A. K. Cho, W. L. Haslett, and D. J. Jenden, J. Pharmacol. Exp. Ther. *138*, 249 (1962).
153. R. Dahlbom, A. Linquist, S. Lindgren, V. Svensson, B. Ringdahl, and M. R. Blair, Jr., Experientia *30*, 1165 (1974).
154. B. Ringdahl and D. J. Jenden, Mol. Pharmacol. *23*, 17 (1983).
155. A. Amstutz, B. Ringdahl, B. Karlen, M. Roth, and D. J. Jensen, J. Med. Chem. *28*, 1760 (1985).
156. F. J. Ehlert and D. J. Jenden, Mol. Pharmacol. *28*, 107 (1985).
157. A. Claesson, A. Asell, S. Bjorkman, and L.-I. Olsson, Acta Pharm. Seuc. *15*, 105 (1978).
158. A. P. Roszkowski, J. Pharmacol. Exp. Ther. *132*, 156 (1961).
159. S. G. Kuznetsov and S. N. Golikov, Synthetic Atropinelike Substances, U.S. Department of Commerce, Joint Publ. Res. Service, *19*, 757 (1963).
160. A. Ladenburg, Justig Liebig's Ann. der Chemie *217*, 82 (1914).
161. B. V. Rama Sastry, in Medicinal Chemistry, 3rd ed. (A. Burger, ed.), Wiley-Interscience, New York, 1970, p. 1544.
162. F. Leonard and L. J. Simet, J. Am. Chem. Soc. *71*, 2855 (1955).
163. D. W. Adamson, P. A. Barrett, and S. Wilkinson, J. Chem. Soc. 52 (1951).
164. R. S. Moffett and D. B. Aspergren, J. Am. Chem. Soc. *79*, 4451 (1957).
165. Roche Products, British Patent 728,579 (1955); Chem. Abstr. *50*, 5773 (1956).
166. J. W. Cusic and R. A. Robinson, J. Org. Chem. *16*, 1921 (1951).
167. F. F. Blicke, Annu. Rev. Biochem. *13*, 549 (1944).
168. F. B. Abramson, R. B. Barlow, M. G. Mustafa, and R. P. Stephenson, Br. J. Pharmacol. *37*, 207 (1969).
169. R. B. Barlow, F. M. Franks, and J. D. M. Pearson, J. Med. Chem. *16*, 439 (1973).
170. R. B. Barlow, J. Med. Chem. *16*, 1037 (1973).
171. K. Fromherz, Naunyn-Schmiedebergs Arch. Exp. Pathol. Pharmakol. *176*, 86 (1933).
172. R. Meier and K. Hoffmann, Helv. Med. Acta *7* (Suppl. 6), 106 (1941).
173. A. Engelhardt and H. Wick, Arzneimittelforsch. *7*, 217 (1957).
174. L. H. Sternbach and S. Kaiser, J. Am. Chem. Soc. *74*, 2215 (1952).
175. M. D. Mashkovsky and L. N. Yakhonotov, Prog. Drug Res. *13*, 293 (1969).
176. L. H. Sternbach and S. Kaiser, J. Am. Chem. Soc. *74*, 2219 (1952).
177. J. P. Robinson, G.-G. Heden, and H. Von Schreeb, CB Weapons Today, Almquist & Wiksell, Stockholm, 1973.
178. W. J. Rzeszotarski, R. E. Gibson, W. C. Eckelman, D. A. Simms, E. M. Jagoda, N. L. Ferreira, and R. C. Reba, J. Med. Chem. *25*, 1103 (1982).

179. R. E. Gibson, W. J. Rzeszotarski, W. C . Eckelman, E. M. Jagoda, D. J. Weckstein, and R. C. Reba, Biochem. Pharmacol. *32*, 1851 (1983).
180. A. Meyerhöffer, J. Med. Chem. *15*, 994 (1972).
181. R. Petenzone, Jr., W. J. Rzeszotarski, W. J. Eckelman, R. E. Gibson, and R. C. Reba, 16th Med. Atl. Reg. Mtg., Am. Chem. Soc. Wilmington, DE, April 1982, Abstr. 98.
182. R. Quirion, Eur. J. Pharmacol. *111*, 287 (1985).
183. H. R. Ing, G. S. Dawes, and J. J. Wajda, J. Pharmacol. Exp. Ther. *85*, 85 (1945).
184. A. M. Lands and C. J. Cavalitto, J. Pharmacol. Exp. Ther *110*, 369 (1954).
185. A. M. Lands and F. P. Luduena, J. Pharmacol. Exp. Ther. *116*, 177, (1956).
186. R. Foster, P. J. Goodford, and H. R. Ing, J. Chem. Soc. *1957*, 3575.
187. L. H. Sternbach and S. Kaiser, J. Am. Chem. Soc. *75*, 6068 (1953).
188. A. B. Funkes and R. F. Rekker, Arzneimittelforsch. *9*, 539 (1959).
189. A. M. Lands, J. O. Hoppe, J. R. Lewis, and E. Ananenko, J. Pharmacol. Exp. Ther. *100*, 19 (1950).
190. J. D. P. Graham and S. Lazarus, J. Pharmacol. Exp. Ther. *68*, 331; *70*, 165 (1940).
191. B. B. Brown, C. R. Thompson, G. R. Klahm, and H. Werner, J. Am. Pharm. Assoc. Sci. Ed. *39*, 305 (1950).
192. F. F. Blicke and C. E. Maxwell, J. Am. Chem. Soc. *64*, 428 (1942).
193. A. A. Goldberg and A. H. Wragg, J. Chem. Soc. *1957*, 4823.
194. C. Melchiorre, A. Cassinelli, and W. Quaglia, J. Med. Chem. *30*, 201 (1987).
195. A. Cassinelli, C. Melchiorre, and W. Quaglia, Program and Abstracts, International Symposium on Muscarinic Cholinergic Mechanisms, (S. Cohen, J. Finberg, and M. Sokolovsky, eds.), The Tel-Aviv University Press, Tel-Aviv, Israel, 1986, p. 29.
196. F. Jellinek, Acta Crystallogr. *10*, 277 (1957).
197. J. G. Cannon and J. P. Long, in Drugs Affecting the Peripheral Nervous System (A. Burger, ed.), Marcel Dekker, New York, 1967, p. 133.
198. K. Fromherz, Naunyn-Schmiedebergs Arch. Exp. Pathol. Pharmakol. *173*, 86 (1933).
199. F. F. Blicke and C. E. Maxwell, J. Am. Chem. Soc. *64*, 431 (1942).
200. P. D. Armstrong and J. G. Cannon, J. Med. Chem. *13*, 1037 (1970).
201. A. R. Cushny, J. Physiol. *30*, 176 (1904).
202. R. S. Cahn, C. K. Ingold, and V. Prelog, Experientia *21*, 81 (1956).
203. J. P. Long, F. P. Luduena, B. F. Tullar, and A. M. Lands, J. Pharmacol. Exp. Ther. *117*, 29 (1956).
204. W. R. Buckett and C. G. Haining, Br. J. Pharmacol. Chemother. *24*, 138 (1965).
205. R. W. Brimblecombe, D. M. Green, T. D. Inch, and P. J. Thompson, J. Pharm. Pharmacol. *23*, 745 (1971).
206. T. D. Inch and R. W. Brimblecombe, J. Pharm. Pharmacol. *23*, 813 (1971).
207. W. M. Duffin and A. F. Green, Br. J. Pharmacol. Chemother. *10*, 383 (1955).
208. R. B. Barlow, J. Pharm. Pharmacol. *23*, 90 (1971).
209. B. W. J. Ellenbrook, R. J. F. Van Rossum, and E. J. Ariëns, J. Pharm. Pharmacol. *17*, 393 (1965).

210. P. A. J. Janssen, C. J. E. Niemegeers, C. J. E. Schellekens, P. Demoen, F. N. Lanaerts, J. M. van Neuten, I. van Wijngaarden, and J. Brugmans, Arzneimittelforsch. *21*, 1365 (1971).
211. I. van Wijngaarden, Life Sci. *9*, 489 (1970).
212. C. Bernard, C. R. Soc. Biol. (Paris) *2*, 195 (1851).
213. C. Bernard, C. R. Scéances Acad. Sci. III *43*, 825 (1856).
214. A. Crum Brown and T. R. Fraser, Trans. R. Soc. Edinburgh *25*, 151 (1868).
215. A. Crum Brown and T. R. Fraser, Proc. R. Soc. Edinburgh *6*, 556 (1869).
216. H. R. Ing and W. M. Wright, Proc. R. Soc. Lond. [Biol.] *109*, 337 (1931).
217. P. E. B. Holmes, D. J. Jenden, and D. B. Taylor, Nature (Lond.) *159*, 86 (1947).
218. J. Thomas and G. A. Starmer, J. Pharm. Pharmacol. *13*, 752 (1961).
219. F. Bergmann and A. Shimoni, Biochim. Biophys. Acta *10*, 49 (1953).
220. J. B. Stenlake, W. D. Williams, N. C. Dhar, and I. G. Marshall, Eur. J. Med. Chem. *9*, 233 (1974).
221. T. O. Soine and J. Naghaway, J. Pharm. Sci. *63*, 1643 (1975).
222. W. A. Varanda, Y. Aracava, S. M. Sherby, W. G. VanMeter, M. E. Eldefrawi, and E. X. Albuquerque, Mol. Pharmacol. *28*, 128 (1985).
223. W. T. Comer and W. L. Matier, in Burger's Medicinal Chemistry, 4th ed., Part III (M. E. Wolff, ed.), Wiley-Interscience, New York, 1981, p. 302.
224. A. J. Everett, L. A. Lowe, and S. Wilkinson, Chem. Commun. 1020 (1970).
225. R. B. Barlow and H. R. Ing, Nature (Lond.) *161*, 718 (1948).
226. D. Bovet, F. Bovet-Nitti, S. Guarino, V. G. Longo, and R. Fusco, Arch. Int. Pharmacodyn. Ther. *88*, 1 (1951).
227. F. Hucho and J. P. Changeux, FEBS Lett. *38*, 11 (1973).
228. S. Loewe and S. C. Harvey, Naunyn-Schmiedebergs Arch. Exp. Pathol. Pharmakol. *214*, 214 (1952).
229. J. J. Lewis, M. Martin-Smith, T. C. Muir, and H. H. Ross, J. Pharm. Pharmacol. *19*, 502 (1967).
230. M. D. Aceto, B. R. Martin, and E. L. May, in Handbook of Stereoisomers: Drugs in Psychopharmacology (D. F. Smith, ed.), CRC Press, Boca Raton, Florida, 1984, pp. 67-78.
231. D. J. Triggle, in Handbook of Stereoisomers: Drugs in Psychopharmacology, (D. F. Smith, ed.), CRC Press, Boca Raton, Florida, 1984, pp. 31-66.
232. C. H. Clothia and P. J. Pauling, Nature (Lond.) *219*, 1156 (1968).
233. C. H. Clothia and P. J. Pauling, Nature (Lond.) *226*, 541 (1970).
234. C. H. Clothia, Nature (Lond.) *225*, 36 (1970).
235. R. W. Baker, C. H. Clothia, P. Pauling, and T. J. Petcher, Nature (Lond.) *230*, 439 (1971).
236. C. J. Cavallito, Prog. Drug Res. *24*, 267 (1980).
237. J. P. Tollenaere, Trends Pharmacol. Sci. *2*, 213 (1982).
238. N. J. M. Birdsall and E. C. Hulme, J. Neurochem. *27*, 7 (1976).
239. R. J. Cushley and H. G. Mautner, Tetrahedron *26*, 2151 (1970).

240. D. Lichtenberg, P. A. Kroon, and S. I. Chan, J. Am. Chem. Soc. *96*, 5934 (1974).
241. P. Pauling and T. J. Petcher, Nature (Lond.) *236*, 112 (1972).
242. D. W. Genson and R. E. Christoffersen, J. Am. Chem. Soc. *95*, 362 (1973).
243. L. B. Kier, Mol. Pharmacol. *3*, 487 (1967).
244. A. Karlin, Fed. Proc. *32*, 1847 (1973).
245. J. M. Schulman, M. L. Savio, and R. L. Disch, J. Med. Chem. *26*, 817 (1983).
246. J. P. Tollenaere, Trends Pharmacol. Sci. *4*, 85 (1984).
247. A. M. Lands and F. P. Luduena, J. Pharmacol. Exp. Ther. *116*, 177, 331 (1956).
248. R. B. Barlow, Introduction to Chemical Pharmacology, Wiley, New York, 1964, p. 211.
249. B. V. Rama Sastry and H. C . Cheng, J. Pharmacol. Exp. Ther. *201*, 105 (1977).
250. R. B. Barlow, K. A. Scott, and R. P. Stephenson, Br.J. Pharmacol. *21*, 509 (1963).
251. J. F. Stubbins, P. M. Hudgins, D. C Murphy, and T. L. Dickerson, J. Pharm. Sci. *61*, 470 (1972).
252. M. T. Flavin, M. C. Lee, E. B. Thompson, and H. N. Bhargava, J. Med. Chem. *30*, 278 (1987).
253. W. J. Rzeszotarski, R. E. Gibson, W. J. Eckelman, and R. C. Reba, U.S. Patent 4,644,003 (1987), and unpublished results.
254. H. Gorissen, G. Aerts, and P. Laduron, FEBS Lett. *96*, 64 (1978).
255. N. J. M. Birdsall, A. S. V. Burgen, and E. C. Hulme, Br. J. Pharmacol. *66*, 337 (1979).
256. T. Haga, FEBS Lett. *113*, 68 (1980).
257. H. Gorissen, G. Aerts, B. Ilien, and P. Laduron, Anal. Biochem. *111*, 33 (1981).
258. C. André, J. P. DeBacker, J. C. Guillet, P. Vanderheyden, G. Vauquelin, and A. D. Strong, EMBO J. *2*, 499 (1983).
259. A. Karlin, Cold Spring Harbor Sym. Quant. Biol. *48*, 1 (1983).
260. G. P. Hess, D. J. Cash, and H. Aoshima, Annu. Rev. Biophys. Bioeng. *12*, 443 (1983).
261. J.-P. Changeux, A. Devillers-Thiéry, and P. Chemoiulli, Science *225*, 1335 (1984).
262. C. F. Stevens, Trends Neurosci. *8*, 1, 336 (1985).
263. R. Miledi and L. T. Potter, Nature (Lond.) *233*, 599 (1971).
264. N. Fujita, N. Nelson, T. D. Fox, T. Claudio, J. Lindstrom, H. Riezman, and G. P. Hess, Science *231*, 1284 (1986).
265. J. L. Marx, Science *219*, 1056 (1983).
266. J.-P. Changeux, T. R. Podleski, and J. C.Meunier, J. Gen. Physiol. *54*, 225 (1969).
267. R. M. Krupka, Biochemistry *6*, 1183 (1967).
268. A. Karlin and E. Bartels, Biochim. Biophys. Acta *126*, 525 (1966).

269. J. Patrick, B. Valeur, L. Monneric, and J.-P. Changeux, J. Membr. Biol. *5*, 102 (1971).
270. J. C. Meunier and J.-P. Changeux, FEBS Lett. *32*, 143 (1973).
271. E. Karlson, D. Eaker, and G. Ponterius, Biochim. Biophys. Acta *257*, 235 (1972).
272. L. Smart, H.-W. Meyers, R. Hilgenfeld, W. Saenger, and A. Maelicke, FEBS Lett. *178*, 64 (1984).
273. W. D. M. Paton and H. P. Rang, Proc. R. Soc. Lond. [Biol.] *163*, 1 (1965).
274. S. H. Snyder, K. J. Chang, M. J. Kuhar, and H. I. Yamamura, Fed. Proc. *34*, 1915 (1975).
275. N. J. M. Birdsall and E. C. Hulme, J. Neurochem. *27*, 7 (1976).
276. F. J. Ehlert, W. Roeske, and H. I. Yamamura, Fed. Proc. *40*, 153 (1981).
277. F. J. Ehlert, Y. Damont, W. R. Roeske, and H. I. Yamamura, Life Sci. *26*, 961 (1980).
278. R. B. Barlow, D. J. Berry, P. A. M. Glenton, N. M. Nikolaou, and K. S. Soh, Br. J. Pharmacol. *58*, 613 (1976).
279. R. Hammer and A. Giachetti, Trends Pharmacol. Sci. *4*, 18 (1984).
280. N. J. M. Birdsall, A. S. V. Burgen, E. C. Hulme, J. M. Stockton, and M. J. Zigmond, Br. J. Pharmacol. *78*, 257 (1983).
281. W. K. Wamsley, M. A. Zarbin, N. J. M. Birdsall, and M. J. Kuhar, Brain Res. *200*, 1 (1980).
282. T. W. Vickroy, M. Watson, H. I. Yamamura, and W. R. Roeske, Fed. Proc. *43*, 2785 (1984).
283. M. Watson, H. I. Yamamura, and W. R. Roeske, Life Sci. *32*, 3001 (1983).
284. L. Noronha-Blob, J. Ferkany, D. Costello, V. Lowe, W. Rzeszotarski, W. Kinnier, and D. U'Prichard, Trends Pharmacol. Sci. (Suppl.) *6*, 84 (1986).
285. E. C. Hulme, N. J. M. Birdsall, A. S. V. Burgen, and P. Mehta, Mol. Pharmacol. *14*, 437 (1978).
286. A. J. Beld and E. J. Ariëns, Eur. J. Pharmacol. *25*, 203 (1976).
287. B. J. Morley, J. F. Lorden, G. B. Brown, G. E. Kemp, and R. J. Bradley, Brain Res. *134*, 161 (1977).
288. B. J. Morley, G. E. Kemp, and P. Salvaterra, Life Sci. *24*, 859 (1979).
289. B. J. Morley and G. E. Kemp, Brain Res. Rev. *3*, 81 (1981).
290. B. J. Morley, Pharmacol. Ther. *15*, 111 (1981).
291. I. Nukina, N. Ogawa, H. Takayama, and Z. Ota, Res. Commun. Chem. Pathol. Pharmacol. *47*, 141 (1985).
292. C. Larsson and A. Nordberg, J. Neurochem. *45*, 24 (1985).
293. S. J. Moss and S. Wonnacott, Biochem. Soc. Trans. *13*, 164 (1985).
294. R. D. Schwartz and K. J. Kellar, Science *220*, 214 (1983).
295. M. J. Marks and A. C. Collins, Pharmacol. Biochem. Behav. *22*, 283 (1985).
296. G. Schrieber, A. Bark, and M. Sokolovsky, J. Cardiovasc. Pharmacol. *7*, 390 (1985).
297. C. R. Houser, G. D. Crawford, R. P. Barber, P. Salveterra, and J. E. Vaughn, Brain Res. *266*, 97 (1983).
298. D. Frambrough and H. C. Hartzell, Science *176*, 189 (1972).
299. C. C. Chang, J. Formosan Med. Assoc. *59*, 416 (1960).

300. C. C. Chang, J. Formosan Med. Assoc. *71*, 315 (1972).
301. T. K. Harden, L. I. Tanner, M. W. Martin, N. Nakahata, A. R. Hughes, J. R. Hepler, T. Evans, S. B. Masters, and J. H. Brown, Trends Pharmacol. Sci. (Suppl.) *6*, 14 (1986).
302. P. M. Laduron and P. F. M. Janssen, J. Neurochem. *33*, 1223 (1979).
303. C. Kaiser and T. Jain, Med. Res. Rev. *5*, 145 (1985).
304. M. Williams and J. L. Robinson, J. Neurosci. *4*, 2906-2911 (1984).

15

Purine Receptors

Michael Williams / CIBA-GEIGY Corporation, Summit, New Jersey

HISTORICAL SURVEY

The current interest (1-4) in adenosine and its related nucleotide, adenosine 5′-triphosphate (ATP), as neuromodulatory agents in mammalian tissues is the culmination of nearly 50 years of skepticism (5). The thought that a cellular component so ubiquitous and so intimately involved in nearly every aspect of tissue function (6) could also act as a chemical messenger was, for a long time, considered heresy, and it was not until Rall suggested that the central stimulant, caffeine, was producing its effects by antagonizing the actions of endogenous adenosine (7) that research efforts intensified to the extent that adenosine agonists and antagonists are now considered potential therapeutic agents (5,8).

In 1929, Drury and Szent-Gyorgi showed (9) that adenosine could affect cardiovascular function, producing hypotension and bradycardia. Unfortunately, subsequent studies that showed that the purine was not useful in the treatment of heart disease (10) failed to take into account the short half-life of the nucleoside (5), which has proven advantageous in the use of adenosine in the treatment of supraventricular tachycardia (11).

Subsequent studies showed that adenosine in the form of its nucleotide ATP was a central sedative (12) with hypnotic activity (13). Following from the discovery of cyclic adenosine 5′-monophosphate (cyclic AMP), it was subsequently found that adenosine was one of the most potent modulators of cyclic nucleotide formation (14), an effect that was susceptible to blockade by methylxanthines, such as caffeine and theophylline (7). In subsequent years, the effects of adenosine on cyclic AMP formation, especially in brain tissue, have been extensively studied (15,16), leading to the description of two adenosine receptor subtypes (17,18), termed A_1 and A_2 based on their opposing actions on adenylate cyclase activity. In 1980, the first of several radioligand binding assays was described (19-21). These enabled the characterization of the distribution and pharmacology of adenosine receptors in a more concise and systematic manner and led to the identification of several classes of nonxanthine compounds with adenosine antagonist activity (22-24) (see also Reference 194). In addition, chemical efforts have led to the synthesis of potent, soluble, and selective xanthine antagonists (25-29). At the same time, a role for the adenine nucleotide ATP as a neuromodulator, especially in peripheral tissues, has emerged (30-32), with two distinct receptor populations described (32).

Despite such progress, the area of purine receptor research is still very much in its infancy. Unlike the more classic neurotransmitters, few chemical entities structurally dissimilar to purines and xanthines can be used as tools for further examining the receptor subtypes. Furthermore, although the actions of adenosine have been described in terms of energy balance changes (33) and cellular homeostatis (34), the only clear picture that exists in relation to endogenous adenosine availability is related to stress situations (34,35). Thus the pathophysiology of the nucleoside is well established, but its physiological role remains relatively obscure.

The findings (36) that the enzyme responsible for adenosine degradation, adenosine deaminase (ADA), could be localized to discrete areas of mammalian brain, together with immunohistochemical studies on adenosine localization (37), has reinforced evidence for a physiological role for the nucleoside but has not resolved the significance of the factors involved in its availability. Adenosine can be formed from several nucleotides (38), and it is generally accepted that the enzyme 5′-nucleotidase (39), which has an anatomic localization similar to that described for adenosine A_1 receptors (40), is the most probable source of extracellular adenosine owing to its actions on AMP.

CLASSIFICATION

Adenosine Receptors

The existence of discrete receptors for adenosine was first suggested by Sattin and Rall (7) and confirmed by the studies of Premont et al. (41) using cell-free brain homogenates. In the 10 years following the description of the adenosine-stimulated adenylate cyclase system, a great deal of work was done in elaborating the role of adenosine and its relationship to other neurotransmitters in central nervous system (CNS) tissue (15,16). An additional body of work related to purinergic neurotransmission has been, and was evolving, from studies conducted by Burnstock (30) and collagues on a phenomenon known as "nonadrenergic, noncholinergic" neurotransmission in peripheral tissues. In 1978, Burn-

Table 1 Purine Receptor Classification

Type	Pharmacology	Effector coupling
Adenosine (P_1) receptors		
A_1 (Ri)	CPA $\geqslant$ CHA $\geqslant$ 2-CADO $\geqslant$ R-PIA > NECA >> S-PIA	Adenylate cyclase inhibition; calcium movement
A_2 (Ra)	NECA $\geqslant$ MECA $\geqslant$ 2-CADO $\geqslant$ R-PIA > CPA > CHA > S-PIA	Adenylate cyclase inhibition
ATP (P_2) receptors		
P_{2x}	α, β-methylene ATP=β, γ-methylene ATP > ATP = 2MeSATP	Prostaglandin formation
P_{2y}	2MeSATP >> ATP > α, β-methylene ATP = β, γ-methylene ATP	Prostaglandin formation Endothelial relaxing factor release

All P_1 receptors are blocked by xanthine adenosine antagonists. The intracellular P adenosine recognition site (1,17) is omitted since it is not a receptor.

stock proposed the first nomenclature for purinoceptors. P-1 receptors were in essence adenosine-sensitive receptors (Table 1), P-2 receptors, as discussed further below, were sensitive to ATP. In the same year, following studies in various cell culture systems, Londos and Wolff in Bethesda (17) and Van Calker and Hamprecht and colleagues (18) in Germany simultaneously described the existence of two separate classes of adenosine receptor. In the schema of Londos and Wolff, the two receptors were termed R, indicating the need for an intact ribose function in the nucleoside nucleus with subscripts i and a to designate whether receptor activation led to an inhibition or stimulation of cyclic AMP formation. Van Calker et al. (18) termed these same two receptors A_1 and A_2, the A_1 being the Ri receptor and the A_2 and Ra receptor. Early criteria indicating that the two classes of receptor could be classified according to their effects on adenylate cyclase activity have since been modified when it became apparent that receptor-mediated phenomena consistent with the activation of either A_1 or A_2 receptors could occur independently of changes in cyclic AMP (42). The current classification of A_1 and A_2 receptors is therefore related to their pharmacological, as opposed to biochemical, effector system-linked effects (43,44). A_1 receptors are generally more sensitive to activation by adenosine analogs substituted in the N^6 position (Table 1 and Figure 1), and A_2 receptors have selectively for agonists substituted in the 2 and 5′ N-positions. Adenosine can also influence calcium fluxes (45,46) and potassium conductance changes (8, 47). Adenosine has pre- and postsynaptic effects in the CNS (48,49). In addi-

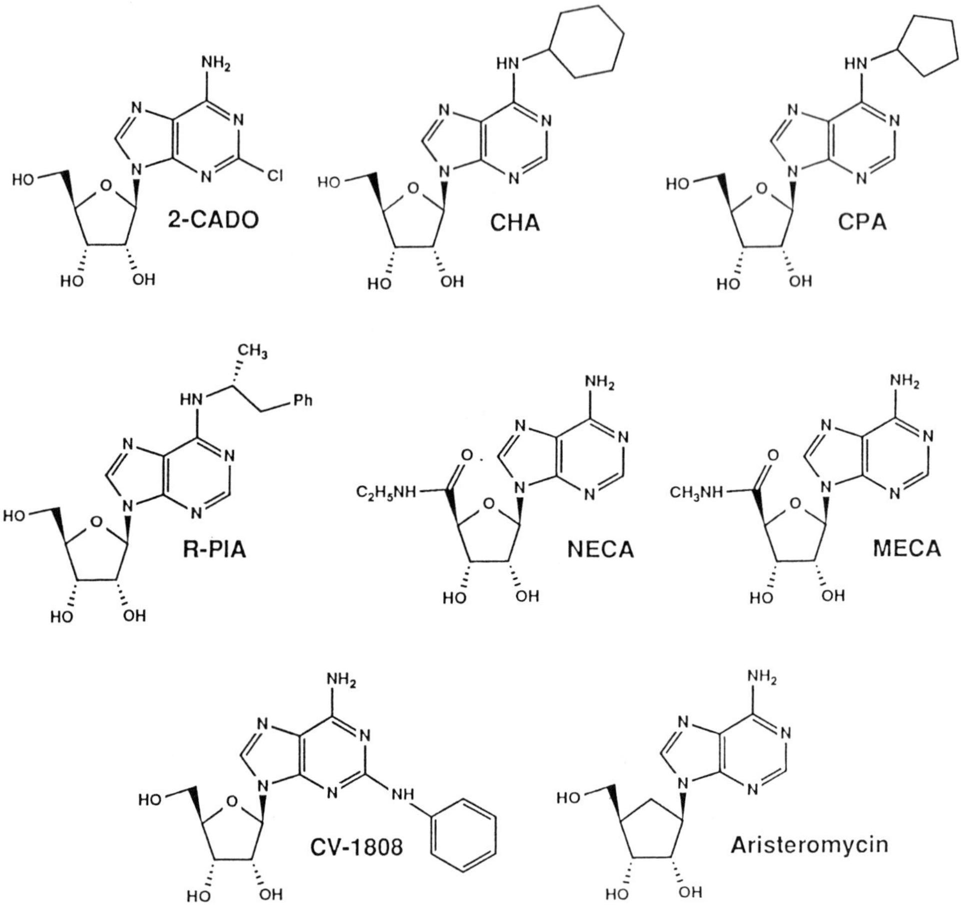

Figure 1 Structures of adenosine receptor ligands.

tion to influencing membrane potentials, the purine can also prolong the time course of calcium-dependent after-hyperpolarization, an effect that can reduce neuronal excitability to long-term stimulatory input (50). In rat hippocampus, the presynaptic effects of the purine on neurotransmitter release appear to be more important than any direct or indirect effects on postsynaptic conductance (49,50). Evidence for subtypes of both A_1 and A_2 receptors has been documented (3,4,51).

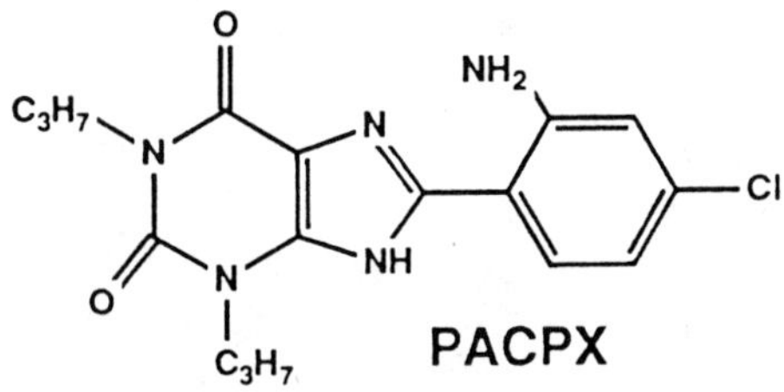

PACPX

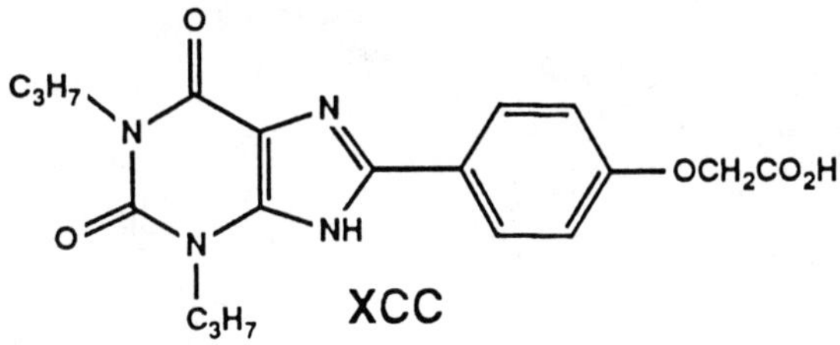

XCC

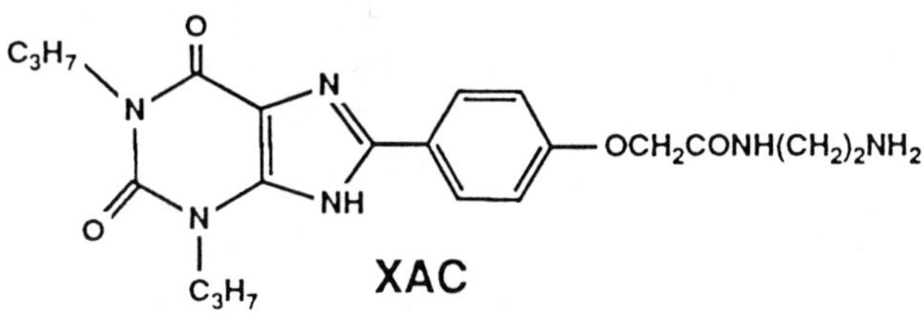

XAC

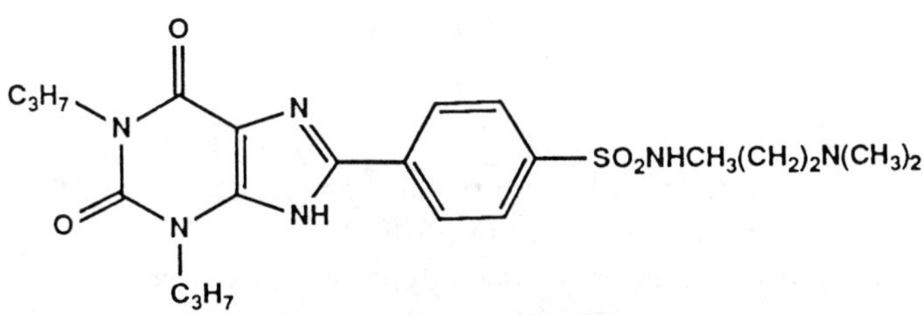

PD 115,199

Figure 1 (Continued)

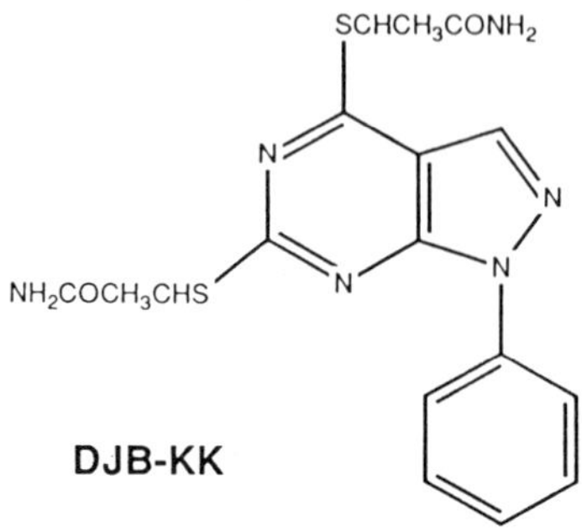

DJB-KK

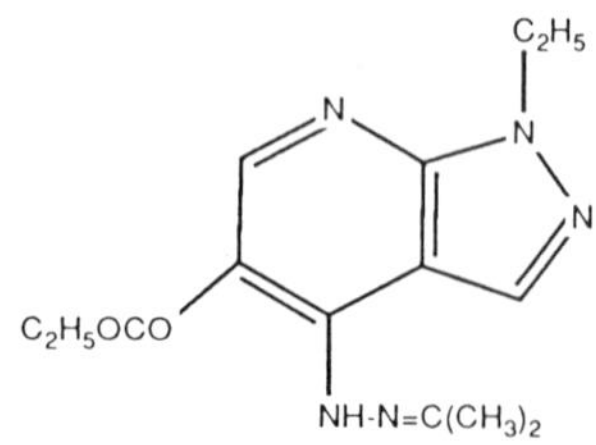

Etazolate

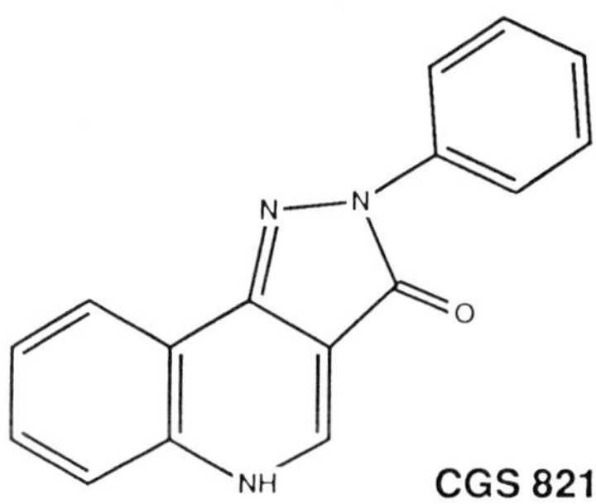

CGS 8216

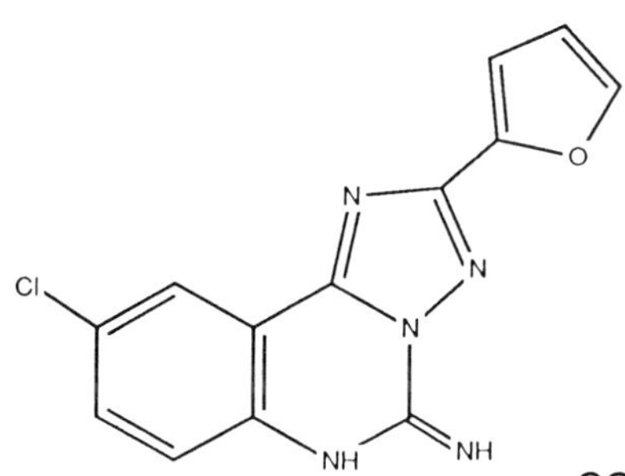

CGS 15943

Figure 1 (Continued)

ATP-Sensitive Purinoceptors

Many of the effects ascribed to actions of the purine nucleotide ATP on cell function have been attributed either to hydrolysis of this labile cellular constituent to adenosine (52) or to the ability of the polyphosphate side chain of the nucleotide to chelate calcium (53) and thus indirectly modulate cell function. Such suggestions, although attractive, are not compatible with the studies performed using nonhydrolyzable analogs of the purine (31) or with the many actions of ATP that are not susceptible to blockade by the xanthines (54,55). The existence of two ATP receptor subtypes termed P_{2x} and P_{2y} has been recently suggesed (Table 1), with the former sensitive to the α,β-methylene isostere of ATP and the latter to the 2-methylthio analog of the purine nucleotide. The major problem in the unequivocal elaboration of a functional role for ATP in mammalian tissues is the lack of availability of selective antagonists. The bee venom apamin, which was originally thought to be a selective ATP antagonist (30), was found to act directly with the potassium channel (56). The azidonitrophenylaminoproprionyl analog of ATP, $ANAPP_3$ (57), is an irreversible ligand that has also received extensive study.

ATP is thought to produce its effects on cell function through the intermediacy of products of the arachidonic acid pathway (30), most notably the prostaglandins.

The anatomic localization of adenosine receptors has not been clearly established. Based on the effects of low concentrations of adenosine ŏn neurotransmitter release (1-4), it has been assumed that A_1 receptors are located presynaptically, although studies on cyclic AMP production (15,16) have suggested that A_2 receptors may occur both pre- and postsynaptically. A number of lesioning studies in brain tissue (58-63) using kainic acid, quinolinate, and 6-hydroxydopamine have yielded equivocal results in relation to A_1 receptor localization, although it is generally agreed that they are localized on cerebellar granule cells (60,61) and on striatal interneurons (58,60,63). Some kainate-resistant receptors are located on corticostriatal projections (63). The inconsistencies reported (58-63) may be related to the extent of lesioning and the time after lesioning at which binding in the issues was studied.

AGONISTS AND ANTAGONISTS OF A_1 and A_2 RECEPTORS

As already indicated, the majority of compounds interacting with adenosine receptors are either purines as agonists or xanthines as antagonists. Selective A_1 agonists include N^6-cyclohexyladenosine (CHA), N^6-cyclopentyladenosine (CPA), and N^6-phenylisopropyladenosine (PIA) (Fig. 1) (19,21,64,65). Althought 2-chloroadenosine (2-CADO) and 5′-*N*-ethylcarboxamidoadenosine (NECA) have been reported as A_1- and A_2-selective agonists, respectively, both are nonselective (64). Thus the effects of NECA, as contrasted to those of an A_1-selective agent, such as CHA, are indicative of additional rather than selective receptor actions. In the presence of 50 nM CPA (64–67) or *N*-ethylmaleimide (68-70), which can selectively titrate out the A_1-receptor population by altering the affinity of A_1 receptors by an action on the Ni coupling protein (69), NECA can be shown to be A_2 selective. In the PC12 pheochromocytoma cell line, which has a population of adenosine receptors that are exclusively A_2 in nature (71,72), binding of the A_1-selective ligands CPA and CHA is negligible but NECA binds in a manner indicative of labeling of an A_2-receptor population (73). The methyl analog of NECA, MECA (Figure 1), has some degree of A_2 selectivity (64-73), as does 2-aminophenyladenosive (CV 1808) and the deaza-adenosine analog aristeromycin (Table 2) (74). None of

Table 2 Receptor Selectivity of Adenosine Receptor Ligands[a]

Compound	Activity (nM) A_1	A_2	A_2-A_1 ratio
Agonists			
CHA	1.8	1001	565
CPA	1.0	685	685
R-PIA	1.6	241	151
NECA	6.3	11.9	1.9
2-CADO	14.0	86	6.1
MECA	260	99	0.4
CV 1808	989	177	0.2
Aristeromycin	2651	821	0.3
S-PIA	29	3540	122
Antagonists			
PACPX	5.4	589	109
DPX	135	5298	39.2
Theophylline	14900	37533	2.5
XAC	15	83	5.5
8PT	176	2056	11.7
CGS 8216	2620	33520	12.8
Etazolate	6100	13800	2.3
DJB-KK	168	736	4.4
CGS 15943A	21	3.3	0.16

[a]For guide to abbreviations, see Figure 1 and text. All data derived by radioligand binding in rat brain tissue.

Source: Data from References 64, 74, 84, and 192.

these compounds is especially potent, however. An A_2-selective adenosine agonist congener has been reported (75).

A similar situation exists for the xanthine adenosine antagonists. Most antagonists are A_1 selective (26,27,76), especially the potent 8-phenyl derivative, PACPX (Figure 1 and Table 2) (27). A_2-selective xanthines have been more difficult to synthesize, although several compounds with weak activity are known (27). The attachment of peptidic side chains to both purines and xanthines, the "functionalized congener" approach, has resulted in a series of interesting compounds with receptor selectivity and affinity (75). Several nonxanthine adenosine antagonists have been reported: the pryazolquinoline, CGS 8216 (22); the pyrazolopyrimidine, DJB-KK (23); the pyrazolpyridine, etazolate (77); and the triazoloquinazoline, CGS 15943A (Figure 1) (see Reference 194). The three former compounds are A_1 selective, with DJB-KK some 20 times more active than either etazolate or CGS 8216 at this receptor (Table 2); CGS 15943A is A_2 selective.

The α_2 agonist clondine has adenosine antagonist activity (78-80), and although a considerable body of evidence supports a relationship between A_1 and α_2 receptors (4), this is still somewhat controversial. It may be noted that both clonidine (81) and adenosine (82) have been found to be useful in the treatment of morphine withdrawal.

PURINERGIC RECEPTOR BINDING

Adenosine Receptors

Although several ligands have been used to label adenosine receptors (1-4), the N^6-substituted analogs CHA, CPA, and R-PIA are most commonly used. To detect high-affinity binding, it is necessary to remove the endogenous adenosine present in both homogenates and tissue sections with the catabolic enzyme ADA (4). Incubations can then be carried out at room temperature using crude cellular homogenates and receptor-ligand complexes isolated by vacuum filtration. Typically, dissociation constants are in the nanomolar range with corresponding apparent B_{max} values in the range 200-900 fmol/mg tissue protein. Some typical values are shown in Table 3. The xanthine antagonist DPX (Figure 1) has also been used to label adenosine receptors. Early studies indicated that both A_1 and A_2 receptors could be labeled with the xanthine (19), but binding of this ligand has not been found to be reproducible. Two functionalized xanthine congeners, XAC and XCC (Figure 1), have given promising binding data in homogenates (83,84) and because of their high specific radioactivity (>140 Ci/mmol) may be of use in determing adenosine receptor characteristics in tissues in which receptor denisty or affinity is low. These factors have tended to preclude the accurate characterization of adenosine receptors in lung (85), heart (86,87), kidney (88), platelets (89), mast cells (90), and other peripheral tissues (91,92). Comparative data on the receptors in tissues other than brain are shown in Table 3.

As already indicated, labeling of A_2 receptors has been limited by the availability of selective ligands. The use of [^{3}H]NECA in the presence of either 50 nM CPA (64-67) or concentrations of NEM below 100 μM (68-70) has allowed A_2 receptors to be studied in adenosine-free tissue, although it may be noted that in lung tissue NECA binding is unaffected by ADA (85). The PC12 pheochromocytoma cell line, on the basis of adenylate cyclase (71,72) and NECA binding (73), appears to be exclusively A_2 in nature and may offer a cell system free of A_1 receptors. A factor of interest in regard to the biochemical, as opposed to the pharmacological, characterization of A_2 receptors has been the concentrations of the purine necessary for receptor activation, usually of the

Table 3 Characteristics of Adenosine Receptors in Mammalian Tissues

Tissue source	K_D (nM)	B_{max} (fmol/mg protein)	References
Rat brain	0.5-2.6	200–800	19,20,65,86
Rat heart	3-6	22	86
Rat adipocyte	3.0	380	86
Bovine myocardium	1.0	31	89
Guinea pig ileum	1.8	140	8
Human platelets	160	8,400	87
	2900	33,400	87
Human neutrophils	230	–	91

order of 10^{-6} M compared with 10^{-9} M for A_1 receptor activation (1). In binding assays, NECA has an affinity around 4 nM. In addition, the treatment of tissues with NEM selectively relates to an interaction with Ni-coupled A_1 receptors (69). Thus, if A_1 receptors not linked to adenylate cyclase are present in a given tissue, these should presumably be unaffected by the sulfhydryl alkylating agent.

ATP Receptors

The labeling of ATP receptors using [^{3}H] ATP has been reported in rabbit bladder (93), and [^{3}H] AppNHp can bind irreversibly to brain membranes with nonomolar affinity (94). $ANAPP_3$ has also been used as a ligand for P-2 receptors (95) in guinea pig vas deferens. Such binding was covalent.

DISTRIBUTION, LOCALIZATION, AND FUNCTION OF ADENOSINE RECEPTORS

Adenosine receptors are widely distributed in mammalian tissues, with the highest denisty in brain tissue (15,16,20). Pharmacological and autoradiographic studies have clearly demonstrated a differential distribution of the A_1 and A_2 receptor subtypes. In the cardiovascular system, A_1 receptors are located in the atrioventricular node and modulate nerve conduction in this region (96), producing negative chronotropic, inotropic, and antiarrhythmic activity. In the coronary vessels, A_2 receptors predominate (97). This differential localization of receptors is also functionally distinct such that compounds with A_2-agonist activity can cause a fall in blood pressure and those with A_1-agonist activity have negative inotropic, chronotropic, and dromotropic actions. Extensive studies on the coronary adenosine receptor have led to the suggestion (98) that it is a hybrid with characteristics of both the A_1 and A_2 receptors (99). The functional significance of this observation may be related to the recent report (100) that the antihypertensive actions of a series of adenosine analogs has no apparent relationship to their binding profiles at CNS adenosine receptors of both subtypes. In addition, adenosine analogs are available that, although showing no receptor subtype selectivity in vitro (64), can reduce blood pressure without affecting heart rate (101,102). This effect has been attributed to an increased sensitivity of A_2 receptors in the intact animal (103). Adenosine may also affect blood pressure by modulating the release of renin from the kidney (104). Such effects would only be seen after chronic administration of adenosine analogs and would be indirect (105).

In elaborating the pharmacological properties of adenosine, many different mammalian species have been used. From such studies, prediction has been made in regard to adenosine receptor pharmacology, which has been found to be species rather than receptor dependent (106). Evaluation of adenosine A_1-receptor binding in a number of species (107,108) has shown that the N^6-substituted analogs of the purine show distinct interspecies differences in activity and the 2- and 5′-substituted analogs, 2-CADO and NECA, had consistent activity in the different animals. For xanthines, the interspecies differences were very marked, with PACPX far more active in cow brain than in guinea pig, rat, or human. Thus although this potent xanthine is 70,000 times more active than theophylline in calf brain (27), in guinea pig brain this difference is nowhere near as marked (108). Similar observations have been made in regard to species differences in the binding of [^{3}H]NECA to caudate A_2 receptors (67). Again, the differences in the activity of agonists at this receptor are more marked for N^6- than for 2- or 5′-substituted analogs. Dramatic differences were also observed for xanthines, especially those substituted at the 8 position with a phenyl moiety, in both binding (67) and inhibition of adenosine-stimulated adenylate cyclase (109). Species differences have also been reported for the enzyme 5′-nucleotidase (110) and for the binding of [^{3}H]nitrobenzylthioinosine (NBI), a putative ligand for the adenosine uptake site (111). In addition to these relatively gross interspecies differences in adenosine systems, subtle intraspecies differences in caffeine responsiveness have been documented in the mouse (112-114).

Functionally, adenosine receptor activation has been linked to a general inhibition of neurotransmitter release (1-4). This effect is distinct from the autoinhibition seen with such receptors as the presynaptic α_2 and dopamine autoreceptors, which selectively modulate norepinephrine and dopamine release, respectively (115,116). Since the purine can modulate the release of acetylcholine (117), γ-aminobutyric acid (GABA) (118), glutamate (119,120), norepinephrine (121), dopamine (121,122), and serontonin (121), it is difficult to relate its physiological actions to any selective action. In addition, compared with the concentrations of neuroeffector agents required to produce autoinhibition of release (115,123), adenosine is required in relatively high concentrations to produce its effects (3). Some resolution of this phenomenon may relate to the role of endogenous adenosine. The effects of the xanthines on cell function, that is, as central stimulants and locomotor stimulants (2), have been described in terms of the existence of a purinergic inhibitory tone (121). Tissue levels of adenosine approximate to 1-2 μM (124), and it seems probable that the availability of the purine in the extracellular milieu subserves a homeostatic function (34) rather than being related to a stimulus-secretion coupling event. The factors governing adenosine availability have yet to be unequivocally defined (4, 125). Stimulus-dependent release has been reported (126-129). However, it is not clear to what extent such release is calcium dependent (126,130). Although adenosine availability may be regulated via a phenomenon described as "neuronal nonexcytotic release" (131), the activity of 5′-nucleotidase has been correlated with A_1-receptor density (40). The purine may also be formed from cyclic AMP and ATP (132,133), the latter of which has been identified as a cotransmitter (134).

Adenosine receptor activation can depress neuronal cell firing (3), due to both direct effects on cell function and indirectly via its effects on neurotransmitter release (2-4). These effects may in turn be related to the behavioral actions of adenosine (2). Blockade of the effects of endogenous adenosine using the methylxanthines has been reported to increase neurotransmitter release (135), cell firing (136), and behavioral activity (2).

PURINERGIC RECEPTOR COUPLING

Mention has already been made of the effects of adenosine on adenylate cyclase activity (15,16). Although having no direct effects on phosphatidylinositol turnover (137), the purine can potentiate the actions of histamine (138), an effect similar to that previously observed in adenylate cyclase interactions (15). As already noted, adenosine can affect transmembrane calcium fluxes (45,46), and direct effects on potassium conductance have been observed (3,47). Despite the profound effects of adenosine on cyclic AMP formation, the physiological, as opposed to pharmacological, role of this second messenger has yet to be determined. Since some effects of adenosine occur independently of cyclic AMP (42), changes in cyclic AMP may be occurring secondarily to more relevant physiological effects, such as ion movement. Again, the lack of availability of agonists and antagonists, other than purines or xanthines, has limited the precise establishment of a cause-and-effect relationship.

In addition to the proposed actions of ATP receptor activation on prostaglandin formation, the recent description of the P_{2x} and P_{2y} receptor subtypes

(32) indicates that endothelial relaxant factor (ERF) may be an important effector of ATP receptor activation.

PHYSIOLOGICAL ROLE: TARGETS FOR DRUGS

The ubiquitous nature of adenosine distribution and its role as an enzyme cofactor and cellular energy constituent (6) has tended to cloud the issues related to specificity in the actions of the purine and, consequently, its potential as a therapeutic candidate. The major impetus in the study of purines as neuromodulatory agents has been the demonstration that the xanthine caffeine, the most widely used psychoactive agent (2,7,139), is an adenosine antagonist. It is somewhat surprising, therefore, given the paucity of selective and effective agonists and antagonists, that agents modifying purine function at the receptor level have been enthusiastically studied as therapeutic entities in many areas.

CNS

Adenosine and its analogs have sedative (12) and hypnotic activity (13) when administered to animals. Conversely, the xanthines are central stimulants (2) with possible nootropic actions (140). An A_1-receptor-mediated inhibition of dopamine release has been associated with R-PIA-induced deep sleep (141). Ethanol-sensitive "long-sleep" mice are more sensitive to the sedative and hypothermic actions of adenosine agonists than the ethanol-insensitive "short-sleep" mice (142). These genetic effects have been related to changes in receptor density and affinity (143).

Purines may play a role in the etiology of anxiety. Caffeine is anxiogenic (144), and the sedative actions of adenosine are similar to those seen with the classic antianxiety agents, the benzodiazepines. The adenosine catabolic product inosine was identified as a ligand, albeit with millimolar activity, for the central benzodiazepine receptor (145). Compounds with antianxiety activity do not in general show any propensity to interact with adenosine receptors (22), although it may be noted that xanthines can antagonize the effects of the benzodiazepine anxiolytics in humans (146) in addition to antagonizing the actions of these compounds on neuronal cell firing and cardiovascular function (147). The benzodiazepine inverse agonist CGS 8216 (22) is an adenosine antagonist. The interrelationship between anxiolytics and adenosine, as with the effects of the purine in relation to other classes of psychoactive agent, has been related to the relatively weak actions of the benzodiazepines as adenosine uptake inhibitors (148).

One of the more tangible roles for adenosine as a CNS effector agent is in the area of epilepsy. Caffeine in high doses is an effective convulsant (144), and adenosine is a potent anticonvulsant (149). Since the endogenous levels of the purine can increase dramatically following seizures (150) and caffeine and theophylline can prolong the duration of kindled seizure activity, it has been suggested (151) that adenosine may function as an endogenous anticonvulsant agent. In this context the purine may function as a "retaliatory metabolite" (33) or homeostatic agent (34). The anticonvulsant carbamazepine has been reported to up-regulate A_1 receptors (152).

Xanthines have been reported to have varying effects on morphine analgesia (153), and as mentioned, adenosine has been found to be effective in the treatment of opiate withdrawal (81). Adenosine itself has been reported both to antagonize the analgesic actions of morphine (154) and to have antinociceptive activity on its own account (153). Caffeine has been described as an analgesic adjuvant (155), a role that may relate to the ability of the xanthines to inhibit cyclooxygenase activity (156).

Excessive doses of caffeine can elicit a self-destructive behavioral phenomenon (157) similar to that seen in Lesch-Nyhan syndrome (158), a disease state due to a genetic lesion in the purine metabolizing enzyme hypoxanthine-guanine phosphoribosyltransferase (HRGPTase). NECA has been found to affect dopamine-related actions on rat rotational behavior (159) and to potentiate apomorphine-induced self-mutilation (160). The relationship between dopamine and adenosine at the level of the striatum (a focal point in Lesch-Nyhan syndrome) is further evidenced by the distinct localization of high-affinity A_2 receptors to this region (66) and by reports related to the effects of adenosine analogs on prolactin release (161) and as putative antipsychotic agents (162).

The possibility that purines may exert their effects on CNS function indirectly by affecting blood pressure and cerebral blood flow has been a topic of concern (2,3). Although many adenosine agonist have been shown to cause a hypotensive effect that can be correlated with effects on neuron firing in the CNS (163), these agents can produce similar effects when administered directly into the brain (164). Evidence does exist, however, to dissociate the hypotensive effects of adenosine agonists from their CNS actions (165).

Cardiovascular System

Much current research in the cardiovascular area is directed toward the development of selective adenosine A_2 agonists as antihypertensives (76). Adenosine can regulate coronary blood flow, atrioventricular node conduction, and sinoatrial node excitation (166). Large increases in endogenous adenosine levels occur in coronary tissue following hypoxia (167) and are thought to protect the heart from excess stress in terms of regulating oxygen supply to demand and may thus subserve a homeostatic role. The distinct distribution of A_1 versus A_2 receptors has made it feasible to search for a selective A_2 agonist that would be hypotensive via its actions on the coronary vasculature without affecting A-V node conduction (167). An additional approach in the use of adenosine in the treatment of hypertension is to elucidate the role of agonists in the inhibition of renin release from the kidney (104,168). The effects of the purine are biphasic, involving the stimulation of A_1 receptors to induce inhibition of renin release and the stimulation of A_2 receptors to potentiate release. The second messenger for these effects appears to be calcium (169).

Adenosine can antagonize the inotropic actions of β-adrenoceptor agonists (170-173), an effect that involves changes at the level of a common pool of cyclic AMP. This interaction is somewhat controversial at the present time (174). Adenosine and its analogs, at low concentrations (10^{-8} M), can evoke vasodilation in cerebral blood vessels (175,176). These vessels are even more sensitive (10^{-11} M) to ATP (176). Such vasodilation and concomitant increases in cerebral blood flow can occur as the result of hypoxia (167).

Adenosine agonists may also be of use in the treatment of atherosclerosis. A component of the etiology of this degenerative disease is the production of free radicals from neutrophils, which can cause endothelial damage in arteries (177). Adenosine can reduce superoxide liberation from neutrophils (178) and may thus prevent some of the damage associated with the liberation of this radical.

The adenosine antagonist aminophylline (theophylline-ethylenediamine) has been widely used as a cardiotonic agent (179), increasing cardiac contractility, although the concentrations required to produce adenosine blockade approximate those necessary for phosphodiesterase inhibition such that the therapeutic use of aminophylline is associated with the production of arrhythmias and CNS side effects.

Lung

Theophylline has long been a drug of choice in the treatment of asthma (179) and is reported (180) to produce its effects by inhibition of lung phosphodiesterase activity and a concomitant increase in tissue cyclic AMP levels, an action compatible with the known effects of selective β_2-receptor agonists on cyclic AMP formation in lung tissue. More recent evidence has indicated that the xanthines act as antiasthmatics by virtue of their ability to block the histamine releasing actions of adenosine on mast cells in asthmatics (181-183). This observation is, however, confounded by the fact that adenosine has either no effect on pulmonary function in normal subjects (181) or is a bronchodilator by virtue of its direct actions on smooth muscle tone (184). However, in asthmatics, the purine is a potent and life-threatening bronchoconstrictor (182).

The contribution of phosphodiesterase inhibitor activity to the therapeutic use of xanthines in asthma is also uncertain. Such agents as the β_2 agonists, which increase cyclic AMP formation, are effective antiasthmatics (180), as is the diterpene forskolin (181), which can elevate tissue cyclic AMP levels independently of receptor activation (182). Since the consequences of asthma are multifaceted, it is possible that both adenosine antagonists and phosphodiesterase inhibitors are therapeutically useful. The combination of these two properties in theophylline is perhaps serendipitous.

Immune Function

Many immune deficiency diseases are characterized by hereditary disorders in purine metabolism (187,188), and adenosine has been reported to inhibit lymphocyte transformation (188), IgE-mediated histamine release (189), and inhibition of platelet aggregation (190). Abnormalities in T-cell function in Hodgkin's disease have been linked to decreases in the activity of ADA and 5′-nucleotidase (191). Adenosine may play an important role in various immunodeficiency and autoimmune disease states (192).

FUTURE PROSPECTS

As already noted, in contrast to many of the other receptors described in the present volume, receptors for adenosine (and to a more limited extent, ATP) are characterized by a paucity of ligands. In the development of new therapeutic modalities and in understanding the systems involved in the physiology and pathophysiology of adenosine function, both agonist and antagonist compounds selective for each of the receptor subtypes are necessary. Furthermore, the possibility of developing such entities, which are neither purines or xanthines, is of crucial importance.

REFERENCES

1. J. W. Daly, J. Med. Chem. *25*, 197 (1972).
2. S. H. Snyder, Annu. Rev. Neurosci. *8*, 103 (1985).
3. T. V. Dunwiddie, Int. Rev. Neurobiol. *27*, 63 (1985).
4. M. Williams, in Receptor Binding in Drug Research (R. A. O'Brien, ed.), Marcel Dekker, New York, 1985, p. 235.
5. Editorial, Lancet, *2* (1985).
6. J. R. S. Arch and E. A. Newsholme, Essays Biochem. *14*, 88 (1978).
7. A. Sattin and T. W. Rall, Mol. Pharmacol. *6*, 13 (1970).
8. M. Williams, in Adenosine: Receptors and Modulation of Cell Function (V. Stefanovich, K. Rudolphi, and P. Schubert, eds.), IRL Press, Oxford, 1985, p. 73.
9. A. N. Drury and A. Szent-Gyorgi, J. Physiol. (Lond.) *68*, 213 (1929).
10. R. M. Honey, W. T. Ritchie, and W. A. R. Thomson, Q. J. Med. *23*, 481 (1930).
11. J. DiMarco, J. D. Sellars, B. B. Lerman, M. L. Greenbert, R. M. Berne, and L. Bellardinelli, J. Am. Coll. Cardiol. *6*, 417 (1985).
12. W. Feldberg and S. L. Sherwood, J. Physiol. (Lond.) *123*, 148 (1954).

13. M. Maitre, L. Cieslielski, A. Lehman, E. Kempf, and P. Mandel, Biochem. Pharmacol. *23*, 2807 (1974).
14. S. Kakuchi, T. W. Rall, and H. McIlwain, J. Neurochem. *16*, 485 (1969).
15. J. W. Daly, Cyclic Nucleotides in the Nervous System, Plenum Press, New York, 1977.
16. G. I. Drummond, Cyclic Nucleotides in the Nervous System, Raven Press, New York, 1984.
17. C. Londos and J. Wolff, Proc. Natl. Acad. Sci. USA *74*, 5482 (1977).
18. D. Van Calker, M. Muller, and B. Hamprecht, J. Neurochem. *33*, 999 (1979).
19. R. F. Bruns, J. W. Daly, and S. H. Snyder, Proc. Natl. Acad. Sci. USA *77*, 5547 (1980).
20. M. Williams and E. A. Risley, Proc. Natl. Acad. Sci. USA *77*, 6892 (1980).
21. U. Schwabe and T. Trost, Naunyn-Schmiedebergs Arch. Pharmacol. *313*, 179 (1981).
22. M. Williams, E. A. Risley, and J. R. Huff, Can. J. Physiol. Pharmacol. *59*, 897 (1981).
23. A. Czernik, B. Petrack, H. J. Kalinsky, S. Psychoyos, W. D. Cash, C. Tsai, R. K. Rinehart, F. R. Granat, R. A. Lovell, D. E. Brundish and R. Wade, Life Sci. *30*, 363 (1982).
24. L. P. Davis, D. J. Brown, S. Chen Chow, and G. A. R. Johnston, Neurosci. Lett. *41*, 189 (1983).
25. F. W. Smellie, C. W. Davis, J. W. Daly, and J. N. Wells, Life Sci. *24*, 2475 (1979).
26. J. W. Ferkany, R. Hicks, J. Connor, and L. Noronha-Blob, Pharmacologist *27*, 292 (1985).
27. R. F. Bruns, J. W. Daly, and S. H. Snyder, Proc. Natl. Acad. Sci. USA *80*, 2077 (1983).
28. J. W. Daly, W. Padgett, M. T. Shamin, P. Butts-Lamb, and J. Waters, J. Med. Chem. *28*, 487 (1985).
29. D. Ukena, J. W. Daly, K. L. Kirk, and K. A. Jacobson, Life Sci. *38*, 797 (1986).
30. G. Burnstock, in Cell Membrane Receptors for Drugs and Hormones (R. W. Straub and L. Bolis, eds.), Raven Press, New York, 1978, p. 107.
31. G. Burnstock, in Regulatory Function of Adenosine (R. M. Berne, T. W. Rall, and R. Rubio, eds.), Martinus Nijhoff, Boston, 1983, p. 49.
32. G. Burnstock and C. V. Kennedy, Gen. Pharmacol. *16*, 433 (1985).
33. A. C. Newby, Trends Biochem. Sci. *9*, 42 (1984).
34. M. Williams, Trends Neurosci. *7*, 164 (1984).
35. J.-P. Boulenger, P. J. Marangos, K. J. Zander, and J. Hanson, Clin. Neuropharmacol. *9*, 79 (1986).
36. J. I, Nagy, F. A. LaBella, M. Buss, and P. E. Dadonna, Science *224*, 166 (1984).
37. K. M. Braas, A. C. Newby, and S. H. Snyder, in Adenosine: Receptors and Modulation of Cell Function (V. Stefanovich, K. Rudolphi, and P. Schubert, eds.), IRL Press, Oxford, 1985, p. 59.

38. H. Nagata, Y. Mimori, S. Nakamura, and M. Kameyama, J. Neurochem. *42*, 1001 (1984).
39. S. Nakamura, Y. Mimori, S. Iijima, H. Nagata, S. Yamao, and M. Kameyama, in Physiology and Pharmacology of Adenosine Derivatives (J. Daly, Y. Kuroda, J. W. Phillis, H. Shimizu, and M. Ui, eds.), Raven Press, New York, 1983, p. 21.
40. R. R. Goodman and S. H. Snyder, J. Neurosci. *2*, 1230 (1982).
41. J. Premont, M. Perez, and J. Bockaert, Mol. Pharmacol. *13*, 662 (1977).
42. T. V. Dunwiddie and B. B. Fredholm, Naunyn-Schmiedebergs Arch. Pharmacol. *326*, 294 (1984).
43. B. Hamprecht and D. Van Calker, Trends Pharmacol. Sci. *6*, 154 (1985).
44. T. W. Stone, in Purines: Pharmacological and Physiological Roles (T. W. Stone, ed.) VCH Publications, Deerfield Beach, Florida, 1985, p. 4.
45. B. A. Henon and D. A. McAfee, J. Physiol. (Lond.) *336*, 607 (1983).
46. J. A. Riberio, A. M. SaAlmedia, and J. M. Namordo, Biochem. Pharmacol. *28*, 1297 (1979).
47. R. W. Greene and H. L. Hass, J. Physiol. (Lond.) *366*, 119 (1985).
48. M. Segal, Eur. J. Pharmacol. *19*, 193 (1982).
49. G. R. Siggins and P. Schubert, Neurosci. Lett. *23*, 55 (1981).
50. T. V. Dunwiddie and H. L. Hass, J. Physiol. (Lond.) *369*, 365 (1985).
51. J. W. Daly, P. Butts-Lamb, and W. Padgett, Cell. Mol. Neurobiol. *3*, 69 (1983).
52. J. W. Phillis and P. H. Wu, Prog. Neurobiol. *16*, 187 (1981).
53. K. Krnjevic, Physiol. Rev. *54*, 418 (1974).
54. C. E. Jahr and T. M. Jessell, Nature (Lond.) *304*, 730 (1983).
55. M. W. Salter and J. L. Henry, Neuroscience *15*, 815 (1985).
56. G. Burnstock, in Adenosine: Receptors and Modulation of Cell Function (V. Stefanovich, K. Rudolphi, and P. Schubert, eds.), IRL Press, Oxford, 1985, p. 3.
57. J. S. Fedan, G. K. Hogaboom, J. P. O'Donnell, J. Colby, and D. P. Westfall, Eur. J. Pharmacol. *69*, 41 (1981).
58. T. F. Murray and D. L. Cheney, Neuropharmacology *21*, 575 (1982).
59. H. G. E. Lloyd and T. W. Stone, Brain Res. *334*, 385 (1985).
60. M. F. Jarvis and M. Williams, Pharmacol. Biochem. Behav. (in press).
61. R. R. Goodman, M. J. Kuhar, L. Heste, and S. H. Snyder, Science *220*, 179 (1983).
62. W. J. Wojcik and N. H. Neff, J. Neurochem. *41*, 759 (1983).
63. J. D. Geiger, Brain Res. *363*, 404 (1985).
64. R. F. Bruns, G. H. Lu, and T. A. Pugsley, Abstr. Soc. Neurosci. *10*, 282 (1984).
65. M. Williams, A. Braunwalder, and T. E. Erickson, Naunyn-Schmiedebergs Arch. Pharmacol. *332*, 179 (1986).
66. E. W. Snowhill and M. Williams, Eur. J. Pharmacol.
67. G. A. Stone, M. A. Sills, B. Weeks, and M. Williams, Drug. Dev. Res. (in press).

68. S.-M. H. Yeung and R. D. Green, Naunyn-Schmiedbergs Arch Pharmacol. *325*, 218 (1984).
69. D. Ukena, E. Poeschla, E. Hutteman, and U. Schwabe, Naunyn-Schmiedbergs Arch. Pharmacol. *327*, 247 (1984).
70. B. B. Fredholm, E. Lindgren,and K. Lindstrom, J. Neurochem. *37*, 1431 (1985).
71. G. Guroff, G. Dickens, D. End, and C. Londos, J. Neurochem. *37*, 1431 (1981).
72. L. Noronha-Blob, R. P. Marshall, W. J. Kinnier, and D. C. U'Prichard, Life Sci. *39*, 1059 (1986).
73. M. Williams, M. Abreu, M. F. Jarvis, and L. Noronha-Blob, J. Neurochem. *48*, 498 (1987).
74. M. Williams and M. F. Jarvis, Pharmacol. Biochem. Behav. *29*, 433 (1988).
75. K. A. Jacobson, J. W. Daly, and K. L. Kirk, 189th American Chemical Society Meeting, Miami Beach, April 1985, MEDI 38 (1985).
76. H. W. Hamilton, D. F. Ortwine, D. F. Worth, E. W. Badger, J. A. Bristol, R. F. Bruns, S. J. Haleen, and R. P. Steffen, J. Med. Chem. *28*, 1071 (1985).
77. T. W. Stone and D. A. Taylor, J. Pharm. Pharmacol. *30*, 792 (1978).
78. S. K. Kulkarni and A. K. Mehta, Life Sci. *34*, 2273 (1984).
79. R. T. Lai, Y. Watanabe, Y. Kamino, and H. Yoshida, Life Sci. *34*, 409 (1984).
80. T. Katsuragi and T. Furukowa, Trends Pharmacol. Sci. *7*, 337 (1985).
81. A. S. Gold, A. C. Pottash, D. R. Sweeney, and H. D. Kleber, JAMA *243*, 343 (1980).
82. H. O. J. Collier, N. T. Plant, J. F. Tucker, and A. Von Uexull, Br. J. Pharmacol. *81*, 131p (1984).
83. K. A. Jacobsen, D. Ukena, J. W. Daly, and K. L. Kirk, J. Labelled Comp. Radiopharm. *23*, 519 (1986).
84. K. A. Jacobson, D. Ukena, K. L. Kirk, and J. W. Daly, Proc. Natl. Acad. Sci. USA *83*, 4089 (1986).
85. D. Ukena, E. Bohme, and U. Schwabe, Naunyn-Schmiedebergs Arch. Pharmacol. *327*, 36 (1984).
86. J. Linden, C. E. Hollen, and A. Patel, Circ. Res. *56*, 728 (1985).
87. M. J. Lohse, D. Ukena, and U. Schwabe, Naunyn-Schmiedebergs Arch. Pharmacol. *328*, 310 (1985).
88. P. H. Wu and P. C. Churchill, Arch. Int. Pharmacodyn. *273*, 83 (1985).
89. E. Hutteman, D. Ukena, V. Lenschow, and U. Schwabe, Naunyn-Schmiedebergs Arch. Pharmacol. *325*, 226 (1984).
90. D. L. Marquardt and S. I. Wasserman, Agents Actions *16*, 453 (1985).
91. B. N. Cronstein, E. D. Rosenstein, S. B. Kramer, G. Weissman, and R. Hirschhorn, J. Immunol. *135*, 1366 (1985).
92. D. Ukena, D. Martens, and U. Schwabe, Naunyn-Schmiedebergs Arch. Pharmacol. *321*, R39 (1982).
93. R. M. Levin, R. Jacoby, and A. J. Wein, Mol. Pharmacol. *23*, 1 (1983).
94. M. Williams and E. A. Risley, Fed. Proc. *39*, 1009 (1980).

95. J. S. Fedan, G. K. Hogaboom, J. P. O'Donnell, C. J. Jeng, and G. Guillory, Eur. J. Pharmacol. *108*, 49 (1985).
96. L. Bellardinelli, J. Shyrock, G. A. West, J. P. Di Marco, and R. M. Berne, Circulation *70*, 1083 (1984).
97. R. M. Berne, in Methods in Pharmacology, Vol. 6, Methods Used in Adenosine Research (D. M. Paton, ed.), Plenum Press, New York, 1985, p. 331.
98. R. A Olssen, Trends Pharmacol.Sci. *5*, 113 (1984).
99. S. Kusachi, R. D. Thompson, W. J. Bugni, N. Yamada, and R. A. Olssen, J. Med. Chem. *28*, 1636 (1985).
100. M. D. Taylor, W. H. Moos, H. W. Hamilton, D. S. Szotek, W. C. Patt, E. W. Badger, J. A. Bristol, R. F. Bruns, T. G. Heffner, and T. E. Mertz, J. Med. Chem. *29*, 346 (1986).
101. M. Collis and C. M. Brown, Eur. J. Pharmacol. *96*, 61 (1983).
102. E. Leung, C. I. Johnston, and E. A. Woodcock, Clin. Exp. Pharmacol. Physiol. *12*, 515 (1985).
103. D. B. Evans, J. A. Schenden, and J. A. Bristol, Life Sci. *31*, 2425 (1983).
104. S. Itoh, D. A. Carretero, and R. D. Murray, J. Clin. Invest. *76*, 1412 (1985).
105. C. S. Sweet and E. H. Blaine, in Cardiovascular Pharmacology, 2nd ed. (M. Antanaccio, ed.), Raven Press, New York, 1984, p. 119.
106. B. B. Fredholm, B. Jonzon, E. Lindgren, and K. Lindstrom, J. Neurochem. *39*, 165 (1983).
107. K. M. M. Murphy and S. H. Snyder, Mol. Pharmacol. *21*, 250 (1982).
108. J. W. Ferkany, H. L. Valentine, G. A. Stone, and M. Williams, Drug Dev. Res. *9*, 85 (1986).
109. U. Schwabe, D. Ukena, and M. J. Lohse, Naunyn-Schmiedebergs Arch. Pharmacol. *330*, 212 (1985).
110. K. S. Lee, P. Schubert, M. Reddington, and G. W. Kreutzberg, Acta Histochem, (Jena) [Suppl.] *31*, 57 (1985).
111. A. Verma and P. J. Marangos, Life Sci. *36*, 283 (1985).
112. T. W. Seale, P. Johnson, J. M. Carney, and O. M. Rennert, Physiol. Biochem. Behav. *20*, 567 (1983).
113. T. W. Seale, P. Johnson, T. H. Roderick, J. M. Carney, and O. M. Rennert, Physiol. Biochem. Behav. *23*, 275 (1984).
114. T. W. seale, T. H. Roderick, P. Johnson, L. Logan, O. M. Rennert, and J. M. Carney, Physiol. Pharmacol. Behav. in press (1986).
115. K. Starke, Annu. Rev. Pharmacol. Toxicol. *21*, 7 (1981).
116. A. Carlsson, in Pre- and Postsynaptic Receptors (E. Usdin and W. E. Bunney, Jr., eds.), Marcel Dekker, New York, 1975, p. 49.
117. J. Sawynok and K. Jhmandas, J. Pharmacol. Exp. Ther. *197*, 379 (1976).
118. C. Hollins and T. W. Stone, Br. J, Pharmacol. *69*, 107 (1980).
119. A. C. Dolphin and E. R. Archer, Neurosci. Lett. *43*, 49 (1983).
120. R. Corradetti, G. LoCentre, F. Moroni, M. B. Possani, and G. Pepeu, Eur. J. Pharmacol. *104*, 19 (1984).
121. H. H. Harms, G. Wardeh, and A. H. Mulder, Eur. J. Pharmacol. *49*, 305 (1978).

122. M. L. Michaelis, E. K. Michaelis, and S. L. Myers, Life Sci. *24*, 2083 (1979).
123. R. H. Roth, Ann. N.Y. Acad. Sci. *430*, 27 (1984).
124. T. Zetterstrom, L. Vernet, U. Ungerstedt, U. Tossman, B. Jonzon, and B. B. Fredholm, Neurosci. Lett. *29*, 11 (1982).
125. M. Williams, in Psychopharmacology: A. Third Generation of Progress (H. Meltzer, ed.), Raven Press, New York, 1987, p. 289.
126. Y. Kuroda and H. McIlwain, J. Neurochem. *22*, 691 (1974).
127. B. B. Fredholm and P. Hedqvist, Biochem. Pharmacol. *29*, 1635 (1980).
128. T. W. Stone, Neuroscience *6*, 523 (1981).
129. C. Barberis, V. Leviel, and J. L. Daval, in Purines: Pharmacology and Physiological Roles (T. W. Stone, ed.), VCH Publications, Deerfield Beach, Florida, 1985, p. 107.
130. M. N. Perkins and T. W. Stone, Br. J. Pharmacol. *80*, 263 (1983).
131. B. Jonzon and B. B. Fredholm, J. Neurochem. *44*, 217 (1985).
132. F. Pons, R. F. Bruns, and J. W. Daly, J. Neurochem. *34*, 1319 (1980).
133. W. F. MacDonald and T. D. White, Prog. Neuropsychopharmacol. Biol. Psychiatry *8*, 487 (1985).
134. M. Israel, B. Lesbats, R. Manaranche, F. M. Meunier, and P. Frachon, J. Neurochem. *34*, 923 (1980).
135. J. W. Phillis, Z. G. Ziang, B. J. Chelack, and P. H. Wu, Pharmacol. Biochem. Behav. *13*, 421 (1980).
136. T. V. Dunwiddie, B. J. Hoffer, and B. B. Fredholm, Naunyn-Schmiedebergs Arch. Pharmacol. *316*, 326 (1981).
137. E. Brown, D. A. Kendall, and S. R. Nahorski, J. Neurochem. *42*, 1379 (1984).
138. E. B. Hollingsworth, R. A. de la Cruz, and J. W. Daly, Eur. J. Pharmacol. *122*, 45 (1986).
139. P. B. Dews, Caffeine, Springer-Verlag, New York, 1984.
140. I. Hindmarch and Z. Subhan, Drug Dev. Res. *5*, 379 (1985).
141. M. Radulovacki, R. M. Virus, M. Djuricic-Nedelson, and R. D. Green, J. Pharmacol. Exp. Ther. *228*, 268 (1984).
142. W. R. Proctor and T. V. Dunwiddie, Science *224*, 519 (1980).
143. B. B. Fredholm, N. R. Zahniser, G. R. Weiner, W. R. Proctor, and T. V. Dunwiddie, Eur. J. Pharmacol. *111*, 133 (1985).
144. S. H. Snyder and P. Sklar, J. Psychiatr. Res. *18*, 91 (1984).
145. P. J. Marangos, S. M. Paul, A. M. Parma, F. K. Goodwin, P. K. Syapin, and P. Skolnick, Life Sci. *24*, 851 (1979).
146. S. E. File, A. J. Bond, and R. G. Lister, J. Clin. Psychopharmacol. *2*, 102 (1982).
147. T. P. Kenakin, J. Pharmacol. Exp. Ther. *222*, 752 (1982).
148. P. H. Wu and J. W. Phillis, Neurochem. Int. *6*, 613 (1984).
149. T. V. Dunwiddie and T. Worth, J. Pharmacol. Exp. Ther. *220*, 70 (1982).
150. H. R. Winn, J. Morii, and R. M. Berne, Ann. Biomed. Eng. *13*, 321 (1985).
151. M. Dragunow, G. V. Goddard, and R. Laverty, Epilepsia *26*, 480 (1985).

152. P. J. Marangos, S. R. B. Weiss, P. Montgomery, J. Patel, P. K. Narang, A. M. Cappabianca, and R. M. Post, Epilepsia *26*, 493 (1985).
153. M. H. Ahiljanian and A. E. Takemori, Eur. J. Pharmacol. *112*, 171 (1985).
154. D. R. H. Gourley and S. K. Beckner, Proc. Soc. Exp. Biol. Med. *144*, 774 (1973).
155. E. M. Laska, A. Sunshine, F. Mueller, W. B. Elven, C. Siegal, and A. Rubin, JAMA *251*, 1711 (1984).
156. A. R. Whorton, J. B. Collawn, M. E. Montgomery, S. L. Young, and R. S. Kent, Biochem. Pharmacol. *34*, 119 (1985).
157. E. M. Boyd, M. Dolman, L. M. Knight, and E. P. Shepard, Can. J. Physiol. Pharmacol. *43*, 995 (1965).
158. W. C. Nyhan, Annu. Rev. Med. *24*, 41 (1973).
159. B. B. Fredholm, M. Herrera-Marschitz, B. Jonzon, and U. Ungerstedt, Pharmacol. Biochem. Behav. *19*, 535 (1983).
160. R. D. Green, H. K. Proudfit, and S.-M. H. Yeung, Science *218*, 58 (1982).
161. S. F. Steward and T. A. Pugsley, Naunyn-Schmiedebergs Arch. Pharmacol. *331*, 140 (1985).
162. T. G. Heffner, D. A. Downs, J. A. Bristol, R. F. Bruns, S. E. Harrigan, W. H. Moos, K. L. Sledge, and J. N. Wiley, Pharmacologist *21*, 293 (1985).
163. J. W. Phillis and P. H. Wu, in Physiology and Pharmacology of Adenosine Derivatives (J. Daly, Y. Kuroda, J. W. Phillis, H. Shimizu, and M. Ui, eds.), Raven Press, New York, 1983, p. 219.
164. R. A. Barraco, A. K. Aggarawal, J. W. Phillis, M. A. Moore, and P. H. Wu, Neurosci. Lett. *48*, 139 (1984).
165. J. J. Katims, Z. Annau, and S. H. Snyder, J. Pharmacol. Exp. Ther. *227*, 167 (1983).
166. R. M. Berne, in Adenosine: Receptors and Modulation of Cell Function (V. Stefanovich, K. Rudolphi, and P. Schubert, eds.), IRL Press, Oxford, 1985, p. 357.
167. H. R. Winn, R. Rubio, and R. M. Berne, Am. J. Physiol. *241*, H235 (1981).
168. R. D. Murray and P. C. Churchill, J. Pharmacol. Exp. Ther. *232*, 189 (1985).
169. P. C. Churchill and M. C. Churchill, J. Pharmacol. Exp. Ther. *232*, 580 (1985).
170. J. G. Dobson, Circ. Res. *52*, 151, (1983).
171. M. Endoh, M. Maruyama, and N. Taira, J. Cardiovasc. Pharmacol. *5*, 131 (1983).
172. R. Seitelberger, W. Schutz, O. Schlappack, and G. Raberger, Naunyn-Schmiedebergs Arch. Pharmacol. *325*, 234 (1984).
173. M. Bohm, R. Bruckner, I. Hackbarth, B. Haubitz, R. Linhart, W. Meyer, B. Schmidt, W. Schmitz, and H. Scholz, J. Pharmacol. Exp. Ther. *230*, 483 (1984).
174. W. Schutz and M. Freissmuth, Trends Pharmacol. Sci. *7*, 310 (1985).
175. M. Wahl and W. Kuschinsky, Pfluegers Arch. *362*, 55 (1976).
176. T. Forrester, A. M. harper, E. T. Mackenzie, and E. M. Thomson, J. Physiol. (Lond.) *296*, 343 (1979).

177. E. L. Bierman, in Harrison's Principles of Internal Medicine, 10th ed. (R. G. Petersdorf, R. D. Adams, E. Braunwald, K. J. Isselbacher, J. B. Martin, and J. D. Wilson, eds.), McGraw-Hill, New York, 1983, p. 1465.
178. B. N. Cronstein, S. B. Kramer, G. Weissman, and R. Hirschhorn, Clin. Immunol. Immunopathol. *30*, 495 (1984).
179. T. Z. Csaky and B. A. Barnes, in Cutting's Handbook of Pharmacology, 7th ed. Appleton-Century-Crofts, Norwalk, Connecticut, 1984, p. 210.
180. Merck Manual, 14th ed. Merck, Sharp and Dohme Research Laboratories, Rahway, New Jersey, 1982, p. 405.
181. M. J. Cushley, A. E. Tattersfield, and S. T. Holgate, Br. J. Clin. Pharmacol. *15*, 161 (1983).
182. J. S. Mann, M. J. Cushley, and S. T. Holgate, Am. Rev. Respir. Dis. *132*, 1 (1985).
183. M. K. Church and S. T. Holgate, Trends Pharmacol. Sci. *7*, 49 (1986).
184. L. E. Gustafsson, N. P. Wiklund, and B. Cederqvist, Eur. J. Pharmacol. *120*, 179 (1986).
185. J. Chang and A. Lewis, Eur. J. Pharmacol. *101*, 217 (1984).
186. K. B. Seamon and J. W. Daly, J. Cyclic Nucleotide Res. *7*, 201 (1981).
187. R. E. Kellems, C.-Y. Yeung, and D. E. Ingolia, Trends Genetics *1*, 278 (1985).
188. G. M. Kramer, Lancet *1*, 50 (1986).
189. R. Hirschhorn, J. Grossman, and G. Weismann, Proc. Soc. Exp. Biol. Med. *133–136* (1970).
190. G. Marone, R. Petracca, and S. Vigorita, Int. Arch. Allergy Appl. Immunol. *77*, 259 (1985).
191. J. L. Murray, R. Perez-Soler, D. Bywaters, and E. M. Herst, Am. J. Hematol. *21*, 57 (1985).
192. B. N. Cronstein and R. Hirschhorn, in The Adenosine Receptors (M. Williams, ed.) Humana, Clifton, New Jersey, 1988 (in press).
193. S. H. Polmar, C. Fernandez-Mejia, and R. E. Birch, in Receptor Biochemistry and Methodology, Vol. 11. Adenosine Receptors (D. M. F. Cooper and C. Londos, eds.), Liss, New York, 1988, p. 97.
194. M. Williams, J. Francis, G. Ghai, A. Braunwalder, S. Psychoyos, G. A. Stone, and W. D. Cash, J. Pharmacol. Exp. Ther. *241*, 415 (1987).

16

Peptide Receptors

Terry W. Moody and David C. Perry / The George Washington University School of Medicine and Health Sciences, Washington, D.C.

HISTORICAL PERSPECTIVE

Many neuropeptides have been found in the mammalian central nervous system (CNS) and periphery; like other endogenous signal molecules, neuropeptides produce their effects by interacting with distinct receptor recognition sites. For example, the neurokinin substance P (SP), an 11 amino acid peptide, mediates pain transmission in primary sensory afferent neurons (1), by interacting with SP receptors that are abundant in the dorsal horn of the spinal cord (2). Cholecystokinin (CCK), a 33-amino-acid peptide, functions as a satiety agent after intraperitoneal injection (3). This reduction in food intake likely results because CCK inhibits gastric emptying (removal of food from the stomach), an action that is mediated at least in part by CCK receptors present in the pyloric sphincter (5).

Neuropeptide receptors have several differences from classic transmitter receptors. They bind endogenous peptides with higher affinity ($K_D \cong 10^{-9}$ M) than transmitter receptors bind their endogenous ligands ($K_D \cong 10^{-6}$ M). This is due primarily to the slow time of dissociation of peptides from their receptors ($k_{-1} = 0.01\ min^{-1}$), which is several orders of magnitude slower than that of classic neurotransmitters (5). Also, the density of peptide receptors is much lower (2 pmol/g wet tissue) than that of classic transmitter receptors (50 pmol/g wet tissue). The analytical problems associated with this low density are partly compensated for by the high specific activity (up to 2200 Ci/mmol) of ^{125}I, the most commonly utilized radiolabel for peptide receptor probes, yielding a greater amount of radioactivity per binding site than tritium.

The characterization of peptide receptors is fraught with numerous technical pitfalls. Most peptides are hydrophobic and thus adhere to filters as well as glass (6). This can be minimized by adding bovine serum albumin (BSA) to the incubation solution and pretreating filters with such agents as polyethyleneimine (6). Also, peptides are readily degraded by proteases present in the tissue preparations. This proteolytic degradation is slowed by the addition of enzyme inhibitors, such as bacitracin. Another problem occurs when the $[^{125}I]$ peptides used as receptor probes undergo radiolysis; this limits the useful life of these probes to approximately 1 month. Also, it is necessary to verify that the bulky iodine atom has not altered the biologic activity of the peptide. Despite these difficulties, over a dozen peptide receptors have been characterized in the past decade, in both the CNS and periphery. This chapter focuses on peptide receptors for bombesin (BN), bradykinin (BK), neurokinins (NK) (including SP), somatostatin (SS), and vasoactive intestinal polypeptide (VIP). Receptors for opioid peptides are considered by Tam (p. 571).

BOMBESIN AND GASTRIN RELEASING PEPTIDE

Bombesin (BN), a 14 amino acid peptide initially isolated from frog skin (7), was demonstrated to be a potent hypothermic agent in cold-exposed rats after injection into the cisterna magnum or lateral ventricles. It is structurally similar to the 27 amino acid peptide, gastrin releasing peptide (GRP), which was initially isolated from the porcine stomach (8). In particular, 9 of the 10 C-terminal amino acid residues are identical (Table 1) and the C-terminal octapeptide is essential for high-affinity receptor binding activity. GRP, like most peptides, is derived from a high-molecular-weight precursor protein. The precursor protein derived from a lung carcinoid tumor is composed of 148 amino acid residues and contains a single copy of GRP near the N terminal (9). The precursor protein is metabolized by trypsinlike enzymes and is then amidated to yield the 27 amino acid peptide GRP, which is biologically active in the gastrointestinal tract. In the CNS and in small cell lung cancer (SCLC), GRP is further metabolized to the 10 amino acid $GRP^{18\text{-}27}$ (neuromedin C), which is biologically active (10,11).

Table 1 Structure of BN-like Peptides[a]

GRP	Ala–Pro–Val–Ser–Val–Gly–Gly–Gly–Thr–Val–Leu–Ala–Lys–Met Tyr–Pro–Arg–Gly–Asn–His–Trp–Ala–Val–Gly–His–Leu–Met–NH_2
BN	Pyr–Gln–Arg–Leu–Gly–Asn–Gln–Trp–Ala–Val–Gly–His–Leu–Met–NH_2

[a]Sequence homologies are underscored.

Initially, biologic assays were performed using BN and related analogs. In particular, nanogram doses of bombesin lowered the core temperature in cold-exposed rats (12). Subsequent studies showed that there was a well-defined structure-activity relationship with the BN-induced hypothermia (13). Deletion of the C-terminal dipeptide amide (Des–Leu^{13}, Met^{14})BN resulted in loss of biologic activity. In contrast, deletion of the N-terminal tetrapeptide (N–Ac–Gly^{5})BN had no effect on biologic activity. These data indicate that the C but not the N terminal is essential for biologic activity. Further conservative amino acid substitutions near the N terminal, such as Tyr for Leu at the 4 position, could be made and had no effect on biologic activity. In contrast, substitutions near the N terminal, such as Tyr for Leu at the 4 position, could be made and had no effect on biologic activity. In contrast, substitutions near the C terminal, such as Tyr for His^{12}, resulted in a dramatic loss of biologic potency. The high potency of BN and the well-defined structure-activity relationship suggested that the BN-induced hypothermia was mediated by a CNS receptor.

Because BN has a blocked N terminal and does not have a lysine amino acid residue, it cannot be iodinated with Bolton-Hunter reagent. Synthetic (Tyr^{4})BN and (Tyr^{12})BN were iodinated using the chloramine-T procedure, (Tyr^{4})BN, a potent BN analog, binds specifically to rat brain membranes (Table 2), whereas (Tyr^{12})BN, which is approximately two orders of magnitude less potent than is BN, did not. Thus the (Tyr^{4})BN that was radiolabeled and purified (it is not commercially available) serves as an excellent receptor probe. High-performance liquid chromatography (HPLC)-purified [^{125}I]GRP is now commercially available (Amersham) and can be used as a receptor probe (Table 2). GRP binds to the receptors, however, with about fourfold lower affinity than does (Tyr^{4})BN.

Data on receptors for BN-like peptides in the CNS, dispersed pancreatic acini, SCLC cells, pituitary cells, and 3T3 cells are collated in Table 3. Radiolabeled (Tyr^{4})BN and GRP have served as receptor probes, and these ligands bind with high affinity (K_D = 0.5-4 nM). In general, the apparent receptor

Table 2 Specific Binding of Radiolabeled BN-like Peptides to Rat Brain Membranes[a]

Peptide	Specific binding (%)
[^{125}I-Tyr^{4}]BN	3.0
[^{125}I-Tyr^{12}]BN	<0.1
[^{125}I]GRP	1.4

[a]Radiolabeled peptide (100,000 cpm) was added to rat brain membranes (1.5 mg) and the percentage specific binding determined in triplicate relative to the total counts added.

Table 3 BN-like Peptide Receptor Binding Properties[a]

Parameter	Brain homogenate	Brain slices	Pancreatic acini	Pituitary cells	3T3 cells	SCLC cells
Radiolabeled probe	(Tyr^4)BN	(Tyr^4)BN	(Tyr^4)BN	(Tyr^4)BN	GRP	(Tyr^4)BN
Buffer	50 mM Tris	Krebs Ringer's	Kreb Ringer's	Ham's F10 medium	DMEM medium	RPMI-1640 medium
Temperature (°C)	4	25	37	37	37	37
K_D (nM)	3	5	2	1	0.5	0.5
Density	80 fmol/mg protein	130 fmol/mg protein	5000 per cell	3700 per cell	100,000 per cell	3000 per cell

[a]Receptors for BN-like peptides were characterized using rat brain homogenate (14), rat striatal slices (20), guinea pig pancreatic acini (27), GH_4C_1 rat pituitary cells (30), Swiss 3T3 cells (32), and human SCLC cells (45).

density is moderate (2000-5000 per cell); however, it is exceedingly high (100,000 per cell) using certain 3T3 clones. These data indicate that BN receptors have been characterized in the CNS and periphery.

The pharmacology of binding has been investigated extensively. Modifications can be made near the N terminal with retention of biologic activity, whereas modifications near the C terminal result in loss of biologic activity. In particular, substitution of D- for the natural L-amino acids at the 8, 10, 12, 13, or 14 position reduces binding affinity and the ability to induce hypothermia by at least two orders of magnitude (14). Similarly, substitution of norleucine or leucine at position 14 for methionine reduces biologic potency by approximately one order of magnitude. If the methionine at position 14 is oxidized to BN-sulfoxide or deamidated, the biologic potency is reduced by two orders of magnitude. Therefore, numerous amino acids near the C terminal, such as the tryptophan at the 8 position, the valine at the 10 position, the histidine at the 12 position, the leucine at the 13 position, and especially the methionine at the 14 position, are essential for biologic activity.

Similar results were obtained using GRP analogs (15). The N terminal of GRP is not essential in that GRP^{1-16} did not have appreciable receptor binding or biologic activity. The C terminal of GRP was essential in that GRP^{18-27} had total receptor binding and biologic activity. Smaller fragments of GRP were also active. In particular, Ac-GRP^{20-27} was approximtely twofold more potent than GRP^{18-27}. GRP^{21-27} and GRP^{18-27} were approximately two and four orders of magnitude less potent than GRP, respectively. These data suggest that the C-terminal octapeptide of GRP or the structurally related BN is the smallest fragment with full biologic activity.

Initially all the GRP and BN analogs synthesized were agonists. Recently, BN receptor antagonists have been identified. Surprisingly, putative SP antagonists, such as spantide [(D–Arg^1, D–$Trp^{7,9}$, Leu^{11})SP], function as BN receptor antagonists. SP and most of the putative antagonists have the same C-terminal dipeptide amide as does BN; however, the other amino acids differ. Nonetheless, spantide inhibits binding of [^{125}I] GRP to SCLC cells with an IC_{50} value of 0.5 μM. Other substance SP antagonists, such as (D–Arg^1, D–$Trp^{7,9}$, Leu^{11})SP, (D–Pro^2, D–Phe^7, D–Trp^9)SP, (D–Pro^2, D–$Trp^{7,9}$)SP, (D–Pro^4, D–$Trp^{7,9}$)SP^{4-11}, and (D–Pro^4, D–$Trp^{7,9,11}$)SP^{4-11}, were slightly less potent (Table 4). Interestingly, SP had no effect on the binding of [^{125}I] GRP to the SCLC cells. Thus although the putative SP receptor antagonists have little sequence homology to BN, their tertiary structure may be such that they bind with moderate affinity to BN as well as SP receptors. These putative SP antagonists have been observed to antagonize BN-induced amylase secretion from guinea pig pancreatic acinar cells (16), BN-induced grooming in rats (17), and BN-induced uptake of [^{3}H] thymidine in 3T3 cells (18). In summary, agonists and antagonists of varying potency have been defined for BN.

Table 4 Pharmacology of BN Receptor Antagonists in SCLC Cells

Peptide	IC_{50} (μM)
(D–Arg^1, D–$Trp^{7,9}$, Leu^{11})SP (spantide)	0.5
(D–Arg^1, D–Pro^2, D–$Trp^{7,9}$, Leu^{11})SP	1
(D–Pro^4, D–$Trp^{7,9,11}$)SP $^{4-11}$	7
(D–Pro^4, D–$Trp^{7,9}$)SP $^{4-11}$	10
Substance P (SP)[a]	>20

[a]The structure of substance P is Arg–Pro–Lys–Pro–Gln–Gln–Phe–Phe–Gly–Leu–Met–NH_2. Thus, substance P has the same C-terminal dipeptide amide as BN.

Initially, receptors for BN-like peptides were characterized in the CNS using a filtration assay (14). This was somewhat difficult as [^{125}I-Tyr^4)] BN readily adhered to glass fiber filters, but this can be minimized by using 1% bovine serum albumin or 0.1% polyethyleneimine (6). Subsequently, centrifugation assays (19) indicated that radiolabeled (Tyr^4)BN bound with high affinity to rat brain homogenate and that these binding sites were discretely distributed in the rat brain. In particular, the density of sites ranged sevenfold and was greatest in the hippocampus and hypothalamus, intermediate in the cortex, striatum, thalamus, and midbrain, and lowest in the medulla-pons and cerebellum.

The distribution of (Tyr^4)BN binding sites was investigated using rat brain slices and autoradiographic techniques (20). High grain densities were observed in the olfactory bulb, especially the granule cell and glomerular layer and the nucleus accumbens and olfactory tubercle (Figure 1). Sone of the receptors in the nucleus accumbens may be present on dopamine-containing neurons (Z. Merali and T. Moody, unpublished). In the hypothalamus, high grain densities were present in the suprachiasmatic, periventricular, and paraventricular nuclei. BN is a potent satiety agent after direct injection in the paraventricular nucleus (21). Moderate receptor densities were also observed in the anterior, ventromedial, and dorsomedial hypothalamus and medial preoptic and arcuate nuclei, as well as the median eminence. BN is a potent hypothermic agent after direct injection into the anterior hypothalamus (22). In the hippocampus, high grain densities were present in the polymorphic layer of the dentate gyrus, subiculum, the oriens layer, stratum radiatum, and the molecular layer. Pressure application of BN causes an excitatory response from pyramidal cells of the CA 1 region of the hippocampus (23). High receptor densities were observed in the paraventricular thalamic, central medial thalamic and rhomboid thalamic nuclei, the nucleus of the solitary tract, and the substantia gelatinosa. Injection

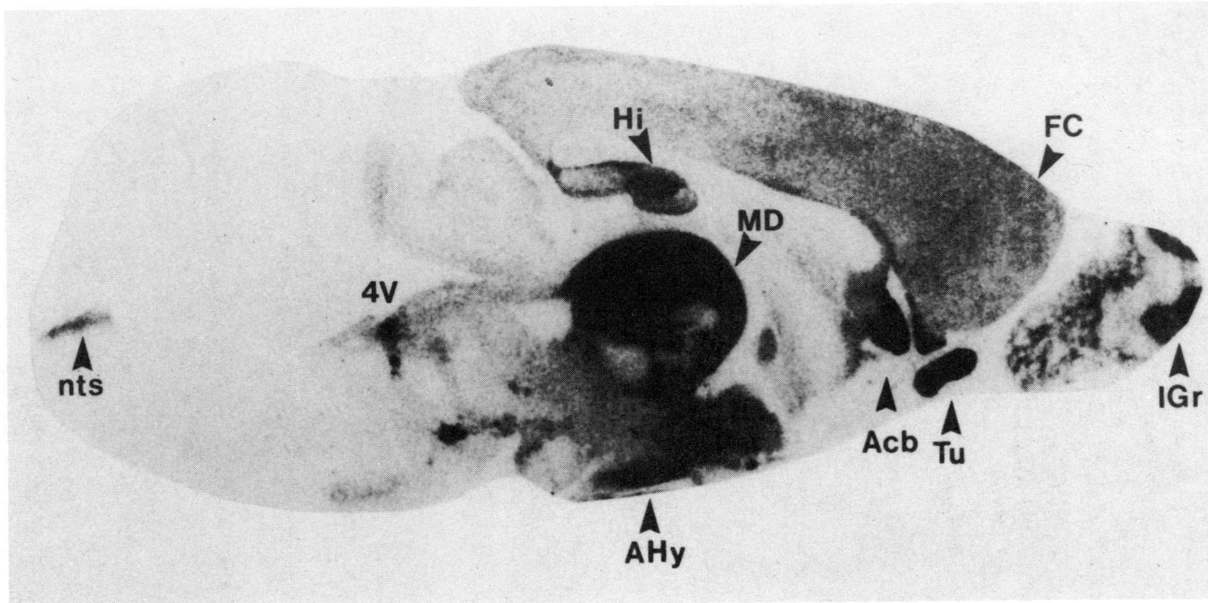

Figure 1 Localization of central receptors for BN-like peptides. The receptors were localized using in vitro autoradiographic techniques. Fresh frozen sections of rat brain were thaw mounted onto glass slides derived from Paxinos and Watson coordinates lateral 0.9 mm and incubated with [^{125}I-Tyr^{4}] BN as described previously (20). Free ligand was removed by washing in buffer, and the tissue-mounted slide, which contained bound peptide, was exposed to LKB Ultrafilm. The dark areas indicate high receptor density. Acb, nucleus accumbens; AHy, anterior hypothalamic area; FC, frontal cortex; Hi, hippocampus; IGr, internal granular layer of the olfactory bulb; MD, mediodorsal thalamic nuclei; nts, nucleus of the solitary tract; Tu, olfactory tubercle; and 4V, fourth ventricle.

of BN into the rat spinal cord causes a biting and scratching response indicative of sensory stimulation (24,25). In summary, receptors for BN-like peptides are discretely distributed in the mammalian brain. When activated these receptors depolarize certain neurons in the hippocampus and spinal cord. In the hypothalamus, striatum, and olfactory tubercle, BN receptors enhance the turnover of the neurotransmitter dopamine (26).

In the pancreas, however, BN stimulates Ca^{2+} flux (27). Using acini derived from the guinea pig pancreas, [^{125}I-Tyr^4] BN bound with high affinity (K_D = 2 nM) to a single class of sites (B_{max} = 5000 per cell). $^{45}Ca^{2+}$ efflux from these cells was stimulated in a dose-dependent manner by BN with an ED_{50} of 0.3 nM. Similarly, BN elevated intracellular cGMP, but not cAMP, levels in the cells and stimulated amylase release from the cells, again with an ED_{50} = 0.3 nM. Because the half-maximal concentration of BN (IC_{50}) to inhibit [^{125}I-Tyr^4] BN was 1.2 nM, it appeared that occupation of only 25% of the receptors was sufficient to induce a full biologic response. Thus there may be spare receptors for BN-like peptides on pancreatic acini. Such receptors desensitize after incubation with 100 nM (Tyr^4)BN for 20 min (28). After desensitization, the dose-response curve for BN to induce amylase release was shifted to the right by approximately an order of magnitude.

When activated, receptors for BN-like peptides stimulate hormone secretion from certain cells. BN-stimulated insulin secretion from HIT-T15 cells derived from hamster islets (29). The ED_{50} for BN was 0.5 nM, and using maximal doses, BN stimulated the insulin secretion rate 10-fold. The effect caused by BN was reversed by somatostatin. BN stimulated growth hormone secretion from GH_4C_1 rat pituitary tumor cells (30) with an ED_{50} = 0.5 nM. Also, [^{125}I-Tyr^4] BN bound with high affinity (K_D = 1.2 nM) to a single class of sites (3600 per cell) using these cells. Some of the radiolabeled (Tyr^4)BN was internalized when the binding assay was performed at 37 but not at 4°C. After internalization, much of the radiolabel was present as [^{125}I] tyrosine. BN-like peptides appear to be degraded by lysozomal enzymes after internalization.

Receptors for BN-like peptides stimulate phosphatidylinositol turnover in certain cells. In particular, the phosphatidylinositol 4,5-bisphosphate generated by BN may be metabolized into diacylglycerol and inositol 1,4,5-trisphosphate (31). The former agent activates protein kinase C, which may phosphorylate certain proteins. In this regard, nanomolar concentrations of BN stimulated phosphorylation of an 80,000 dalton protein within 1 min using a BN-responsive clone of Swiss 3T3 cells (32). These cells bind [^{125}I] GRP (33) with high affinity (K_D = 0.5 nM) to a single class of sites (100,000 per cell). The phosphorylation caused by BN was reversed by the BN receptor antagonist (D–Arg^1, D–Pro^2, D–Trp^{7,9}, Leu^{11})SP. Inositol 1,4,5-trisphosphate may release calcium from intracellular pools, such as the endoplasmic reticulum. In this regard, when 3T3 cells were equilibrated with quin 2, a fluorescent indicator that estimates the intracellular Ca^{2+}, nanomolar concentrations of BN increased the intracellular Ca^{2+}. This increase in calcium caused by BN was reversed by micromolar concentrations of (D–Arg^1, D–Pro^2, D–Trp^{7,9}, Leu^{11})SP. Also, BN stimulated $^{45}Ca^{2+}$ efflux from 3T3 cells (34). Further, nanomolar concentrations of BN in the presence of 1 μg/ml insulin caused a fivefold increase in the uptake of [^{3}H] thymidine by quiescent 3T3 cells (35). BN by itself caused only a 1.5-fold increase in the uptake of [^{3}H] thymidine, whereas insulin by itself had no effect. Thus BN and insulin are synergistic with respect to the growth of 3T3 cells. The increase in [^{3}H] thymidine caused by BN has half-maximal using 0.3 nM BN, and the increase in radioactive thymidine uptake was reversed by (D–Arg^1, D–Pro^2, D–Trp^{7,9}, Leu^{11})SP in a dose-dependent manner. Half-maximal inhibition of [^{3}H] thymidine uptake was caused by 10 μM (D–Arg^1, D–Pro^2, D–Trp^{7,9}, Leu^{11})SP. Besides stimulating the growth of 3T3 cells, BN stimulates the growth of normal human bronchial epithelial cells (36) and small cell lung cancer cells (37).

Although neuropeptides and their receptors have not been shown to be the cause of a central disease, they have been associated with certain peripheral diseases. In particular, high levels of BN-like peptides are associated with SCLC, a disease that comprises about 25% of the 110,000 new cases of malignant lung cancer each year in the United States (38-40). The BN-like peptides are then secreted from the tumor cells into the extracellular fluids (41,42); in patients with extensive disease, the levels of immunoreactive BN in the plasma are elevated (41-43). These BN-like peptides may activate receptors present in certain SCLC cells (46) that bind radiolabeled (Tyr^4)BN with high affinity (K_D = 0.5 nM) to a single class of sites (B_{max} = 2000 per cell). Because trypsin, which degrades proteins enriched in basic amino acid residues, and *N*-ethylmaleimide, which chemically modifies cysteine amino acid residues, greatly reduced specific [^{125}I-Tyr^4] BN binding activity, the BN receptor may be a protein with essential cysteine amino acid residues. Initial biochemical studies indicated that the molecular weight of the SCLC receptor for BN-like peptides may be 78,000 daltons. Currently much effort is being devoted to solubilization, purification, and the genetic composition of the BN receptor. Electron microscopic studies indicate that BN receptors are present on the surface of the SCLC cells (47). Thus if a monoclonal antibody is made against the BN receptor, it may serve as a valuable therapeutic agent. BN functions as an autocrine growth factor in SCLC (37). Thus anti-BN monoclonal antibodies neutralize the endogenous BN-like peptides that are secreted and inhibit tumor growth. It is interesting to note that the monoclonal antibodies, similar to the receptor, recognize the C terminal of BN. Thus, when the BN-like peptides are bound to the antibody, they cannot bind to the receptor and hence they are neutralized.

NEUROKININS

Substance P is an 11 amino acid peptide initially discovered in 1930 and isolated and sequenced from the bovine hypothalamus in 1970 (48). Initially, it was identified based on its ability to contract smooth muscle in the intestine, and because it was not acetylcholine, it was named SP for "preparation" (49). It has a free N terminal, whereas the C terminal is amidated; hence it can be readily degraded by aminopeptidases. It is structurally similar to the 12 amino acid peptide kassinin and the 10 amino acid peptide substance K (SK), each of which was isolated from the porcine spinal cord. Each of the three peptides has a common C-terminal heptapeptide (Phe-X-Gly-Leu-Met-NH_2). Kassinin has 7 of the same 11 amino acid residues as SP, and SP and SK have 5 identical amino acids (Table 5). Nonetheless, there are distinct receptors for each of these three peptides.

This class of peptides is generally referred to as fast-acting kinins or neurokinins (NK). Precursor proteins have been identified in the bovine corpus striatum. One precursor, β-preprotachykinin, contains one copy each of SP and SK. The precursor is synthesized as a 130 amino acid protein and contains an N-terminal hydrophobic sequence of 25 amino acid residues. Because the SP and SK sequences are flanked by basic amino acid residues, they may be cleaved from the precursor protein by trypsin like enzymes and then further processed by amidating enzymes. Also, α-preprotachykinin contains a single copy of SP (50). The precursor for kasinin remains unknown. An additional two tachykinins are physalaemin, which is structurally similar to SP, and eledoisin, which is structurally similar to kassinin. Initially, two types of tachykinin receptors were identified using smooth muscle preparations. The SP-physalaemin (NK-1; SP-P) type of receptor was identified in the guinea pig ileum, guinea pig bladder, and rat bladder. The order of peptide potency for NK-1 receptors was physalaemin = SP > kassinin = eledoisin. Also, the SP-eledoisin (NK-3; SP-E) receptor was identified in the rat vas deferens, rat duodenum, hamster bladder, and mouse

Table 5 Structure of Tachykinins[a]

Substance P	Arg–Pro–Lys–Pro–Gln–Gln–Phe–Phe–Gly–Leu–Met–NH_2
Substance K	His–Lys–Thr–Asp–Ser–Phe–Val–Gly–Leu–Met–NH_2
Eledoisin	Pyr–Pro–Ser–Lys–Asp–Ala–Phe–Ile–Gly–Leu–Met–NH_2
Physalaemin	Pyr–Ala–Asp–Pro–Asn–Lys–Phe–Tyr–Gly–Leu–Met–NH_2
Kassinin	Asp–Val–Pro–Lys–Ser–Asp–Gln–Phe–Val–Gly–Leu–Met–NH_2

[a]Sequence homologies relative to substance P are underscored.

bladder. The order of peptide potency for the NK-3 receptor was eledoisin = kassinin >> physalaemin = SP (51).

The multiplicity of tachykinin receptors has been confirmed by binding studies. For NK-1 receptors, [^{3}H] SP, [^{125}I] Bolton Hunter (BH)-SP, [^{3}H] physalaemin, and [^{125}I] physalaemin have been employed as receptor probes in the rat CNS (52-55). [^{3}H] SP, [^{125}I] BH-SP, and [^{3}H] physalaemin are now commercially available. The rank order of tachykinin potency to inhibit specific binding to NK-1 receptors is SP = physalaemin >> SK = kassinin = eledoisin. For NK-3 receptors, [^{125}I] BH-eledoisin has been used as a receptor probe (56, 57). The NK potency to inhibit specific binding of [^{125}I] BH-eledoisin is eledoisin = kassinin > physalaemin > SK > SP. Also, SK (NK-2) receptors have been identified in the gastrointestinal tract, urinary bladder smooth muscle preparations, and the substantia nigra of the guinea pig (58,59). The rank order of tachykinin potency to inhibit specific [^{125}I] BH-SK binding is SK >> kassinin = eledoisin > SP = physalaemin. Thus the pharmacology of NK-1, NK-3, and NK-2 receptors is distinct.

The NK-1 receptors bind physalaemin and substance P with high affinity, but the density of sites in the brain and exocrine pancreas is very low (Table 6). The entire sequence of SP is required for high-affinity binding activity to rat CNS receptor sites. The affinity of deca(2-11)SP and octa(4-11)SP is reduced by one and two orders of magnitude, respectively, relative to SP (IC_{50} = 1.1 nM). Hepta(5-11)SP and hexa(6-11)SP are 0.6 and 0.2% as potent as SP, respectively. Penta(7-11)SP and Tri(9-11)SP are 0.005 and <0.001% as potent as SP. It is generally accepted that the C-terminal hexapeptide of SP is essential for biologic activity. The pharmacology of the putative SP antagonists is shown in Table 7. Spantide, (D–Arg^1, D–Pro^2, D–Trp^{7,9}, Leu^{11})SP, and (D–Pro^2, D–Trp^{7,9})SP had approximately the same potency (IC_{50} = 2 μM). (D–Pro^2, D–Phe^7, D–Trp^9)SP was less potent (IC_{50} = 10 μM). (D–Pro^4, D–Trp^{7,9})SP

Table 6 Pharmacology of Substance P Antagonists in Rat Brain Slices[a]

Peptide	IC_{50} (μM)
Spantide	2
(D–Arg^1, D–Pro^2, D–Trp^{7,9}, Leu^{11})SP	2
(D–Pro^2, D–Trp^{7,9})SP	2
(D–Pro^4, D–Trp^{7,9,11})SP$^{4-11}$	5
(D–Pro^4, D–Trp^{7,9})SP$^{4-11}$	5
(D–Pro^2, D–Phe^7, D–Trp^9)SP	10

[a]The ability of various SP analogs to inhibit specific [^{125}I] physalaemin binding to rat brain slices was determined.

Table 7 SP-P Receptor Binding Properties[a]

Parameter	Brain homogenate	Brain slices	Brain slices	Pancreatic acini
Radiolabeled probe	$[^3H]$ SP	$[^{125}I]$ physalaemin	$[^3H]$ SP	$[^{125}I]$ physalaemin
Buffer	20 mM HEPES	Krebs-Ringer	50 mM Tris	Krebs-Ringer
Temperature (°C)	4	25	25	37
K_D (nM)	0.4	0.3	0.5	2
Density	27 fmol/mg protein	22 fmol/mg protein	22 fmol/mg protein	500 per cell

[a]Receptors for SP-P receptors were characterized using rat brain membranes (53), rat striatal slices (54,52), and dispersed guinea pig pancreatic acini (66).

functions as a central SP antagonist in that it blocked the ability of SP to depolarize locus ceruleus neurons (60). Because (D–Arg^1, D–Pro^2, D–Trp^{7,9}, Leu^{11})SP antagonized the ability of BN or SP to stimulate amylase release (13), these SP antagonists may also function as BN receptor antagonists.

Regional distribution of NK-1 receptors shows moderate to high grain densities in the nucleus accumbens and caudate putamen. Because SP elevates homovanillic acid and 3,4-dihydroxyphenylacetic acid levels in the striatum, NK-1 receptors may be present on DA-containing neurons (61). Low to moderate grain densities were present in the cortex; SP is an excitatory agent in the cortex and striatum (62,63). Moderate to high grain densities were observed in the olfactory bulb and tubercle, dentate gyrus, hippocampus, and subiculum. Moderate grain densities were present in certain hypothalamic regions, such as the suprachiasmatic nucleus, as well as the central gray. Because moderate levels of SP are present in these regions, the NK-1 receptors may play a role in the regulation of circadian cycles and pain threshold, respectively. Also, moderate to high grain densities were observed in the superior colliculus and inferior colliculus, and these areas are involved in visual and auditory information processing, respectively. High grain densities were present in the amygdala, where SP is an excitatory agent (64). High grain densities were present in the locus ceruleus and dorsal parabrachial nuclei but not the mesencephalic tract of the trigeminal nerve; SP is excitatory in the former two areas but not the latter area (64). High grain densities were observed in the nucleus tractus solitarii, where SP is a potent hypotensive agent (65). High NK-1 receptor density was observed in the dorsal but not ventral horn of the spinal cord, and SP may mediate pain transmission in the spinal cord via primary sensory afferent neurons (2).

SP, like BN, does not stimulate the second messenger cAMP. Rather, it stimulates Ca^{2+} flux in pancreatic acini (66). The ED_{50} for physalaemin is 2 nM, whereas SP, eledoisin, and kassinin were less potent with ED_{50} = 5, 100, and 250 nM, respectively. Similarly, SP elevated intracellular cGMP concentrations in, and caused amylase secretion from, pancreatic acini; SP was more potent than either eledoisin or kassinin. SP, similarly to BN, stimulates phosphatidylinositol turnover (67). Thus it is likely that SP generates diacylglycerol-activating protein kinase C and inositol-1,4,5 trisphosphate, which would raise intracellular Ca^{2+} levels. Recently, it was reported that SP stimulates DNA synthesis in cultured arterial smooth muscle cells and human skin fibroblasts (68). Further, the increase in connective tissue cell growth caused by SP is inhibited by the SP antagonist spantide.

Currently, efforts are focusing on the biochemical characterization of SP receptors by solubilization from rat brain membranes using 0.125% sodium glycodeoxycholate and 1 M NaCl (69). Approximately 50% of the high-affinity binding sites can be recovered in the solubilized fraction. The pharmacology of binding to the solubilized receptor was similar to the membrane-bound receptor in that deca(2-11)SP, octa(4-11)SP, and hepta(5-11)SP were approximately one, two, and three orders of magnitude less potent than SP. Physalaemin and spantide were equipotent and three orders of magnitude less potent than SP. Thus the solubilized receptor has a pharmacology similar to that of the membrane-bound receptor.

Although SP receptors have not been implicated in any particular disease, the endogenous peptide has. SP levels are greatly reduced in the caudate putamen of patients with Huntington's disease (70,71).

VIP AND SECRETIN

Vasoactive intestinal polypeptide, a 28 amino acid peptide, and secretin, a 27 amino acid peptide, were initially isolated from the small intestine and sequenced in 1970 (72,73). VIP decreases blood pressure following intravenous injection in dogs, whereas secretin is a hormone released from the small intestine that causes electrolyte secretion from the pancreas (74,75). They are structurally similar in that they share nine sequence homologies; eight of the nine are in the N-terminal half of the peptide. In spite of these similarities, there are two distinct types of receptors, one that binds VIP with high and secretin with low affinity and the other that binds secretin with high and VIP with low affinity. Peptide histidine isoleucine (PHI) is a 27 amino acid peptide that has 14 sequence homologies with VIP (76). PHI binds with low affinity to both VIP as well as secretin receptors. Gastric inhibitory peptide (GIP), a 42 amino acid peptide with four sequence homologies to VIP (Table 8), and glucagon, a 29 amino acid peptide with six sequence homologies to VIP, do not bind to VIP or secretin receptors.

Table 8 Amino Acid Sequence of the VIP-Related Peptides[a]

	1 — 5 — 10 — 15 — 20
SEC	His–Ser–Asp–Gly–Thr–Phe–Thr–Ser–Glu–Leu–Ser–Arg–Leu–Arg–Asp–Ser–Ala–Arg–Leu–Gln–Arg–Leu–Leu–
VIP	His–Ser–Asp–Ala–Val–Phe–Thr–Asp–Asn–Tyr–Thr–Arg–Leu–Arg–Lys–Gln–Met–Ala–Val–Lys–Tyr–Leu–
PHI	His–Ala–Asp–Gly–Val–Phe–Thr–Ser–Asp–Phe–Ser–Arg–Leu–Leu–Gly–Gln–Leu–Ser–Ala–Lys–Lys–Tyr–Leu–
GLU	His–Ser–Gln–Gly–Thr–Phe–Thr–Ser–Asp–Tyr–Ser–Lys–Tyr–Leu–Asp–Ser–Arg–Arg–Ala–Gln–Asp–Phe–Val–
GIP	Tyr–Ala–Glu–Gly–Thr–Phe–Ile–Ser–Asp–Tyr–Ser–Ile–Ala–Met–Asp–Lys–Ile–Arg–Gln–Gln–Asp–Phe–Val–
SEC	Gln–Gly–Leu–Val–NH_2
VIP	Asn–Ser–Ile–Leu–Asp–NH_2
PHI	Glu–Ser–Leu–Ile–NH_2
GLU	Gln–Trp–Leu–Met–Asn–Thr
GIP	Asn–Trp–Leu–Leu–Ala–Gln–Lys–Gly–Lys–Lys–Ser–Asp–Trp–Lys–His–Asn–Ile–Thr–Gln
	25 — 30 — 35 — 40

[a]Sequence homologies relative to SEC are underscored.

The precursor protein for VIP has been cloned from human neuroblastoma cells (77). Prepro-VIP and pro-VIP have molecular weights of 20,000 and 17,500 daltons, respectively. Pro-VIP contains one VIP and one PHM, a 27 amino acid peptide that has 25 of the same 27 amino acid sequences that PHI has. Pro-VIP is likely to be metabolized by trypsinlike enzymes as dibasic amino acid residues precede and follow amino acid residues 125-153 of pro-VIP. Pro-$VIP^{125-153}$ is then metabolized by amidating enzymes to convert the glycine at position 153 to an amide, resulting in VIP, which has an amidated C terminal. Little is known about the synthesis of secretin.

VIP receptors were first characterized in rat liver plasma membranes and adipocytes using [^{125}I]VIP and centrifugation techniques (78,79). Subsequently, VIP receptors were characterized in guinea pig pancreatic acini (80) and in the CNS using rat and guinea pig brain membranes (81,82). [^{125}I]VIP binds to two classes of sites using rat brain synaptosomes (83) (Table 9), one of high affinity (K_D = 3 nM) and one of moderate affinity (K_D = 40 nM). Pharmacological studies indicated that specific [^{125}I]VIP binding was inhibited with high affinity by VIP (IC_{50} = 4 nM) and low affinity by secretin (IC_{50} = 600 nM) but not glucagon ($IC_{50} > 10,000$ nM).

Secretin receptors have been characterized in the guinea pig pancreas and neuroblastoma cells as well as the rat CNS (84-86) using [^{125}I]secretin and a centrifugation assay (Table 10). [^{125}I]Secretin bound with high affinity (K_D = 0.2 nM) to a single class of sites (B_{max} = 9 fmol/mg protein) using rat brain homogenate (84). In contrast, radiolabeled secretin bound with slightly lower affinity to neuroblastoma cells and exocrine pancreas cells (85,86). Specific

Table 9 VIP Receptor Binding Properties[a]

Parameter	Brain homogenate	Brain slices	Pancreatic acini
Radiolabled probe	[^{125}I]VIP	[^{125}I]VIP	[^{125}I]VIP
Buffer	50 mM HEPES	Krebs-Ringer	Krebs-Ringer
Temperature (°C)	37	25	37
K_D (nM)	3, 62	3, 28	0.9
Density	0.2, 8.4 pmol/mg protein	0.3, 1.5 pmol/mg protein	9000 per cell

[a]Receptors for VIP were characterized using rat brain cerebellar membranes (82), rat striatal slices (91), and dispersed guinea pig pancreatic acini (80).

Table 10 Secretin Receptor Binding Properties[a]

Parameter	Brain homogenate	Neuroblastoma cells	Pancreatic acini
Radiolabeled probe	[^{125}I] Secretin	[^{125}I] Secretin	[^{125}I] Secretin
Buffer	10 mM HEPES	Krebs-Ringer's	Krebs-Ringer's
Temperature (°C)	37	37	37
K_D (nM)	0.2	1	11
Density	9 fmol/mg protein	3200 fmol/mg protein	93,000 per cell

[a]Receptors for secretin were characterized using rat cerebellar membranes (85), c1300 clone of N18T62 neuroblastoma cells (86), and dispersed guinea pig pancreatic acini (84).

[^{125}I] secretin binding to rat brain homogenate was inhibited with high affinity by secretin (IC_{50} = 0.2 nM) and low affinity by VIP (IC_{50} = 150 nM) and PHI ($IC_{50} >$ 10,000 nM).

The intrinsic activity of VIP and secretin resides in the N terminal of the peptide. Secretin$^{1\text{-}14}$ functions as an agonist, whereas secretin $^{5\text{-}27}$ or secretin$^{14\text{-}27}$ does not (80,87,88). Both the N-terminal and C-terminal regions of VIP and secretin, however, are required for high-affinity binding activity. Thus the affinities of secretin$^{1\text{-}14}$ and secretin$^{14\text{-}27}$ are over three orders of magnitude lower than the affinity of secretin for both VIP and secretin receptors. Because secretin$^{5\text{-}27}$ inhibits specific [^{125}I] VIP and [^{125}I] secretin binding with IC_{50} of 100,000 and 400 nM, respectively, it is approximately three orders of magnitude less potent for both VIP and secretin receptors than is secretin.

Recently, the discrete regional distribution of VIP receptors was determined using in vitro autoradiographic techniques (89-91). The density of [^{125}I] VIP receptors was moderate in the cerebral cortex, especially lamina 1 (Figure 2), where VIP can cause excitation (92) and stimulate glycogenolysis (93). Moderate grain densities were present in the caudate putamen but not the globus pallidus. VIP stimulates glucose utilization using 2-deoxyglucose in the striatum and the cingulate cortex (94,95). Moderate to high grain densities were observed in the supraoptic, suprachiasmatic, periventricular, and paraventricular nuclei. Because VIP stimulates prolactin, growth hormone, and leutinizing hormone secretion (96), these hypothalamic VIP receptors may be involved in the regulation of pituitary hormone secretion. Because moderate to high densities of [^{125}I] VIP binding sites were present in the superior colliculus, olfactory bulb, and olfactory tubercle, VIP may play a role in visual information processing and olfaction, respectively. High grain densities were observed in the pineal gland. VIP stimulates cAMP accumulation and serotonin-

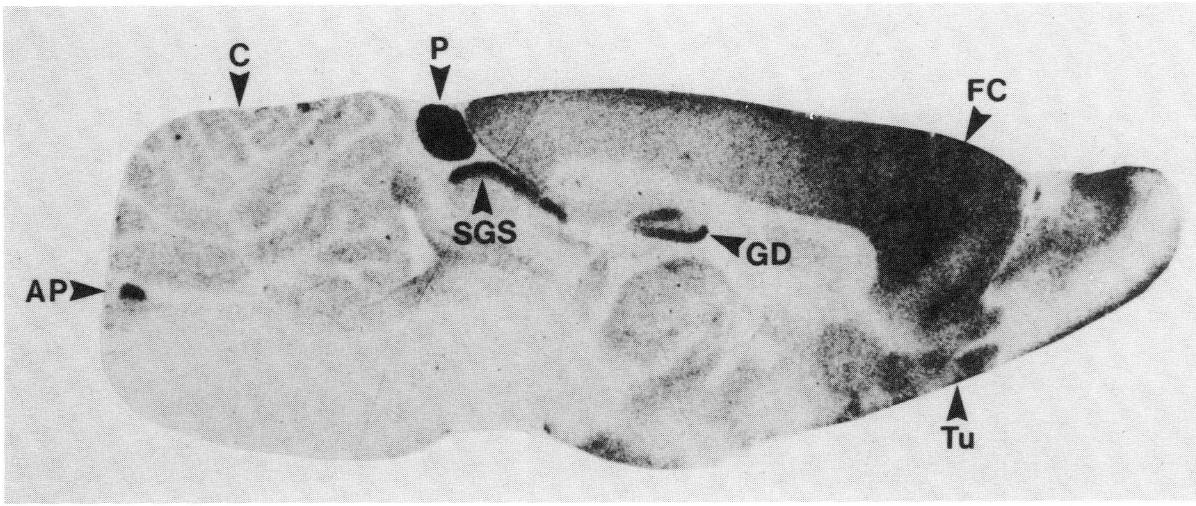

Figure 2 Localization of central VIP receptors. A light field autoradiogram illustrates [^{125}I] VIP binding sites in a sagittal rat brain section derived from Paxinos and Watson (24) coordinates lateral 0.4 mm. AP, area postrema; C, cerebellum; FC, frontal cortex; GD, dentate gyrus; P, pineal; SGS, superficial gray layer of the superior colliculus; and Tu, olfactory tubercle.

N-acetyltransferase activity in the rat pineal, resulting in increased melatonin synthesis (97). Because moderate densities of VIP receptors are present in the area postrema, VIP may regulate cerebral blood flow. VIP causes vasodilation of isolated middle cerebral arteries (98). The densities of VIP receptors in the hippocampus, dentate gyrus, and paraventricular thalamic nucleus were moderate to high. VIP can elevate cAMP levels in the cortex, caudate putamen, hypothalamus, thalamus, and hippocampus (99-101).

The distribution of secretin receptors has not been investigated using autoradiographic techniques, as radiolabeled secretin readily adheres to the glass or plastic slides used to mount the brain sections. In homogenate binding studies, however, specific secretin binding was highest in the cerebellum, intermediate in the cortex, striatum, hippocampus, thalamus, and hypothalamus, and lowest in the midbrain and medulla-pons (85). In contrast, the density of VIP receptors in the cerebellar cortex was low. Therefore there is a different regional distribution of VIP and secretin receptors.

VIP elevates cAMP levels from rat brain slices in a concentration-dependent manner. Micromolar but not nanomolar doses of VIP elevate cAMP levels 9-fold from rat frontal cortex membranes and 100-fold from rat hippocampal membranes (102). In contrast, using dispersed acini derived from the guinea pig pancreas, nanomolar concentrations of VIP (ED_{50} = 0.9 nM) elevate cAMP levels 10-fold. Because three orders of magnitude more VIP is required to elevate cAMP levels in the brain than in the exocrine pancreas, VIP receptors in the exocrine pancreas may be more tightly coupled to adenylate cyclase than are receptors in the CNS.

Secretin can also elevate cAMP levels in rat brain slices, superior cervical ganglia, cultured brain cells, neuroblastoma, and neuroblasoma x glioma cell lines (103-107). In brain slices, micromolar but not nanomolar doses of secretin increased the cAMP levels approximately threefold. Thus in the CNS, the magnitude of the cAMP increase caused by VIP exceeds that of secretin. This may result because the density of VIP receptors is approximately an order of magnitude greater than that of secretin. The increase in cAMP caused by secretin (1 μM) is reversed by micromolar doses (ED_{50} = 2000 nM) of secretin$^{5\text{-}27}$. Also, secretin$^{5\text{-}27}$ inhibited [^{125}I] secretin binding with an IC_{50} value of 600 nM. In contrast, secretin$^{5\text{-}27}$ (25 μM) did not inhibit the increase in cAMP levels caused by VIP (1 μM). These data suggest that, in the CNS, secretin$^{5\text{-}27}$ may function as a secretin but not VIP receptor antagonist.

VIP, unlike secretin, is a potent secretagogue. VIP (ED_{50} = 0.2 nM) causes secretin of the enzyme amylase from guinea pig pancreatic acini. Also, VIP causes secretion of growth hormone and prolactin from the rat pituitary. VIP (1 μM) elevates the intracellular levels of cAMP fourfold and causes secretion of BN-like peptides from SCLC cells (44). Also, the survival of embryonic spinal cord neurons in culture is reduced as a result of electrical blockade of tetrodotoxin, and this loss of spinal cord neurons is reversed by VIP. VIP may enhance neuronal survival by causing secretion of essential growth factors (108). Thus VIP is a secretagogue that causes the release of enzymes, hormones, and growth factors from certain cells. In contrast, secretin does not stimulate amylase release from pancreatic acini. Also, secretin stimulates hypothalamic tyrosine hydroxylase activity, resulting in reduced secretion of prolactin and luteinizing hormone from the pituitary (109,110).

VIP receptors are being characterized biochemically (111). In human lymphoblast [^{125}I] VIP was cross-linked to a single polypeptide of molecular weight 43,000 daltons using disuccinymidyl suberate (DSS). Similarly, using rat liver membranes, [^{125}I] VIP was cross-linked to a major and minor band of molecular weights 48,000 and 86,000, respectively (112). Using guinea pig pancreatic acinar membranes [^{125}I] VIP was cross-linked using DSS to major and minor bands of molecular weights 45,000 and 30,000 daltons (113).

Abnormal production of VIP has been associated with watery diarrhea syndrome (114).

BRADYKININ

The kinins, including the prototype peptide bradykinin, are highly active in a number of different organs. BK is thought to play an important role in several pathophysiological conditions, including inflammation, pain, shock, and hypertension (115-120). BK acts not as a standard blood-borne hormone but rather a local hormone, released at or near the site of action by the action of enzymes called kallikreins on protein precursors called kininogens. The resultant bioactive peptides, including BK, kallidin (LysBK), and several related peptides (see Table 11), share a spectrum of activity: they are potent vasodilators, lowering blood pressure; they increase capillary permeability, causing edema; they are potent algesic agents; they cause contraction of various smooth muscles; and they cause diuresis, natriuresis, and intestinal chloride secretion (115-120).

The kinins also have numerous nervous system interactions; the best studied is the pain response, which is thought to involve direct stimulation of free afferent nerve endings (120-123). Pupillary constriction caused by BK is thought to arise from stimulation of sensory nerves and relase of SP (124,125), and other

Table 11 Amino Acid Sequences for Bradkininlike Peptides

Bradykinin	Arg–Pro–Pro–Gly–Phe–Ser–Pro–Phe–Arg
Kallidin	Lys–Arg–Pro–Pro–Gly–Phe–Ser–Pro–Phe–Arg
(DesArg^9)BK	Arg–Pro–Pro–Gly–Phe–Ser–Pro–Phe
(DPhe^7)BK	Arg–Pro–Pro–Gly–Phe–Ser–DPhe–Phe–Arg

smooth muscle effects of kinins may also be caused at least in part from stimulation of intrinsic nerves. The CNS role of kinins is less well understood despite their presence in brain tissue (126,127). Pharmacologically, injection of kinins intracerebroventricularly (ICV) has numerous effects; the most well documented is a sharp but transient increase in mean arterial blood pressure starting about 1 min after ICV injection of 1-5 μg BK (although some effects have been reported at lower doses) (see Reference 128). The locus for this effect is not clear; distinct effects have been reported following the direct tissue injection into the pars ventralis of the lateral septum (129) and into various hypothalamic nuclei (130). BK may act on sites adjacent to the third ventricle to increase blood pressure (131). Numerous other effects have been reported for ICV BK, including an increase in heart rate, antinociception, antidiuresis, hyperthermia, hyperglycemia, miosis, and a distinct behavioral syndrome of excitation followed by depression (128).

Until recently, most information on the BK receptor came from indirect sources, such as in vitro bioassays. BK has potent effects on a wide variety of smooth muscles, including most blood vessels (primarily vasodilation of small resistance vessels, with some constriction seen in larger veins and arteries), the guinea pig ileum (the characteristic slow contraction of this muscle led to the name BK), and stomach, the cat jejunum, the rat uterus (contraction of these two muscles can be seen at concentrations as low as 0.1 ng/ml), rat duodenum (relaxation), rat vas deferens (contraction), and bronchial muscles (bronchoconstriction seen in most species) (119). Many of these, especially the guinea pig ileum, have been used as bioassay preparations, permitting structure-activity relationship (SAR) studies of kinin analogs. Non-smooth muscle assays have also been employed, including BK stimulation of ion transport in the gut (132). An in vivo assay sometimes employed with kinins uses the blood pressure drop following intra-arterial injection in rats (133).

With one important exception noted below, studies in these preparations have demonstrated a conservative kinin structure activity relationship. The entire BK structure is required for activity; deletion of either terminal arginine destroys activity (Table 12) (116-119). Kallidin is essentially equipotent with BK; small but variable potency increases are sometimes obtained with N-terminal extensions, such as DArg or LysLys (134). There have been some reports of synthetic BK analogs with a somewhat enhanced potency compared with that of BK (e.g., β-thienyl-alanine in positions 5 and 8) (21), but these effects have not been very large and seem to be variable across different tissues (119,132,134-137).

Bioassays are essential for the development of antagonist molecules, a goal that has proved frustratingly elusive for peptide researchers. Recently, Vavrek and Stewart (138) announced the synthesis of a series of BK analogs with a DPhe substitution in the seventh position that demonstrated clear-cut antagonist potency in vitro (guinea pig ileum and rat uterus) and in vivo (rat blood pressure).

Table 12 Kinin Receptor Binding Properties[a]

Parameter	Guinea pig ileum	Bovine uterus	Human fibroblasts	N1E-115 cells
Radiolabeled probe	[^{3}H] BK	[^{125}I-Tyr] kallidin	[^{3}H] BK	[^{3}H] BK
Buffer	25 mM TES, pH 6.8	20 mM PIPES pH 6.3	Hanks' medium	DMEM medium
Temperature (°C)	25	4	37	25
K_D (nM)	0.013, 0.91	0.165	6.1	1.0[b]
Density	8.3,14 pmol/ tissue	165 fmol/mg protein	211 fmol/mg protein	95,000 per cell[b]

[a]Receptors for kinin peptides were characterized using homogenates of guinea pig ileum (132,166,168) or bovine uterine myometrium (136,165), or to whole-cell preparations of human fibroblasts (146,147) or N1E-115 neuroblastoma clones (154).
[b]Three sites detected by computer analysis; numbers are from site correlated with cGMP production (154).

Some of these compounds appear to be pure competitive antagonists with reasonable potency (128,139); this represents an important advance in this field.

Although kinin SAR in most assays are quite similar, Regoli and associates (119) have reported the presence in certain tissues of a kinin receptor that shows a marked preference for BK analogs minus the C-terminal arginine. As these peptides are virtually inactive in standard kinin bioassay, this implies a different receptor; they have termed this desArg^9-BK-preferring site the "B-1" receptor and the classic site the "B-2" receptor (119). A number of compounds, notably desArg^{10} [Leu^8] kallidin, act as competitive inhibitors at the B-1 receptor (but not at B-2 receptors) (119). The prototype tissue used to assay B-1 activity is the rabbit aorta; the receptor has also been detected in several other (mostly vascular) preparations. The level of B-1 sites is normally quite low; however, it can be shown to form de novo in tissues allowed to sit several hours in vitro or following in vivo injections of lipopolysaccharide (119). Thus it may be that the B-1 site is of importance primarily in response to certain pathological situations. B-2 (or classic) kinin receptors are also associated with pathology, of course, as many of the effects of BK noted previously are part of the body's normal response to injury, inflammation, or disease (115-120). There are also some extreme pathological situations in which kinins may be involved, including carcinoid syndrome, postgastrectomy dumping syndrome, and inflammatory bowel disease (132). The actual involvement of kinins and their receptors in any of these conditions has yet to be proven.

A number of studies suggest that many kinin effects are mediated by prostaglandins or other metabolic products of arachidonic acid. BK has been shown to stimulate the release of arachidonate, PGE_2, and $PGF_{2\alpha}$ in several different in vitro preparations (e.g., 118,123). Whether this release is physiologically significant or whether the prostaglandins contribute to or are required for the subsequent kinin effects is not clear and may vary with the tissue. The ability of BK to produce a pain response may be accentuated by prostaglandins, which are thought to "sensitize" the tissue to the painful stimuli of BK (140); the synthesis of the prostaglandins, in turn, may be in part caused by BK (123).

There is also considerable evidence for BK effects on cyclic nucleotide formation, phosphatidylinositol metabolism, and calcium mobilization. BK has been shown to stimulate production of cAMP in a variety of tissues and cell types, including guinea pig lung slices (141), the supraoptic nucleus of the rat hypothalamus (142), rabbit renal medulla (143), and human and mouse fibroblasts (144-147) and mast cells (148). In some cases, this effect was blocked by inhibitors of prostaglandin synthesis (i.e., indomethacin), suggesting that the increase in cAMP was secondary to stimulation of prostaglandin synthesis (141,144,149); in other cases, the rise in cAMP was independent of prostaglandin synthesis (143,145,148). BK *inhibits* adenylate cyclase in NG108-15 neuroblastoma X glioma cells (150).

Others have found BK-stimulated increases in cGMP, in such tissues as rat muscle (151), guinea pig lung (141), and human umbilical artery (152), and cell lines, such as the C6-4-2 glioma (153), the NG108-15 (153), and the N1E-115 neuroblastoma (154). This effect was not found to be secondary to prostaglandin formation in guinea pig lung (141) but was inhibited by lipoxygenase inhibitors in N1E-115 cells (154), suggesting the involvement of leukotrienes.

BK can stimulate the breakdown of inositol phospholipids into inositol phosphates in human fibroblasts (155), bovine endothelial cells, and canine carotid artery segments (156) and clonal cell lines, including NCB-20 (157), NG108-15 (128,150,158-160), and N1E-115 (128,139).

The electrophysiological effects of BK have been measured in several systems. BK causes a change in potential difference across guinea pig ileal mucosal membrane due to active secretion of Cl^- (132,161). This action is blocked by the calcium entry blocker verapamil, indicating a link to calcium flux; it is partially blocked by various inhibitors of arachidonic acid metabolism (132). Other effects of BK have also been shown to be sensitive to verapamil, including a neuronally induced contraction of the rat vas deferens (162). BK stimulates calcium influx in human fibroblasts (144). In rat glioma C6-4-2 cells, BK causes hyperpolarization lasting 10-30 sec, thought to be due to an increased K^+ conductance (163). In neuroblastoma-glioma NG108-15 cells, BK produces a sustained depolarization preceded by a transient hyperpolarization (158). The hyperpolarization phase may be due to breakdown of phosphatidylinositol leading to an eventual increase of intracellular Ca^{2+} that activates calcium-dependent K^+ channels; the depolarization phase appears to be caused by inhibition of M channels, decreasing K^+ efflux (150).

Kinin receptors have been studied using radioligand assays with tritiated or iodinated peptides. Substitution of tyrosine for phenylalanine in positions 5 or 8 or addition of tyrosine to the N terminal (to give [Tyr^1] kallidin, or Tyr^0-BK) allows radioiodination to yield [^{125}I-Tyr] BK ligands (Table 11). These have been widely employed as radioimmunoassay tracers (e.g., References 127 and 164) and also receptor ligands; [^{125}I-Tyr^1] kallidin was found to be optimal for the latter funcion (164,165). Tritiated BK with high specific activity (30-80 Ci/mmol) has become the radioligand of choice for most BK receptor studies.

Traditional kinin pharmacology is primarily concerned with peripheral systems, and most successful BK binding studies have been with peripheral tissues. Reports of BK binding studies in the CNS showed (166) specific binding of [^{3}H] BK to rat brain membranes that was not readily characterizable, although [^{3}H] BK binding in guinea pig ganglia and spinal cord using autoradiographic techniques has been more successful (167).

The first demonstration of BK receptors occurred when [^{125}I-Tyr^1] kallidin was found to label two sites in bovine uterus myometrium, a high-affinity (K_D = 0.1 nM) and a lower affinity (K_D = 20 nM) site (165). Such binding was inhibited by mono- and divalent cations and showed a pH optimum of 6.5. Binding was also seen in bovine esophageal mucosa, ductus arteriosus, left ventricle, lung parenchyma, and bronchi (165). The possibility that the binding was to membrane-bound degradative enzymes (i.e., kallikrein and others) was addressed by examining the ability of an enzyme inhibitor (SQ 20,881; Teprotide) to compete for binding. Saturable binding was reduced by Teprotide in the renal medulla but not the myometrium, leading to the conclusion that binding to the former tissue was to an enzyme rather than a true biologic receptor (165). In a subsequent report, the binding potency of a large number of BK analogs versus [^{125}I-Tyr^1] kallidin in myometrial membrane was tested, showing reasonable correlation with biologic potency (136). A solubilized preparation of this binding activity was prepared using the nonionic detergent CHAPS (136). Similar techniques were used by this group to demonstrate high-affinity binding to rat renomedullary interstitital cells (134).

Examination of [^{3}H] BK binding in guinea pig tissues using relatively low specific activity ligand (12 Ci/mmol) and a centrifugation assay showed highest levels in guinea pig ileum, colon, and duodenum and estrous rat uterus. Binding had high affinity (K_D = 5 nM) and appeared to be to a single class of sites. Binding potencies correlated well with contractile potencies in guinea pig ileum and estrous rat uterus for a series of 32 analogs (166).

Improved results were obtained by using high specific activity ligand (52-76 Ci/mmol), an elaborate cocktail of peptidase inhibitors (which allowed assays at room temperature, improving the ratio of specific to nonspecific binding), and a filtration technique to separate free from bound. Binding of [^{3}H] BK to glass fiber filters, which had previously prevented the use of this technique,was prevented by presoaking filters in 0.1% polyethyleneimine (132, 168). Two binding sites of higher affinity than previously reported were observed in guinea pig ileum with K_D values of 13 and 910 pM (168). Two sites were also detected in guinea pig heart and kidney. Ileal contractile potencies of a series of BK analogs correlated well with potency at the high-affinity site (168). The rank order of analog potency, the sensitivity to inhibition by ions, and most of the general physical characteristics of this binding were very similar to that reported earlier (166). In both studies. DesArg^{9}BK was found to be essentially inactive at inhibiting [^{3}H] BK binding, distinguishing it from the B-1 site (119). [^{3}H] BK binding has also been studied in guinea pig kidney, ureter and bladder (169) and spinal cord and ganglia (167). Binding in the mucosal and muscle layers of the guinea pig ileum correlated well with the ability of BK to stimulate intestinal chloride secretion (132).

[^{3}H] BK can label sites in human fibroblasts (146); significant down-regulation of these sites was seen with prolonged exposure to BK (147). Binding of [^{3}H] BK to isolated nephron segments of rabbit has also been reported (170); highest binding was seen in the collecting ducts. Three binding sites for [^{3}H] BK were detected in N1E-115 neuroblastoma cells using computer analysis; the site with a K_D of 1 nM correlated best with the stimulation of cGMP when a series of BK analogs was examined (154). Comparison of the ability of a series of antagonist analogs of BK to inhibit binding of [^{3}H] BK to membranes from guinea pig ileum and N1E-115 cells and to inhibit BK-stimulated production of phosphatidylinositol (PI) turnover showed similarities (139); however, differential agonist-antagonist activities were noted when the PI assay was compared with the smooth muscle assays (see below) (138,139).

To date there has been only one report on the regulation of kinin receptors: A 75-90% decrease in [^{3}H] BK binding sites in human fibroblasts was seen following 5 min preincubation with 1 μM BK at 37°C (147). Following 15 min exposure to BK, binding could be almost completely restored by 30 min incubation at 37°C in the absence of BK. The extent of down-regulation was BK concentration dependent. The ability of BK to stimulate prostacyclin (PGI_2) release in these cells showed down- and up-regulation that closely paralleled these binding changes (147). Prolonged exposure (15 min to 8 hr) to BK allowed no recovery of binding, as long as bacitracin was added to prevent metabolic destruction of BK. Such rapid autoregulation of BK receptors may be a model for the rapid desensitization that can be seen to many BK responses (119).

Increasingly, multiple forms of receptor sites are being reported for peptides. A number of lines of evidence suggest that this is also true for kinin receptors. The B-1 receptor differs from classic kinin receptors by the high potency of (desArg^9)BK analogs (and relatively low potency of BK) and the specific antagonism elicited by [Leu^8](DesArg^9)BK (119). The B-1 receptor is found primarily in vascular tissue and increases in number following certain pathological manipulations or following prolonged in vitro incubation in artificial media. Evidence for two kinin receptors in smooth muscle comes from observations of biphasic actions, such as relaxation and contraction in the guinea pig tenia coli (171). Others, using guinea pig ileum and rat duodenum, noted that lower concentrations of BK are needed for relaxation than for contraction, that BK-potentiating peptides selectively enhance one effect, and that the contractile response is selectively desensitized (172,173). BK effects in rat vas deferens appear to be mediated by two separate sites located in the smooth muscle (postsynaptic) and adrenergic terminals (presynaptic) (163). These sites differ in their sensitivity to such agents as tetrodotoxin, verapamil, and indomethacin (163). The potency order for centrally mediated antinociception in rabbit was exactly the reverse of that seen with various other peripheral effects of kinins, suggesting a different specificity of brain kinin receptors (174).

Antagonists have traditionally been a preferred tool for the differentiation of multiple receptor types. Initial findings with BK antagonists have already yielded evidence of kinin receptor multiplicity. For example, ($DPhe^7$)BK is a pure (albeit weak) antagonist in the guinea pig ileum contraction assay, whereas it is a weak agonist in the rat uterus contraction assay (138). Further evidence for multiplicity was obtained by comparing antagonist activities at inhibiting BK-stimulated PI turnover in N1E-115 cells to these activities in bioassays in vitro. ($Thi^{5,8}DPhe^7$)BK is an agonist of equivalent potency to BK in N1E-115 cells, failing to block PI turnover in concentrations as high as 1 μM [also true of ($DPhe^7$)BK] (139), whereas this analog acts as a pure antagonist in both the guinea pig ileum assay (pA_2 6.3) and rat uterus assay (pA_2 6.4) (138). Other $DPhe^7$-substituted analogs demonstrate antagonist activity in all three assays (138,139). Thus the receptor populations in the three preparations may well be distinct.

Binding studies have also revealed evidence of different kinin receptors. The two sites of different affinity for [^3H] BK in guinea pig ileum, heart, and kidney were mentioned earlier (168). In these tissues there were twice as many lower affinity sites (K_D = 910 pM) as higher affinity sites (K_D = 13 pM); the lower affinity sites showed a preference for (Lys)BK (kallidin) analogs over BK (168). Two affinities for [^{125}I-Tyr^1] kallidin binding to bovine myometrium have also been detected; however, the significance of these two sites or any difference in analog selectivity was not discussed (165). The inclusion of specific inhibitors for BK degradative enzymes was thought to be sufficient to rule out the possibility that part of the binding was to a degradative enzyme. Computer analysis of [^3H] BK binding to intact N1E-115 cells found three sites with affinities of 0.83, 1.0, and 4.9 nM; only the intermediate affinity site appeared to correlate with production of cGMP (154).

SOMATOSTATIN

Somatostatin is a cyclic 14-amino acid peptide originally isolated and characterized from ovine hypothalami (Table 13) (175). It was originally defined by its ability to inhibit pituitary growth hormone (GH) secretion and was found later to inhibit the secretion of other pituitary hormones [i.e., thyrotropin stimulating hormone and prolactin (176)] . Subsequent work has made it clear that the name somatostatin is too limited; it has been located in and demonstrated to have potent biologic activity (mostly inhibitory in nature) in a wide variety of organs. It has many paracrine and neurocrine effects in the periphery: it is secreted from D cells in pancreatic islets, inhibiting the release of insulin and glucagon; it inhibits protein secretion from the exocrine pancreas; it inhibits secretion of gastric acid, gastrin, and pepsin in the stomach; it inhibits nutrient absorption and motility in the gut; and it blocks the stimulation of adrenal aldosterone secretion by angiotensin II, to list just some of its better studied effects (177-179).

Somatostatin has been localized in a subpopulation of primary afferent neurons and may serve as a neurotransmitter for these sensory nerves (177,178, 180); it has also been localized in enteric neurons and sympathetic ganglia (177, 178). Although very high concentrations of this peptide are found in the median eminence, as befitting a release (inhibiting) factor; it is also widely distributed throughout the rest of the brain, including the cortex, hippocampus, striatum, amygdala, hypothalamus, various nuclei in the pons and medulla, and the

Table 13 Amino Acid Sequences for Somatostatin Peptides

SS[a]	14	13	12	11	10	9	8	7	6	5	4	3	2	1
	Lys–	Arg–	Glu–	Arg–	Pro–	Ala–	Met–	Ala–	Pro–	Asn–	Ser–	Asn–	Ala–	Ser
	|													
	15	16	17	18	19	20	21	22	23	24	25	26	27	28
	Ala–	Gly–	Cys–	Lys–	Asn–	Phe–	Phe–	Trp–	Lys–	Thr–	Phe–	Thr–	Ser–	Cys
	1	2	3	4	5	6	7	8	9	10	11	12	13	14
SMS 201-995	H–Phe–Cys–Phe–DTrp–Lys–Thr–Cys–Thr(ol)													
CGP 23996[b]	(desamino)Cys–Lys–Asn–Phe–Trp–Lys–Thr–Tyr–Thr–Ser–Cys													

[a]Numbers above SS are for SS28; numbers below are for SS14. Cys(3) and Cys(14) are connected by a disulfide link in both peptides.

[b]The two cysteine residues are connected by a dicarba link.

dorsal horn of the spinal cord (181). In many of these areas, somatostatin activity is colocalized with other potential transmitters, notably neuropeptide Y (181). It has a variety of behavioral and electrophysiological activities when administered centrally, although its precise role in CNS function is unknown (177, 180,181). Somatostatin has been suggested to play a role in a number of disease states, including Parkinson's disease, Alzheimer's disease, Huntington's disease, epilepsy, diabetes, duodenal ulcer, and sleep apnea (177,178,180-182).

Somatostatin derives from a high-molecular-weight precursor, prosomatostatin. The sequence of this molecule has been deduced from the nucleotide sequence of cloned somatostatin cDNA from various sources; human somatostatin cDNA codes for a single 116 amino acid precursor of which the final 14 amino acids constitute somatostatin (181). A 28 amino acid peptide containing somatostatin at the C terminal, called somatostatin-28 (SS28; see Table 13), has been described; it is present in many tissues in variable concentrations, sometimes higher than somatostatin (SS14) itself (177,181). SS28 shows a spectrum of activity similar to that of SS14; at present it remains unclear whether SS28 is simply a precursor form of somatostatin or has a physiological role separate from that of SS14.

Because of its broad spectrum of activity, much attention has been devoted to the structure-activity relationships of somatostatin in an attempt to find clinically useful analogs. SAR findings can be summarized as follows: cyclization is not absolutely required, activity resides primarily in the C terminal, and the four residues Phe-7 through Thr-10 appear to be the core of activity (Table 13) (178,183). Several small, potent analogs have been synthesized, including several with oral potency and a relatively long duration of action (184,185). A parentally active analog that antagonizes somatostatin inhibition of GH, insulin, and glucagon release has been synthesized (186).

The biologic function of SS28 has been examined in a variety of tissues. In the pancreas, SS14 was 10 times more potent at inhibiting glucagon release than SS28, but SS28 was 10 times more potent at inhibiting insulin release (187). In CNS-related functions, including inhibition of bombesin-stimulated increases in plasma glucose and epinephrine, SS28 was more potent than SS14 (188); SS28 was also more potent at producing CNS-mediated hyperthermia (189). SS14 appears to be slightly more potent than SS28 at suppression of pituitary GH release (188) and inhibition of gastric acid and pepsin secretion in the stomach (190).

Somatostatin may produce its effects via inhibition of adenylate cyclase activity (191). Hormone-stimulated cyclic AMP production is blocked by somatostatin in cultured brain cells (192), S49 lymphoma cells (193), GH_3 rat pituitary tumor cells (194), GH_4C_1 rat pituitary cells (191), and AtT-20 mouse pituitary tumor cells (195-197). Studies with forskolin stimulation of adenylate cyclase indicate that somatostatin acts via interaction with Gi, the inhibitory guanine nucleotide regulatory unit. Pertussis toxin, which ADP-ribosylates and prevents its activity, blocks the ability of somatostatin to inhibit the release of GH from primary cultures of the pituitary (196) as well as the ability to inhibit ACTH release from AtT-20 cells (195,199) and to inhibit prolactin release from GH_4C_1 cells (191) and GH_3 cells (194). Further evidence that somatostatin receptors are linked to adenylate cyclase is that guanine nucleotides can inhibit binding (e.g., see Reference 191). Although there is considerable evidence for inhibition of adenylate cyclase, the significance of the inhibition of adenylate cyclase to the observed effects of somatostatin remains controversial (177).

Somatostatin can also modulate calcium levels, inhibiting either calcium influx or intracellular mobilization. Certain effects of somatostatin can be reversed with the calcium ionophore A23187 (177). Somatostatin also causes an increased K^+ permeability in islet cells, implying that calcium effects may be secondary to the increased K^+ influx (200). The effects of somatostatin on islet cells are blocked by agents (quinine and tetraethylammonium) that interfere with potassium channels (200). Somatostatin transiently reduces intracellular Ca^{2+} in GH_3 clonal pituitary cells (201); this effect may also be mediated by Gi (191). Somewhat conflicting data exist regarding the electrophysiological effects of somatostatin; both inhibitory and excitatory effects are seen within the same brain region (180) and in guinea pig myenteric plexus neurons (202). Hyperpolarization and inhibition have been reported in hippocampal slices, rat cortex, cerebellum, and hypothalamus, and cat spinal cord; depolarization and excitation have been reported in neurons from mouse spinal cord, hippocampus, and rat and rabbit cortex (177,180). Dose-dependent and somewhat complex excitatory effects of somatostatin have been observed in dispersed rat cortical neuron (203). Somatostatin appears to affect potassium channels (probably indirectly), either opening or closing them depending upon the concentration applied (202).

Various tyrosine-containing analogs of somatostatin have been synthesized for use as ^{125}I-labeled radioligands (Table 13). Most early studies used ligands with the tyrosine in the 1 or in the 0 position (on the N-terminal end). These two analogs were compared with a third, [^{125}I-Tyr^{11}] SS14; this last was found to be much less susceptible to metabolic destruction and demonstrated higher ratios of specific to nonspecific binding (183,204); this ligand has subsequently seen widespread use. Other SS14 analogs used as radioligands include [^{125}I-Tyr^{1}DTrp^{8}] SS14, [^{125}I-Tyr^{0}DTrp^{8}] SS14, [^{125}I] CGP 23996 (131), and [^{125}I] SMS 204-090 (Table 13) (206). Radiolabeled analogs of SS28, including [^{125}I-Tyr^{0}] SS28 and [^{125}I-Leu^{8},DTrp^{22},Tyr^{25}] SS28 (= [^{125}I] LTT-SS28), have also been employed. The latter is relatively resistant to metabolic degradation, and its decrease lipophilicity yields reduced levels of nonspecific binding (207).

The first study on somatostatin receptor binding used [^{125}I-Tyr^1] SS14 to demonstrate a single class of high-affinity (K_D = 0.6 nM) binding sites in membranes from GH_4C_1 pituitary tumor cells (208); somatostatin inhibited prolactin and GH secretion in these will with a dose-response similar to its ability to inhibit radioligand binding. Binding to these receptors is inhibited by guanine nucleotides and by pretreatment with pertussis toxin (191). Subsequent studies have measured somatostatin binding to normal rat pituitary membranes (209-217), as well as mouse pituitary AtT-20 tumor cells (218-219) and human somatotrophic tumors (220). Ultrastructural autoradiography has localized the binding in normal rat pituitary to somatotropes, thyrotropes, and lactotropes, the three known targets for pituitary somatostatin (221). Pituitary binding is apparently to a single population of sites, with reported dissociation constants ranging from 0.1 to 2.4 nM (183). Some differences in pituitary binding from that in other tissues has been reported. For instance, SS28 is generally more potent than SS14 at pituitary receptors, whereas the reverse is true in brain and pancreas (183) (also see below); other tissue differences in binding potency of other analogs have also been reported (see Reference 183). Also, GH_4C_1 receptors show some differences in binding selectivity from normal pituitary receptors (191,212). In general, good correlation exists between binding potency and biologic potency for somatostatin in the pituitary (191,210,213) (although see Reference 212).

Binding sites for somatostatin have been detected in insulinoma cells (primarily pancreatic β cells); SS28 showed somewhat higher affinity than SS14 (222). In vitro autoradiography showed somatostatin binding sites on A, β, and D cells in pancreatic islets (223). Exocrine pancreas also exhibits SS binding: a single class of sites was seen in guinea pig acinar membranes using [^{125}I-Tyr^{11}] SS14, with slightly higher affinity for SS14 than for SS28 (224,225). Binding to these sites was regulated by sodium and calcium ions. Rat pancreatic acinar membranes are similarly labeled by [^{125}I-Tyr^1] SS14 (226,227). With isolated gastric cells, [^{125}I-Tyr^1] SS14 appears to label two sites (K_D = 0.08 and 4.5 nM); the higher affinity binding was ascribed to parietal cells (228). Capsular particles from the glomerulosa zone of the adrenal exhibit very high affinity binding (K_D = 0.06 nM) for [^{125}I-Tyr^0] SS14 (229), and adrenal cortex membranes bind [^{125}I-Tyr^1] SS14 with a K_D of 1.1 nM (183).

Numerous studies have examined somatostatin receptors in rat brain using various radioligands and binding protocols. [^{125}I-Tyr^{11}] SS14 binds to a single class of sites with very high affinity for SS14 (K_D = 0.08 nM) and a B_{max} of 155 fmol/mg protein in cortical membranes (204). Using the SS28 analog [^{125}I] LTT-SS28 and a filtration assay, a single class of high-affinity sites (SS14, K_D = 0.31 nM) was seen (207). However, the SS14 radioligand, compared to total cortical binding was somewhat higher (197 fmol/mg protein), and the regional distribution noticeably different: Binding was highest in hippocampus, followed by amygdala, olfactory tubercle, caudate, cortex, spinal cord, midbrain, and hypothalamus; little was seen in cerebellum (207). A single class of high-affinity sites (K_D = 0.23 nM) was seen using [^{125}I-Tyr^1] SS14 (230), in rat cortical membranes, although with a lower density (B_{max} = 53 fmol/mg protein) (230). This latter study showed a regional distribution unlike either of the previous two; cortex had the highest binding, followed by the preoptic area, olfactory tubercle, nucleus accumbens and amygdala, hypothalamus, striatum, hippocampus, and brain stem; minimal binding was seen in cerebellum or olfactory bulb. The reasons for these discrepancies in regional distribution are unclear. The distribution of binding sites does not completely correspond to the distribution of somatostatin in rat brain: immunoreactivity is highest in hypothalamus, with decreasing levels found in cortex, medulla-pons, thalamus, striatum, and cerebellum (204). Much of the high concentration of somatostatin in hypothalamus is present in cell bodies and fibers for eventual release into the pituitary system, so one might not expect a correspondingly high receptor density here compared with a terminal field, for example.

Subcellular distribution studies showed binding located largely in synaptosomal fractions (204,230). Using rat cortical synaptosomal membranes, the metabolic stability of three ligands was studied (205) and it was found that [^{125}I] CGP 23996 > [^{125}I-Tyr^{11}] SS14 > [^{125}I-Tyr^0] SS14; protection was greatly enhanced by adding 5 mM $MgCl_2$ and 20 μg/ml bacitracin, a peptidase inhibition mixture similar to that employed in earlier studies (204).

A large number of natural and synthetic somatostatin analogs have been employed to examine the SAR for somatostatin receptor binding. The cyclic structure appears to be a requirement, residues 7-11 (Table 13) being essential for ligand recognition (183). Binding was enhanced by substitution of DTrp^8; halogenation of this residue further increasing binding (183). As mentioned, different analog selectivities for different tissues have been detected. For instance, SS28 is more potent than SS14 in pituitary (210,216); the order is reversed in pancreas (183), and in brain, the potency of the competing ligand apparently depends on the labeled tracer, suggesting heterogeneity in brain tissue (see below). Similar differences among tissues have been found with synthetic analogs (183). Binding potency has been compared to potency in various bioassays; in general, excellent correlations have been obtained (204, 205,210,211,215,230).

Various factors serve to regulate somatostatin receptors. Binding to most tissues is enhanced by 5 mM Mg^{2+} (although see below), whereas Ca^{2+} augments binding only in acinar membranes (224). Guanine nucleotides inhibit somatostatin binding in different tissues (e.g., 191,214), implying linkage to adenylate cyclase. Chronic exposure of AtT-20 mouse pituitary tumor cells to somatostatin enhanced sensitivity of these cells to forskolin stimulation of cAMP and release of ACTH (125). Chronic exposure of these cells to either SS14 or

SS28 down-regulates binding sites measured by $[^{125}\text{I-Tyr}^{11}]$SS14 (218,219). The accompanying increased forskolin sensitivity is apparently *not* causally linked to the down-regulation; this could mean that somatostatin is linked to more than one effector system in these cells (219). Up-regulation has also been reported in cerebral cortex following in vivo administration of cysteamine to deplete brain somatostatin (231). Rat pancreatic acinal somatostatin receptors are apparently down-regulated by cholecystokinin-8; this effect is not seen in brain tissue (226). Phorbol esters and synthetic diacylglycerol also down-regulate acinar receptors; this effect is increased by cotreatment with A23187 (227). This finding suggests a link to the phosphatidylinositol second messenger system. In rat pituitary, somatostatin binding decreases at 2 and 6 hr following a bolus of exogenous GH to rats; at 24 hr, binding has increased to 48% above controls (232). Thyrotropin release hormone can regulate somatostatin binding in GH_4C_1 cells (233) but not normal pituitary (210). Finally, lesion of the rat nucleus basalis lowers somatostatin binding in frontal cortex, implying presynaptic localization of these sites in cholinergic terminals in this region (234).

Somatostatin receptor distribution has been measured using quantitative autoradiography. Similar distributions were found using $[^{125}\text{I-Tyr}^{11}]$SS14 and $[^{125}\text{I}]$LTT-SS28 (235). A subsequent and more detailed study utilized primarily $[^{125}\text{I}]$LTT-SS28, which yielded specific-nonspecific binding in ratios from 2:1 to 3:1 (236). Binding in gray matter regions showed striking heterogeneity, with a predominance in rostral structures; little was seen in most white matter regions. In the forebrain, very dense cortical binding was seen in infragranular layers, with a relatively receptor-poor zone in the granular layer (lamina IV) and moderately dense binding in deeper layers. Other forebrain regions that showed high levels of binding include the olfactory bulb, anterior olfactory area, lateral olfactory nucleus, lateral septum, subfornical organ, bed nucleus of the stria terminalis, and amygdala (especially lateral nuclei) (235,236). The hippocampus showed striking region- and lamina-dependent heterogeneity: low binding in the pyramidal layer contrasted with high binding in the strata oriens and radiata; binding was high in CA1 and subiculum, moderate in dentate gyrus, and low in CA2-4. In the diencephalon, thalamic binding was mostly low to moderate, but several areas of the hypothalamus showed higher levels, including periventricular and preoptic regions. Substantial levels of binding wer were also seen in the medial habenula and zona incerta. The highest midbrain binding was seen in the interpeduncular nucleus, with more modest levels in the substantia nigra, periaqueductal gray, and superior colliculus (235,236). Hindbrain structures with moderate to high binding were seen along the floor of the fourth ventricle, including the locus ceruleus, dorsal tegmental nucleus, and vestibular nuclei; other labeled regions include parabrachial, hypoglossal, and gracile nuclei. Cerebellum showed minimal binding; however, heavy labeling was seen in the deep cerebellar nuclei. Modest binding was seen in the spinal cord dorsal horn (235,236).

This distribution corresponds well with the distribution of immunoreactive somatostatin (181), with some exceptions. As noted, the highest somatostatin levels are found in the hypothalamus, which showed only modest levels of receptors. Also, high levels of binding were seen in the medial habenula and interpeduncular nucleus, regions relatively poor in immunoreactive somatostatin (181). Binding in rat brain and pituitary using sections wiped from slides showed a saturation analysis indicating a single class of sites for each ligand in both brain and pituitary, with ligand affinities between 0.2 and 0.5 nM for each ligand and each tissue (237). Distribution was similar to that already described (235,236), although differences were noted between the SS14 and SS28 ligands (see below).

[^{125}I] CGP 3996 has also been used for somatostatin autoradiography; this ligand labels an apparently homogeneous population of sites, with K_D values of 2.7-4.0 nM (238,239). The distinctive layering in cortex and hippocampus seen with earlier studies was also seen with this ligand; the highest binding was found in the strata oriens and radiatum of CA1-2, with high binding also detected in the subiculum, laminae V and VI of the temporal cortex, the cingulate cortex, the basolateral nucleus of the amygdala, and the medial habenular nucleus (238). In another study (239), a much more homogeneous pattern was reported; the highest binding was in the anterior temporal cortex and ventral cochlear nucleus; hippocampus was also high, especially CA1-2, but concentrated in strata pyramidale and radiatum; cortical binding was highest in laminae I-II. Other regions of moderate to high binding include anterior olfatory nucleus, anterior cingulate cortex, claustrum, olfactory tubercle, periaqueductal gray, trigeminal nucleus, superior olive, and cerebellar granular layer. There were several differences in regional distribution in these studies; of particular interest was the relatively high level of binding seen in the cerebellum, especially the granular and molecular layers (239), since this structure shows little or no binding in most other studies. [^{125}I] SMS 204-090 has also been employed for somatostatin autoradiography (239,240); this ligand demonstrates considerably higher affinity (K_D = 0.19 nM in cortical homogenates) and a very low level of nonspecific binding; it apparently labels a subpopulation of somatostatin receptors (see below). The highest binding was seen in superior colliculus, locus ceruleus, periaqueductal gray, entorhinal and frontoparietal cortex Ilamina IV), dorsal lateral septal nucleus, medial habenula, and amygdala-hippocampal area (239,240).

[^{125}I] LTT-SS28 has been used to label sites in human and monkey brain (180,236). In human brain, the highest binding was seen in cortex; cingulate cortex was the highest region, followed by inferior temporal gyrus, frontal cortex, medial temporal gyrus, premotor cortex, superior temporal gyrus, and the temporal pole (241). Binding was higher in laminae V and VI, with a receptor-poor zone in laminae IV (182,236). Binding to other human brain regions ranked: amygdala, substantia nigra pars compacta, hippocampus, caudate, globus pallidus, nucleus accumbens, putamen, and substantia innominata (241). Binding was low in hypothalamus and brain stem. A similar distribution was seen in monkey brain, although the nucleus accumbens, substantia innominata, and caudate were relatively higher in density (241). The presence of somatostatin receptors in a number of afferent sensory pathways suggested a role in sensory functions; dense cortical and hippocampal binding implied a role in cognitive processes; parabrachial binding underscored a probable function in respiration; and binding in the striatum and substantia nigra suggested somatostatin's possible role in the extrapyramidal system (180). A depletion of nigral somatostatin receptors was also found in Parkinson's disease, implying a location on nigrostriatal neurons (180). Somatostatin receptors are also decreased 40-50% in selected cortical and hippocampal regions in brain from patients with Alzheimer's disease, with a corresponding decrease in somatostatin immunoreactivity (182).

Although most binding studies have reported an apparently homogeneous population of binding sites, there is increasing evidence for heterogeneity. One such possibility, that different organs contain different receptors, was addressed previously. Although largely similar in binding characteristics, different tissues can be distinguished by differences in binding potencies and biologic potencies, both with natural analogs (SS14 and SS28) and with synthetic analogs (183).

A second possibility is the existence of different sites for SS14 and SS28. Studies using radioligand analogs of SS14 and SS28 in homogenate binding have found a number of differences but have failed to conclusively demonstrate the presence of separate receptors for these two peptides. Differences in sensitivity to ions and guanine nucleotides between [^{125}I-Tyr^{11}] SS14 and [^{125}I] LTT-SS28, have been reported (242,243), as well as biphasic displacement curves versus each radioligand (242), but no differences in displacement potencies using a large number of analogs. SS28 was however, more potent than SS14 at inhibiting binding of [^{125}I-Tyr^{0}] SS28 (in both rat brain and pituitary), but SS14 and SS28 were equipotent at inhibiting binding of [^{125}I-Tyr^{0}] SS14 (192). No other differences were found between the two peptides, including regional distribution. Others have confirmed that SS28 has higher affinity in brain than SS14 (using a different SS28 ligand, [^{125}I] LTT-SS28) (207), but in most studies SS14 has been found to be more potent than SS28 versus [^{125}I-Tyr^{11}] SS14 binding (210,244). In CNS-based bioassays, SS28 is more potent than SS14 (188,189). Comparison of regional differences in binding capacity using two radioligands have yielded only small differences, and none strongly indicating separate populations. The DTrp^{22} substitution in LTT-SS28 may confer more "SS14-like" properties, making this ligand a poor choice for differentiating distinct sites for SS14 and SS28 (183), although similar results have been reported using [^{125}I-Tyr^{0}] SS28 (216).

Receptor autoradiography may prove to be a more sensitive tool for the purposes of delineating two binding site populations. Differences in the distribution of binding of [^{125}I-Tyr^{0}, DTrp^{8}] SS14 and [^{125}I] LTT-SS28 using in vitro autoradiographic techniques have been reported (237,245). In wipe studies, the SS28 analog had a B_{max} value 2.7 times that of the SS14 ligand, although complete displacement of each radioligand was obtained with both SS14 and SS28. Autoradiograms showed that the SS28 ligand labeled all sites labeled by the SS14 ligand, as well as additional sites, found almost entirely in caudal structures. Thus, several deep cerebellar nuclei, the cochlear nuclei, the medial vestibular nucleus, and substantia nigra, the lateral geniculate, and the hypoglossal nucleus all showed substantial binding with the SS28 radioligand but none with the SS14 radioligand; other regions were labeled by both radioligands but showed relatively more labeling with the 28 than the 14 radioligand (237,245). No differences were detected in pituitary binding between the two ligands. These results support the existence of two subtypes of somatostatin receptors in rat brain, an SS28-preferring form in caudal structures and a nondiscriminating form in all other regions. [^{125}I] LTT-SS28 perferentially labeled sites in the median eminence (183).

The somatostatin analogs SMS 201-995 and cyclo [AlaCysPheDTrpLysThr-Cys] distinguish between two subtypes of receptors in rat brain labeled by both [^{125}I Tyr^{11}] SS14 and [^{125}I] LTT-SS28 (217,244,246). These have been termed SS_A, with high affinity, and SS_B with low (micromolar) affinity. The proportion of the two sites varied, with cortex, hippocampus, and striatum exhibiting 45, 30, and 63% SS_A binding, respectively (244). Only SS_A binding was seen in pituitary and pancreas.

SUMMARY

Neuropeptides are synthesized as high-molecular-weight prohormones, which are then metabolized by enzymes to smaller biologically active peptides. Like classic neurotransmitters, most neuropeptides are stored in intracellular granules, are released upon appropriate stimulation, and produce their effects by binding to specific, high-affinity receptors on target cells. The activated receptors then induce second messenger production and, ultimately, a biologic response. Although this chapter has specifically focused on receptors for bombesin, neurokinins, vasoactive intestinal peptide, bradykinin, and somatostatin and their properties and potential physiological roles, many more peptides with putative neuromodulatory actions are known to exist and more are being discovered regularly. The ability to measure their receptors is an important contribution toward understanding their role in mammalian tissue function. Future progress in this field will depend greatly on recent innovations in the molecular neurobiology of these peptides and their receptors, and the increased availability of selective and potent antagonists as well as novel nonpeptiditic ligands.

ACKNOWLEDGMENTS

During the preparation of this manuscript the authors were supported by NSF grant BNS-8500552 and NCI grants CA-33767 and CA-42306 to Moody and AHA grant 84-368 to Perry.

REFERENCES

1. J. L. Henry, Brain Res. *114*, 439 (1976).
2. S. S. Wolf, T. W. Moody, R. Quirion, and T. L. O'Donohue, Brain Res. *332*, 299 (1985).
3. J. Gibbs, R. C. Young, and G. P. Smith, Nature *245*, 323 (1973).
4. T. H. Moran, P. H. Robinson, and P. R. McHugh, Ann. N.Y. Acad. Sci. *448*, 621 (1985).
5. J. L. Bennett, in Neurotransmitter Receptor Binding (H. I. Yamamura, S. J. Enna, and M. J. Kuhar, eds.), Raven Press, New York, 1978, p. 57.
6. Bruns, R. F., Biochem. Pharmacol.
7. A. Anastasi, V. Erspamer, and M. Bucci, Arch. Biochem. Biophys. *148*, 443 (1973).
8. T. J. McDonald, J. Jornvall, G. Nilsson, M. Vagne, M. Ghatei, S. R. Bloom, and V. Mutt, Biochem. Biophys. Res. Commun. *90*, 227 (1979).
9. E. R. Spindel, W. W. Chin, J. Price, L. H. Rees, G. M. Besser, and J. F. Habener, Proc. Natl. Acad. Sci. USA *81*, 5699 (1984).
10. N. Minamino, K. Kangawa, and H. Matsuo, Biochem. Biophys. Res. Commun. *119*, 14 (1984).
11. T. W. Moody, L. Y. Korman, and D. N. Carney, in Neural and Endocrine Peptides and Receptors (T. W. Moody, ed.), Plenum Press, New York, 1986, p. 421.
12. M. Brown, J. Rivier, and W. Vale, Science *196*, 998 (1977).
13. J. Rivier and M. Brown, Biochemistry *17*, 1766 (1978).
14. T. W. Moody, C. B. Pert, J. Rivier, and M. R. Brown, Proc. Natl. Acad. Sci. USA *75*, 5372 (1978).
15. S. S. Wolf and T. W. Moody, Peptides *6*, 111 (1985).
16. R. T. Jensen, S. W. Jones, K. Folkers, and J. D. Gardner, Nature *308*, 61 (1984).
17. A. T. Yachnis, J. N. Crawley, R. T. Jensen, M. M. McCrane, and T. W. Moody, Life Sci. *35*, 1963 (1984).
18. A. N. Corps, L. H. Rees, and K. D. Brown, Biochem. J. *231*, 781 (1985).
19. A. Pert, T. W. Moody, C. B. Pert, L. A. DeWald, and J. E. Rivier, Brain Res. *193*, 209 (1980).
20. M. A. Zarbin, M. J. Kuhar, T. L. O'Donohue, S. S. Wolf, and T. W. Moody, J. Neurosci. *5*, 429 (1985).
21. G. L. Willis, J. Hansky, and G. C. Smith, Peptides *9*, 87 (1984).
22. Q. J. Pittman, Y. Tache, and M. Brown, Life Sci. *26*, 725 (1980).
23. J. Dodd and J. S. Kelly, Brain Res. *205*, 337 (1981).

24. G. Paxinos and C. Watson, The Rat Brain in Stereotaxic Coordinates, Academic Press, Sydney, 1982.
25. T. L. O'Donohue, V. J. Massari, C. J. Pazoles, B. M. Chronwall, C. W. Shults, R. Quirion, T. N. Chase, and T. W. Moody, J. Neurosci. *4*, 2956 (1984).
26. E. Widerlov, D. Muller, A. Frye, and G. Breese, Peptides *5*, 523 (1984).
27. R. T. Jensen, T. Moody, C. Pert, J. E. Rivier, and J. D. Gardner, Proc. Natl. Acad. Sci. USA *75*, 6139 (1978).
28. S. J. Pandol, R. T. Jensen, and J. D. Gardner, J. Biol. Chem. *257*, 12024 (1982).
29. S. L. Swope and A. Schonbrunn, Proc. Natl. Acad. Sci. USA *81*, 1822 (1984).
30. J. Westendorf and A. Schonbrunn, J. Biol. Chem. *258*, 7527 (1983).
31. J. P. Heslop, D. M. Blakeley, K. D. Brown, R. F. Irvine, and M. J. Berridge, Cell *4*, 703 (1986).
32. I. Zachary, J. W. Sinnett-Smith, and E. Rozengurt, J. Cell Biol. *102*, 2211 (1986).
33. I. Zachary and E. Rozengurt, Proc. Natl. Acad. Sci. USA *82*, 7616 (1985).
34. S. A. Mendoz, J. A. Schneider, A. Lopez-Rivas, J. W. Sinnet-Smith, and E. Rozengurt, J. Cell Biol. *102*, 2223 (1986).
35. E. Rozengurt and J. Sinnett-Smith, Proc. Natl. Acad. Sci. USA *80*, 2936 (1983).
36. J. C. Willey, J. F. Lechner, and C. C. Harris, Exp. Cell Res. *153*, 245 (1984).
37. F. Cuttitta, D. N. Carney, J. Mulshine, T. W. Moody, J. Fedorko, A. Fischler, and J. D. Minna, Nature *316*, 823 (1985).
38. T. Moody, C. Pert, A. Gazdar, D. Carney, and J. Minna, Science *214*, 1246 (1981).
39. S. Wood, J. Wood, M. Ghatei, U. Lee, D. Shaughnessy, and S. Bloom, J. Clin. Endocrinol. *53*, 1310 (1981).
40. M. R. Erisman, O. Linnoila, R. Hernandez, R. DiAugustine, and L. Lazarus, Proc. Natl. Acad. Sci. USA *79*, 2379 (1982).
41. G. D. Sorenson, O. S. Pettengill, C. C. Cate, M. A. Ghatei, K. E. Molyneux, E. J. Gosselin, and S. R. Bloom, Life Sci. *33*, 1939 (1982).
42. T. W. Moody, E. K. Russel, T. L. O'Donohue, C. D. Linden, and A. Gazdar, Life Sci. *32*, 487 (1983).
43. C. B. Pert and U. K. Schumacher, Lancet *1*, 509 (1982).
44. S. M. Wood, J. R. Wood, M. A. Ghatei, G. D. Sorenson, and S. R. Bloom, Lancet *1*, 690 (1982).
45. L. Y. Korman, D. N. Carney, M. L. Citron, and T. W. Moody, Cancer Res. *46*, 1214 (1986).
46. T. W. Moody, D. N. Carney, F. Cuttitta, K. Quatrocchi, and J. D. Minna, Life Sci. *36*, 105 (1985).
47. P. M. Lackie, F. Cuttitta, J. D. Minna, S. R. Bloom, and J. M. Polak, Histochemistry *83*, 56 (1985).
48. M. M. Chang and S. E. Leeman, J. Biol. Chem. *245*, 4784 (1970).
49. U. S. Von Euler and J. H. Gaddum, J. Physiol. (Lond.) *72*, 74 (1931).

50. H. Nawa, T. Hirose, H. Takashima, S. Inayama, and S. Nakanishi, Nature *306*, 32 (1983).
51. S. H. Buck and E. Burcher, Trends Pharmacol. Sci. 65-68 (1986).
52. R. Quirion, C. S. Shults, T. W. Moody, C. B. Pert, T. N. Chase, and T. L. O'Donohue, Nature *303*, 714 (1983).
53. M. R. Hanley, B. E. B. Sandburg, C. M. Lee, L. L. Iversen, D. E. Brundish, and R. Wade, Nature *286*, 810 (1980).
54. S. S. Wolf, T. W. Moody, R. Quirion, and T. L. O'Donohue, Brain Res. *332*, 299 (1985).
55. P. Mohini, S. W. Bahouth, C. E. Brundish, and J. M. Musacchio, J. Neurosci. *5*, 2078 (1985).
56. M. A. Cascieri, G. G. Chicchi, and T. Liang, J. Biol. Chem. *260*, 1501 (1985).
57. J. C. Beaujouan, Y. Torrens, A. Viger, and J. Glowinski, Mol. Pharmacol. *26*, 248 (1984).
58. S. H. Buck, E. Burcher, C. W. Shults, W. Lovenberg, and T. L. O'Donohue, Science *226*, 987 (1984).
59. R. Quirion and T. V. Dam, Neuropeptides *6*, 191 (1985).
60. G. Enberg, T. H. Srensson, S. Rosell, and K. Folkers, Nature *293*, 222 (1981).
61. P. C. Waldmeir, R. Kam, and K. Stocklin, Brain Res. *159*, 223 (1978).
62. J. W. Phillis and J. J. Limacher, Brain Res. *69*, 158 (1974).
63. L. Le Gal Da Salle and D. Dentri, Brain Res. *135*, 174 (1977).
64. P. G. Guyenet and G. H. Aghajanian, Brain Res. *136*, 178 (1977).
65. C. J. Helke, T. L. O'Donohue, and D. M. Jacobowitz, Peptides *1*, 1 (1981).
66. R. T. Jensen and J. D. Gardner, Proc. Natl. Acad. Sci. USA *75*, 6139 (1979).
67. M. J. Berridge, C. P. Downes, and M. R. Hanley, Biochem. J. *206*, 587 (1982).
68. J. Nilsson, A. M. von Euler, and C. J. Calsgaard, Nature *315*, 61 (1985).
69. M. Maruyama, Brain Res. *370*, 186 (1986).
70. S. H. Buck, T. F. Burk, M. R. Brown, and H. I. Yamamura, Brain Res. *209*, 464 (1981).
71. P. C. Emson, A. Arregui, V. Clement Jones, B. F. P. Sandberg, and M. Rosser, Brain Res. *209*, 464 (1981).
72. V. Mutt, J. E. Jorpes, and S. Magnuson, Eur. J. Biochem. *15*, 513 (1970).
73. S. I. Said and V. Mutt, Science *69*, 1217 (1970).
74. W. M. Bayliss and E. H. Starling, Proc. R. Soc. Lond. [Biol.] *73*, 310 (1904).
75. M. Bodanszky, Y. S. Klausner, and S. I. Said, Proc. Natl. Acad. Sci. USA *70*, 382 (1973).
76. K. Tatemoto and V. Mutt, Proc. Natl. Acad. Sci. USA *78*, 6603 (1981).
77. N. Itoh, K. Obata, N. Yanaihara, and H. Okamoto, Nature *304*, 547 (1983).
78. D. Bataille, P. Freychet, and G. Rosselin, Endocrinology *95*, 713 (1974).
79. B. Desbuquois, Eur. J. Biochem. *46*, 439 (1974).

80. J. P. Christophe, T. P. Conlon, and J. D. Gardner, J. Biol. Chem. *251*, 4629 (1976).
81. P. Robberecht, P. DeNeef, M. Lammens, M. Deschodt-Lanckman, and J. P. Christophe, Eur. J. Biochem. *90*, 147 (1978).
82. D. P. Taylor and C. B. Pert, Proc. Natl. Acad. Sci. USA *76*, 660 (1979).
83. P. Staun-Olsen, B. Ottesen, J. Gammeltoft, and J. Fahrenkrug, Brain Res. *330*, 317 (1985).
84. R. T. Jensen, C. G. Charlton, H. Adachi, S. W. Jones, T. L. O'Donohue, and J. D. Gardner, Am. J. Physiol. *245*, G186 (1983).
85. R. T. Fremeau, R. T. Jensen, C. G. Charlton, R. L. Miller, T. L. O'Donohue, and T. W. Moody, J. Neurosci. *3*, 1620 (1983).
86. B. L. Roth, M. C. Beinfeld, and A. C. Howlett, J. Neurochem. *42*, 1145 (1984).
87. J. D. Gardner, A. J. Rottman, S. Natarajan, and M. Bodanszky, Biochim. Biophys. Acta *583*, 491 (1979).
88. R. Robberecht, T. P. Conlon, and J. D. Gardner, J. Biol. Chem. *251*, 4635 (1976).
89. J. Besson, M. Dussailant, J. C. Marie, W. Rostene, and G. Rosselin, Peptides *5*, 339 (1984).
90. E. B. DeSouza, H. Seifert, and M. J. Kuhar, Neurosci. Lett. *56*, 113 (1985).
91. M. M. Shaffer and T. W. Moody, Peptides *7*, 283 (1986).
92. J. W. Phyllis, J. R. Kirk, and S. I. Said, Can. J. Physiol. Pharmacol. *56*, 337 (1978).
93. P. J. Magistratti, J. H. Morrison, W. J. Shoemaker, V. Sapin, and F. E. Bloom, Proc. Natl. Acad. Sci. USA *78*, 6530 (1981).
94. J. McCulloch and P. A. Kelly, Nature *304*, 438 (1983).
95. J. McCulloch, P. A. T. Kelly, R. Uddman, and L. Edvisson, Proc. Natl. Acad. Sci. USA *80*, 1472 (1983).
96. E. Vijayan, W. K. Samson, S. I. Said, and S. M. McCann, Endocrinology *104*, 53 (1977).
97. T. Kaneko, P. Cheng, H. Oka, T. Oka, N. Yanaihara, and C. Yanaihara, Biomed. Res. *1*, 84 (1980).
98. S. P. Duckles and S. I. Said, Eur. J. Pharmacol. *78*, 371 (1982).
99. M. Deschodt-Lanckman, P. Robberecht, and J. Christophe, FEBS Lett. *83*, 76 (1977).
100. R. W. Kerwin, S. Pay, K. D. Bhoola, and C. J. Paycock, J. Pharm. Pharmacol. *32*, 561 (1980).
101. M. Quick, L. L. Iversen, and S. R. Bloom, Biochem. Pharmacol. *27*, 2209 (1978).
102. A. M. Etgen and E. T. Browning, J. Neurosci. *3*, 2487 (1983).
103. R. T. Fremeau, R. T. Jensen, C. G. Charlton, R. L. Miller, T. L. O'Donohue, and T. W. Moody, J. Neurosci. *3*, 1620 (1983).
104. O. Van Calker, M. Muller, and B. Hamprecht, Proc. Natl. Acad. Sci. USA *77*, 6907 (1980).
105. F. Propst, L. Morader, E. Wuncsh, and B. Hamprecht, J. Neurochem. *32*, 1495 (1979).

106. N. Y. Ip, C. K. Ho, and R. E.. Zigmond, Proc. Natl. Acad. Sci. USA *79*, 7566 (1982).
107. B. L. Roth, M. C. Beinfeld, and A. C. Howlett, J. Neurochem. *42*, 1145 (1984).
108. D. E. Brenneman and L. E. Eiden, Proc. Natl. Acad. Sci. USA *85*, 1159 (1986).
109. G. Babu and E. Vijayan, Brain Res. Bull. *11*, 25 (1983).
110. W. K. Samson, M. D. Lumpkin, and S. M. McCann, Life Sci. *34*, 155 (1984).
111. C. L. Wood and M. S. O'Dorisio, J. Biol. Chem. *260*, 1243 (1985).
112. A. Couvineau and M. Laburthe, Biochem. J. *225*, 473 (1985).
113. K. E. McArthur, C. L. Wood, M. S. O'Dorisio, Z.-C. Zhou, J. D. Gardner, and R. T. Jensen, J. Biol. Chem., in press.
114. S. R. Bloom and J. M. Polak, VIPomas, in Advances in Peptide Hormone Research Studies (S. I. Said, ed.), Raven Press, New York, 1982, p. 457.
115. R. W. Kellermeyer and R. C. Graham, N. Engl. J. Med. *279*, 757, 802, 859 (1968).
116. E. G. Erdos (ed.), Bradykinin, Kallidin and Kallikrein. Handbook of Experimental Pharmacology, Vol. 25, Springer-Verlag, Berlin, 1970.
117. J. J. Pisano and K. F. Austen, Chemistry and Biology of the Kallikrein-Kinin System in Health and Disease, DHEW Publication NIH76-791, 1974.
118. E. G. Erdos (ed.), Bradykinin, Kallidin and Kallikrein. Handbook of Experimental Pharmacology, Vol. 75 (Suppl.), Springer-Verlag, Berlin, 1979.
119. D. Regoli and J. Barabe, Pharmacol. Rev. *32*, 1 (1980).
120. W. W. Douglas, in The Pharmacological Basis of Therapeutics, 7th ed. (A. Gilman, L. S. Goodman, T. W. Rall, and F. Murad, eds.), MacMillan, New York, 1985, p. 639.
121. R. K. S. Lim, Annu. Rev. Physiol. *32*, 269 (1970).
122. D. Armstrong, in Bradykinin, Kallidin and Kallikrein. Handbook of Experimental Pharmacology (E. G. Erdos, ed.), Vol. 25, Springer-Verlag, Berlin, 1970, p. 434.
123. F. Lembeck, H. Popper, and H. Juan, Naunyn-Schmeidebergs Arch. Pharmacol. *294*, 69 (1976).
124. G. Bynke, R. Hakanson, J. Hong, and S. Leander, Eur. J. Pharmacol. *91*, 469 (1983).
125. N. Neda, I. Muramatsu, and M. Fujiwara, J. Pharmacol. Exp. Ther. *230*, 469 (1984).
126. F. M. A. Correa, R. B. Innis, G. R. Uhl, and S. H. Snyder, Proc. Natl. Acad. Sci. USA *76*, 1489 (1979).
127. D. C. Perry and S. H. Snyder, J. Neurochem. *43*, 1072 (1984).
128. D. C. Perry, in Neural and Endocrine Peptides and Receptors (T. W. Moody, ed.), Plenum Press, New York, 1986, p. 367.
129. F. M. A. Correa and E. G. Graeff, J. Pharmacol. Exp. Ther. *192*, 670 (1975).
130. D. I. Diz and D. M. Jacobowitz, Brain Res. Bull. *12*, 409 (1984).

131. R. E. Lewis and M. I. Phillips, Am. J. Physiol. *247*, R63 (1984).
132. D. C. Manning, S. H. Snyder, J. F. Kachur, R. J. Miller, and M. Field, Nature *299*, 256 (1982).
133. J. Roblero, J. R. Ryan, and J. M. Stewart, Res. Commun. Chem. Pathol. Pharmacol. *6*, 207 (1973).
134. E. Schroder, in Bradykinins, Kallidin and Kallikrein. Handbook of Experimental Pharmacology, Vol. 25 (E. G. Erdos, ed.), Springer-Verlag, Berlin, 1970, p. 324.
135. F. W. Dunn and J. M. Stewart, J. Med. Chem. *14*, 779 (1971).
136. M. J. Frederick, R. J. Vavrek, J. M. Stewart, and C. E. Odya, Biochem. Pharmacol. *33*, 2887 (1984).
137. M. J. Frederick, F. C. Abel, W. A. Rightsel, E. E. Muirhead, and C. E. Odya, Life Sci. *37*, 331 (1985).
138. R. J. Vavrek and J. M. Stewart, Peptides *6*, 161 (1985).
139. K. M. Braas, D. C. Manning, D. C. Perry, and S. H. Snyder, Br. J. Pharmacol. *94*, 3 (1988).
140. L. A. Chahl and A. Iggo, Br. J. Pharmacol. *59*, 343 (1977).
141. J. Stoner, V. C. Manganiello, and M. Vaughan, Proc. Natl. Acad. Sci. USA *70*, 3830 (1973).
142. E. S. Redgate, J. D. Dupress, and J. Axelrod, Brain Res. *365*, 61 (1986).
143. T. V. Senser, N. S. Rapp, L. A. Spry, and B. B. Davis, Endocrinology *144*, 541 (1984).
144. D. L. Bareis, V. C. Manganiello, F. Hirata, M. Vaughan, and J. Axelrod, Proc. Natl. Acad. Sci. USA *80*, 2514 (1983).
145. M. Fujimoto and T. O. Kabayashi, Jpn. J. Pharmacol. *41*, 223 (1986).
146. A. A. Roscher, V. C. Manganiello, C. L. Jelsema, and J. Moss, J. Clin. Invest. *72*, 626 (1983).
147. A. A. Roscher, V. C. Manganiello, C. L. Jelsema, and J. Moss, J. Clin. Invest. *74*, 552 (1984).
148. T. Ishizaka, M. Iwata, and K. Ishizaka, J. Immunol. *134*, 1880 (1985).
149. J. V. Fahey, C. P. Ciosek, Jr., and D. S. Newcombe, Agents Actions *7*, 255 (1977).
150. H. Higashida, R. A. Streaty, W. Klee, and M. Nirenberg, Proc. Natl. Acad. Sci. USA *83*, 942 (1986).
151. H. Frucht, G. Lilling, and R. Bertner, Int. J. Biochem. *16*, 397 (1984).
152. R. I. Clyman, A. S. Blacksin, V. C. Manganiello, and M. Vaughan, Proc. Natl. Acad. Sci. USA *72*, 3883 (1975).
153. G. Reiser, U. Walter, and B. Hamprecht, Brain Res. *290*, 367 (1984).
154. R. M. Snider and E. Richelson, J. Neurochem. *43*, 1749 (1984).
155. L. M. Vicentini and M. L. Villareal, Biochem. Biophys. Res. Commun. *123*, 663 (1984).
156. C. K. Derian and M. A. Moskowitz, J. Biol. Chem. *261*, 3831 (1986).
157. P. C. Francel and G. Dawson, Biochem. Biophys. Res. Commun. *135*, 507 (1986).
158. K. Yano, H. Higashida, R. Inoue, and Y. Nozawa, J. Biol. Chem. *59*, 10201 (1984).

159. K. Yano, H. Higashida, H. Hattori, and Y. Nozawa, FEBS Lett. *181*, 403 (1985).
160. T. Osugi, S. Uchida, T. Imaizumi, and H. Yoshida, Brain Res.*379*, 84 (1986).
161. S. W. Cuthbert and H. S. Margolius, Br. J. Pharmacol. *75*, 587 (1982).
162. J. P. Huidodro-Toro, R. Herreros, and A. Pinto-Corrado, Eur. J. Pharmacol. *121*, 305 (1986).
163. G. Reiser and B. Hamprecht, Brain Res. *239*, 191 (1982).
164. C. E. Odya and T. L. Goodfriend, in Bradykinin, Kallidin and Kallikrein. Handbook of Experimental Pharmacology, Vol. 25 (Suppl.), (E. G. Erdos, ed.), Springer-Verlag, Berlin, 1979.
165. C. E. Odya, T. L. Goodfriend, and C. Peña, Biochem. Pharmacol. *29*, 175 (1980).
166. R. B. Innis, D. C. Manning, J. M. Stewart, and S. H. Snyder, Proc. Natl. Acad. Sci. USA *78*, 2630 (1981).
167. D. C. Manning and S. H. Snyder, Neurosci. Abstr. *9*, 4590 (1983).
168. D. C. Manning, R. Vavrek, J. M. Stewart, and S. H. Snyder, J. Pharmacol. Exp. Ther. *237*, 504 (1986).
169. D. C. Manning and S. H. Snyder, in Kinins IV, Part A (L. M. Greenbaum and H. S. Margolius, eds.), Plenum Press, New York, 1986, p. 563.
170. K. Tomita and J. J. Pisano, Am. J. Physiol. *246*, F732 (1984).
171. P. N. Aarsen and M. von Caspel-de Bruyn, Eur. J. Pharmacol. *12*, 348 (1970).
172. P. Borschov, A. C. M. Paiva, T. B. Paiva, and S. I. Shimuta, Br. J. Pharmacol. *83*, 591 (1984).
173. A. Camargo and S. H. Ferreira, Br. J. Pharmacol. *42*, 305 (1971).
174. S. A. Ribeiro and M. Rocha e Silva, Br. J. Pharmacol. *47*, 517 (1973).
175. P. Brazeau, W. Vale, P. Burgus, N. Ling, M. Butcher, J. Rivier, and R. Guillemin, Science *179*, 77 (1973).
176. W. Vale, C. Rivier, P. Brazeau, and R. Guillemin, Endocrinology *95*, 968 (1974).
177. S. Reichlen, in Brain Peptides (D. T. Krieger, M. J. Brownstein, and J. B. Martin, eds.), John Wiley and Sons, New York, 1983, p. 711.
178. S. Reichlen, N Engl. J. Med. *309*, 1495, 1556 (1983).
179. Y. C. Patel and C. B. Srikant, Annu. Rev. Physiol. *48*, 551 (1986).
180. M. F. Beal, G. Uhl, M. F. Mazurek, N. Kowall, and J. B. Martin, in Neuropeptides in Neurologic and Psychiatric Disease (J. B. Martin and J. D. Barchas, eds.), Raven Press, New York, 1986, p. 215.
181. J. C. W. Finely, J. L. Maderut, L. J. Rogers, and P. Petrusz, Neuroscience *6*, 2173 (1981).
182. M. F. Beal, M. F. Mazurek, V. T. Tran, G. Chattha, E. D. Bird, and J. B. Martin, Science *229*, 289 (1985).
183. C. B. Srikant and Y. C. Patel, Adv. Exp. Biol. Med. *188*, 291 (1986).
184. W. A. Murphy, C. A. Myers, and D. H. Coy, Endocrinology *109*, 491 (1981).

185. D. F. Veber, F. W. Holly, R. F. Nutt, S. L. Brogstrand, S. F. Brady, R. Hirschmann, M. S. Glitzer, and R. Saperstein, Nature *280*, 512 (1979).
186. J. L. Fries, W. A. Murphy, J. Sueiras-Dias, and D. H. Coy, Peptides *3*, 811 (1982).
187. L. Mandarino, D. Sterner, W. Blanchard, S. Nissen, J. Gerich, N. Ling, P. Brazeau, P. Bohlen, F. Esch, and R. Guillemin, Nature *291*, 76 (1981).
188. M. Brown, J. Rivier, and W. Vale, Endocrinology, *108*, 2391 (1981).
189. M. Brown, N. Ling, and J. Rivier, Brain Res. *214*, 127 (1981).
190. B. H. Hirst, J. M. Conlon, D. H. Coy, J. Holland, and B. Show, Regul. Pept. *4*, 227 (1982).
191. A. Schonbrunn, L. J. Dorflinger, and B. D. Koch, Adv. Exp. Biol. Med. *188*, 305 (1986).
192. D. Van Calker, M. Muller, and B. Hamprecht, Proc. Natl. Acad. Sci. USA *77*, 6907 (1980).
193. K. H. Jakobs, K. Aktories, and G. Schultz, Nature *303*, 177 (1983).
194. Y. Yajima, Y. Yakita, and T. Saito, J. Biol. Chem. *261*, 2684 (1986).
195. J. Axelrod and T. D. Reisine, Science *224*, 452 (1984).
196. M. J. Cronin, A. D. Rogel, G. A. Myers, and E. L. Hewlett, Endocrinology *113*, 209 (1983).
197. S. Heisler, T. D. Reisine, V. Y. H. Hook, and J. Axelrod, Proc. Natl. Acad. Sci. USA *79*, 6502 (1982).
198. K. H. Jakobs and G. Schultz, Proc. Natl. Acad. Sci. USA *80*, 3899 (1983).
199. T. Reisine, J. Pharmacol. Exp. Ther. *229*, 14 (1984).
200. C. S. Pace and J. T. Tarvin, Diabetes *30*, 836 (1981).
201. W. Schlegel, F. Wuarin, C. B. Wollheim, and G. R. Zahnd, Cell Calcium *5*, 223 (1984).
202. R. A. North, in Neuropeptides in Neurologic and Psychiatric Disease (J. B. Martin and J. B. Barchas, eds.), Raven Press, New York, 1986, p. 71.
203. J. R. Delfs and M. A. Dichter, J. Neurosci. *3*, 1176 (1983).
204. C. B. Srikant and Y. C. Patel, Proc. Natl. Acad. Sci. USA *78*, 3930 (1981).
205. A. J. Czernik and B. Petrack, J. Biol. Chem. *258*, 5525 (1983).
206. J. C. Reubi, Life Sci. *36*, 1829 (1985).
207. J. C. Reubi, M. H. Perrin, J. E. Rivier, and W. Vale, Life Sci. *28*, 2191 (1981).
208. A. Schonbrunn and A. H. Tashjian, Jr., J. Biol. Chem. *253*, 6473 (1978).
209. C. B. Srikant and Y. C. Patel, Nature *294*, 259 (1981).
210. C. B. Srikant and Y. C. Patel, Endocrinology *110*, 2138 (1982).
211. J. C. Reubi, M. Perrin, J. Rivier, and W. Vale, Biochem. Biophys. Res. Commun. *105*, 1538 (1982).
212. J. C. Reubi, M. Perrin, J. Rivier, and W. Vale, Regul. Pept. *4*, 141 (1982).
213. G. Aguilera and D. S. Parker, J. Biol. Chem. *257*, 1134 (1982).
214. A. Enjalbert, R. Rasolonjanahary, E. Moyse, C. Kordon, and J. Epelbaum, Endocrinology *113*, 822 (1983).
215. C. B. Srikant, D. H. Coy, and Y. C. Patel, Life Sci. *33*, 201 (1983).
216. E. Moyse, R. Benoit, A. Enjalbert, J. P. Gautron, C. Koredon, N. Ling, and J. Epelbaum, Regul. Pept. *9*, 129 (1984).

217. J. C. Reubi, Neurosci. Lett. *49*, 259 (1984).
218. S. Heisler and C. B. Srikant, Endocrinology *117*, 217 (1985).
219. C. B. Srikant and S. Heisler, Endocrinology *117*, 271 (1985).
220. J. C. Reubi and A. M. Laudalt, J. Clin. Endocrinol. Metab. *59*, 1148 (1984).
221. G. Morel, P. Leroux, and G. Pelletier, Endocrinology *116*, 1615 (1985).
222. J. C. Reubi, J. Rivier, M. Perrin, M. Brown, and W. Vale, Endocrinology *110*, 1049 (1982).
223. Y. C. Patel, M. Amherdt, and L. Orci, Science *217*, 1155 (1982).
224. D. Taparel, J. P. Esteve, C. Susini, N. Vaysse, D. Balas, G. Berthon, E. Wunsch, and A. Ribet, Biochem. Biophys. Res. Commun. *115*, 827 (1983).
225. J. P. Esteve, C. Sassini, N. Vaysse, H. Antoniotti, E. Wansch, G. Berthon, and A. Ribet, Am. J. Physiol. *247*, G62 (1984).
226. C. Sakamoto, I. D. Goldfine, and J. A. Williams, J. Biol. Chem. *259*, 9623 (1984).
227. T. Matozaki, C. Sakamoto, M. Nagao, and S. Baba, J. Biol. Chem. *261*, 1414 (1986).
228. F. Reyl, C. Silve, and M. J. M. Lewin, in Hormone Receptors in Digestion and Nutrition (G. Rosselin, P. Fromageot, and S. Bonfil, eds.), Elsevier, Amsterdam, 1979, p. 391.
229. G. Aguilera, D. S. Parker, and K. J. Catt, Endocrinology *111*, 1376 (1982).
230. J. Epelbaum, L. Tapia Arancibia, C. Kordon, and A. Enjalbert, J. Neurochem. *38*, 1515 (1982).
231. C. B. Srikant and Y. C. Patel, Endocrinology *115*, 990 (1984).
232. H. Katakami, M. Berelowitz, M. Marbach, and L. A Frohman, Endocrinology *117*, 557 (1985).
233. A. Schonbrum and A. H. Tashjian, Jr., J. Biol. Chem. *255*, 190 (1980).
234. J. Epelbaum, Y. Lamour, A. Enjalbert, M. Hamon, P. Dutar, and C. Kordon, Brain Res. *371*, 376 (1986).
235. V. T. Tran, G. R. Uhl, D. C. Perry, D. C. Manning, W. W. Vale, M. H. Perrin, J. E. Rivier, J. B. Martin, and S. H. Snyder, Eur. J. Pharmacol. *301*, 307 (1984).
236. G. R. Uhl, V. Tran, S. H. Snyder, and J. B. Martin, J. Comp. Neurol. *240*, 288 (1985).
237. P. Leroux and G. Pelletier, Peptides *5*, 503 (1984).
238. K. Gulya, J. K. Wamsley, D. Gehlert, J. P. Pelton, S. P. Duckles, V. J. Hruby, and H. I. Yamamura, J. Pharmacol. Exp. Ther. *235*, 254 (1985).
239. J. Epelbaum, M. Dusscullant, A. Enjalbert, C. Kordon, and W. Rostene, Peptides *6*, 713 (1985).
240. J. C. Reubi and R. Maurer, Neuroscience *15*, 1183 (1985).
241. M. F. Beal, V. T. Tran, M. F. Mazurek, G. Chattha, and J. B. Martin, J. Neurochem. *46*, 359 (1986).
242. D. C. Perry, D. C. Manning, J. E. Rivier, W. Yale, and S. H. Snyder, Neurosci. Abstr. *9*, 171 (1983).

243. D. C. Perry, D. C. Manning, and S. H. Snyder, J. Steroid Biochem. *20*, 1614 (1984).
244. V. T. Tran, M. F. Beal, and J. B. Martin, Science *228*, 492 (1985).
245. G. Pelletier, P. Leroux, and G. Morel, Methods Enzymol. *124*, 607 (1986).
246. J. C. Reubi and R. Maurer, Regul. Pept. *14*, 301 (1986).

17

Opioid Receptors

S. William Tam / E. I. du Pont de Nemours & Company, Inc., Wilmington, Delaware

HISTORICAL PERSPECTIVES

The concept of multiple opioid receptors is now well established. The recognition of these receptors was made possible by the synthesis of a large number of morphine analogs. Some of the most important early compounds include the mixed agonist-antagonists nalorphine, cyclazocine, pentazocine, ketazocine, SKF 10,047 (*N*-allylnormetazocine), and the antagonist naloxone. The early evidence of multiple opioid receptors could be found in the studies of various combinations of morphine and nalorphine in which lower doses of nalorphine antagonized the analgesic effects of morphine, larger doses of nalorphine produced a lesser degree of antagonism, but nalorphine is analgesic in its own right (1,2). Detailed analysis of structure-activity relationships among analgesics led to the first suggestion of the existence of multiple opioid receptors (3). This concept was further supported by early pharmacological studies that indicated that opioid receptors in the gut are different from those in the CNS (4).

The analgesic activity of opioids was explained by their interaction with a stereospecific receptor binding site (5,6). However, analysis of the different syndromes produced by morphine and a number of mixed agonist-antagonists in the nondependent chronic spinal dog led Martin and associates (7,8) to designate μ, κ, and σ receptors to account for the agonistic activities of morphine, ketocyclazocine, and SKF 10,047, respectively. The κ benzomorphans behaved as agonists in the guinea pig ileum and mouse vas deferens but required more naloxone for antagonism than morphine (9). The presence of a distinct ethylketocyclazocine site that correlates with the κ receptor was demonstrated by receptor binding studies (10).

Shortly after the discovery of the endogenous opioid pentapeptides methionine-enkephalin and leucine-enkephalin (11), the δ opioid receptor was identified by receptor binding and isolated mouse vas deferens studies (12,13). β Endorphin was found to be more potent than morphine in the rat vas deferens, and thus the ϵ (β endorphin) receptor was designated (14,15). Receptor binding studies confirmed the presence of distinct morphine (μ) and enkephalin (δ) binding sites that correlate with μ and δ receptors (13,16). The presence of a distinct binding site for ϵ receptor has also been confirmed (17).

Such σ opiates as cyclazocine and SKF 10,047 were found to inhibit phencyclidine (PCP) binding (18,19), and this led to the suggestion that the PCP receptor is the σ receptor. However, it is not clear why some σ opiates, such as pentazocine (20), have only very low affinity for the PCP site. Recently, a naloxone-inaccessible σ-binding site for (+)-SKF 10,047 and (+)-ethylketocyclazocine that binds σ opiates with high affinity has been identified in rat brain and spinal cord (21). This σ-binding site differs from the PCP receptor in drug selectivity, stereospecificity, and topography (21). A similar etorphine-inaccessible site was identified independently in guinea pig brains using tritiated racemic SKF 10,047 (22). More recent studies also indicate that a large number of antipsychotic drugs also bind to this σ-binding site (23) and that PCP binds to this σ site in addition to the PCP receptor (24).

BINDING SITES AND RADIOLIGANDS

The μ, δ, κ, σ, and ϵ receptors represent the five opioid receptor types. There is controversy about whether the σ receptor should be called an opioid receptor since it is not blocked by naloxone. Subtypes of some of these receptor types have been described, such as the μ_1 and μ_2 receptor subtypes. Selective binding conditions or selective ligands for these receptors have been determined. However, it must be pointed out that these binding assays are performed under in vitro conditions, in most cases with disrupted cell homogenates in nonphysiological buffers. Thus, the drug selectivity of a type of binding site or the receptor selectivity of a ligand is defined in a particular set of experimental conditions, and therefore, the selectivity and potency obtained under such conditions do not necessarily generalize to those measured under in vivo conditions or in isolated tissue bioassays.

The Scatchard plots for many radiolabeled opioid ligands are nonlinear, indicating their binding to multiple sites (13,16,21,25). In competition experiments, morphine was shown to be more potent in inhibiting binding of [^{3}H] dihydromorphine, and the enkephalins are more potent in inhibiting the binding of [^{3}H] D-Ala^2, D-Leu^5] enkephalin (DADLE).

The μ binding site can be labeled by [^{3}H] D-Ala^2, *N*-methyl-Phe^4, Glyol^5]-enkephalin (DAGO), [^{3}H] sufentanyl, [^{3}H] dihydromorphine, [^{125}I] FK 33,824

[(Tyr, D-Ala^2, Gly, *N*-methyl-Phe^4, Met (*O*)ol)enkephalin], and [^{3}H] naloxone when these ligands are used at concentrations below 1 nM. At higher concentrations, these ligands also bind to the δ-binding sites.

The δ-binding site can be selectively labeled by using [^{3}H] DADLE, [^{125}I]-DADLE, [^{3}H]-D-Ser^2, Thr^6] leu-enkephalin (DSTLET), [^{3}H] leu-enkephalin, and [^{3}H] met-enkephalin, provided that appropriate μ-selective blocking agents, such as DAGO, sufentanil, or morphiceptin, are used to block the μ-binding sites.

The δ binding site can be selectively labeled by [^{3}H] dynorphin A (26) or [^{3}H] tifluadom (27). It can also be labeled using the nonselective ligands (–)-[^{3}H] ethylketocyclazocine (24) or (–)-[^{3}H] bremazocine (28) in the presence of appropriate unlabeled μ-, and δ-selective blocking agents, such as DAGO and DADLE.

The ϵ-binding site can be labeled with [^{3}H] diprenorphine in the presence of morphiceptin and DADLE to suppress μ and δ receptor binding (29). The ϵ binding site can also be labeled with [^{3}H] β_h-endorphin (30-32).

Recent studies by Tam (21,24), Tam and Cook (23), and Martin et al. (33) have demonstrated a high-affinity binding site for the σ agonist [^{3}H] (+)-SKF 10,047. This site has binding properties predictive of a σ site and demonstrates regional distribution in the central nervous system (CNS) of rats (21) and guinea pigs (24). The σ-binding site can be selectively labeled with [^{3}H] (+)-SKF 10,047 at a concentration of 4 nM or less. At high concentrations, [^{3}H] (+)-SKF 10,047 also interacts with the PCP receptors. [^{3}H] Haloperidol has been used to selectively label the σ-binding site provided that 100 nM (+)-butaclamol or 25 nM spiperone is present to prevent binding of the ligand to dopamine D_2 receptors (23,25). The putative dopamine autoreceptor agonist [^{3}H] (+)-3-(3-hydroxyphenyl)-*N*-(1-propyl)piperidine (3-PPP), has also been shown to selectively label the σ-binding site (34). This σ-binding site has also been shown to be selectively labeled by [^{3}H] (+)-ethylketocyclazocine (21,24).

Subtypes of μ binding sites have been reported. A common high-affinity binding site for opiates, enkephalins, and endorphin that was irreversibly blocked by naloxazone (35,36) and naloxonazine (37) has been reported and termed the μ_1 site. It was proposed that the μ_1 sites mediate supraspinal analgesia and prolactin release but not growth hormone release. The lower affinity morphine binding sites have been termed μ_2 sites. Another opioid receptor subypte (λ) has been identified by in vitro and in vivo labeling of brain membranes (38,39). The λ binding sites have high affinity for 4,5-epoxymorphinans, such as naloxone and nalorphine, but have low affinity for diprenorphine, etorphine, and buprenorphine. The physiological and pharmacological functions of this site are not known. Anomalous binding of opiates in kidney and liver membranes has been described (40). These binding sites have a reversed stereoselectivity for opiates compared with opioid receptor binding. However, the drug selectivity of these sites is different from that of the σ (+)-SKF 10,047 binding sites. Naloxone and morphine bind to these anomalous binding sites, but not to the (+)-SKF 10,047 binding sites.

SELECTIVITY AND SAR

It has been recognized that pharmacological activity correlates better with μ opioid receptor binding when performed in the presence of NaCl (41). It has also been determined that NaCl decreases the affinity of agonists for the μ receptor (41). Thus, the sodium ratio, defined as the ratio of the IC_{50} obtained in the presence of 100 mM NaCl and the IC_{50} obtained in the absence of NaCl in displacement studies, has been used quite successfully to define agonist, mixed agonist-antagonists, and antagonists. However, we have found that some of the most potent and selective μ agonists have small sodium ratios similar to those of antagonists (Table 1). Therefore, the use of sodium ratios has its limitations.

We have found that although NaCl decreases agonist affinity for the μ binding sites, it improves the correlation between μ agonist binding and analgesic activity (S. W. Tam and W. K. Schmidt, unpublished observation). A good correlation also exists between analgesia and κ agonist binding to the κ receptor. Therefore, we include 100 nM NaCl in the incubation mixture when we perform opioid receptor binding assays. The results are shown in Table 2. Under these conditions, sufentanil, morphine, fentanyl, PL017, naltrexone, and naloxone are relatively μ selective. Dynorphin A 1-13, tifluadom (42), and U50,488H (43) are κ selective. DADLE and met-enkephalin are δ selective, and (+)-SKF 10,047 is σ selective. Sodium ions not only decrease the affinity of agonists for the μ-binding sites, they also decrease the relative selectivity of the drugs for the μ binding sites. For example, (N-methyl-Phe^3, D-Pro^4)morphiceptin is 215 times more potent in binding to μ than to δ binding sites in the absence of NaCl but is

Table 1 Relative Potency Ratio in the Presence and Absence of 100 mM NaCl in Displacement of 0.5 nM [^{3}H] Naloxone Binding in Rat Brain Membranes

μ Agonists	Sodium ratios
Morphine	32
Oxymorphone	23
Fentanyl	20
Sufentanil	4.0
Etonitazene	1.5
Lofentanil	0.8

Table 2 Affinities of Ligands for the Different Types of Opioid Binding Sites[a]

Ligands	K_i (nM)			
	μ	κ	δ	σ
μ Agonists				
Etorphine	0.87	2.0	3.2	> 10,000
Sufentanil	5.3	140	41	94
Oxymorphone	15	720	140	>100,000
Morphine	38	1,900	510	>100,000
Fentanyl	53	1,900	420	590
Methadone	170	4,400	760	ND
PL017	228	>10,000	7,300	ND
Partial μ agonist				
Buprenorphine	0.77	1.1	2.2	>100,000
κ Agonists				
Dynorphine A 1–13[b]	31	0.98	12	> 10,000
Tifluadom	32	2.1	190	6,600
U50,488H	820	15	21,000	700
δ Agonists				
DADLE	150	>10,000	1.8	> 10,000
Met-enkephalin	130	>10,000	3.2	> 10,000
Mixed agonist-antagonists				
Bremazocine	0.90	0.67	2.9	200
Mr 2034	1.9	2.0	9.4	160
(–)-Ethylketocyclazocine	2.3	2.2	5.2	19,000
(–)-SKF 10,047	3.0	4.7	15	1,800
Cyclazocine	0.45	5.9	6.3	36
Butorphanol	1.7	7.4	13	2,300
Nalbuphine	6.3	61	160	>100,000
Pentazocine	39	87	470	18
σ Agonist				
(+)-SKF 10,047	1,900	1,600	19,000	48
Antagonists				
Naltrexone	0.37	4.8	9.4	>100,000
Naloxone	1.2	12	19	>100,000
Mr 2266	1.3	2.1	4.4	940

[a]The apparent affinities at the binding sites were determined in the presence of 100 mM NaCl according to the method of Tam (24). Some data are from Tam (24).
[b]Dynorphin A 1-13 is Tyr-Gly-Gly-Phe-Leu-Arg-Arg-Ile-Arg-Pro-Lys-Leu-Lys.

only 32 times more potent as a μ ligand than a δ ligand in the presence of 100 mM NaCl. Again, the selectivity of a compound depends on the conditions under which the assay is performed. In other studies, other μ-selective ligands, such as morphiceptin (44), DAGO (45), FK 33,824 (46), and syndyphalin (47), and other δ-selective ligands, such as (D-Ser^2, Leu^5, Thr^6)enkephalin (DSTLE) (48), (D-Pen^2, L-Pen^5)enkephalin (49), and (D-Pen^2, D-Pen^5)enkephalin (49), have been reported. A N-diallyl derivative of leu-enkephalin, ICI 154,129, has been shown to be a weak but very selective δ antagonist (50). Opioid peptides and their selectivity for different opioid receptor types is given in Table 3.

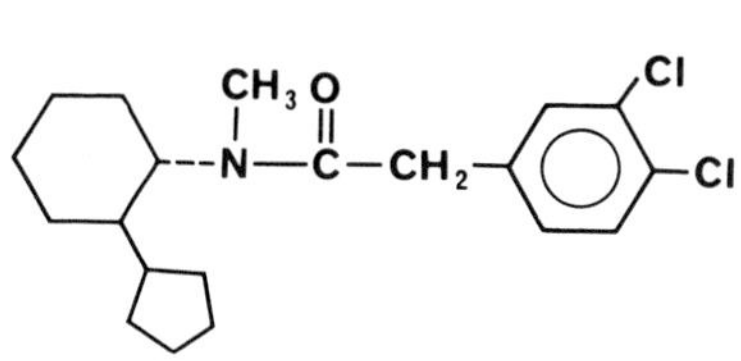

U50,488H

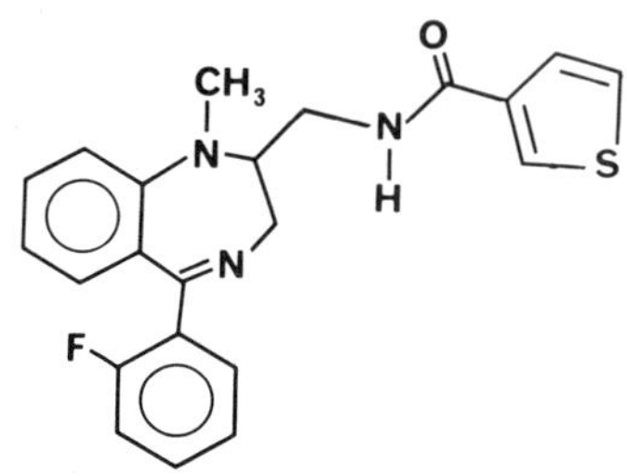

Tifluadom

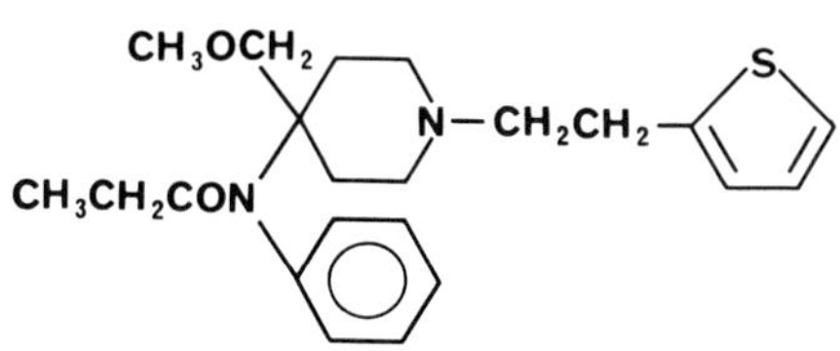

Sufentanil

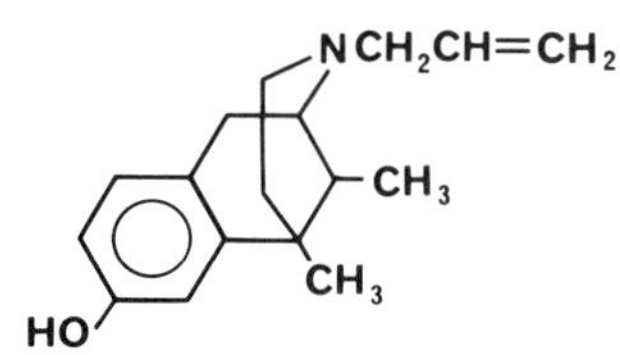

SKF 10,047

The σ-binding site was discovered very recently (21-24,33). Although the binding of σ opiates, such as pentazocine, cyclazocine, and SKF 10,047, the stereoselectivity, and the regional distribution of the σ binding site are consistent with the hypothesis that the σ-binding site is pharmacologically relevant, the physiological function of this site is not known. Recently, Tam and Cook (23) noticed a striking difference among some antipsychotic drugs. Haloperidol and the phenothiazines demonstrated high affinity for σ binding site, whereas spiperone, (+)-butaclamol, thiothixene, loxapine, and clozapine demonstrated low affinity for the σ site (Table 4). All of these antipsychotic drugs have high affinity for the dopamine D_2 receptor site. These findings led to the hypothesis that some antipsychotic drugs may act therapeutically by antagonizing both this site and dopamine receptors, whereas others may specifically antagonize only dopamine receptors.

Table 3 Opioid Peptides

Peptides	Receptor selectivity
β Endorphin	μ, ε
Leu-enkephalin	δ, μ
Met-enkephalin	δ, μ
Dynorphin A 1–13	κ
Dynorphin A 1–17	κ
Morphiceptin	μ
PL017	μ
DAGO	μ
FK 33824	μ
DSTLET	δ
DADLE	δ
DPDPE	δ
ICI 154,129	δ

Most receptor binding assays are done with membrane preparations of cell homogenates. It is possible that the use of nonphysiological buffer and the loss of second messengers and ions in these membrane preparations may affect the coupling of the receptor to effectors and result in changes in the conformation of the receptor. For this reason, Rogers and El-Fakahany (51) studied opioid receptor binding in dissociated intact rat brain cells (52,53) and demonstrated μ-receptor binding using [^{3}H] naloxone and [^{3}H] dihydromorphine in a physiological buffer. These studies have been extended to demonstrate the existence of μ, δ, κ, σ, and PCP binding sites in viable intact guinea pig brain cells in a physiological buffer. In general, the ligands are two to eight times less potent in binding to opioid binding sites in intact brain cells in physiological buffer (Table 5) than in membranes in nonphysiological buffer (Table 2). The affinity of morphine for μ binding sites is three times lower in the intact cells than in the membranes, and its affinity for κ binding sites is two times higher in the intact cells (Table 5) than in membranes in Tris buffer containing 100 mM NaCl (Table 2). Both morphine and naloxone are less μ selective in intact cells in physiological buffer than in membranes in nonphysiological buffer. On the other hand, DADLE is more δ selective in the intact cells, and there is no significant change in the selectivity of U50,488H in the intact cells. However, the affinities of

Table 4 Affinities of Antipsychotic Drugs for the σ Binding Site and the Dopamine D_2 Receptor

Drugs	IC_{50} (nM)	
	σ	D_2
Haloperidol	4.2	1.9
Perphenazine	13	4.0
Fluphenazine	18	6.5
Acetophenazine	57	ND[a]
Trifluoperazine	70	ND
Molindone	130	ND
Pimozide	150	ND
Thioridazine	180	16
Chlorpromazine	190	9.1
Triflupromazine	220	ND
Spiperone	1,100	0.8
(+)-Butaclamol	1,400	2.7
Thiothexene	1,500	3.6
Loxapine	1,700	12
Clozapine	12,000	ND

[a]Not determined.
Source: Data from Tam and Cook (23).

Table 5 Affinities of Ligands for the Different Types of Opioid Binding Sites in Intact Guinea Pig Brain Cells in Physiological Buffer[a]

Ligands	K_i (nM)		
	μ	κ	δ
Morphine	110	950	1,500
U50,488H	5,600	120	>100,000
DADLE	780	>100,000	5.4
Naloxone	2.3	13	46

[a]The binding buffer contains 110 mM NaCl, 5.3 mM KCl, 1.8 mM $CaCl_2$, 1 mM $MgSO_4$, 25 mM glucose, 62.8-70.1 mM sucrose, and 20 mM HEPES, pH 7.4, with osmolarity adjusted to 336-340 mOsM with sucrose (41). The following binding conditions were used: μ binding, 1 nM [^{3}H]naloxone with 20 nM DADLE and 300 nM U50,488H; κ binding, 4 nM (–)-[^{3}H]EKC with 100 nM DAGO and 500 nM DADLE; δ binding, 2 nM [^{3}H]DADLE with 100 nM DAGO and 300 nM U50,488H.

U50,488H for the μ and κ binding sites are decreased seven to eight times compared with that in membranes in Tris buffer with NaCl. In another study using membranes from guinea pig cortex and rat brains, Werling et al. (54) have demonstrated μ, δ, and κ receptor binding in a physiological buffer and that DAGO, DSLET, and U50,488H are high-affinity and selective ligands for μ, δ, and κ sites, respectively.

REGIONAL LOCALIZATION IN THE CENTRAL NERVOUS SYSTEM

Receptor binding studies with homogenates of tissues indicate marked differential distribution of μ and δ sites in the central nervous system of rats (21,46), guinea pigs (24), and humans (55). The μ binding sites in rat brain are most dense in the nucleus accumbens, caudate putamen, hypothalamus, substantia nigra, and ventral tegmentum. The δ sites are most dense in the olfactory tubercle, caudate putamenm, and nucleus accumbens (Table 6) (21). These findings are in general agreement with the findings in autoradiographic studies (56,57). In the guinea pig brain, the μ-binding sites are most dense in the hypothalamus and midbrain, followed by the striatum, thalamus, and pons and medulla; the δ sites are most dense in the striatum (24).

The concentration of κ sites is low in rat brains but high in guinea pig brains. In guinea pig brain homogenates, κ sites are quite dense in cortex, striatum, hypothalamus, midbrain, hippocampus, cerebellum, and pons and medulla (24). An autoradiographic study found κ sites in the gray matter of the rat spinal cord (58). It is a concentrated band of sites within the dorsal part of the substantia gelatinosa (58), and dynorphin is localized in the same area (59). The guinea pig cerebellum contains κ sites but is essentially lacking μ and δ sites (60), thus making it an ideal source for κ receptor isolation. Autoradiographic studies with the guinea pig central nervous system showed κ sites in the layers V and VI of the cerebral cortex (57), the substantia nigra and bulbus olfactorius (61), and the distal region of the molecular layer of the dentate gyrus (62). Moderate concentrations of κ sites are also found in the molecular layer of the cerebral cortex, lamina VI of the neocortex, and the stratum lacunosum moleculare of the hippocampus of the human brain (63).

The σ-binding site has been found in brains of rats (21), mice (33,64), guinea pigs (22-24,34), and gerbils (64). In the rat CNS, the σ sites are most concentrated in the spinal cord, hypothalamus, hippocampus, pons and medulla, and cerebellum (Table 6). This pattern of regional distribution is unlike any other receptors for neurotransmitters. In the guinea pig brain, the highest concentrations of σ sites are in the hypothalamus, anterior pituitary, midbrain, and pons and medulla (24). Specifically, some σ sites are localized in the pyramidal layer of the hippocampus, the granular layer of the dentate gyrus, and the cranial nerve large motor neurons (65). This pattern of regional distribution is clearly different from that of the phencyclidine binding sites (21,65-68).

Table 6 Regional Distribution of Multiple Binding Sites in Rat CNS

Regions	μ	δ	σ	PCP
Olfactory bulb	++	++	+	ND[a]
Frontal cortex	++	+	+	++
Olfactory tubercle	+++	+++	–	ND
Nucleus accumbens	++++	++	–	ND
Septum	++	+	–	ND
Caudate putamen	++++	+++	+	ND
Remaining cortex	++	++	+	++
Globus pallidus	+++	++	+	ND
Amygdala	++	+	–	++
Hypothalamus	++++	+	++	ND
Anterior pituitary	+	–	+	ND
Posterior pituitary	+	–	–	ND
Hippocampus	+	–	++	+++
Thalamus	+++	++	+	ND
Substantia nigra	++++	+	–	ND
Ventral tegmentum	++++	+	+	ND
Cerebellum	+	–	++	+
Pons and medulla	+++	+	++	+
Spinal cord	++	+	++	+

[a]Not determined.
Source: Data are from Tam (21).

IRREVERSIBLE LIGANDS

A number of irreversible ligands have been described for opioid receptors. They have been extremely useful in the determination of multiple receptor types or subtypes. The use of the irreversible antagonists naloxazone and its azine derivative naloxonazine has led to the postulation of two μ receptor subtypes, μ_1 and μ_2. Naloxazone and naloxonazine selectively and irreversibly inhibit a high-affinity common binding site for opioids, or μ_1 site and block opioid analgesia with little effect on other classes of binding sites (35-37,69). This μ_1 site accounts for about 40% of the high-affinity binding of [^{3}H] D-Ala^2-Met^5 enkephalinamide (37) and is proposed as a mediator of supraspinal analgesia. The lower affinity binding site for morphine, μ_2, was thought to account for respiratory depression and represents the morphine receptor in the guinea pig ileum. These compounds provide useful tools for opioid receptor research.

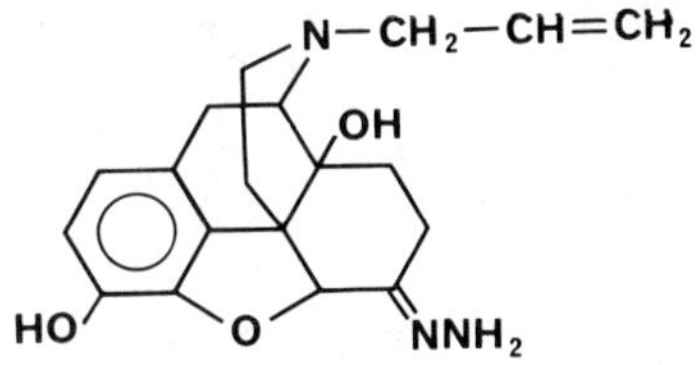

Naloxazone

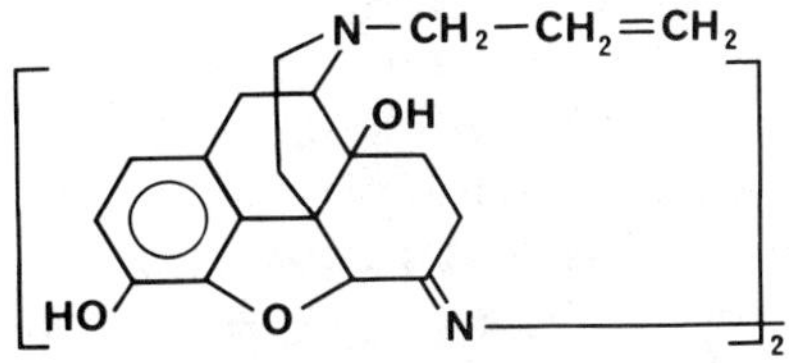

Naloxonazine

In another effort to obtain irreversible ligands for opioid receptors, the isothiocyanate derivatives of fentanyl (FIT) and etonitazene (BIT) and the fumaramido derivative of endoethenotetrahydrooripavine (FAO) have been synthesized (70). In vitro incubation of these ligands with rat brain membranes and NG108-15 hybrid cells indiated that BIT is a selective alkylator for μ receptors and FIT and FAO are selective alkylators for δ receptors. The reactivity of these ligands with κ-receptors has not been determined. The δ receptor selective affinity ligand [^{3}H] FIT has been used for identification and partial purification of a M_r 58,000 glycoprotein subunit of the opioid receptor of NG108-15 cells (71). More recently, a more potent δ-receptor-selective acylating agent, 3-[^{3}H] methylfentanylisothiocyanate (super-FIT), has been used to purify the δ receptor in NG108-15 cells is apparent homogeneity (72). This purified portein has a MW near 58,000 and it has a strong tendency to dimerize.

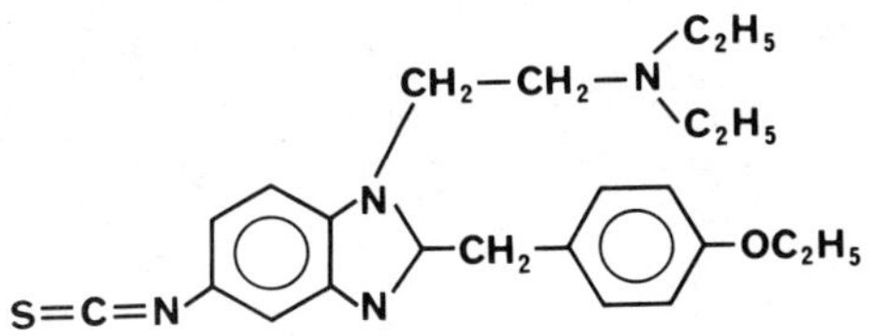

BIT

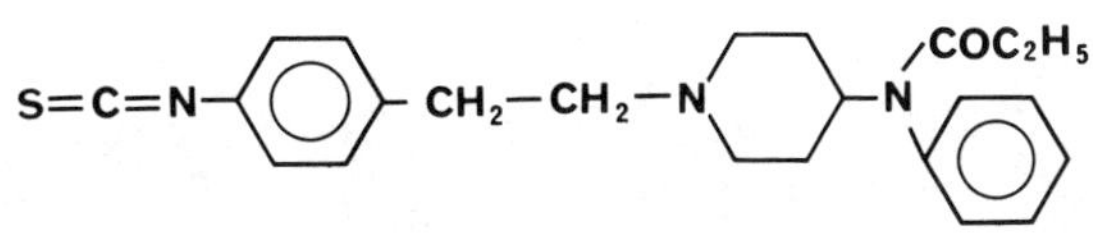

FIT

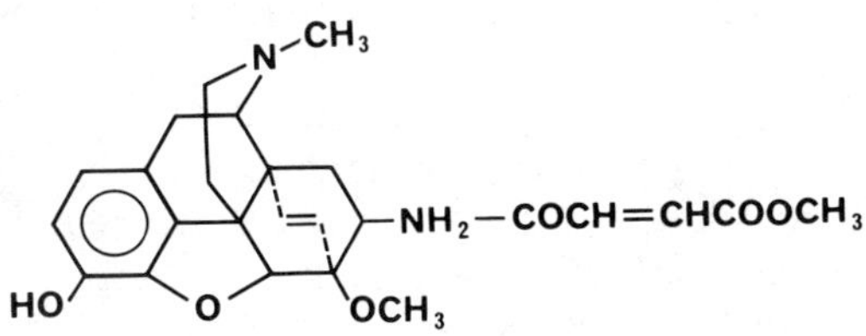

FAO

The first useful antagonist affinity label for opioid receptors was a nitrogen mustard derivative of naltrexone, β-chlornaltrexamine (β-CNA) (73,74). β-CNA is a potent irreversible ligand for opioid receptors in vitro and in vivo. β-CNA produced irreversible antagonism in both the guinea pig ileum (GPI) and mouse vas deferens (MVD) preparations (75,76). β-CNA is not a very selective opioid irreversible ligand. It alkylates opioid receptors with potency in the order $\mu > \kappa > \delta$. The most widely used irreversible antagonist selective for opioid receptor is the fumaramate methyl ester derivative of naltrexone, β-funaltrexamine (β-FNA) (77). β-FNA has been used very successfully to deplete the GPI of functional μ receptors (78), thereby affording a near homogeneous population of κ receptors. This preparation is very useful in the evaluation of κ activity of agonists and is particularly useful in evaluating those agonists that have high binding affinity for both the μ and κ receptors, such as the mixed agonist-antagonists. In vivo, β-FNA demonstrated analgesic activity in mice of short duration (79). This analgesic effect is mediated by κ receptors. β-FNA is a reversible κ-receptor agonist and an irreversible μ receptor antagonist. In contrast, the antagonist activity of β-FNA is μ selective and can last for days.

Some binding studies have also been performed with β-FNA (80-83). In these studies, a high concentration of β-FNA (1 μM) was used to show irreversible inhibition of μ-receptor binding (80-82). In GPI, pretreatment with 1 μM β-FNA did not alter μ, κ, or δ receptor binding (83). However, pretreatment of guinea pig brain membranes with β-FNA irreversibly inhibited μ and δ binding only when the β-FNA concentration was greater than 250 and 62 nM, respectively (83). Since high concentrations of β-FNA were required to obtain irreversible inhibition, it is not clear whether the irreversible inhibition was due to nonspecific alkylation or specific receptor-related reaction.

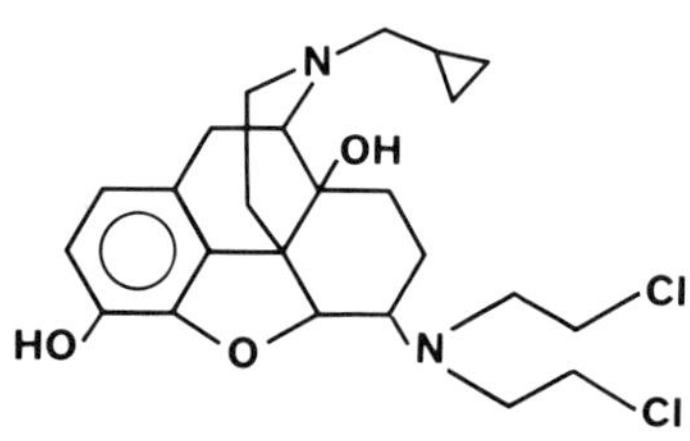

β-CNA

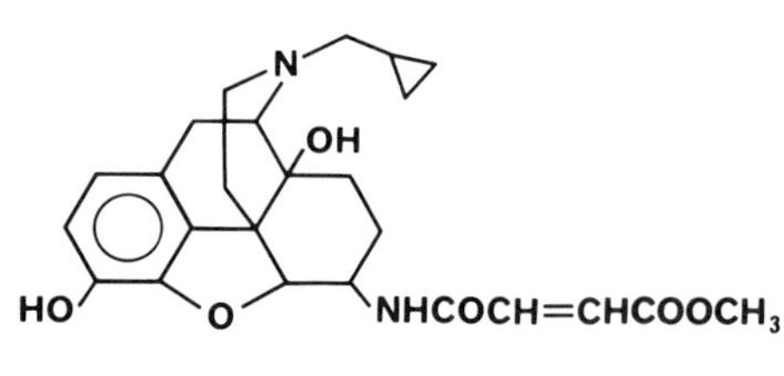

β-FNA

In a more detailed binding study with β-FNA, it was found that it binds reversibly to μ, κ, and δ receptors with K_i of 2.2, 14, and 78 nM, respectively (84). Thus, β-FNA is μ selective. In addition, it was found that pretreatment of guinea pig brain membranes with 500 nM β-FNA in Tris buffer caused a 25% irreversible inhibition of μ binding and a 15% irreversible inhibition of δ binding. The κ binding was not affected irreversibly. Interestingly, β-FNA became much more potent in irreversibly inhibiting μ binding and δ binding when the pretreatment with brain membranes was performed in Tris buffer containing 100 mM NaCl (Figure 1). However, the irreversible inhibition of μ binding plateaued at about 50% at a wide concentration range of β-FNA. This result may be interpreted as evidence for μ receptor subtypes: one that irreversibly reacts with β-FNA, and one that does not. However, since NaCl caused such a dramatic shift toward greater potency and extent of irreversible reactivity with β-FNA, it is possible that NaCl simply shifts the equilibrium between two or more conformational states of the μ receptor. This latter hypothesis was supported by the finding that the extent of irreversible inhibition was increased

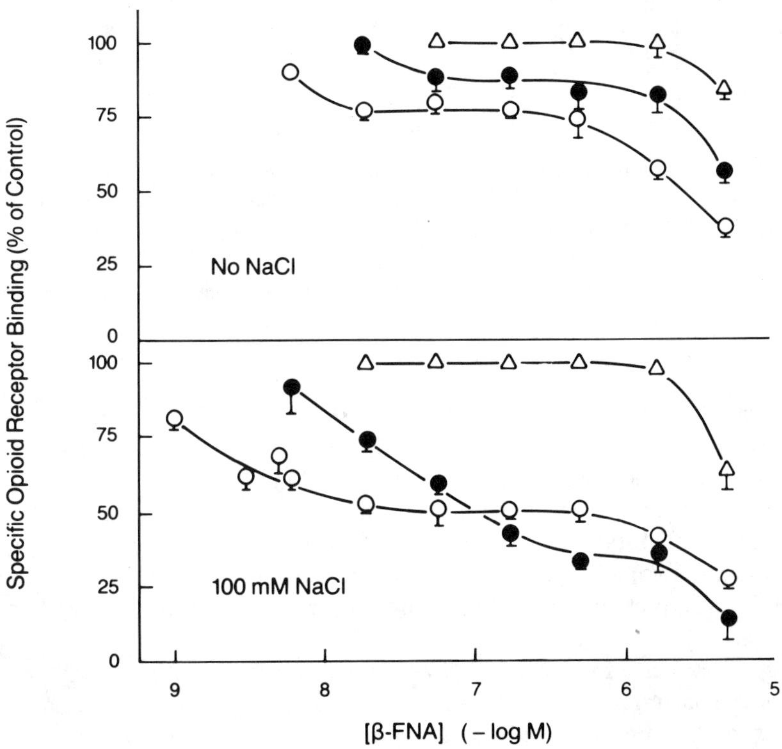

Figure 1 Irreversible inhibition of opioid receptors binding by pretreatment of guinea pig brain membranes with various concentrations of β-FNA. Membranes were treated with β-FNA at 37°C for 1 hr in the absence or presence of 100 mM NaCl and washed five times. Specific μ (○), δ(●), and κ(△) opioid receptor binding was performed in the presence of 100 mM NaCl.

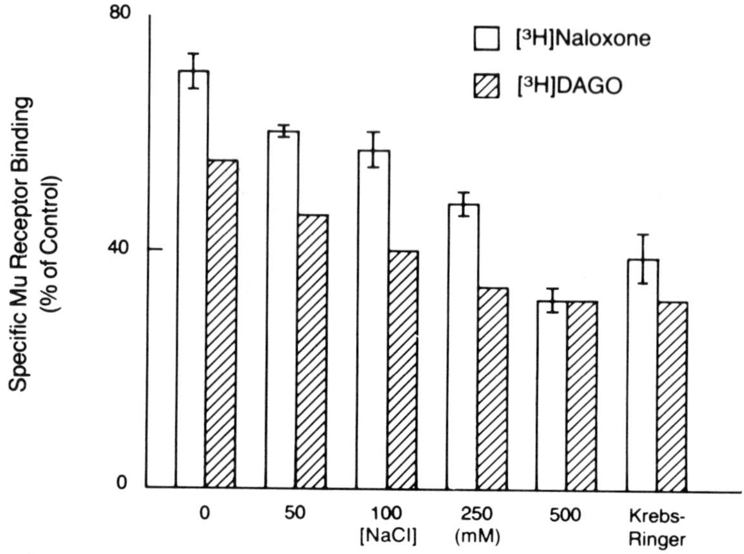

Figure 2 Effect of concentration of NaCl on irreversible inhibition of μ-opioid receptor binding by pretreatment of membranes with β-FNA. Membranes were pretreated with 100 nM β-FNA at 37°C for 1 hr in the presence of various concentrations of NaCl or in Krebs-Ringer solution and washed extensively. [^{3}H]Naloxone (0.5 nM) binding was performed in the presence of 100 mM NaCl; [^{3}H]DAGO (1 nM) binding was performed in the absence of NaCl.

with increasing concentrations of NaCl in the pretreatment buffer (Figure 2). Importantly, very significant irreversible inhibition of μ binding was also observed when pretreatment with β-FNA was performed in physiological Krebs-Ringer solution. Thus the results suggest that the μ receptor can exist in different conformational states and that NaCl favors a conformation of the receptor that has a nucleophile close enough to react irreversibly with β-FNA (85). Again, it is critical to choose the right medium to study receptor-ligand interaction.

A direct irreversible binding of β-FNA to the μ receptor has been shown with very low concentrations of [^{3}H]β-FNA (Figure 3) (85). This irreversible binding is stereospecific and is a two-step process. First β-FNA binds reversibly to the μ receptor, and then the irreversible reaction occurs as a second step. [^{3}H]β-FNA is a useful tool for μ receptor isolation. [^{3}H]β-FNA specifically and irreversibly labels only one major brain membrane glycoprotein, which shows up as a broad band of molecular weight 68-97 (kilodaltons) in sodium dodecyl sulfate polyacrylamide gel electrophoresis (86).

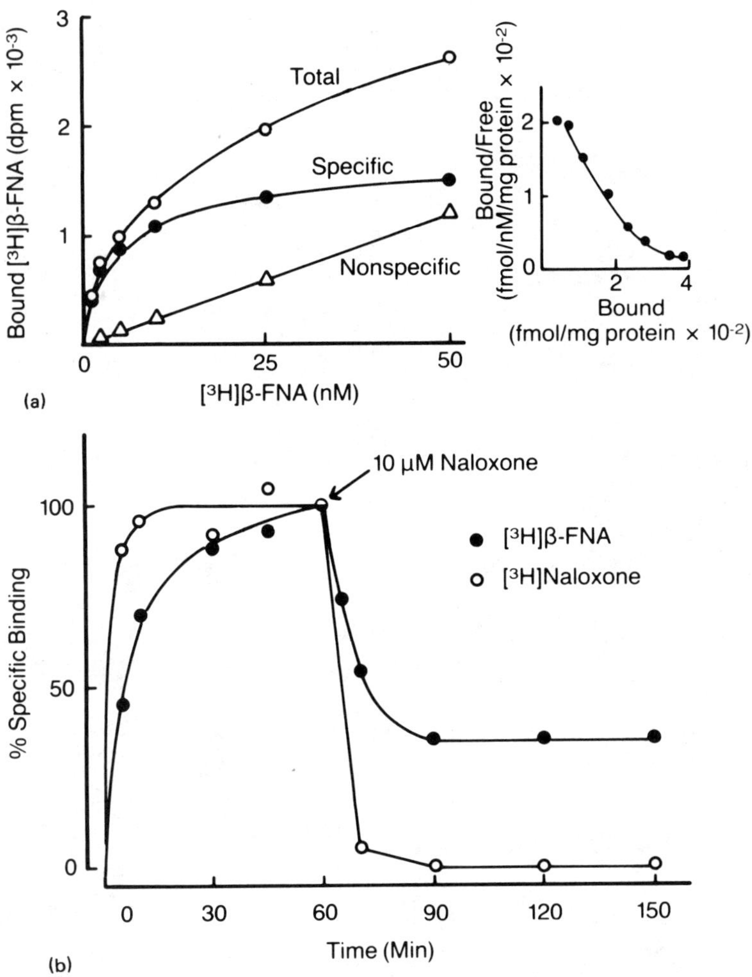

Figure 3 (a) Saturation studies of [^{3}H]β-FNA binding to bovine striatal membranes at 25°C for 60 min. (b) Time course of association and dissociation of 5 nM [^{3}H]β-FNA and [^{3}H] naloxone (control). Excess unlabeled naloxone was added after a 60 min incubation of ligands with guinea pig brain membranes at 25°C to begin displacement.

Table 7 Bivalent Ligands for Opioid Receptors

Structures	Selective receptor activity
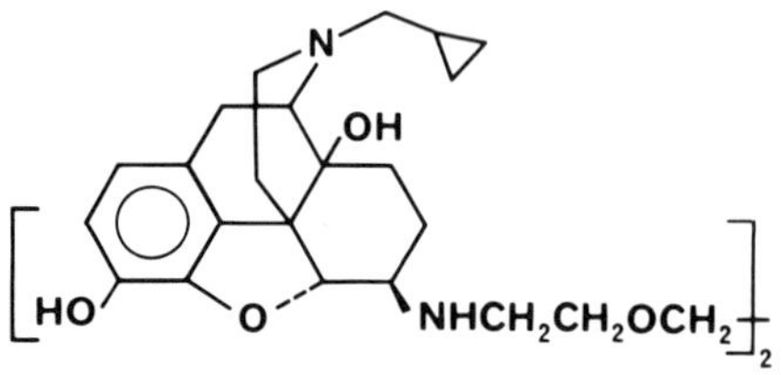	κ Antagonist
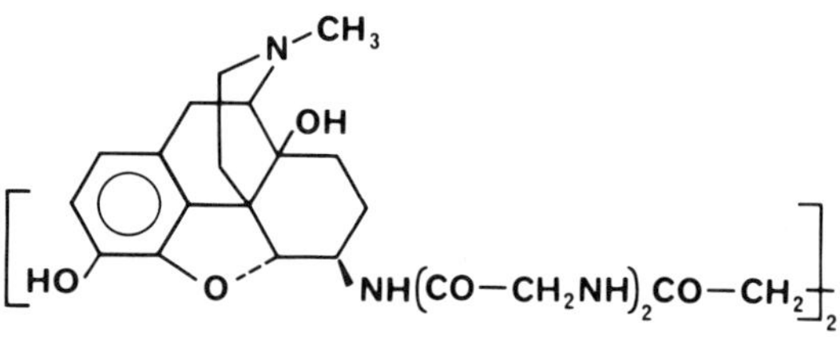	μ Agonist
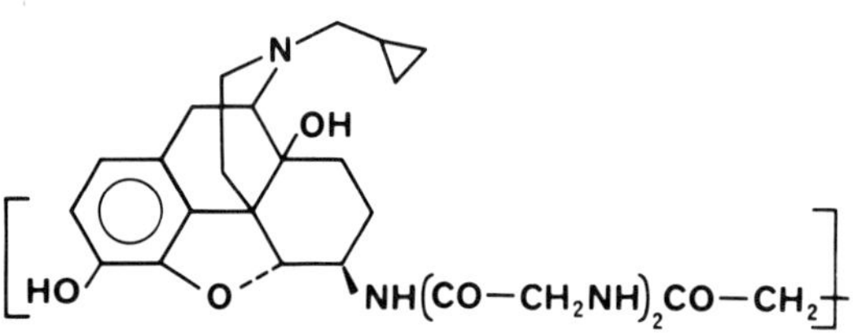	μ Antagonist

BIVALENT LIGANDS

Ligands having two pharmacophores connected by a spacer (bivalent ligands or dimers) have the potential for bridging vicinal receptors. Such bridging should be manifested by a substantial increase in potency due to the high local concentration of the free pharmacophore in the vicinity of the proximal recognition site when the bivalent ligand is bound in a monovalent mode. Indeed, bivalent ligands containing opiate or peptide pharmacophores have been reported to possess enhanced opioid agonist or antagonist potencies at a specific spacer length (87-96). Bivalent ligands derived from enkephalins (87-90), tetrapeptide enkephalin analogs (91), naltrexamine (91-95), and oxymorphamine (87, 94,95) have been studied. However, such enhanced potencies can also be achieved by bridging between an opioid receptor and a vicinal nonopioid recognition site. The requirement for these bivalent ligands to enhance potencies by bridging vicinal opioid receptors has been established by the preparation of bivalent ligands containing one active (–)-opiate pharmacophore and one inactive (+)-opiate pharmacophore (94). Such meso isomers have either no or only a small enhancement in potency compared with the monovalent ligands. The optimal spacer length varies with receptor types (95). In the guinea pig ileum, the most potent and selective κ antagonist is TENA, a naltrexamine dimer with a triethylenedioxy bridge (Table 7) (94). The very potent μ agonist and μ antagonist in the guinea pig ileum are dimers of oxymorphamine and naltrexamine, respectively, with spacer containing four glycines and a succinyl bridge (Table 7) (94). Selective bridging of δ receptors occurs with a naltrexamine dimer with a spacer containing six ethylene glycol units (92).

REGULATION AND FUNCTIONS

The classic study by Martin et al. (8) indicates that prototype agonists of different types of opioid receptors mediate different behavioral responses in chronic spinal dogs (Table 8). Other behavioral studies indicate that although both μ and κ agonists mediate analgesia, unlike μ agonists, κ agonists are not self-administered by monkeys (96), do not suppress abstinence in morphine-dependent dogs (8) or monkeys (96), and do not express cross-tolerance with morphine (7,97). The properties of opioid receptors are listed in Table 9. μ Agonists mediate analgesia, depress respiration, slow food transit, and cause

Table 8 Pharmacological Actions of the μ-, κ-, and σ-Receptor Agonists in Chronic Spinal Dogs[a]

	μ	κ	σ
Pulse rate	–	0	+
Pupil diameter	–	–	+
Respiratory rate	–	0	+
Temperature	–	0	+
Flexor reflex	–	–	–
Skin twitch reflex	–	–	0
Morphine abstinence	Suppress	No	No
Cyclazocine abstinence	Suppress	Suppress	
Behavioral effects	Indifference	Sedation	Delirium

[a]+, Increase; –, decrease; 0, no change.
Source: From Martin et al. (8).

Table 9 General Functions of Opioid Receptors

Type	Prototype agonists	Effects
μ	Morphine, morphiceptin	Analgesia, respiratory regulation, constipation, physical dependence
κ	Ethylketocyclazocine, dynorphin	Analgesia, regulation of appetite, sedation
δ	Leu-enkephalin, DADLE	Emotional behavior mediation, epileptic, analgesia
σ	SKF 10,047, possibly PCP	Mydriasis, tachypnea, tachycardia, mania

physical dependence. κ Agonists mediate analgesia and sedation, stimulate feeding, and increase urination. The κ opioid analgesia in the hot plate and tail-flick tests is dependent on serotonergic mechanisms (98). The diuretic effect of κ agonists is probably mediated by the suppression of vasopressin (99). δ Agonists mediate satisfaction and reward (100), seizures in the dorsomedial thalamus (101,102), and, to a lesser extent, analgesia. The σ agonists produce psychotomimetic effects, such as depersonalization, dysphoria, suspiciousness, and hallucinations (103).

Opioid receptors appear to modulate specific cholinergic pathways. Morphine depresses acetylcholine turnover in the parietal and occipital cortices, hippocampus, and nucleus accumbens but not in the frontal cortex, striatum, amygdala, substantia nigra, dorsal raphe, septum, or nucleus interpeduncularis (104-109) [for review see Wood (110)]. These neocortical and hippocampal cholinergic neurons are modulated by μ, δ, and ϵ receptor regulations (109, 111-113). Studies of the septal-hippocampal pathway indicate that these receptor sites are at the level of the cholinergic cell bodies or on afferent terminals or interneurons innervating these cells (111-113).

The μ and δ receptor agonists activate presynaptic opioid receptors in the nigrostriatal dopaminergic pathway of the rat (114-117) to increase dopamine synthesis but not release (105,118-123). The activation of presynaptic opioid receptors results in depressed dopamine transmission (123-126) and elicits changes in feedback activity to cause increases in substantia nigra cell firing after morphine is given parenterally (127-129). On the other hand, κ agonists do not alter dopamine metabolism in the nigrostriatal pathway (124) but can antagonize contralateral head turning evoked by electrical stimulation of the striatum (130). In contrast to their action in the rat, μ and δ agonists increase dopamine release in the mouse (123,131,132) and cause behavioral excitation (133).

κ and δ Agonists have been reported to inhibit the depolarization-induced efflux of γ-aminobutyrate (GABA) in rat substantial nigra slices (134). However, this effect requires very high concentrations of opioids (10--400 μM) and naloxone at these concentrations also has similar inhibitory effects. Thus, it is not clear whether the inhibitory effects of opioids on GABA release is a specific receptor-mediated effect.

Opioids also regulate hormonal release. An inhibitory enkephalinergic system exists in the hypothalamic dopaminergic neurons (135). The μ, δ, and ϵ receptor agonists act at the hypothalamic level to increase prolactin and growth hormone release in rats and humans (136-143). Blockade of the putative μ_1 sites by the irreversible antagonist naloxazone decreases morphine-induced release of prolactin but increases morphine-induced release of growth hormone in the rats (143).

Opioids modulate the activity of adenylate cyclase in the brain. Opioids have been reported to inhibit increased cAMP levels in brain or brain regions stimulated by prostaglandin (144), norepinephrine (145), or dopamine (146-150). The effects of opioids on adenylate cyclase activity are best understood by using the neuroblastoma cell line that contains only δ receptors. Opioids reduce the basal and prostaglandin E_1-stimulated adenylate cyclase activity in whole cells (151-153), and cell homogenates (154-157). Koski and Klee (158) demonstrated that opioids stimulate a membrane-associated GTPase in NG108-15 neuroblastoma cells and, further, provided evidence that opioids inhibit adenylate cyclase through stimulation of GTP hydrolysis (159).

ELECTROPHYSIOLOGY

The electrophysiological effects of opioids have been studied extensively in primary cultures of neuronal cells and brain slices (see reviews 160–162). In primary cell cultures of dissociated dorsal root ganglion neurons, several enkephalin analogs decrease both the amplitude and duration of the calcium spike triggered by depolarization current injection; the effects can be blocked by naloxone (163,164). Both μ and δ receptors are believed to mediate the decrease in calcium-dependent action potential (165). Opioids depress the firing rate of locus coeruleus neurons (166). Opioids hyperpolarize locus coeruleus neurons by increasing potassium conductance (167-169). This effect is mediated by μ receptors but not δ or κ receptors.

Opioid peptides have been studied in the hippocampal slice preparation (170-173). Opioids are excitatory in these preparations by reducing dendritic and/or somatic inhibitory postsynaptic potentials. Both μ and δ receptors produce the excitatory response on CA1 pyramidal cells (174). The κ agonist ethylketocyclazocine has no effect in this preparation.

The presence of both μ and δ receptors in the same enteric neuron has been demonstrated (175). Both μ and δ agonists produce inhibitory effects in these neurons. A small portion of these neurons may have only μ or δ receptors.

SPECULATIONS AND FUTURE DIRECTION

The opioid system is very complex and appears to modulate a wide variety of biologic functions. Thus, it is difficult to assess which aspect of opioid regulation is most important. However, several important issues are particularly worth mentioning. Opioids are very important analgesics, although the mechanisms of tolerance and physical dependence are still poorly understood. It is hoped that development of highly selective κ agonists will produce moderately strong analgesics without the physical dependence properties. However, κ agonists still produce tolerance. Thus, a clear understanding of the mechanism of analgesic tolerance and dependence will be extremely useful in the future development of potent analgesics with less undesirable effects.

It has been proposed that the interconversion can occur between the μ and δ forms of the opioid receptor in rat striatal patches (176). Although cross-protection studies, specific alkylation and lack of cross-tolerance among the different receptor types provide strong evidence that these receptor types are separate entities, the ultimate proof requires the purification and sequencing of each of these receptor types. The δ receptor from NG108-15 neuroblastoma-glioma hybrid cells has been purified to homogeneity (72), and the μ receptor from bovine striatum has been purified to near homogeneity (177). These receptors have very similar molecular weights. However, not enough quantities have been purified to allow amino acid sequencing. The question of in vivo interaction between μ, κ, and δ receptors is also important (178).

The discovery of a σ binding site that is different from the PCP site is very exiciting (21-24). That psychotomimetics and a number of antipsychotic drugs bind to this site with high affinity led to the hypothesis that in addition to blockade of dopamine receptors, some antipsychotic drugs exert their therapeutic action through the σ site (23). Consistent with this hypothesis, a non-dopaminergic potential antipsychotic agent rimcazole that has shown antipsychotic activity in open clinical trial (179) has been shown recently to bind to the σ site in vitro and in vivo (180). It is hoped that the development of a selective antagonist will lead to a new class of antipsychotic drugs without extrapyramidal symptoms. If the σ receptor is important for the regulation of physiological function, then one would expect there is a specific neurotransmitter or hormone for the σ receptor. Indeed, 5 years ago we have observed potent σ receptor binding activity from a low-molecular-weight fraction of rat brain extracts (S. W. Tam and L. Davis, unpublished observation). However, the purification of this active component has been difficult. This finding has recently been confirmed in guinea pig (181) and human brain extracts (182). It will be very important to purify and identify this endogenous substance.

REFERENCES

1. R. W. Houde and S. L. Wallenstein, Fed. Proc. *15*, 440 (1956).
2. W. R. Martin, Pharmacol. Rev. *19*, 463 (1967).
3. P. S. Portoghese, J. Med. Chem. *8*, 609 (1965).
4. A. E. Takemori, H. J. Kupferberg, and J. W. Miller, J. Pharmacol. Exp. Ther. *169*, 39 (1969).
5. C. B. Pert and S. H. Snyder, Proc. Natl. Acad. Sci. USA *70*, 2243 (1973).
6. E. J. Simon, J. M. Hiller, and I. Edelman, Proc. Natl. Acad. Sci. USA *70*, 1947 (1973).
7. P. E. Gilbert and W. R. Martin, J. Pharmacol. Exp. Ther. *198*, 66 (1976).
8. W. R. Martin, C. G. Eades, J. A. Thompson, R. E. Huppler, and P. E. Gilbert, J. Pharmacol. Exp. Ther. *197*, 517 (1976).
9. M. Hutchinson, H. W. Kosterlitz, F. M. Leslie, A. A. Waterfield, and L. Terenius, Br. J. Pharmacol. *55*, 541 (1975).
10. H. W. Kosterlitz, S. J. Paterson, and L. E. Robson, Br. J. Pharmacol. *73*, 939 (1981).
11. J. Hughes, T. W. Smith, H. W. Kosterlitz, L. A. Fothergill, B. A. Morgan, and H. R. Morris, Nature *258*, 577 (1975).
12. J. A. H. Lord, A. A. Waterfield, J. Hughes, and H. W. Kosterlitz, in Opiates and Endogenous Opioid Peptides (H. W. Kosterlitz, ed.), North-Holland, Amsterdam, 1976, p. 275.
13. J. A. H. Lord, A. A. Waterfield, J. Hughes, and H. W. Kosterlitz, Nature *267*, 495 (1977).
14. S. Lemaire, J. Magnan, and D. Regoli, Br. J. Pharmacol. *64*, 327 (1978).
15. R. Schulz, E. Faase, M. Wuster, and A. Herz, Life Sci. *24*, 843 (1979).
16. K.-J. Chang and P. Cuatrecasas, J. Biol. Chem. *254*, 2610 (1979).
17. K.-J. Chang, S. G. Blanchard, and P. Cuatrecasas, Mol. Pharmacol. *26*, 484 (1984).
18. R. S. Zukin and S. R. Zukin, Mol. Pharmacol. *20*, 246 (1981).
19. R. Quirion, R. P. Hammer, Jr., M. Herkenham, and C. B. Pert, Proc. Natl. Acad. Sci. USA *78*, 5881 (1981).
20. J. H. Jaffe and W. R. Martin, in The Pharmacological Basis of Therapeutics (A. G. Gilman, L. S. Goodman, and A. Gilman, eds.), Macmillan, New York, 1980, p. 494.
21. S. W. Tam, Proc. Natl. Acad. Sci. USA *80*, 6703 (1983).
22. T.-P. Su, J. Pharmacol. Exp. Ther. *223*, 284 (1982).
23. S. W. Tam and L. Cook, Proc. Natl. Acad. Sci. USA *81*, 5618 (1984).
24. S. W. Tam, Eur. J. Pharmacol. *109*, 33 (1985).
25. G. W. Pasternak and S. H. Snyder, Nature *253*, 563 (1975).
26. E. A. Young, J. M. Walker, R. Houghten, and H. Akil, Eur. J. Pharmacol. *91*, 327 (1983).

27. W. P. Burkard, Eur. J. Pharmacol. *97*, 337 (1984).
28. A. D. Corbett, S. J. Paterson, A. T. McKnight, J. Magnan, and H. W. Kosterlitz, Nature *299*, 79 (1982).
29. K.-J. Chang, E. Hazum, and P. Cuatrecasas, Proc. Natl. Acad. Sci. USA *78*, 4141 (1981).
30. P. Fenara and C. H. Li, Proc. Natl. Acad. Sci. USA *77*, 5746 (1980).
31. R. G. Hammonds, Jr., N. Ling, and D. Puett, Anal. Biochem. *114*, 75 (1981).
32. P. Y. Law, H. H. Lo, and C. H. Li, Proc. Natl. Acad. Sci. USA *76*, 5455 (1979).
33. B. R. Martin, J. S. Katzen, J. A. Woods, H. L. Tripathi, L. S. Harris, and E. L. May, J. Pharmacol. Exp. Ther. *231*, 539 (1984).
34. B. L. Largent, A. L. Gundlach, and S. H. Snyder, Proc. Natl. Acad. Sci. USA *81*, 4983 (1984).
35. G. W. Pasternak, S. R. Childers, and S. H. Snyder, Science *208*, 514 (1980).
36. G. W. Pasternak, Neurology *31*, 1311 (1981).
37. E. F. Hahn, M. Carrol-Buatti, and G. W. Pasternak, J. Neurosci. *2*, 572 (1982).
38. J. Grevel and W. Sadee, Science *221*, 1198 (1983).
39. W. Sadee, M. L. Richards, J. Grevel, and J. S. Rosenbaum, Life Sci. *33*, Sup. I, 187 (1983).
40. R. Simantov, S. R. Childers, and S. H. Snyder, Mol. Pharmacol. *14*, 69 (1978).
41. C. B. Pert and S. H. Snyder, Mol. Pharmacol. *10*, 868 (1974).
42. D. Romer, H. H. Buscher, R. C. Hill, R. Maurer, and T. J. Petcher, Nature *298*, 759 (1982).
43. P. F. VonVoigtlander, R. A. Lahti, and J. H. Ludens, J. Pharmacol. Exp. Ther. *224*, 7 (1983).
44. K.-J. Chang, T. Killian, E. Hazum, P. Cuatrecasas, and J. K. Chang, Science *212*, 75 (1981).
45. H. W. Kosterlitz and S. J. Paterson, Br. J. Pharmacol. *73* (Suppl) 299 (1981).
46. K.-J. Chang, B. R. Cooper, E. Hazum, and P. Cuatrecasas, Mol. Pharmacol. *16*, 91 (1979).
47. R. Quirion, Y. Kiso, and C. B. Pert, FEBS Lett. *141*, 203 (1982).
48. G. Gacel, M. C. Fournie-Zaluski, and B. P. Roques, FEBS Lett. *118*, 245 (1980).
49. H. I. Mosberg, R. Hurst, V. J. Hruby, K. Gee, K. Akiyama, H. I. Yamamura, J. J. Galligan, and T. F. Burks, Life Sci. *33* (Suppl. I), 447 (1983).
50. J. J. Gormley, J. S. Morley, T. Priestly, J. S. Shaw, M. J. Turnbull, and H. Wheeler, Life Sci. *31*, 1263 (1982).
51. N. F. Rogers and E. E. El-Fakahany, Life Sci. *37*, 307 (1985).
52. P. Honegger and E. Richelson, Brain Res. *109*, 335 (1976).
53. P. Honegger and E. Richelson, Brain Res. *162*, 89 (1979).
54. L. L. Werling, G. D. Zarr, S. R. Brown, and B. M. Cox, J. Pharmacol. Exp. Ther. *233*, 722 (1985).
55. K. A. Bonnet, J. Groth, T. Gioannini, M. Cortes, and E. J. Simon, Brain Res. *221*, 437 (1981).

56. R. Quirion, J. M. Zajac, J. L. Morgat, and B. P. Roques, Life Sci. *33*, Sup. I, 227 (1983).
57. R. R. Goodman and S. H. Snyder, Proc. Natl. Acad. Sci. USA *79*, 5703 (1982).
58. P. Slater and S. Patel, Eur. J. Pharmacol. *92*, 159 (1983).
59. L. J. Botticelli, B. M. Cox, and A. Goldstein, Proc. Natl. Acad. Sci. USA *78*, 7783 (1981).
60. L. E. Robson, R. W. Foote, R. Maurer, and H. W. Kosterlitz, Neuroscience *12*, 621 (1984).
61. R. W. Foote and R. Maurer, Life Sci. *33* (Suppl. I), 243 (1983).
62. R. W. Foote and R. Maurer, Eur. J. Pharmacol. *85*, 99 (1982).
63. R. Maurer, R. Cortes, A. Probst, and J. M. Palacios, Life Sci. *33* (Suppl. I), 231 (1983).
64. S. W. Tam, G. F. Steinfels, and L. Cook, in Sigma and Phencyclidine-like Compounds as Molecular Probes in Biology (E. Domino and J. J. Kamenka, eds.), NPP Book, Ann Arbor, Michigan, 1988, p. 383.
65. A. L. Gundlach, B. L. Largent, and S. H. Snyder, Eur. J. Pharmacol. *113*, 465 (1985).
66. S. R. Zukin and R. S. Zukin, Proc. Natl. Acad. Sci. USA *76*, 5372 (1979).
67. J. P. Vincent, B. Kartalovski, P. Geneste, J. M. Kamneka, and M. Lazdunski, Proc. Natl. Acad. Sci. USA *76*, 4678 (1979).
68. R. Quirion, R. P. Hammer, Jr., M. Herkenham, and C. B. Pert, Proc. Natl. Acad. Sci. USA *78*, 5881 (1981).
69. A. Z. Zhang and G. W. Pasternak, Life Sci. *29*, 843 (1981).
70. K. C. Rice, A. E. Jacobson, T. R. Burke, Jr., B. S. Bajwa, R. A. Streaty, and W. A. Klee, Science *220*, 314 (1983).
71. W. A. Klee, W. F. Simonds, W. F. Sweat, T. R. Burke, Jr., A. E. Jacobson, and W. A. Klee, FEBS Lett. *150*, 125 (1982).
72. W. F. Simonds, T. R. Burke, Jr., K. C. Rice, A. E. Jacobson, and W. A. Klee, Proc. Natl. Acad. Sci. USA *82*, 4974 (1985).
73. P. S. Portoghese, D. L. Larson, J. B. Jiang, A. E. Takemori, and T. P. Caruso, J. Med. Chem. *21*, 598 (1978).
74. P. S. Portoghese, D. L. Larson, J. B. Jiang, T. P. Caruso, and A. E. Takemori, J. Med. Chem. *22*, 168 (1979).
75. T. P. Caruso, A. E. Takemori, D. L. Larson, and P. S. Portoghese, Science *204*, 316 (1979).
76. S. J. Ward, P. S. Portoghese, and A. E. Takemori, Eur. J. Pharmacol. *80*, 377 (1982).
77. P. S. Portoghese, D. L. Larson, L. M. Sayre, D. S. Fries, and A. E. Takemori, J. Med. Chem. *23*, 233 (1980).
78. S. J. Ward, P. S. Portoghese, and A. E. Takemori, Eur. J. Pharmacol. *85*, 163 (1982).
79. S. J. Ward, P. S. Portoghese, and A. E. Takemori, J. Pharmacol. Exp. Ther. *220*, 494 (1982).
80. R. B. Rothman, W. D. Bowen, U. K. Schumacher, and C. B. Pert, Eur. J. Pharmacol. *95*, 147 (1983).

81. R. B. Rothman, U. K. Schumacher, and C. B. Pert, J. Neurochem. *43*, 1197 (1984).
82. S. J. Ward, P. S. Portoghese, and A. E. Takemori, Eur. J. Pharmacol. *80*, 377 (1982).
83. A. D. Corbett, H. W. Kosterlitz, A. T. McKnight, S. J. Paterson, and L. E. Robson, Br. J. Pharmacol. *85*, 665 (1985).
84. S. W. Tam and L.-Y. Liu-Chen, J. Pharmacol. Exp. Ther. *239*, 351 (1986).
85. S. W. Tam, V. J. Nickolson, and L.-Y. Liu-Chen, Eur. J. Pharmacol. *119*, 259 (1985).
86. L.-Y. Liu-Chen, C. A. Phillips, and S. W. Tam, Fed. Proc. *45*, 322 (1986).
87. E. Hazum, K.-J. Chang, H. J. Leighton, O. W. Lever, and P. Cuatrecasas, Biochem. Biophys. Res. Commun. *104*, 347 (1982).
88. S. A. Krumins, T. Costa, and D. Rodbard, Life Sci. *32*, 511 (1983).
89. S. A. Krumins, T. Costa, Y. Shimohigashi, P. J. Munson, and D. Rodbard, Biochem. Biophys. Res. Commun. *108*, 406 (1982).
90. Y. Shimohigashi, T. Costa, S. Matsuura, H.-C. Chen, and D. Rodbard, Mol. Pharmacol. *21*, 558 (1982).
91. Y. Shimohigashi, T. Costa, H.-C. Chen, and D. Rodbard, Nature *297*, 333 (1982).
92. M. Erez, A. E. Takemori, and P. S. Portoghese, J. Med. Chem. *25*, 847 (1982).
93. P. S. Portoghese and A. E. Takemori, Life Sci. *36*, 801 (1985).
94. P. S. Portoghese, D. L. Larson, C. B. Yim, L. M. Sayre, G. Ronsisvalle, A. W. Lipkowski, A. E. Takemori, K. C. Rice, and S. W. Tam, J. Med. Chem. *28*, 1140 (1985).
95. P. S. Portoghese, D. L. Larson, L. M. Sayre, C. B. Yim, G. Ronsisvalle, S. W. Tam, and A. E. Takemori, J. Med. Chem. *29*, 1855 (1986).
96. J. H. Woods, C. L. Fly, and H. H. Swain, Dev. Neurosci. *4*, 403 (1978).
97. P. Tepper and J. H. Woods, Psychopharmacology *58*, 125 (1978).
98. P. F. VonVoigtlander, R. A. Lewis, and G. L. Neff, J. Pharmacol. Exp. Ther. *231*, 270 (1984).
99. J. D. Leander, R. L. Zerbe, and J. C. Hart, J. Pharmacol. Exp. Ther. *234*, 463 (1985).
100. L. Stein and J. D. Belluzzi, Adv. Biochem. Psychopharmacol. *18*, 299 (1978).
101. H. Frenk, G. Urca, and J. C. Liebeskind, Brain Res. *147*, 327 (1978).
102. H. Frenk, B. G. McCarty, and J. C. Liebeskind, Science *200*, 335 (1978).
103. C. A. Haertzen, Psychopharmacologia *18*, 366 (1970).
104. D. L. Cheney, M. Trabucchi, G. Racagni, C. Wang, and E. Costa, Life Sci. *15*, 1977 (1974).
105. E. Costa, D. L. Cheney, G. Racagni, and G. Zsilla, Life Sci. *17*, 1 (1975).
106. G. Zsilla, D. L. Cheney, G. Racagni, and E. Costa, J. Pharmacol. Exp. Ther. *199*, 662 (1976).
107. G. Zsilla, G. Racagni, D. L. Cheney, and E. Costa, Neuropharmacology *16*, 25 (1977).
108. D. E. Schmidt and D. M. Buxbaum, Brain Res. *147*, 194 (1978).

109. P. L. Wood and L. M. Stotland, Neuropharmacology *19*, 975 (1980).
110. P. L. Wood, Neuropharmacology *21*, 487 (1982).
111. F. Moroni, D. L. Cheney, and E. Costa, Naunyn-Schmiedebergs Arch. Pharmacol. *299*, 149 (1977).
112. F. Moroni, D. L. Cheney, and E. Costa, Neuropharmacology *17*, 191 (1978).
113. P. L. Wood, D. L. Cheney, and E. Costa, Neuroscience *4*, 1479 (1979).
114. H. Pollard, C. Llorens, and J. C. Schwartz, Nature *268*, 745 (1977).
115. H. Pollard, C. Llorens, J. C. Schwartz, C. Gros, and F. Dray, Brain Res. *151*, 392 (1978).
116. E. L. Gardner, R. S. Zukin, and M. H. Makman, Brain Res. *194*, 232 (1980).
117. L. C. Murrin, J. T. Coyle, and M. J. Kuhar, Life Sci. *27*, 1175 (1980).
118. L. Ahtee and I. Kaariainen, Eur. J. Pharmacol. *22*, 206 (1973).
119. C. Gauchy, Y. Agid, J. Glowinski, and A. Cheramy, Eur. J. Pharmacol. *22*, 311 (1973).
120. M. F. Sugrue, Br. J. Pharmacol. *52*, 159 (1974).
121. A. Guidotti, F. Moroni, K. Gale, and K. Kumakura, in Catecholamines: Basic and Clinical Frontiers (E. Usdin, I. J. Kopin, and J. Burchas, eds.), Pergamon Press, New York, 1978, p. 1035.
122. F. Moroni, E. Peralta, D. L. Cheney, and E. Costa, J. Pharmacol. Exp. Ther. *208*, 190 (1979).
123. P. L. Wood, M. Stotland, J. W. Richard, and A. Rackham, J. Pharmacol. Exp. Ther. *215* 697 (1980).
124. A. Algeri, N. Brunello, G. Calderini, and A. Consolazione, Adv. Biochem. Psychopharmacol. *18*, 199 (1978).
125. P. Slater and C. Blundell, Neuropharmacology *18*, 705 (1979).
126. P. Moleman and J. Bruinvels, J. Neural. Trans. *46*, 225 (1979).
127. K. Iwatsuba and D. H. Clouet, J. Pharmacol. Exp. Ther. *202*, 429 (1977).
128. M. C. Nowycky, J. R. Walters, and R. H. Roth, J. Neural Trans. *49*, 99 (1978).
129. E. P. Finnerty and S. H. H. Chan, Eur. J. Pharmacol. *59*, 307 (1979).
130. P. Slater and D. A. Longman, Neuropharmacology *19*, 1153 (1980).
131. K. Kuschinsky and O. Hornykiewicz, Eur. J. Pharmacol. *26*, 41 (1974).
132. G. Racagni, F. Cattabeni, and R. Paoletti, in Animal Models in Psychiatry and Neurology (I. Hain and E. Usdin, eds.), Pergamon Press, New York, 1977, p. 315.
133. R. J. Katz and B. J. Carroll, Prog. Neuropsychopharmacol. *3*, 395 (1979).
134. M. S. Starr, Life Sci. *37*, 2249 (1985).
135. T. Okajima, T. Motomatsu, K. Kato, and H. Ibayashi, Life Sci. *26*, 699 (1980).
136. L. Cusan, A. Dupont, G. S. Kledzik, F. Labrie, D. H. Coy, and A. V. Schally, Nature *268*, 544 (1977).
137. C. J. Shaar, R. C. A. Frederickson, N. B. Dininger, and L. Jackson, Life Sci. *21*, 853 (1977).

138. V. Locatelli, D. Cocchi, S. Bajusz, S. Spampinato, and E. E. Muller, Experientia *34*, 1650 (1978).
139. B. V. Graffenfried, E. D. Pozo, J. Roubicek, E. Krebs, W. Poldinger, P. Burmeister, and L. Kerp, Nature *272*, 729 (1978).
140. H. Y. Meltzer, R. J. Miller, R. G. Fessler, M. Simonovic, and V. S. Fang, Life Sci. *22*, 1931 (1979).
141. K. M. Foley, I. A. Kourides, C. E. Inturrisi, R. F. Kaikor, C. G. Zaroulis, J. B. Posner, R. W. Houde, and C. H. Li, Proc. Natl. Acad. Sci. USA *76*, 5377 (1979).
142. L. Grandison, W. Fratta, and A. Guidotti, Life Sci. *20*, 1633 (1980).
143. K. Spiegel, I. A. Kourides, and G. W. Pasternak, Science *217*, 745 (1982).
144. H. O. J. Collier and A. C. Ray, Nature *248*, 24 (1974).
145. D. Tsang, A. T. Tan, J. C. Henry, and S. Lal, Brain Res. *152*, 521 (1978).
146. D. Wilkening, R. K. Mishra, and M. H. Makman, Life Sci. *19*, 1129 (1976).
147. K. P. Minneman, Br. J. Pharmacol. *59* (1977).
148. U. T. Motonatsu, M. Lis, N. Seidah, and M. Chretien, Biochem. Biophys. Res. Commun. *77*, 442 (1977).
149. U. Havemann and K. Kuschinsky, Pharmacology *16*, 295 (1978).
150. S. A. Walczak, D. Wilkening, and M. H. Mokman, Brain Res. *160*, 105 (1979).
151. J. Traber, K. Fischer, S. Latzin, and B. Hamprecht, Nature *253*, 120 (1975).
152. M. Brandt, K. Fischer, L. Moroder, E. Wunsch, and B. Hamprecht, FEBS Lett. *68*, 38 (1976).
153. M. Brandt, C. Buchen, and B. Hamprecht, FEBS Lett. *80*, 251 (1977).
154. S. K. Sharma, M. Nirenberg, and W. A. Klee, Proc. Natl. Acad. Sci. USA *72*, 590 (1975).
155. W. A. Klee and M. Nirenberg, Nature, *263*, 609 (1976).
156. A. Lampert, M. Nirenberg, and W. A. Klee, Proc. Natl. Acad. Sci. USA *73*, 3165 (1976).
157. A. Goldstein, B. M. Cox, W. A. Klee, and M. Nirenberg, Nature *265*, 362 (1977).
158. G. Koski and W. A. Klee, Proc. Natl. Acad. Sci. USA *78*, 4185 (1981).
159. G. Koski, R. A. Streaty, and W. A. Klee, J. Biol.. Chem. *257*, 14035 (1982).
160. G. Henderson, Br. Med. Bull. *39*, 59 (1983).
161. A. W. Duggan, Br. Med. Bull. *39*, 65 (1983).
162. R. A. North and T. M. Egan, Br. Med. Bull. *39*, 71 (1983).
163. A. W. Mudge, S. E. Leeman, and G. D. Fischbach, Proc. Natl. Acad. Sci. USA *76*, 526 (1979).
164. M. A. Werz and R. L. MacDonald, J. Pharmacol. Exp. Ther. *227*, 394 (1983).
165. M. A. Werz and R. L. MacDonald, Nature *299*, 730 (1982).
166. J. Korf, B. S. Bunney, and G. K. Aghajanian, Eur. J. Pharmacol. *25*, 165 (1974).
167. C. M. Pepper and G. Henderson, Science, *209*, 394 (1980).

168. J. T. Williams, T. M. Egan, and R. A. North, Nature *299*, 74 (1982).
169. J. T. Williams, G. Henderson, and R. A. North, in Brain Slices (R. Dingledine, ed.), Plenum Press, New York, 1984.
170. R. A. Nicoll, B. E. Algher, and C. E. Jahr, Nature *287*, 22 (1980).
171. H. K. Lee, T. Dunwiddie, and B. Hoffer, Brain Res. *184*, 331 (1980).
172. H. L. Haas and R. W. Ryall, J. Physiol. (Lond.) *308*, 315 (1980).
173. R. Dingledine, J. Neurosci. *1*, 1022 (1981).
174. R. J. Valentino and R. Dingledine, J. Pharmacol. Exp. Ther. *223*, 502 (1982).
175. T. M. Egan and R. A. North, Science *214*, 923 (1981).
176. W. D. Bowen, S. Gentlemen, M. Herkenham, and C. B. Pert, Proc. Natl. Acad. Sci. USA *78*, 4818 (1981).
177. T. L. Gioannini, A. D. Howard, J. M. Hiller, and E. J. Simon, J. Biol. Chem. *260*, 15117 (1985).
178. J. W. Holaday, J. B. Long, and F. C. Tortella, Fed. Proc. *44*, 2860 (1985).
179. G. Chouinard, and L. Annable, Psychopharmacology *84*, 282 (1984).
180. R. M. Ferris, F. L. M. Tang, K.-J. Chang, and A. Russell, Life Sci. *38*, 2329 (1986).
181. T.-P. Su, A. D. Weissman, and S.-Y. Yeh, Life Sci. *38*, 2199 (1986).
182. A.-Z. Zhang, K. N. Mitchell, L. Cook, and S. W. Tam, in Sigma and Phencyclidine-like Compounds as Molecular Probes in Biology (E. Domino and J. M. Kamenka, eds.), NPP Book, Ann Arbor, Michigan, 1988, p. 335.

18

Eicosanoid Receptors

Stanley T. Crooke, Seymour Mong, Mike Clark,* G. Kurt Hogaboom, Monica Lewis, and John Gleason / Smith Kline and French Laboratories, Philadelphia, Pennsylvania

HISTORICAL PERSPECTIVE

Several lines of evidence suggested that specific receptors for leukotrienes must exist and that they may display significant heterogeneity. The multiplicity of the peptidyl leukotrienes, their potency, and variations in rank order potencies in various tissues among the peptidyl leukotrienes, and the effects of antagonists in various pharmacological studies suggested strongly that multiple peptidyl leukotriene receptors must exist (1,2). For example, in guinea pig myocardium, the effects of the leukotrienes (LK) were blocked by the putative leukotriene end organ antagonist, FPL 55712 (3). In guinea pig trachea, when LTC_4 was incubated in the presence of serine-borate to prevent conversion to LTD_4 and LTE_4, it produced contractions that were not antagonized by FPL 55712, suggesting that the effects induced by the leukotrienes were mediated via receptors, that there are distinct receptors for LTC_4, and that FPL 55712 antagonizes the LTD_4 receptors (4). Furthermore, indirect evidence suggested that three specific receptors for LTC_4, LTD_4, and LTE_4 may exist in guinea pig lung (5). In human endothelial cells, LTC_4 appeared to induce specific desensitization of prostacyclin production, suggesting that the LTC_4 receptors were coupled to prostacyclin production and that the receptors were regulated by excess LTC_4 (6). Moreover, indirect evidence for the existence of LTB_4 receptors mediating neutrophil chemotaxis and other activities has also been reported (7,8).

Although many studies indirectly suggested the existence of heterogeneity and the importance of leukotriene receptors, only with direct radioligand bind-

**Present affiliation*: Washington University Medical School, St. Louis, Missouri

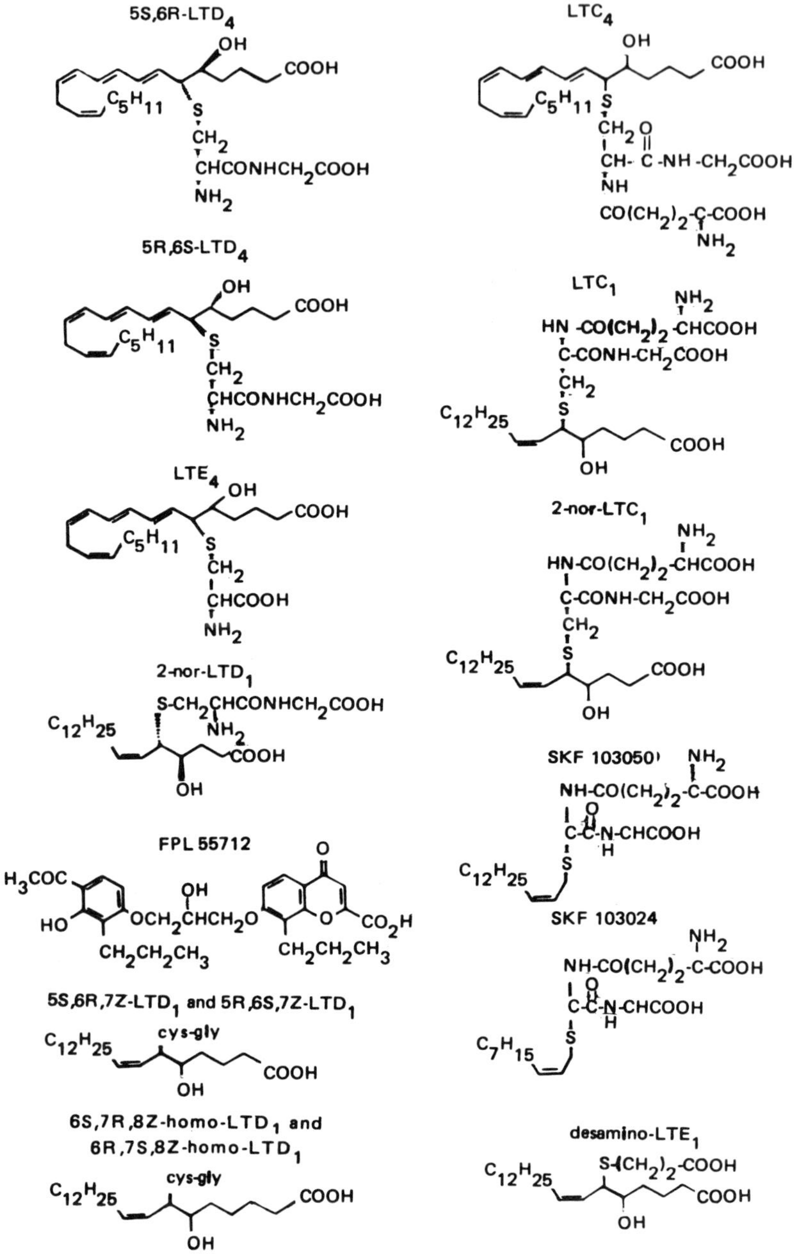

Figure 1 Peptidyl leukotriene analogs employed in this study.

ing studies have they been conclusively demonstrated and characterized. In the 5 years since the initial receptor binding studies were reported, our understanding of the leukotriene receptors has rapidly developed. Similarly, in the past 2 years substantial progress has been made in the elucidation of signal transduction processes. It is to these topics that the remainder of this review is directed.

LIGANDS EMPLOYED TO IDENTIFY AND CHARACTERIZE THE LEUKOTRIENE RECEPTORS

Figure 1 shows the structures of a number of ligands that have been employed to study the peptidyl leukotriene receptors. Numerous agonists and antagonists have been synthesized and employed to develop rank order potencies in receptor binding assays and physiological studies. Isotopically labeled [^{3}H] 14,15-5S,6R-LTC_4 ([^{3}H] LTC_4), and [^{3}H] 14,15-5S,6R-LTD_4 ([^{3}H] LTD_4) were prepared by Dupont NEN (Boston) with specific activities of 35-45 Ci/mmol. These radiolabeled agonists have been employed in competitive binding assays. However, a critical requirement that has not yet been met is the synthesis of radiolabeled high-affinity, high-specific activity antagonists of the LTC_4 and LTD_4 receptors. Although [^{3}H] LTB_4 has been synthesized and employed to study LTB_4 receptors, a much more limited array of analogs has been available to study these receptors.

PEPTIDYL LEUKOTRIENE RECEPTORS

LTD_4 Receptors

Binding Characteristics

Several laboratories have reported the presence of saturable stereoselective, high-affinity binding sites for LTD_4 in membranes prepared from several tissues. The ligand binding characteristics of LTD_4 receptors in guinea pig and human lung have been characterized extensively in our laboratory. Figure 2 shows a saturation binding isotherm for LTD_4 in guinea pig lung. From these and other studies, it was shown that LTD_4 rapidly and reversibly associated with LTD_4 specific receptors in guinea pig lung membranes. The K_D and B_{max} were 0.21 ± 0.05 nM and 1100 ± 200 mol/mg protein, respectively, (9-11). The dissociation rate constant was 0.181 min^{-1}. That the binding sites are specific for LTD_4 was demonstrated by competition binding analyses. These studies also clearly demonstrated that LTC_4 bound to LTD_4 receptors with much lower affinity than did LTD_4. Similar results were reported by other laboratories (12). Figure 3 shows that a number of mono- and divalent cations affect binding of LTD_4 to its guinea pig lung receptors. Furthermore, binding was shown to be optimal in

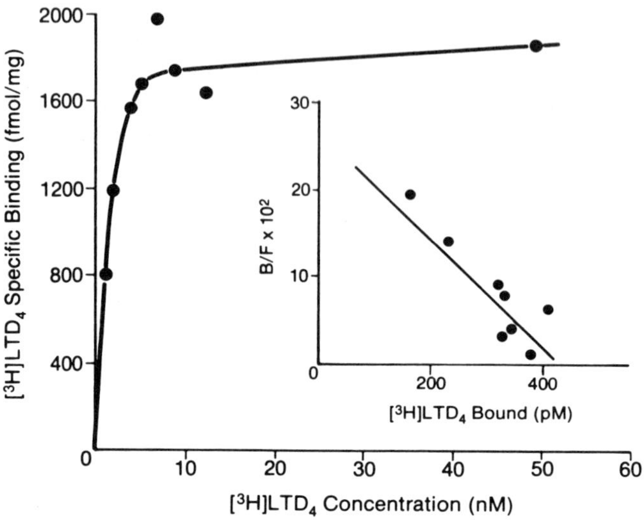

Figure 2 Saturation and Scatchard analyses of $[^3H]LTD_4$ receptor sites. Saturation analysis. The crude membrane preparation isolated from guinea pig lung was incubated with $[^3H]LTD_4$ (from 1 to 50 nM) in 10 mM PIPES buffer (pH 6.5) containing 5 mM cysteine, 5 mM glycine, 5 mM $CaCl_2$, and 5 mM $MgCl_2$. The nonspecific binding was estimated by including the unlabeled LTD_4 at concentrations 1000-fold that of the $[^3H]LTD_4$ under identical conditions. Specific binding at each concentration of $[^3H]LTD_4$ was determined after equilibrium was achieved. Each point represents the mean derived from triplicate determinations. Inset, Scatchard analysis of the data obtained from saturation analysis. (From Reference 13.)

PIPES buffer at pH 6.5, and guanine nucleotides were shown to inhibit binding (9,10,13-15).

LTD_4 receptors in adult and fetal human lung have been shown to be remarkably similar to those found in guinea pig lung. A single class of binding sites with K_D values of 0.15 ± 0.09 and 0.12 ± 0.003 nM and B_{max} values of 68 ± 29 and 62 ± 14 fmol/mg protein was observed in adult and fetal human lung, respectively. Binding to the LTD_4 receptors was affected by sodium ions and guanine nucleotides in a fashion entirely analogous to that observed in guinea pig lung (16). Thus from these studies we may conclude that receptors for LTD_4 are present in guinea pig and human lung and that they are comparable, although the density of the receptors may vary in different species. Moreover, we can conclude that there is no evidence of developmental variation in either density or other characteristics in LTD_4 receptors in humans.

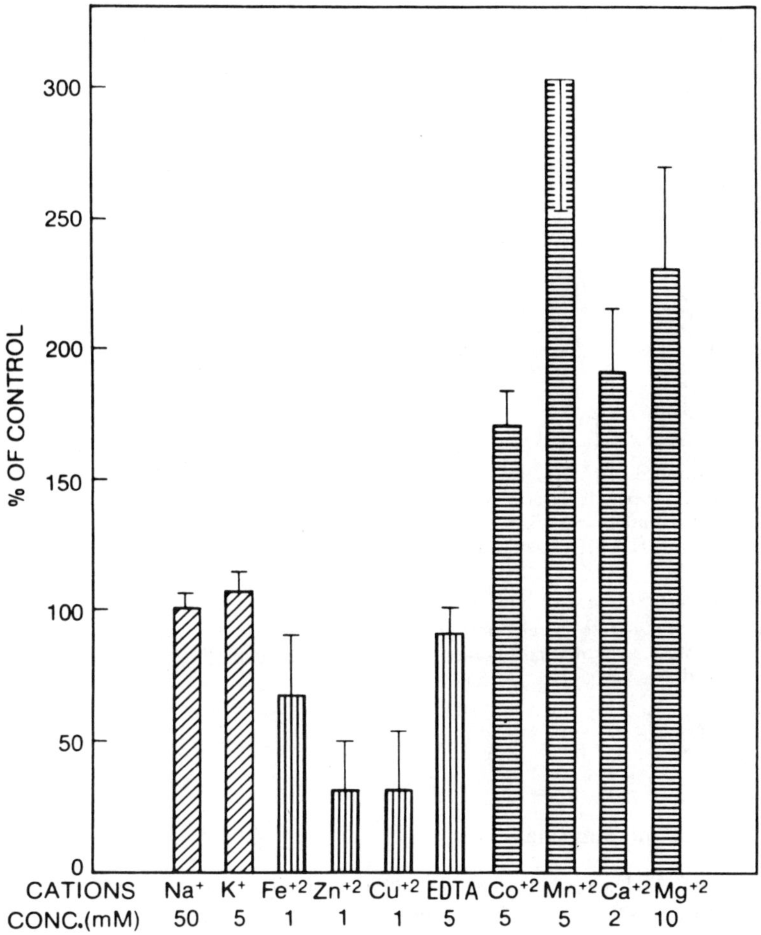

Figure 3 Stimulatory and inhibitory effect of monovalent and divalent cations on [^{3}H] LTD_4 specific binding. The [^{3}H] LTD_4 (3.5 nM) was incubated with the membrane preparations in the 10 mM PIPES (pH 6.5) containing 50 mM NaCl, in the presence of 3.5 mM LTD_4 (to determine the nonspecific binding) or absence of LTD_4 (to determine the total binding) at 20°C from 0 to 40 min. The specific binding at the equilibrium was taken as 100% of control. The effects of various cations were determined by inclusion of millimolar concentrations of $CaCl_2$, $MgCl_2$, $ZnCl_2$, $FeSO_4$, $CuSO_4$, and KCl in the incubation mixtures. The percentage stimulation or inhibition was calculated by taking the ratio of the amount of the specific binding to that of the control (50 mM of NaCl). (From Reference 14.)

Table 1 Tissue and Species Variability of the [^{3}H] LTD_4-Specific Binding Sites[a]

Tissue	Specifically bound [^{3}H] LTD_4 (fmol)	
	Rabbit[b]	Guinea pig
Lung	ND[c]	191 ± 60
Trachea	ND	65
Ileum	ND	ND
Uterus	ND	44 ± 11
Brain	NT[d]	ND
Heart	136 ± 30	166 ± 30
Liver	ND	ND

[a]The presence or absence of [^{3}H] LTD_4-specific binding sites were determined by kinetic experiments using the methods described in the legend of Figure 3. [^{3}H] LTD_4 (3.5 nM) was incubated with the membranes isolated from rabbit or guinea pig tissues in the presence of 10 mM Tris buffer (pH 7.5), 150 mM NaCl, and 2 mg/ml of protein from 0 to 40 min at 37°C.

[b]Female Flemish giant rabbits with body weight 1.5 kg were used in this study.

[c]Not detected; the steady-state level of nonspecific binding was either greater than or equivalent to the steady-state level of total binding. Hence, no specific binding of [^{3}H] LTD_4 could be determined.

[d]Not tested.

Source: From Reference 17.

Tissue Distribution and Subtypes

Table 1 shows that LTD_4 receptors are detected in several guinea pig and rabbit tissues. However conditions for LTD_4 binding must be optimized for each tissue. If standard conditions for binding to membranes from lung are employed, no binding to other tissues is observed (9,10,14,17,18). The distribution of LTD_4 receptor correlates well with the sensitivity of various tissues to LTD_4.

Evidence for tissue subtypes of LTD_4 receptors derives primarily from comparisons of receptors in guinea pig heart and lung. Table 2 compares some of the properties of LTD_4 receptors derived from these tissues. Although the lack of cation sensitivity of LTD_4 receptors from heart may possibly be explained by variations in the amount of adventitious cations present in that tissue, the differences in the effects of guanine nucleotides and sulfhydryl reagents strongly suggest that cardiac and lung LTD_4 receptor subtypes may exist (19). Thus, these studies suggest the possibility of developing tissue-selective antagonists.

Although LTD_4 receptors have been identified in the membranes of smooth muscle and endothelial cells maintained in tissue culture, the specific binding

Table 2 Comparative Properties of LTD_4 Receptors in Guinea Pig Lung and Heart

	Lung	Heart
Affinity (K_D, nM)	0.2 ± 0.05	3.4 ± 2.1
Density (B_{max}, pmol/mg protein)	1-2	0.85 ± 0.09
Cation modulation		
Ca^{2+}, Mg^{2+}	Stimulation	None
Na^+	Inhibition	None
Guanine nucleotide modulation	Transition of affinity state	None
Sensitivity to		
NEM, pHMB	Sensitive	Insensitive
DTT	Resistant	Sensitive

component is not adequate to characterize the binding more fully (20). To develop improved binding conditions that allow further characterization of LTD_4 receptors in smooth muscle and endothelial cells remains an important objective as it is of considerable interest to determine their subtype and other characteristics.

Chemical Characteristics

Preliminary observations relative to the chemistry of LTD_4 receptors were derived from studies employing sulfhydryl- and disulfide-containing reagents. In guinea pig myocardium, dithiothreitol (DTT), and sulfhydryl reducing agent, resulted in a decrease in the B_{max} without a change in K_D. Pretreatment with agonist did not prevent this effect. These results suggest that LTD_4 receptors in guinea pig myocardium may have one or more essential disulfides and that they are not in the agonist binding site (19). In contrast, LTD_4 receptors in membranes derived from guinea pig lung were insensitive to DTT but sensitive to such reagents as *N*-ethylmaleimide (9). We have solubilized, partially purified, and characterized LTD_4 receptors from guinea pig lung membranes after labeling with $[^3H]LTD_4$. Of the detergents tested, digitonin was the most effective at 0.5-1.0%. Following solubilization, the receptors had characteristics analogous to their characteristics in the membranes. The dissociation of $[^3H]LTD_4$ from the receptors was increased by sodium ions and guanine nucleotides, and the receptors were sensitive to heat, trypsin, and chymotrypsin. Table 3 shows the recoveries and partial purification of the receptors from membranes employing a procedure involving size-exclusion chromatography.

Figure 4 shows the elution profile of the LTD_4 receptors off several lectin columns, and Figure 5 shows the elution profile of the receptors from a size-

Table 3 Partial Purification of LTD_4 Receptor Complex for Guinea Pig Lung Membranes

Procedures	Protein concentration (μg/ml)	Total protein (μg)	Receptor activity (fmol)	Yield (%)	Specific activity (fmol/mg)	Enrichment (-fold)
Crude membrane protein[a]	1750	4725	2948	100	504	1
Solubilization[b]	450	450	1500	51	3330	5.6
Size-exclusion chromatography[c]	40	120	594	20	4950	8.3
WGL column[d]	7	7	85	3	12080	20.3

[a]Crude guinea pig lung membranes were employed.
[b]The LTD_4 receptors were labeled with 2 nM [^{3}H] LTD_4, under standard conditions, and solubilized with 0.5% digitonin.
[c]TSK 3000 and TSK 4000 size-exclusion HPLC columns were connected in series to fractionate the solubilized receptor complex.
[d]Wheat germ lectin column was employed to fractionate the solubilized [^{3}H] LTD_4 receptor complex. The concentration of protein, the specific activity of receptor, degree of enrichment, and percentage yield were calculated as described in the text.
Source: From Reference 18.

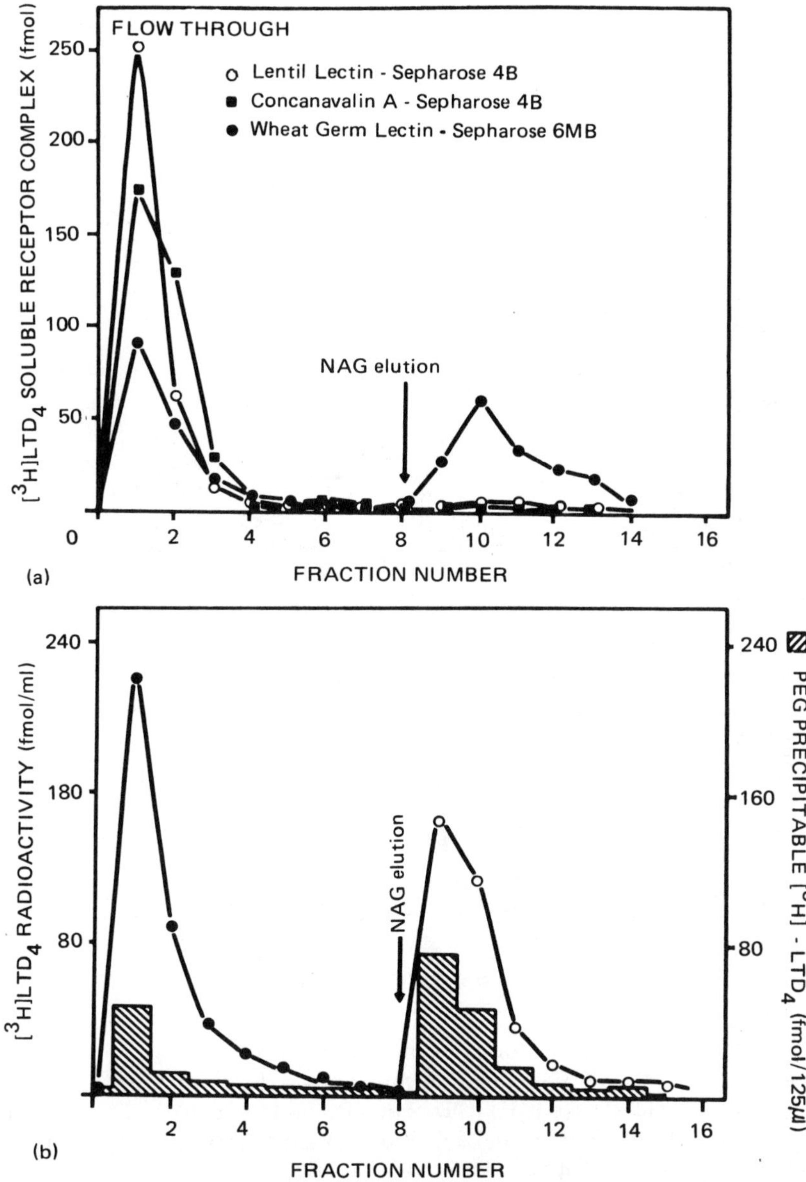

Figure 4 Lectin affinity chromatography of solubilized receptor complex. (a) [^{3}H] LTD_4-labeled soluble receptor complex was prepared and loaded on Lentil lectin-Sepharose, conconavalin A-Sepharose 4B, or wheat germ lectin-Sepharose 6MB columns. The columns were then washed stepwise eight times as flow-through using 1 ml buffer for each wash. *N*-acetylglucosamine (0.1 M) was used to elute the radioactivity bound to the column in a stepwise manner with 1 ml NAG elution in each fraction. (b) Radioactivity of the flow-through and eluted fraction was monitored by the polyethylene glycol precipitation method or direct scintillation spectrometry of the soluble fractions. (From Reference 18.)

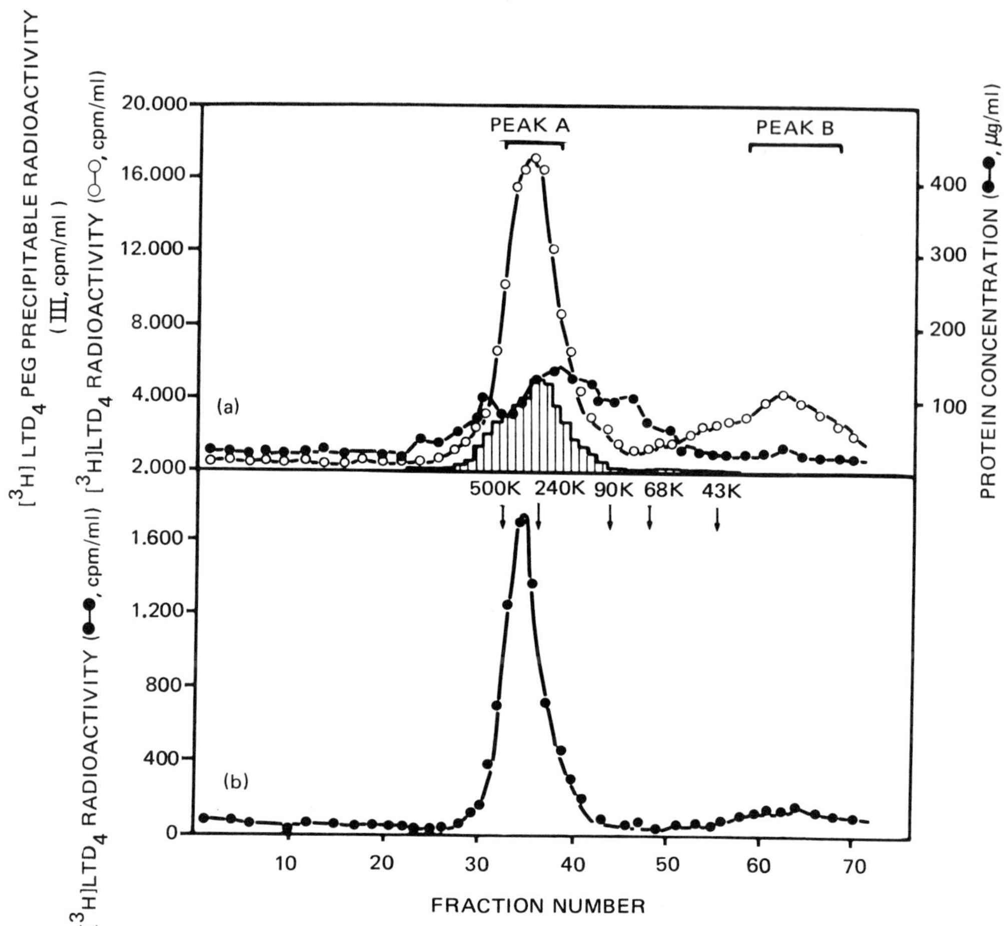

Figure 5 Size-exclusion chromatography of solubilized receptor-ligand complex. A TSK 4000 and TSK 3000 size-exclusion chromatography column was connected in series to fraction the [^{3}H] LTD$_4$-bound soluble receptors. The column was eluted with buffer containing 0.5% digitonin at a flow rate of 0.5 ml/min. (a) The radioactivity of PEG-precipitable material in 200 μl each fraction was determined (▥). Alternatively, the radioactivity in 100 μl of each fraction was determined directly (○–○). The protein concentration in each fraction was determined by the Bradford method (●–●) as the presence of digitonin interfered with ultraviolet absorption spectra or the Lowry method for the determination of protein concentration. (b) The fractions 32-38 of the chromatogram in (a) were pooled and injected into the same columns together with the molecular weight protein markers under conditions identical to those described above. The radioactivity in fraction 58-68 (peak B) area corresponds to free [^{3}H] LTD$_4$. (From Reference 18.)

exclusion column. These studies demonstrate that the receptor can be solubilized from guinea pig lung membranes as a multimeric complex with an apparent molecular weight of 240,000-500,000 daltons in the presence of digitonin (17, 18,21,22).

Physiologic Role

To characterize the physiological role of LTD_4 receptors, we compared the affinities of a series of LTD_4 receptor ligands with their activities in smooth muscle contraction assays employing guinea pig lung parenchyma. Figure 6 shows an excellent correlation between affinity for receptors and contractile activities for these analogs. Thus, we conclude that most, if not all, LTD_4 binding sites in guinea pig lung membranes are coupled to smooth muscle contractile activities (9).

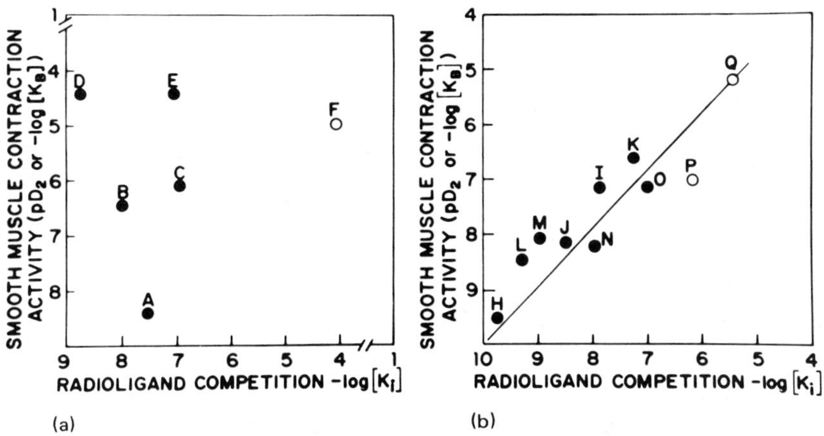

Figure 6 Correlation of smooth muscle contraction and ligand binding activities of LT analogs. The agonist-induced smooth muscle contraction activities were determined from organ-bath systems using guinea pig lung parenchymal strips. The pD_2 and K_1 were calculated. K_a values for FPL 55712 and 2-nor-LTD_1 were determined by pretreating tracheal tissue with 100 mM FPL 55712 in the presence of 45 mM serine-borate followed by LTC_4 treatment. (a) LTC analogs: A, LTC_4, B, LTC_1, C, 2-nor-LTC_1; D, SKF 103024; and F, antagonist FPL 55712. (b) LTD analogs: H, 5S-6R-LTD_4; I, 5R,6S-LTD_4; J, 5S,6R-LTD_1; K, 5R,6S-LTD_1; L, desamino-LTE_1; M, LTE_4; N, 6S,7R-8Z-homo-LTD_1; O, 6R,7S-8Z-homo-LTD_1; P, antagonist FPL 55712; Q, 2-nor-LTD_1. The antagonist activity of FPL 55712 and 2-nor-LTD_1 was determined in the absence of serine-borate. pD_2, K_B, and K_1 for each of the agents was determined from one to four experiments. The SE of these values were smaller than the symbols used. (From Reference 9.)

LTC_4 Receptors

Binding Characteristics

Binding sites for LTC_4 have been observed in a number of tissues in several laboratories. LTC_4 binding sites were observed in rat lung (12), guinea pig uterus (23), in guinea pig ileum (24), and a cultured smooth muscle cell line (25,26).

In our laboratory, we have also studied LTC_4 binding sites in a number of tissues. In the presence of serine-borate to inhibit conversion to LTD_4, LTC_4 specifically bound to sites in membranes prepared from guinea pig heart. The binding was saturable, yielding a K_D of 27.5 ± 5 nM and a B_{max} of 19.9 ± 3 pmol/mg protein. Competition binding curves for LTC_4 and LTD_4 binding sites in guinea pig lung demonstrated that the binding sites for the two leukotrienes differ. LTC_4 binding sites in guinea pig lung were sensitive to *N*-ethylmaleimide, stimulated by Ca^{2+}, Mg^{2+}, and Na^+, and unaffected by guanine nucleotides. They were also relatively more resistant to heating than LTD_4 receptors in guinea pig lung (9,11). Similar results were obtained in human fetal and adult lung. A K_D of 26 ± 6 nM and a B_{max} of 84 ± 18 pmol/mg protein were observed. Moreover, the characteristics of binding and effects of various monovalent and divalent cations, guanine nucleotides, and various competing ligands were comparable to those observed in guinea pig lung. The guinea pig appears to be an appropriate model for humans (27).

We have also demonstrated LTC_4 binding sites in guinea pig myocardium (28). The binding sites were saturable, displayed high affinity for LTC_4, and were highly specific for LTC_4. The K_D and B_{max} values were 27.5 ± 6.0 nM and 19.9 ± 5.2 pmol/mg protein, respectively. In the myocardium, LTC_4 binding was enhanced by mono- and divalent cations, and this appeared to be due to the increase in B_{max} with no significant effect on the K_D. *N*-ethylmaleimide induced a concentration-dependent reduction in LTC_4 binding. At 30 mM, *N*-ethylmaleimide reduced the B_{max} of the $[^3H]LTC_4$ binding site from 19.9 ± 5.2 to 8.2 ± 3.1 pmol/mg protein without significant effects on the K_D. Pretreatment with LTC_4 protected the binding sites from inactivation by *N*-ethylmaleimide and suggested that a sulfhydryl group was located at or near the agonist binding site.

LTC_4 binding has been demonstrated in membrane fractions from two smooth muscle cell lines, the hamster line DDT1MF2, and the murine line BC3H1 (20,26). Moreover, similar results were obtained in studies on the bovine endothelial cell line, CPAE (29). In the DDT1MF2 cells the K_D and B_{max} were 2.1 ± 4 nM and 55 ± 5 pmol/mg protein, respectively, values comparable to those reported by Krilis et al. (25). In the murine smooth muscle cell line, BC3H1, the K_D and B_{max} were 33 ± 16 nM and 25 ± 6 pmol/mg protein, respectively. Furthermore, competition binding analyses demonstrated that the binding sites were specific for LTC_4 and different from LTD_4 binding sites in the membranes of these cells.

Tissue Distribution and Subtypes

Table 4 shows the B_{max} for LTC_4 and LTD_4 binding to guinea pig tissues employing conditions optimal for binding to membranes derived from guinea pig lung. Obviously, specific LTC_4 binding to membranes of various guinea pig tissues was observed; however, binding of LTC_4 was considerably less fastidious than that of LTD_4. Moreover, a number of tissues not typically considered responsive to leukotrienes displayed LTC_4 binding sites (9,17,18,21,22).

Less substantial evidence for tissue subytpes of LTC_4 binding sites existed than for LTD_4 sites. Table 5 compares the properties of LTC_4 binding sites in guinea pig lung to those in guinea pig ventricular myocardium (9). No significant difference in mono- or divalent cation sensitivity, effects of guanine nucleo-

Table 4 Specific Binding of [^{3}H] LT (fmol/mg of Protein Specifically Bound) to Guinea Pig Tissues

Tissue	LTD_4[a]	LTC_4[b]
Liver	ND[c]	11736 ± 73
Brain	ND	8456 ± 260
Ileum	ND	7512 ± 808
Uterus	ND	7128 ± 832
Heart	ND	7024 ± 1048
Kidney	ND	7024 ± 444
RBC	ND	6448 ± 200
Lung	971 ± 108	5312 ± 328
Skeletal muscles (musculus gastrocnemius)	ND	4616 ± 560
Spleen	ND	4392 ± 76
Stomach	ND	2400 ± 620

[a]The [^{3}H] LTD_4 specific binding experiment was conducted under standard conditiions in kinetic experiment as described in note b.

[b]The number of [^{3}H] LTC_4 specific binding was obtained from kinetic binding experiments. [^{3}H] LTC_4 binding was initiated by adding membrane protein to 1 ml incubation mixtures (final protein concentration of 100 μg/ml) that contained 10 mM Tris-HCl (pH 7.5). 80 mM serine-borate 1 mM $CaCl_2$, 10 mM $MgCl_2$, 5 mM cysteine, and 5 mM glycine (standard condition). The level of [^{3}H] LTC_4 specific binding was determined after [^{3}H] LTC_4 binding has reached equilibrium (30 min incubation at 20°C).

[c]Not detected under these experimental conditions.

Source: From Reference 14.

Table 5 Comparative Properties of LTC_4-Specific Binding in Guinea Pig Lung and Heart[a]

Property	Lung	Heart
Affinity (K_D, nM)	15 ± 5	27.5 ± 6.0
Density (B_{max}, pmol/mg)	50-80	19.9 ± 5.2
Cation modulation		
Mg^{2+}, Ca^{2+}	Stimulation	Stimulation
Na^+	Stimulation	Stimulation
Guanine nucleotide modulation	No effect	No effect
Sensitivity to		
NEM	Decrease	Decrease
pHMB	Density of specific binding	Density

[a]The conditions employed in this study are identical to those indicated in Figures 5 and 6.
Source: From References 9 and 28.

tides, or sensitivity to sulfhydryl alkylating agents were observed in binding sites in the myocardium and lung of the guinea pig. To determine more definitively if tissue-specific subtypes of LTC_4 binding sites exist, more detailed competitive binding studies employing a larger number of LTC_4 analogs are required.

Physiological Role

The physiological role of the LTC_4 binding site was less clearly defined than that of LTD_4 receptors. Figure 6 shows that, in contrast to LTD_4 receptors (Figure 6b), minimal correlation binding assays and lung parenchymal contractile activity between the binding of LTC_4 analogs in competitive was observed (9). Moreover, the broad tissue distribution of LTC_4 binding sites, the relative lack of evidence for tissue-specific subtypes of binding sites, and the lack of effects of guanine nucleotides on binding provided no additional circumstantial evidence that the binding sites observed were physiologically coupled receptors.

Nevertheless, substantial evidence suggested that at least a fraction of the LTC_4 binding sites detected are physiologically coupled receptors. Clearly, LTC_4 induced contraction of smooth muscle cells, lung parenchyma, and other tissue and the contraction occurred at concentrations too low to be explained by interactions with LTD_4 receptors. Furthermore, in cell lines that respond to LTC_4, LTC_4 binding sites were observed and the response correlated temporally and relative to potency with the K_D and B_{max} of the LTC_4 receptors (20,26). We concluded that although LTC_4 binding site K_D and B_{max} values may not correlate directly with contraction of guinea pig lung, in other tissues and cells a significant fraction of LTC_4 binding sites may be physiologically relevant and at least a fraction of LTC_4 binding sites in guinea pig lung are coupled to contractile responses in that tissue.

LTE_4 Binding

Pharmacological studies suggested that specific receptors for LTE_4 exist in guinea pig trachea (5). These studies showed that LTE_4 enhanced the contractions induced in guinea pig tracheal spirals by histamine but had no effect on the contractions induced by histamine in parenchyma. This effect was not induced by LTC_4 or LTD_4. However, direct radioligand binding studies in our laboratory and others have failed to identify LTE_4 binding sites in guinea pig lung. Rather, LTE_4 appeared to bind to LTD_4 receptors and its effects were explicable on that basis (10). Clearly, additional studies are required to determine if specific LTE_4 receptors exist in some tissues and to define their physiologic and pathophysiologic roles.

Comparison of LTC_4 and LTD_4 Binding Sites

Substantial evidence suggests that LTC_4 and LTD_4 and their analogs bind to distinctly different sites in the tissues studied. Table 6 compares the properties

Table 6 Comparative Properties of LTC_4 Receptors and LTD_4 Specific Binding Sites in Guinea Pig Lung Membranes

Property	$[^3H]LTC_4$ specific binding	$[^3H]LTD_4$ receptors
Affinity (K_D, nM)	10-20	0.2 ± 0.05
Density (B_{max}, pmol/mg protein	50-80	1-2
Sensitivity to		
NEM (10^{-4} M)	Sensitive	Resistant
Heating (80°C)	Sensitive	Sensitive
Heating (60°C)	Resistant	Sensitive
Modulated by		
Ca^{2+}, Mg^{2+} (1-10 mM)	Stimulation	Stimulation
Na^+ (10-200 mM)	Stimulation	Inhibition
Guanine nucleotides	No effect	Transition of affinity states

Source: From Reference 9.

Table 7 Biochemical Properties of Myocardial Leukotriene Receptors

Properties	LTC_4	LTD_4
K_D (nM)	25	6
B_{max} (pmol/mg)	42	0.4
Regulation cations	Yes	No
Sulfhydryl reagents	NEM	DTT
Antagonist competition	SK& F 102081 > SK&F 101132 ⩾ SK&F 88046 >> FPL 55712	SK&F 101132 >> SK&F 102081 >> SK&F 88046 = FPL 55712
Agonist competition	LTD_4 >>> LTE_4 ⩾ LTD_4	LTD_4 > LTC_4 > LTE_4

Source: From References 19 and 28.

of LTC_4 and LTD_4 binding sites in guinea pig lung. Clearly, these binding sites differ in a number of important properties. Similar conclusions derive from a comparison of LTC_4 and LTD_4 binding sites in guinea pig myocardium (Table 7).

LTB_4 Receptors

Studies on LTB_4-induced chemotaxis and aggregation of neutrophils demonstrated that the effects of LTB_4 analogs were stereoselective and structurally specific and suggested that these effects may be mediated by interactions with LTB_4 receptors (8,30). More recently, direct radioligand binding studies have demonstrated LTB_4 receptors in human, rat, and rabbit polymorphonuclear leukocytes (PMN) (31-34).

That the binding sites observed were physiologically coupled receptors has been inferred by comparison of rank order potency for binding to the potencies in chemotactic and aggregation assays. However, human and rat PMN may differ in several regards. In human cells, high- and low-affinity states of LTB_4 receptors were observed. In rat PMN, only the high-affinity state was observed. Furthermore, rat PMN failed to respond chemotactically to LTB_4, suggesting that the low-affinity state-specific binding site for LTB_4 may be necessary for chemotaxis in PMN (32,34). Additional support for the concept that the LTB_4 binding sites in PMN are receptors derived from studies showing that LTB_4 selectively induced mobilization of calcium in these cells (35). Moreover, recent data showing that pertussis toxin inhibits LTB_4-induced chemotaxis and binding of LTB_4 to its receptors suggest that, in rat, rabbit, and human neurotrophils, LTB_4 receptors may be coupled to guanine nucleotide binding proteins (18,36).

SIGNAL TRANSDUCTION SYSTEMS

Peptidyl Leukotrienes

The most direct data concerning mechanisms of signal transduction systems for the peptidyl leukotrienes derives from studies in our laboratory employing smooth muscle and endothelial cells in tissue culture. By screening a number of cell lines, we identified several cell lines that contracted in response to LTC_4 and LTD_4, some that relaxed, and others that appeared to be unaffected. LTC_4 and LTD_4 induced contractions in a smooth muscle cell line, BC3H1, grown on a latex substrate. These effects were stereospecific and inhibited by inhibition of thromboxane synthesis and leukotriene receptor antagonists (20). Moreover, contraction was shown to be dependent on thromboxane synthesis and require RNA and protein synthesis. Thus from these and other studies, we concluded that LTC_4 and LTD_4 receptors were present in these cells, they both induced contraction, and contraction was mediated by thromboxane synthesis that was dependent on RNA and protein synthesis (20,26).

Similarly, in an endothelial cell line (CPAE), LTC_4 and LTD_4 induced an increase in arachidonic acid metabolism resulting in an increase in biosynthesis of prostacyclin. The effects were stereospecific, blocked by antagonists, and required RNA and protein synthesis. The calcium ionophore A23187 and bradykinin also induced prostacyclin synthesis in these cells but more rapidly than LTD_4 or LTC_4. Furthermore, as shown in Table 8, the activities of the calcium ionophore and bradykinin were unaffected by cycloheximide, and only bradykinin-induced increases in prostacyclin production were blocked by actinomycin D. From these data we concluded that in CPAE cells, LTC_4 and LTD_4 bound to their respective receptors and induced an increase in arachidonic acid utilization that resulted in an increase of prostacyclin production. This increase required RNA and protein synthesis, and the mechanisms by which increased arachidonic acid utilization was induced by bradykinin and the calcium ionophore were different (29).

Additional studies suggest that a fraction of the LTC_4 and LTD_4 receptors in some smooth muscle and endothelial cells were coupled to the inhibitory guanine nucleotide binding protein (Gi). Incubation of CPAE cells with islet activating protein pertussis toxin (IAP) resulted in ADP ribosylation of Gi and inhibition of LTD_4-induced prostacyclin production (37). IAP had no effect on bradykinin- or ionophore-induced increases in prostacyclin production. Similar results were observed in the guinea pig lung (21).

LTD_4-induced contraction in lung parenchyma appeared to be mediated via LTD_4 receptors inducing an increase in arachidonic acid utilization resulting in increases in all prostanoids, but thromboxane A_2 synthesis appeared to be greatest. Figure 7a shows the correlation between thromboxane synthesis and contractile activity of a number of analogs. Figure 7b shows the correlation

Table 8 Effects of Protein and RNA Synthesis Inhibitors on Prostacyclin Synthesis in CPAE Cells[a]

Culture conditions	None	Cycloheximide[b]	Inhibitor: actinomycin D[b]
Control	2.4 ± 0.6	1.4 ± 0.5	2.1 ± 0.3
+LTD_4 (1 μM)	4.6 ± 0.2	1.9 ± 0.3**	1.9 ± 0.2**
+LTC_4 (1 μM)	4.1 ± 0.4	1.4 ± 0.5**	1.5 ± 0.5**
+Arachidonic acid (0.1 mM)	3.7 ± 0.2	3.2 ± 0.3	3.5 ± 0.2
+A23187 (1 μM)	4.2 ± 0.1	3.9 ± 0.2	4.6 ± 0.2
+Bradykinin (20 pM)	4.3 ± 0.2	2.0 ± 0.2	4.3 ± 0.2

[a]Values expressed as ng/ml per 10^5 cells per 10 min of 6-keto-prostaglandin $F_{1\alpha}$. Values significantly different (**$p > 0.05$) from the control samples that were treated with LTC_4, LTD_4, A23187, BK, or arachidonic acid but not cycloheximide or actinomycin D.

[b]The effects of cycloheximide (100 μg/ml) and/or actinomycin D (10 μg/ml) or prostacyclin synthesis. Cells were pretreated with these inhibitors for 10 min, rinsed with saline, and then treated with LTC_4 (1 μM), LTD_4 (10 μM), BK (20 pM), arachidonic acid (0.1 mM), or A23187 (1 M) in saline containing inhibitor for an additional 20 min before the supernatants were assayed for 6-keto-PGF_1 by RIA. The data shown (mean ± SD) were obtained from three separate experiments.

Source: From Reference 29.

Figure 7 (a) Correlation of receptor binding affinity to thromboxane B_2 (TxB_2) synthesis activity. The LTD_4 receptor binding affinity K_i of the agonists was determined from radioligand competition studies. The agonists induced TxB_2 synthesis activity were expressed as pD_2. The standard deviations, when not shown, are smaller than the symbols employed. (b) Correlation of smooth muscle contraction and prostanoid synthesis activities of LT agonists. Smooth muscle contractile activities for the agonists were determined using guinea pig lung parenchyma strips and expressed as pD_2. The pD_2 were correlated with TxB_2 synthesis activities and for agonists in guinea pig lung.

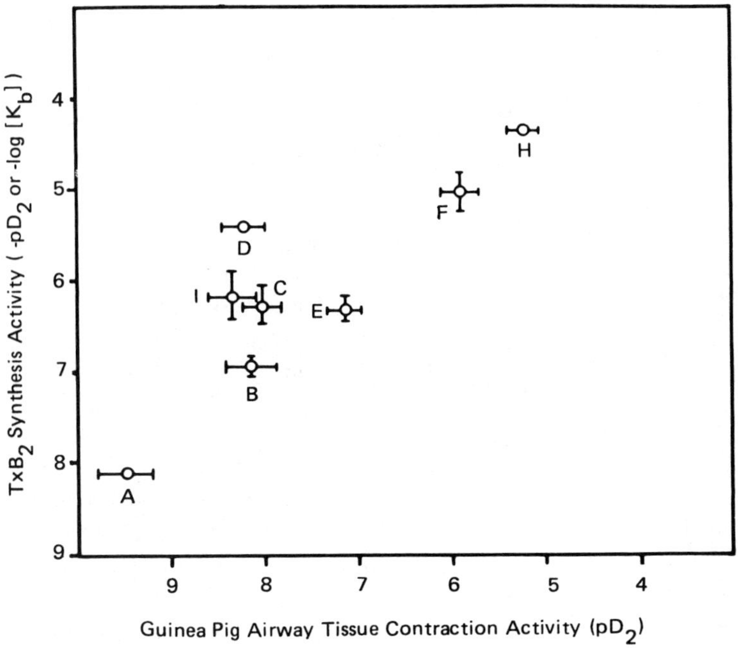

TxB2 Synthesis Activity (-pD2 or -log [Kb])
Guinea Pig Airway Tissue Contraction Activity (pD2)
A
B
C
D
E
F
H
I

(a)

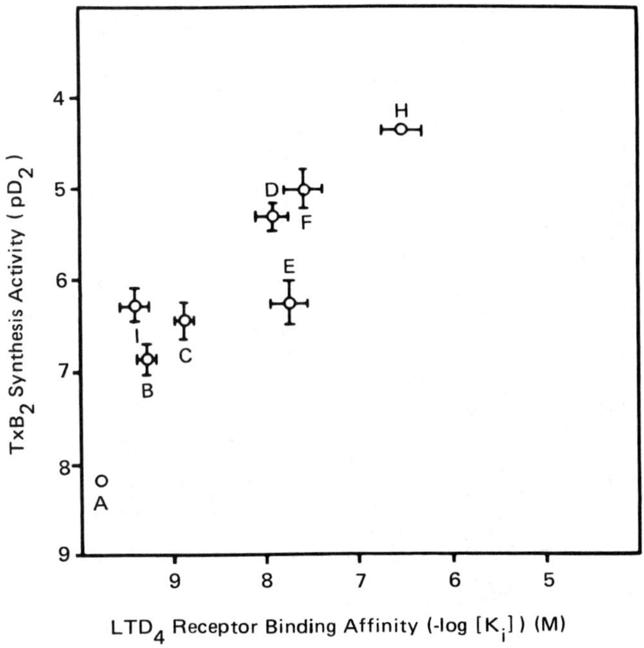

TxB2 Synthesis Activity (pD2)
LTD4 Receptor Binding Affinity (-log [Ki]) (M)
A
B
C
D
E
F
H
I

(b)

between receptor binding affinity and thromboxane synthesis. The effects were highly specific, blocked by LTD_4 receptor antagonists, and partially blocked by IAP pretreatment (21).

To induce an increase in arachidonic acid utilization, LTC_4 and LTD_4 interacted with their Gi or other guanine nucleotide binding protein-coupled receptors and induced transcription and translation of a protein we have called phospholipase A_2 activating protein. PLA_2 activating protein interacted with phospholipase A_2, causing an increase in the number of active phospholipase A_2 molecules, that is, an increase in V_{max} (29). We are currently purifying and charactizing this protein. These data allowed us to propose a model for LTC_4 and LTD_4 signal transduction (Figure 8).

Clearly, a critical question relative to the model proposed is, how generic is it? In our studies on a variety of cell lines and tissues, we have seen no effect of LTC_4 and LTD_4 that cannot be explained via an increase in arachidonic acid metabolism mediated by specific interactions with LTC_4 and LTD_4 receptors. However, several studies have suggested that there are effects of LTC_4 and LTD_4 that are independent of cyclooxygenase products. Therefore, at present we can conclude that the signal transduction process that we have defined is of importance in a large fraction of the cells and tissues in which these agents are active, but we cannot exclude other important mechanisms of signal transduction in some tissues.

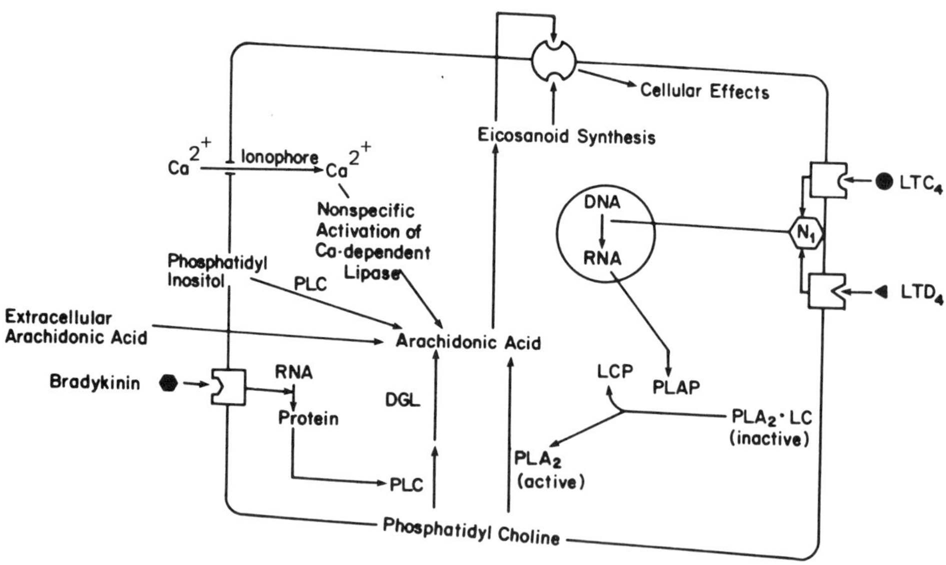

Figure 8 A model of the mechanism of action for peptidyl leukotrienes.

LTB_4

The signal transduction process for LTB_4 has been studied in PMN. LTB_4 appeared to bind to specific LTB_4 receptors that were coupled to Gi (36,38,39). LTB_4 induced calcium mobilization in the same cells, and this effect appeared to be dependent on coupling to Gi (32,36). Studies in our laboratory showed that in rat peritoneal PMN, LTB_4 and analogs induced a rapid increase in inositol phospholipid metabolism that was correlated with LTB_4 agonist potency and was inhibited by IAP. Thus, we concluded that in these cells LTB_4 receptors were coupled to Gi and mediated an increase in inositol turnover, and this was perhaps responsible for the calcium mobilization (17). However, no substantial changes in inositol metabolism were observed in human neutrophils that displayed substantial mobilization of calcium (35). It is therefore possible that the receptor transduction systems for LTB_4 may vary in different cells or species.

CONCLUSIONS AND FUTURE DIRECTIONS

Clearly, during the past several years very substantial progress has been made in understanding the receptors and signal transduction processes for the leukotrienes. These studies provide new insights into the mechanism of action of the leukotriene and a variety of opportunities to develop novel therapeutic agents. Equally clearly, much is left to be learned.

For example, we must resolve the questions about LTC_4 binding sites. What fraction of these sites are physiologically coupled receptors? If a portion of the binding sites is not receptors, what are they? What is the subcellular localization of LTD_4 and LTC_4 receptors? Are they all located in the plasma membrane? Are all leukotriene receptors coupled to Gi? What are the chemical characteristics of the leukotriene receptors, and what chemistry is practiced by agonists at the receptor?

For peptidyl leukotrienes are there signal transduction processes that do not involve cyclooxygenase products? What are the characteristics of the phospholipase activating proteins? How are interactions with membrane localized receptors transduced to result in selective gene activation? What are the relevant genetic mechanisms to this process? How is the increase in phospholipase A_2 activity terminated, and what are relationships of various inhibitors of phospholipase A_2 and the phospholipase activating protein?

Moreover, we know essentially nothing about the regulation of the leukotriene receptors and their signal transduction processes. Are the leukotriene receptors subject to homologous and heterologous desensitization? How is phopholipase C and its regulation in LTB_4-induced effects?

These and many other questions await additional studies. However, the progress made in a very few recent years, increasing number of synthetic agonists, partial agonists, and antagonists as well as the cellular behavior pharmacological model systems suggest that we shall have answers to these questions. The answers undoubtedly will be exciting and provide fascinating insights into cellular behavior, the roles of the leukotrienes in pathophysiology, and offer new opportunities for the discovery and development of novel therapeutics.

ACKNOWLEDGMENTS

The authors would like to thank Hsiao-Ling Wu, Gloria Chi-Rosso, Miller O. Scott, Mike Cook, Terry Conway, James Foley, Henry S. Sarau, and David Littlejohn for excellent technical assistance and Denise Granata for excellent typographical assistance.

REFERENCES

1. R. A. Lewis, K. F. Austen, J. M. Drazen, N. A. Soter, J. C. Figueiredo, and E. J. Corey, in Advances in Prostaglandin, Thromboxane and Leukotriene Research, Vol. 9 (B. Samuelssön and R. Paoletti, eds.), Raven Press, New York, 1982, p. 137.
2. R. A. Lewis, C. W. Lee, L. Levine, R. A. Morgan, J. W. Weiss, J. M. Drazen, H. Oh, D. Hoover, E. J. Corey, and K. F. Austen, in Advances in Prostaglandin, Thromboxane and Leukotriene Research, Vol. 11 (B. Samuelssön, R. Paoletti, and P. W. Ramwell, eds.), Raven Press, New York, 1983, p. 15.
3. Y. Hattori and R. Levi, J. Pharmacol. Exp. Ther. *230*, 646 (1984).
4. D. W. Snyder and R. D. Krell, J. Pharmacol. Exp. Ther. *231*, 616 (1984).
5. T. H. Lee, K. F. Austen, E. J. Corey, and J. M. Drazen, Proc. Natl. Acad. Sci. USA *81*, 4922 (1984).
6. W. Benjamin, N. K. Hopkins, T. D. Olgeesby, and R. R. Gorman, Biochem. Biophys. Res. Commun. *117*, 780 (1983).
7. A. W. Ford-Hutchinson, M. A. Bray, M. V. Doig, M. E. Shipley, and M. J. H. Smith, Nature *286*, 264 (1980).
8. R. A. Lewis, E. J. Goetzl, J. M. Drazen, N. A. Soter, K. F. Austen, and E. J. Corey, J. Exp. Med. *154*, 1243 (1981).
9. S. Mong, H. L. Wu, M. O. Scott, M. A. Lewis, M. A. Clark, B. M. Weichman, C. M. Kinzig, J. G. Gleason, and S. T. Crooke, J. Pharmacol. Exp. Ther. *234*, 316 (1985).
10. S. Mong, M. O. Scott, M. A. Lewis, H. L. Wu, G. K. Hogaboom, M. A. Clark, and S. T. Crooke, Eur. J. Pharmacol. *109*, 183 (1985).
11. G. K. Hogaboom, S. Mong, H. L. Wu, and S. T. Crooke, Biochem. Biophys. Res. Commun. *116*, 1136 (1983).
12. S. S. Pong and R. N. DeHaven, Proc. Natl. Acad. Sci. USA *80*, 7415 (1983).
13. S. Mong, H. L. Wu, G. K. Hogaboom, M. A. Clark, and S. T. Crooke, Eur. J. Pharmacol. *102*, 1 (1984).
14. S. Mong, H. L. Wu, G. K. Hogaboom, M. A. Clark, J. M. Stadel, and S. T. Crooke, Eur. J. Pharmacol. *106*, 241 (1985).

15. R. Bruns, W. J. Thomsen, and T. A. Pugsley, Life Sci. *33*, 645 (1983).
16. M. A. Lewis, S. Mong, R. L. Vessella, and S. T. Crooke, Biochem. Pharmacol. *34*, 4311 (1985).
17. S. Mong, G. K. Hogaboom, M. A. Clark, and S. T. Crooke, in Receptors in Drug Research (R. O'Brien, ed.), Marcel Dekker, New York, 205-224 (1986).
18. S. Mong, H. L. Wu, M. J. Stadel, and S. T. Crooke, Mol. Pharmacol, in press (1986).
19. G. K. Hogaboom, S. Mong, J. M. Stadel, and S. T. Crooke, J. Pharmacol. Exp. Ther. *233*, 686 (1985).
20. M. A. Clark, M. Cook, S. Mong, and S. T. Crooke, Eur. J. Pharmacol. *116*, 207 (1985).
21. S. Mong, H. L. Wu, M. A. Clark, J. G. Gleason, and S. T. Crooke, ms. submitted J. Pharmacol. Exp. Ther. *239*, 35-41 (1986).
22. S. Mong, G. Chi-Rosso, P. Bender, and S. T. Crooke, Mol. Pharmacol. *30*, 235 (1986).
23. S. L. Levinson, Prostaglandins *28*, 229 (1984).
24. S. Nicosia, H. J. Crowley, D. Oliva, and A. F. Welton, Prostaglandins, *27*, 483 (1983).
25. S. Krilis, R. A. Lewis, E. J. Corey, and K. F. Austen, J. Clin. Invest. *72*, 1516 (1983).
26. M. A. Clark, M. Cook, S. Mong, G. K. Hogaboom, R. Shorr, J. Stadel, and S. T. Crooke, Life Sci. *35*, 441 (1984).
27. M. A. Lewis, S. Mong, R. L. Vessella, G. K. Hogaboom, H. L. Wu, and S. T. Crooke, Prostaglandins *27*, 961 (1984).
28. G. K. Hogaboom, S. Mong, J. M. Stadel, and S. T. Crooke, Mol. Pharmacol. *27*, 236 (1985).
29. M. A. Clark, T. M. Conway, E. F. Bennett, S. T. Crooke, and J. M. Stadel, Proc. Natl. Acad. Sci. *83*, 7320 (1986).
30. A. W. Ford-Hutchinson, P. J. Piper, and M. N. Samhoun, Br. J. Pharmacol. *76*, 215 (1982).
31. A. H. Lin, P. L. Ruppel, and R. R. Gorman, Prostaglandins *28*, 837 (1984).
32. R. A. Kreisle, C. W. Parker, G. L. Griffin, R. M. Senior, and W. F. Stenson, J. Immunol. *134*, 3356 (1985).
33. D. W. Goldman and E. J. Goetzl, J. Immunol. *129*, 1600 (1982).
34. D. W. Goldman and E. J. Goetzl, J. Exp. Med. *159*, 1027 (1984).
35. M. Volpi, R. Yassin, W. Tao, T. F. P. Molski, P. H. Naccache, and R. I. Sha'afi, Proc. Natl. Acad. Sci. USA *81*, 5966 (1984).
36. D. W. Goldman, D. M. Olson, L. A. Gifford, and E. J. Goetzl, Fed. Proc. *44*, abstract 1898 (1985).
37. M. A. Clark, P. Littlejohn, S. Mong, and S. T. Crooke, Prostaglandins *31*, 157-166 (1986).
38. E. L. Becker, J. C. Kermode, P. H. Naccache, R. Yassin, M. L. Marash, J. J. Munoz, and R. I. Sha'afi, J. Cell Biol. *100*, 1641 (1985).
39. J. T. Rosenbaum, H. Enkel, D. E. Chenoweth, and D. W. Goldman, Fed. Proc. *44*, 580, abstract 994 (1985).

19

Δ^9-Tetrahydrocannabinol, Phencyclidine, and Tricyclic Antidepressant Receptors

Billy R. Martin / Medical College of Virginia, Virginia Commonwealth University, Richmond, Virginia

The explosive growth in the past few years of receptor research has led to the characterization of binding sites for almost all centrally acting drugs. Although central receptors for opioids, neurotransmitters, and others are relatively well defined, many of the receptors for numerous drugs either are just beginning to be characterized or remain controversial. In this chapter the current status of putative receptors for psychoactive cannabinoids, phencyclidine, and related psychotomimetics, as well as tricyclic antidepressants, is reviewed. It is obvious that these drug classes differ appreciably in their pharmacological effects on the central nervous system. In addition, the type of information available for supporting a receptor for each of these drugs is also varied. Therefore, the review of the putative receptors for these different classes of drugs is not uniform. Throughout this chapter, the terms "specific binding sites" and "receptors" are sometimes used interchangeably, but generally in vitro binding that appears to be selective for a specific ligand as defined solely by competition studies with appropriate agonists and antagonists is referred to as specific binding. The term receptor is reserved for when there is more definitive evidence, such as identification of an endogenous ligand or identification of the second messenger, in addition to the in vitro binding data and the in vivo pharmacological characterization.

There is a great deal of speculation regarding the mechanisms of action of Δ^9-tetrahydrocannabinol (THC), the psychoactive cannabinoid of marihuana. However, there are relatively few in vitro binding data available that support a cannabinoid receptor. The primary evidence for a cannabinoid receptor is based on structure-activity relationships derived from in vivo experiments, which are discussed in the cannabinoid section.

In contrast to the cannabinoid series, there are appreciable in vitro binding data to support a specific binding site for phencyclidine (PCP). In some ways, the in vitro data are more supportive of a PCP binding site than the in vivo data. For example, only recently have successes been obtained in the development of a PCP antagonist, and even they are modest achievements. Therefore, the emphasis in the PCP section is on the in vitro characterization rather than on in vivo evidence. In addition, the interrelationship between PCP binding sites and those for psychotomimetic opioids is discussed. Aside from the postulate by some investigators that these sites are one and the same, the intriguing prospect that dysphoria is a receptor-mediated event can be supported by the observations that receptors exist for numerous psychotomimetics. It is for this reason that the evidence for specific binding sites for psychotomimetic opioids is included in this section.

The review of the specific binding sites for tricyclic antidepressants differs from those for the cannabinoids and PCP primarily in that numerous antidepressants can serve as ligands for in vitro bindings. In addition, many of these antidepressants have different modes of action, which suggests different receptor populations. Therefore, the review of antidepressant receptors involves in vitro binding studies of several ligands that produce a common pharmacological effect through different receptor populations.

A common link between these classes of compounds is that identification of a receptor for these compounds may provide important new insights in understanding the etiologies of mental abberations.

CANNABINOIDS

Historical Perspective

The recreational use of marihuana is a centuries-old practice that has become a persistent worldwide problem. Achieving an understanding of the way in which Δ^9-THC produces its unique behavioral effects may provide some insight into how its abuse may be controlled. Undoubtedly, elucidation of its mechanism of action would provide valuable insight into how the central nervous system processes sensory information. There have been numerous attempts to determine the mechanism of action of Δ^9-THC. Although evidence for the possible existence of cannabinoid receptors has emerged slowly over the past several years, some of the criteria (most notably structure-activity relationships and specific antagonists) for cannabinoid receptors have been fulfilled primarily through in vivo pharmacological studies. The most supportive data are the in vivo structure-activity relationships (SAR) that exist for cannabinoids.

Structure-Activity Relationships

The SAR for cannabinoids have been reviewed extensively elsewhere (1), so that only a summary of some of the structural requirements for cannabinoid activity, as they relate to possible receptor function, is presented here. It should be pointed out that two nomenclatural systems are currently in use for the cannabinoids. Typically, the monoterpene system is used in Europe and the dibenzopyran system is preferred in North America. The latter system is used throughout this review. It is worthwhile to note that Δ^9-THC, 11-OH-Δ^9-THC, and Δ^8- THC in the dibenzopyran system are Δ^1-THC, 7-OH-Δ^1-THC, and Δ^6-THC, respectively, in the monoterpene system.

It has been clearly established that modest structural modifications of the THC molecule, depicted in Table 1, can result in profound changes in cannabinoid behavioral activity. Alterations in any of the three rings or in the side chain of THC change its behavioral activity. The effect of hydroxylation on cannabinoid activity is presented in Table 1. It can be seen that cannabinoid behavioral activity can be either increased or decreased depending upon the location of the hydroxyl group (2-9). For example, hydroxylation in the 3′ position enhances activity twofold, whereas hydroxylation in another position of the side chain may reduce activity as much as fivefold. It is also clear that incorporation of a hydroxyl group in the 11 position of Δ^9-THC tends to increase activity, whereas a hydroxy group at position 8 decreases activity. It was known from the pioneering work of Adams (10) that the position of the double bond in the A ring was important for central activity. The examples in Table 1 demonstrate the profound effect of changes in the position of the double bond on central activity (11,12). It has been firmly established that optimum activity is obtained when the double bond is in the 9-10 position. In addition, saturation of the A ring reduces activity (11). Removal of the 9-methyl group (13) or addition of a methyl, methoxy, amino, or acetamido group to the 9-methyl group attenuates behavioral activity (13,14). As stated in Table 1, hydroxylation of the 9-methyl results in either no change or an increase in potency (2,3). Replacement of the 9-methyl group with a hydroxyl group enhances potency (15). It has generally been assumed that an intact B ring is essential for activity because cannabidiol lacks psychoactivity (16,17). However, Melvin et al. (18) synthesized a derivative of 9-*nor*-9β-hydroxyhexahydrocannabinol that lacked the B ring entirely and had a dimethylheptyl side chain at the 3 position. It proved to have a pharmacological profile similar to that of Δ^9-THC, although it was more potent than Δ^9-THC (19). The spatial relationship between the A and B rings appears to be critical. In the naturally occurring psychoactive cannabinoids, the A and B rings are fused through a trans junction, and synthetic cis analogs lack effects on the central nervous system (20,21), as described in Table 1. Alterations in the gem dimethyls on the B ring, such as replacement with a ketone, also reduce centrally mediated effects (22,23). Expansion of the B ring to a 7-membered ring eliminates the effects in the dog static-ataxia test but not general depressant effects in mice (22).

As far as the C ring is concerned, early work indicated that behavioral activity was dependent upon the presence of a free or esterified phenolic hydroxyl

Table 1 Structural Requirements for Cannabinoid Activity

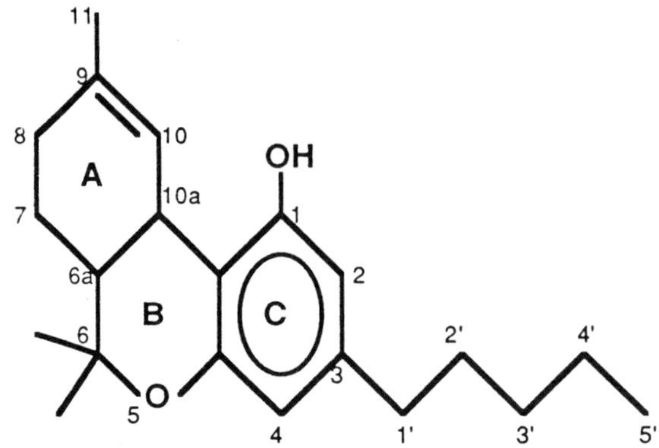

Cannabinoid	Species	Activity	Relative potency[a]	Reference
Hydroxylation				
Δ^9-THC	–	–	1.0	
11-OH-Δ^9-THC	Human	Psychoactivity	1.2	2, 3
8α-OH-Δ^9-THC			⩾ 0.25	2, 4
8β-OH-Δ^9-THC			⩽ 0.2	2, 4
12β-OH-Δ^9-THC	Dog	Static ataxia	1.0	5
1′-OH-Δ^8-THC	Mouse	Catalepsy	0.25	6
2′-OH-Δ^8-THC			> 0.2	6
3′-OH-Δ^8-THC			2.0	6
4′-OH-Δ^8-THC			≈ 0.5	6
5′-OH-Δ^8-THC			1.0	6
R-3′-OH-Δ^9-THC	Rat	Drug discrimination	0.15	7
S-3′-OH-Δ^9-THC			1.0	7
7α,11-DiOH-Δ^8-THC	Mouse	Overt behavior	0.04	8
7β,11-DiOH-Δ^8-THC			0.06	8
3′,11-DiOH-Δ^9-THC	Dog	Static ataxia	0.33	9
Unsaturation				
Δ^8-THC	Human	Psychoactivity	0.75	2
$\Delta^{6a\text{-}10a}$-THC			0.15-0.30	2
Δ^7-THC	Monkey	Overt behavior	Inactive	11
$\Delta^{9\text{-}11}$-THC			0.01	12
	Rat	Drug discrimination	0.15	12

Table 1 (Continued)

Cannabinoid	Species	Activity	Relative potency[a]	Reference
Stereoselectivity				
(+)-*trans*-Δ^9-THC	Monkey	Scheduled responding	0.01	20
		Overt behavior	< 0.05	32
	Dog	Static ataxia	< 0.1	22
	Mouse	Hypothermia	0.1	20
		Spontaneous activity	0.18	20
		Ring test	0.08	31
(±)-*cis*-Δ^9-THC	Mouse	Lethality	> 2.0	21
		Ataxia	0.05	21
		Analgesia	< 0.2	21
		Spontaneous activity	0.05	21
(+)-*cis*-Δ^9-THC	Dog	Static ataxia	< 0.04	20

[a]Compared to the corresponding parent compound, except the unsaturated compounds, which are compared to Δ^9-THC.

group (11,25). A morpholinobutyrate of Δ^9-THC has been prepared, but its behavioral activity is thought to be due to hydrolysis in vivo to Δ^9-THC (26). Mechoulam and Edery (11) also noted that activity was eliminated by the addition of electronegative groups at positions 2 and 4. Hydroxylation at position 2 eliminates activity as well (22). It has also been demonstrated that transposing the phenolic hydroxyl and the side chain also results in loss of activity (27).

Adams (28) first demonstrated the importance of the alkyl group by showing in the $\Delta^{6a\text{-}10a}$-THC series that increasing the length of the side chain increased potency until a maximum was reached with n-hexyl. Branching of the side chain has an unusually profound effect on potency regarding centrally mediated effects, the most notable example being dimethyheptylpyran, which is $\Delta^{6a\text{-}10a}$-THC with a dimethyl heptyl side chain. Hardman et al. (29) have shown that this cannabinoid is more potent than Δ^9-THC in addition to having a longer duration of action. Hydroxylation of the side chain represents an important metabolic process for the cannabinoids (30). Addition of a hydroxyl group to the side chain can have dramatic effects on potency, as mentioned briefly. In the Δ^8-THC series, hydroxylation at the 5′ position did not alter behavioral activity, hydroxylation at the 3′ position increased activity at least twofold, and hydroxylation at 2′ and 1′ decreased activity fivefold or greater (6). Introduction of a hydroxy in positions 1′, 2′, 3′, or 4′ results in a chiral center. The geometric isomers for only 3′-OH-Δ^9-THC have been synthesized, and the behavioral activity appears to reside in the S isomer rather than the R isomer (7). This latter observation suggests that the conformation of the side chain is important for activity.

Stereoselectivity is an important criterion for drug-receptor interactions because stereoisomers share the same physicochemical characteristics. Indeed, Jones et al. (31) showed that mouse brain levels of [^{3}H](+)-Δ^9-THC and [^{3}H](–)-Δ^9-THC were comparable following administration of similar doses of the respective compounds. It has been firmly established that central cannabinoid activity is stereoselective. Edery et al. (32) demonstrated that (–)-*trans*-Δ^9-THC was at least 20 times more potent than (+)-*trans*-Δ^9-THC in monkeys, and Jones et al. (31) found the (–)-*trans* isomer to be 13 times more potent than the (+)-*trans* isomer in the mouse ring test. Based upon the observations that (–)-*trans*-Δ^9-THC was 6, 9, ≈10, and 100 times more effective than (+)-*trans*-Δ^9-THC in producing motor hypoactivity in mice, hypothermia in mice, static ataxia in dogs, and depression of schedule-induced responding in monkeys (20), respectively, it appears that the magnitude of cannabinoid stereoselectivity is dependent upon either the species or the pharmacological test or a combination of both.

The structural modifications discussed here represent only a small portion of those that have been made, but they serve to illustrate that the pharmacological activity of cannabinoids is very dependent upon structure. Of course, SAR itself does not prove that a receptor exists. There can be biochemical actions not involving receptors that are affected by structural changes in the ligand. Anesthetics serve as one of the most noteworthy examples. It is generally assumed that anesthetics produce narcosis by disordering membranes rather than by interacting with receptors and that alterations in potency occur when structural changes alter lipophilicity. Goldstein (33) has reviewed the stereoselectivity of numerous compounds that cause membrane perturbation. Several investigators have demonstrated that cannabinoid-induced changes in membrane order can be related to psychoactivity (34,35). However, there are currently insufficient data to support the notion that cannabinoid psychoactivity arises from membrane perturbation. Nonetheless, the SAR in these biochemical systems illustrate that structural requirements are not restricted to drug-receptor interactions.

Another issue of SAR, alluded to earlier, is whether alterations in centrally mediated cannabinoid effects arising from these structural modifications are due to changes in receptor interactions or to differences in penetration, distribution, and elimination of the cannabinoids from the central nervous system.

There is evidence that the biodisposition of some of the metabolites of THC may differ from that of the parent compound. For example, there are reports that 11-OH-Δ^9-THC is at least as potent as, or more potent than Δ^9-THC, yet the increased potency of 11-OH-Δ^9-THC may be due, at least in part, to a decrease in plasma protein binding and concomitant increase in brain penetrability. Schou et al. (36) showed that when Δ^9-THC was administered into the carotid artery of mice, its ability to penetrate brain was diminished when plasma was used as the vehicle, whereas the penetrability of 11-OH-Δ^9-THC was essentially the same regardless of whether the vehicle was saline or plasma. These findings suggested that the greater potency of 11-OH-Δ^9-THC was due to greater brain penetrability as a result of its lesser affinity for plasma proteins. Perez-Reyes et al. (37) extended the observations of Schou and coworkers by comparing the brain concentrations of 11-OH-Δ^9-THC and Δ^9-THC in mice by administering equivalent doses of each compound. The brain levels of 11-OH–Δ^9-THC were almost five times higher than those of Δ^9-THC. In contrast to the notion that 11-OH-Δ^9-THC is equipotent with or more potent than Δ^9-THC (3,8), it appears 11-OH-Δ^9-THC may be less potent when brain concentrations are considered (37). The findings of Gough and Olley (38) also support this suggestion in that intrastriatal injections of Δ^9-THC were more effective than those of 11-OH-Δ^9-THC in producing catalepsy in rats.

Ohlsson et al. (39) carried out studies in which they compared the biodisposition of Δ^8-THC to that of metabolites hydroxylated in the side chain. They found that the hydroxylated metabolites penetrated brain more readily than did Δ^8-THC. However, it did not appear that pharmacokinetic features were totally responsible for determining the central activity of these cannabinoid metabolites. For example, 1′-OH- and 4′-OH-Δ^8-THC reached the highest concentrations in brain but the most potent cannabinoids were Δ^8-THC, 11-OH-Δ^8-THC, and 3′-OH-Δ^8-THC. Also, differences in their lipophilicity could not account for their differences in pharmacological potency. It was concluded that the differences in the potency of these compounds depended upon inherent pharmacodynamic properties rather than pharmacokinetics. Binder et al. (40) have also shown that the pharmacokinetics of the inactive Δ^{9-11}-THC are the same as those of Δ^9-THC. The important question is whether the structure-activity relationships already discussed implicate the involvement of a receptor.

Cannabinoid Antagonists

Antagonists have played major roles in the characterization of most receptors. Unfortunately, a specific antagonist has not been found for Δ^9-THC. Numerous drugs have been used to attenuate the effects of Δ^9-THC; however, it is more likely that these studies characterize drug interactions as opposed to specific antagonism. One of the first putative cannabinoid antagonists was cannabidiol (CBD), a natural constituent of marihuana, that interfered with the effects of

Δ^9-THC (41,42). Human studies conducted by Karniol et al. (43) demonstrated that oral ingestion of comparable doses of CBD and Δ^9-THC resulted in an attenuation of Δ^9-THC-induced tachycardia as well as its psychological effects. However, considerable effort by other investigators has failed to provide convincing evidence for the antagonistic properties of CBD. Bird et al. (44) found no influence of CBD on the pharmacological effects of THC after oral administration, but Hollister (16) found the effects of Δ^9-THC were slightly intensified when it was coadministered with CBD. Others reported that CBD did not influence Δ^9-THC-induced tachycardia when they were smoked together and that it exerted only minimal effects on the psychotropic activity of THC (45). Animal studies also provide conflicting results in that CBD has been shown to block Δ^9-THC induced catatonia in mice, corneal areflexia, bradycardia, and hypothermia in rabbits and motor hypoactivity and hypothermia in rats (46,47). Doses of CBD that were 8 and 50 times greater than those of Δ^9-THC provided a partial reversal of Δ^9-THC food-reinforced operant performance in rats and pigeons, respectively. CBD failed to block THC discrimination in rats in studies carried out by Browne and Weissman (35). Also, studies carried out in our laboratories have not shown CBD to block Δ^9-THC-induced analgesia (48). When the CBD antagonism studies are viewed collectively, there does not appear to be sufficient evidence to support CBD as a specific antagonistic of Δ^9-THC.

Although CBD may not be a cannabinoid antagonist, other constituents of marihuana may interfere with the effects of Δ^9-THC. Cannabinol (CBN) is another substance in marihuana that has been reported to block the effects of Δ^9-THC, in this case, potentiation of pentobarbital sleeping time (49). However, Hollister and Gillespie (50) found that CBN exerted no influence on the subjective effects of Δ^9-THC in humans. Musty et al. (51) found that CBN tended to potentiate rather than block the effects of THC in humans, which was consistent with the observations of Perez-Reyes et al. (3) that CBN exhibits weak cannabinoid activity in humans. Cannabichromene, another constituent of marihuana that has been evaluated for synergistic and antagonistic properties, was found to have little antagonistic effect on several pharmacological properties of Δ^9-THC in mice (52,53).

Kudrin and Davydova (54) reported that phenitrone was capable of preventing as well as reversing all the symptoms of hashish intoxication in dogs. However, subsequent studies carried out by numerous investigators failed to show any cannabinoid antagonist activity of phenitrone in dogs, rats, mice, and pigeons (55-58). Miscellaneous drugs reportedly alter the action of cannabinoids. For example, the central nervous system (CNS) depressant effects of Δ^9-THC were blocked by amphetamine in rats (59), and the stimulant properties of amphetamine were enhanced by Δ^9-THC in mice (60). Thyrotropin-releasing hormone (TRH), which has been shown to stimulate the CNS at pharmacological doses, has been shown to antagonize decreased locomotor activity induced by Δ^9-THC (61). Lastly, Δ^9-THC-induced muricide has been blocked by imipramine (62). These examples serve to illustrate the lack of a specific cannabinoid antagonist. It will be difficult to prove the existence of THC receptors until a specific antagonist is found.

Binding Sites and Radioligands

Limited information regarding the characterization of cannabinoid binding sites is available. It is important to realize that the physicochemical properties of the cannabinoids have made it virtually impossible to conduct in vitro binding studies because of a large degree of nonspecific, nonsaturable binding due to their high lipophilicity (63,64). However, Harris et al. (65) were able to demonstrate high-affinity, saturable binding of [^{3}H] Δ^8-THC to hepatoma cells in tissue culture. Despite that the relevance of this binding has not been established, a significant aspect of this study was the demonstration that saturable binding could be measured in an in vitro system. Harris et al. (65) also investigated cannabinoid binding to brain homogenates, which obviously has more relevance to psychoactivity. Unfortunately, the heterogeneous, lipophilic nature of brain tissue precluded the quantitation of saturable binding. A small percentage of binding was displaced by unlabeled Δ^8-THC, and the binding was dependent upon temperature. Roth and Williams (63) studied the binding of [^{3}H] Δ^9-THC to crude and purified rat synaptosomal membranes. They reported a linear relationship between the quantity of [^{3}H] Δ^9-THC bound and the amount added, which they interpreted as a single binding site. However, no effort was made to establish the specificity or saturability of this site since it appeared to represent nonspecific binding.

Recently, Seltzman et al. (66) synthesized 5′-trimethylammonium-Δ^8-THC (Figure 1), which should be much less lipophilic than Δ^8- and Δ^9-THC. Nye et al. (67,68) studied the binding of this radiolabeled ligand to rat neuronal membranes and found a saturable, reversible binding site to which 5′-trimethylammonium-Δ^8-THC bound with high affinity (K_D = 89 nM) and large B_{max} (1.2 nmol/mg protein). A Hill plot of these data suggested a single binding site. There was an uneven distribution of these binding sites throughout brain, with the highest density in the hippocampus and the lowest in the medulla-pons. These investigators also demonstrated that 5′-trimethylammonium-Δ^8-THC would bind to a solubilized protein from rat brain that was in the molecular weight range of 60,000 daltons.

It may be that other cannabinoids with reduced lipophilicity will be useful receptor probes. Zitko et al. (69) made a water-soluble derivative of Δ^9-THC by addition of a morpholinobutyrate at the phenolic hydroxyl (Figure 1). This ester retained cannabinoid activity, which could be due to either the parent compound or to enzymatic hydrolysis to Δ^9-THC. Likewise, Yoshimura et al. (70),

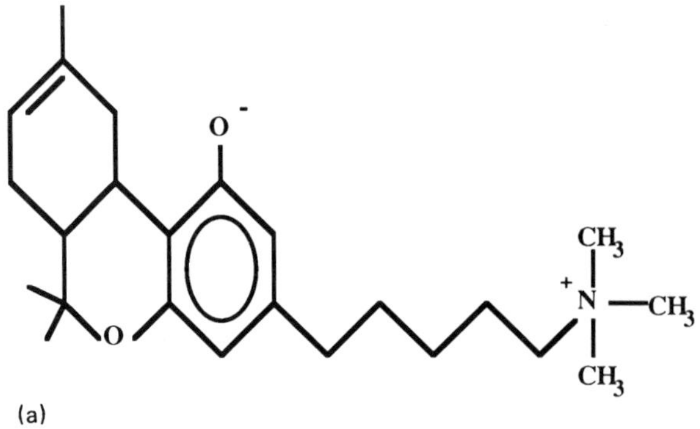

(a)

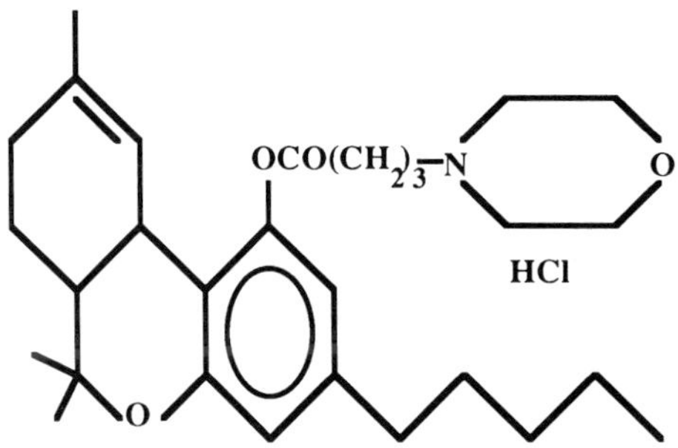

(b)

Figure 1 Structure of 5′-trimethylammonium-Δ^8-THC phenolate (a) and Δ^9-THC morpholinobutyrate (b).

prepared a water-soluble derivative by addition of a phosphate to the phenolic hydroxyl of Δ^8-THC. This phosphate was considerably less potent than Δ^8-THC and had a longer onset of action. The apparent difference in pharmacological profiles of the morpholinobutyrate and the phosphate may be due to the latter being hydrolyzed to a lesser extent.

[^{3}H]-Δ^8-THC-1-morpholinobutyrate (80 Ci/mmol) has recently been prepared, and the characteristics of its binding to rat brain membranes were studied. Using conditions similar to those described by Nye et al. (67,68), saturable binding in the picomolar to micromolar concentration range was not observed (unpublished observations). It should be noted that although Δ^8-THC-1-morpholinobutyrate is considered a water-soluble derivative, it is only sparingly soluble in water. Its relative hydrophobicity may have resulted in nonspecific binding, masking the specific binding much in the same was as postulated for Δ^8- or Δ^9-THC.

Selectivity and SAR

The SAR for cannabinoid psychotomimetic effects has been established as discussed earlier. Attempts to duplicate this SAR for in vitro binding have been hampered by the obvious lack of a good assay for a saturable binding site. In the studies carried out by Nye et al. (67,68), the site labeled by 5′-trimethylammonium-Δ^8-THC appeared to be selective for cannabinoids in that agents that interact with receptors for dopamine, serotonin, catecholamine, histamine, γ-aminobutyric acid (GABA), benzodiazepine, adenosine, and calcium channels did not compete for binding. Numerous cannabinoids competed effectively for this binding site. However, there was not a good correlation between the binding affinities and psychoactivity of the cannabinoids, which raises the question of the biologic significance of this site. Δ^9-THC produces many effects on the central nervous system, and it may well be that this site is associated with a nonpsychoactive effect of cannabinoids. Recent studies in our laboratory have been directed toward characterizing the central effects of 5′-trimethylammonium-Δ^8-THC. It was found to be more effective than Δ^8-THC in producing general CNS depression in mice. After intravenous administration, 5′-trimethylammonium-Δ^8-THC was apporximately five times more potent than Δ^8-THC in producing motor hypoactivity, hypothermia, and lethality. When 5′-trimethylammonium-Δ^8-THC and Δ^8-THC were also compared in the dog static ataxia test, which is highly selective for cannabinoid psychoactivity (1), 5′-trimethylammonium-Δ^8-THC was ineffective at doses 5-10 times greater than active doses of Δ^8-THC (B. R. Martin, unpublished observations). The pharmacological activity of $\Delta^{9\text{-}11}$-THC, which was one of the most potent analogs competing for [^{3}H] 5′-trimethylammonium-Δ^8-THC binding, produced motor hypoactivity and hypothermia in mice but was less potent than Δ^9-THC. However, it failed to produce the classic static ataxia in dogs (71). These observations suggest that 5′-trimethylammonium-Δ^8-THC may be labeling a nonpsychoactive cannabinoid site.

Function and Regulation

A problem that is confounding the elucidation of a cannabinoid mechanism of action is the multiplicity of centrally mediated effects of cannabinoids as alluded to earlier and discussed elsewhere (72). It would appear that a single action of cannabinoids is not responsible for producing the myriad behavioral effects of cannabinoids, in addition to alterations in cardiovascular function, body temperature, and others. It has been suggested that cannabinoids alter adenylate cyclase activity through a receptor-mediated event (73-75).

Speculation and Future Direction

There has been considerable speculation regarding the mechanism(s) through which Δ^9-THC produces its effects on the central nervous system. One of the most prominent suggestions is that cannabinoids produce their effects by fluidizing membranes, much in the same way as general anesthetics are thought to act. Numerous investigators have shown that cannabinoids perturb membranes and that the SAR is consistent with psychoactivity (76-80). Conversely, there is the other notion discussed here, that cannabinoids interact with a specific receptor to produce their psychoactivity. It seems plausible that cannabinoids may act through both mechanisms, in addition to others, so that the behavioral effects of cannabinoids represent a composite of numerous actions.

It is obvious that continued progress in the area of in vitro binding is essential if the concept of a cannabinoid receptor is to be accepted. The primary interest should be the characterization of a binding site associated with psychoactivity. The ratio of specific to nonspecific binding to brain membranes must be improved appreciably before specific binding can be studied in a reliable manner. Several options can be utilized in order to decrease the nonspecific binding component. Possible future directions include synthesis of radiolabeled ligands that are less lipophilic and yet retain psychoactivity. Another obvious way to increase the ratio of specific to nonspecific binding sites is to utilize some type of sample purification. Attempts in our laboratory to purify specific binding sites by subcellular fractionation techniques and column chromatography have produced disappointing results. There have been suggestions that preparations of affinity columns may be feasible for the cannabinoids (81). The affinity column depicted in Figure 2 has been prepared by coupling Δ^9-THC to Affinose 102 (aminoalkyl agarose, 80-108 μm, Bio-Rad Laboratories) using dicyclohexylcarbodiimide (B. R. Martin, unpublished results). However, it remains to be determined whether affinity column chromatography will be beneficial in isolating cannabinoid binding sites.

Another important area of research that needs attention is the development of a cannabinoid antagonist. There appear to have been few systematic attempts to synthesize such an antagonist. A major stumbling block has been the lack of a simple assay suitable for evaluation of compounds for antagonism of cannabinoid behavioral activity. However, a specific antagonist of Δ^9-THC would serve as an important research tool and at the same time lend credence to the suggestion of a cannabinoid receptor.

PHENCYCLIDINE

Historical Perspective

Phencyclidine has a unique pharmacological profile that has attracted considerable attention. PCP was found to be a surgical anesthetic that did not markedly

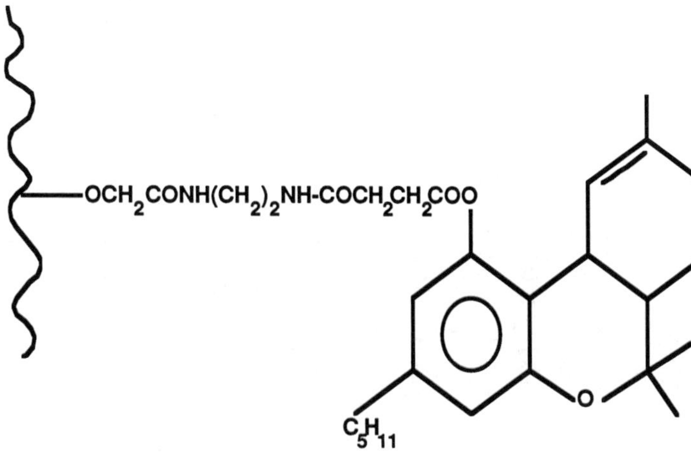

Figure 2 Affinose-Δ^9-THC affinity column.

affect the respiratory, cardiovascular, or gastrointestinal system (82). Enthusiasm for the use of PCP as a surgical anesthetic was short-lived owing to the thought disturbances, disorientation, hallucination, and agitation that it produced postoperatively (82). Although these properties relegated PCP to the ranks of veterinary anesthesiology, they appealed to some members of the drug culture, which led to illicit use of PCP beginning in the mid 1960s (84). The most troubling problems of PCP abuse are the wide range of effects that can appear in the user and the unpredictability of their intensity from episode to episode. The incidence of bizarre and violent behavior that is sometimes exhibited by the user has served to focus attention on the misuse of PCP.

One of the most intriguing aspects of the pharmacology of PCP is the schizophrenialike effects that it sometimes produces. At present, it is not known whether PCP acts directly to produce schizophrenia or acts indirectly by precipitating a latent mental disorder. There is obvious interest in elucidating the mechanism(s) by which PCP produces its central effects since they are unique. Considerable effort has been expended in determining the effects PCP exerts on neurotransmitter homeostasis as well as other biochemical processes in the central nervous system. A discussion of these findings is not germane to

Table 2 Summary of the Characteristics of the PCP Receptor

Ligand	Tissue	KD[a]	B_{max}[b]	Buffer[c]	Reference
$[^3H]$PCP	Rat brain	150	1.5	Tris-HCl (50 mM)	85
$[^3H]$PCP	Rat brain	250	2.4	Tris-HCl (50 mM)	86
$[^3H]$PCP	Rat brain	63	ND	Tris-HCl (5 mM)	87
$[^3H]$PCP	*Torpedo* electric organ	7,500	290.0	NaCl (250 mM)	88
$[^3H]$ 3-NH_2-PCP	*Torpedo* electric organ	15,500	320.0	NaCl (250 mM)	88
$[^3H]$PCP	Rat brain	250[d]	ND[e]	Tris-HCl (50 mM)	89
$[^3H]$PCP	Rat liver	230[d]	ND	Tris-HCl (50 mM)	89
$[^3H]$PCP	Rat kidney	130[d]	ND	Tris-HCl (50 mM)	89
$[^3H]$PCP	Glass fiber filters	240	1.1[f]	Tris-HCl (50 mM)	90
$[^3H]$PCP	Rat brain	300	1.1[f]	Tris-HCl (50 mM)	90
$[^3H]$PCP	Denatured rat brain	360	2.4[f]	Tris-HCl (50 mM)	90
$[^3H]$PCP	Rat olfactory bulb slices	46	10.5	Tris-HCl(5 mM)	91
$[^3H]$PCP	Rat brain	85	2.0	Tris-HCl (5 mM)	92
	Rat brain	195	3.5	Tris-HCl (50 mM)	
[3H]PCP	Guinea pig muscle	36[g]	ND	Tris-HCl (5 mM)	93
	Myenteric plexus	2,000[g]	ND	Tris-HCl (5 mM)	93
$[^3H]$PCP	Rat brain	ND	ND	Tris-HCl (50 mM)	94
$[^3H]$PCP	Pigeon brain	ND	ND	Tris-HCl (50 mM)	94
$[^3H]$PCP	Rat brain	27	0.9	HEPES (10 mM)	95
$[^3H]$PCP	Rat brain	24	2.8	Tris-HCl (5 mM)	96
	Rat brain	7,600	96.0	Tris-HCl (5 mM)	96
	Pigeon brain	40	2.5	Tris-HCl (5 mM)	96

	Pigeon brain	5,000	126.0	Tris-HCl (5 mM)	96
[^{3}H]TCP	Rat brain	7.4	1.0	Tris-HCl (5 mM)	99
[^{3}H]Dexoxadrol	Rat brain	1.7	0.4	Tris-HCl (50 mM)	107
[^{3}H]Dexoxadrol	Rat brain	8-15	ND	Tris-HCl (5 mM)	108
[^{3}H](±)-NANM	Rat brain	0.35	0.8[h]		117
	Rat brain	7.5	20.9		
[^{3}H](±)-NANM	Rat brain	6	ND		118
[^{3}H](±)-NANM	Rat brain	252	663.0	Tris-HCl (100 mM)	119
[^{3}H](±)-NANM	Rat brain	2.4	0.2	Na^+/K^+PO_4 (5 mM)	120
	Rat brain	9.3	0.6	Na^+/K^+PO_4 (5 mM)	120
	Rat brain	245	6.2	Na^+/K^+PO_4 (5 nM)	120
[^{3}H](+)-NANM	Mouse brain	12	0.4	Tris-HCl (50 mM)	121
[^{3}H](+)-NANM	Rat brain	20	0.3	HEPES-KOH (10 mM)	122
[^{3}H](+)-NANM	Rat brain	150	2.9	Tris-HCl (5 mM)	123
[^{3}H]Cyclazocine	Rat brain	0.2	0.1	Tris-HCl (50 mM)	126
	Rat brain	70	1.0	Tris-HCl (50 mM)	126
[^{3}H](+)-3-PPP	Rat brain	30	31.0[i]	Tris-HCl (50 mM)	127

[a]Nanomolar.

[b]Picomoles per milligram protein.

[c]Concentration of buffer in parentheses.

[d]Concentration that reduced binding of 1 nM [^{3}H]PCP by 50%.

[e]Not determined.

[f]Picomoles per assay.

[g]Values obtained from an unresolved curvilinear Scatchard plot.

[h]Fentomoles per milligram tissue.

[i]Picomoles per gram wet weight.

this review. A notion that has generated much interest is the possibility that receptors specific for PCP exist in the central nervous system and that these receptors are involved in the expression of psychotomimetic activity.

Binding and Radioligands

The first evidence to support PCP receptors emerged in the late 1970s, and research efforts to characterize and define these receptors continues at the present time. A summary of some of the PCP binding studies is presented in Table 2. A large number of studies have been carried out with [^{3}H]PCP as the radiolabeled ligand (85-96). In most cases, binding characteristics were determined for rat brain membranes. The data in Table 2 shows that [^{3}H]PCP has a relatively low affinity for its binding site, with the K_D in the range of 150-300 nM. This rather weak affinity coupled with the observation that [^{3}H]PCP binds to glass fiber filters under conditions that were used for measuring bound [^{3}H]PCP (90) cast doubt on the validity of the PCP binding. The technical aspects of the binding procedure were addressed by several investigators who demonstrated that pretreatment of the filters with various surfactants eliminated binding to the filters but not to brain membranes (92,94,95). [^{3}H]PCP has also been reported to bind in a saturable manner to brain slices, a procedure that does not involve filtration (91). Vincent et al. (87) made an important observation when they discovered that [^{3}H]PCP binding was greatly influenced by the ionic strength of the buffer (87). Reducing the buffer from 50 to 5 mM Tris-HCl increased the binding affinity of [^{3}H]PCP fourfold. When 5 mM Tris-HCl was used, the dissociation constants for [^{3}H]PCP binding to rat brain ranged from 24 to 86 nM. Binding to the PCP receptor in rat brain is glutamate dependent (see Reference 202).

An important analog of PCP is *N*-(1-[2-thienyl] cyclohexyl)piperidine (TCP), which is also a drug of abuse (97) that produces behavioral effects qualitatively similar to those of PCP (98). Vignon et al. (99) demonstrated that [^{3}H]TCP binds to rat brain homogenates with higher affinity (K_D = 7.4 nM) than PCP. These investigators pointed out that [^{3}H]TCP was a preferred ligand because of this high affinity, as well as because it slowly dissociated from the binding site and produces less nonspecific binding. Although this ligand has improve the in vitro binding assay, its most significant contribution is that its binding affinity is in the low nanomolar range, which is consistent with that of other centrally active drugs for their binding sites.

Some nonarylcycloalkylamine compounds, such as the 1,3-substituted dioxolanes, exhibit PCP-like properties. Both dexoxadrol and etoxadrol, depicted in Figure 3, have been shown to produce behavioral effects in animals similar to those of PCP (100-102). They also exhibit psychotomimetic properties in humans. In vitro binding studies have also been conducted with [^{3}H]-

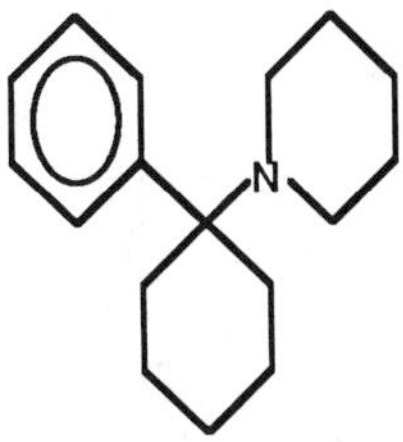

Phencyclidine (PCP)

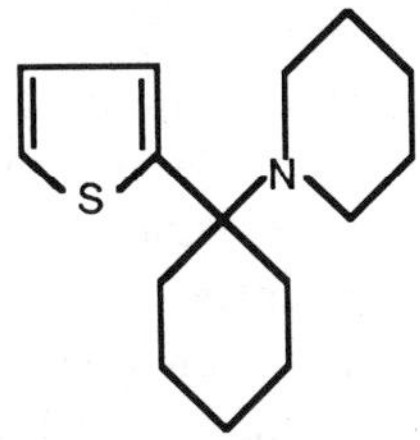

N-(1-[2-Thienyl]cyclohexyl)piperidine (TCP)

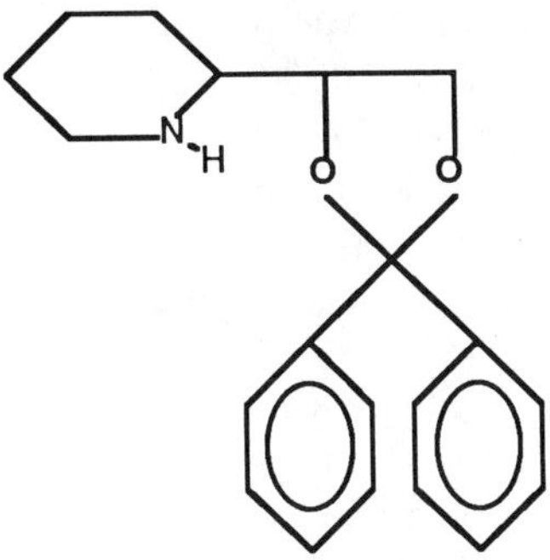

Dexoxadrol

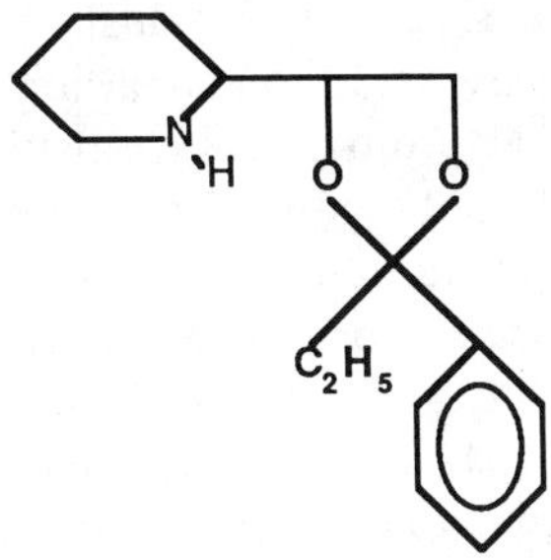

Etoxadrol

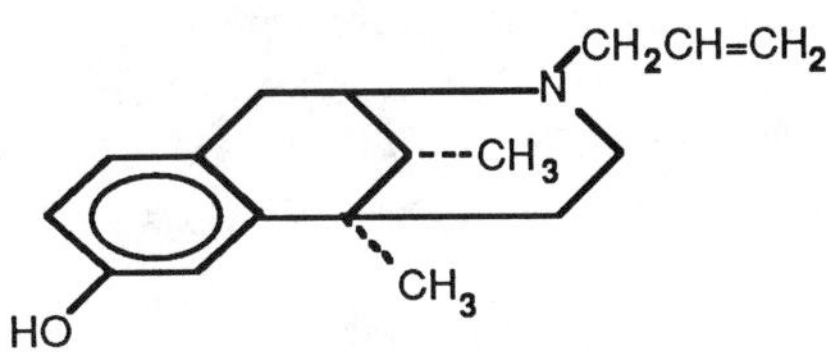

N-Allylnormetazocine (NANM)

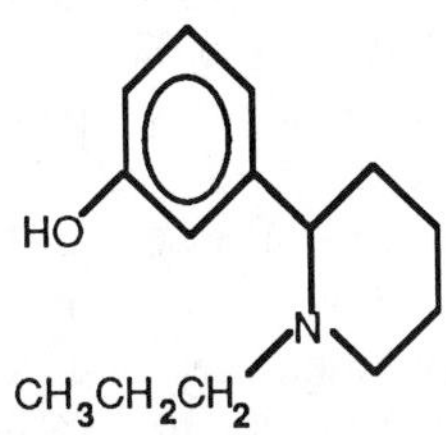

3-(3-Hydroxyphenyl)-N-(1-propyl)piperidine (3-PPP)

Figure 3 Structures of NAMN, 3-PPP, and PCP-related compounds.

dexoxadrol, and it binds reversibly (103-106) but with high affinity to rat brain membranes to what appears to be the PCP-specific site (107,108). Therefore, [^{3}H] dexoxadrol provided further evidence that a high-affinity site could be found in brain tissue for PCP and similar psychotomimetic compounds.

The observation in 1979 by Vaupel and Jasinski (109) that the pharmacological properties of PCP were similar to those of *N*-allylnormetazocine (NANM) in the chronic spinal dog preparation has generated considerable interest. It can be seen in Figure 3 that NANM is a benzomorphan, whereas PCP is an arylcyclohexylamine. In 1976, Martin et al. (110) presented the concept of multiple opiate receptor subtypes in an attempt to explain the myriad opioid effects. His classification scheme included the μ receptor, which was responsible for opioid-induced miosis, bradycardia, hypothermia, and antinociception, with morphine the prototypical agonist; the κ receptor, which was primarily responsible for sedative properties, with ketocyclazocine (another benzomorphan) the prototypical agonist; and finally the σ receptor, which was associated with mydriasis, tachycaradia, and delirium (see Chapter 16 for further discussion of σ sites). The prototypical agonist for the σ receptor was NANM. Since it had been reported that NANM produced psychotomimetic activity in clinical trials (111), in addition to its PCP-like properties in animals, it was only natural that an extensive effort to use NANM as a tool for probing the psychotomimetic action mechanism of PCP would follow. The psychotomimetic activity of NANM was established using a racemic mixture rather than the individual stereoisomers. Brady et al. (112) and Shannon (113) demonstrated that the (+)isomer of NANM is more PCP-like than the (–)isomer in drug discrimination procedures. This difference is probably attributable to (+)-NANM lacking μ and κ properties, whereas the (–)isomer has these opiate properties, which mask any σ properties it may contain. (–)-NANM produced PCP-like effects in rats trained under a schedule of electric shock termination and in pigeons trained under a schedule of food presentations (113,114) but not in rats, squirrel monkeys, or rhesus monkeys trained under schedules of food presentation (112,115). Both (+)- and (–)-NANM produced effects similar to those of PCP in the chronic spinal dog preparation (116). Thus it seems that (+)-NANM is similar to PCP in all test systems, whereas (–)-NANM is similar to PCP in selected species and/or procedures.

Initial binding studies conducted with $[^3H]$(±)-NANM revealed the presence of two binding sites, and it was presumed that the high-affinity site was the μ or κ receptor and the low-affinity site was the σ receptor (117). Wolozin et al. (118) blocked the μ-binding component of $[^3H]$(±)-NANM with naloxazone and found a resulting binding site with a K_D of 6 nM. Su (119) used a similar approach with $[^3H]$(±)-NANM by blocking the μ-binding component with etorphine, but the K_D of the remaining binding component was 252 nM, which was much greater than that reported by Wolozin et al. (118). The difference in these results may be due to the high ionic strength of the buffer used by Su (119). Johnson (120) found three binding sites when he examined the binding of $[^3H]$(±)-NANM to rat brain membranes, which was consistent with the two isomers binding to different receptors. The binding characteristics of the individual isomers of [^{3}H] NANM were evaluated in our laboratory (121) and found to be distinctly different from each other, which supported the binding data obtained indirectly from blocking experiments. The K_D of [^{3}H] (–)-NANM was 2.1 nM, whereas that for [^{3}H] (+)-NANM was 12 nM. As discussed below, competition studies suggested the (–)isomer was associated with μ and κ receptors, whereas the (+) isomer was not. At approximately the same time, Itzhak et al. (122) obtained a K_D of 20 nM when they examined the binding of [^{3}H]-(+)-NANM using low ionic strength buffer. However, under similar conditions, Mendelsohn et al. (123) obtained a much higher K_D (150 nM) for [^{3}H] (+)-NANM binding.

(±)-Cyclazocine is another benzomorphan that has been used in attempts to validate the PCP receptor. Martin et al. (110) described (±)-cyclazocine as having μ, κ, and σ properties, and it was subsequently shown that PCP would generalize to (±)-cyclazocine in rats trained to discriminate (±)-cyclazocine from saline (124) and that (±)-cyclazocine would generalize to the PCP cue in PCP-trained rats (125). Zukin and Zukin (126) investigated the binding of [^{3}H]-(±)-cyclazocine to rat brain membranes and reported three binding sites. The high- and low-affinity sites were characterized as the μ and σ binding sites, respectively.

One of the more interesting compounds to be used as a potential PCP receptor probe is (+)-3-(3-hydroxyphenyl)-*N*-(1-propyl)piperidine (3-PPP) (Figure 3). Considerable evidence supports a role of 3-PPP as a dopamine autoreceptor agonist, although this point is somewhat controversial (127). There is evidence to suggest that the dopaminergic system is involved in some of the actions of PCP (128). In addition, it has been shown that haloperidol, a dopaminergic antagonist, has very high affinity for the (+)NANM binding site. Largent et al. (127) described a specific high-affinity binding site for [^{3}H] (+)-3-PPP that does not appear to be independent of the dopaminergic system.

An anti-ischemic anticonvulsant, MK 801, binds to PCP-like sites in brain tissue (203) and, like PCP, acts as an indirect, *N*-methyl-D-aspartate (NMDA) antagonist (204). Considerable evidence has accumulated for an NMDA/PCP receptor complex (205).

Selectivity and SAR

The site labeled with [^{3}H] PCP under high ionic conditions is selective for PCP analogs and (±)-NANM (site 3 in Figure 4). Drugs that did not displace at 100 μM included numerous agonists and antagonists of CNS transmitters, neuropeptides, amino acids, naloxone, ketocyclazocine, levorphanol, haloperidol, and diazepam. Although this site was selective for PCP and related compounds, these agents exhibited low affinity, with IC_{50} values in the range of 160-2400 nM (85,86). Vincent et al. (86) reported a correlation coefficient of 0.84 be-

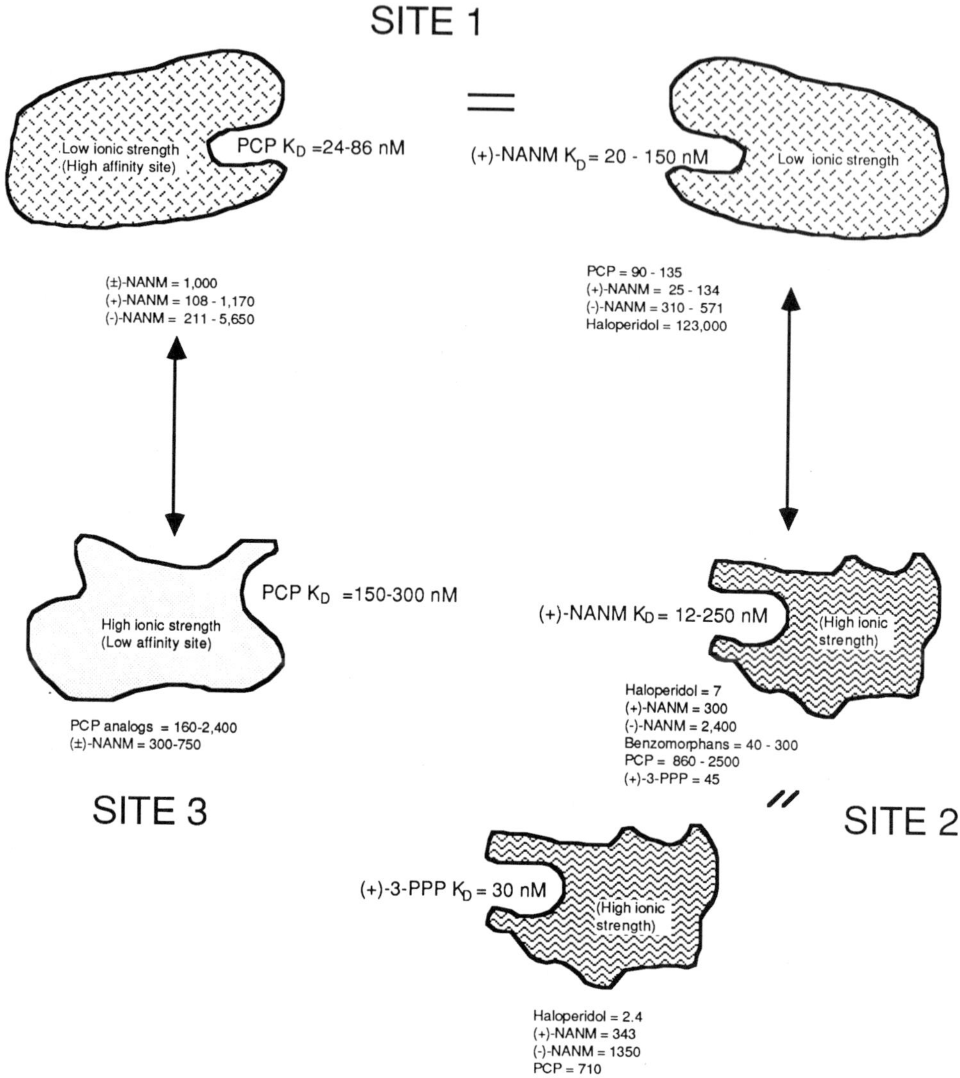

Figure 4 Configuration of hypothetical PCP and σ receptors. IC_{50} values are indicated for selected compounds listed under each receptor. Equals signs are used to point out receptors that are the same, whereas the arrows are used to suggest receptor interconversions.

tween behavioral potency (rotarod activity) and displacement of [^{3}H]PCP by 9 PCP-related analogs. Similar results were obtained with [^{3}H]PCP binding to slices of rat olfactory bulb (91). Hampton et al. (94) examined several stereoisomer pairs for their ability to compete for [^{3}H]PCP binding to rat brain. They found that dexoxadrol was 80 times more effective than levoxadrol in displacing [^{3}H]PCP, which was consistent with the behavioral activity of these isomers (129,130). (+)-NANM was approximately three times more effective than (–)-NANM in competing for binding. However, it was perplexing that the IC_{50} values for ketamine, (–)-NANM, and levoxadrol were essentially the same despite their differing pharmacological potencies.

Lowering the ionic strength of the buffer altered the affinity of PCP for its binding site without exerting appreciable effects on selectivity (referred to as site 1 in Figure 4). Zukin et al. (92) demonstrated that numerous benzomorphans competed for [^{3}H]PCP binding. Murray and Leid (95) reported similar results with benzomorphans and PCP-related compounds. The (+) isomer of NANM was five times more potent than the (–) isomer in displacing [^{3}H]PCP. However, Mendelsohn et al. (96) reported much higher affinities for PCP-related compounds competing for this binding site. The IC_{50} for PCP itself was 27 nM, but those for (+)- and (–)-NANM were 108 and 211 nM, respectively. Haloperidol did not bind to this site.

Although it is clear that PCP and (+)-NANM share some common behavioral properties, binding studies with the stereoisomers of [^{3}H]NANM have served to illustrate the complex relationship between PCP and σ receptors. The studies conducted with the racemic mixture of [^{3}H]NANM are somewhat more difficult to interpret owing to the interaction of the two isomers to different subsets of opioid receptors. Attempts to block the high-affinity μ binding with various agonists and antagonists have been quite successful. In this regard, Su (119) demonstrated that there was an etorphine-inaccessible binding site of [^{3}H]-NANM that appeared to be the σ receptor (site 2 in Figure 4). Numerous psychotomimetic benzomorphans competed for this binding site with IC_{50} in the range 40-300 nM. (+)-NANM was approximately 10 times more potent than (–)-NANM in binding to this site. PCP had weak affinity for this site (IC_{50} = 2540 nM). Other results with [^{3}H](+)-NANM confirmed these observations, with the exception that the affinity of [^{3}H](+)-NANM for its binding site under conditions of high ionic strength was 12 nM (121). Haloperidol was the most potent displacer, followed by the (+) isomers of the benzomorphans. The IC_{50} for PCP was 860 nM. The results from these studies suggest that this site is not responsible for PCP-related psychoactivity because numerous agents (such as naloxone, (+)-methadone, and (+)-morphine) that failed to produce PCP-like effects were equally effective as PCP in competing for binding. The most noteworthy exception between the PCP sites (in the presence of either low or high ionic strength buffer) and the (+)-NANM site (under high ionic strength) is haloperidol's high affinity for the latter. However, recent studies in which the ionic strength was lowered for [^{3}H] (+)-NANM binding revealed similarities between this binding site and that of PCP under low ionic strength. The IC_{50} of PCP, (+)-NANM, and (–)-NANM were similar for these two sites in addition to the fact that they were haloperidol insensitive (122,123).

As mentioned earlier, attempts have also been made to label the PCP σ site with (+)-3-PPP. Displacement studies, which have been carried out under high ionic strength, indicate that this site is closely related to the site labeled with (+)-NANM under high ionic strength buffer (127,131). Both sites are labeled by (+)-3-PPP, (+)-NANM, and haloperidol, to a lesser extent with (–)-NANM, and only very weakly with PCP and its structural analogs (site 2 in Figure 4). This site does not appear to be dopamine related since potent dopamine agonists and antagonists compete weakly for binding and the localization of these sites does not parallel those of dopamine receptors (127).

To summarize these rather complex findings, it appears that either the PCP σ receptor is able to assume more than one conformation or multiple receptor subtypes exist, as suggested in Figure 4. At low ionic strength, the PCP and (+)-NANM binding sites appear to be more closely related to each other and may in fact be one and the same (termed site 1). However, increasing the ionic strength serves to accentuate differences between PCP and (+)-NANM binding such that two distinct binding sites emerge. These sites could represent different interconversions of site 1, or they could be distinctly different sites that only bind their ligands under conditions of higher ionic strength. There is good evidence that PCP produces psychotomimetic activity by interaction with a receptor, and there has even been some preliminary evidence of an endogenous ligand for this site (132). It also appears that PCP and σ opioids are capable of sharing some form of a common receptor. Clearly, there is sufficient evidence to show that σ opioids can interact at a distinctly different site. It remains to be established which site is responsible for σ-opioid behavioral effects.

TRICYCLIC ANTIDEPRESSANTS

Historical Perspective

One of the first plausible explanations for the biochemical basis of effective disorders was provided by Schildkraut (133) and Bunney and Davis (134), who proposed that mania essentially resulted from excessive catecholamine output and depression from a deficient catecholamine output. A similar indoleamine hypothesis was also advanced (135). The observation that tricyclic antidepressants inhibited the neuronal uptake of monoamines and serotonin led to the suggestion that this action was responsible for their antidepressant effects, which was consistent with the neurotransmitter hypothesis. However, there were numerous unanswered questions regarding neuronal uptake as the mechanism of antidepressant action, not the least of which was that antidepressant drugs could be shown to act quickly on neuronal systems in animals but clinical effects were notoriously delayed. Neurotransmitter receptor desensitization was advanced as an explanation for this discrepancy between biochemical changes and clinical improvement, which was supported by such findings as thyroid hormone acceleration of the onset of imipramine-induced clinical improvement (136).

It now appears that the antidepressant action of the tricyclic compounds is more complex than mere inhibition of neuronal uptake of neurotransmitters. For example, some postsynaptic effects of tricyclic antidepressants on norepinephrine, serotonin, and acetylcholine responses appear to be independent of inhibition of uptake. The emergence of effective antidepressant drugs, such as mianserin, viloxazine and iprindol, that are weak inhibitors of neurotransmitter uptake weakens the argument of antidepressant activity being due to this action (137-140). In addition, some neuronal uptake inhibitors, such as cocaine, are devoid of antidepressant properties (141). The discovery of Langer and his colleagues of specific, high-affinity binding sites for [^{3}H] imipramine (142-144) led to a new concept regarding the etiology of mental depression. It is now thought that antidepressants produce their effects by interacting with their own receptors, which results in down-regulation of neurotransmitter receptors (see References 145 and 146 for reviews). The purpose of this section is to review the emergence of information that established that these antidepressant receptors exist in the central nervous system. As pointed out in earlier sections of this review, the receptor concept is strengthened by in vitro binding data that are congruent with in vivo observations, most notably of which are SAR and the existence of an antagonist.

Structure-Activity Relationships

Several tricyclic ring systems possess antidepressant properties when they are appropriately substituted (see an earlier review, Reference 147). Imipramine (Figure 5) is a derivative of 10,11-dihydro-5H-dibenz[*b,f*] azepine. Activity is restricted to such compounds as imipramine that have two or three carbons in the side chain; activity is lost in compounds lacking the side chain or with branched side chains containing more than four carbonds. Removal of an N-methyl results in desipramine, which retains clinical efficacy comparable to that of imipramine (148). Replacement of the ring heteroatom by an sp^2- or sp^3-hybridized carbon atom does not eliminate antidepressant activity as evidenced by clinical effects of amitriptyline and nortriptyline. Various alterations can be made to the 10,11 bridge with little effect on potency. For example, dihydrodibenzazepines and dibenzazepines are similar in potency. Antidepressant activity is retained with substitutions in the ring (as in the case with doxepin) or when the bridge is eliminated altogether. Also, the cis isomer of doxepin is more potent than the trans isomer. Substitutions on the aromatic rings at

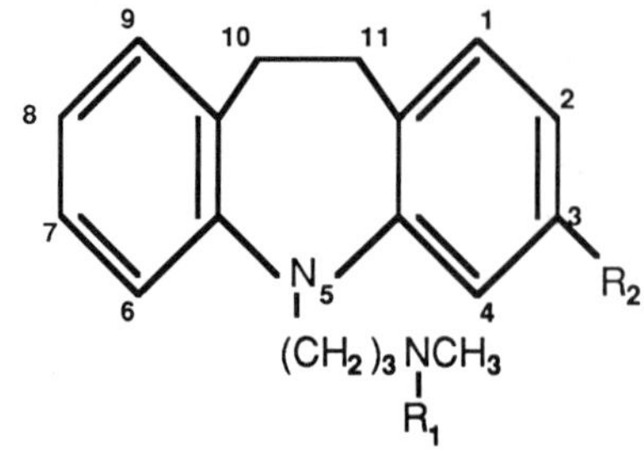

	R_1	R_2
Imipramine	CH_3	H
Desipramine	H	H
Ro 11-2465	CH_3	CN

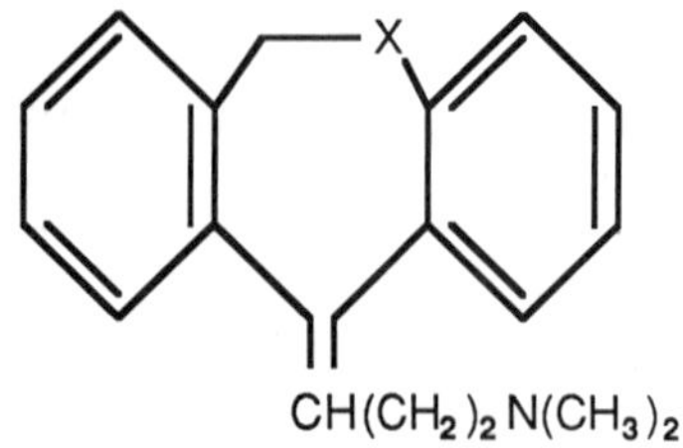

Amitriptyline	$X = CH_2$
Doxepin	$X = O$

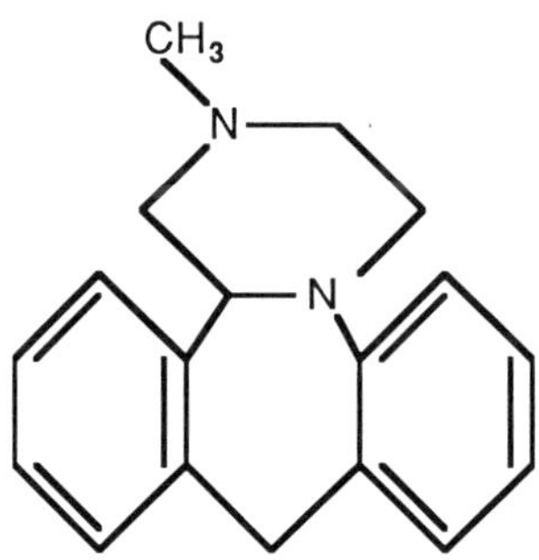

Mianserin

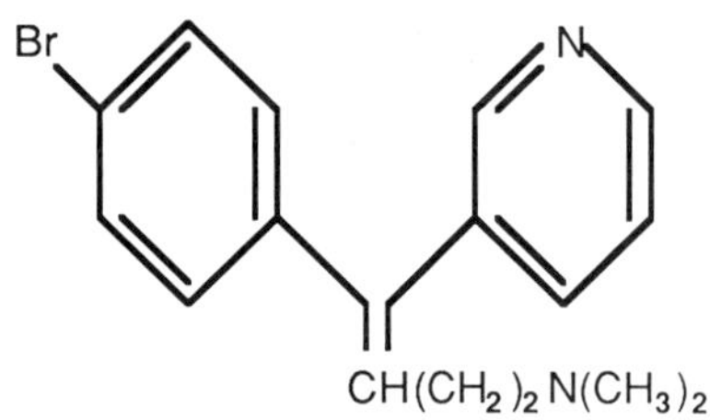

Zimelidine

Figure 5 Structures of tricyclic antidepressants.

positions 3(chloro), 10 (methyl), and 10,11 (dimethyls) do not necessarily eliminate activity. Also, 3-cyanoimipramine, Ro 11-2465, exhibits a pharmacological profile similar to that of imipramine (149).

Antagonists

It is obviously not possible to test compounds in the clinic for their potential to antagonize tricyclic antidepressant effects on affective disorders. Hence, there are no compounds known to specifically antagonize the antidepressant properties of the tricyclics. On the other hand, there is in vitro evidence that some derivatives may bind in an irreversible fashion. There is considerable evidence implicating the involvement of serotonin uptake in the action of some tricyclic antidepressants. 2-Nitroimipramine, which competes for [^{3}H]-imipramine binding and blocks serotonin uptake, has been reported to bind to platelets almost irreversibly at low temperatures (150). Likewise, [^{3}H] 3-cyanoimipramine binds irreversibly to [^{3}H] imipramine binding sites in platelets in the presence of Na^+ at 4°C (151). It would appear that these derivatives could fulfill the role of specific antagonists for antidepressants.

Binding and Radioligands

The major portion of evidence for antidepressant receptors has been derived from in vitro binding studies with [^{3}H] imipramine, as summarized in Table 3 (142-144,152-177). These studies have demonstrated that there are saturable binding sites through the central nervous system in several species, including the human. The K_D of the high affinity [^{3}H] imipramine site in rat cortex ranges from 2.7 to 7.5 nM and the B_{max} ranges from 0.22 to 0.51 pmol/mg protein. A low-affinity, high-capacity site has also been reported in rat cortex (160,165). The two binding sites that have been characterized in synaptosomes prepared from rat cortex correspond to these high- and low-affinity sites (161, 163). The high-affinity binding has been demonstrated in hypothalamus (K_D = 4.1-6.8 nM), striatum (2.8 nM), hippocampus (3.3-8.3 nM), midbrain (2.4 nM), pons-medulla (7.8 nM), cerebellum (8.0 nM), and spinal cord (4.0 nM). The density of the high-affinity [^{3}H] imipramine binding site was found to be similar in rat cortex, hypothalamus, and hippocampus, somewhat lower in midbrain and striatum, and very low in cerebellum, as summarized in Table 3. [^{3}H] imipramine binding has also been examined extensively in human brain, and it agrees closely with that reported for the rat. There is some disagreement among the results of experiments with human brains, which is to be expected, but in general the highest densities of binding sites were in cortex and hypothalamus, with lower concentrations in hippocampus, caudate, and cerebellum.

Desipramine, the N-demethylated metabolite of imipramine that is in itself a clinically potent antidepressant, has also been used extensively in characterizing the tricyclic antidepressant receptor (162,178-185). It has a somewhat higher affinity than imipramine in rat cortex, with a K_D in the range of 0.8-4.5 nM. High-affinity [^{3}H] desipramine binding was also found in rat hypothalamus (K_D = 0.9-2.6 nM), striatum (3.4), medulla (1.0), and cerebellum 1.1-1.5), which exhibited higher affinity than the corresponding [^{3}H] imipramine binding except in the case of striatum. The densities of [^{3}H] desipramine binding sites were 0.07-0.68 pmol/mg protein (B_{max}) for rat cortex, 0.16-0.22 pmol/mg protein for hypothalamus, 0.03 pmol/mg protein for striatum, and

Table 3 Summary of the Binding Characteristics of the Tricyclic Antidepressants in Central and Peripheral Tissues

Ligand	Species	Tissue	K_D[a]	B_{max}[b]	Ref.
[^{3}H]imipramine	Rat	Cortex	6.8	22.70[c]	142
			4.0	13.80[c]	143
			7.5	0.33	154
			2.7	0.39	155
			3.2	36.50[c]	156
			9.5[d]	0.45[d]	160
			4.0	0.25	162
			3.6	0.22	164
			6.9	0.24	165
			6.6	0.49	167
			4.0	0.25	169
			5.3	0.51	186
			133	1.87	160
			292	2.46	165
		Cortex	6.8	0.43	161
		synaptosomes			
			14	3.10	163
			99	14.20	163
		Hypothalamus	5.2	0.28	162
			4.1	0.32	169
			6.8	0.64	186
		Striatum	2.8	0.16	162
			2.8	0.16	169
		Hippocampus	4.7	0.26	164
			8.3	0.80	167
			3.3	0.14	169
			4.5	0.35	170
			4.1	0.36	171
		Midbrain	2.4	0.16	169
		Pons-medulla	7.8	0.60	170
		Cerebellum	8.0	0.06	169
			8.0	0.06	162
		Spinal cord	4.0	0.20	169
		Vas deferens	ND[e]	ND	160
			ND	ND	162
		Heart	ND	ND	162
		Submaxillary gland	ND	ND	162
		Platelets	2.3	1.75	162
	Cat	Hypothalamus	2.7	0.40	158
		Platelets	2.4	4.10	158
	Horse	Striatum (neuronal)	6.4	0.19	166
		Striatum (glia)	2.6	0.03	166

Table 3 (continued)

Ligand	Species	Tissue	K_D [a]	B_{max} [b]	Ref.
	Cow	Retina	6.4	0.78	175
	Human	Cortex	7.9	0.38	168
		Cortex (normals)	6.4	0.59	159
		Cortex (suicides)	8.0	0.33	159
		Frontal cortex	1.7	7.00[c]	152
		Hypothalamus	6.3	0.50	168
			1.7	16.2[c]	152
		Hippocampus	4.7	0.27	168
		Caudate nucleus	4.6	0.24	168
		Cerebellar cortex	1.7	4.00[c]	152
		Platelets (normals)	6.0	ND	153
			1.6	0.56	162
		Platelets (depressed)	2.4	0.32	162
		Platelets (normal males)	2.0	0.76	172
		Platelets (depressed males)	2.2	0.61	172
		Platelets (normal females)	1.8	0.65	172
		Platelets (depressed (females)	1.8	0.52	172
		Platelets (normals)	3.0	0.66	17
		Platelets (depressed)	3.0	0.56	173
		Platelets (manic-depressed)	3.2	0.53	173
		Platelets (normals)	0.6	1.01	174
			4.6	0.69	176
			0.7	0.64	177
[³H] desipramine	Rat	Cortex	2.4	0.10	162
			4.5	0.07	178
			1.5	0.24	179
			0.8	6.48[c]	181
			2.0	6.30[c]	182
			1.4	16.00[c]	184
			2.4	0.10	185
			2.3	0.11	180
			80	0.68	178
		Hypothalamus	2.6	0.16	162
			0.9	0.22	179
			2.6	0.16	185

Table 3 (continued)

Ligand	Species	Tissue	K_D[a]	B_{max}[b]	Ref.
		Striatum	3.4	0.03	162
			ND	ND	179
			3.4	0.03	185
		Medulla	1.0	0.10	179
		Septum	1.7	0.32	179
		Cerebellum	1.5	0.08	162
			1.5	0.08	185
			1.1	0.10	179
		Brain slices	6.6	0.36	183
		Vas deferens	2.6	1.02	162
			1.2	0.67	180
			2.6	1.02	185
		Heart	1.5	0.06	162
			1.1	0.18	180
			1.5	0.06	185
		Submaxillary gland	2.3	0.31	162
			2.3	0.31	185
		Platelets	ND	ND	162
[^{3}H] mianserin	Rat	Cortex	1.4	0.25	154
			0.8	0.16	157
			0.8	0.15	186
		Hypothalamus	1.5	0.27	186
		Hippocampus	0.9	0.24	172
		Cortex	1.4	0.25	187
			0.6	ND	188
	Guinea pig	Cerebellum	2.4	23.10[c]	188
	Calf	Caudate	0.7	0.14	189
[^{3}H] Ro 11-2465	Rat	Cortex	0.9	0.29	190
			0.3	0.27	154
			12.7	1.90	187
			195	4.47	154
[^{3}H] 2-nitroimipramine	Human	Platelets	1.3	0.81	177
[^{3}H] doxepin	Rat	Cortex	0.02	0.01	188
			3.6	0.74	188

[a]Nanomolar.
[b]Picomoles per milligram protein unless indicated otherwise.
[c]Picomoles per gram tissue.
[d]Calculated from data provided in the article.
[e]Not detected.

0.08-0.10 pmol/mg protein for cerebellum, which were approximately 50, 50, and 80% less than those found for [^{3}H] imipramine in cortex, hypothalamus, and striatum, respectively.

Antidepressant receptors have been identified with [^{3}H] mianserin (154, 157,171,186-189). This compound is considered an atypical antidepressant in that it does not block neurotransmitter reuptake. The affinity of [^{3}H] mianserin binding is in the range 0.6-1.5 nM (K_D), which is comparable to that of [^{3}H]-desipramine. However, evidence that mianserin and imipramine are capable of binding to different sites is discussed later.

3-Cyanoimipramine (Ro 11-2465) is a specific serotonin uptake inhibitor that has been used as a receptor probe (154,190). It was found to bind to both a high-affinity and low-affinity site in rat cortex. The K_D and B_{max} of the high-affinity site were in the same range as those of the other radiolabeled ligands. Another serotonin uptake blocker, [^{3}H] 2-nitroimipramine, has been used successfully to label a high-affinity binding site in human platelets (177). The antidepressant doxepin also binds to rat cortex, and two high-affinity binding sites have been characterized (191).

Peripheral tissues have been examined for high-affinity [^{3}H] imipramine binding, and platelets were the only tissue to contain these sites (Table 3). The K_D of [^{3}H] imipramine were similar for platelets from rat (2.3 nM) and normal human subjects (1.6-4.6 nM). [^{3}H] Desipramine binding was not detectable in platelets but was found in vas deferens, heart, and salivary glands.

Selectivity and SAR

The high-affinity binding site of [^{3}H] imipramine was shown to be selective for typical tricyclic antidepressants in that the IC_{50} for imipramine, desipramine, protriptyline, chlorimipramine, and amitriptyline were in the range 7-25 nM but that of doxepin was 300 nM (143). The monoamine uptake blockers (fluoxetine, nisoxetine, quipazine, and cocaine) displaced [^{3}H] imipramine in the concentration range 200-900 nM. The IC_{50} for the atypical antidepressants iprindole (5500 nM), viloxazine (11,500), and mianserin (20,000) demonstrated that they do not bind to this site. Serotonin, norepinephrine, and dopamine, along with their agonists and antagonists, competed very weakly or not at all for [^{3}H] imipramine binding. Despite that tricyclic antidepressants act directly on cholinergic (192), histaminergic (193), α-adrenergic (194), and serotonergic receptors (195), this evidence suggested that the imipramine binding was distinguishable from these sites. Dumbrille-Ross et al. (154) demonstrated that [^{3}H] imipramine binding was effectively displaced by serotonin uptake inhibitors, was Na^+ dependent, and was decreased by midbrain raphe lesions. They suggested that [^{3}H] imipramine was binding to a presynaptic site associated with serotonin uptake, particularly in light of the fact that serotonin uptake was Na^+ dependent.

Gross et al. (156) demonstrated that [^{3}H] imipramine binding sites in rat brain were located on presynaptic serotonergic nerve terminals rather than on noradrenergic neurons in that 5,7-dihydroxytryptamine (a serotonergic neurotoxin) treatment reduced the B_{max} by 96%, whereas 6-hydroxydopamine (a noradrenergic neurotoxin) treatment decreased the B_{max} by only 20%. Brunello et al. (157) also found [^{3}H] imipramine binding to be attenuated, but not completely eliminated, by 5,7-dihydroxytryptamine, which led them to suggest that these binding sites were also associated with other synaptic systems. However, Briley et al. (158) demonstrated that cats treated chronically with imipramine exhibited a down-regulation of β receptors as well as [^{3}H] imipramine binding sites but the rate of recovery from this down-regulation differed, which led them to conclude that the [^{3}H] imipramine binding site was not the same as the β adrenoceptor. Kinnier et al. (167) also found down-regulation of [^{3}H] dihydroalprenolol and [^{3}H] imipramine binding sites in rats treated chronically with imipramine; however, the location within the brain of the sites that were down-regulated did not coincide, which added additional evidence for differences between the two sites. [^{3}H] Spiperone (5-HT_2) binding was unaltered by chronic imipramine treatment, which again distinguished 5-HT_2 receptors from imipramine binding sites. Hrdina (164) found that chronic treatment with the typical antidepressant nortriptyline would down-regulate [^{3}H] imipramine binding sites in rat brain but, surprisingly, so did chronic treatment with haloperidol. This latter observation raised questions about whether these sites actually play a role in affective disorders since haloperidol lacks antidepressant efficacy.

As for subcellular distribution, Rehavi et al. (161) found [^{3}H] imipramine binding sites in the synaptosomal fraction and in microsomes but not in myelin and nuclear fractions; Stockert et al. (1984) found [^{3}H] imipramine binding both pre- and postsynaptically in synaptosomes, as well as in a purified nuclear fraction. Palkovits et al. (169) have established the regional distribution of [^{3}H] imipramine binding sites, which was highly correlated with the localization of endogenous serotonin levels, but only weakly so with that of norepinephrine and not with that of dopamine.

Although [^{3}H] desipramine binding was similar to that of [^{3}H] imipramine with regard to K_D and B_{max}, competition studies made it clear that these two drugs were labeling different sites in rat brain (178). For example, nortriptyline and nisoxetine were the most potent displacers of [^{3}H] desipramine bindings, but chlorimipramine and fluoxetine were the most potent in competing for [^{3}H] imipramine binding. Hdrina (178) established a significant correlation between the ability of various drugs to block [^{3}H] desipramine binding and their ability to block the neuronal uptake of norepinephrine in rat brain. Rehavi et al. (179) also found that norepinephrine uptake inhibitors were the most effective displacers of [^{3}H] desipramine binding. In addition, they found a marked decrease in [^{3}H] desipramine binding, but not [^{3}H] imipramine binding, in 6-hydroxydopamine-lesioned rats. Further support for an adrenergic link was provided by Langer et al. (180), who found that inhibition of norepinephrine uptake in both central and peripheral rat tissues by several compounds was correlated with their ability to inhibit [^{3}H] desipramine binding. Lee et al. (182) also found norepinephrine uptake blockers to be effective displacers of [^{3}H] desipramine binding and that *N*-(2-chloroethyl)-*N*-ethyl-2-bromobenzylamine, a specific neurotoxin for norepinephrine neurons, selectively abolished [^{3}H] desipramine binding sites. Using quantitative autoradiography, high concentrations of [^{3}H] desipramine binding sites were found in the locus ceruleus, the anterior ventral thalamus, the ventral portion of the bed nucleur of the stria terminalis, and the paraventricular and the dorsomedial nuclei of the hypothalamus, a distribution pattern that coincided with that of norepinephrine. 6-Hydroxydopamine lesions decreased [^{3}H] desipramine binding by 60-90% (183).

Raisman et al. (143) showed that mianserin did not compete for [^{3}H] imipramine binding, which suggested that it binds to a different site. The observation that [^{3}H] imipramine and [^{3}H] mianserin may bind to different receptors was explored further by Dumbrille-Ross et al. (154). They demonstrated that [^{3}H] imipramine binding was effectively displaced by serotonin uptake inhibitors, was Na^+ dependent, and was decreased by midbrain raphe lesions, and binding of [^{3}H] mianserin was displaced by serotonin 5-HT_2 antagonists, such as spiperone, and by antihistaminics (H_1) when spiperone was present and was Na^+ independent. They suggested that [^{3}H] mianserin was binding to postsynaptic 5-HT_2 (see Chapter 11) and histamine (H_1) (see Chapter 12) sites in contrast to [^{3}H] imipramine presynaptic sites. Brunello et al. (157,186) reported an increase in [^{3}H] mianserin binding following 5,7-dihydroxytryptamine treatment, which supported postsynaptic localization. Hrdina (164) reported that chronic treatment with mianserin failed to down-regulate [^{3}H] imipramine binding sites. In addition, Kinnier et al. (167) found that iprindole, another atypical antidepressant, also failed to down-regulate [^{3}H] imipramine binding sites but did down-regulate the β-adrenergic receptor labeled with [^{3}H] dihydroalprenolol. Although there were suggestions that mianserin may bind to a yet-to-be-identified site, Peroutka and Snyder (188) only found [^{3}H] mianserin binding at H_1 and 5-HT_2 receptors.

Doxepin binding represents the least characterized of the antidepressant binding sites. It is clear that the high-affinity site is related to H_1 sites, but it remains to be established unequivocally whether they are one and the same (191). H_1 antagonists have a high affinity for both [^{3}H] doxepin and [^{3}H]-pyrilamine (an H_1 antagonist) sites, but tricyclic antidepressants were more effective in competing for [^{3}H] doxepin than for [^{3}H] pyrilamine.

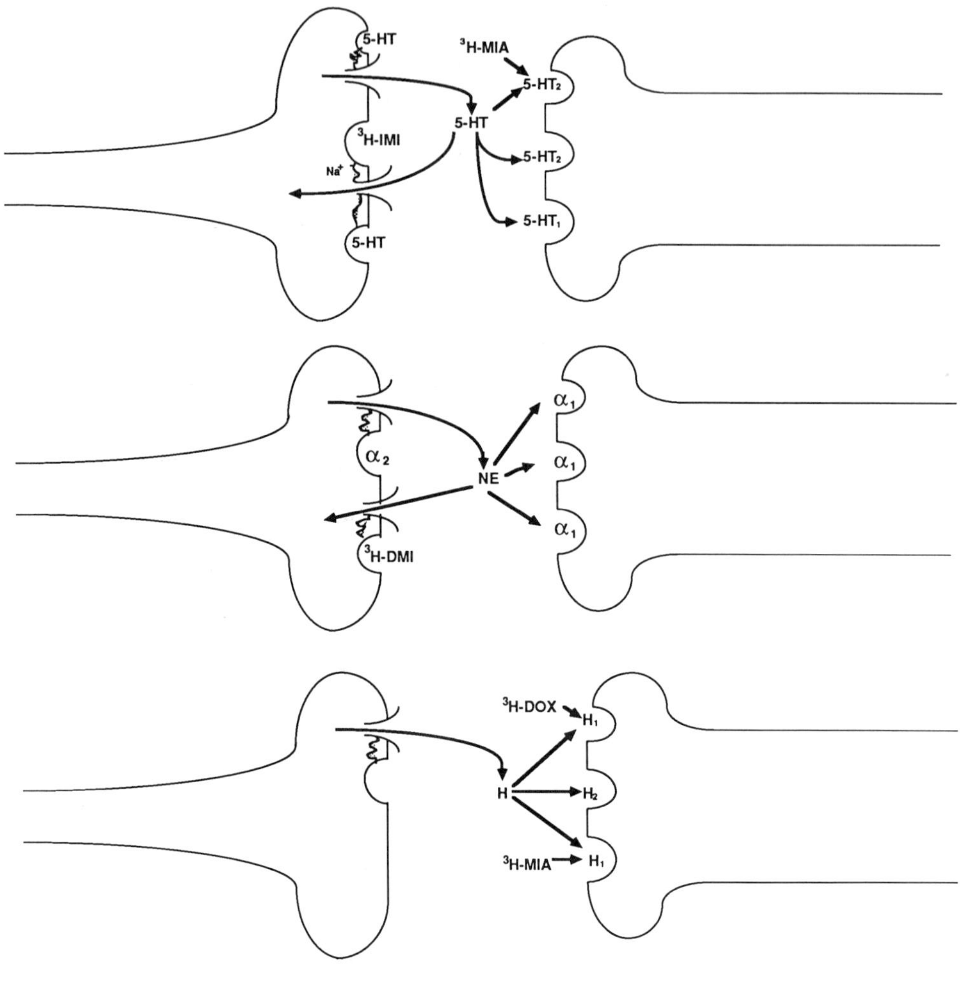

Figure 6 Localization of antidepressant receptors for imipramine (IMI), desipramine (DMI), mianserin (MIA), and doxepin (DOX) and their relationship to norepinephrine (NE), serotonin (5-HT), and histamine (H) receptors.

Function and Regulation

Considerable emphasis has been placed on establishing the functional significance of the [^{3}H] imipramine binding sites. The evidence discussed here establishes a strong link between the serotonergic system and the imipramine binding site despite that several questions remain unanswered. One of the most important questions deals with the relationship between serotonin uptake and the imipramine binding site. Costa et al. (196) concluded that imipramine does not actually bind to the macromolecules directly involved in serotonin uptake but rather binds to an allosteric site that may serve a regulatory role in serotonin uptake (Figure 6). As to the molecular mechanism of action of imipramine, two crucial elements are the delayed responsiveness to antidepressant therapy and the down-regulation of β-adrenergic stimulation of adenylate cyclase. These observations form the basis of Costa's premise that imipramine and desipramine relieve the symptoms of depression by down-regulating an increased responsiveness of β-adrenergic receptors that arise during depression. Langer and Raisman (162) also concluded that the presynaptic [^{3}H] imipramine site was different from the substrate recognition site for serotonin and that it was the presynaptic serotonin site that regulated serotonin release and the imipramine site that regulated serotonin uptake. If serotonin acts at a presynaptic site to block its own uptake, this site would presumably be distinguishable from the imipramine site given serotonin's low affinity for the latter. That chronic treatment with imipramine down-regulates its recognition site suggests that there is a concomitant increase in uptake of serotonin and a resultant decrease in serotonergic transmission. One may therefore speculate that depression results from decreased concentrations of an endogenous imipraminelike substance or from a depletion of these sites, either of which would produce increased serotonergic transmission. In this light, it is particularly important that Brusov et al. (197) have identified substances in plasma that inhibit serotonin uptake as well as [^{3}H] imipramine binding. Although it can be argued that decreased serotonergic transmission elicited by chronic treatment with imipramine attenuates β-adrenergic synaptic function and adenylate cyclase the question remains as to why ablation of serotonergic neurons does not produce the same effects (196).

Another reason for the emphasis on establishing the relevance of [^{3}H] imipramine binding sites is that it is possible to select a population of individuals who should exhibit a permanently altered biochemical system, a clear advantage over studying a temporarily induced drug state as with Δ^9-THC and PCP. Although the rat experiments have provided convincing evidence for the existence of specific binding sites for antidepressants, it was the discovery of [^{3}H]-imipramine binding sites in human brain that greatly enhanced their significance (152,159,168). Of course, it is particularly important that this binding site appears to be identical in both human and rat brain (168). Numerous arguments can be presented for and against the role that [^{3}H] imipramine binding sites play in affective disorders. However, it is noteworthy that Stanley et al. (159) found a 44% decrease in the number of [^{3}H] imipramine binding sites in frontal cortex of suicides, which is consistent with the theory presented earlier. Another aspect that has contributed to the attention of establishing the significance of antidepressant binding sites has been the long-term objective of clinicians to establish biologic markers for affective disorders. It is unnecessary to detail the difficulties that clinicians have in classifying and diagnosing mental disorders in order to appreciate the value of a reliable marker that could be readily assessed in peripheral tissue. It should not be forgotten that monoamine oxidase activity in platelets was once considered a possible marker for bipolar depression. After the observations that platelets contained [^{3}H] imipramine binding sites that exhibited the same characteristics as those in the central nervous system, special attention has focused on determining whether they might serve as a biologic marker for depression. It has been established that [^{3}H] imipramine binding sites are decreased in number in depressed patients, but there does not appear to be a relationship between the severity of depression and the number of sites (198). Also, [^{3}H] imipramine binding sites do not rise to within the normal range after patients improve following imipramine treatment. It is for these reasons that Langer et al. (198) have suggested that [^{3}H] imipramine binding may reflect a vulnerability factor or a genetic susceptibility to depression. The utility of this biologic marker would be in confirming depression rather than assessing the status of clinical improvement.

The localization of [^{3}H] desipramine binding sites in tissues and brain areas with high noradrenergic innervation, along with the competition studies, strongly suggests that these sites are involved in modulating the neuronal uptake of norepinephrine much in the same way that imipramine has been proposed to regulate serotonin uptake (Figure 6).

Mianserin is a tetracyclic antidepressant that has little effect on norepinephrine and serotonin uptake (199,200). It differs from both imipramine and desipramine in that it binds to postsynaptic rather than presynaptic receptors and it also binds to histamine (H_1) receptors (Figure 6). The main commonality between the typical antidepressants and the atypical antidepressants, which include mianserin, is that they down-regulate recognition sites for neurotransmitters (200,201).

Speculation and Future Direction

The discovery of specific binding sites for the antidepressants has provided an exciting means for elucidating the biologic etiology of affective disorders. It appears that the binding sites for the different antidepressants are associated with distinctively different biologic events, although a common thread may be β-adrenergic receptor desensitization. A reasonable assumption would be that some of these binding sites are not involved in the causation of depression, particularly the histamine binding site. However, it is important to remember that aberrations in numerous neurotransmitter systems could result in similar symptomatology that is collectively described in depression. Establishment of the biologic significance of these binding sites may well depend upon the identification of the endogenous substance that interacts with them.

ACKNOWLEDGMENTS

Portions of the research described in the cannabinoid and phencyclidine sections were supported by National Institute on Drug Abuse grants DA-03672 and DA-02396, respectively. The author is grateful for the critical review of this article by Dr. W. L. Dewey and the assistance of T. J. Martin and P. J. Little.

REFERENCES

1. R. K. Razdan, Pharmacol. Rev. *38*, 75 (1986).
2. L. E. Hollister, Pharmacology *121*, 3 (1974).
3. M. Perez-Reyes, M. C. Timmons, M. A. Lipton, K. H. Davis, and M. E. Wall, Science *177*, 633 (1972).
4. M. Perez-Reyes, M. C. Timmons, M. A. Lipton, H. D. Christensen, K. H. Davis, and M. E. Wall, Experientia *29*, 1009 (1973).
5. R. K. Razdan, J. F. Howes, D. B. Uliss, H. C. Dalzell, G. R. Handrick, and W. L. Dewey, Experientia *32*, 416 (1976).
6. A. Ohlsson, M. Widman, S. Carlsson, T. Ryman, and C. Strid, Acta Pharmacol. Toxicol. (Copenh.) *47*, 308 (1980).
7. B. Martin, M. J. Kallman, G. F. Kaempf, L. S. Harris, W. L. Dewey, and R. K. Razdan, Pharmacol. Biochem. Behav. *21*, 61 (1984).
8. H. D. Christensen, R. I. Freudenthal, J. T. Gidley, R. Rosenfield, G. Boegli, L. Testino, D. R. Brine, C. G. Pitt, and M. E. Wall, Science *172*, 165 (1971).
9. G. R. Handrick, R. P. Duffley, G. Lambert, J. G. Murphy, H. C. Dalzell, J. F. Howes, R. K. Razdan, B. R. Martin, L. S. Harris, and W. L. Dewey, J. Med. Chem. *25*, 1447 (1982).
10. R. Adams, Harvey Lect. *37*, 168 (1942).
11. R. Mechoulam and H. Edery, in Marihuana, Chemistry, Pharmacology, Metabolism and Clinical Effects (R. Mechoulam, ed.), Academic Press, New York, 1973, p. 101.
12. A. Semjonow and M. Binder, Psychopharmacology (Berlin) *85*, 178 (1985).
13. B. R. Martin, W. L. Dewey, L. S. Harris, and E. L. May, Pharmacol. Biochem. Behav. *3*, 849 (1975).
14. R. S. Wilson, B. R. Martin, and W. L. Dewey, J. Med. Chem. *22*, 879 (1979).
15. R. S. Wilson, E. L. May, B. R. Martin, and W. L. Dewey, J. Med. Chem. *19*, 1165 (1976).
16. L. E. Hollister, Experientia *29*, 825 (1973).
17. M. Perez-Reyes, M. C. Timmons, K. H. Davis, and M. E. Wall, Experientia *29*, 1368 (1973).
18. L. S. Melvin, M. R. Johnson, C. A. Harbert, G. M. Milne, and A. Weissman, J. Med. Chem. *27*, 67 (1984).
19. A. Weissman, G. M. Milne, and L. S. Melvin, Jr., J. Pharmacol. Exp. Ther. *223*, 516 (1982).
20. B. R. Martin, R. L. Balster, R. K. Razdan, L. S. Harris, and W. L. Dewey, Life Sci. *29*, 565 (1981).

21. D. B. Uliss, H. C. Dalzell, G. R. Handrick, J. F. Howes, and R. K. Razdan, J. Med. Chem. *18*, 213 (1975).
22. B. R. Martin, L. S. Harris, and W. L. Dewey, in The Cannabinoids: Chemical, Pharmacologic, and Therapeutic Aspects (S. Agurell, W. L. Dewey, and R. E. Willette, eds.), Academic Press, New York, 1983, p. 503.
23. A. Loev, P. E. Bender, F. Dowalo, E. Macko, and P. J. Fowler, J. Med. Chem. *16*, 1200 (1973).
24. T. C. U. Jarbe and G. C. Ohlin, Psychopharmacology (Berlin) *54*, 193 (1977).
25. H. Edery, Y. Grunfeld, Z. Ben-Zvi, and R. Mechoulam, Ann. N.Y. Acad. Sci. *191*, 40 (1971).
26. B. A. Zitko, J. F. Howes, R. K. Razdan, B. C. Dalzell, H. C. Dalzell, J. C. Sheehan, H. G. Pars, W. L. Dewey, and L. S. Harris, Science *177*, 442 (1972).
27. M. D. Adams, J. T. Earnhardt, B. R. Martin, L. S. Harris, W. L. Dewey, and R. K. Razdan, Experientia *33*, 1204 (1977).
28. R. Adams, Harvey Lect. *37*, 168 (1942).
29. H. F. Hardman, E. F. Domino, and M. H. Seevers, Pharmacol. Rev. *23*, 295 (1971).
30. S. Agurell, M. Binder, K. Fonseka, J.-E. Lindgren, K. Leander, B. Martin, I. M. Nilsson, M. Nordqvist, A. Ohlsson, and W. Widman, in Marihuana: Chemistry, Biochemistry and Cellular Effects (G. G. Nahas, ed.) Springer-Verlag, New York, 1976, p. 141.
31. G. Jones, R. G. Pertwee, E. W. Gill, W. D. M. Paton, I. M. Nilsson, M. Widman, and S. Agurell, Biochem. Pharmacol. *23*, 439 (1974).
32. H. Edery, Y. Greenfield, Z. Ben-Zvi, and R. Mechoulam, Ann. N.Y. Acad. Sci. *191*, 40, (1971).
33. D. B. Goldstein, Annu. Rev. Pharmacol. Toxicol. *24*, 43 (1984).
34. D. K. Lawrence and E. W. Gill, Mol. Pharmacol. *11*, 595 (1975).
35. A. S. Bloom and C. J. Hillard, in Marihuana '84 (D. J. Harvey, ed.), IRL Press, Oxford, 1985, p. 217.
36. J. Schou, L. D. Prockop, G. Dahlstrom, and C. Rohde, Acta Pharmacol. Toxicol. (Copenh.) *41*, 33 (1977).
37. M. Perez-Reyes, J. Simmons, D. Brine, G. L. Kimmel, J. H. Davis, and M. E. Wall, in Marihuana: Chemistry, Biochemistry and Cellular Effects (G. G. Nahas, ed.), Springer-Verlag, New York, 1976, p. 179.
38. A. L. Gough and J. E. Olley, Neuropharmacology, *17*, 137 (1978).
39. A. Ohlsson, M. Widman, S. Carlsson, T. Ryman, and C. Strid, Acta Pharmacol. Toxicol. (Copenh.) *47*, 308 (1980).
40. M. Binder, F.-J. Witteler, B. Schmidt, I. Franke, E. Bohnenberger, and H. Sanderman, Jr., in The Cannabinoids: Chemical, Pharmacologic, and Therapeutic Aspects (S. Agurell, W. L. Dewey, and R. E. Willette, eds.), Academic Press, New York, 1983, p. 709.
41. I. G. Karniol and E. A. Carlini, J. Pharm. Pharmacol. *24*, 833 (1972).
42. E. A. Carlini, I. G. Karniol, R. F. Renault, and C. R. Schuster, Br. J. Pharmacol. *50*, 299 (1974).

43. I. G. Karniol, I. Shirakawa, N. Kasinski, A. Pfeferman, and E. A. Carlini, Eur. J. Pharmacol. *28*, 172 (1974).
44. K. D. Bird, T. Boleyn, G. B. Chesher, D. M. Jackson, G. A. Starmer, and R. K. C. Teo, Psychopharmacology (Berlin) *71*, 181 (1980).
45. L. Lemberger, B. Dalton, R. Martz, B. Rodda, and R. Farney, Ann. N.Y. Acad. Sci. *281*, 219 (1976).
46. L. A. Borgen and W. M. Davis, Res. Commun. Chem. Pathol. Pharmacol. *7*, 663 (1974).
47. I. G. Karniol and E. A. Carlini, Psychopharmacology (Berlin) *33*, 53 (1970).
48. V. S. Jorapur, Z. H. Khalil, R. P. Duffley, R. K. Razdan, B. R. Martin, L. S. Harris, and W. L. Dewey, J. Med. Chem. *28*, 783 (1985).
49. J. C. Krantz, Jr., H. J. Berger, and B. L. Welch, Am. J. Pharmacol. *143*, 149 (1971).
50. L. Hollister and H. Gillespie, Clin. Pharmacol. Ther. *18*, 80 (1975).
51. R. Musty, T. Karniol, I. Shirakawa, R. Takahashi, and E. Kolbel, in The Pharmacology of Marihuana (M. C. Braude and S. Szara, eds.), Raven Press, New York, 1976, p. 559.
52. W. M. Davis and N. S. Hatoum, Gen. Pharmacol. *14*, 247 (1983).
53. N. S. Hatoum, W. M. Davis, M. A. Elsohly, and C. E. Turner, Gen. Pharmacol. *12*, 357 (1981).
54. A. N. Kudrin and O. N. Davydova, Farmakol. J. Toksikol. *31*, 549 (1968).
55. H. J. Berger and J. C. Krantz, Jr., J. Pharm. Pharmacol. *24*, 492 (1972).
56. R. G. Browne and A. Weissman, J. Clin. Pharmacol. *21*, 227S (1981).
57. P. Lomax and C. Campbell, Experientia, *27*, 1191 (1971).
58. T. C. Spaulding, R. D. Ford, W. L. Dewey, D. E. McMillan, and L. S. Harris, Eur. J. Pharmacol. *19*, 310 (1972).
59. E. O. H. Lew and J. S. Richardson, Drug Alcohol Depend. *8*, 93 (1981).
60. J. C. Garriott, L. J. King, R. B. Forney, and F. W. Hughes, Life Sci. *6*, 2119 (1967).
61. H. N. Bhargava and G. A. Matwyshyn, Eur. J. Pharmacol. *68*, 147 (1980).
62. M. Fujiwara, N. Ibii, Y. Kataoka, and U. Showa, Psychopharmacology (Berlin) *68*, 7 (1980).
63. S. H. Roth and P. J. Williams, J. Pharm. Pharmacol. *31*, 224 (1979).
64. P. Seeman, M. Chau-Wong, and S. Moyyen, Can. J. Physiol. Pharmacol. *50*, 1193 (1972).
65. L. S. Harris, R. A. Carchman, and B. R. Martin, Life Sci. *22*, 1131 (1978).
66. H. H. Seltzman, S. R. Setzer, D. L. Williams, I. Demian, C. Wyrick, and C. G. Pitt, in Marihuana '84 (D. J. Harvey, ed.), IRL Press, Oxford, 1985, p. 183.
67. J. S. Nye, H. H. Seltzman, C. G. Pitt, and S. H. Snyder, in Marihuana '84 (D. J. Harvey, ed.), IRL Press, Oxford, 1985, p. 253.
68. J. S. Nye, H. H. Seltzman, C. G. Pitt, and S. H. Snyder, J. Pharmacol. Exp. Ther. *234*, 784 (1985).
69. B. A. Zitko, J. F. Howes, R. K. Razdan, B. C. Dalzell, H. C. Dalzell, J. C. Sheehan, H. G. Pars, W. L. Dewey, and L. S. Harris, Science *177*, 442 (1972).

70. H. Yoshimura, K. Watanabe, K. Oguri, M. Fujiwara, and S. Ueki, J. Med. Chem. *21*, 1079 (1978).
71. B. R. Martin, J. A. Scimeca, B. R. Thomas, J. T. Martin, U. Estrada, and W. L. Dewey, Fed. Proc. *45*, 1058 (1986).
72. B. R. Martin, Pharmacol. Rev. *38*, 45 (1986).
73. A. C. Howlett, Life Sci. *35*, 1803 (1984).
74. A. C. Howlett and R. M. Fleming, Mol. Pharmacol. *26*, 532 (1984).
75. A. C. Howlett, Mol. Pharmacol. *27*, 429 (1985).
76. D. K. Lawrence and E. W. Gill, Mol. Pharmacol. *11*, 595 (1975).
77. I. Tamir and D. Lichtenberg, J. Pharm. Sci. *72*, 458 (1983).
78. D. Bach, A. Raz, and R. Goldman, Biochem. Pharmacol. *25*, 1241 (1976).
79. C. J. Hillard, R. A. Harris, and A. S. Bloom, J. Pharmacol. Exp. Ther. *232*, 579 (1985).
80. D. R. Wing, J. T. A. Leuschner, G. A. Brent, D. J. Harvey, and W. D. M. Paton, in Marihuana '84 (D. J. Harvey, ed.), IRL Press, Oxford, 1985, p. 85.
81. I. Franke, A. Grothey, S. Koch, F.-J. Witteler, M. Binder, and H. Sandermann, Jr., Biochem. Pharmacol. *34*, 3983 (1985).
82. G. Chen, C. R. Ensor, D. Russel, and B. Bohner, J. Pharmacol. Exp. Ther. *127*, 241 (1959).
83. F. E. Greifenstein, M. DeVault, J. Yoshitake, and J. E. Gajewski, Anesth. Anal. *37*, 283 (1958).
84. S. E. Lerner and R. S. Burns, in Phencyclidine Abuse: An Appraisal (R. C. Petersen and R. C. Stillman, eds.), NIDA Research Monograph 21, U.S. Government Printing Office, Washington, D.C., 1978, p. 66.
85. S. R. Zukin and R. S. Zukin, Proc. Natl. Acad. Sci. USA *76*, 5372 (1979).
86. J. P. Vincent, B. Kartalovski, P. Geneste, J. M. Kamenka, and M. Lazdunski, Proc. Natl. Acad. Sci. USA *76*, 4678 (1979).
87. J.-P. Vincent, J. Vignon, B. Kartalovski, and M. Lazdunski, Eur. J. Pharmacol. *68*, 73 (1980).
88. Y. Kloog, A. Kalir, O. Buchman, and M. Sokolovsky, FEBS Lett. *109*, 125 (1980).
89. J.-P. Vincent, J. Vignon, B. Kartalovski, and M. Lazdunski, Eur. J. Pharmacol. *68*, 79 (1980).
90. S. Maayani and H. Weinstein, Life Sci. *26*, 2011 (1980).
91. R. Quirion, R. P. Hammer, Jr., M. Herkenham, and C. B. Pert, Proc. Natl. Acad. Sci. USA *78*, 5881 (1981).
92. S. R. Zukin, M. L. Fitz-Syage, R. Nichtenhauser, and R. S. Zukin, Brain Res. *258*, 277 (1983).
93. A. R. Gintzler, R. S. Zukin, and S. R. Zukin, Br. J. Pharmacol. *75*, 261 (1982).
94. R. Y. Hampton, F. Medzihradsky, J. H. Woods, and P. J. Dahlstrom, Life Sci. *30*, 2147 (1982).
95. T. F. Murray and M. E. Leid, Life Sci. *34*, 1899 (1984).
96. L. G. Mendelsohn, G. A. Kerchner, V. Kalra, D. M. Zimmerman, and J. D. Leander, Biochem. Pharmacol. *33*, 3529 (1984).

97. E. F. Domino (ed.), in PCP (Phencyclidine): Historical and Current Perspectives, NPP Books, Ann Arbor, Michigan, 1981, p. 515.
98. K. T. Brady, R. L. Balster, L. T Meltzer, and D. Schwertz, Pharmacol. Biochem. Behav. *12*, 67 (1980).
99. L. Vignon, R. Chicheportiche, M. Chicheportiche, J.-M. Kamenka, P. Geneste, and M. Lazdunski, Brain Res. *280*, 194 (1983).
100. K. T. Brady, W. L. Woolverton, and R. L. Balster, J. Pharmacol. Exp. Ther. *220*, 56 (1982).
101. H. E. Shannon, in Phencyclidine and Related Arylcyclohexylamines: Present and Future Applications (J. M. Kamenka, E. F. Domino, and P. Geneste, eds.), NPP Books, Ann Arbor, Michigan, 1983, p. 311.
102. E. J. Cone, R. L. McQuinn, and H. E. Shannon, J. Pharmacol. Exp. Ther. *228*, 147 (1984).
103. L. Lasagna and J. W. Pearson, Proc. Soc. Exp. Biol. Med. *118*, 352 (1965).
104. A. M. Simopoulos, A. Pinto, P. W. Babikow, A. Kurkland, and C. Savage, Dis. Nerv. Syst. *31*, 203 (1970).
105. M. W. Williams, E. A. Williford, C. S. Williams, and J. C. Towne, Arch. Int. Pharmacodyn. Ther. *178*, 26 (1969).
106. R. D. Wilson, D. L. Traber, E. Barratt, D. Creson, R. Schmitt, and C. R. Allen, Anesth. Analg. Curr. Res. *49*, 236 (1970).
107. V. H. Sethy and J. M. McCall, Drug. Dev. Res. *4*, 635 (1984).
108. C. Pilaph, P. Contreras, T. L. O'Donohue, and R. Quirion, Neurosci. Lett. *56*, 1 (1985).
109. D. B. Vaupel and D. R. Jasinski, Fed. Proc. *38*, 1094 (1979).
110. W. R. Martin, C. G. Eades, J. A. Thompson, R. E. Huppler, and F. E. Gilbert, J. Pharmacol. Exp. Ther. *197*, 517 (1976).
111. A. S. Keats and J. Telford, Adv. Chem. Ser. *45*, 170 (1964).
112. K. T. Brady, R. L. Balster, and E. L. May, Science *215*, 178 (1982).
113. H. E. Shannon, Eur. J. Pharmacol. *84*, 225 (1982).
114. J. F. Grippo, S. Herling, R. E. Solomon, and J. H. Woods, Pharmacologist *23*, 152 (1981).
115. R. E. Solomon, S. Herling, and J. H. Woods, in Advances in Endogenous and Exogenous Opioids (H. Takagi and E. J. Simon, eds.). Elsevier, New York, 1981, p. 484.
116. D. B. Vaupel, Pharmacologist *24*, 117 (1982).
117. G. W. Pasternak, M. Carroll-Buatti, and K. Spiegel, J. Pharmacol. Exp. Ther. *219*, 192 (1981).
118. B. L. Wolozin, S. Nishimura, and G. W. Pasternak, J. Neurosci. *2*, 708 (1982).
119. T.-P. Su, J. Pharmacol. Exp. Ther. *223*, 284 (1982).
120. K. M. Johnson, Res. Commun. Subst. Abuse *4*, 33 (1983).
121. B. R. Martin, J. S. Katzen, J. A. Woods, H. L. Tripathi, L. S. Harris, and E. L. May, J. Pharmacol. Exp. Ther. *231*, 539 (1984).
122. Y. Itzhak, J. M. Hiller, and E. J. Simon, Mol. Pharmacol. *27*, 46 (1985).
123. L. G. Mendelsohn, V. Kalra, B. G. Johnson, and G. A. Kerchner, J. Pharmacol. Exp. Ther. *233*, 597 (1985).

124. J. J. Teal and S. G. Holtzman, J. Pharmacol. Exp. Ther. *212*, 368 (1980).
125. S. G. Holtzman, J. Pharmacol. Exp. Ther. *214*, 614 (1980).
126. R. S. Zukin and S. R. Zukin, Mol. Pharmacol. *20*, 246 (1981).
127. B. L. Largent, A. L. Gundlach, and S. H. Snyder, Proc. Natl. Acad. Sci. USA 81, 4983 (1984).
128. K. M. Johnson, T. W. Vickory, S. M. Leventer, and L. S. Mok, in Phencyclidine and Related Arylcyclohexylamines: Present and Future Applications (J. M. Kamenka, E. F. Domino, and P. Geneste, eds.), NPP Books, Ann Arbor, Michigan, 1983, p. 505.
129. S. Herling and J. H. Woods, Life Sci. *28*, 1571 (1981).
130. B. L. Slifer and R. L. Balster, Subst. Alcohol Actions/Misuse *5*, 273 (1985).
131. A. L. Gundlach, B. L. Largent, and S. H. Snyder, Eur. J. Pharmacol. *113*, 465 (1985).
132. R. Quirion, D. A. DiMaggio, E. D. French, P. C. Contreras, J. Shiloach, C. B. Pert, H. Everist, A. Pert, and T. L. O'Donohue, Peptides *5*, 967 (1984).
133. J. J. Schildkraut, Am. J. Psychiatry *122*, 509 (1965).
134. W. E. Bunney and J. M. Davis, Arch. Gen. Psychiatry *13*, 483 (1965).
135. S. O. Ogren, K. Fuxe, L. F. Agnati, J. A. Gustafsson, G. Jonsson, and A. C. Holm, J. Neural Transm. *46*, 85 (1979).
136. A. J. Prange, Jr., I. C. Wilson, A. E. Knox, T. K. McClane, G. R. Greese, B. R. Martin, L. B. Alltop, and M. A. Lipton, J. Psychiatr. Res. *9*, 187 (1972).
137. B. N. Rosloff and J. M. Davis, Psychopharmacology (Berlin) *40*, 53, (1974).
138. P. Bevan, C. M. Bradshaw, and E. Szabadi, Br. J. Pharmacol. *55*, 17 (1975).
139. W. Lippman and T. A. Pugsley, Can. J. Physiol. Pharmacol. *54*, 494 (1976).
140. P. A. Baumann and L. Maitre, Naunyn-Schmiedebergs Arch. Pharmacol. *300*, 31 (1977).
141. R. M. Post, J. Kotin, and F. K. Goodwin, Am. J. Psychiatr. *131*, 511 (1974).
142. R. Raisman, M. Briley, and S. Z. Langer, Eur. J. Pharmacol. *54*, 307 (1979).
143. R. Raisman, M. Briley, and S. Z. Langer, Nature *281*, 148 (1979).
144. M. S. Briley, R. Raisman, and S. Z. Langer, Eur. J. Pharmacol. *58*, 347 (1979).
145. E. Costa, D. M. Chuang, M. L. Barbaccia, and O. Gandolfi, Experientia *39*, 855 (1983).
146. S. M. Stahl, J. Clin. Psychiatry *45*, 37 (1984).
147. T. C. Daniels and E. C. Jorgensen, in Wilson and Gisvold's Textbook of Organic Medicinal and Pharmaceutical Chemistry (R. F. Doerge, ed.), J. B. Lippincott, Philadelphia, 1982, p. 392.
148. D. S. Janowsky and B. Byerley, J. Clin Psychiatry *45*, 3 (1984).
149. A. Dumbrille-Ross and S. W. Tang, Mol. Pharmacol. *23*, 607 (1982).
150. M. Rehavi, Y. Ittah, P. Skolnick, K. C. Rice, and S. M. Paul, Naunyn-Schmiedebergs Arch. Pharmacol. *320*, 45 (1982).

151. A. Davis, J. M. Morris, and S. W. Tang, Eur. J. Pharmacol. *109*, 97 (1985).
152. M. Rehavi, S. M. Paul, P. Skolnick, and F. K. Goodwin, Life Sci. *26*, 2273 (1980).
153. S. M. Paul, M. Rehavi, K. C. Rice, Y. Ittah, and P. Skolnick, Life Sci. *28*, 2753 (1981).
154. A. Dumbrille-Ross, S. W. Tang, and D. V. Coscina, Life Sci. *29*, 2049 (1981).
155. M. Briley and S. Z. Langer, Eur. J. Pharmacol. *72*, 377 (1981).
156. G. Gross, M. Gothert, H.-P. Ender, and H.-J. Schumann, Naunyn-Schmiedebergs Arch. Pharmacol. *317*, 310 (1981).
157. N. Brunello, D. M. Chuang, and E. Costa, in Typical and Atypical Antidepressants: Molecular Mechanisms (E. Costa and G. Racagni, eds.), Raven Press, New York, 1982, p. 179.
158. M. Briley, R. Raisman, S. Arbilla, M. Casadamont, and S. Z. Langer, Eur. J. Pharmacol. *81*, 309 (1982).
159. M. Stanley, J. Virgilio, and S. Gershon, Science *216*, 1337 (1982).
160. P. D. Hrdina, Prog. Neuropsychopharmacol. Biol. Psychiatr. *7*, 457 (1983).
161. M. Rehavi, P. Skolnick, and S. M. Paul, Eur. J. Pharmacol. *87*, 335 (1983).
162. S. Z. Langer and R. Raisman, Neuropharmacology *22*, 407 (1983).
163. M. Stockert, A. Veiga, S. Butman, and E. DeRobertis, J. Recept. Res. *4*, 755 (1984).
164. P. D. Hrdina, Psychiatr. Res. *11*, 271 (1984).
165. P. D. Hrdina, Eur. J. Pharmacol. *102*, 481 (1984).
166. M. S. Briley, G. Fillion, D. Beaudoin, M.-P. Fillion, and S. Z. Langer, Eur. J. Pharmacol. *64*, 191 (1980).
167. W. J. Kinnier, D.-M. Chuang, and E. Costa, Eur. J. Pharmacol. *67*, 289 (1980).
168. S. Z. Langer, F. Javoy-Agid, R. Raisman, M. Briley, and J. Agid, J. Neurochem. *37*, 267 (1981).
169. M. Palkovits, R. Raisman, M. Briley, and S. Z. Langer, Brain Res. *210*, 493 (1981).
170. P. D. Hrdina, B. A. Pappas, R. J. Bialik, and C. L. Ryan, Eur. J. Pharmacol. *83*, 343 (1982).
171. N. Brunello, D.-M. Chuang, and E. Costa, Eur. J. Pharmacol. *78*, 383 (1982).
172. K. B. Asarch, J. C. Shih, and A. Kulcsar, Commun. Psychopharmacol. *4*, 425 (1980).
173. P. L. Wood, B. Suranyi-Cadotte, G. Schwartz, and N. P. V. Nair, Biol. Psychiatr. *18*, 715 (1983).
174. E. T. Mellerup, P. Plenge, and R. Rosenberg, Psychiatr. Res. *7*, 221 (1982).
175. H. O. Borbe and I. Zube, Brain Res. *264*, 178 (1983).
176. A. Davis, J. Morris, and S. W. Tang, Eur. J. Pharmacol. *86*, 353 (1983).
177. M. Rehavi, H. Tracer, K. Rice, P. Skolnick, and S. M. Paul, Life Sci. *32*, 645 (1983).
178. P. D. Hrdina, Prog. Neuropsychopharmacology *5*, 553 (1981).

179. M. Rehavi, P. Skolnick, M. J. Brownstein, and S. M. Paul, J. Neurochem. *38*, 889 (1982).
180. S. Z. Langer, R. Raisman, and M. Briley, Eur. J. Pharmacol. *72*, 423 (1981).
181. M Radulovacki and N. Micovic, Brain Res. *235*, 393 (1982).
182. C. M. Lee, J. A. Javitch, and S. H. Snyder, J. Neurosci. *2*, 1515 (1982).
183. A. Biegon and T. C. Rainbow, J. Neurosci. *3*, 1069 (1983).
184. M. Rehavi, P. Skolnick, B. Hulihan, and S. M. Paul, Eur. J. Pharmacol. *70*, 597 (1981).
185. R. Raisman, M. Sette, C. Pimoule, M. Briley, and S. Z. Langer, Eur. J. Pharmacol. *78*, 345 (1982).
186. N. Brunello, D. M. Chuang, and E. Costa, Science *215*, 1112 (1982).
187. A. Dumbrille-Ross, S. W. Tang, and P. Seeman, Eur. J. Pharmacol. *68*, 395 (1980).
188. S. J. Peroutka and S. H. Snyder, J. Pharmacol. Exp. Ther. *216*, 142 (1981).
189. P. M. Whitaker and A. J. Cross, Biochem. Pharmacol. *29*, 2709 (1980).
190. W. P. Burkard, Eur. J. Pharmacol. *61*, 409 (1980).
191. J. E. Taylor and E. Richelson, Eur. J. Pharmacol. *78*, 279 (1982).
192. S. H. Snyder and H. I. Yamamura, Arch. Gen. Psychiatr. *24*, 236 (1977).
193. J. P. Green and S. Maayani, Nature *269*, 163 (1977).
194. D. C. U'Prichard, D. A. Greenberg, P. P. Sheenan, and S. H. Snyder, Science *199*, 197 (1978).
195. J. L. Bennett and G. K. Aghajanian, Life Sci. *15*, 1935 (1975).
196. E. Costa, D. M. Chuang, M. L. Barbaccia, and O. Gandolfi, Experientia *39*, 855 (1983).
197. O. S. Brusov, A. M. Fomenko, and A. B. Katasonov, Biol. Psychiatr. *20*, 235 (1985).
198. S. Z. Langer, R. Raisman, D. Sechter, C. Gay, H. Loo, and E. Zarifian, in Frontiers in Biochemical and Pharmacological Research in Depression (E. Usdin, ed.), Raven Press, New York, 1984, p. 113.
199. W. F. Kofoe and B. E. Leonard, Arch. Int. Pharmacodyn. Ther. *206*, 389 (1973).
200. A. P. Zis and F. K. Goodwin, Arch. Gen. Psychiatr. *36*, 1097 (1979).
201. F. E. Soroko, N. B. Mehta, R. A. Maxwell, R. M. Ferris, and D. H. Schroeder, J. Pharm. Pharmacol. *29*, 767 (1977).
202. P. Loo, A. Braunwalder, J. Lehmann, and M. Williams, Eur. J. Pharmacol. *123*, 467 (1986).
203. E. H. K. Wong, J. A. Kemp, T. Priestley, A. R. Knight, and G. N. Woodruff, Proc. Natl. Acad. Sci. USA *83*, 7104 (1986).
204. J. A. Kemp, A. C. Foster, and E. H. F. Wong, Trends Neurosci. *10*, 294 (1987).
205. M. Williams, P. A. Loo, D. E. Murphy, A. F. Braunwalde, M. F. Jarvis, and M. A. Sills, J. Rec. Res. *8*, 195 (1988).

20

Calcium Channels and Their Receptor Sites

D. J. Triggle / State University of New York, Buffalo, New York

R. A. Janis / Miles Institute for Preclinical Pharmacology, West Haven, Connecticut

HISTORICAL PERSPECTIVE

The fundamental role for Ca^{2+} as a cellular messenger is made possible by the highly asymmetric distribution of free Ca^{2+} across the plasma membrane and by the existence of specific intracellular Ca^{2+} binding proteins. These binding proteins serve to translate the informational content of the cytoplasmic Ca^{2+} concentration, elevated during excitation from $<10^{-8}$ M to approximately 10^{-6} M, into biologic response. Thus Ca^{2+} plays a critical role in stimulus-response coupling processes (1-3).

That Ca^{2+} fulfills an intermediary role in many processes of cellular regulation demands that its movements and concentrations be constantly controlled. At the cellular level this is made possible by a variety of transport and sequestration processes localized at both plasma and intracellular membranes (Figure 1). In principle, a more perfect understanding of the various pathways regulating Ca^{2+} mobilization and Ca^{2+} function should lead to discrete classes of drugs that are capable of specifically modulating these pathways (4). One such class of compounds has achieved both therapeutic and pharmacological significance in recent years.

The group of compounds variously referred to as Ca^{2+} entry blockers, Ca^{2+} channel blockers, or Ca^{2+} channel antagonists is of substantial therapeutic significance particularly for disorders of the cardiovascular system (5-8). This chemically heterogeneous group of compounds, including the clinically available nifedipine, verapamil, and diltiazem, is currently used in the treatment of various anginal conditions, certain cardiac arrhythmias, hypertension, a number of peripheral vascular disorders, and myocardial preservation. How-

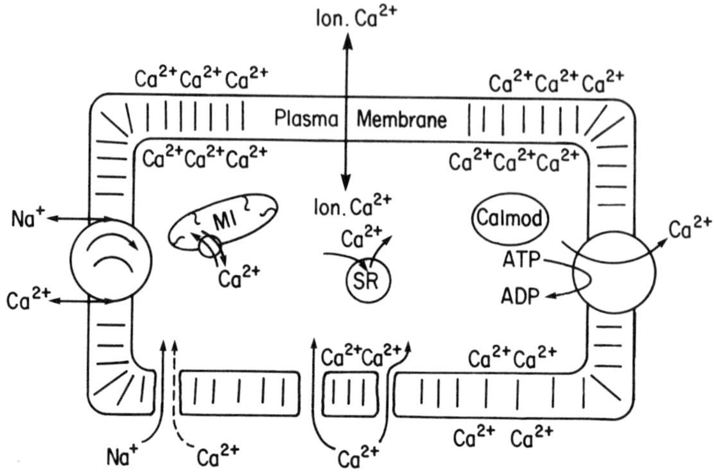

Figure 1 Schematic representation of cellular Ca^{2+} regulation depicting Ca^{2+} entry through the voltage-dependent Na^+ channel, the Ca^{2+} channels, Na^+-Ca^{2+} exchange, and ionophore-mediated processes. Intercellular sequestration release processes are depicted in mitochondria and sarcoplasmic reticulum, and Ca^{2+} targets are depicted as calmodulin. Ca^{2+} exit processes are depicted as Na^+-Ca^{2+} exchange and Ca^{2+}-ATPase.

ever, their potential use is wider and should, in principle, apply to the control of Ca^{2+} currents in all tissues. That these drugs exert their primary action at the potential-dependent Ca^{2+} channel of the plasma membrane has been demonstrated by direct electrophysiologic observations in several tissue categories (9-11). Evidence for a specific, rather than a nonspecific, mode of action was early recognized by the existence of distinct structure-activity relationships described for the major classes of agents (4,12-15). Nonetheless, the chemical and pharmacological heterogeneity of these agents and the absence of an inclusive structure-activity relationship indicates that they may share the common property of Ca^{2+} channel antagonism but likely act at different sites and by different mechanisms. Recently, a further dimension of interest has been added by the discovery of 1,4-dihydropyridine analogs of nifedipine that serve as potent activators of the Ca^{2+} channel. Bay K 8644 [Figure 2; 1,6-dimethyl-3-carbomethoxy-5-nitro-2-(trifluoromethylphenyl)-1,4-dihydropyridine] is a potent example of this type of agent (16). Of particular interest are the observations that in this type of compound, opposing pharmacological activities, activation, and antagonism are associated with enantiomeric species (Figure 2) (17,18).

In the past several years these ligands have proved to be valuable molecular probes of the voltage-dependent Ca^{2+} channel. The availability of an increasing

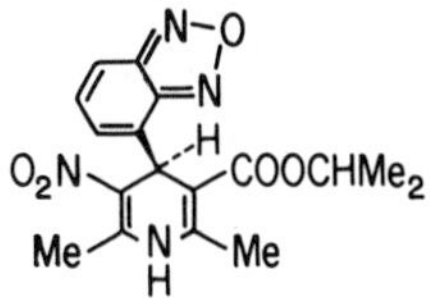

(+) S 202-791

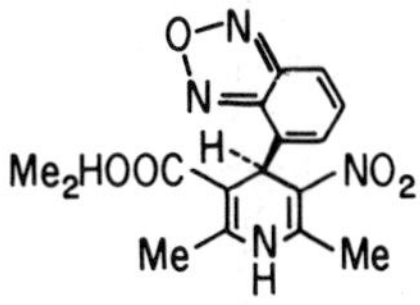

(−) R 202-791

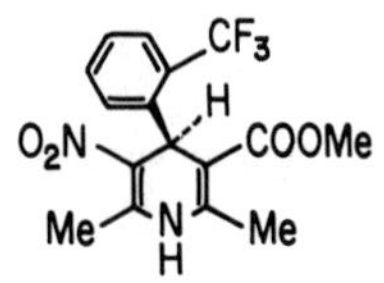

(−) S Bay k 8644

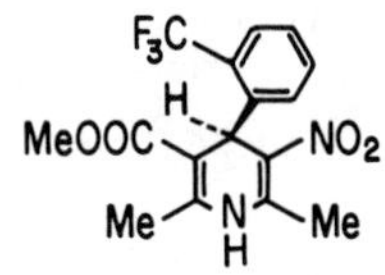

(+)R Bay k 8644

Figure 2 Stereoisomeric pairs of 1,4-dihydropyridine activator-antagonists.

number of these compounds as tritium- or iodine-labeled species of high specific activity (Figure 3) has been a major factor in the rapidly developing interest in Ca^{2+} channels. The availability of these radioligands has permitted approaches to the following questions:

1. Do Ca^{2+} channel ligands have specific sites of action?
2. What is the localization of these sites?
3. Are the sites different for each ligand class, and what are their mutual interactions?
4. What are the relationships of the binding sites to Ca^{2+} channel function and to the process of Ca^{2+} permeation?
5. What is the basis for the apparent tissue selectivity of action exhibited by available Ca^{2+} channel ligands?

RADIOLIGAND BINDING

Definition of Specific Binding

Several [^{3}H] 1,4-dihydropyridines, including both antagonist and activator structures, [^{125}I] 1,4-dihydropyridine and [^{3}H] verapamil, [^{3}H] desmethoxyverapamil, and [^{3}H] diltiazem have been used (Figure 3) in radioligand binding studies of the Ca^{2+} channel in excitable tissues (for general reviews, see References 19-21). The basic technique employed has been a rapid filtration over glass fiber filters, employing in the majority of cases tissue homogenates or microsomal fractions. Most studies have employed a tris(hydroxymethyl)amino-

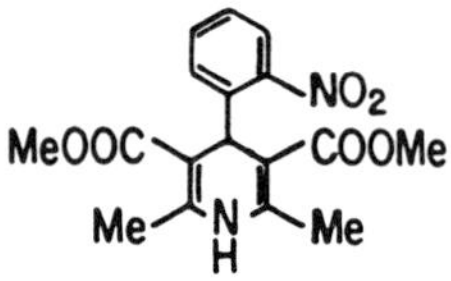

Nifedipine

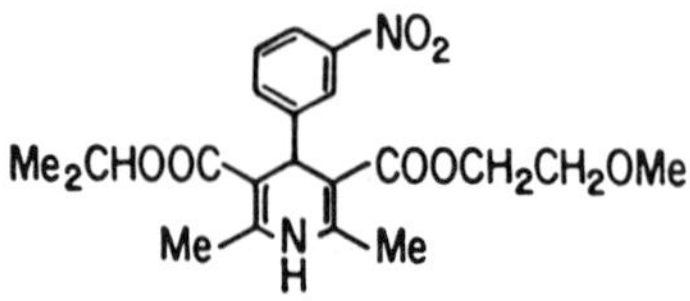

Nimodipine

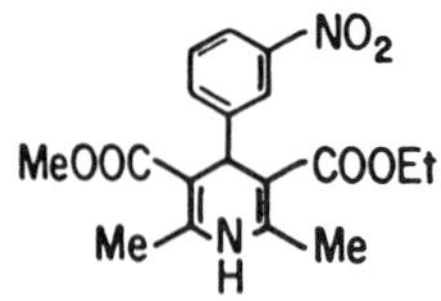

Nitrendipine

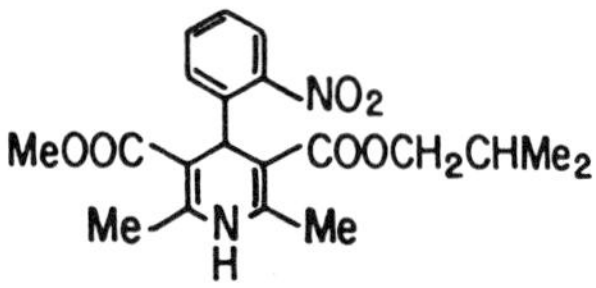

Nisoldipine

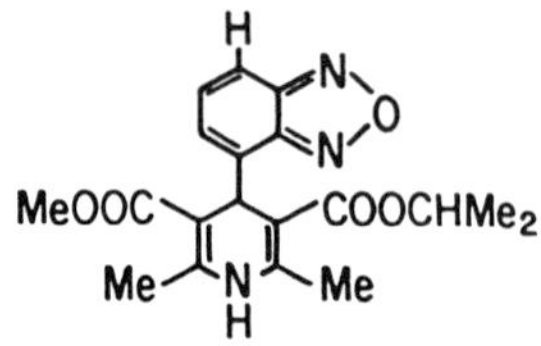

PN 200 110

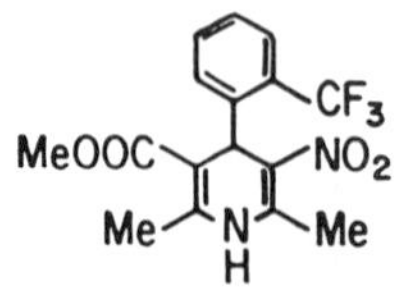

Bay K 8644

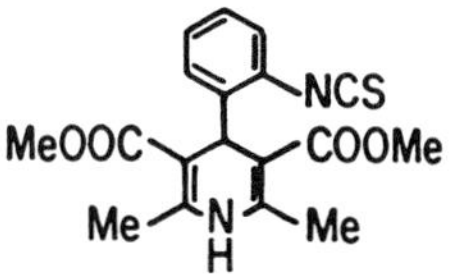

O-NCS-Nifedipine

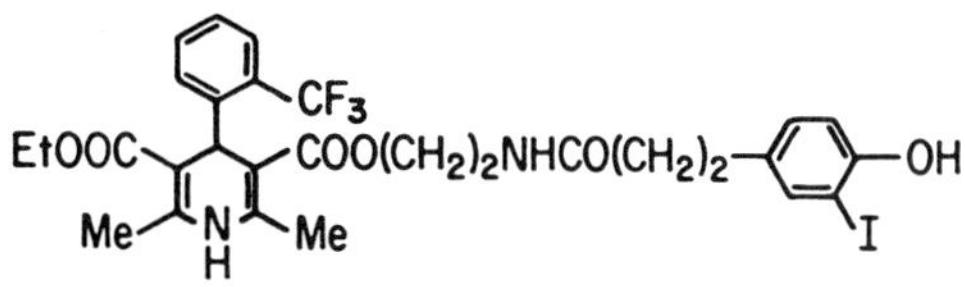

Iodipine

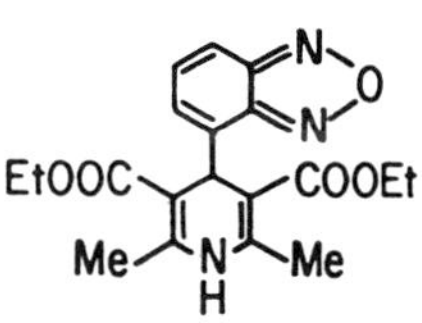

PY108 068

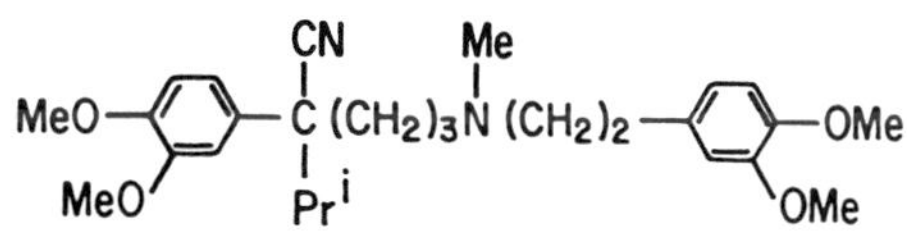

Verapamil

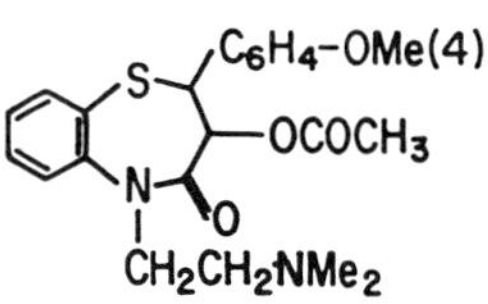

Diltiazem

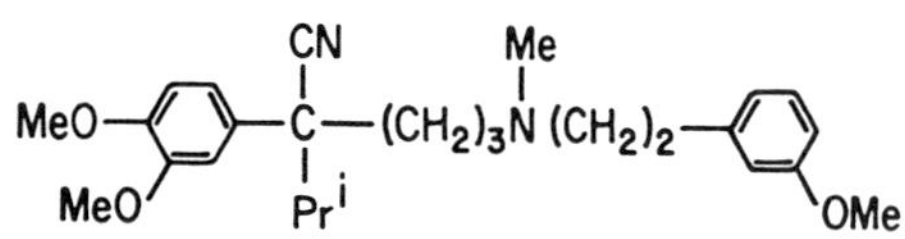

Desmethoxyverapamil

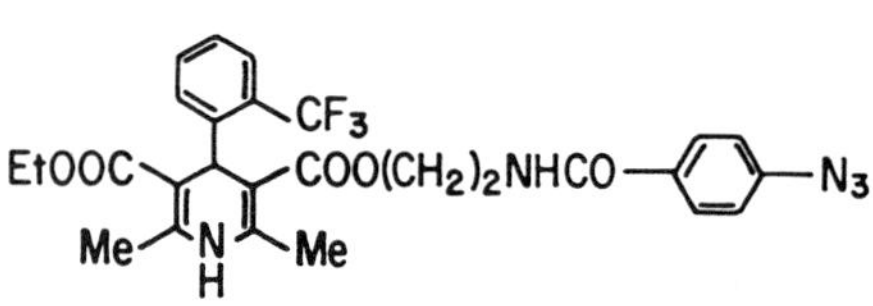

Azidopine

Figure 3 Structural formulas of radiolabeled Ca^{2+} channel ligands. With the exception of Bay K 8644, these are all Ca^{2+} channel antagonists.

methane (Tris) buffer at pH 7.0-7.4, at 20-37°C with a dilute tissue suspension and 10^{-7}-10^{-5} M unlabeled ligand with which to define nonspecific binding. [^{3}H] 1,4-dihydropyridines have been most often employed, particularly [^{3}H] nitrendipine, the first commercially available ligand. Only a comparatively few studies have employed the non-1,4-dihydropyridine ligands verapamil, desmethoxyverapamil, and diltiazem.

Equilibrium binding data have been summarized previously (4,19-21), and these references should be consulted for listings of specific details. Most binding studies have employed [^{3}H] 1,4-dihydropyridines (22-90). Saturable high-affinity binding found with the [^{3}H] 1,4-dihydropyridines in smooth, neuronal, and cardiac preparations is of high affinity, 10^{-9}-10^{-10} M, save in a number of instances in which very high protein concentrations were used in the binding assay. Most reports describe only one category of binding site ($n_H \cong 1.0$), but in many instances only a limited concentration range of radioligand has been employed. Binding densities range from approximately 10 to 1500 fmol/mg protein for homogenate and microsomal preparations: the lower densities are found in vascular smooth and some neuronal cell lines. [^{3}H] 1,4-dihydropyridine binding to skeletal muscle preparations is consistently of lower affinity (1.5-8.0×10^{-9} M) but of much higher density, approximately 50 to 1000fold, than that observed in smooth and cardiac muscle and neuronal preparations. Skeletal muscle thus differs both in possessing lower affinity sites and in possessing an extremely high density of binding sites. Binding of the activator ligand [^{3}H] Bay K 8644 is also of significantly lower affinity. Only limited studies of [^{3}H] verapamil (91-98), [^{3}H] desmethoxyverapamil (40,69,95,98), and [^{3}H]-diltiazem (99,100) binding are available: these ligands appear to be of generally lower affinity than the 1,4-dihydropyridines.

The specificity of the binding process has been examined in many, but not all, studies. Quite generally, specific binding is sensitive to pharmacologically defined Ca^{2+} channel ligands of various structural categories and is insensitive to or weakly sensitive to a variety of receptor-directed ligands and ligands for Na^+ and K^+ channels.

Reversibility of Binding

Reversibility of ligand binding has been demonstrated in many preparations, but detailed rate measurements are frequently unavailable. However, the majority of available studies (Table 1) indicate single association and dissociation processes, and binding constants calculated from these rate studies are generally in good agreement with those derived from equilibrium studies.

However, more detailed analyses of the kinetics of binding suggest that the simple bimolecular association model may be an oversimplification. A detailed analysis of the kinetics of binding of (±)-nitrendipine and (+)-PN 200

Table 1 Rate Constants for Radioligand Binding to Ca^{2+} Channels

Ligand	System	Temperature (°C)	k_1 (M/min^{-1})	k_{-1} (min^{-1})	Refs.
[^{3}H]Nitrendipine	Guinea pig ileum	25	6.0×10^8	0.06	43
	Bovine aorta	25	1.84×10^8	0.018	50
	Rabbit uterus	4	5.0×10^7	0.06	46
	Rabbit heart	25	4.4×10^8	0.048	26, 52
	Canine heart	25	2.44×10^8	0.035	52
	Rat heart	37	6.5×10^8	0.333	86
	Guinea pig heart	37	$1.0\text{-}1.5 \times 10^7$	0.0074–0.028	22
	Guinea pig heart	15	–	0.01	103
	Rabbit muscle,	37	–	0.289	57
	t tubules	10	1.0×10^8	0.10	
	Rabbit muscle, t tubules (soluble)	4	7.8×10^6	0.066	104
	Rat brain, synaptosomes	25	–	0.032	61
	Rat brain, synaptosomes	25	4.8×10^8	0.034	64
	Rat brain, membranes	25	Biphasic	Biphasic	65[a]
[^{3}H]Nimodipine	Rat heart	37	3.7×10^8	0.23	86
	Human heart	37	8.2×10^8	0.40	70
	Guinea pig brain	37	3.2×10^8	0.18	75
	Rat brain	37	–	0.099	79
	Guinea pig, skeletal microsomes	37	–	0.343	72
[^{3}H]PN 200 110	Rat brain	25	Biphasic	0.004-0.006	65[a]
	Rabbit muscle, t tubules (soluble)	4	1.2×10^8	0.015	105
[^{3}H]Nifedipine	Rabbit heart	25	7.7×10^7	0.052	84
	Rabbit heart	37	3.1×10^8	0.49	86
[^{3}H]Verapamil	Cardiac sarcolemma	25	1×10^8	5.8	93
	Rabbit muscle, t tubules	10	1.98×10^6	0.096	95
	Guinea pig, skeletal microsomes	2	$\sim 2 \times 10^6$	Biphasic	96[a]
[^{3}H]Desmethoxy-verapamil	Guinea pig, skeletal microsomes	25	5×10^7	0.032	98
	Bovine heart, sarcolemma	4	2.1×10^9	1.08	69
[^{3}H]Diltiazem	Guinea pig, skeletal microsomes	2	5.3×10^5	0.02	100
[^{125}I]Iodipine	Guinea pig, Skeletal microsomes	25	1.0×10^8	0.06	101, 102

[a]See text for discussion.

110 reveals that both ligands label an apparently identical high-affinity population of sites in brain membranes but that the onset kinetics are biphasic (65). Additionally, in the case of nitrendipine the dissociation kinetics are biphasic, consistent with the two optical isomers binding with the same, or closely similar, association rates but having different dissociation rates. Several kinetic schemes can accommodate ligand binding with biphasic and monophasic association and dissociation rates, respectively: in particular, schemes involving isomerization of the binding protein before and after binding are of interest (Figure 4). The best data fit was obtained with scheme C, in which the binding protein exists in two states with high (R′) and low (R) affinity for the ligands under the experimental conditions employed. Rhodes et al. (106) have noted the high partition coefficients of 1,4-dihydropyridine ligands (107) and suggest that these ligands likely interact with their binding protein via a membrane approach rather than via a three-dimensional diffusion from the aqueous phase.

Inadequate data are available for ligands of other structural classes. However, the binding of [^{3}H] verapamil to skeletal muscle is reported to have monophasic association and biphasic dissociation kinetics (96). This may reflect the different dissociation rates of the stereoisomers of verapamil.

Stoichiometry of Binding

Two major issues need to be resolved. Does the binding site capacity in excitable tissues differ among ligands within a given structural class, and does binding

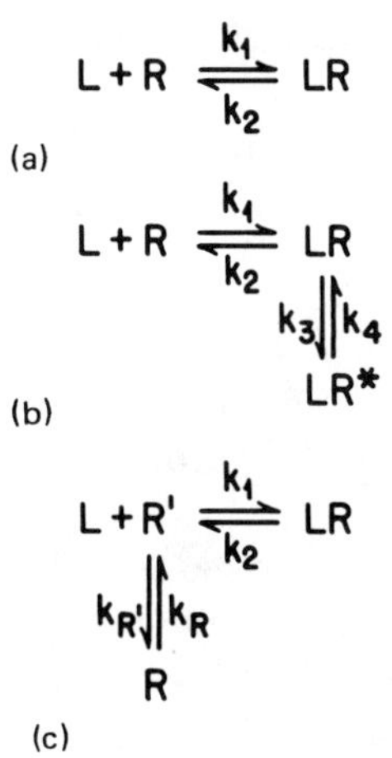

Figure 4 Representation of ligand binding to channel proteins. (a) Simple bimolecular binding. (b) Isomerization ($R \rightleftharpoons R^*$) after binding. (c) Isomerization ($R \rightleftharpoons R'$) prior to binding. [After Weiland and Oswald (65).]

capacity differ according to structural classes? These are important issues for they reflect on the one hand the relative ability of Ca^{2+} channel ligands to stabilize different states of the Ca^{2+} channel and, on the other hand, the relative stoichiometry of the binding sites for the different structural classes of Ca^{2+} channel ligands.

Unfortunately, given the variation in protocols between laboratories this information is not to be obtained by a simple literature survey (4,19-21,108, 109,111) as truly comparative studies are rare. The available data suggest that, with the possible exception of skeletal muscle, the 1,4-dihydropyridines label equivalent numbers of sites in membrane preparations, and it is probable that the ratio of verapamil, diltiazem, and dihydropyridine sites is 1:1:1. Whether this is true for all tissues remains to be established, and in skeletal muscle Glossmann et al. argue (96,102) that the relative stoichiometry is 1,4-dihydropyridine-phenylalkylamine-diltiazem is 2:4:1.

Distribution of Binding Sites

Binding sites for Ca^{2+} channel ligands enjoy a very widespread distribution: they are found in virtually all excitable tissues from animal species and, very recently, have been characterized in some plant tissues (122,123). The latter observations are of interest because calcium channel antagonists block Ca^{2+}-dependent cytokinin-induced responses (124).

The subcellular localization of binding sites has been studied quite extensively in smooth, cardiac, and skeletal muscle. The distribution of [^{3}H] 1,4-dihydropyridine binding sites has been examined in three smooth muscle systems, rat uterus and stomach and guinea pig ileum (45,48,125). In rat uterus and stomach fundus smooth muscles there was an excellent correlation between [^{3}H] nitrendipine binding site distribution and the occurrence of the plasma membrane markers K^+-activated ouabain-sensitive *p*-nitrophenylphosphatase, 5′-nucleotidase, phosphodiesterase, and Mg-ATPase but not between mitochondrial or endoplasmic reticulum markers. A similar localization of [^{3}H] nitrendipine binding sites is observed in guinea pig ileal longitudinal muscle, as measured either by [^{3}H] nitrendipine or by labeling with the putative irreversible ligand 2,6-dimethyl-3,5-dicarbomethoxy-4-(2-isothiocyanophenyl)-1,4-dihydropyridine (125).

In cardiac muscle, binding of [^{3}H] 1,4-dihydropyridines has been reported as preferentially associated with sarcolemma (114), but subsequent work (31, 68) indicated that the binding sites did not copurify with solely plasmalemmal markers. It is likely that the binding sites for nitrendipine and other Ca^{2+} channel ligands in cardiac tissue are asymmetrically localized on plasmalemmal regions physically linked to junctional sarcoplasmic reticulum (28). Two fractions of ventricular microsomes that bound nitrendipine were found, with the major population following the distribution of markers for junctional sarcoplasmic reticulum. This distribution does not parallel that for muscarinic receptors, which are preferentially associated with light microsomes (28).

Electrophysiologic experiments indicate that Ca^{2+} channels in skeletal muscle are restricted to the transverse tubules (126-128). Consistent with this, the transverse tubules are also the site of preferential localization of Ca^{2+} channel ligands as judged by a variety of biochemical and immunohistochemical markers (57,113,129,130). The extremely high density of binding sites found in skeletal muscle t-tubule preparations and their comparatively ready availability makes this tissue fraction an attractive candidate for Ca^{2+} channel purification studies.

Verapamil and diltiazem binding is also associated with the t tubules of skeletal muscle (95,96,108). However, verapamil binding to cardiac preparations seems to indicate a dual location (69,93): with crude homogenates the ratio of verapamil to nitrendipine binding sites is 4:1, but in purified membranes the ratio becomes 1:1 (93).

Interactions Between Binding Sites

It has long been suspected that the chemical, pharmacological, and therapeutic heterogeneity of the major Ca^{2+} channel antagonist structures is consistent with their interaction at different sites at which to regulate Ca^{2+} channel function. This suggestion has been considerably strengthened by the radioligand binding studies, which have led to the generally accepted conclusion that these structures occupy three discrete but allosterically linked binding sites at, or in close association with, the Ca^{2+} channel (for general reviews, see References 15 and 19-21).

Verapamil, desmethoxyverapamil, and D600, as representatives of the phenylalkylamine class of antagonist, interact with [^{3}H] 1,4-dihydropyridine binding sites in negative heterotropic fashion in cardiac muscle (26,34,37,70, 103), smooth muscle (43), neuronal tissue (35,37,59,64,75,83,131), and skeletal muscle (72,104,112,113). From these studies a general picture of site-site interactions has been established. Verapamil and structurally related phenylalkylamines serve as stereoselective partial inhibitors of 1,4-dihydropyridine binding and function by reducing the affinity of the 1,4-dihydropyridine binding site by increasing k_{-1} without change in B_{max} (Table 2). However, the detailed interpretation of these data presents a richly complex picture involving the interactions of high- and low-affinity binding states for phenylalkylamines with high- and low-affinity binding states for 1,4-dihydropyridines (21,76,102). In smooth and cardiac muscle, the stereoselectivity [(–) > (+)] of verapamil, desmethoxyverapamil, and D600 activity in displacing bound 1,4-dihydropyridines parallels the stereoselectivity of pharmacological activities and the stereoselectivity of displacement of bound desmethoxyverapamil. For neuronal tissue and skeletal

Table 2 Effects of Verapamil (Phenylalkylamines) and Diltiazem on 1,4-Dihydropyridine Binding

	Control				Verapamil			Diltiazem			
Preparation and ligand	B_{max} (fmol/mg)	K_D (x 10^{-9} M)	k_{-1} (min^{-1})	Temperature (°C)	B_{max} (fmol/mg)	K_D (x 10^{-9} M)	k_{-1} (min^{-1})	B_{max} (fmol/mg)	K_D (x 10^{-9} M)	k_{-1} (min^{-1})	Ref.
Brain, [^{3}H] Nitr[a]	114	0.17	0.036	25	–	–	–	111	0.12	0.012	61
	107	0.12	–	0	116	0.34	–	119	0.22	–	64
	122	0.47	–	37	240	3.12	–	258	0.12	–	64
Heart, [^{3}H] Nitr	250	0.19	0.059	37	250	0.75	0.125	260	0.16	0.036	26
	120	0.23	–	25	120	∿0.90	–	–	–	–	34
	170	1.2	0.047	25	–	–	–	170	0.70	0.025	110
	95	0.28	0.37	37	–	–	2.66	132	0.23	–	70
	300	0.35	0.34	–	–	–	–	300	0.35	0.017	69
Smooth muscle, [^{3}H] Nitr	1140	0.16	–	25	–	–	–	1600	0.17	–	
	1340	0.20	–		1300	0.48	–				43
Skeletal muscle, [^{3}H] Nitr	2100	1.5	1.5	37	–	–	–	10200	0.99	0.34	72
	11700	1.0	1.36	37	–	–	–	57900	1.3	0.38	113
	2100	1.5	1.5	37	–	–	–	10000	1.0	0.34	112

[a]Nitr = nitrendipine.

muscle more complex pictures have been suggested in which the stereoselectivity of inhibition of 1,4-dihydropyridine binding may depend upon the experimental conditions. Thus, in skeletal muscle (72,104,112,113) (+)-D600 was much more potent ($>10^4$) than (–)-D600 in displacing [^{3}H] nimodipine, but stereoselectivity was inverted in the presence of diltiazem.

Diltiazem serves as a stereoselective positive heterotropic effector of 1,4-dihydropyridine binding, serving to increase either B_{max} or to increase the affinity of binding by reducing k_{-1} (Table 2). These effects have been observed in cardiac muscle (26,69,103,110), smooth muscle (30,43), neuronal tissue (37,61,64, 131), and skeletal muscle (72-74,112,113). Diltiazem thus serves to increase the high-affinity 1,4-dihydropyridine binding population. This action is stereoselective, *d-cis*-diltiazem > *l-cis*-diltiazem, consistent with the pharmacological activities (132); furthermore, the effects of diltiazem are highly temperature dependent, being more apparent at 37°C than at room temperature (21,76,102).

The heterotropic interactions of diltiazem and phenylaklylamines have been analyzed primarily from the perspective of 1,4-dihydropyridine binding. A smaller number of studies have measured the effects of 1,4-dihydropyridines on diltiazem and phenylalkylamine binding. The results are generally confirmatory of the conclusions drawn from [^{3}H] 1,4-dihydropyridine binding. In cardiac membranes, nimodipine serves as a negative allosteric regulator of [^{3}H] desmethoxyverapamil binding (69). In skeletal muscle and brain, binding of [^{3}H]-desmethoxyverapamil is inhibited by 1,4-dihydropyridines, and diltiazem by a reduction in B_{max} and an increase in K_D, respectively (98,133). Binding of [^{3}H] *d-cis*-diltiazem is regulated in both skeletal muscle and brain membranes (99,100). In skeletal muscle (100), binding is enhanced by 1,4-dihydropyridines by a B_{max} effect at 37°C, but at 2°C inhibition is observed: at 2°C more sites are labeled by diltiazem than at 37°C, and this may reflect a temperature-dependent equilibration between two affinity states. In brain (99), dihydropyridines also regulate diltiazem binding in a temperature-dependent manner, inhibiting binding at 0°C and enhancing binding at 30 and 37°C. These stimulant effects are mediated by an increase in affinity of diltiazem. In this preparation the effects of the Ca^{2+} channel activator Bay K 8644 were of particular interest, since at 0°C it exhibited high-potency inhibition (IC_{50} = 3.8 nM) but at 37°C it exhibited low-potency stimulation (IC_{50} = 300 nM) of [^{3}H] diltiazem binding. In this respect, Bay K 8644 contrasted with other 1,4-dihydropyridines.

SELECTIVITY AND STRUCTURE-ACTIVITY RELATIONSHIPS

Most studies of Ca^{2+} channel radioligand binding have made some effort to determine the specificity of the radioligand binding process by measuring the sensitivity of this process to ligands of the various structural classes. Comprehensive studies of the structure-activity relationship of radioligand binding are rare as studies of the relationship between radioligand binding and pharmacological activities (4,19-21,52). To date, the most comprehensive of these studies are with [^{3}H] 1,4-dihydropyridines.

Examples of correlations between binding and pharmacological activities for 1,4-dihydropyridine series in smooth and cardiac muscle and neuronal membranes are shown in Figure 5 (26,43,52,79). Several important conclusions are permissible. There is an excellent linear and virtually 1:1 correlation between binding (inhibition of [^{3}H] nitrendipine) and pharmacology (inhibition of depolarization-induced tension response) in guinea pig ileal smooth muscle (43). Similar correlations, but on a more limited basis, are available in other smooth muscle systems (32,46,52,53). There is a similarly good linear correlation for cardiac tissue, but the relationship is not 1:1, the apparent dissociation reflecting high-affinity binding and the low-affinity cardiac pharmacology that underlie the smooth muscle selectivity expressed by the 1,4-dihydropyridines. However, the rank order correlation is virtually identical in smooth and cardiac muscle, and a similar high-affinity binding is expressed in both tissues (Figure 5c). This suggests that the same strucutre-activity relationship is being observed in binding and pharmacological studies in both cardiac and smooth muscle. Very similar binding sites for 1,4-dihydropyridines are found in brain preparations, the affinities of which correlate excellently with functional properties in smooth muscle (Figure 5d).

Ca^{2+} channel ligands are frequently observed in both clinical and experimental situations to be weakly active in neuronal preparations (58). The available data indicate that an apparently common (or very similar) high-affinity binding site for 1,4-dihydropyridines is expressed in membranes from excitable cells of cardiac, smooth, and neuronal tissues and that the extent to which this binding correlates with observed pharmacological properties is tissue dependent and may well depend upon factors that couple binding state to channel function. Skeletal muscle generally binds 1,4-dihydropyridines with lower affinity, but the same structure-activity relationship is expressed (19-21,72,102, 112,113). Stereoselectivity of 1,4-dihydropyridine interaction observed in pharmacological experiments (17,18,116,117) is maintained in binding states (Table 3).

As yet comparatively few studies characterize in detail the binding of verapamil, desmethoxyverapamil, and diltiazem. Binding of verapamil and [^{3}H] desmethoxyverapamil is sensitive to phenylalkylamine derivatives in a stereoselective fashion (Table 4). The stereoselectivity observed in cardiac muscle, (−)-verapamil > (+)-verapamil, (−)-D600 > (+)-D600, and (−)-desmethoxyverapamil > (+)-desmethoxyverapamil, accords with the stereoselectivity observed in working hearts, and there is also good agreement with the

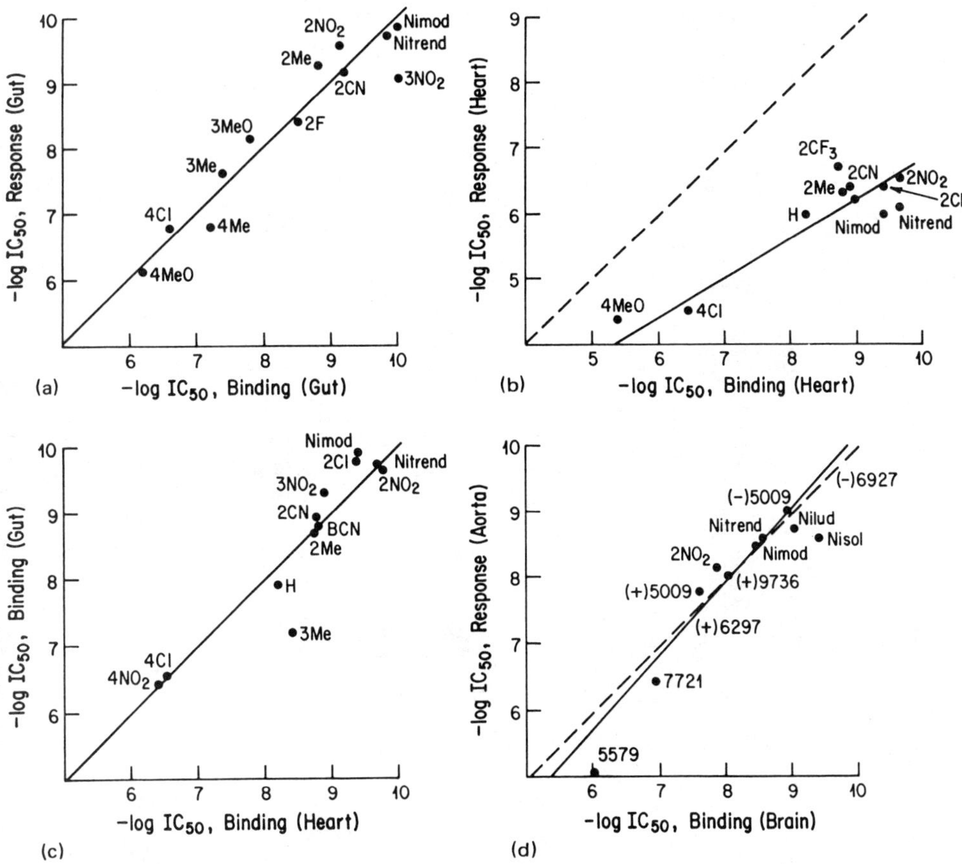

Figure 5 Correlations between binding and pharmacological behavior of 1,4-dihydropyridines. (a) Binding of a series of 1,4-dihydropyridines (substituted nifedipine analogs; 2-NO_2 = nifedipine) to guinea pig intestinal smooth muscle membranes and inhibition of response (tonic component) to K^+ depolarization. (From Reference 43.) (b) Binding to rabbit heart membranes and inhibition of response of stimulated cat papillary muscle. (From References 13 and 26.) (c) Binding to rabbit heart and guinea pig intestine membranes. (From References 13, 26, and 43.) (d) Binding to rat brain membranes and inhibition of response of rabbit aorta to K^+ depolarization. (From Reference 79.) Filled lines are regression and dashed lines indicate unit slope.

Table 3 Competition for [^{3}H] 1,4-Dihydropyridine Binding by Enantiomeric 1,4-Dihydropyridines: K_1 (or IC_{50}) Value (X 10^{-9} M)

		Nitrendipine		Nimodipine		Bay e 6927		Nicardipine		Bay K 8644		PN 202-791		PN 200 110		
Species and tissue	[^{3}H] ligand[a]	(–)	(+)	(–)	(+)	(–)	(+)	(–)	(+)	(-)	(+)	(–)	(+)	(–)	(+)	Refs.
Rat brain	[^{3}H] Nim	0.43	8.8	1.04	2.4	0.09	27.5									79
Pig brain	[^{3}H] Nim	2.0	30.0	1.0	30.0	0.1	30.0									85
Rat brain	[^{3}H] PN 200-110							2.90	0.37					14.9	0.06	37
Rat brain	[^{3}H] PN 200-110													10.9	0.09	83
Guinea pig brain	[^{3}H] Nim							28	2.3							21
Rat brain	[^{3}H] Nim					0.01	6.9									32
Rabbit heart	[^{3}H] Nit					0.006	2.4									26
Canine heart	[^{3}H] Nit					0.03	2.8									114
Bovine aorta	[^{3}H] Nit					0.02	15									50
Guinea pig ileum	[^{3}H] Nit					0.01	1.1									43
Rat tail artery	[^{3}H] Nit									7.2	1.2	0.14	55			115
Rabbit skeletal muscle	[^{3}H] Nim													206	2.2	72, 112, 113

[a]Nim = Nimodipine; Nit = Nitrendipine.

Table 4 Inhibition of [^{3}H] Verapamil or [^{3}H] Desmethoxyverapamil Binding by Phenylalkylalamine Derivatives

Ligand	Skeletal muscle t tubules ($IC_{50} \times 10^{-9}$ M)				Cardiac muscle ($IC_{50} \times 10^{-9}$ M)		
Temperature (°C)	10	2	10	25	37	25	4
(–)-Verapamil		77.0	40				21
(+)-Verapamil		30	10				556
(±)-Verapamil	25	–			50		
(–)-D600		15.0	20	12		36	21
(+)-D600		67.0	60	55		7000	560
(±)-D600	70	–					
Tiapamil		221.0					
D890							Inactive
(–)-Desmethoxy-verapamil				3			10
(+)-Desmethoxy-verapamil				17			280
(±)-D619			5000	2855			
Bepridil	40					2300	
d-cis-Diltiazem	200	140	200	215			
d-cis-Diltiazem		14000	4000	6900			
Reference	95	96	97	98	88	93	69

absolute activities [IC_{50} values for (–)verapamil and (–)-D600 are approximately 10^{-8} and 5×10^{-9} M] with stereoselectivity ratios, that is, (–) to (+) of 5-100 (70,118,121).

Few studies of binding of [^{3}H] diltiazem are available. In brain and skeletal muscle (99,100), binding is stereospecific (*d-cis*-diltiazem > *l-cis*-diltiazem) and is sensitive to other structural classes of Ca^{2+} channel ligands. In skeletal muscle, binding is highly temperature dependent, an approximately fivefold decrease in B_{max}, with no change in K_D, being measured as temperature goes from 2 to 37°C.

RADIOLIGAND BINDING AND Ca^{2+} CHANNEL FUNCTION AND REGULATION

Function of Ligand Binding Sites

There can be little doubt but that the binding sites identified in the majority of studies represent a component of the voltage-dependent Ca^{2+} channel. The location of these binding sites and the state(s) of these binding sites in relationship to the functional Ca^{2+} channel machinery remain, however, to be determined.

Cation Dependence of Ligand Binding

That [^{3}H] 1,4-dihydropyridine binding was dependent upon low concentrations of divalent cation was an early observation of substantial interest (60,134). High-affinity binding of [^{3}H] 1,4-dihydropyridines is dependent upon the presence of micromolar concentrations of Ca^{2+} or Mg^{2+}. Monovalent cations are without major effect on binding (25,60,134). Thus in the incubation media commonly employed there is sufficient contaminating M^{2+} to ensure maximum or nearly maximum binding. The dependence of 1,4-dihydropyridine binding has been studied extensively in brain (34,60), smooth muscle (135), and heart (40,135) with broadly similar results. In M^{2+}-depleted membrane preparations, [^{3}H] nitrendipine binding is restored in the sequence $M^{2+} > M^{3+}$, with a general trend among divalent cations for restoration of maximum binding to be optimum at ionic radii similar to that of Ca^{2+} and for binding to decrease with increasing charge density (Figure 6). The majority of investigations of the cation dependence of 1,4-dihydropyridine binding have focused only upon the availability of the high-affinity binding site, with the general assumption that an interconversion from a low-affinity to a high-affinity binding site is dependent upon M^{2+}. In chick heart membranes this interconversion has been quantitated (Table 5).

The relationship of this high-affinity divalent cation stabilization of ligand binding to channel function remains to be determined. High-affinity 1,4-dihydropyridine binding is blocked by calmodulin antagonists (Figure 7) (135) with the same potency sequence for inhibiting calmodulin-dependent phosphodiesterase activity. Possibly the 1,4-dihydropyridine binding site is associated with a cation binding protein that resembles calmodulin. Recent descriptions of a permeation mechanisms of Ca^{2+} channels indicate that they involve high-affinity M^{2+} binding sites (136,137). However, [^{3}H] 1,4-dihydropyridine binding to skeletal muscle t tubules does not depend upon the presence of divalent cations (57,108), although cation binding sites would be assumed to be a general

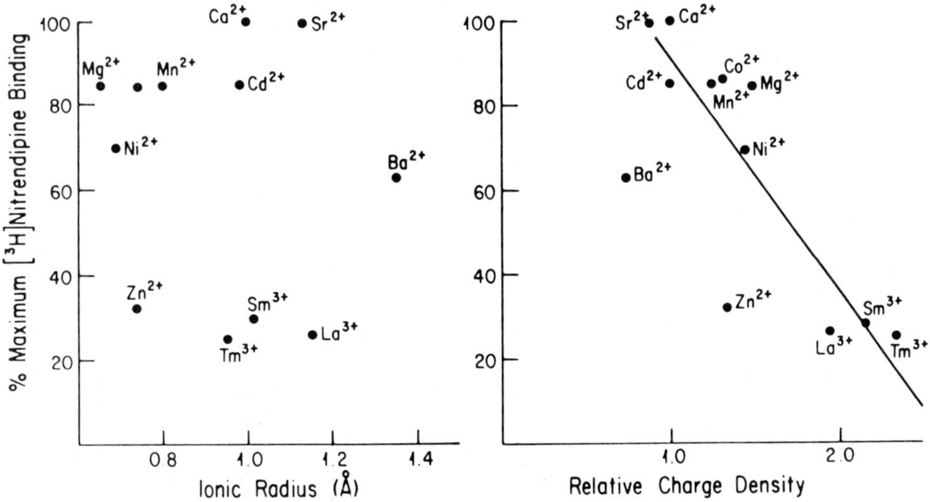

Figure 6 Dependence of maximum restoration of [^{3}H] nitrendipine binding in intestinal smooth muscle membranes upon (left) ionic radius of added cation and (right) relative charge density (Ca^{2+} = 1.0). [Data from Luchowski et al. (135).]

Table 5 Cation Dependence of 1,4-Dihydropyridine Binding in Chick Heart[a]

	M^{2+} depleted[a]				M^{2+} added			
Radioligand	K_H	K_L	B_{max}(H)	B_{max}(L)	K_H	K_L	B_{max}(H)	B_{max}(L)
(+)-PN 200 110	0.1	2.70	0.19	0.70	0.1[b]	–	1.4[b]	–
Nitrendipine	0.1[c]	5.2[c]	0.27[c]	1.56[c]	0.18[d]	–	1.6[d]	–

[a]Cation depleted with EDTA (10^{-3} M). Values for K_H and K_L are expressed × 10^{-9} M. B_{max}(H) and (L) are in picomoles per milligram protein.
[b]Mg^{2+}, 50 μM.
[c]Ca^{2+}, 10 μM.
[d]Ca^{2+}, 250 μM.

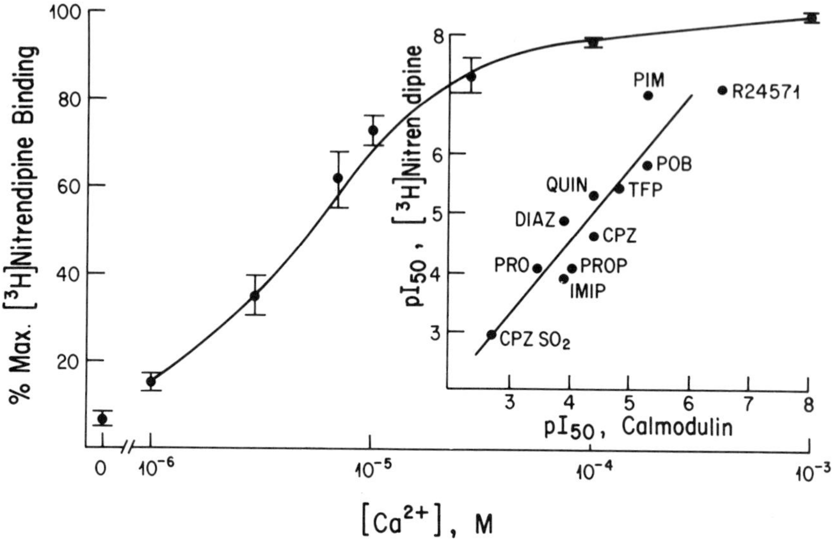

Figure 7 Dependence in guinea pig intestinal smooth muscle membranes of maximum [^{3}H] nitrendipine binding on added Ca^{2+}. Inset, inhibition of [^{3}H]-nitrendipine binding by a series of calmodulin antagonists as a function of their ability to inhibit calmodulin-dependent phosphodiesterase (PIM, pimozide; POB, phenoxybenzamine; TFP, trifluoperazine; QUIN, quinacrine; DIAZ, diazoxide; CPZ, chlorpromazine, PROP, propranolol; PRO, promethazine, IMIP, imipramine; $CPZSO_2$, chlorpromazine sulfoxide). [Data from Luchowski et al. (135).]

component of Ca^{2+} channels. A requirement for divalent cations is, however, observed in the solubilized state (112).

A similar contrast between skeletal and cardiac muscle binding is produced by the cation dependence of verapamil, desmethoxyverapamil, and diltiazem binding. In skeletal muscle t tubule preparations, binding is inhibited by divalent cations (97,108), whereas in cardiac muscle binding of desmethoxyverapamil is, similarly to 1,4-dihydropyridine binding, dependent upon the presence of divalent cations (40,69).

A possible resolution of these findings is that skeletal muscle lacks the divalent cation-controlled high-affinity binding state for Ca^{2+} channel ligands found in cardiac, neuronal, and smooth muscle membranes, consistent with the lower affinity binding found in skeletal muscle, but remains sensitive to the inhibitory actions of di- and trivalent cations.

Relationship of Ligand Binding to Ca^{2+} Channel Permeation States

It is quite clear that the Ca^{2+} channel antagonists do not function merely as channel occluders. Electrophysiologic data of increasing sophistication show very clearly that channel ligand interaction, both antagonist and activator, is frequency and voltage dependent (11,138-142). Thus, both the quantitative and qualitative expressions of ligand activity are dependent upon the state of the channel that it encounters or that it can stabilize. Voltage-dependent channels exist minimally in three states , or families of states, resting, open, and inactivated; the transitions between these states are voltage and time dependent (138, 143). Ca^{2+} channel ligands may access these states via different pathways (hydrophilic or hydrophobic) and will have different affinities for the several states. Thus, nifedipine and other 1,4-dihydropyridine antagonists have a higher affinity ($\cong 10^3$-fold) for the inactivated state than for the resting state (11,141, 142), whereas Bay K 8644 and activator ligands show preferential stabilization of the open state (11,144). Similarly, verapamil and related structures that show prominent frequency dependence of interaction likely access the channel via the open state.

There are important consequences from these electrophysiologic studies for radioligand binding. These studies suggest that several binding states of different affinities should be measurable and that with cellular preparations it should be possible via membrane potential changes to alter binding parameters. Furthermore, these electrophysiologic experiments suggest that binding to membrane fragments may represent, largely or even exclusively, binding to the inactivated state of the Ca^{2+} channel. Whole-cell binding studies are, therefore, highly desirable.

From the data already reviewed it is very clear that multiple binding sites for Ca^{2+} channel ligands have been detected. These studies, however, usually document the allosteric regulation of radioligand binding by metal cations, guanine nucleotides, or various structural categories of channel ligand. In contrast, very few studies document different ligand binding states in the absence of such manipulation. Furthermore, in the few such available studies (22,38, 51,97,97,145,146), for both 1,4-dihydropyridines and verapamil, the putative low-affinity binding site is now well characterized. Of interest is the report that the number of $[^3H]$nitrendipine binding sites in myocytes increases in reversible fashion with K^+ depolarization (145): this would be consistent with production of a high-affinity binding state by depolarization. However, the low-affinity binding site for $[^3H]$nimodipine ($K_D = 3\text{-}5 \times 10^{-8}$ M) detected in erythrocytes (147) and cardiac membranes (69) is blocked by the adenosine transport inhibitors, nitrobenzylthioinosine, dilazep, and dipyridamole, suggesting that it does not represent a low-affinity state of the Ca^{2+} channel. Careful whole-cell binding studies under a variety of stimulated conditions are necessary to resolve this situation.

Regulation of Ca^{2+} Channels

The availability of the radioligand binding assay as a marker for Ca^{2+} channels, both functional and putative, has made possible a variety of studies detailing binding site changes under several regulating influences, including aging, chronic drug treatment, hypertension, heavy metal intoxication, and disease states.

There have been a number of functional studies of Ca^{2+} channel ontogeny, particularly in cardiac tissue (148-150): binding studies reveal complementaary details. Thus, in chick hearts (39), binding capacity for [^{3}H]nitrendipine increases during embryonic development from day 3 to day 14 and undergoes a second increase after hatching. During embryonic development the slow action potentials are originally insensitive to nitrendipine, suggesting that binding sites may be uncoupled from channel function. However, development of chick brain binding sites appears to be slower than cardiac development (151). Biphasic development of [^{3}H]nitrendipine binding sites also occurs in chick skeletal muscle cells in culture (152): the binding site is absent in myoblasts and appears at the fusion stage to myotubes, where the sarcoplasmic reticulum and the t tubules are formed. Upon denervation of skeletal muscle there is an initial increase of nitrendipine binding sites but a decrease in the number of tetrodotoxin and ouabain binding sites (152,153).

In a comparative study (154) in rat tissues of the development of binding sites in brain and skeletal and cardiac muscle, brain binding sites follow a pattern of development similar to that established for other receptors. Skeletal muscle binding sites appear in concert with the differentiation of sarcoplasmic reticulum and t tubules, whereas cardiac binding sites reach a maximum density only at 7-10 days postnatal. This latter development may parallel that for cardiac β-adrenoceptor responsiveness. Little is known of Ca^{2+} channel behavior during aging, but a decline in neuronal [^{3}H]nitrendipine binding sites has been reported for aged rats (155).

A number of studies indicate that Ca^{2+} channel ligand binding sites are regulated during chronic drug exposure (Table 6): the corresponding functional studies in general are not available. Chronic lead exposure to rats results in an increase in brain nitrendipine binding in the striatum, but not hippocampus, an effect apparently due to lead accumulation (161).

Ca^{2+} channel binding sites are reported to be altered during hypertension and myocardial ischemia. In spontaneously hypertensive rats (SHR), nitrendipine binding to brain, but not to heart, membranes is increased (41), but in desoxycorticosterone-hypertensive rats binding to the brain is decreased (162). The relevance of these changes to hypertension remains to be established. In ischemic hearts there is a decrease in the binding capacity of both high-affinity ($K_D = 2.4 \times 10^{-10}$ M) and low-affinity (K_D 4.8×10^{-8} M) binding sites (146).

Table 6 Effect of Chronic Drug Action on Calcium Channel Ligand Binding

System	Chronic exposure	Ligand binding site	Effect K_D	Effect B_{max}	Ref.
Mice, brain (in vivo)	Chlorpromazine	Nimodipine	↑	–	156
Mice, brain (in vivo)	Morphine	Nitrendipine	–	↑	157
Rats, smooth muscle (in vivo)	Reserpine	Nitrendipine	–	↑	158
Rats, myotubules (in vitro)	Isoproterenol	Nitrendipine	↑	↑	159
	Alprenolol	Nitrendipine	↑	↓	159
	Reserpine	Nitrendipine	↑	↓	159
Mice, brain (in vivo)	Nifedipine	Nitrendipine	–	↓	160
	Verapamil	Nitrendipine	–	↓	160

CURRENT AND FUTURE DIRECTIONS

The radioligand binding assay with Ca^{2+} channel ligand has proved to be extremely useful in providing a molecular approach to Ca^{2+} channel quantitation and localization that complements the approaches made by electrophysiology and pharmacology. In the few years since the first publications of Ca^{2+} channel ligand binding, an impressive amount of ligand and channel characterization has been possible. Indeed, individual workers in the field are frequently more competitive and antagonistic than the ligands that they study.

The Ca^{2+} channel ligands are currently proving to be of considerable value in attempts to purify and reconstitute the Ca^{2+} channel, primarily from skeletal muscle (104,112,113,163-166), and have been employed by a number of workers to obtain estimates of molecular weights of the membrane-associated channel by the technique of radiation inactivation (96,98,167-171). Additionally, the Ca^{2+} channel ligands have generated several affinity labels of both the conventional and photosensitive varieties (164,167,172,173). Collectively, these studies have led to the conclusion that the Ca^{2+} channel is of subunit organization consisting of two large units (M_r 175 and 170 kD) and at least two smaller subunits (M_r 30-50 kD). The largest subunit, α_1, carries the drug binding sites (174-176). These ligands will continue to be of value in efforts to isolate the channel and the ongoing work on appropriate gene characterization and expression (177).

From a pharmacological perspective, the current radioligand binding assays will continue to be employed in drug and tissue screening protocols, but more sophisticated understanding of ligand action and of the basis for ligand selectivity will demand whole-cell binding. Additionally, the realization from electrophysiologic studies that there are several different categories of Ca^{2+} channel will provoke new efforts to define protocols and lead structures with which to analyze other categories of voltage-dependent Ca^{2+} channel. Such pharmacological studies will complement electrophysiologic investigations and will collectively serve to define both the molecular characteristics and functions of these several Ca^{2+} channel classes.

These approaches should ensure that by the end of this decade the primary structure and membrane organization of at least one category of Ca^{2+} channel will have been elucidated and that substantial progress should have been made toward defining defects in Ca^{2+} channel properties or expression that may underlie a number of disease states. The contribution of the Ca^{2+} channel radioligand binding assay to these successes will have been considerable.

ACKNOWLEDGMENTS

This work was supported in part by grants from the National Institutes of Health (HL 16003, HL/AI 31178).

REFERENCES

1. R. H. Kretsinger, Coord. Chem. Rev. *18*, 29-124 (1976).
2. A. K. Campbell, Intracellular Calcium, John Wiley and Sons, New York, 1984.
3. R. P. Rubin, G. B. Weiss, and J. W. Putney (eds.), Calcium in Biological Systems, Plenum Press, New York, 1985.
4. R. A. Janis, P. J. Silver, and D. J. Triggle, Adv. Drug. Res., *16*, 309-591 (1987).
5. A. Fleckenstein, Annu. Rev. Pharmacol. Toxicol. *17*, 149-166 (1977).
6. A. Fleckenstein, Calcium Antagonism in Heart and Smooth Muscle: Experimental Facts and Therapeutic Prospects, John Wiley and Sons, New York, 1983.
7. S. F. Flaim and R. A. Zelis (eds.), Calcium Blockers, Mechanisms of Action and Clinical Implications, Urban and Schwarzenberg, Baltimore, 1982.
8. L. H. Opie (ed.), Calcium Antagonists and Cardiovascular Disease, Raven Press, New York, 1984.
9. R. W. Tsien, Annu. Rev. Physiol. *45*, 341-358 (1983).
10. R. Bayer, R. Kaufman, R. Mannhold, and R. Rodenkirchen, Prog. Pharmacol. *5*, 53-86 (1982).
11. P. Hess, J. B. Lansman, and R. W. Tsien, Nature *311*, 538-544 (1984).
12. L. B. Rosenberger and D. J. Triggle, in Calcium and Drug Action (G. B. Weiss, ed.), Plenum Press, New York, 1978, pp. 3-31.
13. R. Mannhold, R. Rodenkirchen, and R. Bayer, Prog. Pharmacol. *5*, 25-52 (1982).
14. D. J. Triggle and V. C. Swamy, Circ. Res. *52*(Suppl. I), 17-28 (1983).

15. R. A. Janis and D. J. Triggle, J. Med. Chem. *26*, 775-785 (1983).
16. M. Schramm, G. Thomas, R. Towart, and G. Franckowiak, Nature *303*, 535-537 (1983).
17. R. P. Hof, U. T. Ruegg, A. Hof, and A. Vogel, J. Cardiovasc. Pharmacol. *7*, 689–693 (1985).
18. G. Franckowiak, M. Bechem, M. Schramm, and G. Thomas, Eur. J. Pharmacol. *114*, 223-226 (1985).
19. R. A. Janis and D. J. Triggle, Drug. Dev. Res. *4*, 257-274 (1984).
20. D. J. Triggle and R. A. Janis, in Modern Methods in Pharmacology (N. Back and S. Spector, eds.), Alan R. Liss, New York, 1984, pp. 1-28.
21. H. Glossmann and D. R. Ferry, Methods Enzymol *109*, 513-551 (1985).
22. P. Bellemann, D. Ferry, F. Lübbecke, and H. Glossmann, Arzeimittelforsch *31*, 2064-2067 (1981).
23. P. Bellemann, D. Ferry, F. Lübbecke, and H. Glossmann, Arzeimittelforsch. *32*, 361-363 (1982).
24. A. B. Fawzi and J. H. McNeill, Eur. J. Pharmacol. *104*, 357-362 (1984).
25. J. Schwartz and J. Vells, Br. J. Pharmacol. *84*, 511-515 (1985).
26. R. A. Janis, J. G. Sarmiento, S. C. Maurer, G. T. Bolger, and D. J. Triggle, J. Pharmacol. Exp. Ther. *231*, 8-15 (1984).
27. W. McBride, A. Mukherjee, Z. Haghani, E. Wheeler-Clark, J. Brady, T. Gandler, L. Bush, L. M. Buja, and J. T. Willerson, Am. J. Physiol. *247*, H775-783 (1984).
28. N. Brandt, Arch. Biochem. Biophys. *242*, 306-319 (1985).
29. L. T. Williams and P. Tremble, J. Clin. Invest. *70*, 209-212 (1982).
30. A. DePover, M. A. Matlib, S. W. Lee, G. P. Dube, I. L. Grupp, G. Grupp, and A. Schwartz, Biochem. Biophys. Res. Commun. *108*, 110–117 (1982).
31. L. T. Williams and L. R. Jones, J. Biol. Chem. *258*, 5344-5347 (1983).
32. R. A. Janis, S. C. Maurer, J. G. Sarmiento, G. T. Bolger, and D. J. Triggle, Eur. J. Pharmacol. *82*, 191-194 (1982).
33. K. M. M. Murphy and S. H. Snyder, Eur. J. Pharmacol. *77*, 201-202 (1982).
34. F. J. Ehlert, W. R. Roeske, E. Itoga, and H. I. Yamamura, Life Sci. *30*, 2191-2202 (1982).
35. P. J. Marangos, J. Patel, C. Miller, and A. M. Martino, Life Sci. *31*, 1575-1585 (1982).
36. H. Schoemaker, H. R. Lee, W. R. Roeske, and H. I. Yamamura, Eur. J. Pharmacol. *88*, 275–276 (1983).
37. H. R. Lee, W. R. Roeske, and H. I. Yamamura, Life Sci. *35*, 721-732 (1984).
38. J. D. Marsh, E. Loh, D. Lachance, W. H. Barry, and T. W. Smith, *Circ. Res.* *53*, 539-543 (1983).
39. J.-F. Renaud, T. Kazazoglou, A. Schmid, G. Romey, and M. Lazdunski, Eur. J. Biochem. *139*, 673-681 (1984).
40. J. Ptasienski, K. K. McMahon, and M. M. Hosey, Biochem. Biophys. Res. Commun. *129*, 910-917 (1985).
41. K. Ishii, T. Kano, Y. Kurobe, and J. Ando, Eur. J. Pharmacol. *88*, 277-278 (1983).

42. G. T. Bolger, P. J. Gengo, E. M. Luchowski, H. Siegel, D. J. Triggle, and R. A. Janis, Biochem. Biophys. Res. Commun. *104*, 1604-1609 (1982).
43. G. T. Bolger, P. Gengo, R. Klockowski, H. Siegel, R. A. Janis, A. M. Triggle, and D. J. Triggle, J. Pharmacol. Exp. Ther. *225*, 291-309 (1983).
44. P. J. Gengo, F. Yousif, R. A. Janis, and D. J. Triggle, Biochem. Pharmacol. *33*, 3445-3449 (1984).
45. T. Godfraind and M. Wibo, Br. J. Pharmacol. *85*, 335-340 (1985).
46. W. C. Miller and J. B. Moore, Life Sci. *34*, 1717-1724 (1984).
47. S. Batra, Br. J. Pharmacol. *85*, 767-774 (1985).
48. A. K. Grover, C.-Y. Kwan, E. Luchowski, E. E. Daniel, and D. J. Triggle, J. Biol. Chem. *259*, 2223-2226 (1984).
49. J. B. Cheng, A. Bewtra, and R. G. Townley, Experientia *40*, 267-269 (1984).
50. J. G. Sarmiento, R. A. Janis, D. Jenkins, A. M. Katz, and D. J. Triggle, in Nitrendipine (A. Scriabine, S. Vanov, and K. Deck, eds.), Urban and Schwarzenberg, Baltimore, 1984, pp. 153-160.
51. C. R. Triggle, D. K. Agrawal, G. T. Bolger, E. E. Daniel, C. Y. Kwan, E. M. Luchowski, and D. J. Triggle, Can. J. Physiol. Pharmacol. *60*, 1738-1741 (1982).
52. D. J. Triggle and R. A. Janis, in Nitrendipine (A. Scriabine, S. Vanov, and K. Deck, eds.), Urban and Schwarzenberg, Baltimore, 1984, pp. 33-52.
53. F. B. Yousif, G. T. Bolger, A. Ruzycky, and D. J. Triggle, Can. J. Physiol. Pharmacol. *63*, 453-462 (1985).
54. M. Fosset, E. Jaimovich, E. Delpont, and M. Lazdunski, Eur. J. Pharmacol. *86*, 141-142 (1983).
55. A. S. Fairhurst, S. A.Thayer, J. E. Colker, and D. A. Beatty, Life Sci. *32*, 1331-1339 (1983).
56. D. R. Ferry, A. Goll, and H. Glossmann, Naunyn-Schmiedebergs Arch. Pharmacol. *323*, 276-277 (1983).
57. M. Fossett, E. Jaimovich, E. Delpont, and M. Lazdunski, J. Biol. Chem. *258*, 6086-6092 (1983).
58. R. J. Miller and S. B. Freedman, Life Sci. *34*, 1205-1221 (1984).
59. F. J. Ehlert, E. Itoga, W. R. Roeske, and H. I. Yamamura, Biochem. Biophys. Res. Commun. *104*, 937-943 (1982).
60. R. J. Gould, K. M. M. Murphy, and S. H. Snyder, Proc. Natl. Acad. Sci. USA *79*, 3656-3660 (1982).
61. H. I. Yamamura, H. Schoemaker, R. G. Boles, and W. R. Roeske, Biochem. Biophys. Res. Commun. *108*, 640-646 (1982).
62. M. Takahashi and A. Ogura, FEBS Lett. *152*, 191-194 (1983).
63. L. Toll, J. Biol. Chem. *257*, 13189-13192 (1982).
64. R. G. Boles, H. I. Yamamura, H. Schoemaker, and W. R. Roeske, J. Pharmacol. Exp. Ther. *229*, 333-339 (1984).
65. G. A. Weiland and R. E. Oswald, J. Biol. Chem. *260*, 8456-8464 (1985).
66. D. A. Greenberg and E. C. Cooper, Alcoholism Clin. Exp. Res. *8*, 568-571, (1984).
67. K. Ishii, T. Kane, Y. Kurobe, and J. Ando, Eur. J. Pharmacol. *88*, 277-278, (1983).

68. A. DePover, S. W. Lee, M. A. Matlib, K. Whitmer, B. A. Davis, T. Powell, and A. Schwartz, Biochem. Biophys. Res. Commun. *113*, 185-191 (1983).
69. P. Ruth, V. Flockerzi, E. von Nettelbladt, J. Oeken, and F. Hofmann, Eur. J. Biochem. *150*, 313-322 (1985).
70. D. R. Ferry, H. Glossmann, and A. J. Kaumann, Br. J. Pharmacol. *84*, 811-824 (1985).
71. D. Rampe, E. Luchowski, A. Rutlege, R. A. Janis, and D. J. Triggle, Can. J. Physiol. Pharmacol. *65*, 1452-1460 (1987).
72. D. R. Ferry and H. Glossman, FEBS Lett. *148*, 331-337 (1982).
73. A. Goll, D. R. Ferry, and H. Glossmann, FEBS Lett. *157*, 63-69 (1983).
74. D. R. Ferry, A. Goll, and H. Glossmann, Naunyn-Schmiedebergs Arch. Pharmacol. *323*, 276-277 (1983).
75. D. R. Ferry and H. Glossmann, Naunyn-Schmiedebergs Arch. Pharmacol. *321*, 80-83 (1983).
76. H. Glossmann, D. R. Ferry, F. Lübbecke, R. Mewes, and F. Hofmann, J. Recept. Res. *3*, 177-190 (1983).
77. P. Bellemann, D. R. Ferry, F. Lübbecke, and H. Glossmann, Arzeimittelforsch. *32*, 361-363 (1982).
78. J. Peroutka and G. S. Allen, J. Neurosurg. *59*, 933-977 (1983).
79. P. Bellemann, A. Schade, and R. Towart, Proc. Natl. Acad. Sci. USA *80*, 2356-2360 (1983).
80. H. R. Lee, W. R. Roeske, and H. I. Yamamura, Life Sci. *35*, 721-732 (1984).
81. D. R. Ferry, A. Goll, and H. Glossmann, Naunyn-Schmiedebergs Arch. Pharmacol. *323*, 276-277 (1983).
82. L. M. Schwartz, E. W. McCleskey, and W. Almers, Nature *314*, 747-751 (1985).
83. P. Supavilai and M. Karobath, J. Neural. Trans. *60*, 49-67 (1984).
84. M. Holck, S, Thorens, and G. Haeusler, Eur. J. Pharmacol. *85*, 305-315 (1982).
85. M. Holck, S. Thorens, and G. Haeusler, J. Recept. Res. *3*, 191-198 (1983).
86. M. Pan, R. A. Janis, and D. J. Triggle, Pharmacologist *25*, 523 abstract (1983).
87. P. Bellemann, FEBS Lett. *167a*, 88-92 (1984).
88. P. Bellemann, J. Recept. Res. *4*, 571-585 (1984).
89. R. A. Janis, D. Rampe, J. G. Sarmiento, and D. J. Triggle, Biochem. Biophys. Res. Commun. *121*, 317-323 (1984).
90. C. M. Su and D. J. Triggle, unpublished data (1986).
91. U. L. Hulthen, R. Landmann, E. Bürgissar, and F. R. Bühler, J. Cardiovasc. Pharmacol. *4*, S291-293 (1982).
92. P. Bellemann, J. Recept. Res. *4*, 571-585 (1984).
93. M. L. Garcia, M. J. Trumble, J. P. Reuben, and G. J. Kaczorowski, J. Biol. Chem. *259*, 15013-15016 (1984).
94. I. J. Reynolds, R. J. Gould, and S. H. Snyder, Eur. J. Pharmacol. *95*, 319-321 (1983).

95. J.-P. Galizzi, M. Fosset, and M. Lazdunski, Biochem. Biophys. Res. Commun. *118*, 239-245 (1984).
96. A. Goll, D. R. Ferry, and H. Glossmann, Eur. J. Biochem. *141*, 177-186 (1984).
97. J.-P. Galizzi, M. Fosset, and M. Lazdunski, Eur. J. Biochem. *144*, 211-215 (1984).
98. A. Goll, D. R. Ferry, J. Striessnig, M. Schober, and H. Glossmann, FEBS Lett. *176*, 371-377 (1984).
99. H. Schoemaker and S. Z. Langer, Eur. J. Pharmacol. *111*, 273-277 (1985).
100. H. Glossmann, T. Linn, M. Rombusch, and D. R. Ferry, FEBS Lett. *160*, 226-232 (1983).
101. D. R. Ferry and H. Glossmann, Naunyn-Schmiedebergs Arch. Pharmacol. *325*, 186-189 (1984).
102. H. Glossmann, D. R. Ferry, A. Goll, and M. Rombusch, J. Cardiovasc. Pharmacol. *6*, S608-621 (1984).
103. M. Holck, W. Fischli, and U. Hengartner, J. Recept. Res. *4*, 557-569 (1984).
104. M. Borsotto, R. I. Norman, M. Fosset, and M. Lazdunski, Eur. J. Biochem. *142*, 449-455 (1984).
105. M. Borsotto, J. Barhanin, R. I. Norman, and M. Lazdunski, Biochem. Biophys. Res. Commun. *122*, 1357-1366 (1984).
106. D. G. Rhodes, J. G. Sarmiento, and L. G. Herbette, Mol. Pharmacol. *27*, 612-623 (1985).
107. L. G. Herbette, D. G. Rhodes, D. W. Chester, R. A. Colvin, J. G. Sarmiento, Y. Vant Erve, and A. M. Katz, in Pathobiology of Cardiovascular Injury, Martinus Nijhoff, Hingham, Massachusetts, 1985.
108. J.-P. Galizzi, M. Fosset, and M. Lazdunski, Biochem. Biophys. Res. Commun. *132*, 49-55 (1985).
109. D. Rampe, E. Luchowski, A. Rutledge, R. A. Janis, and D. J. Triggle, Can. J. Physiol. Pharmacol. *65*, 1452-1460 (1987).
110. G. Thorgeirsson and S. A. Rudolph, Biochem. Biophys. Res. Commun. *121*, 657-663 (1984).
111. J. L. Balwierczak and A. Schwartz, Eur. J. Pharmacol. *116*, 193-194 (1985).
112. H. Glossmann and D. R. Ferry, Naunyn-Schmiedebergs Arch. Pharmacol. *232*, 279-291 (1983).
113. H. Glossmann, D. R. Ferry, and C. B. Boschek, Naunyn-Schmiedebergs Arch. Pharmacol. *323*, 1-11 (1983).
114. J. G. Sarmiento, R. A. Janis, R. A. Colvin, D. J. Triggle, and A. M. Katz, J. Mol. Cell. Cardiol. *15*, 135–137 (1983).
115. D. J. Triggle, unpublished data (1986).
116. R. Towart, E. Wehinger, and H. Meyer, Naunyn-Schmiedebergs Arch. Pharmacol. *317*, 183-185 (1981).
117. T. Shibanuma, M. Iwanani, K. Okuda, T. Takenaba, and M. Murakami, Chem. Pharm. Bull. *28*, 2809-2812 (1980).

118. A. J. Kaumann and J. R. Serur, Naunyn-Schmiedebergs Arch. Pharmacol. *291*, 307-258 (1975).
119. C. Ludwig and C. Nawroth, Br. J. Pharmacol. *59*, 411-417 (1977).
120. H. Nawrath, I. Blei, R. Gengen, C. Ludwig, and V.-G. Zang, in Calcium Antagonism in Cardiovascular Therapy (A. Zanchetti, ed.), Excerpta Medica, Amsterdam, 1981, pp. 52-63.
121. K. Satoh, T. Yanagisawa, and N. Taira, J. Cardiovasc. Pharmacol. *2*, 309–318 (1980).
122. A. M. Hetherington and A. J. Trewavas, Plant. Sci. Lett. *35*, 109-113 (1984).
123. E. Andrejaukcas, R. Hertel, and D. Marme, J. Biol. Chem. *260*, 5411-5414 (1985).
124. M. J. Saunders and P. K. Hepler, Dev. Biol. *99*, 41-49 (1983).
125. A. K. Grover, C. Y. Kwan, E. E. Daniel, S. Ahmad, T. Ramlal, P. Oakes, and D. J. Triggle, Arch. Int. Pharmacodyn. *273*, 74-82 (1985).
126. L. N. Siri, J. A. Sanchez, and E. Stefani, J.Physiol. (Lond.) *305*, 87-96 (1980).
127. W. Almers, R. Fink, and P. T. Palade, J. Physiol. (Lond.) *312*, 177-207 (1981).
128. E. Stefani and D. J. Chiarandini, Annu. Rev. Physiol. *44*, 357–372 (1982).
129. M. Rosemblatt, C. Hidalgo, C. Vergara, and N. Ikemoto, J. Biol. Chem. *256*, 8140-8148 (1981).
130. N. R. Brandt, R. M. Kawamoto, and A. H. Caswell, J. Recept. Res. *5*, 155-170 (1985).
131. K. M. M. Murphy, R. J. Gould, B. L. Largent, and S. H. Snyder, Proc. Natl. Acad. Sci. USA *80*, 860-864 (1983).
132. T. Nagao, M. Sato, Y. Iwasawa, T. Takada, R. Ischida, H. Nakajimos, and A. Kiyomota, Jpn. J. Pharmacol. *22*, 467-478 (1972).
133. D. R. Ferry, A. Goll, C. Gadow, and H. Glossmann, Naunyn-Schmiedebergs Arch. Pharmacol. *327*, 183-187 (1984).
134. H. Glossmann, D. R. Ferry, F. Lübbecke, R. Mewes, and F. Hofmann, Trends Pharmacol. Sci. *2*, 431-437 (1982).
135. E. M. Luchowski, F. Yousif, D. J. Triggle, S. C. Maurer, J. G. Sarmiento, and R. A. Janis, J. Pharmacol. Exp. Ther. *230*, 607-613 (1984).
136. P. Hess and R. W. Tsien, Nature *309*, 453-455 (1984).
137. W. Almers, E. W. McCleskey, and P. T. Palade, J. Physiol. (Lond.) *353*, 565-583 (1984).
138. L. M. Hondeghem and B. G. Katzung, Annu. Rev. Pharmacol. Toxicol. *24*, 387-423 (1984).
139. T. T. McDonald, D. Pelzer, and W. Trautwein, J. Physiol. (Lond.) *352*, 217-244 (1984).
140. M. Kohlhardt, Naunyn-Schmiedebergs Arch. Pharmacol. *323*, 251-260 (1983).
141. M. C. Sanguinetti and R. S. Kass, Circ. Res. *55*, 336-348 (1984).
142. B. P. Bean, Proc. Natl. Acad. Sci. USA *81*, 6388-6392 (1984).

143. B. Hille, Ionic Channels of Excitable Membranes, Sinauer Associates, Sunderland, Massachusetts, 1984.
144. M. C. Sanguinetti and R. S. Kass, J. Mol. Cell Cardiol. *16*, 667-670 (1984).
145. F. J. Green, B. B. Farmer, G. L. Wiseman, M. J. L. Jose, and A. M. Watanabe, Circ. Res. *56*, 576-585 (1985).
146. W. G. Nayler, J. S. Dillon, J. S. Elz, and M. McKelvie, Eur. J. Pharmacol. *115*, 81-89 (1985).
147. J. Striessnig, G. Zernig, and H. Glossmann, Eur. J. Pharmacol. *108*, 329-330 (1985).
148. N. Sperelakis and K. Shigenobu, J. Gen. Physiol. *60*, 430-453 (1972).
149. N. Sperelakis and K. Shigenobu, J. Mol. Cell Cardiol. *6*, 449-471 (1974).
150. M. Kojima and N. Sperelakis, J. Dev. Physiol. *7*, 355-363 (1985).
151. P. J. Marangos, N. Sperelakis, and J. Patel, J. Neurochem. *42*, 1338-1342 (1984).
152. A. Schmid, J.-F. Renaud, M. Fosset, J.-P. Neaux, and M. Lazdunski, J. Biol. Chem. *259*, 11366-11372 (1984).
153. A. Schmid, T. Kazazoglou, J.-F. Renaud, and M. Lazdunski, FEBS Lett. *172*, 114-118 (1984).
154. T. Kazazoglou, A. Schmid, J.-F. Renaud, and M. Lazdunski, FEBS Lett. *164*, 75-79 (1983).
155. S. Govoni, R. A. Rius, F. Battaini, A. Bianchi, and M. Trabucchi, Brain Res. *333*, 374-377 (1985).
156. V. Ramkumar and E. E. El-Fakahany, Res. Commun. Chem. Pathol. Pharmacol. *48*, 463-466 (1985).
157. V. Ramkumar and E. E. El-Fakahany, Eur. J. Pharmacol. *102*, 371-372 (1984).
158. R. E. Powers and W. S. Colucci, Biochem. Biophys. Res. Commun. *132*, 844-849 (1985).
159. A. Schmid, J.-F. Renaud, and M. Lazdunski, J. Biol. Chem. *260*, 13041-13046 (1985).
160. G. Panza, J. A. Grebb, R. C. Shelton, and I. Hanbauer, Fed. Proc. 2765 abstract (1985).
161. S. Govoni, R. A. Rius, F. Battaini, and M. Trabucchi, Soc. Neurosci. Abstr. *11*, 1597 (1985).
162. H. R. Lee, M. Watson, H. I. Yamamura, and W. R. Roeske, Life Sci. *37*, 971-977 (1985).
163. B. M. Curtis and W. A. Catterall, Biochemistry *23*, 2113-2118 (1984).
164. T. L. Kirley and A. Schwartz, Biochem. Biophys. Res. Commun. *123*, 41-49 (1984).
165. B. M. Curtis and W. A. Catterall, J. Biol. Chem. *258*, 7280-7283 (1983).
166. A. Rengasamy, J. Ptasienski, and M. M. Hosey, Biochem. Biophys. Res. Commun. *126*, 1-7 (1985).
167. J. C. Venter, C. M. Fraser, J. S. Schaber, C. Y. Jung, G. Bolger, and D. J. Triggle, J. Biol. Chem. *258*, 9344-9348 (1983).
168. D. R. Ferry, A. Goll, and H. Glossmann, Naunyn-Schmiedebergs Arch. Pharmacol. *323*, 292-297 (983).

169. A. Goll, D. R. Ferry, and H. Glossmann, FEBS Lett. *157*, 63-69 (1983).
170. R. I. Norman, M. Borsotto, M. Fosset, M. Lazdunski, and J. C. Ellory, Biochem. Biophys. Res. Commun. *111*, 878-883 (1983).
171. D. R. Ferry, A. Goll, and H. Glossmann, EMBO J. *2*, 1729-1732 (1983).
172. D. R. Ferry, M. Rombusch, A. Goll, and H. Glossmann, FEBS Lett. *169*, 112-118 (1984).
173. K. P. Campbell, G. M. Lipshutz, and G. H. Denney, J. Biol. Chem. *259*, 5384-5387 (1984).
174. A. T. Leung, T. Imagawa, B. Block, C. Franzini-Armstrong, and K. P. Campbell, J. Biol. Chem. *263*, 994-1001 (1988).
175. H. Glossmann and J. Striessnig, ISI Atlas of Science: Pharmacology *2*, 202-210 (1988).
176. M. Takahashi and W. A. Catterall, Science *236*, 88-91 (1987).
177. J. P. Leonard, J. Nargeot, T. P. Snutch, N. Davidson, and H. A. Lester, J. Neurosci. *7*, 875-881 (1987).

21

Intact Cells: A Model for Studying Receptor Binding and Function

Esam E. El-Fakahany / University of Maryland School of Pharmacy, Baltimore, Maryland

HISTORICAL PERSPECTIVE

Radioligand-neurotransmitter receptor binding techniques have revolutionized the prevailing approaches to our understanding of molecular pharmacology. These techniques have been extremely helpful in providing a host of information regarding one of the earliest steps in the cascade of drug action, that is, the interaction of a drug with its specific receptors. Owing to the simplicity of performing receptor binding experiments, this approach has been gaining ground over many other classic pharmacological techniques used to study neurotransmitter receptors through the quantitation of a drug-induced response in a certain tissue or organ, for example, a change in blood pressure, muscle contraction, or gland secretion.

Although combinations of receptor binding studies and drug-induced pharmacological responses represent a very important, if not a mandatory, approach in studying drug action, this fruitful marriage seldom appears in the literature. Further, the majority of receptor binding experiments are usually conducted under conditions that fail to mimic the in vivo situation, for example in buffers of a composition that is far from physiological. Most importantly, broken-cell preparations have been used most extensively in receptor binding studies.

In order to increase relevant correlations between studying a receptor as a "drug binding site" or a "response-triggering unit," the conditions for studying receptor binding and receptor-mediated responses should be approximated as much as possible. Intact cell systems therefore present an excellent model for such an ideal combination. The use of intact cell systems offers several advantages, including the following:

1. They bear a much closer resemblance to intact tissues in vivo, in terms of having both their integrity unharmed and their full complement of the cytoplasmic milieu.
2. They can be used to study biochemical responses consequent to receptor activation, especially those responses that cannot be elicited in broken-cell preparations, such as ion fluxes, activation of guanylate cyclase, or alteration of phosphoinositide turnover.
3. These cells contain an undisrupted receptor-effector unit, which makes them ideal for studying mutual allosteric interactions between these subunits.
4. Detailed studies of receptor translocation upon different manipulations, for example, desensitizing conditions, require an intact cell system.

Because of the vast applicability of intact cells in receptor studies, this chapter concentrates only on their use in elucidating mechanisms related to muscarinic acetylcholine receptors. Most radioligand binding studies identifying the properties of muscarinic receptors have been performed using tissue homogenates. The popularity of this latter preparation stems from the ease of obtaining a large quantity of tissue in a homogeneous form, although consisting of different cell types. In addition, their use guarantees, in most instances, very low levels of nonspecific binding. Such binding studies have demonstrated that classic muscarinic receptor antagonists bind with high affinity to a homogeneous population of the receptors (1,2). However, heterogeneity of binding sites has been shown for certain antagonists, such as pirenzepine (3-6) and *N*-methylscopolamine (7,8), and for others under certain experimental conditions (9-12). On the other hand, agonists are capable of discriminating between multiple receptor conformations comprising high- and low-affinity binding sites (2,13), and occasionally super high-affinity sites (14). Certain muscarinic agonist-mediated responses have been associated with activation of the high-affinity sites (2,15,16); others have been attributed to the ability of agonists to interact with the low-affinity receptor species (13,17-19). In spite of the useful information gathered over the years using brain homogenates to study muscarinic cholinergic receptors in the central nervous system, this preparation suffers from several drawbacks, resulting mainly from disrupting the cell membrane integrity. This may lead to uncoupling of muscarinic receptors and their effectors or other membrane components that may modulate receptor conformation. Disrupting the cell membrane also exposes its cytoplasmic side to a nonphysiological environment while depriving it of its natural intracellular milieu. Added to this is the common use of nonphysiological buffers in most muscarinic receptor binding assays in brain homogenates. Although purified brain synaptosomes suffer less from these pitfalls compared with brain homogenates, they are quite tedious and time consuming to prepare. Furthermore, synaptosomes can actually be considered sealed vesicles that re-form from broken-cell preparations under certain conditions.

The strength of the approach of using intact cells to perform receptor binding experiments lies in the fact that agonist binding can be correlated with agonist-induced responses in the same cells and under the same experimental conditions. Throughout this chapter, I concentrate on discussing the usefulness of using intact cells to characterize the properties and regulation of muscarinic acetylcholine receptors, in addition to the various biochemical and physiological responses mediated by activation of these receptors. The preparations that are discussed include tissue slices, clonal cell lines, and dissociated cells in primary culture.

RECEPTOR BINDING STUDIES

Brain slices represent a simple approach to study muscarinic receptors. Gilbert et al. (20) compared the binding of the potent muscarinic receptor antagonist [^{3}H] quinuclidinyl benzilate ([^{3}H] QNB) in rat brain slices and homogenate preparations. Although binding in slices retained its stereospecificity, it was consistently of a lower affinity than that in crude synaptic membranes. Similar results were obtained in slices and homogenates obtained from rabbit iris (21). The discrepancy of the affinity constants in these two types of preparations may be due to the presence of diffusion barriers in the slices or to a conformational change in the receptor sites induced by tissue disruption.

Some binding studies have been performed in intact neurons in primary culture prepared from fetal rats (22,23). The specific binding of [^{3}H] QNB in this preparation demonstrates saturability and high affinity. Although the rank order of potency of several muscarinic receptor agonists and antagonists in displacing [^{3}H] QNB binding in this preparation correlated well with that obtained in brain homogenates, about 10 times higher concentrations of the cholinergic drugs were required for the displacement in intact cells compared with broken-cell preparations (23). A similar difference in potencies to displace [^{3}H] QNB binding by drugs was observed when binding in intact rat pituitary cells (24) was compared with that in membrane preparations from the same cells (25) or from sheep pituitary glands (26). Primary cultures of neurons from the brain and the spinal cord have also been very useful in demonstrating the ontogeny of muscarinic receptors and its relationship to the development of other cholinergic markers (22,27,28).

Recently, freshly prepared dissociated brain cell aggregates obtained from adult rats have been used to study ligand binding to muscarinic cholinergic receptors in the central nervous system (29). Such cells are prepared simply by mechanical dissociation of brain tissue in the absence of calcium and magnesium ions, without using enzymatic treatment, for example, with trypsin, hyaluronidase, or collagenase. The dissociated cells are sieved through microfilament nylon meshes of different pore size and washed repeatedly using low-speed centrifugation. Using these cells, we found that the specific binding of [^{3}H]*N*-methylscopolamine ([^{3}H]NMS) is related linearly to the amount of cell protein included in the binding assay, up to 400 μg protein per assay tube. Increasing the concentration of [^{3}H]NMS in the range 0.01-2 nM results in increased total binding, and specific binding of [^{3}H]NMS to muscarinic acetylcholine receptors demonstrates saturability at higher ligand concentrations (Figure 1a). On the other hand, nonspecific binding measured in the presence of 2 μM atropine increases linearly with the ligand concentration (Figure 1a). It is noteworthy that the level of nonspecific binding is reasonably low compared with that of total binding. Scatchard analyses of the saturation isotherms indicate a maximum binding capacity of 450 fmol/mg protein and an average equilibrium dissociation constant of 0.17 nM (Figure 1b). Scatchard plots are linear, and the Hill coefficients are about 1.0 (Figure 1b), suggesting that [^{3}H]NMS binds to a single homogeneous population of muscarinic acetylcholine receptors in rat brain cell aggregates with no evidence for cooperative interactions.

Displacement of the specific binding of 0.2 nM [^{3}H]NMS in intact brain cells by muscarinic receptor antagonists results in steep displacement curves, with Hill coefficients close to unity (Figure 2). Atropine, scopolamine, and (±)-QNB displace the specific binding with average median inhibitory concentrations of 0.96, 2.75, and 1.09 nM, respectively, which translate into corresponding inhibition constants of 0.42, 1.27, and 0.25 nM (Table 1).

On the other hand, muscarinic receptor agonists displace specific [^{3}H]NMS binding in intact brain cell aggregates with very shallow inhibition curves extending over three logarithmic units (Figure 3). Hence, the Hill coefficients for carbamylcholine, oxotremorine, and pilocarpine are 0.64, 0.65, and 0.78, respectively (Table 1). Nonlinear least-squares regression analysis of displacement of [^{3}H]NMS-specific binding in brain cells by agonists resolved the displacement curves into high- and low-affinity components. The equilibrium dissociation constants of the high-affinity sites for carbamylcholine, oxotremorine, and pilocarpine are 4.3 μM, 90 nM, and 1.6 μM, respectively, and the respective values

---

Figure 1 Saturation isotherm of [^{3}H]NMS binding in adult rat brain cell aggregates. (a) Rat brain cells (100 μg protein) were incubated in triplicate with increasing concentrations of [^{3}H]NMS in the absence (○, total binding) and in the presence (▲, nonspecific binding) of 2 μM atropine. Specific binding (●) was obtained by subtracting nonspecific binding from total binding. The data presented are the average of 14 independent experiments. (b) Scatchard plot of [^{3}H]NMS specific binding presented in a (r = 0.992). Inset: Hill plot of the saturation isotherm (n_H = 0.96).

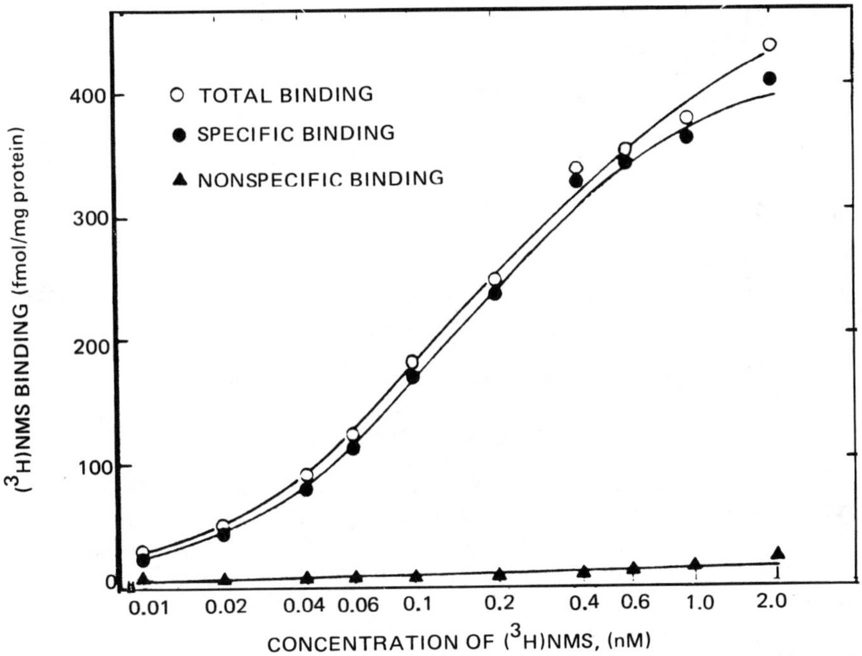

TOTAL BINDING
SPECIFIC BINDING
NONSPECIFIC BINDING
(^{3}H)NMS BINDING (fmol/mg protein)
0
100
200
300
400
0.01
0.02
0.04
0.06
0.1
0.2
0.4
0.6
1.0
2.0
CONCENTRATION OF (^{3}H)NMS, (nM)

(a)

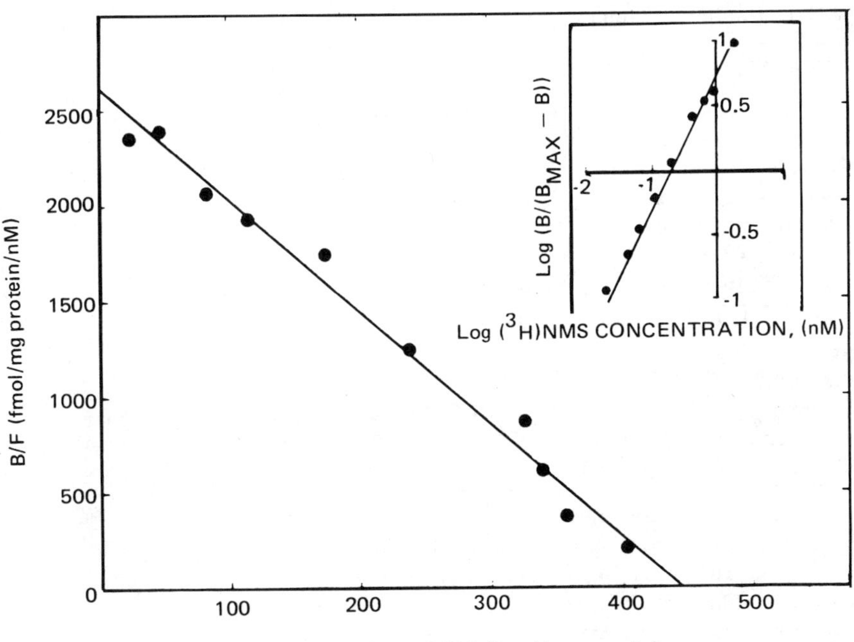

B/F (fmol/mg protein/nM)
0
500
1000
1500
2000
2500
100
200
300
400
500
SPECIFIC BINDING (fmol/mg protein)
Log (B/(B$_{MAX}$ − B))
1
0.5
-0.5
-1
-2
-1
Log (^{3}H)NMS CONCENTRATION, (nM)

(b)

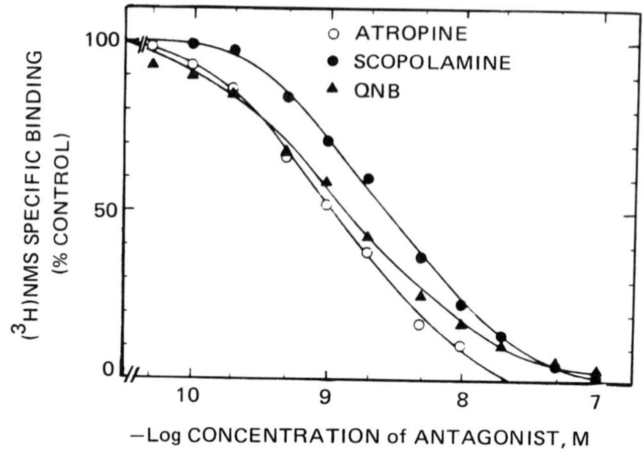

Figure 2 Displacement of the specific binding of [^{3}H]NMS by muscarinic receptor antagonists in adult intact rat brain cells. Cells were incubated in triplicate with 0.2 nM [^{3}H]NMS in the presence of increasing concentrations of atropine (○), scopolamine (●), or (±)-quinuclidinyl benzilate (▲). Data are presented as a percentage of specific binding obtained in the absence of displacers (with a mean of 18 fmol per assay tube) by averaging the results of three to five independent experiments.

Table 1 Binding Parameters of [^{3}H]NMS in Intact Adult Rat Brain Cell Aggregates[a]

	Parameter				
Displacer	K_H[b] (M)	K_L[c] (M)	$\%R_H$[d]	$\%R_L$[e]	n_H[f]
Antagonists					
Atropine (3)[g]	$0.42 \pm 0.08 \times 10^{-9}$	–	100	–	0.96
Scopolamine (4)	$1.27 \pm 0.12 \times 10^{-9}$	–	100	–	0.94
(–)-QNB (5)	$0.25 \pm 0.01 \times 10^{-9}$	–	100	–	0.81
Agonists					
Carbamyl-choline (3)	$4.3 \pm 1.8 \times 10^{-6}$	$8.9 \pm 1.3 \times 10^{-5}$	18.5	81.5	0.64
Oxotremorine (2)	8.98×10^{-8}	1.80×10^{-6}	18.9	81.1	0.65
Pilocarpine (2)	1.60×10^{-6}	1.81×10^{-5}	34.5	65.5	0.78

[a]Values expressed as mean ± SEM.
[b]Equilibrium dissociation constant of high-affinity sites.
[c]Equilibrium dissociation constant of low-affinity sites.
[d]Percentage of high-affinity sites.
[e]Percentage of low-affinity sites.
[f]Hill coefficient.
[g]Number of independent experiments.

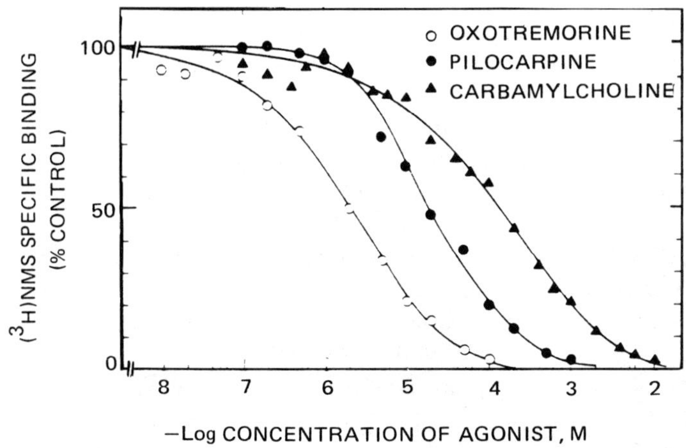

Figure 3 Displacement of specific [^{3}H]NMS binding by muscarinic receptor agonists in intact adult rat brain cell aggregates. Cells were incubated in triplicate with 0.2 nM [^{3}H]NMS in the presence of increasing concentrations of carbamylcholine (▲), oxotremorine (○), or pilocarpine (●). Data are shown as a percentage of control binding in absence of displacers, which averaged 14 fmol per assay tube and are presented as the mean of two to three independent experiments.

for the low-affinity sites are 89, 1.8, and 18.1 μM (Table 1). In addition, the proportion of the density of high-affinity sites to total receptor concentration is 18.5, 18.9, and 34.5% for carbamylcholine, oxotremorine, and pilocarpine, respectively (Table 1).

Dissociated adult rat brain cell aggregates may thus provide a very useful and physiologically relevant model to study muscarinic acetylcholine receptor binding in vitro. This technique offers an easy method to obtain intact differentiated brain cells with minimal diffusion barriers. As mentioned earlier, fetal rat brain cell aggregates in primary culture have also been used for muscarinic receptor binding experiments. However, using this method requires a large number of animals and a sterile environment, in addition the the expense of growth media. Moreover, these fetal cells lack a full complement of muscarinic receptors. In addition to the usefulness of intact cells prepared from adult rat brain in receptor binding experiments, they also have a great potential as a tool to study muscarinic receptor-mediated biochemical responses that require intact cells, such as cyclic GMP formation and phosphoinositide turnover.

Clonal cell lines have also provided an excellent model of the muscarinic acetylcholine receptors in the central nervous system (2). These cells offer several advantages, including providing a large number of homogeneous cells in monolayer, where diffusion barriers are minimal. In addition, it is possible to precisely control the environment around the cells in order to study the effect of different pharmacological manipulations on cellular function. Mouse neuroblastoma (clone N1E-115) cells so far represent the most widely used clone in studies of muscarinic receptor binding and function (30). Binding studies in these cells in our laboratory and in Richelson's laboratory have shown that radiolabeled muscarinic receptor antagonists demonstrate saturable and high-affinity binding to the receptors, and agonists bind to multiple receptor sites. Data from other laboratories using other neuronal clones and cell hybrids have demonstrated similar findings. Although the number of antagonist binding sites in neuroblastoma N1E-115 cells is not related to the growth cycle of the cells in culture (31), the receptor density declines as the cells are passed (15). In addition, there is a significant preferential loss of the low-affinity agonist binding sites with continuous subculturing (15). The multiple affinity states for agonists in these cells are believed to be coupled to different biochemical responses, as is discussed later.

A serious problem that is sometimes encountered when using intact cells to study muscarinic receptor binding is the relatively high levels of nonspecific binding, especially when a lipophilic ligand, such as [^{3}H]QNB, is used to label the receptors. This may be due to the contribution of ligand trapping in some intracellular compartment, probably the lysosomes (32,33). This would result in an overestimation of receptor density by including a nonreceptor-related component. In attributing all the apparent "displaceable" bound radioactivity to "specific" binding to receptor sites, serious erroneous conclusions may be reached. Such ligand trapping can be avoided if a hydrophilic ligand, such a [^{3}H]NMS, is used in the binding experiments. However, very low levels of nonspecific [^{3}H]QNB binding have also been shown in dissociated intact adult rat brain cells, making this system suitable for studying the binding of both hydrophilic and lipophilic ligands (29). In addition, ligand trapping could also be avoided or minimized if binding is studied in homogenized cultured cells. However, it should be emphasized that this manipulation totally defeats the purpose of the expense and the effort spent in growing intact cells in culture, if disrupted cells are used in binding experiments.

Cellular components of human blood have been used to study muscarinic receptors, with the aim of developing a model to reflect the changes in the properties of these receptors in pathological conditions and after prolonged drug treatment. These include erythrocytes (34), neutrophils (35), and fibroblasts (36). However, contrary to preliminary reports, muscarinic receptor binding in human fibroblasts could not be replicated by other laboratories (37). In addition, [^{3}H]QNB binding demonstrated in intact human neutrophils has a very low affinity (35), which makes the properties of muscarinic receptors on these cells quite different from those found in other human tissues.

PHARMACOLOGICAL RESPONSES

Despite the usefulness of radioligand binding studies, they do not provide complete information regarding how neurotransmitter receptors are involved in mediating a biologic response. Intact cells avail a means of combining receptor binding experiments with those demonstrating agonist-induced effects. Most importantly, they have been a very strong tool in gradually unfolding the mystery of multiple binding sites for agonists, the significance of which is not yet completely understood.

In N1E-115 cells, muscarinic receptor agonists elicit a large, rapid, and transient increase in cyclic GMP formation (Figure 4) (31,38-40). Muscarinic receptor-mediated elevation of cyclic GMP levels has the pharmacology expected for a typical muscarinic receptor-mediated effect in the central nervous system, since the ability of various antagonists to block this response closely parallels their affinities for human brain muscarinic receptors (41). It has been suggested that the cyclic GMP formed may be involved in mediating muscarinic agonist-induced hyperpolarization in these cells (42). This cyclic GMP response is also dependent on temperature (Figure 4) (38) and extracellular calcium (43). For example, it is abolished at temperatures below 20°C or in the absence of calcium in the incubation medium. In addition, calcium ionophores increase cyclic GMP levels (40) and calcium channel antagonists are capable of blocking this muscarinic receptor-mediated effect (Figure 5) in a noncompetitive fashion (44). However, muscarinic receptor agonists fail to increase calcium influx in these cells (45), but they do accelerate the rate of calcium efflux (46). The importance of extracellular calcium for cyclic GMP responses has been linked to muscarinic receptor-mediated activation of phospholipase A_2 and release of arachidonic acid, which are calcium-dependent processes (45). In addition, cyclic GMP responses to muscarinic receptor activation are related to the cell cycle. Thus, the ability of muscarinic receptor agonists to increase cyclic GMP levels in mouse neuroblastoma cells increases as the cells enter the stationary phase (31). This can be accounted for by an increase in the activity of guanylate cyclase, in the absence of any significant changes in the concentration of muscarinic receptors (Figure 6) (31). It has been suggested that agonist low-affinity binding sites are responsible for cyclic GMP formation in mouse neuroblastoma cells (19). This hypothesis is supported by recent findings by McKinney et al. (15) that continuous subculturing of these cells results in a parallel loss of both agonist low-affinity sites and the ability of agonists to increase cyclic GMP synthesis.

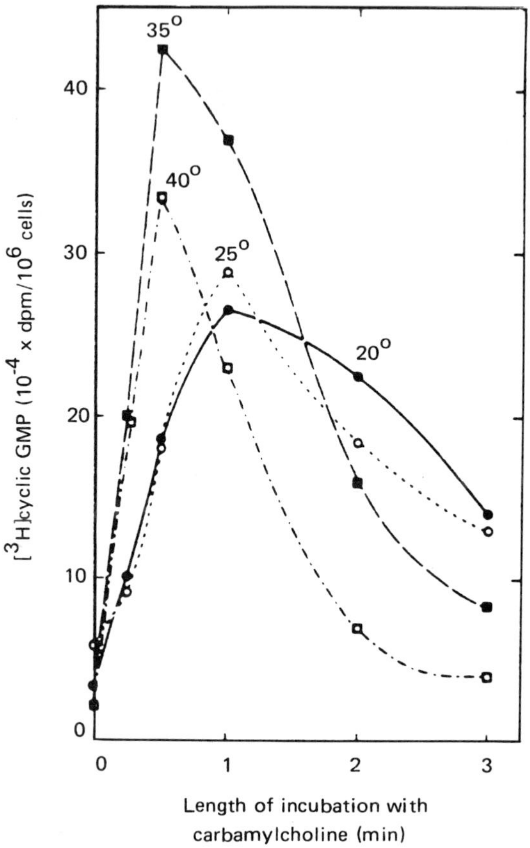

Figure 4 Agonist-stimulated cyclic GMP formation in mouse neuroblastoma cells (clone N1E-115). Cells were prelabeled with $[^3H]$guanine for subsequent measurements of $[^3H]$cyclic GMP. Prelabeled cells were incubated with 1 mM carbamylcholine at different temperatures for the indicated time periods.

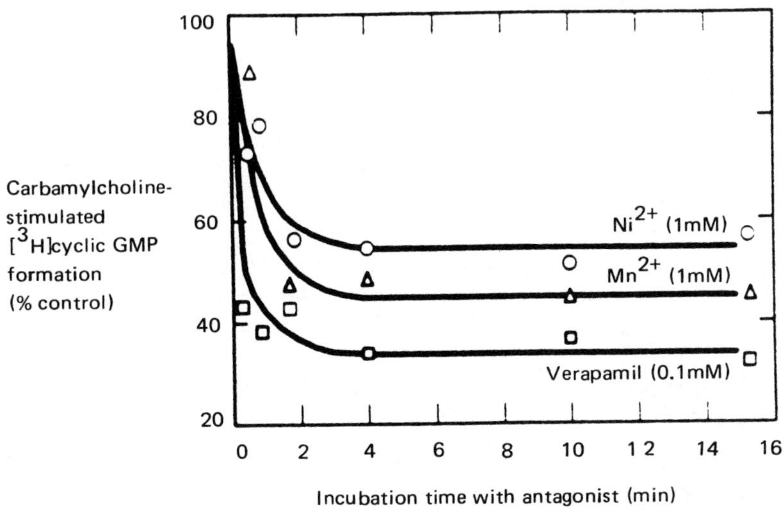

Figure 5 Effect of some calcium channel antagonists on carbamylcholine-induced cyclic GMP formation in mouse neuroblastoma (N1E-115) cells. Cells prelabeled with [^{3}H]guanine were preincubated with the calcium antagonists for different time periods at 37°C and then stimulated with 1 mM carbamylcholine for 30 sec at 37°C.

Another response that is mediated by muscarinic receptors in N1E-115 cells is the inhibition of prostaglandin E_1-induced elevations in cyclic AMP levels (47). This response is stable over several subcultures of the cells, which is different from the decline observed in cyclic GMP responses (15). An elegant study by Richelson and coworkers has provided strong evidence that the high- and low-affinity receptor conformations for agonists mediate the inhibition of cyclic AMP formation and the increase in cyclic GMP levels, respectively (16). Muscarinic receptor-mediated inhibition of cyclic AMP formation has also been reported in other cell types, such as neuroblastoma-glioma hybrids (NG108-15) (48) an human astrocytoma cells (1321N1) (49,50). It is interesting that the mechanisms implicated in the inhibition of cyclic AMP synthesis by activation of muscarinic receptors are different in various cell clones. For example, although this response involves activation of phosphodiesterase in 1321N1 cells, it is mediated by inhibition of adenyl cyclase in NG108-15 cells (51).

Activation of muscarinic acetylcholine receptors in intact cells also results in an enhancement of the hydrolysis of phosphoinositides, a response that has been implicated in the regulation of intracellular calcium concentration (2). This response has been demonstrated in mouse neuroblastoma cells (clone N1E-115)

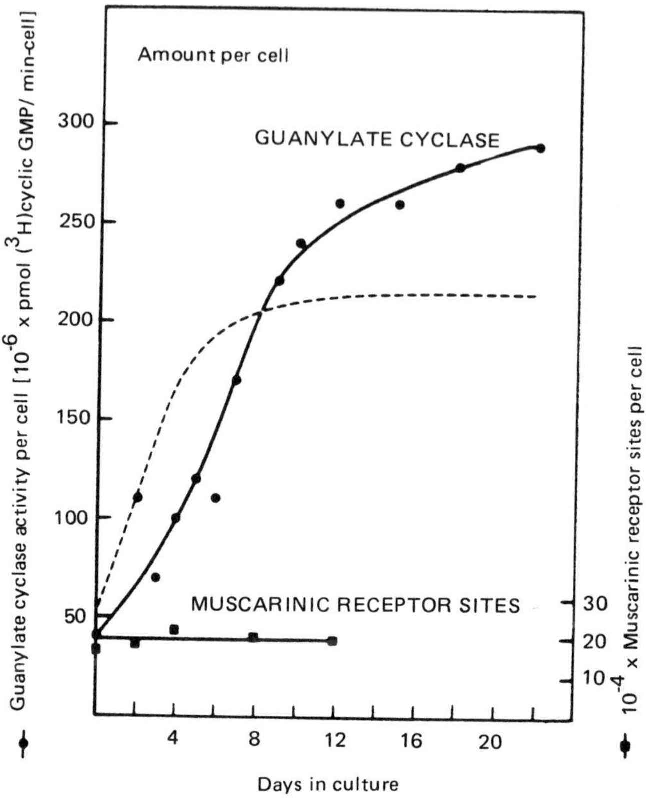

Figure 6 Relationship between days in culture, guanylate cyclase activity, and muscarinic receptor density in mouse neuroblastoma N1E-115 cells. Muscarinic receptor density was measured using 1 nM [^{3}H]QNB. The interrupted line represents cell number per culture flask.

(52), neuroblastoma glioma hybrids (53), and human astrocytoma 1321N1 cells (54,55). It is noteworthy that clonal cell lines are very useful in this regard, since phosphoinositide turnover can be elicited only in an intact cell system.

Cultured cells have also been used to demonstrate the relationship between the ability of muscarinic receptor agonists to bind to the receptors and to cause certain mechanical and ionic responses. For example, in cardiac cells in primary culture, muscarinic receptor agonists induce a negative chronotropic response (56,57), increase the rate of potassium efflux (17), and affect calcium mobilization (58). In addition, they increase phosphatidylinositol turnover and inhibit cyclic AMP formation (59). The ability of muscarinic agonists to slow the heart rate and increase potassium permeability has been correlated with their high-affinity, guanine nucleotide-sensitive binding site (60). This receptor conformation also seems to be involved in the mediation of calcium mobilization (58).

REGULATION OF MUSCARINIC RECEPTOR DENSITY AND RESPONSES

Muscarinic acetylcholine receptors in the central nervous system and in the periphery are subject to regulation upon prolonged contact with their own agonists. For example, repeated injections of acetylcholinesterase inhibitors in intact animals, a treatment that produces a sustained elevation of synaptic acetylcholine levels, result in a significant reduction in the concentration of muscarinic receptors in the brain (61-63). Similar results have been obtained upon chronic injections of muscarinic receptor agonists (64,65). Investigations of the possible molecular mechanisms underlying receptor regulation require the use of intact cells. Hence, cultured cells have been widely used to study the regulation of muscarinic receptor function and concentration. For example, brief exposure of mouse neuroblastoma (clone N1E-115) cells to muscarinic receptor agonists results in a specific and reversible desensitization of receptor-mediated increases in cyclic GMP formation (38,39,66). This desensitization is temperature, concentration, and time dependent (38). The time course of this desensitization is quite rapid, with a half-time of about 4 min at maximal agonist concentrations, and is not accompanied by any significant loss in the number of receptor sites as measured by [^{3}H] QNB binding (Figure 7) (67). However, more prolonged incubations (>30 min) of these cells with muscarinic receptor agonists cause a gradual decrease in receptor density (Figure 7), with a half-time of 4 hr (67,68). Reversal of this receptor down-regulation is also slow and requires synthesis of new receptors (67). Similar findings have been repored in neuroblastoma-glioma hybrid cells (69), in addition to cardiac cells in primary culture (70). In these latter cells, however, a biphasic decrease in [^{3}H] QNB binding has been detected (70). In cases in which detailed investigations of desensitization of muscarinic receptor-mediated functions have been carried out in an attempt to correlate the loss of the ability of agonists to activate the receptor and the loss of [^{3}H]-QNB receptor binding sites, it has been generally concluded that these two phenomena have a poor temporal association (66,67,71,72). However, better correlations have been obtained recently when [^{3}H] NMS was used to monitor the changes in receptor concentration (71,72). This ligand is very hydrophilic and is presumably able to bind to muscarinic receptors only if they exist on the cell surface. Using this ligand, we have demonstrated that brief incubation of mouse neuroblastoma clone N1E-115 cells with carbamylcholine leads to a rapid and reversible decline in [^{3}H] NMS binding, reaching a steady state within 30 min at

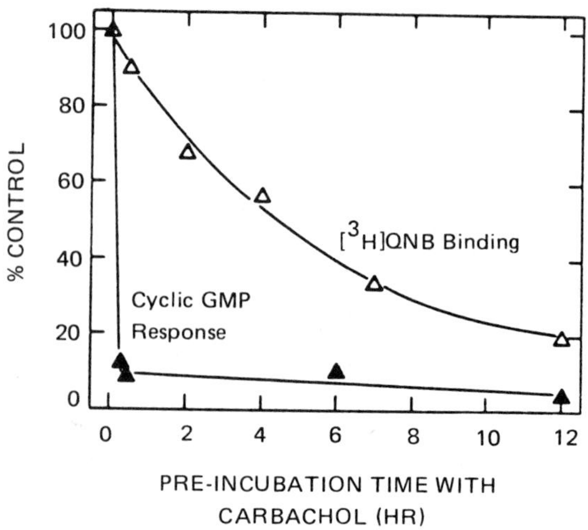

Figure 7 Time course of desensitization of muscarinic receptor-mediated cyclic GMP formation and loss of specific [^{3}H]QNB binding. Mouse neuroblastoma cells (clone N1E-115) were preincubated without (control) or with (desensitized) 1 mM carbamylcholine. The cells were then washed and used for cyclic GMP experiments (1 mM carbamylcholine, 30 sec at 37°C) or binding assays (1 nM [^{3}H]QNB, 45 min at 37°C).

37°C (Figure 8). This decline in [^{3}H]NMS binding is due solely to a decrease in the maximum binding capacity, without any significant change in the receptor affinity (72). On the other hand, similar parallel incubations do not cause significant change in [^{3}H]QNB binding. However, it should be noted that it is difficult to reach absolutely concrete conclusions regarding the regulation of [^{3}H-QNB binding sites, owing to its high and erratic nonspecific binding in this system. It has also been reported that preincubation of neuroblastoma-glioma hybrid cells (clone 108CC15) and cerebellar granule cells with carbamylcholine for very short time periods leads to a rapid disappearance of muscarinic receptors labeled with [^{3}H]NMS in intact cells, with a half-time of 4.5 min (73). In addition, Galper et al. (74) have reported that in intact heart cells in primary culture, muscarinic receptor agonists cause a much faster decline in [^{3}H]NMS binding than in [^{3}H]QNB binding. In this latter system, down-regulation of [^{3}H]NMS binding sites was compared to desensitization of the negative chronotropic effects and the increase in potassium efflux induced by muscarinic agonists. It was found that receptor down-regulation and desensitization of receptor function share a quite similar time course, suggesting a cause-and-effect relationship (74). The ability of agonists to reduce subsequent [^{3}H]NMS binding in neuroblastoma N1E-115 cells increases with increasing the agonist concen-

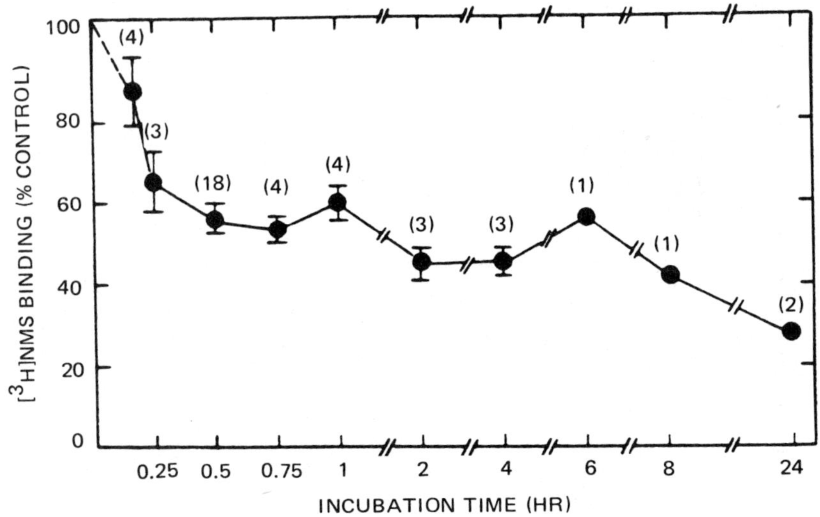

Figure 8 Time course of agonist-induced decrease of specific [^{3}H]NMS binding in mouse neuroblastoma cells (clone N1E-115). Cells in monolayer were incubated with or without 1 mM carbamylcholine for different time periods at 37°C and then harvested and washed, followed by incubation with 2 nM [^{3}H]NMS for 90 min at 15°C. Data shown are the mean ± SEM (% of control without carbamylcholine), and the numbers in parentheses represent the number of independent experiments, each determined in quadruplicate.

tration, and it reaches a maximum of 50% reduction (Figure 9) (72). The dose-response relationship for the regulation of [^{3}H]NMS binding is very similar to that for stimulation of cyclic GMP formation and desensitization of this response (2,72). In addition, the potency of carbamylcholine in this regard correlates well with its equilibrium dissociation constant for the low-affinity binding sites in these cells (2), again suggesting that activation of these sites is responsible for cyclic GMP formation. The decrease in [^{3}H]NMS binding is directly proportional to the incubation temperature, and it does not take place below 20°C. Furthermore, this effect is abolished by pretreatment of cells with atropine (10 nM). Atropine itself does not cause effects similar to those observed with agonists, suggesting that receptor activation is necessary for this process. It has been suggested that, upon short-term desensitization, the receptors are translocated either intracellularly or in a lipophilic domain in the cell membrane, where they can be labeled only by a lipophilic ligand, such as [^{3}H]QNB, while being inaccessible to [^{3}H]NMS or muscarinic receptor agonists (72). However, another possibility that should be taken into consideration is that desensitiza-

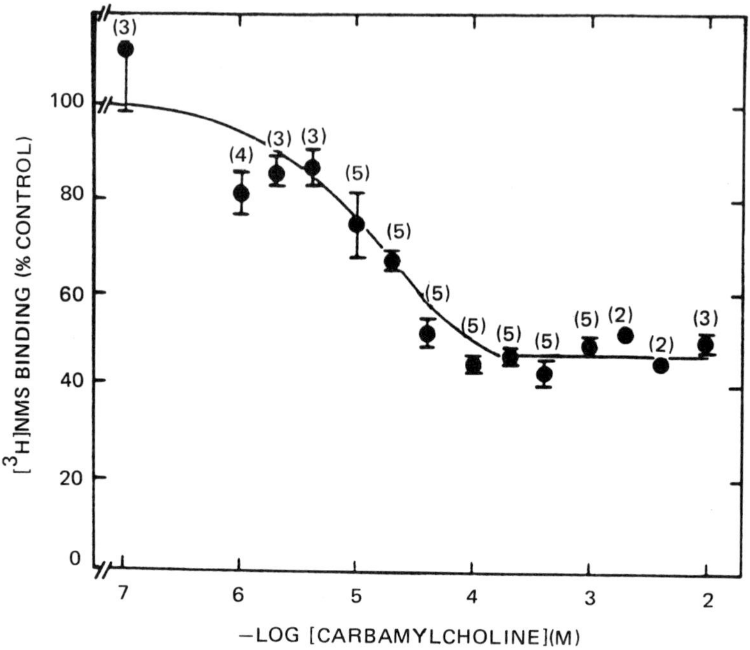

Figure 9 Dose-response relationship of carbamylcholine-induced reduction of $[^3H]$NMS binding in mouse neuroblastoma cells. Cells were incubated with different carbamylcholine concentrations for 30 min at 37°C and then harvested and washed, and the binding of 2 nM $[^3H]$NMS was determined. The data are shown as mean ± SEM, and the number of experiments is shown in parentheses.

tion may result in a receptor conformation able to bind only $[^3H]$QNB (74). The existence of internalized receptors upon desensitization still needs direct experimental evidence.

Taken together, our data suggest that the observed agonist-induced reduction in $[^3H]$NMS binding is physiologically relevant since it bears several similarities to agonist-induced desensitization of cyclic GMP formation, such as time course, dependence on agonist concentration, temperature sensitivity, and reversibility.

Comparative studies have been performed in intact cells dissociated from adult rat brain. In this system, we have found that preincubation of brain cells with carbamylcholine results in a significant reduction in the specific binding of $[^3H]$NMS to the receptors after the agonist is washed away (75). This agonist-mediated receptor down-regulation increases in magnitude with prolongation of the exposure period to the agonist, and a steady state is reached after 3 hr at

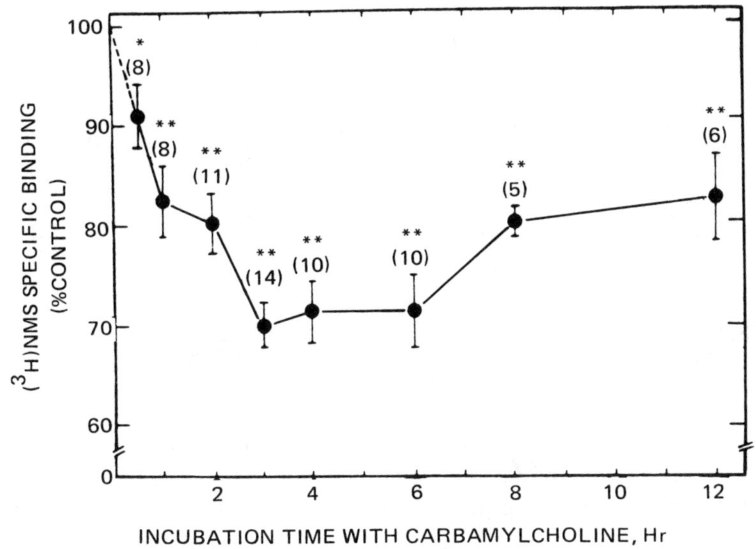

Figure 10 Time course of carbamylcholine-induced receptor down-regulation in intact rat brain cells. Cells were preincubated with 1 mM carbamylcholine for different time periods at 37°C and washed and assayed for the specific binding of 0.2 nM [^{3}H]NMS at 15°C. Specific binding is expressed as the percentage of control cells incubated under identical conditions in the absence of carbamylcholine, which averaged 237 ± 22.2 fmol/mg protein. The data are presented as the mean ± SEM of n independent experiments, each determined in quadruplicate. (*) Significantly different from control, $p < 0.025$. (**) Significantly different from control, $p < 0.005$.

37°C (Figure 10) (75). As in N1E-115 cells, carbamylcholine-induced receptor down-regulation is not apparent when exposure to agonist is performed at 15°C. In addition, it is abolished when the receptors are blocked by atropine. As in the case of mouse neuroblastoma cells, the decline in [^{3}H]NMS binding induced by agonist is reflected as a significant reduction in the receptor density with no change in receptor affinity (Table 2) (72,75). However, differences in the mode of regulation of [^{3}H]NMS binding exist between dissociated brain cells and clonal cell lines. For example, the time course of this phenomenon is slower, and the magnitude of the reduction of binding is lower in the former than the latter (72,75). The physiological significance of the regulation of muscarinic receptors in intact brain cells awaits further studies related to desensitization of muscarinic receptor-mediated biochemical responses in these cells.

In summary, intact cells provide a very useful means to investigate the correlation between binding of ligands to muscarinic receptors and their ability to

Table 2 Effect of Desensitization on the Maximum Binding B_{max} and Equilibrium Dissociation Constant K_D of [^{3}H]NMS in Intact Rat Brain Cells[a]

Treatment	B_{max} (fmol/mg protein)	K_D (nM)
Control (8)[b]	436 ± 32	0.156 ± 0.02
Desensitized (4)[c]	278 ± 30[d]	0.136 ± 0.11

[a]Values are reported as the mean ± SEM. Ligand is [^{3}H]NMS.
[b]Numer of independent determinations.
[c]Cells were preincubated with 1 mM carbamylcholine for 3 hr at 37°C.
[d]Significantly different from control ($p < 0.01$).

either activate or inhibit receptor-mediated changes in the cell function. In addition, they are extremely helpful in comparing the regulation of receptor concentration with actual loss of receptor function.

ACKNOWLEDGMENTS

The author would like to thank Mrs. Pat Tretter for her excellent secretarial help. The research from the author's laboratory cited in this chapter was supported by grants from the National Science Foundation (BNS-84-06357) and NIH (NS24158 and AG07118) and a contract from the U. S. Army Research Office (DAAG-29-85-K-0123). The author is a recipient of a Research Career Development Award from the NIH (AG-00344).

REFERENCES

1. E. C. Hulme, N. J. M. Birdsall, A. S. V. Burgen, and P. Mehta, Mol. Pharmacol. *14*, 737 (1978).
2. M. McKinney and E. Richelson, Annu. Rev. Pharmacol. Toxicol. *24*, 121 (1984).
3. N. J. M. Birdsall and C. Hulme, Trends Pharmacol. Sci. *4*, 459 (1983).
4. R. Hammer and A. Giachetti, Life Sci. *31*, 2991 (1982).
5. R. Hammer, C.P. Berrie, N. J. M. Birdsall, A. S. V. Burgen, and E. C. Hulme, Nature *283*, 90 (1980).
6. G. R. Luthin and B. B. Wolfe, J. Pharmacol. Exp. Ther. *228*, 648 (1984).
7. J.-H. Lee and E. E. El-Fakahany, Eur. J. Pharmacol. *110*, 263 (1985).
8. J.-H. Lee and E. E. El-Fakahany, J. Pharmacol. Exp. Ther. *233*, 707 (1985).
9. F. J. Ehlert, W. R. Roeske, and H. I, Yamamura, Life Sci. *28*, 2441 (1981).
10. E. C. Hulme, C. P. Berrie, N. J. M. Birdsall, and A. S. V. Burgen, Eur. J. Pharmacol. *73*, 137 (1981).

11. J. S. Aguilar, P. J. C. Salas, and E. DeRobertis, Mol. Pharmacol. *22*, 304 (1982).
12. M. W. Martin, M. M. Smith, and T. K. Harden, J. Pharmacol. Exp. Ther. *230*, 424 (1984).
13. N. J. M. Birdsall, A. S. V. Burgen, and E. C. Hulme, Mol. Pharmacol. *14*, 723 (1978).
14. N. J. M. Birdsall, E. C. Hulme, and A. S. V. Burgen, Proc. R. Soc. Lond. [Biol.] *207*, 1 (1980).
15. M. McKinney, S. Stenstrom, and E. Richelson, Mol. Pharmacol. *26*, 156 (1984).
16. M. McKinney, S. Stenstrom, and E. Richelson, Mol. Pharmacol. *27*, 223 (1985).
17. J. B. Galper, L. C. Dziekan, D. S. Miura, and T. W. Smith, J. Gen. Physiol. *80*, 231 (1982).
18. S. W. Halvorsen and N. M. Nathanson, J. Biol. Chem. *256*, 7941 (1981).
19. P. G. Strange, N. J. M. Birdsall, and A. S. V. Burgen, Biochem. Soc. Trans. *5*, 189 (1977).
20. R. F. T. Gilbert, M. R Hanley, and L. L. Iversen, Br. J. Pharmacol. *65*, 451 (1979).
21. W. C. Taft, A. A. Abdel-Latif, and R. A. Akhtar, Biochem. Pharmacol. *29*, 2713 (1980).
22. C. K. Atterwill, A. Kingsbury, J. Nicholls, and A. Prince, Br. J. Pharmacol. *83*, 89 (1984).
23. W. Frick, F. Hefti, K. Citherlet, A. Dravid, and G. Gmelin, Neurosci. Lett. *40*, 45 (1983).
24. A. Mukherjee, G. Snyder, and S. M. McCann, Life Sci. *27*, 475 (1980).
25. R. L. Taylor and D. R. Burt, Eur. J. Pharmacol. *65*, 305 (1980).
26. D. R. Burt and R. L. Taylor, Neuroendocrinology *30*, 344 (1980).
27. N. Brooks and D. R. Burt, Dev. Neurosci. *3*, 118 (1980).
28. Y. Dudai and E. Yavin, Brain Res. *155*, 368 (1978).
29. J.-H. Lee and E. E. El-Fakahany, Biochem. Pharmacol, *34*, 4299 (1985).
30. E. Richelson, in International Review of Biochemistry, Vol. 26 (K. F. Tipton, ed.), University Press, Baltimore, 1979, pp. 81.
31. E. E. El-Fakahany and E. Richelson, J. Neurochem. *35*, 941 (1980).
32. J.-M. Maloteaux, A. Gossuin, P. J. Pauwels, and P. M. Laduron, FEBS Lett. *156*, 103 (1983).
33. A. Gossuin, J.-M. Maloteaux, A. Trouet, and P. Laduron, Biochim. Biophys. Acta *804*, 100 (1984).
34. L. C. Tang, E. Schoomaker, and W. P. Weismann, Biochim. Biophys. Acta *772*, 235 (1984).
35. B. H. Dulis, M. A. Gordon, and I. B. Wilson, Mol. Pharmacol. *15*, 28 (1979).
36. N. S. Nadi, J. I, Numberger, Jr., and E. S. Gershon, N. Engl. J. Med. *311*, 225 (1984).
37. R. H. Lenox, R. J. Hitzemann, E. Richelson, and J. R. Kelsoe, N. Engl. J. Med. *312*, 861 (1985).
38. E. E. El-Fakahany and E. Richelson, J. Neurochem. *34*, 1288 (1980).

39. E. E. El-Fakahany and E. Richelson, Proc. Natl. Acad. Sci. USA *77*, 6897 (1980).
40. E. Richelson and E. E. El-Fakahany, Biochem. Pharmacol. *30*, 2887 (1981).
41. E. E. El-Fakahany and E. Richelson, Br. J. Pharmacol. *78*, 97 (1983).
42. G. J. Wastek, J. R. Lopez, and E. Richelson, Mol. Pharmacol. *19*, 15 (1981).
43. E. Richelson, F. G. Prendergast, and S. Divinetz-Romero, Biochem. Pharmacol. *27*, 2039 (1978).
44. E. E. El-Fakahany and E. Richelson, J. Neurochem. *40*, 705 (1983).
45. R. M. Snider, M. McKinney, C. Forray, and E. Richelson, Proc. Natl. Acad. Sci. USA *81*, 3905 (1984).
46. T. Deguchi, S. Ohsako, M. Nakane, M. Ichikawa, and M. Yoshida, J. Neural Transm. (Suppl.) *18*, 369 (1983).
47. H. Matsuzawa and M. Nirenberg, Proc. Natl. Acad. Sci. USA *72*, 3472 (1975).
48. D. A. Green and R. B. Clark, J. Neurochem. *39*, 1125 (1982).
49. R. B. Meeker and T. K. Harden, Mol. Pharmacol. *22*, 310 (1982).
50. R. B. Meeker and T. K. Harden, Mol. Pharmacol. *23*, 384 (1983).
51. T. Evans, M. M. Smith, L. I. Tanner, and T. K. Harden, Mol. Pharmacol. *26*, 395 (1984).
52. N. M. Cohen, D. M. Schmidt, R. C. McGlennen, and W. L. Klein, J. Neurochem. *40*, 547 (1983).
53. R. G. Siman and W. L. Klein, J. Neurochem. *37*, 1099 (1981).
54. S. B. Masters, T. K. Harden, and J. H. Brown, Mol. Pharmacol. *26*, 149 (1984).
55. S. B. Masters, M. T. Quinn, and J. H. Brown, Mol. Pharmacol. *27*, 325 (1984).
56. J. B. Galper, W. Klein, and W. A. Catterall, J. Biol. Chem. *252*, 8692 (1977).
57. J. B. Galper and T. W. Smith, Proc. Natl. Acad. Sci. USA *75*, 5831 (1978).
58. G. Oettling, H. Schmidt, and U. Drews, J. Cell Biol. *100*, 1073 (1985).
59. S. A. Orellana and J. H. Brown, Biochem. Pharmacol. *34*, 1321 (1985).
60. J. B. Galper, L. C. Dziekan, and T. W. Smith, J. Biol. Chem. *259*, 7382 (1984).
61. J. Ben-Barak, H. Grazit, J. Silman, and Y. Dudai, Eur. J. Pharmacol. *74*, 73 (1981).
62. L. G. Costa, B. W. Schwab, and S. D. Murphy, Biochem. Pharmacol. *31*, 3407 (1982).
63. S. P. Sivam, J. C. Norris, D. K. Lim, B. Hoskins, and I. K. Ho, J. Neurochem. *40*, 1414 (1983).
64. M. J. Marks, L. D. Artman, D. M. Patinkin, and A. C. Collins, J. Pharmacol. Exp. Ther. *218*, 337 (1981).
65. M. R. Meyer, M. W. Grainer, and N. M. Nathanson, Mol. Pharmacol. *21*, 280 (1982).
66. E. Richelson, Nature *272*, 366 (1978).
67. J. E. Taylor, E. E. El-Fakahany, and E. Richelson, Life Sci. *25*, 2187 (1979).
68. R. G. Siman and W. L. Klein, Proc. Natl. Acad. Sci. USA *76*, 4141 (1979).

69. W. L. Klein, N. Nathanson, and M. Nirenberg, Biochem. Biophys. Res. Commun. *90*, 506 (1979).
70. J. B. Galper and T. W. Smith, J. Biol. Chem. *255*, 9579 (1980).
71. P. Feigenbaum and E. E. El-Fakahany, Res. Commun. Chem. Pathol. Pharmacol. *43*, 519 (1984).
72. P. Feigenbaum and E. E. El-Fakahany, J. Pharmacol. Exp. Ther. *233*, 134 (1984).
73. J.-M. Maloteaux, A. Gossuin, C. Waterkeyn, and P.-M. Laduron, Biochem. Pharmacol. *32*, 2543 (1983).
74. J. B. Galper, L. C. Dziekan, D. S. O'Hara, and T. W. Smith, J. Biol. Chem. *257*, 10344 (1982).
75. J.-H. Lee and E. E. El-Fakahany, Life Sci. *37*, 515 (1985).

22

Dopamine Receptors

H. B. Niznik and K. R. Jarvie / University of Toronto, Toronto, Ontario, Canada

HISTORICAL PERSPECTIVE

The contention that the primary site of action of neuroleptic drugs is the blockade of postsynaptic receptors for the neurotransmitter dopamine (DA) was introduced in the 1960s (1,2), soon after the successful use of neuroleptics in the treatment of psychotic symptoms. At that time, however, there was no biochemical assay for DA receptors and consequently no direct evidence for this hypothesis. In the early 1970s, two major developments occurred. The first was the identification in neural tissue of a DA-sensitive adenylate cyclase (3,4) the activity of which was competitively inhibited by the phenothiazine class of neuroleptics (5-7) (see Figure 1). In general, there was a strong parallel between the pharmacological potencies for phenothiazines as antagonists of DA-stimulated adenylate cyclase and antipsychotic activity in the human, which in essence supported the hypothesis that neuroleptic action is mediated via the blockade of dopamine receptors. However, there were discrepancies that were initially overlooked, particularly with the butyrophenone class of neuroleptics (see Figure 1). For instance, spiperone, which is much more potent than chlorpromazine in controlling schizophrenia, was virtually inactive in inhibiting DA-sensitive adenylate cyclase. Similar results were obtained with potent substituted benzamide neuroleptics (see Figure 2). This raised the possibility that there was more than one type of DA receptor; that is, butyrophenones and substituted benzamides exhibit low affinity for DA receptors coupled to the stimulation of adenylate cyclase while displaying high affinity for dopamine receptors mediating most of the clinical affects of neuroleptics. In 1975, the single methodological breakthrough that provided direct in vitro evidence for the site of action of neuro-

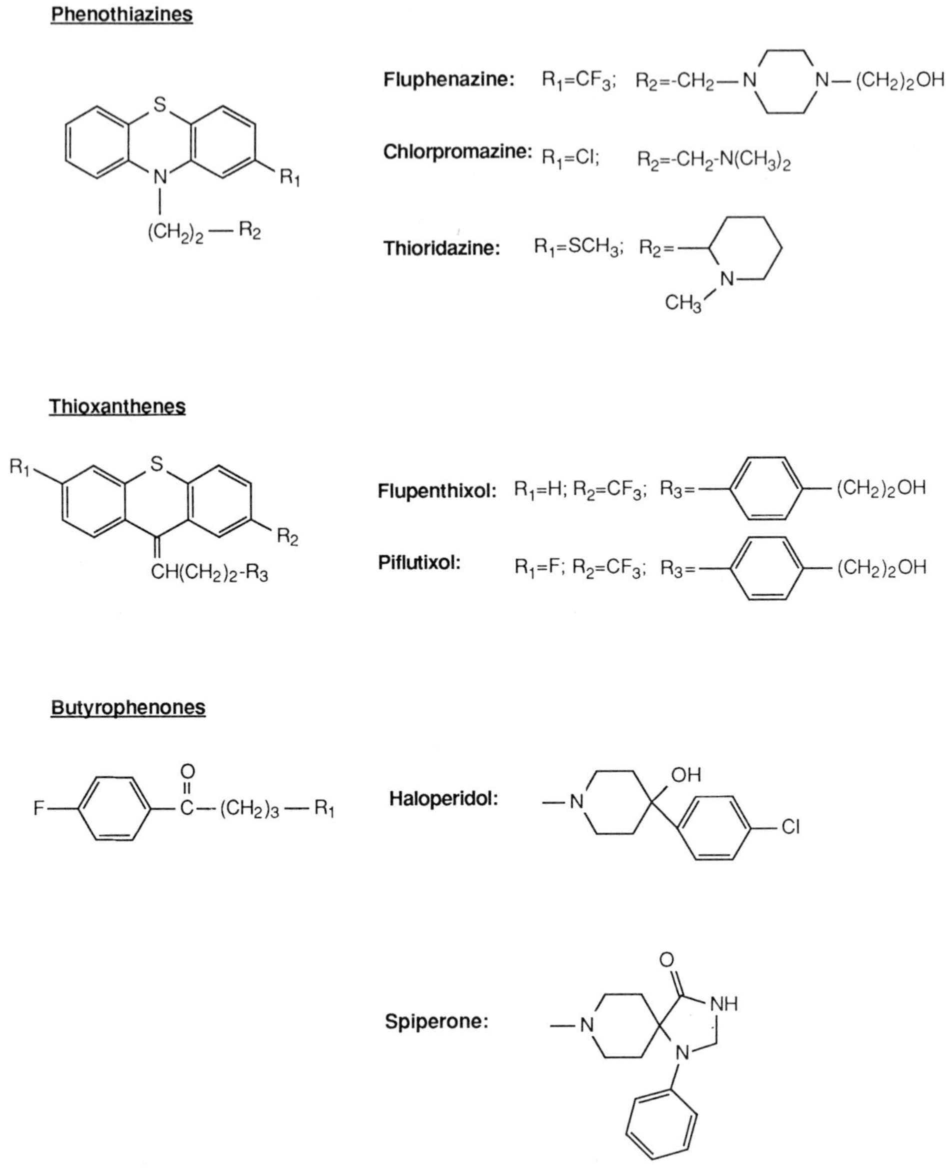

Figure 1 Chemical structures of some neuroleptic drugs.

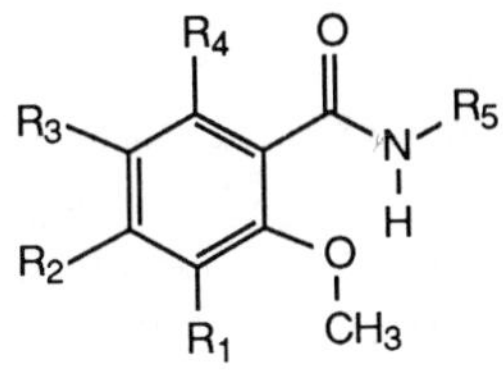

	R_1	R_2	R_3	R_4	R_5
sulpiride	-H	-H	$-SO_2NH_2$	-H	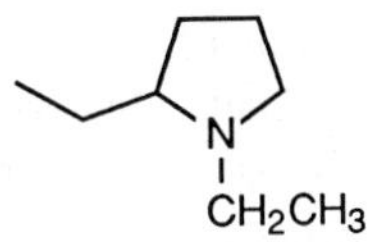
YM-09151-2	-H	$-NHCH_3$	-Cl	-H	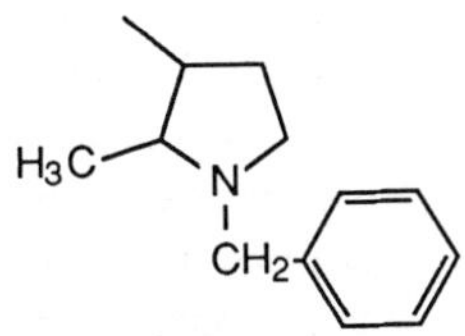
sultopride	-H	-H	$-SO_2C_2H_5$	-H	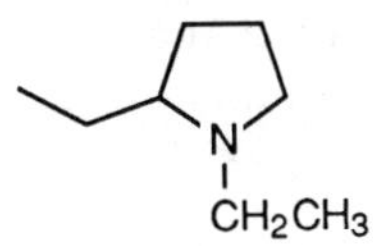
raclopride	-Cl	-H	-Cl	-OH	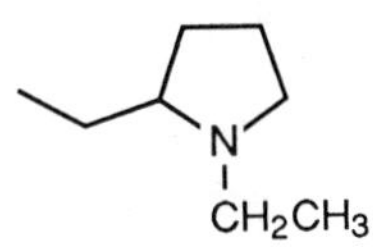
eticlopride	-Cl	-H	$-CH_2CH_3$	-OH	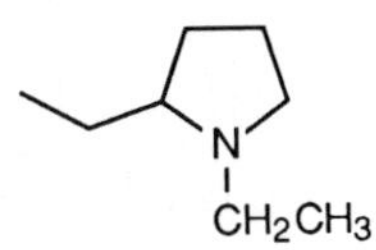

Figure 2 Substituted benzamide D_2 antagonists.

leptics and given rise to a wealth of pharmacological and biochemical data on DA receptors was the development of direct radioligand binding methodology with the use of [^{3}H]butyrophenone neuroleptics (8,9). It was soon found that the affinities of various structural classes of neuroleptics for DA receptors identified by either [^{3}H]haloperidol or [^{3}H]spiperone correlated closely with their clinical potencies in the human (10,11). A couple of years later, a high-affinity binding site labeled by the [^{3}H]thioxanthenes, [^{3}H]flupenthixol (12-15) and [^{3}H]piflutixol (see Figure 1) (16), was identified in neural tissue and appeared to represent the DA receptor coupled to the stimulation of adenylate cyclase. As expected, this receptor displays poor affinity for neuroleptics of the butyrophenone and substituted benzamide class. The potency of phenothiazine and thioxanthene neuroleptics in inhibiting DA-sensitive cyclase activity correlates well with their affinity in displacing [^{3}H]flupenthixol but not [^{3}H]spiperone binding (17).

Following considerable controversy, the present consensus is that there are two distinct DA receptor moieties that are classified on the basis of pharmacological and biochemical criteria (reviewed in References 18-24). These receptors have been termed D_1 and D_2 and are defined biochemically by their ability to stimulate (D_1) or inhibit (D_2) adenylate cyclase activity. In addition to the mode of receptor-effector coupling, DA D_1 and D_2 receptors are discriminated pharmacologically by their ability to respond to and bind specific dopaminergic agonists and antagonists. Indeed, few developments in the past 5 years have made a greater impact on the biochemical identification of DA D_1 and D_2 receptors than the synthesis of agonists and antagonists selective for each of these receptors. Thus, the benzazepine, SKF-38393 (see Figure 3), and the partial

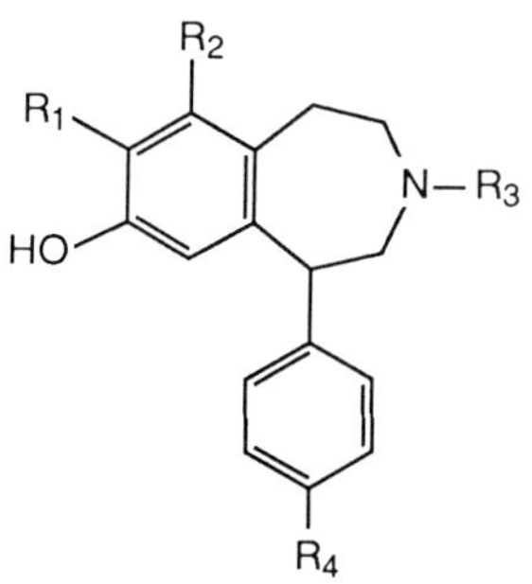

	R_1	R_2	R_3	R_4
SK&F 38393	OH	H	H	H
SK&F 82526	OH	Cl	H	OH
SCH 23390	Cl	H	CH_3	H
SK&F 75670	OH	H	CH_3	H
SK&F 83509	Cl	H	H	H
SK&F 83566	Br	H	CH_3	H
SCH 23982	I	H	CH_3	H

Figure 3 Benzazepines selective for the D_1 receptor.

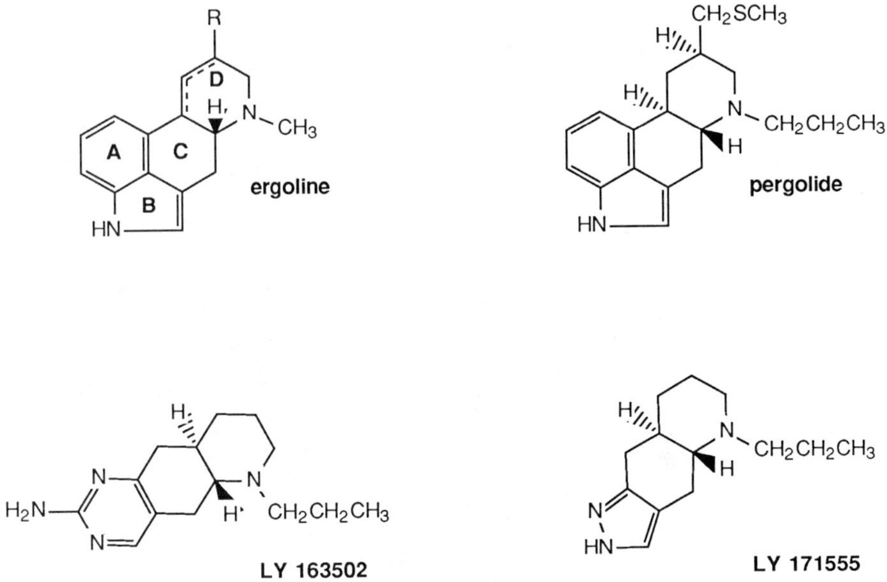

Figure 4 Ergoline-derived D_2-selective agonists.

ergolines, LY 171555 (see Figure 4), and (+)PHNO (MK458) (see Figure 5), are the most selective D_1 and D_2 receptor agonists, respectively (21,22,24-26). Although selective DA D_2 receptor antagonists, such as the substituted benzamides, sulpiride and YM-09151-2, and the potent butyrophenones like spiperone, have been available for some time, only recently has a selective D_1 receptor antagonist of the benzazepine class, SCH-23390, been developed (27). Moreover, the availability of these ligands in radiolabeled form has made it feasible to characterize both D_1 and D_2 receptors with radioligand binding techniques.

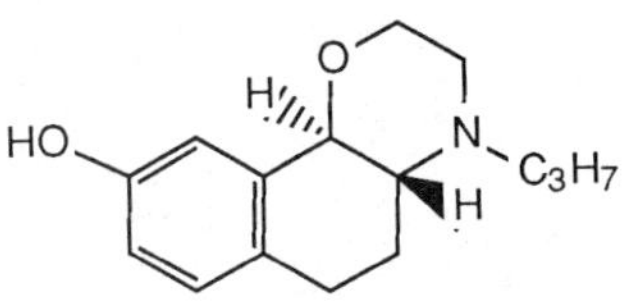

Figure 5 Diagram of 1,2,3,4a,5,6-hexahydro-9hydroxy-4-n-propyl-4H-napth-[1,2-b] [1,4]-oxane [(+)PHNO] : D_2-selective hexahydronaphthoxazine agonist.

Table 1 Radioligands for Dopamine Receptors[a]

Radioligand	Receptor labeled	K values (nM) of nonradioactive molecule at other neurotransmitter receptors				
		D_1	D_2	$5HT_2$	α_1	α_2
Antagonists						
[^{3}H](–)DO-710 (28)	D_2	1000	3	∼1000		
[^{3}H]domperidone (29-33)	D_2	∼1000 (23)	0.6	78 (34)		
[^{3}H]eticlopride (35)	D_2	>1000	0.09 (24)	288 (24)	110	
[^{3}H]haloperidol (36,82)	D_2	71[u]	1.5	∼ 25 (33)	∼ 12 (37,38)	m2000(39)
[^{3}H]metoclopramide (40)	D_2	>1000[u]	50 (24)	2500 (35)		900(39)
[^{3}H]*N*-methyl-spiperone (41,42)	D_2		0.23			
[^{3}H]raclopride (43)	D_2	>1000	3.4 (24)	2200 (24)	> 30	
[^{3}H]spiperone (33,44,45)	D_2	∼ 400[u]	0.08 (24)	∼ 1 (24,35)	∼ 8 (37,46)	∼ 700(39)
[^{3}H]sulpiride (47)	D_2	>1000[u] (57)	11 (24)	>10,000 (24,46)	>10,000 (46)	>1000(39)
[^{3}H]sultopride (48)	D_2	>40,000 (57)	5.8			
[^{3}H]YM-09151-2 (49)	D_2	> 1000[u] (57)	0.05 (24,49)	18 (24)		
[^{125}I]iodospiperone (50)	D_2		0.3	> 1000		
[^{125}I]sulpiride (51,52)	D_2	> 1000	1-6	> 1000	∼ 3000	
[^{3}H]flupentixol (12,13, 53,54,58)	$D_1 < D_2$	18[u]	∼ 1.0 (46,55)	6 (45)	∼ 10 (37,46)	∼ 150(39)
[^{3}H]fluphenazine (55,56)	D_1-D_2	∼ 1 (57)	∼ 1.0 (46)		15 (46)	∼ 900(39)

[^{3}H] piflutixol (16,59,92)	D_1-D_2	∿	0.3 (57)						
[^{3}H] SCH-23390 (57,60-66)	D_1		0.09	∿ 800u (46)	∿	15 (46,61)	>	600 (47.71)	∿100(67)
[^{3}H] SKF-83566 (68)	D_1	∿	1.1 (38)	>1000			>	1000 (69)	
[^{125}I] IODO-SCH-38548 (70)	D_1		0.3						
[^{125}I] IODO-SCH-83692 (71,72)	D_1	∿	0.3						
Agonists									
[^{3}H] ADTN(6,7-diol-2-amino tetralin) (73-75)	$D_1 > D_2$		∿ 4-60[u](57)[b]	∿10 (25,76)					
[^{3}H] apomorphine (74)	$D_1 \geqslant D_2$		∿ 1-10[u] (57)	0.8 (76)		935 (33)		2300 (77)	
[^{3}H] dihydroergocriptine (78,79)	$D_1 > D_1$			∿ 0.6 (76,79)					
[^{3}H] dopamine (36,80-82)	$D_1 > D_2$		0.8-10[u] (62)	8.0 (76)		93,000 (33)		44,000 (77)	
[^{3}H] fenoldopam (SKF 82526) (83)	$D_1 > D_2$		2.3	>100			>	1000	
[^{3}H] *N*-propylnorapomorphine (NPA) (84,85)	$D_1 > D_2$		∿ 2.0[u] (25)	0.4 (76)					
[^{3}H] pergolide (86)	$D_2 > D_1$		∿ 1.0[u] (25)	0.14 (76)		150		100	
[^{3}H] Sandoz-205-501 (87)	$D_2 > D_1$	>	1000	1.0		>10,000		10,000	
[^{3}H] SKF-38393 (88-91)	$D_1 > D_2$		∿ 1.5 (25,57)	157					

[a]Numbers in parentheses refer to reference source; u = Niznik, Fogel and Seeman, unpublished data.
[b]K_d = dissociation constant for high-affinity state.

GENERAL CHARACTERISTICS OF DOPAMINE RECEPTOR CLASSIFICATION AND FUNCTION

Radiologands for DA Receptors

Numerous radiolabeled antagonist and agonist ligands have been developed to identify dopamine D_1 and D_2 receptors; a number of these are summarized in Table 1. Iodinated ligands possess the advantage of increased sensitivity of receptor detection because of their much higher specific activity (∿2200 Ci/mmol). Unfortunately, most iodinated probes available for the study of D_1 and D_2 receptors possess lower affinity than the [^{3}H]-labeled parent molecule.

Several [^{3}H]-labeled agonist ligands have been developed to study DA receptors. Among these, N-propylapomorphine ([^{3}H]NPA) is the most commonly used because of its fairly high affinity. In most systems, [^{3}H]agonist ligands bind only to a portion of the total DA receptor population, that is, the receptor pool capable of forming a stable high-affinity guanine nucleotide-sensitive complex with the agonist ligand. The one major drawback with most [^{3}H]agonists for DA receptors is their poor selectivity in labeling D_1 or D_2 receptors. For a more complete listing of available radioligands, see Reference 23.

Classification of DA Receptors by Radioligand Binding

By definition, DA receptors are classified not only by the [^{3}H]ligand used, but also by the rank order of potency of various dopaminergic agonists and antagonists in inhibiting the binding of the radioactive probe. This is simply because most available [^{3}H]ligands are not absolutely selective for either D_1 or D_2 DA receptors but exhibit considerable affinity for other neurotransmitter receptors as well (see Table 1). [^{3}H]spiperone, for example, which has become the ligand of choice for labeling D_2 receptors because of its high affinity (∿80 pM) also displays high affinity (∿1 nM) for serotonin - 2 receptors (45,93, 94). Similarly, [^{3}H]SCH-23390, the most selective D_1 receptor antagonist available to date, displays fairly high affinity (∿7 nM) for $5HT_2$ receptors (61,95-99). The most selective [^{3}H]ligands available for D_2 receptors are of the substituted benzamide class, DO-710, eticlopride, and sulpiride (Table 1).

The rank order of potencies obtained from radioligand bindin on D_1 receptors in striatal membranes is, for agonists, SKF-38393 > (−)NPA > dopamine 6,7,dihydroxy-2-amino-tetralin [(+)ADTN] > norepinephrine > serotonin > LY-171555 and, for antagonists, SCH-23390 > (+)butaclamol > α-flupenthixol > haloperidol > spiperone > ketanserin >> substituted benzamides. For D_2 receptors the agonist rank order is (−)NPA > (+)ADTN > apomorphine > LY-171555 > dopamine > norepinephrine > SKF-38393 > serotonin and, for antagonists, spiperone > (+)butaclamol > haloperidol > (−)sulpiride > ketanserin > SCH-23390. The rank order of potencies of some agonists and antagonists for D_1 and D_2 receptors identified by [^{3}H]SCH-23390 and [^{3}H]spiperone in competition binding experiments are illustrated in Figure 6. Similar rank orders of potency for dopaminergic agonists and antagonists are obtained in assays of DA-stimulated (D_1) (14,27, 57,60,82,100), or inhibited (D_2) (101) adenylate cyclase activity and in numerous pharmacological and behavioral assay systems (27,102-105).

Using either DA-sensitive cyclase or direct [^{3}H] ligand binding, dopamine D_1 and/or D_2 receptors have been identified in numerous tissues and species, a sampling of which includes the central and peripheral nervous systems of both vertebrates (see above references and 51 and 106-116) and invertebrates (117-120); the retina (see References 21,66,121-124 and references therein), where dopamine modulates the electrical coupling between horizontal cells (125); and in pancreatic acinar cells (126), adrenal chromaffin cells (42), the adrenal glomerulosa (see Reference 127 and references therein), clonal human astrocytoma cell lines (128), the vas deferens (129), human term placenta (130), portal veins (131) and smooth muscle of *Mytilus* (132,133).

Classification of Cardiovascular DA Receptors

Two types of DA receptor have independently been identified in the cardiovascular system, designated DA_1 and DA_2. In general, it appears that DA_1 and DA_2 receptors are similar in their pharmacological characteristics to D_1 and D_2 receptors defined in either the brain or other peripheral tissues (see below) (21, 23,134-136). DA_1 receptors have been identified in renal, mesenteric, splenic, coronary, cerebral, and pulmonary arteries and vascular beds (see References 137-149 and references therein), where dopamine elicits the relaxation of vascular smooth muscle, and possibly in cardiac (150) and renal tissue (151,152) as well. It is surprising to note that although the existence of dopamine receptors in the cardiac and renal vasculature is supported by numerous pharmacological and biochemical data (23,153), the positive identification of these sites as D_1 or D_2 receptors, as indexed by radioligand binding of selective antagonist ligands, has been attempted but has not been successfully demonstrated (154). These proteins may be of a slightly different molecular constitution than those observed in either the brain or other tissues. In any event, activation of cardiovascular DA_1 receptors appears to stimulate adenylate cyclase activity (154). DA_2 receptors appear to be localized on preganglionic sympathetic nerve terminals that mediate the in vitro and in vivo inhibition of norepinephrine release (23, 106,153). The molecular form of these receptors has yet to be extensively characterized.

As originally defined by Kebabian and Calne (20), the prototypical locations of dopamine D_1 and D_2 receptors are the parathyroid and pituitary glands, respectively. The general pharmacological and molecular characteristics of these receptors are summarized in Table 2.

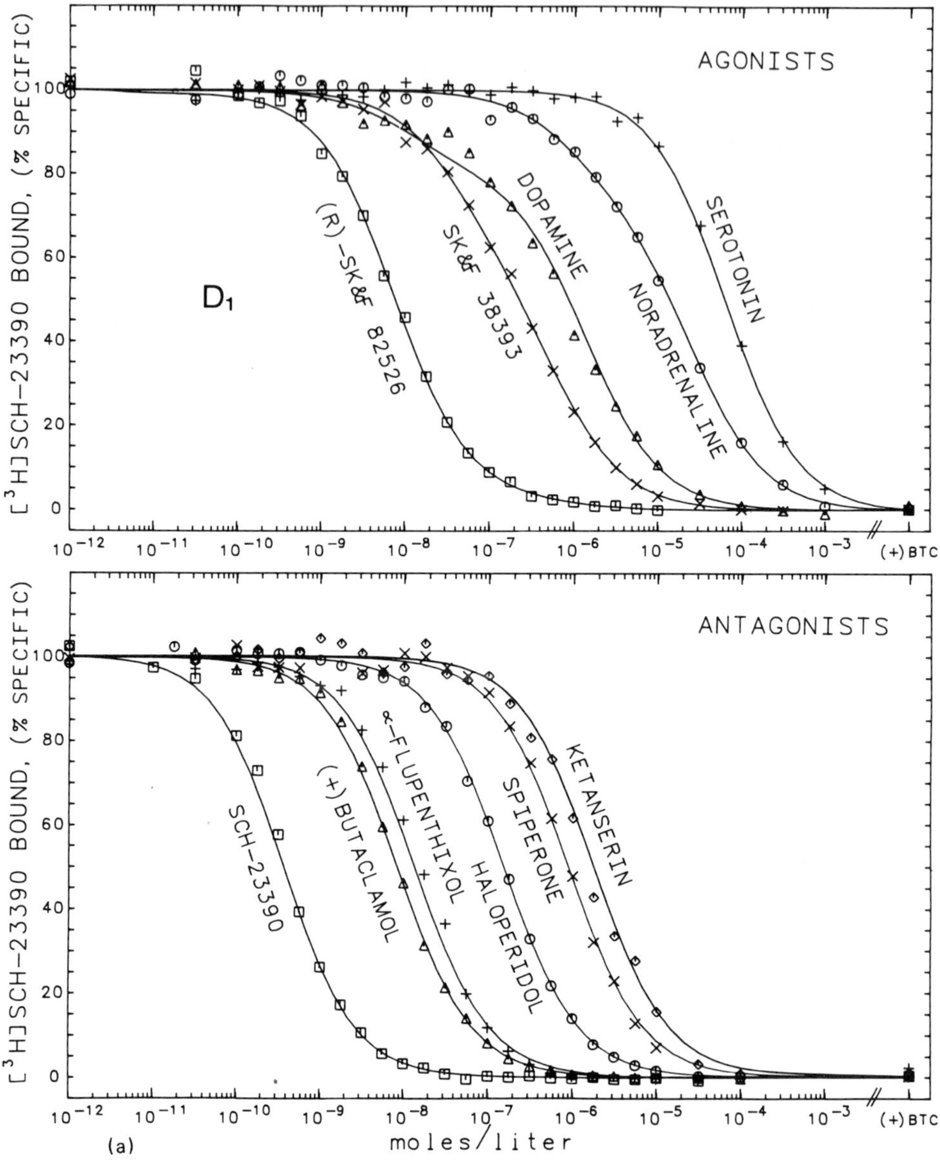

Figure 6 Rank order of potencies of dopaminergic agonists and antagonists on striatal D_1 or D_2 receptors. Increasing concentrations of agonists or antagonists were incubated with receptors and [^{3}H] SCH-23390 (∼100 pM) or [^{3}H] spiperone (∼80 pM) and assayed for D_1 and D_2 receptor activity, respectively, using standard rapid filtration techniques. Data were analyzed for one- and two-site fits by the computer program LIGAND.

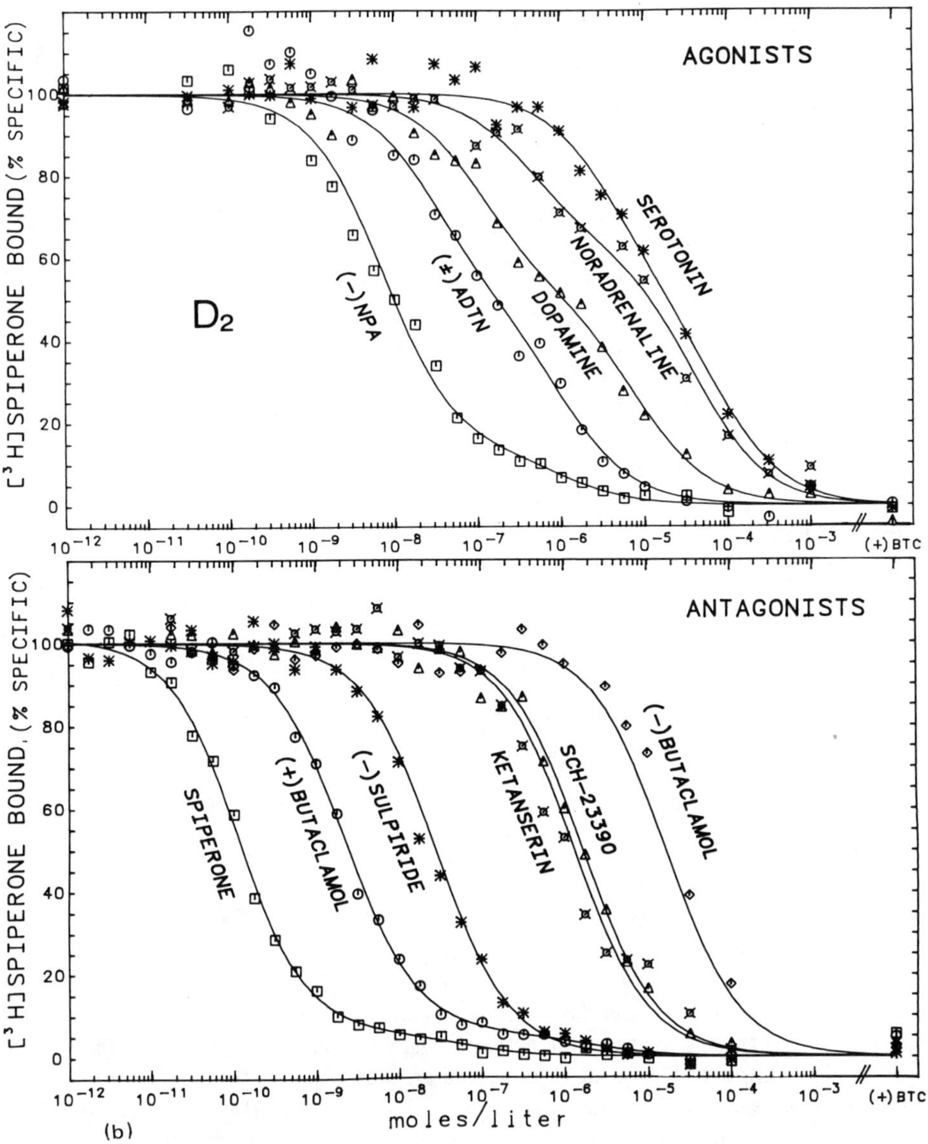

Figure 6 (Continued)

Table 2 Dopamine Receptor Subtypes

Receptor archetype	D_1	D_2
Location	Parathyroid gland	Anterior or intermediate pituitary lobes
Agonist potency series	DA $\geqslant$ (±)-ADTN > SKF-38393 > APO > NE	NPA > (±)-ADTN > APO > DA > NE > 5-HT
Specific agonists	SKF-38393	LY-171555
Specific antagonists	SCH-23390 SKF-83692 SKF-83566	Sulpiride Eticlopride YM-09151-2 Spiperone
Physiological response	Stimulation of parathyroid hormone release	Inhibition of prolactin and MSH release
Biochemical response	Stimulation of adenylate cyclase, phosphorylation of cAMP-dependent protein kinase; phosphorylation of DARPP-32 (?)	Inhibition of adenylate cyclase, inhibition of phosphatidylinositol metabolism; Ca^{2+} mobilization
Transducer units	Guanine nucleotide binding protein, G_s	Guanine nucleotide binding protein(s) G_i (G_o ?)

D_1 Dopamine Receptors

In the parathyroid, DA as well as other dopaminergic agents (SKF-38393 and ADTN) causes a transient increase in the rate of bovine parathyroid hormone (PTH) release both in vitro (155-159) and in vivo (160,161). DA-mediated PTH release appears to be independent of that stimulated by β-adrenergic agents or low Ca^{2+} but is dependent on the activation of adenylate cyclase activity (162-165). The effects of DA in the parathyroid are stereoselectively blocked by dopaminergic antagonists, such as flupenthixol, but not by selective D_2 receptor antagonists, such as YM-09151-2 (see Table 2). The radioligand binding characteristics of [^{3}H] piflutixol, a nonselective D_1 and D_2 receptor antagonist, in crude homogenates of human parathyroid appears to confirm the existence of a D_1 dopamine receptor subtype in this tissue (166). Moreover, DA-stimulated adenylate cyclase activity and PTH release are sensitive to both GTP and cholera toxin and appear to be due to the association of the receptor with a guanine nucleotide binding protein, presumably G_s (coupled to the stimulation of adenylate cyclase). Cholera toxin ADP-ribosylation of the α subunit of G_s (M_r = 45,000) leads to a persistent activation of G_s and increased cAMP production (167,168) by reducing intrinsic GTPase activity.

The molecular nature of the parathyroid D_1 receptor and the mechanism(s) by which DA mediates PTH release are still unknown. Possibly, DA-mediated cAMP-dependent protein kinase activity leads to enhanced PTH release (169, 170). Whether D_1 receptors in the parathyroid are associated with the phosphorylation of a dopamine-adenosine monophosphate-regulated phosphoprotein (M_r = 32,000) or DARPP-32 (see Reference 171 and references therein), as in the central nervous system (see below), is unknown at present. Ca^{2+} inhibition of PTH release, however, appears to be mediated by protein kinase C(172) and Ca^{2+}/calmodulin-dependent protein kinase C activity (173), possibly stimulated by a pertussis toxin-sensitive calcium channel (174,175). Further characterization of the parathyroid D_1 receptor in terms of its ligand binding properties and subunit composition should provide significant insight on the role of this protein in regulating PTH activity and provide a model system in which to study receptor-effector coupling and receptor desensitization.

Unlike DA receptors in the parathyroid, D_1 receptors in the central nervous system have been extensively characterized by radioligand binding techniques using the selective D_1 receptor antagonist probes [^{3}H] SCH-23390 (57,60,111, 113-115) and [^{125}I] SCH-23982 (see Figure 3) (71,72,107,112,116). These ligands bind to striatal membranes in a saturable and reversible manner with high affinity ($K_D \sim 0.1$-0.3 nM). Dopaminergic agonists and antagonists inhibit the binding of these probes in a stereoselective fashion with a typical D_1 receptor profile. Thus, thioxanthenes (e.g., fluphentixol) and benzazepines (e.g., SCH-23390 and SKF-38393) display nanomolar potency for D_1 sites but classic neuroleptics of the butyrophenone (e.g., spiperone) or substituted benzamide class (e.g., sulpiride and YM-09151-2) are virtually inactive (500-10,000 nM).

The association of D_1 receptors with a guanine nucleotide binding protein, presumably G_s, has been implied by virtue of the fact that guanine nucleotides modulate agonist high-affinity binding interactions. This is reflected in the rather shallow (>2 log units) agonist/[^{3}H] SCH-23390 competition curves (illustrated in Figures 6 and 7) where DA, for example, recognizes two affinity forms of the receptor, one with high affinity, termed D_1^{high}, and one with low affinity, D_1^{low}. Similarly, guanine nucleotides inhibit [^{3}H] agonist binding to the high-affinity form of D_1 receptors (82,83,88-91). In contrast, D_1 receptor antagonists recognize only one form of the D_1 receptor as indexed by steep (∼2 log units) competition curves (Figure 7) and their insensitivity to added guanine nucleotides. As with the β-adrenergic receptor (176), which is linked in a stimulatory fashion to adenylate cyclase, the agonist high-affinity form of

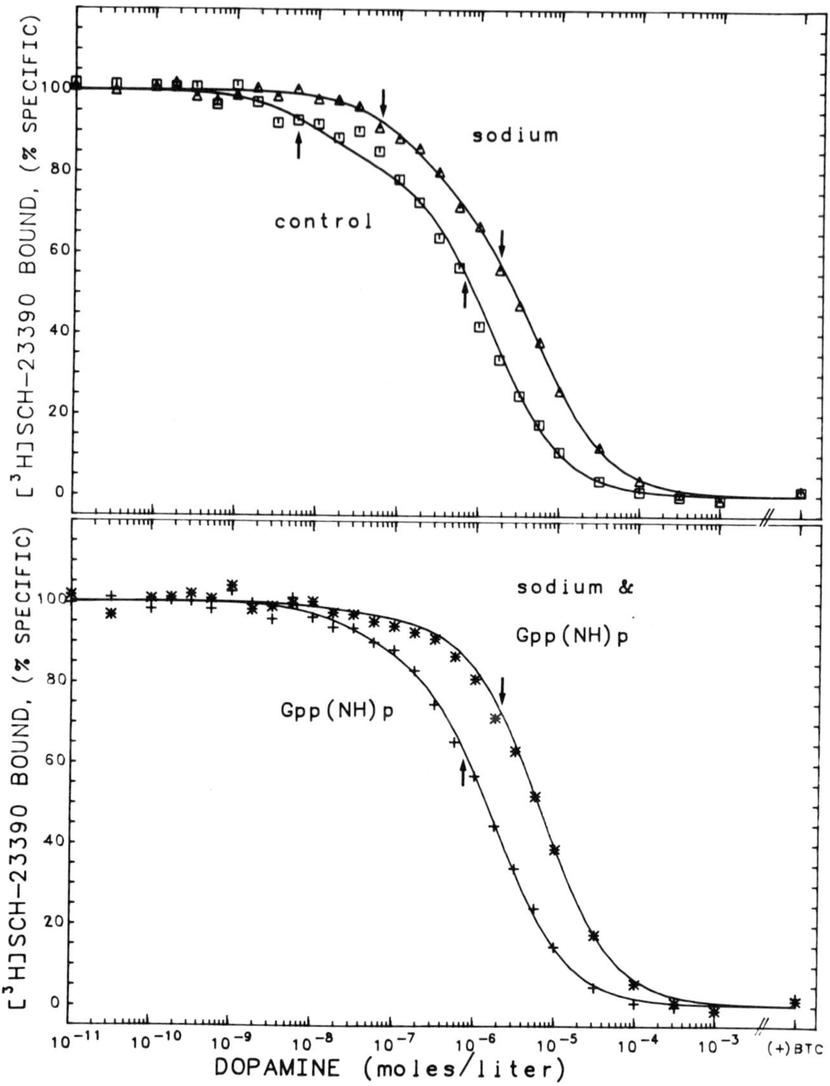

Figure 7 Modulation of agonist high-affinity binding interactions at the striatal D_1 dopamine receptor. As seen in the top panel, dopamine-[^{3}H] SCH-23390 competition curves are biphasic and can best be described as comprised of two sites using LIGAND, with the K_D for dopamine (indicated by arrows) at D_1^{high} ≅ 7 nM and at D_1^{low} ≅ 800 nM. The addition of Gpp(NH)p (bottom panel) converts the biphasic curve into a monophasic pattern, suggesting the presence of a single homogeneous receptor population displaying only low affinity (D_1^{low}) for agonists (K_D ≅ 800 nM). Sodium appears to shift the potency of dopamine to the right (i.e., less potent) without affecting the major pattern of the competition curve. (From Niznik, Fogel, and Seeman, unpublished.)

the D_1 receptor is presumed to reflect the stable association of the receptor with a guanine nucleotide binding protein, G_s. Dissociation of the receptor-G_s complex by the addition of the guanine nucleotide, Gpp(NH)p, mediates the transition of the agonist high-affinity form of the receptor to one displaying only low affinity for agonists. Moreover, DA-stimulated adenylate cyclase is cholera toxin sensitive (177), suggestive of D_1 receptor-G_s coupling. Although G_s is assumed to be coupled to D_1 receptors, the exact molcular form of this protein, of which there are at least two (reviewed in Reference 168), is still unknown.

As already mentioned, the stimulation of adenylate cyclase activity and its antagonism by SCH-23390 is characteristic of D_1 dopamine receptors. In the brain and retina, dopamine-stimulated cyclase has been extensively examined (57,60,100,178). Some concern, however, has been raised about whether D_1 receptors labeled by [^{3}H] SCH-23390 are identical to those sites conferring dopamine-stimulated adenylate cyclase activity. Evidence suggesting the lack of a 1:1 correspondence between [^{3}H] SCH-23390 binding and DA-stimulated cyclase activity are primarily based on the observations that, under certain assay conditions, (1) the K_D of dopaminergic antagonists at inhibiting DA-sitimulated cyclase is some 10 to 40-fold higher than that observed in [^{3}H] SCH-23390 binding assays (100); and (2) the rate of D_1 receptor inactivation by various membrane-perturbating and alkylating agents is not similar to that of DA-stimulated cyclase (179-181). Various models have been proposed to account for these observations (179,180), but the exact molecular nature of the D_1 receptor and of dopamine-stimulated adenylate cyclase remains to be determined. In any event, DA-stimulated cyclase activity is associated with the induction of cAMP-dependent protein kinases (182) and the phosphorylation of DARPP-32 (183,184), a protein that shares structural homology with protein phosphatase inhibitor 1 (171). DARPP-32 is a potent inhibitor of protein phosphatase-1 and appears to be localized in neurons containing D_1 receptors (185). The pharmacological profile of DARPP-32 phosphorylation by D_1 receptor agonists or its attenuation (if any) by selective D_1 antagonists has not been reported (however, see Reference 186). At least in some neuronal systems, DA D_1 receptors appear to mediate a decrease in cAMP-dependent K^+ (120,187) conductance and phosphorylation of specific proteins (188) that may be G-protein mediated (189). Unfortunately, except for those neurons defined by Stoof et al. (120), the nature of the DA receptors (D_1 or D_2) mediating these effects is unknown.

At a behavioral level, little functional significance has been ascribed to D_1 DA receptors. For years, the D_2 DA receptor in the central nervous system has been considered the sole mediator of many of the behavioral actions of DA and by nonselective DA congeners (18). Since the introduction of SKF-38393 and SCH-23390, this concept has been radically altered (reviewed in Reference 190). In essence, two points have emerged. One is that the activation of central

D_1 receptors with selective agonists induced behavioral responses distinct from that mediated by the D_2 DA receptor (see References 191 and 192 and references therein). The second point is that D_1 receptors appear to facilitate the expression of activities mediated by DA D_2 receptors at both the neurophysiological and behavioral level (see References 193-196 and references therein). As yet the significance of these findings at a molecular level is unknown.

D_2 Dopamine Receptors

The DA receptor in both the anterior and neurointermediate lobes of pituitary gland has long been considered the prototypical D_2 receptor (Table 2) (20). On mammotrophs of the anterior pituitary, dopamine and other dopaminergic agonists, including the selective D_2 receptor agonist LY-171555 (quinpirole), inhibit both prolactin release (21,197,198) and adenylate cyclase activity (see References 199 and 200 and references therein). Moreover, the K_D of agonists to inhibit prolactin release, adenylate cyclase activity, and [^{3}H] spiperone binding are all virtually identical. Selective D_2 receptor antagonists, such as YM-09151-2 and (−)sulpiride, block the effects of dopaminergic agonists on both prolactin release and adenylate cyclase.

The effects of DA were first assumed to be mediated solely by the inhibition of adenylate cyclase since pertussis toxin abolishes both dopaminergic inhibition of prolactin release and cAMP (201). Pertussis toxin catalyzes the ADP ribosylation of G_i ($M_r \cong 41{,}000$) and G_o ($M_r \cong 39{,}000$) with the uncoupling of the receptor from G (168). Recent evidence, however, has questioned the role of cAMP in transducing the effects of dopamine in the anterior pituitary, since dopamine-mediated inhibition of prolactin release is independent of that mediated by cAMP (202,203). Moreover, dopaminergic inhibition of prolactin release is associated with Ca^{2+} mobilization (see References 202-205 and references therein), an event possibly mediated by dopaminergic inhibition of phosphatidylinositol (PI) turnover in pituitary tissues (206-209). The inhibitory effects of dopamine on adenylate cyclase may be due to its primary action on Ca^+ mobilization (210,211), since Ca^{2+} itself has been shown to inhibit adenylate cyclase activity in the anterior pituitary (212). Receptor stimulation of phospholipase C leads to the formation of Ca^{2+} mobilizing second messengers (IP_3) and diacylglycerol, an activator of protein kinase C (see References 213-217 for recent reviews). Thus, D_2 receptors in the anterior pituitary appear to be functionally coupled to the inhibition of PI metabolism and possibly the attenuation of phospholipase C. The molecular nature of the G-protein coupling the D_2 receptor to phospholipase C (218), if any, is unknown. G_o is a possible candidate, since it copurifies with anterior pituitary D_2 receptors (219) and has been identified as a pertussis toxin substrate in the pituitary gland (220). G_o appears to regulate the activity of Ca^{2+} channels (221). Alternatively, there are other pertussis toxin-sensitive (222), and insensitive G proteins

(168) that may mediate these events. The D_2 receptor on melanotrophs in the intermediate lobe is associated with the inhibition of propiomelanocortin (POMC) synthesis (223-227), α-MSH (melanocyte stimulating hormone) release, and inhibition of adenylate cyclase activity. Dopaminergic inhibition of adenylate cyclase activity appears to be responsible for the inhibition of peptide hormone release, possibly by attenuating POMC (mRNA) gene expression (223-227). The mechanism responsible for the effect is unknown.

The D_2 dopamine receptor in both the pituitary and brain has been defined by radioligand binding techniques and by its ability to inhibit adenylate cyclase. Thus, $[^3H]$ spiperone, a potent butyrophenone neuroleptic, binds with high affinity (∼100 pM) to both pituitary (see References 228–231 for review and references therein) and brain striatal membranes (reviewed in References 19, 23, and 232). The binding of $[^3H]$ spiperone is competitively inhibited by dopamin-

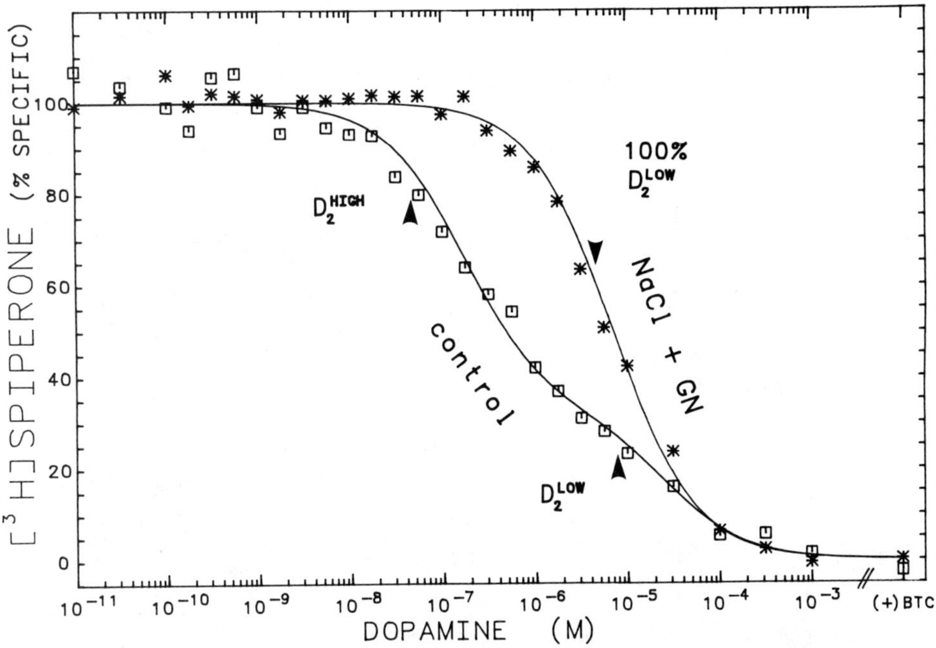

Figure 8 Modulation of agonist high-affinity binding interactions at the striatal D_2 dopamine receptor. DA-$[^3H]$ spiperone competition curves are biphasic with computer-generated estimates of a 61 nM K_D at $D_2{}^{high}$ and ∼7 μM at $D_2{}^{low}$. Guanine nucleotides and sodium affect the transition of the agonist high-affinity form of the receptor to one displaying only low affinity for agonists (indicated by arrow). (From Grigoriadis, Niznik, and Seeman, unpublished).

Table 3 Photoaffinity Ligands for Dopamine Receptors

Compound	Structure	Apparent K_d (nM)	Specific activity (Ci/mmol)	References
Substituted benzamides	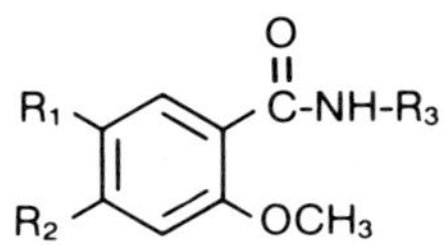			
Azidoclebopride	$R_1 = Cl$; $R_2 = N_3$; $R_3 =$ 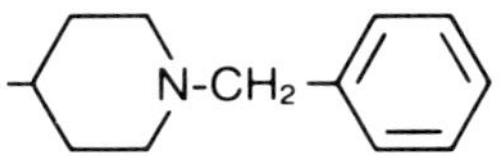	20	–	250–253
Iodoazidoclebopride	$R_1 = I$; $R_2 = N_3$; $R_3 =$ 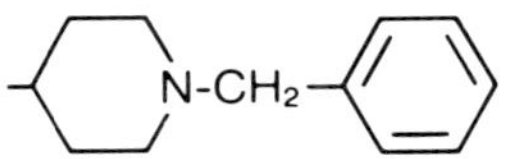	15 8	 14	254 255
Azidosulpride	$R_1 = CH_3NHSO_2$; $R_2 = N_3$; $R_3 = CH_2$- 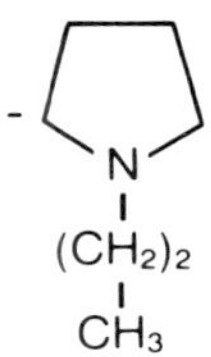	∼ 3	58	244

Butyrophenone	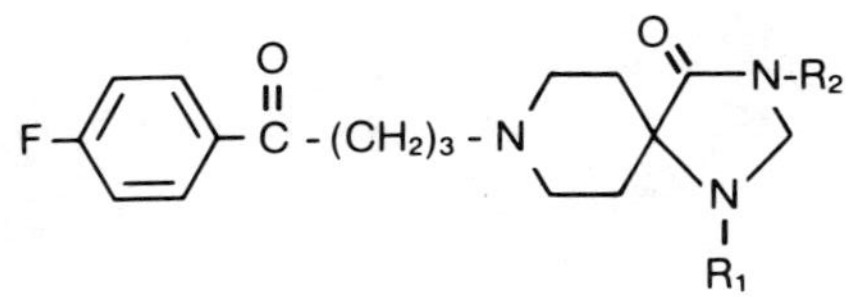			
Iodoazido-*N*-phenethylspiperone	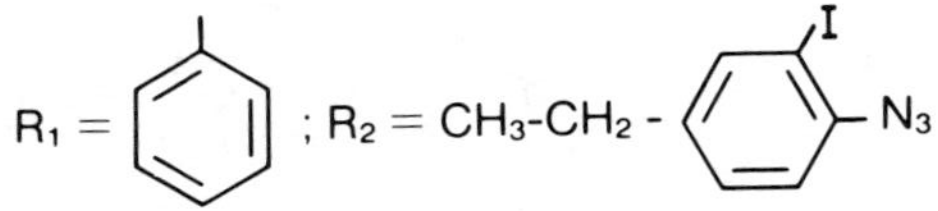	0.02	2200	243,246
Azido-*N*-methyl-spiperone	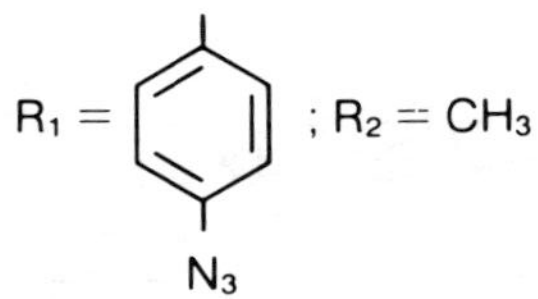	0.06	76	248,249
Phenothiazine	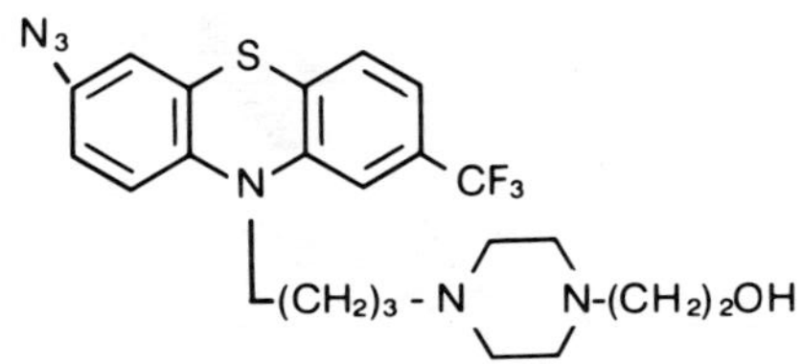			
Azidofluphenazine		–	40	247

Table 3 (Continued)

Compound	Structure	Apparent K_d (nM)	Specific activity (Ci/mmol)	References
Benzazepine SCH-38548	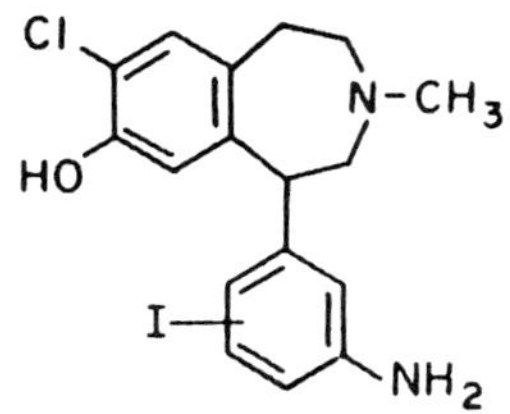	∼ 0.2	<2200	70

ergic agonists (e.g., NPA, ADTN, and LY-171555) and antagonists of the butyrophenone (e.g., spiperone and haloperidol) and substituted benzamide class [YM-09151-2 and (−)sulpiride] with high affinity (76). As with D_1 dopamine receptors, the binding of agonists to D_2 receptors is regulated by both sodium ions and guanine nucleotides (Figure 8) and is pertussis toxin sensitive, suggestive of either receptor-G_i or receptor-G_o coupling (224,232-235). Although it has been known for some time that the D_2 receptor mediates the inhibition of cAMP in the pituitary, linkage of the neuronal D_2 receptor to the inhibition of adenylate cyclase activity has only recently been demonstrated (101,236,237). Not all neuronal D_2 receptors, however, appear to be coupled to the inhibition of adenylate cyclase. Presynaptic D_2 receptors (238) and D_2 receptors in some neuronal systems are cAMP independent (120) and, like anterior pituitary D_2 receptors, appear to mobilize Ca^{2+} and inhibit PI turnover (239) and as such may be in need of reclassification.

PROBES OF DOPAMINE RECEPTOR STRUCTURE

To understand the action of DA or dopaminergic agents at the molecular level, it is essential to characterize the structure and biochemical properties of dopamine receptors. Two powerful approaches that have yielded a wealth of information concerning the structure of minor membrane proteins, such as hormone and neurotransmitter receptors, are photoaffinity labeling and affinity chromatography. These techniques are based on modifications of ligand binding methodology and provide a means by which the ligand binding subunit(s) of a molecule can be identified, characterized, and purified. Photoaffinity labels are ligands that possess inherent affinity for binding sites of proteins but also contain either electrophilic or photosensitive moieties that are capable of forming covalent interactions at or near the ligand binding domain of the receptor. These ligands, under the appropriate conditions, can identify a macromolecular receptor in a complex mixture of biologic membrane proteins and/or isolate the region within a macromolecule responsible for ligand binding and receptor activation. One of the major problems associated with the introduction of functional reactive groups in the generation of photoaffinity labels and the preparation of ligands suitable for linkage to affinity-chromatographic supports (e.g., Sepharose) is the loss of ligand specificity and biologic activity. Only recently has there been some success.

Photoaffinity Ligands

Until the end of 1984, suitable probes for the identification of dopamine receptors were restricted to electrophilic reagents that could not be successfully incorporated into dopamine receptors (reviewed in References 240 and 241). By the end of 1986, there were six photoaffinity ligands capable of identifying the subunit structure of dopamine D_1 and D_2 receptors. The structures of these probes are illustrated in Table 3.

The ligand binding subunit of striatal D_1 DA receptors has recently been identified (70). [^{125}I] SCH-38548, a derivative of the potent and selective D_1 receptors antagonist SCH-23390, covalently incorporates into striatal membranes following photoaffinity cross-linking into a broad band of M_r = 80,000-65,000 (centered at M_r = 72,000) as assessed by SDS-PAGE (sodium dodecyl sulfate-polyacryamide gel electrophoresis) and autoradiography. Photoincorporation of this probe into the diffuse band of M_r = 72,000 is specifically antagonized by dopaminergic agents with an appropriate D_1 receptor pharmacological specificity. Thus, the pharmacological specificity that defines the D_1 DA receptor in radioligand binding assays is retained during photoaffinity cross-linking, validating the use of this probe for the molecular characterization of D_1 dopamine receptor structure. Whether the diffuse pattern of labeling represents marked proteolytic degradation of a polypeptide at M_r = 80,000 or the interference of carbohydrate moieties on the D_1 receptor protein (65,242) on protein mobility on SDS-polyacrylamide gels awaits further clarification.

The identification of the ligand binding subunit of D_2 DA receptor by specific photoaffinity ligands has been extensively described (240). As outlined in Table 4, the major binding subunit of the D_2 DA receptor resides on a polypeptide of M_r = 94,000-92,000 in a variety of tissues. Photoincorporation of these probes into the protein at M_r = 92,000-94,000 is stereoselectively blocked by dopaminergic agonists and antagonists with an appropriate D_2 pharmacological profile, but not by adrenergic or serotonergic agents (see Figure 9). As can be seen in Figure 9, a number of other specifically labeled bands appear at M_r = 80,000-32,000. Whether these "lower" molecular weight peptides represent (1) proteolytic degradation products or (2) multiple subunits of the receptor protein is unknown at present. Of particular interest is the observation of a higher molecular weight form ($M_r \cong$ 120,000) of the D_2 receptor in both the rat neurointermediate lobe and in porcine tissue. The peptide at $M_r \cong$ 120,000 also displays the appropriate pharmacological specificity for D_2 dopaminergic agents. Whether the high-molecular-weight form of the receptor represents a

---

Figure 9 Photoaffinity labeling and pharmacological specificity of [^{3}H] N_3-methylspiperone (upper panel) and [^{125}I] N_3-*N*-phenethylspiperone (lower panel) incorporation into striatal dopamine receptors. Autoradiographs depicting the selective labeling of the ligand binding subunit of the D_2 dopamine receptor, $M_r \cong$ 93,000, in the absence or presence of the indicated concentrations of various agents: BTC, butaclamol; ETIC, eticlopride; SPIP, spiperone; SULP, sulpiride; KET, ketanserin (a S_2 serotonergic antagonist); NPA, *N*-propylnorapomorphine. (From Reference 248 and Grigoriadis, Niznik, and Seeman.)

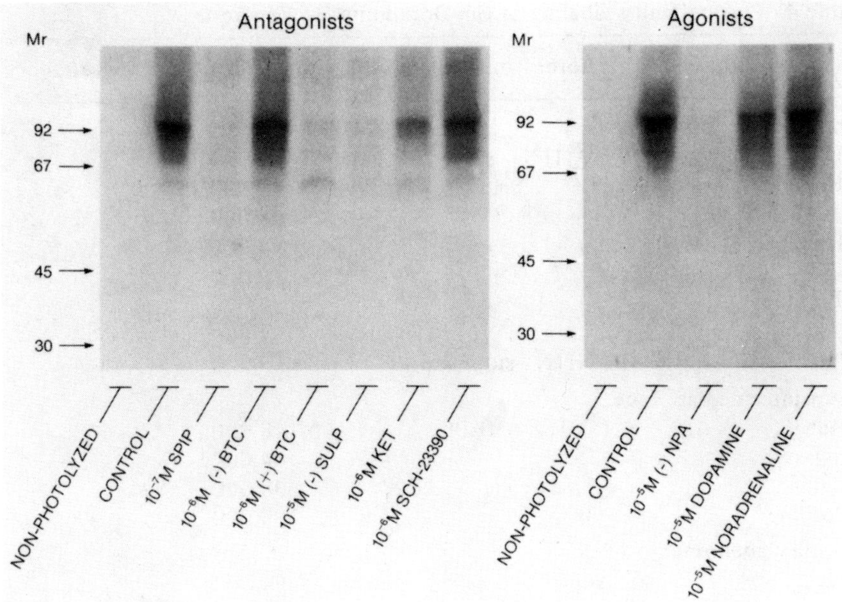

Antagonists
Agonists
Mr
92
67
45
30
NON-PHOTOLYZED
CONTROL
10^{-7}M SPIP
10^{-6}M (-) BTC
10^{-6}M (+) BTC
10^{-5}M (-) SULP
10^{-6}M KET
10^{-6}M SCH-23390
NON-PHOTOLYZED
CONTROL
10^{-5}M (-) NPA
10^{-5}M DOPAMINE
10^{-5}M NORADRENALINE

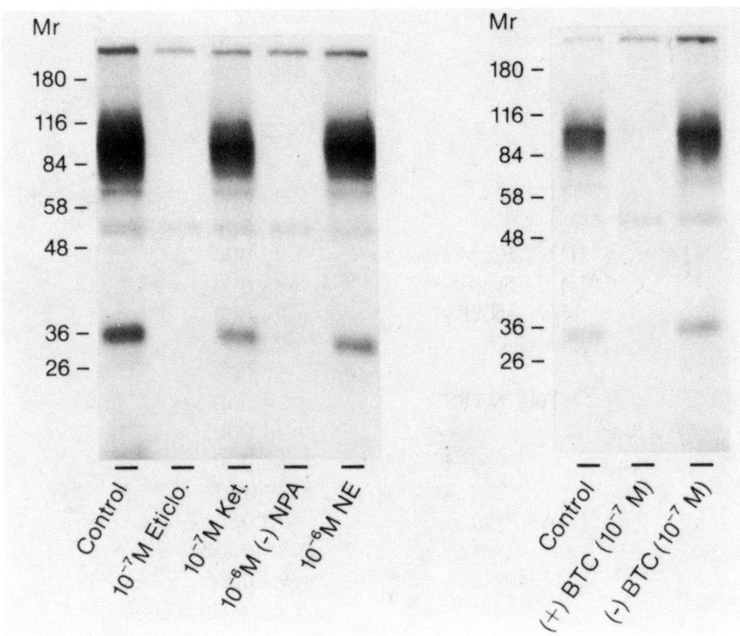

Mr
180
116
84
58
48
36
26
Control
10^{-7}M Eticlo.
10^{-7}M Ket.
10^{-6}M (-) NPA
10^{-6}M NE
Control
(+) BTC (10^{-7} M)
(-) BTC (10^{-7} M)

Table 4 Photoaffinity labeling of the Dopamine D_2 Receptor

Tissue and source	Photoaffinity ligand	M_r[a]	Reference
Anterior pituitary			
Rat	[^{125}I] N_3-NAPS[b]	94,000	243
Bovine		94,000	
		38,000	
Hamster		94,000	
		38,000	
Porcine		130,000	—[c]
		94,000	
Rat	[^{3}H] N_3 sulpiride	∿85,000	244
Neurointermediate lobe			
Rat	[^{125}I] N_3-NAPS	∿118,000	243
		94,000[d]	
		38,000	
Porcine		130,000	—[c]
Pituitary adenoma			
Rat		94,000	245
Retina			
Bovine	[^{125}I] N_3-NAPS	94,000	243
		∿80,000	
		32,000	
Rabbit		94,000	
		∿70,000	
		∿38,000	
		∿36,000	
Striatum			
Rat	[^{125}I] NAPS	94,000	243,246
		32,000	
	[^{3}H] N_3-Fluphenazine	∿92,000[e]	247
	[^{3}H] N_3-Sulpiride	∿85,000	244
Canine	[^{125}I] N_3-NAPS	94,000	243
		∿73,000	
		32,000	
	[^{125}I] N_3-NAPS	94,000	—[f]
		∿73,000	
		32,000	
		<20,000	
	[^{3}H] AMS[g]	92,000	248
		∿75,000[d]	
		93,000[e]	249

Table 4 (Continued)

Tissue and source	Photoaffinity ligand	M_r[a]	Reference
Striatum (Cont)			
Porcine	$[^{125}I]N_3$-NAPS	∿120,000 94,000 32,000[d]	—[c]
Bovine		94,000	243
Rabbit		94,000	
Guinea pig		94,000	
Hamster		94,000 32,000	
Olfactory bulb			
Rat	$[^{3}H]N_3$-Sulpiride	∿85,000	244

[a]As determined by SDS-PAGE and autoradiography unless indicated otherwise.
[b]$[^{125}I]N_3$-NAPS, *N*-azidophenethylspiperone.
[c]K. Jarvie, H. B. Niznik, and P. Seeman, unpublished.
[d]Minor labeled peptide.
[e]M_r determined by counting gel slices.
[f]D. Grigoriadis, H. B. Niznik, and P. Seeman, unpublished.
[g]$[^{3}H]$AMS, azido-*N*-methylspiperone.

differentially processed precursor of the protein at M_r = 94,000, a distinct molecular form of the receptor, or a heavily glycosylated receptor moiety is unknown at present.

It is difficult to ascertain whether the photolabeled D_2 protein at M_r = 120,000-94,000 represents the binding subunit of D_2 receptors alone or of the functional receptor. Previous estimates of the size of D_2 receptors by radiation inactivation of membrane preparations from various tissues and different species have ranged from 110,000 to 191,000 daltons and are summarized in Table 5. These are well within the range of the estimated molecular mass of D_2 receptors by photoaffinity labeling.

Dopamine Receptor Purification via Affinity Chromatography

DA D_1 and D_2 receptors have been successfully solubilized by a variety of detergents (reviewed in References 240 and 241). Initial attempts to purify solubilized receptors have been limited to the use of conventional purification techniques, such as lectin-affinity, ion-exchange, size-exclusion chromatography and isoelectric focusing. These methods have generally yielded estimates of 2 to 16-fold purification of D_2 or D_1 receptors and are inadequate as a means of purifying dopamine receptor proteins.

Table 5 Radiation Inactivation of Dopamine Receptors

Tissue and species	Receptor site	Ligand	Functional target size (daltons)	Ref.
Antagonists				
Striatum				
Canine	D_2	$[^3H]$spiperone	123,000	256
Human	D_2	$[^3H]$spiperone	123,000	256
Rat	D_2	$[^3H]$spiperone	136,000	257
Rat	D_2	$[^3H]$spiperone	191,000	258
Bovine	D_2	$[^3H]$spiperone	110,000 (60%) 1,500,000 (40%)	259
Pituitary adenoma				
Rat	D_2	$[^3H]$spiperone	143,000	245
Rat	D_2	$[^3H]$spiperone	106,000[a]	
Striatum				
Rat	D_1	$[^3H]$piflutixol	79,500	257
Rat	D_1	$[^3H]$SCH-23390	78,000	91
Agonists				
Striatum				
Rat	D_2	$[^3H]$NPA	82,000	260
Bovine	D_2	$[^3H]$NPA	1,500,000[b]	259
Rat	D_1	$[^3H]$SKF-38393	132,000	91
Rat	Dopamine-stimulated cyclase[c]		202,000	257

[a]Following solubilization with the detergent (3-[3-cholamidopropyl-dimethylamino]-1-propanesulfonate, CHAPS).
[b]In the presence or absence of guanine nucleotides.
[c]Estimated target size for adenylate cyclase, 122,500 (257).

Recently, specific affinity chromatographic supports for D_2 receptors (based on structural analogs of spiperone and haloperidol) have been synthesized (261-264). At best, D_2 receptors are purified by ∿1000-fold over solubilized receptor preparations (261,262). Specific affinity chromatographic supports for the purification of D_1 receptors have not been reported to date. It is of interest to note that affinity-purified preparations of D_2 receptors require reconstitution into phospholipid vesicles for full expression of $[^3H]$spiperone binding activity (261) and that anterior pituitary D_2 receptors copurify with a guanine nucleotide binding protein, tentatively identified as G_o (219). The purification of D_1 and D_2 receptors and subsequent reconstitution with purified transducer-effector units should aid in the elucidation of the molecular mechanism(s) responsible for receptor-mediated adenylate cyclase activation and phosphatidylinositol metabolism.

LIGANDS FOR THE IDENTIFICATION OF DOPAMINE RECEPTOR SUBTYPES

Dopamine D_1 Receptor Agonists

SKF 38393 (2,3,4,5-Tetrahydro-7,8-dihydroxy-1-phenyl-1H-3-benzazepine)

As a starting point for the generation of selective dopamine D_1 receptor-active compounds, the tetrahydro-3-benzazepine nucleus has been shown to be particularly useful (see Figure 3). Of the several benzazepine-derived agents with agonist activity, SKF 38393 has been shown to be particularly potent and specific (24). SKF 38393 is a partial agonist at the DA D_1 receptor linked to the stimulation of adenylate cyclase (265,266). SKF 38393 produces half-maximal adenylate cyclase stimulation at concentrations that are 50-fold less [EC_{50} of 0.075 μM for SKF 38393 versus 3.5 μM for DA; rat caudate homogenate (265)] or equivalent [EC_{50} of approximately 1 μM; teleost retina (266); EC_{50} of 3 μM; dispersed bovine parathyroid cells (157)] to that of dopamine.

In intact animals, SKF 38393 does not reduce serum prolactin levels or produce stereotyped movements (265), responses typically associated with agonist activation of dopamine D_2 receptors (18). In rats with DA-depleting unilateral 6-OHDA lesions of the substantia nigra, SKF 38393 produces the stereotyped behavior and contralateral rotation (265) associated with the activation of central DA receptors (18,267). In the intact rat, administration of the active R enantiomer of SKF 38393 is associated with dose-dependent increases in nonstereotyped sniffing, locomotion, and grooming (191). The grooming response to R-SKF 38393 has been shown to be inhibited by coadministration of the specific D_1 receptor antagonist SCH 23390 but not the selective D_2 receptor antagonist metoclopramide (268).

At the peripheral DA DA_1 receptor, SKF 38393 acts as a partial agonist (269), producing a specific dilation of the renal vascular resistance vessels without producing changes in cardiac output or mean arterial blood pressure (270). These effects are produced with a potency of 0.1 (270) to 0.01 (269) that of dopamine itself.

SKF 82526 [Fenoldopam: 6-chloro-7,8-dihydroxy-1-(p-hydroxyphenyl)-2,3,4,5-tetrahydro-(1H)-3-benzazepine]

The interaction of SKF 82526 (fenoldopam) (see Figure 3) with central and peripheral DA receptors has also been well characterized. SKF 82526 is a 6-chloro, 4′-hydroxy analog of SKF 38393. SKF 82526 is a potent agonist at the

DA D_1 receptor associated with the stimulation of the enzyme adenylate cyclase. SKF 82526 has been observed to stimulate adenylate cyclase with an EC_{50} of approximately 20 nM (18 nM) (271,272). This is in contrast to the EC_{50} of DA that has been observed as approximately 3 μM [3.5 μM (271) and 3 μM (272)]. SKF 82526 maximally stimulates the adenylate cyclase to 80% of the maximal level observed with DA (271,272) and as such is a partial agonist at this receptor. Similar results have been obtained (83) using the active R enantiomer (273) of SKF 82526. Relative to the resolved R enantiomer of its analog SKF 38393, (R)-SKF 82526 is more potent at both the stimulation of adenylate cyclase (EC_{50} of 40 versus 63 nM) and the inhibition of [^{3}H] SKF 82526 binding (2.4 versus 5.1 nM) (83). The binding of [^{3}H] SKF 82526 to rat striatal preparations has been examined. [^{3}H] SKF 82526 binds saturably with a dissociation constant K_D of 2.3 ± 0.1 nM and a maximal binding capacity B_{max} of 590 ± 40 fmol/mg protein (83). The site labeled by [^{3}H] SKF 82526 is the D_1 receptor associated with the enzyme adenylate cyclase. This inference is drawn from the observation that both the ability of antagonists to inhibit rat striatal DA-stimulated adenylate cyclase and the ability of agonists to stimulate the adenylate cyclase correlate highly with the abilities of these compounds to inhibit the binding of [^{3}H] SKF 82526 to rat striatal preparations (83). The D_1 versus D_2 selectivity of SKF 82526, as assessed by in vitro binding studies, has been estimated to be from 17-fold (274) to more than 100-fold selective (83) for the D_1 receptor.

In intact animals, SKF 82526 does not readily cross the blood-brain barrier (271,272). The central activity of SKF 82526 is, however, apparent after intracaudal administration, where it produces contralateral rotation in caudate-lesioned rats (272) with a potency one-fifth of that of intracaudal DA (271). No change in serum prolactin levels is apparent after systemic administration (272).

Peripherally, the (R) enantiomer of SKF 82526 appears to act as a full agonist in producing renal vasodilation in anesthetized dogs (269). In this regard, SKF 82526 is nine times more potent than SKF 38393 (271). Both the 6-chloro and 4′-hydroxy groups are important for renal vasodilatory activity (271). Recently, SKF 82526 has been administered to healthy male volunteers and observed to cause systemic arteriolar vasodilation accompanied by renal vasodilation and increased sodium excretion (141).

Tetrahydro-3-benzazepine analogs

Other 3-benzazepine derivatives (see Figure 3) have been synthesized and their activity has been assessed. The N-allyl ($R = CH_2CH{=}CH_2$) derivative of SKF 82526 (SKF 85174) exhibits a slight increase in its agonist potency at the DA-stimulated adenylate cyclase (273). The increased potency of SKF 85174 is, however, associated with a decrease in the D_1 versus D_2 agonist (273,275) selectively associated with an increase in D_2 agonist potency. The N-n-propyl-substituted analog of SKF 82526 exhibits decreases in both D_1 and D_2 potency, but N-methyl substitution is associated with the maintenance of D_2 agonist potency and the conversion of D_1 agonist activity to antagonist activity (275). Experiments conducted with analogs of the N-allyl compound SKF 85174 modified in the 6 position have demonstrated that although replacement of the chloro with a bromo little alters activity, both DA_1 and DA_2 activities decrease upon replacement of chloro with a hydrogen. Introduction of a 6-methyl group decreases DA_1 agonist potency and abolishes DA_2 agonist activity (276).

3′,4′-Dihydroxynomifensine (8-Amino-1,2,3,4-tetrahydro-4-(3,4-dihydroxyphenyl)-2-methylisoquinoline)

3′,4′-Dihydroxynomifensine (Figure 10) is a nonbenzazepine DA D_1-selective full agonist that exhibits in rat striatal preparations a potency against the DA-sensitive adenylate cyclase slightly less than that of dopamine itself (EC_{50} rat striatum: 3′,4′-dihydroxynomifensine, 6 μM; dopamine, 3 μM) (277). Peripherally, as assessed by vasodilation of the renal vasculature, 3′,4′-dihydroxynomifensine acts as a partial agonist (269,278), exhibiting a potency approximately 0.01 that of dopamine (278). The enantiomers of 3′,4′-dihydroxynomifensine have been resolved, and it has been shown that the S enantiomer is most potent at the D_1 receptor (279). Similar results are obtained with the teleost retina adenylate cyclase, where 3′,4′-dihydroxynomifensine is a partial agonist with a potency 10-fold lower than that of DA (266).

Dopamine D_1 Antagonists

SCH 23390 ((R)-(+)-8-chloro-2,3,4,5-tetrahydro-3-methyl-5-phenyl-1H-3-benzazepine-7-ol)

The tetrahydro-3-benzazepine (see Figure 3) nucleus has also been an extremely useful starting point for the synthesis of DA D_1 and DA_1 receptor antagonists.

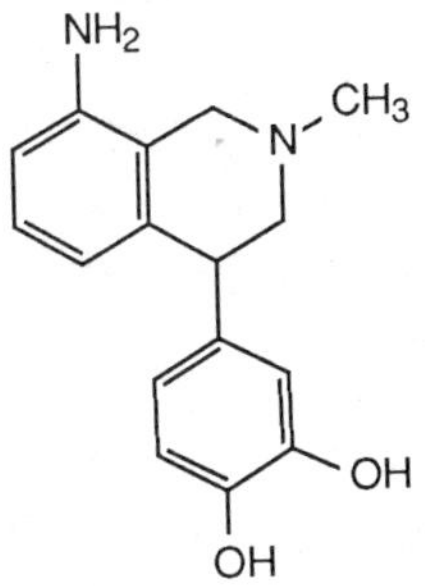

Figure 10 3′,4′-Dihydroxynomifensine.

SCH 23390 is a benzazepine that has displayed in assays in vitro an extraordinary selectivity for the DA D_1 receptor linked to the stimulation of adenylate cyclase (27,103) and the DA DA_1 receptor associated with renal vasodilation (148,280). SCH 23390 has been shown to inhibit rat striatal adenylate cyclase with K_i values in the low nanomolar range [K_i = 4 nM (103) and 10 nM (27)], but micromolar concentrations are required to inhibit the D_2-specific rat striatal homogenate binding of [^{3}H] spiperone (K_i = 24 μM) (27). Studies conducted with radiolabeled [^{3}H] SCH 23390 have provided further support for the selective nature of SCH 23390 binding. [^{3}H] SCH 23390 binds to one site in striatal homogenates with a subnanomolar affinity [K_D = 0.53 nM (60); 0.18 nM (57)] and a density of binding sites in rat and mouse striata (57) that is two to three times that of the number of D_2 sites. (R)-SCH 23390 is the most potent competitor for sites labeled by [^{3}H] SCH 23390, but the (S) isomer (SCH 23388) competes with a 600-fold lesser potency (60). Other potent competitors are *cis*-(*Z*)-piflutixol and *cis*-(*Z*)-flupenthixol (the trans isomers are less potent) (60). The DA D_2 receptor-selective ligands spiperone and sulpiride are not potent competitors for those sites labeled by [^{3}H] SCH 23390 (60). The IC_{50} values for the inhibition of [^{3}H] SCH 23390 binding by various neuroleptics are highly correlated with the corresponding values for the inhibition of DA-sensitive adenylate cyclase, and the inhibition of the nonselective neuroleptics [^{3}H]-piflutixol and [^{3}H]-*cis*-(*Z*)-flupenthixol (57), indicating that these radioligands most likely interact with the binding site associated with the dopaminergic stimulation of adenylate cyclase. The selectivity of SCH 23390 for D_1 receptors has also been confirmed using functional neurochemical model systems (103, 281). As such, SCH 23390, which potently and competitively inhibits the DA-sensitive adenylate cyclase, failed to reverse the apomorphine-induced inhibition of the electrically evoked release of [^{3}H] DA (a test for presynaptic D_2 antagonism) or [^{3}H] acetylcholine (a test for postsynaptic D_2 antagonism) from rabbit striatal slices preloaded with the respective radioligands (103). Interestingly, even though the results obtained in a different D_2 receptor test system (inhibition of the K^+-evoked release of [^{3}H] acetylcholine from superfused rat striatal slices) indicated that the potency of SCH 23390 on the D_2 receptor was only 1 log unit less than the potency of SCH 23390 on the D_1 receptor linked to adenylate cyclase, SCH 23390 was still the most D_1-selective compound assessed (281). Similarly, SCH 23390 has been observed to compete for rat frontal cortex 5-HT-2 receptors in vivo with a potency greater than would be expected from in vitro studies (99).

Analogs of SCH 23390

Studies that have assessed the properties of SCH 23390 analogs (see Figure 3) have greatly aided in the understanding of the structural requirements necessary for D_1 receptor antagonism. Replacement of the 7-chloro substituent with a hydroxyl moiety yields the partial D_1 agonist SKF 75670. Thus it can be assumed that chloro substitution in the seventh position contributes to D_1 antagonist activity (282). The desmethyl analog of SCH 23390 (SKF 83509) is a D_1-selective antagonist of relatively low potency. This indicates that although the N-methyl of SCH 23390 may not be important in determining efficacy, its presence increases D_1 receptor binding affinity (282). The affinity enhancement provided by 3-methylation is apparent for tetrahydro-3-benzazepines with both 7-OH and 7-H substituents (283). The rank order of potency conferred by different 7 substituents has been shown to be Cl = Br >> CH_3 > OH ≥ H (283). Substitution of a bromo substituent for the chloro substituent of SCH 23390 yields SKF 83566, a potent DA_1-selective antagonist of the dopaminergic relaxation of the renal vasculature (284). Recently, a radioiodinated analog of SCH 23390 ([^{125}I] SCH 23982), a 7-[^{125}I] iodo analog has been synthesized and characterized as the ligand of choice for the identification of the D_1 receptor (72).

Dopamine D_2 Agonists

Ergolines

A large series of compounds can be grouped together on the basis of their similarity to the structural nucleus of the ergot alkaloids (see Figure 4).

LY 141-B (pergolide: (8β)-8-[(methylthio)methyl]-6-propylergoline) LY 141-B was one of several synthetic ergoline derivatives initially assessed for potency as a selective D_2 agonist. LY 141-B, the N-propyl analog, was found to be consecutively more potent than the N-ethyl or N-methyl derivatives in both in vivo tests (rotation in rats with 6-OHDA lesions of the nigrostriatal dopamine pathway and suppression of surgically induced tremors in monkeys) and in competing for [^{3}H] DA and [^{3}H] spiperone binding (K_i = 12.8 and 34.2 nM, respectively) (285). More recently it has been shown that pergolide competes for those sites labeled by [^{3}H] spiperone in a biphasic and sodium-dependent manner (no Na^+, $K_D{}^{high}$ = 0.75 nM and $K_D{}^{low}$ = 60 nM; 100 mM Na^+, $K_D{}^{high}$ = 0.14 nM and $K_D{}^{low}$ = 109 nM) (76).

LY 141865 and LY 171555 (respectively, the racemic and (−)-enantiomer quinpirole of trans-4,4a,5,6,7,8,8a,9-octahydro-5-propyl-1H-pyrazolo[3,4-g]-quinoline) The high potency (comparable to pergolide) of the linear tricyclic pyrazole LY 141865 was first demonstrated in prolactin inhibition and lesioned rat turning assays (286). This result was taken as evidence in favor of the pyrrole ethylamine moiety of the ergolines being the DA agonist-active portion of the ergoline nucleus (286). The selective and potent agonism of LY 141865 decreased the basal release of α-melanocyte stimulating hormone from dispersed neurointermediate lobe cells with an IC_{50} ≅ 30 nM. At higher concentrations

(IC_{50} = 30 μM), LY 141865 half-maximally inhibited intermediate lobe basal or (−)isoprenaline-stimulated adenylate cyclase activity and the adenylate cyclase activity in homogenates of cholera toxin-treated intermediate lobe tissue. In contrast, LY 141865 displayed neither agonist or antagonist activity at the D_1 receptor linked to the stimulation of adenylate cyclase in carp retina (287). Peripherally, LY 141865 lowers the mean arterial blood pressure and cardiac rate in anesthetized and conscious spontaneously hypertensive rats. This selective antagonism of neurogenic vasoconstriction was suggested to be due to the agonist action of LY 141865 at presynaptic dopamine receptors that mediate inhibition of peripheral adrenergic neurotransmission (288). Of the resolved enantiomers of LY 141865, the (−)enantiomer (LY 171555) was shown to be the biologically active isomer (289). LY 171555, although displaying no agonist activity at the D_1 receptor of carp retina, evoked a DA D_2 receptor-mediated (290) inhibition of the release of immunoreactive α-MSH-like molecules from enzymatically dispersed melantrophs of the intermediate lobe of the rat pituitary (IC_{50} = 27 nM) (291). LY 171555 inhibits the binding of [^{3}H] spiperone to pig anterior pituitary homogenates in a biphasic manner with a K_D^{high} = 4.8 nM and K_D^{low} = 3680 nM (76).

LY 1613502 [trans-(−)-5,5a,6,7,8,9,9a,10-octahydro-6-propylpyrimido(4,5-g)-quinoline-2-amino · 2HCl] is another partial ergoline reputed to be a selective agonist of dopamine D_2 receptors (292) with a potency greater than that of either pergolide or quinpirole (293).

Aminotetralins

2-Aminotetralin is capable of competing for the binding of [^{3}H] spiperone to the DA D_2 receptor of porcine anterior pituitary membranes. The addition of a hydroxyl group (−OH) corresponding to the meta-hydroxyl group of DA increases potency by an order of magnitude; addition of a hydroxyl group in the position corresponding to the para-hydroxyl group of dopamine does not affect potency (76). In general, the addition of N-n-dipropyl groups enhances potency [e.g., (±)-5,6-dihydroxy-2-aminotetralin, K_D^{high} = 22.3 nM, versus (±)-5,6-dihydroxy-2-dipropylaminotetralin, K_D^{high} = 0.82 nM], although this is not the case in all instances [e.g., (±)-6,7-dihydroxy-2-aminotetralin (ADTN) K_D^{high} = 1.7 nM, versus (±)-6,7-dihydroxy-2-dipropylaminotetralin, K_D^{high} = 12 nM] (76). The potent aminotetralin ADTN is a nonsubtype-selective agonist of DA receptors (24). In a series of subtype-nonselective phenolic hydroxy-2-aminotetralins, N,N-dipropylation was observed to affect D_2 activity more profoundly (294). This tendency for D_2 agonists to exhibit enhanced activity as tertiary amines has been dubbed the "n-propyl phenomenon" (295). A comparison of the resolved enantiomers of a series of primary and tertiary 2-aminotetralin derivatives indicated that although the 2R enantiomer of the primary amine was most active, the (2S) enantiomer was the most active of the resolved enantiomers of the tertiary amine. On the basis of this observation it was hypothesized that N,N-dipropylation causes a change in the preferred conformation from the β-rotameric to the α-rotameric form of dopamine (294). Methylation at C-1 on the aminotetralin nucleus [to yield *cis*-5-hydroxyl-1-methyl-2-(di-*n*-propyl-(amino)tetralin] has been observed to yield compounds that display a stereochemically dependent agonism or antagonism selective for dopaminergic autoreceptors (296). Recently, additional compounds displaying stereochemically dependent agonism or antagonism have been characterized. From these studies, it appears that the orientation of the ammonium hydrogen (or lone pair electrons) may be important for determining dopaminergic agonist and antagonist activity (297).

N-0437 [2-(N-n-propyl-N-thenylethylamino)-5-hydroxytetralin] Of a homologous series of N,N-disubstituted 2-aminotetralins studied, the 5-hydroxy, N,N-n-propylthenylethyl derivative (N-0437) (see Figure 11) was the most potent agonist at the dopamine D_2 receptor. N-0437 inhibited prolactin release from cultured anterior pituitary cells (the concentration producing a 50% decrease in the forskolin-stimulated prolactin release was approximately 20 pM) and inhibited the adenylate cyclase activity in cell-free homogenates of cholera toxin-treated intermediate lobe tissue (IC_{50} = 51 nM). The high degree of D_2 selectivity of this compound was typical of the 5-hydroxy-substituted 2-aminotetralins (298). In various in vivo test systems, N-0437 was observed to display a high D_2 receptor potency, lack of significant α_2-adrengergic receptor-mediated effects, long duration of action, and oral activity (299). Analysis of the binding characteristics of [^{3}H]N-0437 have verified that this ligand interacts specifically and potently with the dopamine D_2 receptor. [^{3}H]N-0437 appears to bind to a single site (calf striatal homogenates) with high affinity (K_D from equilibrium binding studies, 1.6 nM; K_D from kinetic binding studies, 0.17 nM) (300). These results are similar to the $K_D{}^{high}$ value obtained for N-0437 by displacement of [^{3}H] spiperone binding to porcine anterior pituitary homogenates ($K_D{}^{high}$ = 0.14 nM; $K_D{}^{low}$ = 38 nM) (76).

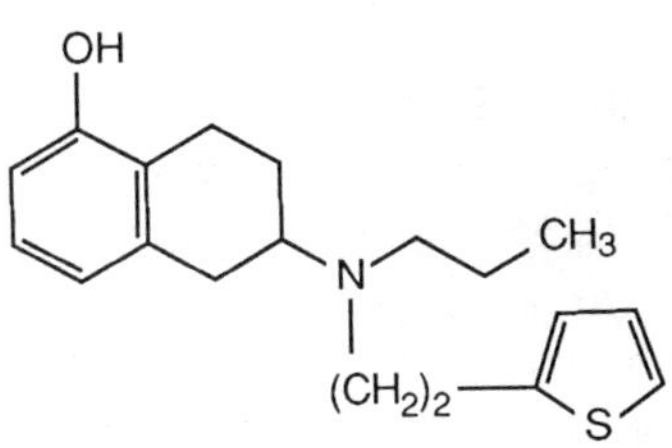

Figure 11 N-0437: a D_2-selective aminotetralin derivative.

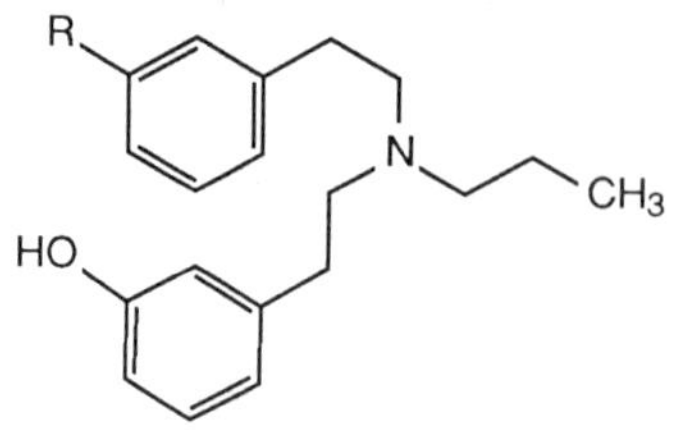

RU 24213 R=H
RU 24926 R=OH

Figure 12 Diphenylethylamine D_2 receptor agonists.

RU 24213 [N-n-propyl-N-phenylethyl-p-(3-hydroxyphenyl)ethylamine hydrochloride] and RU 24926 [N-n-propyl-di-β-(3-hydroxyphenyl)ethylamine hydrochloride] RU 24213 and RU24926 (see Figure 12) are *N*-diphenylethylamine derivatives that have been classed as orally active, selective agonists of the dopamine D_2 receptor. RU 24213 and RU 24926 inhibit prolactin release from primary cultures of rat anterior pituitary cells (IC_{50} of 5 and 3 nM, respectively) and displace the binding of [^{3}H] dihydroergocryptine to bovine anterior pituitary membranes (K_i values of 150 and 100 nM, respectively). In several in vivo test systems (inhibition of prolactin secretion, decreasing striatal DA turnover, and increasing striatal acetylcholine levels in intact animals and animals with unilateral 6-OHDA-induced lesions of the nigrostriatal pathway), these two compounds were both more potent than and their effects were longer lasting than those of apomorphine. Both compounds were inactive on either basal or dopamine-stimulated striatal adenylate cyclase activity (301). Structurally, the hydrated disecoquinoline skeletons of RU 24926 and RU 24213 are similar to the hydrated quinoline structures of the selective D_2 agonists quinpirole and LY-149632 (302).

(+)PHNO ((+)-1,2,3,4a,5,6-hexahydro-9-hydroxy-4-n-propyl-4H-napth[1,2-b][1,4]-oxazine) Within a series of N-n-propylhexahydronaphthoxazines, the 9-hydroxy derivative (Figure 5) has been shown to have the highest affinity for the dopamine D_2 receptor (26,303). Examination of the in vitro activity of the (+)enantiomer of the 9-hydroxy compound [(+)-PHNO] showed that it inhibited the binding of [^{3}H] apomorphine and [^{3}H] spiperone at nanomolar concentrations (IC_{50} = 23.4 and 55 nM, respectively). The rank order for the inhibition of in vitro [^{3}H] spiperone binding was lisuride > *N*-propylnorapomorphine > pergolide > bromocriptine > (+)PHNO. This relative lack of affinity in studies in vitro is, however, in contrast to the results in studies in vivo. In vivo, after intraperitoneal administration, (+)PHNO and *N*-propylnorapomorphine were the

most potent agonists tested (303). When administered orally, (+)PHNO was the most potent inducer of contralateral rotation in 6-OHDA-lesioned rats. (+)PHNO fails to stimulate the dopamine-sensitive adenylate cyclase and as such should be regarded as a selective D_2 agonist with a high in vivo potency (303).

Dopamine D_2 Receptor Antagonists

A large number of compounds from several chemical classes have been observed to inhibit neuroleptic activity (304). However, although neurolepsis is associated with the blockade of DA D_2 receptors (24), many neuroleptics also bind with an appreciable affinity to other receptors (18). It is only a smaller subset of this large number of compounds that display a notable selectivity toward the DA D_2 receptor. (See Table 1.)

Butyrophenones

Haloperidol (4-[4-(p-chlorophenyl)-4-hydroxypiperidino]-4'-fluorobutyrophenone) [^{3}H] Haloperidol (see Figure 1) was the first ligand employed for the selective labeling of the DA neuroleptic receptor (8,57). The ability of a number of neuroleptics from several drug classes to compete for the binding of [^{3}H]-haloperidol correlated well with the potency of these compounds in reversing the apomorphine-induced suppression of prolactin secretion from cultured rat pituitary cells (305). The ability of neuroleptics from several drug classes to inhibit the binding of [^{3}H] haloperidol did not, however, correlate well with the abilities of these compounds to inhibit the binding of the nonselective neuroleptic [^{3}H] flupenthixol or to inhibit the dopamine-stimulated adenylate cyclase of rat striatum (13). Butyrophenones and diphenylbutylpiperidines were, in general, particularly weak in this regard (13). The butyrophenones haloperidol and spiperone were also shown to be relatively weak inhibitors of the dopamine-stimulated adenylate cyclase of carp retina (266). The predominantly D_2 dopaminergic binding of [^{3}H] haloperidol occurs with a dissociation constant of approximately 1 nM (306).

Spiperone (spiroperidol; 8-[4-(fluorophenyl)-4-oxobutyl]-1-phenyl-1,3,8-triazospiro[4.5]decane-4-one) [^{3}H] Haloperidol was largely supplanted as the ligand of choice for the characterization of the "neuroleptic receptor" upon the synthesis and characterization of the binding properties of the related butyrophenone, [^{3}H] spiperone (see Figure 1). This outcome was the result of [^{3}H]-spiperone being a higher affinity ligand (K_D = 143 pM) that exhibited a slower rate of dissociation and lower nonspecific binding (306). The high potency of spiperone was evident in functional tests of antagonist potency in which spiperone was the most potent antagonist of the dopamine-mediated inhibition of α-melanocyte stimulating hormone release from a primary culture of pars intermedia cells (K_D = 20 pM) (307). That the receptors with which [^{3}H] spiperone interacted were in fact DA D_2 receptors was inferred from the observation that antagonists produced similar effects upon both [^{3}H] spiperone binding and upon the D_2 receptor of the rat intermediate lobe that mediates the inhibition of adenylate cyclase (308). Conversely, spiperone was a comparatively weak antagonist of the DA-stimulated adenylate cyclase of carp retina (266). Although the D_2-D_1 selectivity of [^{3}H] spiperone is very good, it should be recognized that [^{3}H] spiperone is also capable of binding with high affinity to the serotonin - 2 receptor (D_2/5HT_2 selectivity ratio of 17) (24).

Benzimidazolones

Domperidone (5-chloro-1-[1-[3-(2-oxo-1-benzimidazolinyl)propyl]-4-piperidyl]-2-benzimidazolinone) Domperidone (Figure 13), as assessed by in vitro experiments, is a potent and specific dopamine D_2 receptor antagonist. The absence of central effects upon peripheral administration indicate, however, that domperidone does not readily cross the blood-brain barrier (309). Additionally, domperidone is inactive as an antagonist of the DA-stimulated cyclase of both carp retina and rat striatum (266). Peripherally, domperidone is highly selective for the DA_2 receptor (269,310). In vitro, [^{3}H] domperidone binds to the rat brain striatal dopamine D_2 receptor with high selectivity and high affinity (from equilibrium binding studies, K_D = 83 pM) (311).

Substituted Benzamides

The substituted benzamides (see Figure 2) are derivatives of procainamide (311). As antagonists of the dopamine D_2 receptor, they share with other neuroleptics the ability to inhibit apomorphine-induced vomiting, to increase serum prolactin levels, and to induce some extrapyramidal disturbances. Substituted benzamides differ from classic neuroleptics, however, in that they appear less cataleptogenic in animal behavioral models, are not potent inhibitors of DA-sensitive adenylate cyclase (313), and their binding to DA D_2 receptors is dependent on the presence of the sodium cation (313).

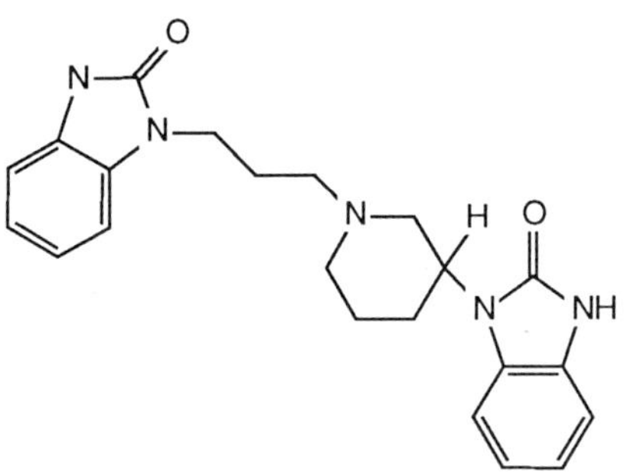

Figure 13 Domperidone: Benzimidazolone D_2 antagonist.

Sulpiride [N-(1-ethyl-2-pyrrolidylmethyl)-2-methoxy-5-aminosulfonyl-benzamide] Early observations of the actions of sulpiride led to the designation of substituted benzamides as atypical neuroleptics. In a manner similar to that of classic neuroleptics, sulpiride produced neurolepticlike changes in behavior in vivo and elevated serum prolactin levels in vivo (314) and in vitro (305). In a manner differing from that of classic neuroleptics, however, sulpiride failed to inhibit the DA-stimulated adenylate cyclase (266,314,315). This atypical profile of activity was an important part of the evidence in favor of the hypothesis of two DA receptors (20).

Sulpiride is a potent inhibitor of the D_2 receptor-specific binding of [^{3}H]-spiperone (IC_{50} = 570 nM; [^{3}H] spiperone concentration was 300 pM) and [^{3}H] sulpiride (IC_{50} = 25 nM; [^{3}H] sulpiride concentration was 10 nM). The displacement of [^{3}H] sulpiride is stereospecific, with (—)-sulpiride the active enantiomer in rat striatal preparations (314). Sulpiride also blocks the inhibitory effect of DA on the D_2 receptor-mediated inhibition of (*l*)-isoproterenol-stimulated accumulation of cAMP in dispersed rat intermediate lobe cells (290). In contrast, sulpiride was not a potent inhibitor of the binding of the nonselective neuroleptic [^{3}H] piflutixol to the dopamine D_1 receptor (314).

The in vitro receptor binding of the substituted benzamides is Na^+-dependent. In the absence of Na^+, the (—)enantiomer of sulpiride is 200-fold less potent in competing for the binding of [^{3}H] spiperone (313). Only Li^+ appears to be able to partially compensate for the absence of Na^+ (314). [^{3}H] sulpiride has been observed to bind to two sites in rat striatal preparations ($K_D{}^{high}$ = 12.7 nM; $K_D{}^{low}$ = 55.4 nM) (316). The addition of Na^+ has been observed to increase the number of [^{3}H] sulpiride binding sites without changing the affinity with which [^{3}H] sulpiride binds. On this basis, it has been hypothesized that [^{3}H] sulpiride may interact with Na^+-dependent sites that are distinct from those sites labeled by [^{3}H] spiperone (317). The question of whether these sodium-dependent and sodium-independent sites are indeed separable remains, as other comparisons have indicated that [^{3}H] sulpiride and [^{3}H]-spiperone (when nonspecific binding is determined by S-sulpiride) label the same set of sites (318).

Peripherally, sulpiride acts as an antagonist at both DA_1 and DA_2 receptors. However, although (S)-sulpiride is a much more active DA_2 antagonist than (R)-sulpiride [this preference is identical to that observed for the striatal D_2 receptor (318)], (R)-sulpiride is slightly more active than (S)-sulpiride as a DA_1 antagonist (269).

YM-09151-2 [cis-N-(benzyl-2-methyl-3-pyrrolidyl)-5-chloro-2-methoxy-4-methylaminobenzamide] Initial assessment of the in vivo and behavioral activity of YM-09151-2 indicated that YM-09151-2 possessed a unique profile of neuroleptic action. YM-09151-2 was unique in that it appeared to be a potent, noncataleptogenic neuroleptic that produced only slight central nervous system depression (319). YM-09151-2 was also observed to inhibit the in vivo dopamine-induced inhibition of rat prolactin secretion with a potency similar to that of spiperone (320). Biochemical studies verified that YM-09151-2 is a highly potent and selective ligand for the D_2 receptor. YM-09151-2 potently displaced [^{3}H] spiperone binding from pituitary and striatal membranes (IC_{50} = 1.0 and 0.4 nM, respectively) (320,321) and antagonized (K_i = 10.5 nM) the dopamine-induced inhibition of adenylate cyclase in homogenates of cholera toxin-treated rat intermediate lobe tissue (322). Conversely, YM-09151-2 was not a potent inhibitor of [^{3}H] piflutixol binding to D_1 receptors (IC_{50} = 22 μM) (321) or the D_1 receptor-mediated stimulation of striatal (IC_{50} = 20 μM) (321) or carp retina adenylate cyclase (322). YM-09151-2 displayed a similar lack of potency at other putative neurotransmitter receptors (323). The saturable binding of [^{3}H]-YM-09151-2 has been characterized. [^{3}H] YM-09151-2 binds to striatal D_2 receptors with high affinity (K_D = 57 pM) and selectivity (relative to [^{3}H] spiperone, the affinity of [^{3}H] YM-09151-2 for the serotonin - 2 receptor was 210- to 18-fold lower) (24,49). As has been observed in the case of other substituted benzamides, the binding of [^{3}H] YM-09151-2 to the D_2 receptor was Na^+ sensitive. In contrast to that observed with [^{3}H] sulpiride, however, the removal of Na^+ produced a sixfold increase in the K_D with no change observed in the maximal binding capacity (49).

Other Substituted Benzamides Of the large number of benzamides that have been synthesized and characterized, several others are notable for their specificity and high affinity. All share the characteristic benzamide D_2 selectivity and Na^+ dependency, although they differ in their affinity: [^{3}H] sultopride, K_D = 5.8 nM (48); [^{3}H] raclopride, K_D = 1.2 nM (43); [^{3}H] eticlopride, K_D = 0.17 pM) (35). Recently, eticlopride has been characterized as having the highest ratio of affinities for the D_2 receptor over the serotonin - 2 receptor of a large number of benzamide and nonbenzamide neuroleptics (24).

Pyrroloisoquinolines

Ro 22-1319 (2,6-dimethyl-3-ethyl-4,4a,5,6,7,8,8a,9-octahydro-4a,8a-*trans*-1H-pyrrolo [2,3-g] isoquinolin-4-one) (see Figure 14), as assessed by behavioral test (discrete avoidance procedure), is a neuroleptic with a potency greater than that of chlorpromazine and approximately equivalent to that of haloperidol. In comparison with haloperidol, however, Ro 22-1319 was less cataleptogenic and had less antistereotypic activity (324). A comparison of the ability of neuroleptics to compete for those sites labeled by [^{3}H] Ro 22-1319 and [^{3}H] spiperone supports the interaction of Ro 22-1319 with dopamine D_2 receptors (325). Ro 22-1319 binds to rat striatal D_2 receptors stereospecifically, with the (−)enantiomer the biologically active enantiomer. The dissociation

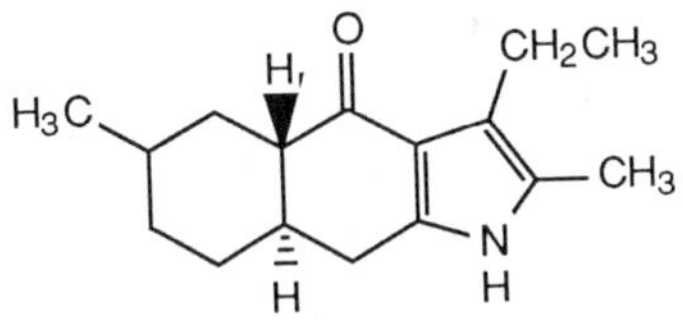

Figure 14 Ro 22-1319: Pyrrolisoquinoline D_2-selective antagonist.

constant for the binding of Ro 22-1319 is temperature dependent (K_D = 3.1 nM at 0°C; 17.6 nM at 37°C). Ro 22-1319 and several other pyrroloisoquinoline derivatives exhibit a dependence on the presence of Na^+ (326,327). Ro 22-1319 has been used to validate a model of the DA D_2 receptor (328).

FUTURE DIRECTIONS

The accumulation of data regarding the molecular structure and function of DA receptors is still in its preliminary stages. The purification to homogeneity of both D_1 and D_2 receptors and the isolation of the genes dictating their synthesis will greatly aid our understanding of dopamine receptor structure and function. Meanwhile, studies on the reconstitution of purified receptor-effector units, as in the case of the β-adrenergic (329) and muscarinic receptor (330,331), may yield some insight into the mechanism(s) responsible for receptor activation of second messenger systems. Additionally, functional studies on the role of DA receptor phosphorylation and receptor desensitization (if any), as has been documented for the β-adrenergic (332,333) receptor, may now be attempted, as well as the significance of D_1 and D_2 receptor-effector cross-talk for the expression of DA-mediated events.

Another major area of uncertainty and an area in need of extensive study is the molecular characterization of DA receptors in both the cardiovascular system and in invertebrate species. Moreover, the functional significance of D_1 and D_2 receptor "interactions" at the behavioral, neurophysiological, and molecular level need to be assessed. Finally, a new generation of selective D_1 and possibly D_2 receptor agents should be developed for both clinical and basic science applications.

ACKNOWLEDGMENTS

The authors are grateful to Marie Botelho for typing the manuscript and to Evan Fogel for preparing some of the figures. H. B. Niznik was supported by a fellowship from the Medical Research Council of Canada.

REFERENCES

1. A. Carlsson and M. Lindqvist, Acta Pharmacol. Toxicol. *20*, 140 (1963).
2. J. M. Van Rossum, Arch. Int. Pharmacodyn. Ther. *160*, 492 (1966).
3. J. W. Kebabian, G. L. Petzold, and P. Greengard, Proc. Natl. Acad. Sci. USA *69*, 2145 (1972).
4. J. H. Brown and M. H. Makman, Proc. Natl. Acad. Sci. USA *69*, 539 (1972).
5. Y. C. Clement-Cormier, J. W. Kebabian, G. L. Petzold, and P. Greengard, Proc. Natl. Acad. Sci. USA *71*, 1113 (1974).
6. L. L. Iversen, Science *188*, 1084 (1975).
7. R. J. Miller, A. S. Horn, and L. L. Iversen, Mol. Pharmacol. *10*, 759 (1974).
8. P. Seeman, M. Chau-Wong, J. Tedesco, and K. Wong, Proc. Natl. Acad. Sci. USA *72*, 4376 (1974).
9. I. Creese, D. R. Burt, and S. M. Snyder, Life Sci. *27*, 993 (1975).
10. P. Seeman, T. Lee, M. Chau-Wong, and K. Wong, Nature *261*, 717 (1976).
11. I. Creese, D. R. Burt, and S. H. Snyder, Science *197*, 596 (1977).
12. J. Hyttel, Prog. Neuropsychopharmacol. *2*, 329 (1978).
13. J. Hyttel, Life Sci. *23*, 551 (1978).
14. J. Hyttel, Psychopharmacology (Berlin) *67*, 107 (1978).
15. A. J. Cross and F. Owen, Eur. J. Pharmacol. *65*, 341 (1980).
16. J. Hyttel, Life Sci. *28*, 563 (1981).
17. J. E. Leysen, C. J. E. Niemegeers, J. P. Tollanaere, and P. M. Laduron, Biochem. Pharmacol. *27*, 307 (1978).
18. P. Seeman, Pharmacol. Rev. *32*, 230 (1980).
19. I. Creese, M. W. Hamblin, and S. E. Leff, in *Handbook of Psychopharma cology*, Vol. 17 (L. L. Iversen, S. D. Iversen, S. H.Snyder, eds.), Plenum Press, N.Y., 1983, p. 81.
20. J. W. Kebabian and D. B. Calne, Nature *277*, 93 (1979).
21. J. C. Stoof and J. W. Kebabian, Life Sci. *35*, 2281 (1984).
22. C. Kaiser and J. Jain, Med. Res. Rev. *5*, 145 (1985).
23. P. Seeman and D. Grigoriadis, Neurochem. Int. *10*, 1 (1987).
24. P. Seeman, D. Grigoriadis, and H. B. Niznik, Drug. Dev. Res. *9*, 63 (1986).
25. P. Seeman and H. B. Niznik, ISI Atlas of Sci.: Pharmacology *2*, 161 (1988).
26. B. Hazelhoff, J. B. de Vries, D. Dijkstra, T. B. A. Mulder, P. B. M. W. M. Timmermans, H. Wynberg, and A. S. Horn, Eur. J. Pharmacol. *124*, 93 (1986).
27. L. C. Iorio, A. Barnett, F. H. Leitz, V. P. Houssen, and C. A. Korduba, J. Pharmacol. Exp. Ther. *226*, 462 (1983).
28. P. Sokoloff, K. Redouane, M. Brann, M. P. Martres, and J.-C. Schwartz, Naunyn-Schmiedebergs Arch, Pharmacol. *329*, 236 (1985).
29. M. Baudry, M. P. Martres, and J.-C. Schwartz, Naunyn-Schmiedebergs Arch. Pharmacol. *332*, 21 (1986).
30. D. Grigoriadis and P. Seeman, Naunyn-Schmiedebergs Arch. Pharmacol. *332*, 21 (1986).
31. M. P. Martres, M. Baudry, and J.-C. Schwartz, Life Sci. *23*, 1781 (1978).
32. S. Lazareno, J. Rec. Res. *3*, 163 (1983).

33. J. E. Leysen, C. J. E. Niemegeers, J. M. Van Nueten, and P. M. Laduron, Mol. Pharmacol. *21*, 301 (1982).
34. M. W. Hamblin, S. E. Leff, and I. Creese, Biochem. Pharmacol. *33*, 877 (1984).
35. H. Hall, C. Kohler, and L. Gawell, Eur. J. Pharmacol. *111*, 191 (1985).
36. D. R. Burt, I. Creese, and S. H. Snyder, Mol. Pharmacol. *12*, 800 (1976).
37. S. J. Peroutka, D. C. U'Prichard, D. A. Greenberg, and S. H. Snyder, Neruopharmacology *16*, 549 (1977).
38. E. Richelson and A. Nelson, Eur. J. Pharmacol. *103*, 197 (1984).
39. B. D. Perry, P. R. Simon, and D. C. U'Prichard, Eur. J. Pharmacol. *95*, 315 (1983).
40. I. Pri-Bar and O. Buchman, J. Labeled Comp. Radiopharmaceut. *20*, 1153 (1983).
41. R. A. Lyon, M. Titeler, J. J. Frost, P. J. Whitehouse, H. N. Wagner, D. F. Wong, R. F. Dannals, J. M. Links, and M. J. Kuhar, J. Neurosci. *6*, 2941 (1986).
42. R. A. Lyon, M. Titeler, L. Bigornia, and A. S. Schneider, J. Neurochem. *48*, 631 (1987).
43. C. Kohler, H. Hall, S. O. Ogren, and L. Gawell, Biochem. Pharmacol. *34*, 2251 (1985).
44. J. L. Fields, T. D. Reisine, and H. I. Yamamura, Brain Res. *136*, 578 (1978).
45. S. J. List and P. Seeman, Proc. Natl. Acad. Sci. USA *78*, 2620 (1981).
46. J. Hyttel, Eur. J. Pharmacol. *91*, 153 (1983).
47. A. Theodorou, M. Crockett, P. Jenner, and C. D. Marsden, J. Pharm. Pharmacol. *89*, 51 (1982).
48. A. Mizuchi, N. Kitagawa, S. Sarata, and Y. Miyachi, Eur. J. Pharmacol. *89*, 51 (1982).
49. H. B. Niznik, D. E. Grigoriadis, I. Pri-Bar, O. Buchman, and P. Seeman, Naunyn-Schmiedebergs Arch. Pharmacol. *329*, 333 (1985).
50. A. L. Gundlach, B. L. Largent, and S. H. Snyder, Life Sci. *35*, 1981 (1984).
51. M. P. Martres, M. L. Bouthenet, N. Sales, P. Sokoloff, and J. C. Schwartz, Science *228*, 752 (1985).
52. M. P. Martres, N. Sales, M. L. Bouthenet, and J. C. Schwartz, Eur. J. Pharmacol. *118*, 211 (1985).
53. S. E. Leff, M. W. Hamblin, and I. Creese, Mol. Pharmacol. *27*, 171 (1985).
54. R. M. Huff and P. B. Molinoff, J. Pharmacol. Exp. Ther. *232*, 57 (1985).
55. D. G. Huff and P. B. Molinoff, J. Pharmacol. Exp. Ther. *232*, 57 (1985).
56. M. Cincotta, D. J. deVries, and P. M. Beart, Neurosci. Lett. *61*, (1985).
57. J. Hyttel and J. Arnt, J. Neural Transm. *68*, 771 (1987).
58. L. C. Murrin, Life Sci. *33*, 2179 (1983).
59. K. M. Mir and S. R. Nahorski, J. Neurochem. *41*, 578 (1983).
60. W. Billard, W. Ruperto, G. Crosby, L. C. Iorio, and A. Barnett, Life Sci. *35*, 1185 (1984).
61. E. J. Hess, G. Battaglia, A. B. Norman, L. C. Iorio, and I. Creese, Eur. J. Pharmacol. *121*, 31 (1986).
62. P. Seeman, C. Ulpian, D. Grigoriadis, I. Pri-Bar, and O. Buchman, Biochem. Pharmacol. *34*, 151 (1985).

63. G. J. Kilpatrick, P. Jenner, and C. D. Marsden, J. Pharm. Pharmacol. *38*, 907 (1986).
64. D. W. Schultz, E. S. Stanford, S. W. Wyrick, and R. B. Mailman, J. Neurochem. *45*, 1601 (1985).
65. H. B. Niznik, N. Y. Otsuka, A. Dumbrille-Ross, D. Grigoriadis, A. Tirpak, and P. Seeman, J. Biol. Chem. *261*, 8397 (1986).
66. M. H. Makman and B. Dvorkin, Mol. Brain Res. *1*, 261 (1986).
67. A. J. Cross, R. D. Mashal, J. A. Johnson, and F. Owen, Neuropharmacology *22*, 1327 (1983).
68. K. E. Flaim, G. W. Gessner, S. T. Crooke, J. R. Heys, and J. Weinstock, Life Sci. *38*, 2087 (1986).
69. E. H. Ohlstein, B. Zabko-Potapovich, and B. A. Berkowitz, Eur. J. Pharmacol. *118*, 321 (1985).
70. N. Amlaiky, J. G. Berger, W. Chang, R. J. McQuade, and M. G. Caron, Mol. Pharmacol. *31*, 129 (1987).
71. A. Sidhu and J. W. Kebabian, Eur. J. Pharmacol. *113*, 437 (1985).
72. A. Sidhu, J. C. van Oene, P. Dandridge, C. Kaiser, and J. W. Kebabian, Eur. J. Pharmacol. *128*, 213 (1986).
73. S. J. List, K. A. Wreggett, and P. Seeman, J. Neurosci. *2*, 895 (1982).
74. P. Seeman, G. N. Woodruff, and J. A. Poat, Eur. J. Pharmacol. *55*, 137 (1979).
75. S. Urwyler and R. Markstein, J. Neurochem. *46*, 1058 (1986).
76. P. Seeman, M. Watanabe, D. Grigoriadis, J. Tedesco, S. R. George, U. Svensson, J. L. G. Nilsson, J. L. Neumeyer, Mol. Pharmacol. *28*, 391 (1985).
77. D. C. U'Prichard, D. A. Greenberg, and S. H. Snyder, Mol. Pharmacol. *13*, 454 (1977).
78. M. Titeler, P. Weinreich, and P. Seeman, Proc. Natl. Acad. Sci. USA *74*, 3750 (1977).
79. C. W. Lin, S. Maayani, and S. Wilk, Biochem. Pharmacol. *30*, 1305 (1981).
80. N. G. Bacopoulos, Life Sci. *34*, 307 (1984).
81. S. J. List and P. Seeman, J. Neurochem. *39*, 1363 (1982).
82. S. E. Leff and I. Creese, Mol. Pharmacol. *27*, 184 (1985).
83. K. E. Flaim, G. W. Gessner, S. T. Crooke, H. M. Sarau, and J. Weinstock, Life Sci. *36*, 1427 (1985).
84. I. Creese, L. Padgett, E. Fazzini, and F. Lopez, Eur. J. Pharmacol. *56*, 411 (1979).
85. M. Titeler and P. Seeman, Eur. J. Pharmacol. *56*, 291 (1979).
86. D. T. Wong, F. P. Bymaster, P. T. Lane, D. Kau, N. J. Bach, and E. C. Kornfeld, Res. Commun. Chem. Pathol. Pharmacol. *30*, 195 (1980).
87. A. Closse, W. Frick, R. Markstein, R. Maurer, and R. Nordmann, J. Neural Transm. *62*, 231 (1985).
88. J. J. Foley, H. M. Sarau, and S. T. Crooke, Pharmacology *27*, 240 (1985).
89. B. Scatton and A. Dubois, Eur. J. Pharmacol. *111*, 145 (1985).
90. A. Dubois, M. Sauasta, O. Curet, and B. Scatton, Neuroscience *19*, 125 (1986).
91. O. Gredal and M. Nielsen, J. Neurochem. *48*, 370 (1987).

92. G. J. Kilpatrick, P. Jenner, and C. D. Marsden, J. Neurochem. *46*, 413 (1986).
93. J. E. Leysen, C. J. E. Niemegeers, J. P. Tollanaere, and P. M. Laduron, Nature *272*, 169 (1978).
94. D. Grigoriadis and P. Seeman, J. Neurochem. *44*, 1925 (1985).
95. D. E. Hicks, H. Schoemaker, and S. Z. Langer, Eur. J. Pharmacol. *105*, 339 (1984).
96. A. Hilditch, G. M. Drew, and R. J. Naylor, Eur. J. Pharmacol. *97*, 333 (1984).
97. E. H. Ohlstein and B. A. Berkowitz, Eur. J. Pharmacol. *108*, 205 (1985).
98. S. J. Gunning and R. J. Naylor, J. Pharm. Pharmacol. *37*, 78 (1985).
99. S. Bischoff, M. Heinrich, J. M. Sonntag, and J. Krauss, Eur. J. Pharmacol. *129*, 367 (1986).
100. P. H. Anderson, F. C. Gronvald, and J. A. Jansen, Life Sci. *37*, 1971 (1985).
101. P. Onali, M. C. Olianas, and G. L. Gessa, Mol. Pharmacol. *28*, 138 (1985).
102. A. V. Christensen, J. Arnt, J. Hyttel, J. J. Larsen, and O. Svendsen, Life Sci. *34*, 1529 (1984).
103. J. Hyttel, Neuropharmacology *23*, 1395 (1984).
104. S. Boyce, E. Kelley, A. Davis, S. Flemminger, P. Jenner, and C. D. Marsden, Biochem. Pharmacol. *34*, 1665 (1985).
105. J. G. Hensler and M. L. Dubocovich, Brain Res. *398*, 407 (1986).
106. J. L. Willems, W. A. Buylaert, R. A. Lefebvre, and M. G. Bogaert, Pharmacol. Rev. *37*, 166 (1985).
107. J. W. Kebabian, T. Augi, J. C. Van Oene, K. Shigematsu, and J. M. Saavedra, Trends Pharmacol. Sci. *7*, 96 (1986).
108. M. H. Sabouni, K. A. Alkadhi, and M. F. Lokhandwala, J. Pharmacol. Exp. Ther. *236*, 65 (1986).
109. K. A. Alkadi, M. H. Sabouni, A. F. Ansari, and R. F. Lokhandwala, J. Pharmacol. Exp. Ther. *238*, 547 (1986).
110. A. G. Fernandez and R. Masshingham, Life Sci. *36*, 1 (1985).
111. S. J. Boyson, P. McGonigle, and P. B. Molinoff, J. Neurosci. *6*, 3177 (1986).
112. C. A. Alter and M. R. Marien, J. Neurosci. *7*, 213 (1987).
113. A. Dubois, M. Savasta, O. Curet, and B. Scatton, Neuroscience *19*, 125 (1986).
114. T. M. Dawson, D. R. Gehlert, R. T. McCabe, A. Bannett, and J. K. Wamsley, J. Neurosci. *6*, 2352 (1986).
115. M. Savasta, A. Dubois, and B. Scatton, Brain Res. *375*, 291 (1986).
116. M. Aiso, K. Shigematsu, J. W. Kebabian, W. Z. Potter, R. A. Cruciani, and J. M. Saavadra, Brain Res. *408*, 291 (1986).
117. S. M. Gospe, Life Sci. *33*, 1945 (1983).
118. P. Deterre, D. Paupardin-Tritsch, and J. Bockaert, Mol. Brain Res. *1*, 101 (1986).
119. S. Pratt and S. C. Pryor, Cell. Mol. Neurobiol. *6*, 325 (1986).

120. J. C. Stoof, T. R. Werkman, J. C. Lodder, and T. A. De Vliegen, Trends Pharmacol. Sci. *7*, 7 (1986).
121. M. Hadjiconstantinou and N. H. Neff, Life Sci. *35*, 1135 (1984).
122. M. L. Dubocovich and N. Weiner, J. Pharmacol. Exp. Ther. *233*, 747 (1985).
123. S. C. Massey and D. A. Redburn, Prog. Neurobiol. *28*, 55 (1987).
124. O. Gredal, D. Parkinson, and M. Nielsen, Eur. J. Pharmacol. *137*, 241 (1987).
125. E. M. Laster and J. E. Dowling, Proc. Natl. Acad. Sci. USA *82*, 3025 (1985).
126. J. Vayssette, N. Vaysse, and A. Ribet, Eur. J. Pharmacol. *122*, 321 (1986).
127. C. Missale, P. Liberini, M. Memo, M. O. Carruba, and P. F. Spano, Endocrinology *119*, 227 (1986).
128. A. J. Balmforth, S. G. Ball, R. I. Freshney, D. I. Graham, H. B. McNamee, and P. F. T. Vaughan, J. Neurochem. *47*, 715 (1986).
129. M. Relja, Z. Lackovic, and N. H. Neff, Life Sci. *31*, 2571 (1982).
130. F. Ferre, Life Sci. *39*, 1893 (1986).
131. F. Amenta, Neurosci. Lett. *56*, 279 (1985).
132. H. Murakami, T. Ishikawa, and M. Sano, Comp. Biochem. Physiol. *84C*, 51 (1986).
133. A. Gies, Comp. Biochem. Physiol. *84*, 61 (1986).
134. J. N. Joyce and J. F. Marshall, Neuroscience *20*, 773 (1987).
135. E. K. Richfield, A. B. Young, and J. B. Penney, Brain Res. *383*, 121 (1986).
136. M. Memo, C. Missale, M. O. Carruba, and P. F. Spano, J. Neural Trans. *22*(Suppl.)., 19 (1986).
137. L. I. Goldberg, Clin. Physiol. Biochem. *3*, 120 (1985).
138. A. Hilditch and G. M. Drew, Trends Pharmacol. Sci. *6*, 396 (1985).
139. L. I. Goldberg, J. D. Kohli, and D. Glock, in *Dopaminergic Systems and Their Regulation* (G. N. Woodruff, J. A. Poat, and P. J. Roberts, eds.), Macmillan Press, New York, 1986, p. 195.
140. M. Schmidt, J. L. Imbs, and J. Schwartz, in *Dopaminergic Systems and Their Regulation* (G. N. Woodruff, J. A. Poat, and P. J. Roberts, eds.), Macmillan Press, New York, 1986, p. 181.
141. N. L. Allison, J. W. Dubb, J. A. Ziemniak, F. Alexander, and R. M. Stote, Clin. Pharmacol. Ther. *41*, 282 (1987).
142.. K. T. Nakamura, R. A. Felder, P. A. Jose, and J. E. Robillard, Am. J. Physiol. *252*, R490 (1987).
143. C. Missale, M. Pizzi, M. Memo, G. B. Picotti, M. O. Carruba, and P. F. Spano, Neurosci. Lett. *61*, 207 (1985).
144. Y. Hoshino, H. Obara, and S. Iwai, Life Sci. *39*, 2525 (1986).
145. S. Dallas, G. M. Drew, and A. Hilditch, J. Cardiovasc. Pharmacol. *8*, 116 (1986).
146. A. L. Blumberg, J. W. Wilson, and J. P. Hieble, J. Cardiovasc. Pharmacol. *7*, 723 (1985).

147. A. G. Dupont, P. Vanderniepen, R A. Lefebvre, and M. G. Bogaert, J. Auton. Pharmacol. *5*, 305 (1986).
148. L. I. Goldberg, D. Glock, J. D. Kohli, and A. Barnett, Hypertension *6*, I-25 (1984).
149. A. Barnett, H. Ahn, W. Billard, E. H. Gold, J. D. Kohli, D. Glock, and L. I. Goldberg, Eur. J. Pharmacol. *128*, 249 (1986).
150. M. Sanderini, A. Benelli, and M. Baraldi, Pharmacol. Res. Commun. *18*, 1151 (1986).
151. R. A. Felder, A. M. Blecher, P. L. Calangno, and P. A. Jose, Am. J. Physiol. *247*, F499 (1984).
152. R. A. Felder, A. M. Blecher, G. M. Eisner, and P. A. Jose, Am. J. Physiol. *247*, F557 (1984).
153. H. A. J. Struyker-Boudier, Prog. Pharmacol. *6*, 81 (1986).
154. O. E. Broode, Life Sci. *31*, 289 (1982).
155. E. M. Brown, R. J. Carrol, and G. D. Aurbach, Proc. Natl. Acad. Sci. USA *74*, 4210 (1977).
156. E. M. Brown, M. F. Attie, S. Reen, D. G. Gardner, J. W. Kebabian, and G. D. Aurbach, Mol. Pharmacol. *18*, 335 (1980).
157. E. M. Brown and B. Dawson-Hughes, in *Dopamine Receptors* (C. Kaiser and J. W. Kebabian, eds.), ACS Symposium Series, Vol. 224, American Chemical Society, Washington, D.C., 1983, p. 1.
158. E. M. Brown, R. Leombruno, J. Thatcher, and M. Burrows, Endocrinology *116*, 1123 (1985).
159. D. A. Hanley and P. G. Wellings, Can. J. Physiol. Pharmacol. *63*, 1139 (1985).
160. J. W. Blum, P. Kunz, J. A. Fischer, V. Binswanger, W. Lichtensteker, and M. DaPrada, Am. J. Physiol. *239*, E255 (1980).
161. J. W. Blum and J. A. Fischer, Life Sci. *35*, 1469 (1984).
162. E. M. Brown, D. G. Gardner, R. A. Windeck, and G. D. Aurbach, Endocrinology *103*, 2323 (1978).
163. E. M. Brown, E. J. Watson, R. Leombruno, and R. H. Underwood, Metabolism *32*, 1038 (1983).
164. M. F. Attie, E. M. Brown, D. G. Gardner, A. M. Spiegel, and G. D. Aurbach, Endocrinology *107*, 1776 (1980).
165. D. M. Shoback, J. Thatcher, R. Leombruno, and E. M. Brown, Proc. Natl. Acad. Sci. USA *81*, 3113 (1984).
166. A. J. Cross, I. N. Rerrier, and C. A. Bloxham, Life Sci. *33*, 743 (1983).
167. A. G. Gilman, Cell *36*, 577 (1984).
168. A. M. Spiegel, Mol. Cell. Endocrinol. *49*, 1 (1987).
169. J. G. Thatcher, D. G. Gardner, and E. M. Brown, Endocrinology *110*, 1367 (1982).
170. E. M. Brown, and J. G. Thatcher, Endocrinology *110*, 1374 (1982).
171. K. R. Williams, H. C. Hemmings, M. B. LoPresti, W. H. Konigsberg, and P. Greengard, J. Biol. Chem. *261*, 1890 (1986).
172. E. M. Brown, J. Redgrave, and J. G. Thatcher, FEBS Lett. *175*, 72 (1984).

173. B. K. Kinder, N. G. Delahunt, J. D. Jamieson, and F. S. Gorelick, Endocrinology *120*, 170 (1987).
174. L. A. Fitzpatrick, M. L. Brani, and G. D. Aurbach, Endocrinology *119*, 2700 (1986).
175. L. A. Fitzpatrick, M. L. Brani, and G. D. Aurbach, Biochem. Biophys. Res. Commun. *138*, 2087 (1986).
176. G. L. Stiles, M. G. Caron, and R. J. Lefkowitz, Physiol. Rev. *64*, 661 (1984).
177. C. K. Mickevicius, J. K. Harrison, and M. E. Gnegy, Mol. Pharmacol. *30*, 469 (1986).
178. G. Battaglia, M. B. Norman, E. J. Hess, and I. Creese, J. Neurochem. *46*, 1180 (1986).
179. P. H. Anderson and C. Braestrup, J. Neurochem. *47*, 1822 (1987).
180. E. J. Hess, G. Battaglia, A. B. Norman, and I. Creese, Mol. Pharmacol. *31*, 50 (1987).
181. A. Sidhu, S. Kassis, J. W. Kebabian, and P. H. Fishman, Biochemistry *25*, 6695 (1986).
182. E. J. Nestler, S. I. Walaas, and P. Greengard, Science *225*, 1357 (1984).
183. S. I. Walaas and P. Greengard, J. Neurosci. *4*, 84 (1984).
184. H. C. Hemmings, A. C. Nairn, D. W. Aswad, and P. Greengard, J. Neurosci. *4*, 99 (1984).
185. C. C. Ouimet, P. E. Miller, H. C. Hemmings, S. I. Walaas, and P. Greengard, J. Neurosci. *4*, 111 (1984).
186. M. Treiman and P. Greengard, Neuroscience *15*, 713 (1985).
187. D. Paupardin-Tritsch, L. Colombaioni, P. Deterre, and H. M. Gerschenfeld, J. Neurosci. *5*, 2522 (1985).
188. J. R. Lemos, I. Novak-Hoffer, and I. B. Levitan, J. Biol. Chem. *260*, 3207 (1985).
189. K. Sasaki and M. Sato, Nature *325*, 259 (1987).
190. J. L. Waddington, Biochem. Pharmacol. *35*, 3661 (1986).
191. A. G. Molloy and J. L. Waddington, Psychopharmacology *92*, 164 (1987).
192. G. H. Fletcher and M. S. Starr, Neuroscience *20*, 589 (1987).
193. J. R. Walters, D. A. Bergstrom, J. H. Carlson, T. N. Chase, and A. R. Braun, Science *236*, 719 (1987).
194. B. G. Weick and J. R. Walters, Brain Res. *405*, 234 (1987).
195. J. Arnt, J. Hyttel, and J. Perregaard, Eur. J. Pharmacol. *133*, 137 (1987).
196. F. J. White, Eur. J. Pharmacol. *135*, 101 (1987).
197. S. R. George, M. Watanabe, T. Di Palo, P. Falardeau, F. Labrie, and P. Seeman, Endocrinology *117*, 690 (1985).
198. N. Ben-Jonathan, Endocrinol. Rev. *6*, 564 (1985).
199. W. M. McDonald, D. R. Sibley, B. F. Kilpatrick, and M. G. Caron, Mol. Cell. Endocrinol. *36*, 201 (1984).
200. B. Borgundvaag and S. R. George, Life Sci. *37*, 379 (1985).
201. M. J. Cronin, G. A. Myers, R. M. MacLeod, and E. L. Hewlett, Am. J. Physiol. *44*, E-499 (1983).

202. D. Delebeke, J. G. Scammell, A. Martinez-Campos, and P. S. Dannies, Endocrinology *118*, 1271 (1986).
203. M. Memo, L. Castelletti, C. Missale, A. Valerio, M. O. Carruba, and P. F. Spano, J. Neurochem. *47*, 1689 (1986).
204. D. Delebeke and P. S. Dannies, Endocrinology *117*, 439 (1985).
205. A. M. Judd, K. Koike, G. Schettini, I. S. Login, E. L. Hewlett, T. Yasumoto, and R. M. MacLeod, Endocrinology *117*, 1215 (1985).
206. P. L. Canonico, C. A. Valdenegro, and R. M. MacLeod, Endocrinology *113*, 7 (1983).
207. P. L. Canonico, C. A. Valdenegro, and R. M. MacLeod, Endocrinology *111*, 347 (1982).
208. A. Enjalbert, F. Sladeczek, G. Guillon, P. Bertrand, C. Shu, J. Epelbaum, A. Garcia-Sainz, S. Jard, C. Lombard, C. Kordon, and J. Bockeart, J. Biol. Chem. *261*, 4071 (1986).
209. S. H. Simmonds and P. G. Strange, Neurosci. Lett. *60*, 267 (1985).
210. P. L. Canonico, W. D. Jarvis, A. M. Judd, and R. M. MacLeod, J. Endocrinol. *110*, 389 (1986).
211. A. B. Cubitt, B. L. Brown, and P. R. M. Dobson, J. Neurochem. *49*, 183 (1987).
212. G. Giannattasio, R. Bianchi, A. Spada, and L. Vallar, Endocrinology *120*, 2611 (1987).
213. M. J.Berridge, I. S. I. Atlas of Science, Pharmacology *1*, 91 (1987).
214. A. A. Abel-Latif, Pharmacol. Rev. *38*, 227 (1986).
215. L. E. Hokin, Annu. Rev. Biochem. *54*, 205 (1985).
216. S. K. Fischer and B. W. Agranoff, J. Neurochem. *48*, 999 (1987).
217. Y. Nishizuka, Science *233*, 305 (1986).
218. J. R. Williamson, Hypertension *8II*, 140 (1986).
219. S. E. Senogles, J. L. Benovic, N. Amlaiky, C. Unson, G. Milligan, R. Vinitsky, A. M. Spiegel, and M. G. Caron, J. Biol. Chem. *262*, 4860 (1987).
220. J. R. Zysk, B. F. Posiner, E. L. Jewlett, J. C. Garrison, and M. J. Cronin, Endocr. Res. *12*, 157 (1986).
221. J. Hescheler, W. Rosenthal, W. Trautwein, and G. Schultz, Nature *325*, 445 (1987).
222. T. Katada, M. Oinuma, K. Kusakabe, and M. Ui, FEBS Lett. *213*, 353 (1987).
223. Y. P. Loh, B. Meyers, B. Wong, D. C. Parish, M. Lang, and M. E. Goldman, J. Biol. Chem. *260*, 8956 (1985).
224. T. E. Cote, E. A. Frey, and R. D. Sekura, J. Biol. Chem. *259*, 8693 (1984).
225. T. E. Cote, R. Folder, J. W. Kebabian, R. D. Sekura, T. Reisine, and H. U. Affolter, J. Biol. Chem. *261*, 4555 (1986).
226. M. Beaulieu, R. Felder, and J. W. Kebabian, Endocrinology *118*, 1032 (1986).
227. T. Yamamoto, Y. Furuki, S. Guild, and J. W. Kebabian, Biochem. Biophys. Res. Commun. *143*, 1076 (1987).
228. S. R. George, M. Watanabe, and P. Seeman, J. Neurochem. *44*, 1168 (1985).

229. M. Watanabe, S. R. George, and P. Seeman, Biochem. Pharmacol. *34*, 2459 (1985).
230. M. Watanabe, S. R. George, and P. Seeman, J. Neurochem. *45*, 1842 (1985).
231. S. H. Simmonds, P. G. Strange, A. W. Hall, and R. J. K. Taylor, Biochem. Pharmacol. *35*, 731 (1986).
232. E. J. Hess and I. Creese, in *Receptor Biochemistry and Methodology*, Vol. 8 (I. Creese and C. M. Fraser, eds.), Alan R. Liss, New York, 1987, p. 1.
233. T. Kuno, O. Shirakawa, and C. Tanaka, Biochem. Biophys. Res. Commun. *115*, 325 (1983).
234. N. Fujita, M. Nakahiro, I. Fukuchi, K. Saito, and H. Yoshida, Brain Res. *333*, 231 (1985).
235. D. J. DeVries and P. M. Beart, Mol. Brain Res. *1*, 29 (1986).
236. S. Weiss, M. Sebben, J. A. Garcia-Sainz, and T. J. Bockaert, Mol. Pharmacol. *27*, 595 (1986).
237. D. M. F. Cooper, C. M. Bier-Laning, M. K. Halford, M. D. Ahlijanian, and N. R. Zahniser, Mol. Pharmacol. *29*, 113 (1986).
238. M. Memo, C. Missale, M. O. Carruba, and P. F. Spano, Neurosci. Lett. *71*, 192 (1986).
239. M. Pizzi, F. D'Agostini, M. DaPrada, P. F. Spano, and W. E. Haefely, Eur. J. Pharmacol. *136*, 263 (1987).
240. H. B. Niznik, Mol. Cell. Endocrinol. *54*, 1 (1987).
241. P. G. Strange, in *Receptor Biochemistry and Methodology* (I. Creese and C. M. Fraser, eds.), Alan R. Liss, New York, 1987, p. 29.
242. H. B. Niznik, D. Grigoriadis, N. Y. Otsuka, A. Dumbrille-Ross, and P. Seeman, Biochem. Pharmacol. *35*, 2974 (1986).
243. N. Amlaiky and M. G. Caron, J. Neurochem. *47*, 196 (1986).
244. K. Redouane, P. Sokoloff, J. C. Schwartz, P. Hamdi, A. Mann, C. G. Wermuth, J. Roy, and J. L. Morgat, Biochem. Biophys. Res. Commun. *130*, 1086 (1985).
245. C. Bouvier, M. Potier, G. Beaugard, J. Lafond, N. Amlaiky, M. G. Caron, and R. Collu, J. Neurochem. *47*, 1653 (1986).
246. N. Amlaiky and M. G. Caron, J. Biol. Chem. *260*, 1983 (1985).
247. J. Y. Lew, E. Meller, and M. Goldstein, Eur. J. Pharmacol. *113*, 145 (1985).
248. H. B. Niznik, D. Grigoriadis, and P. Seeman, FEBS Lett. *209*, 71 (1986).
249. P. Seeman and H. B. Niznik, Eur. J. Pharmacol. *127*, 297 (1986).
250. H. B. Niznik, J. H. Guan, J. L. Neumeyer, and P. Seeman, Eur. J. Pharmacol. *104*, 389 (1984).
251. H. B. Niznik, J. H. Guan, J. L. Neumeyer, and P. Seeman, Mol. Pharmacol. *27*, 193 (1985).
252. W. Wouters, J. Van Dun, and P. M. Laduron, Eur. J. Biochem. *145*, 273 (1984).
253. W. Wouters, J. Van Dun, and P. M. Laduron, Biochem. Pharmacol. *33*, 3517 (1984).
254. H. B. Niznik, A. Dumbrille-Ross, J. H. Guan, J. L. Neumeyer, and P. Seeman, Neurosci. Lett. *55*, 267 (1985).

255. J. L. Neumeyer, J. G. Guan, H. B. Niznik, A. Dumbrille-Ross, P. Seeman, S. Padmanabhan, and D. R. Elmaleh, J. Med. Chem. *28*, 405 (1985).
256. L. Lilly, C. M. Fraser, C. Y. Jung, P. Seeman, and J. C. Venter, Mol. Pharmacol. *24*, 10 (1983).
257. M. Nielsen, V. Klimek, and J. Hyttel, Life Sci. *35*, 325 (1984).
258. M. R. Braun, Biochem. Biophys. Res. Commun. *133*, 1181 (1985).
259. T. Kuno and C. Tanaka, Biochem. Biophys. Res. Commun. *117*, 65 (1983).
260. O. Gredal and M. Nielson, J. Neurochem. *48*, 364 (1987).
261. S. F. Senogles, N. Amlaiky, A. L. Johnson, and M. G. Caron, Biochemistry *25*, 749 (1986).
262. J. Ramwani and R. K. Mishra, J. Biol. Chem. *261*, 8894 (1986).
263. L. Antonian, E. Antonian, R. B. Murphy, and D. I. Schuster, Life Sci. *38*, 1847 (1986).
264. V. Soskic and J. Petrovic, Iugoslav. Physiol. Pharmacol. Acta *22*, 329 (1986).
265. P. E. Setler, H. M. Sarau, C. L. Zirkle, and H. L. Saunders, Eur. J. Pharmacol. *50*, 419 (1978).
266. K. J. Watling and J. E. Dowling, J. Neurochem. *36*, 559 (1981).
267. U. Ungerstedt, Acta Physiol. Scand. *82*(Suppl. 367), 69 (1971).
268. A. G. Molloy and J. L. Waddington, Psychopharmacology *82*, 409 (1984).
269. L. I. Goldberg and J. D. Kohli, in *Dopamine Receptors* (C. Kaiser and J. W. Kebabian, eds.), ACS Symposium Series, Vol. 224, American Chemical Society, Washington, D.C., 1983, p. 101.
270. R. G. Pendleton, L. Samler, C. Kaiser, and P. T. Ridley, Eur. J. Pharmacol. *51*, 19 (1978).
271. J. Weinstock, J. W. Wilson, D. L. Ladd, C. K. Brush, F. R. Pfeiffer, G. Y. Kuo, K. G. Holden, N. C. F. Yim, R. A. Hahn, J. R. Wardell, Jr., A. J. Tobia, P. E. Setler, H. M. Sarau, and P. T. Ridley, J. Med. Chem. *23*(9), 973 (1980).
272. R. A. Hahn, J. R. Wardell, Jr., H. M. Sarau, and P. T. Ridley, J. Pharmacol. Exp. Ther. *232*(2), 305 (1982).
273. C. Kaiser, in *Dopamine Receptors* (C. Kaiser and J. W. Kebabian, eds.), ACS Symposium Series, Vol. 224, American Chemical Society, Washington, D.C., 1983, p. 223.
274. D. R. Sibley, S. E. Leff, and I. Creese, Life Sci. *31*, 637 (1982).
275. S. T. Ross, R. G. Franz, J. W. Wilson, M. Brenner, R. M. DeMarinis, J. P. Hieble, and H. Sarau, J. Med. Chem. *29*, 733 (1986).
276. S. T. Ross, R. G. Franz, G. Gallagher, Jr., M. Brenner, J. W. Wilson, R. M. De Marinis, J. P. Hieble, and H. M.Sarau, J. Med. Chem. *30*, 35 (1987).
277. J. A. Poat, G. N. Woodruff, and K. J. Watling, J. Pharm. Pharmacol. *30*, 495 (1978).
278. J. D. Kohli and L. I. Goldberg, J. Pharm. Pharmacol. *32*, 225 (1980).
279. P. A. Dandridge, C. Kaiser, M. Brenner, D. Gaitanopoulos, L. D. Davis, R. L. Webb, J. J. Foley, and H. M. Sarau, J. Med. Chem. *27*, 28 (1984).
280. L. I. Goldberg, D. Glock, J. D. Kohli, and J. Berger, Clin. Res. *31*(4), 716A (1983).

281. J. F. Plantje, F. J. Daus, H. A. Hansen, and J. C. Stoof, Naunyn-Schmiedebergs Arch. Pharmacol. *327*, 180 (1984).
282. Y. Itoh, M. Beaulieu, and J. W. Kebabian, Eur. J. Pharmacol. *100*, 119 (1984).
283. K. M. O'Boyle and J. L. Waddington, J. Neurochem. *48*, 1039 (1987).
284. B. A. Berkowitz, B. Zabko-Potapovich, S. Sherman, J. P. Hieble, J. Weinstock,and E. H. Ohlstein, Fed. Proc. *43*(3), 743 (1984).
285. J. Y. Lew, S. Nakamora, A. F. Battista, and M. Goldstein, Commun. Psychopharmacol. *3*, 179 (1979).
286. N. J. Bach, E. C. Kornfeld, N. D. Jones, M. O. Chaney, D. E. Dorman, J. W. Paschal, J. A. Clemens, and E. B. Smalstia, J. Med. Chem. *23*, 481 (1980).
287. K. Tsuruta, E. A. Frey, C. W. Grewe, T. E. Cote, R. L. Eskay, and J. W. Kebabian, Nature *292*, 463 (1981).
288. R. A. Hahn, B. R. MacDonald, and M. A. Martin, J. Pharmacol. Exp. Ther. *224*, 206 (1983).
289. R. D. Titus, E. C. Kornfield, N. D. Jones, J. A. Clemens, E. B. Smalotig, R. W. Fuller, R. A. Hahn, M. D. Hynes, N. R. Mason, D. T. Wong, and M. W. Foreman, J. Med. Chem. *26*,1112 (1983).
290. M. Munemura, T. E. Cote, K. Tsuruta, R. L. Eskay, and J. W. Kebabian, Endocrinology *107*(6), 1676 (1980).
291. Y. Itoh, M. E. Goldman, and J. W. Kebabian, Eur. J. Pharmacol. *108*, 99 (1985).
292. F. P. Bymaster, L. R. Reid, C. L. Nichols, E. C. Kornfeld, and D. T. Wong, Life. Sci. *38*, 317 (1986).
293. E. B. Smalstig and J. A. Clemens, Pharmacologist *26*(3), 136 (1984).
294. M. P. Seiler and R. Markstein, Mol. Pharmacol. *26*, 452 (1984).
295. D. E. Nichols, in *Dopamine Receptors* (C. Kaiser and J. W. Kebabian, eds.), ACS Symposium Series, Vol. 224, American Chemical Society, Washington, D.C., 1983, p. 201.
296. A. M. Johansson, L.-E. Arridsson, V. Hacksell, J. L. G. Nilsson, K. Svensson, S. Hjorth, D. Clark, A. Carlsson, D. Sanchez, B. Anderson, and H. Wikstrom, J. Med. Chem. *28*, 1049 (1985).
297. M. Froimowitz, J. L. Neumeyer, and R. J. Baldessarini, J. Med. Chem. *29*, 1570 (1986).
298. M. Beaulieu, Y. Itoh, P. Tepper, A. S. Horn, and J. W. Kebabian, Eur. J. Pharmacol. *105*, 15 (1984).
299. J. Vander Weide, J. B. de Vries, P. G. Tepper, and A. S. Horn, Eur. J. Pharmacol. *125*, 273 (1986).
300. J. Vander Weide, J. B. de Vries, P. G. Tepper, and A. S. Horn, Eur. J. Pharmacol. *134*, 211 (1987).
301. C. Euvard, L. Ferland, T. Di Paolo, M. Beaulieu, F. Labrie, C. Oberlander, J. P. Raynaud, and J. R. Boissier, Neuropharmacology *19*, 379 (1980).
302. M. E. Goldman and J. W. Kebabian, in *Clinical Pharmacology in Psychiatry* (S. G. Dahl, L. F. Gram, S. M. Paul, and W. Z. Potter, eds.), Springer-Verlag, New York, 1987, p. 201.

303. G. E. Martin, M. Williams, D. J. Pettibone, G. G. Yarbrough, B. V. Clineschmidt and J. H. Jones, J. Pharmacol. Exp. Ther. *230*, 569 (1984).
304. P. A. J. Janssen, in *Antipsychotic Drugs: Pharmacodynamics and Pharmacokinetics* (G. Sedvall, B. Urnas, and Y. Zotterman, eds.), Pergamon Press, New York, 1976, p. 5.
305. C. Denel, and J.-J. Folleboucht, Life Sci. *23*, 431 (1978).
306. J. E. Leysen, W. Gommeron, and P. M. Laduron, Biochem. Pharmacol. *27*, 307 (1978).
307. H. Meunier and F. Labrie, Can. J. Biochem. Cell Biol. *61*, 516 (1983).
308. E. A. Frey, T. E. Cote, C. W. Grewe, and J. W. Kebabian, Endocrinology *110*(6), 1897 (1982).
309. P. M. Laduron and J. E. Leysen, Biochem. Pharmacol. *28*, 2161 (1979).
310. D. Glock, J. D. Kohli, and L. I. Goldberg, Fed. Proc. *41*, 1651 (1982).
311. S. Lazareno and S. R. Nahorski, Eur. J. Pharmacol. *81*, 273 (1982).
312. B. M. Angrist, in *The Benzamides: Pharmacology, Neurobiology and Clinical Aspects* (J. Rotrosen and M. Stanley, eds.), Raven Press, New York, 1982, p. 1.
313. P. Jenner and C. D. Marsden, Life Sci. *25*, 479 (1979).
314. E. Stefanini, A. M. Marchisio, P. DeVoto, F. Vernaleone, R. Collu, and P. F. Spano, Brain Res. *198*, 229 (1980).
315. P. Jenner and C. D. Marsden, Neuropharmacology *20*, 1285 (1981).
316. B. D. Roufogalis, M. Thornton, and D. N. Wade, Life Sci. *19*, 927 (1976).
317. P. Jenner, A. Theodorou, and C. D. Marsden, in *The Benzamides: Pharmacology, Neurobiology and Clinical Aspects* (J. Rotrosen and M. Stanley, eds.), Raven Press, New York, 1982, p. 109.
318. A. E. Theodorou, P. Jenner, and C. D. Marsden, Life Sci. *32*, 1243 (1983).
319. N. R. Zahniser and M. L. Dubocovich, J. Pharmacol. Exp. Ther. *227*(3), 592 (1983).
320. M. Yamamoto, S. Usuda, S. Tachikawa, and H. Maeno, Neuropharmacology *21*, 945 (1982).
321. H. Y. Meltzer, M. Mikuni, M. Simonoric, and G. L. Gudelsky, Life Sci. *32*, 1015 (1983).
322. S. Fleminger, H. Van de Waterbeemd, N. M. J. Rupniak, C. Reavill, B. Testa, P. Jenner, and C. D. Marsden, J. Pharm. Pharmacol. *33*, 363 (1983).
323. C. W. Grewe, E. A. Frey, T. E. Cote, and J. W. Kebabian, Eur. J. Pharmacol. *81*, 149 (1982).
324. M. Terai, S. Usuda, I. Kuroiwa, O. Noshiro, and H. Maeno, Jpn. J. Pharmacol. *33*, 749 (1983).
325. A. B. Davidson, E. Boff, D.A. MacNeil, J. Wenger, and L. Cook, Psychopharmacology *79*, 32 (1983).
326. T. Nakajima, K. Iwata, I. Kuruma, and K. Nakamura, Jpn. J. Pharmacol. *33*(Suppl.) 197p. (1983).
327. T. Nakajima and K. Iwata, Mol. Pharmacol. *26*, 430 (1984).

328. G. L. Olson, H.-C. Cheung, E. Chiang, and L. Berger, in *Dopamine Receptors* (C. Kaiser and J. W. Kebabian, eds.), ACS Symposium Series, Vol. 224, American Chemical Society, Washington, D.C., 1983, p. 251.
329. R.J. Lefkowitz, R. A. Cerione, J. Codina, L. Birnbaumer, and M. G. Caron, J. Membr. Biol. *87*, 1 (1985).
330. M. Kurose, T. Katada, T. Haga, K. Haga, A. Ichiyama, and M. Ui, J. Biol. Chem. *261*, 6423 (1986).
331. K. Haga, T. Haga, and A. Ichiyama, J. Biol. Chem. *261*, 10133 (1986).
332. D. R. Sibley and R. J.Lefkowitz, Nature *317*, 124 (1985).
333. R. J. Lefkowitz and M. G. Caron, J. Mol. Cell. Cardiol. *18*, 1150 (1986).

Index *

*This index was prepared using Apple® Hypercard®.

D

E

F

O

P

T

About the Editors

MICHAEL WILLIAMS is Assistant Director in the Drug Discovery Division at CIBA-GEIGY Pharmaceuticals in Summit, New Jersey. The author or coauthor of over 130 articles and book chapters, Dr. Williams is an editorial board member of the *Journal of Pharmacology and Experimental Therapeutics*. He is a member of the Society for Neuroscience, American Society for Pharmacology and Experimental Therapeutics, British Pharmacological Society, International Society for Neurochemistry, American Chemical Society (Medicinal Chemistry Division), and American Association for the Advancement of Science. Dr. Williams received the Ph.D. degree (1974) in neurochemistry from the University of London's Institute of Psychiatry and the D.Sc. degree (1987) in pharmacology, also from the University of London.

RICHARD A. GLENNON is Professor of Medicinal Chemistry in the School of Pharmacy at Virginia Commonwealth University's Medical College of Virginia in Richmond. The author or coauthor of over 150 papers and book chapters, he is a member of the American Chemical Society's Medicinal Chemistry Division and the European Behavioral Pharmacology Society. Professor Glennon received the B.S. degree (1967) in pharmacy and the M.S. degree (1969) in medicinal chemistry from Northeastern University, and a Ph.D. degree (1973) in medicinal chemistry from the State University of New York at Buffalo.

PIETER B. M. W. M. TIMMERMANS is Research Manager for Cardiovascular Diseases at E. I. du Pont de Nemours & Company, Inc., in Wilmington, Delaware. He is the author or coauthor of over 350 scientific publications and has given numerous presentations on various aspects of cardiovascular and ocular pharmacology, emphasizing the autonomic nervous system. Dr. Timmermans is an editorial board member of the *Journal of Pharmacology and Experimental Therapeutics*. He is a member of the American Society for Pharmacology and Experimental Therapeutics, British, Dutch, and German Pharmacological Societies, and the International College of Angiology. Dr. Timmermans received the B.S. (1971) and doctoral (1973) degrees in synthetic organic chemistry from the State University of Leiden, and Ph.D. degree (1976) in pharmacology from the University of Amsterdam in The Netherlands.